Recent Advances in Graphene and Graphene-Based Technologies

Online at: https://doi.org/10.1088/978-0-7503-3999-5

Recent Advances in Graphene and Graphene-Based Technologies

Edited by
Anoop Chandran
St. Cyril's College, Department of Physics, Adoor, Kerala, India

N V Unnikrishnan
Mahatma Gandhi University, School of Pure and Applied Physics, Kottayam, Kerala, India

M K Jayaraj
University of Calicut, Kerala, India

Reenu Elizabeth John
Saintgits College of Engineering, Kottayam, Kerala, India

Justin George
University of Auckland, Department of Mechanical Engineering, New Zealand

IOP Publishing, Bristol, UK

ISBN 978-0-7503-3999-5 (ebook)
ISBN 978-0-7503-3997-1 (print)
ISBN 978-0-7503-4000-7 (myPrint)
ISBN 978-0-7503-3998-8 (mobi)

DOI 10.1088/978-0-7503-3999-5

Version: 20230601

IOP ebooks

British Library Cataloguing-in-Publication Data: A catalogue record for this book is available from the British Library.

Published by IOP Publishing, wholly owned by The Institute of Physics, London

IOP Publishing, No.2 The Distillery, Glassfields, Avon Street, Bristol, BS2 0GR, UK

US Office: IOP Publishing, Inc., 190 North Independence Mall West, Suite 601, Philadelphia, PA 19106, USA

Contents

Preface

Graphene, hailed as the wonder material of the 21st century, from right after its discovery in 2004, has opened up a wide range of possibilities in terms of technological and industrial uses. The diverse and superior properties of graphene like high thermal stability, exorbitant mechanical strength, elevated conductivity, large surface area, improved optical transparency and impermeability have led to a strong pursuit of graphene commercialization. Hitherto, commercialization of graphene is still in its start-up phase. Developing a proper pathway for product-based graphene devices requires substantial advancements in scientific as well as technological innovations. Considering the pressing need for the emergence of graphene-based innovative technologies and their commercialization, global efforts have initiated certain national and international consortiums in addition to government agencies, to provide long term funding and support for bridging the gap between academic research and a graphene-based device market. Due to these collective efforts, research on graphene and graphene-based technologies is advancing at a rapid pace, keeping track of which is extremely important for the scientific community.

This book, spread across 18 chapters, provides a precise overview of the fundamentals as well as the state-of-the-art in the field of graphene technology. A comprehensive review of various forms of graphene like reduced graphene oxide, graphene quantum dots, graphene nanoribbons and functionalized graphene is included. Separate chapters have been dedicated to discussing recent developments in various sensing, electronic, photonic and optoelectronic applications of graphene. The uniqueness of this book is that it covers hot topics like graphene metamaterials and graphene-magnetism. The final chapter provides a comprehensive review of the global market opportunities of graphene and the challenges it faces. A discussion on the technology and innovation road map for graphene-related materials (GRMs) by compiling the most recent status and prospects for GRM industrialization is also included in this chapter.

Aimed for industries exploring graphene use, the book's tone makes it accessible to readers with a general, undergraduate background in the sciences or engineering. Furthermore, the broad variety of topics relating to graphene and its applications presented in a cogent form can be readily used by academicians and PhD students pursuing their research in this field. The reader will be well equipped after finishing this book, to advance independently in graphene research and development and to put what they have learned there to use. The book will be self-contained, but each chapter offers a large number of references to help readers study, research, and explore the themes in greater depth.

Editor biographies

Dr Anoop Chandran, MSc, PhD

Dr Anoop Chandran was awarded a PhD in 2015 for his work on 1D semiconductor nanomaterials. The very same year, he was awarded Dr D S Kothari Post-doctoral Fellowship by UGC for his proposal on semiconductor–CNT/graphene hybrids. He is currently an assistant professor and research supervisor in the Department of Physics, St. Cyril's college, Adoor, Kerala, India. He has numerous articles published in reputed international journals like *Journal of Applied Physics*, *Applied Surface Science*, *Physical Chemistry Chemical Physics*, *ACS Omega*, to mention a few. Additionally, he has co-authored a number of book chapters and worked as a referee for many foreign publications. He is an affiliate member of the Royals Society of Chemistry. His current research interest includes 1D nanomaterials, dielectrics, hybrid conducting polymer nanocomposites, self-healing composites and photovoltaics.

Professor N V Unnikrishnan (MSc, PhD)

Professor N V Unnikrishnan obtained his doctoral degree from Rohtak University, Delhi for work on mechanism in optical materials under pulsed LASER excitations. From 1985 to 1987, he worked as a research associate in BARC, Mumbai. In 1987 he assumed the position of a lecturer in Mahatma Gandhi University, Kottayam and was eventually elevated to the professor level in 2003. After his retirement in 2014, he served as an emeritus professor in the same university till 2019. Currently, he is working as a visiting Professor in Mahatma Gandhi University. Professor Unnikrishnan, during his thirty five year-long teaching career, has supervised 33 PhDs and six post-doctoral fellows. He has completed ten major research projects funded by agencies like UGC, CSIR, DAE-DRNS, TWAS etc and published about 282 research papers in various international journals. He is a life member of Asian Crystallographic Association and also referee of many international journals. His areas of interest include photonic devices, optical wave guides, 1D and 2D materials-based sensors and supercapacitors, self-cleaning materials and non-linear optics.

Prof. M K Jayaraj (MSc, PhD)

Prof. M K Jayaraj received PhD degree in Physics from CUSAT in 1990. From 1990 to 1991 he worked at the Physical Research Laboratory. He was provided the SERC young scientist award by the department of science and technology in 1991. Dr Jayaraj was a visiting scientist to ENEA, Italy during 1994–95 and received the European commission fellowship to work at University of Pierre-et-Marie-Curie during 1995–96. From 1999 to 2001 he worked at Oregon State University. He also worked as a visiting professor at Tokyo Institute of Technology, Japan. He was a Professor in CUSAT till 2020 and guided many research scholars. At present, he is the vice chancellor of the University of Calicut, Kerala. He has more than 220 publications to his credit. His research focus is in the areas of nanostructures devices, transparent conducting oxides, electroluminescent devices, ferroelectric thin films, lasers, graphene, few-layered transition metal dichalcogenides etc.

Dr Reenu Elizabeth John (MSc, MPhil, PhD)

Dr John obtained her PhD in 2019 for her work on 1D transition metal oxides. Her areas of interest are ferroelectric nanomaterials and conducting polymer nanocomposites for EMI shielding. Presently, she is working as an assistant professor in the Department of Physics, Saintgits College of Engineering (Autonomous), Kottayam, Kerala, India.

Mr Justin George (MSc)

Mr George is currently doing research in the University of Auckland under the supervision of distinguished professor Debes Bhattacharyya. He is working on a project to fabricate novel carbon-based composites for sensing and energy applications.

List of contributors

Dr Amritroop Achari
National Graphene Institute, University of Manchester, Manchester, UK

Mr Souren Adhikary
Department of Physics, Indian Institute of Science Education and Research, Tirupati, Andhra Pradesh, India

Dr Arpana Agrawal
Department of Physics, Shri Neelkantheshwar Government Post-Graduate College, Khandwa, Madhya Pradesh, India

Dr Prashant S Alegaonkar
Department of Physics, School of Basic and Applied Sciences, Central University of Punjab, Ghudda, Punjab, India

Mr Augusto D Alvarenga
Nanotechnology National Laboratory for Agriculture (LNNA), Embrapa Instrumentação, Centro, São Carlos, SP, Brazil
PPGQ, Department of Chemistry, Center for Exact Sciences and Technology, Federal University of São Carlos (UFSCar), São Carlos, SP, Brazil

Ms K S Anju
Department of Physics, Cochin University of Science and Technology, Kochi, Kerala, India

Distinguished Professor Debes Bhattacharyya
Centre for Advanced Composite Materials, Department of mechanical engineering, The University of Auckland, New Market, Auckland, New Zealand

Dr Zainal Arif Burhanudin
Department of Electrical and Electronic Engineering, Nanotechnology Research Laboratory, Center of Innovative Nanostructures and Nanodevices (COINN), Universiti Teknologi PETRONAS, Perak, Malaysia

Dr Avishek Chanda
Composite Materials and Engineering Center, Washington State University, Pullman, WA, USA

Dr Anoop Chandran
Department of Physics, St. Cyrils College, Adoor, Kerala, India

Mr L Chandrasekar
Department of Electronics and Communication Engineering, Indian institute of information technology design and manufacturing, Vandalur, Chennai, India

Dr Daniel Correa
Nanotechnology National Laboratory for Agriculture (LNNA), Embrapa Instrumentação, Centro, São Carlos, SP, Brazil
and
PPGQ, Department of Chemistry, Center for Exact Sciences and Technology, Federal University of São Carlos (UFSCar), São Carlos, SP, Brazil
and
PPG-Biotec, Center for Exact Sciences and Technology, Federal University of Sao Carlos (UFSCar), Sao Carlos, SP, Brazil

Dr Neena D
Centre for Advanced Composite Materials, Department of Mechanical Engineering, The University of Auckland, New Zealand

Mr Soumya Ranjan Das
Department of Physics, Indian Institute of Science Education and Research, Tirupati, Andhra Pradesh, India

Dr Sudipta Dutta
Department of Physics, Indian Institute of Science Education and Research, Tirupati, Andhra Pradesh, India

Mr Murilo Henrique Moreira Facure
Nanotechnology National Laboratory for Agriculture (LNNA), Embrapa Instrumentação, Centro, São Carlos, SP, Brazil
and
PPGQ, Department of Chemistry, Center for Exact Sciences and Technology, Federal University of São Carlos (UFSCar), São Carlos, SP, Brazil

Mr Justin George
Centre for Advanced Composite Materials, Department of mechanical engineering, The University of Auckland, New Market, Auckland, New Zealand

Ms Monika Gupta
Department of Electrical and Electronic Engineering, Nanotechnology Research Laboratory, Center of Innovative Nanostructures and Nanodevices (COINN), Universiti Teknologi PETRONAS, Perak, Malaysia

Dr Huzein Fahmi Hawari
Department of Electrical and Electronic Engineering, Nanotechnology Research Laboratory, Center of Innovative Nanostructures and Nanodevices (COINN), Universiti Teknologi PETRONAS, Perak, Malaysia

Professor M K Jayaraj
University of Calicut, Kerala, India

Dr Reenu Elizabeth John
Department of Physics, Saintgits college of Engineering, Kottayam, Kerala, India

Mr Dohun Kim
Department of Energy Science and Engineering, Daegu Gyeongbuk Institute of Science and Technology, Daegu, Republic of Korea

Dr Joon Young Kim
Hydrogen Energy Technology Laboratory Korea Institute of Energy Technology (KENTECH), Jeollanamdo, Republic of Korea
and
Research Institute NEEL Sciences, INC., Jeollanamdo, Republic of Korea

Dr Jung Kyu Kim
School of Chemical Engineering, Sungkyunkwan University (SKKU), Suwon, Republic of Korea

Dr Hasna Kudilatt
Department of Physics, Govt. Arts and Science College, Calicut, Kerala, India

Dr Pradeep Kumar
Nanotechnology Research Laboratory, Center of Innovative Nanostructures and Nanodevices (COINN), Universiti Teknologi PETRONAS, Perak, Malaysia

Dr Aman Mahajan
Department of Physics, Material Science Laboratory/Centre for Sustainable Habitat, Guru Nanak Dev University, Amritsar, India

Dr Sesha Manuguri
Department of neuroscience and biomedical engineering, school of Science, Aalto, Finland

Dr Luiza A Mercante
Institute of Chemistry, Federal University of Bahia (UFBA), Salvador, BA, Brazil

Mr Avishek Mishra
Centre for Advanced Composite Materials, Department of Mechanical Engineering, Auckland, New Zealand

Professor Rahul Raveendran Nair
Department of Chemical Engineering and Analytical Science (CEAS), Sackville Street, Manchester, UK

Dr Badri Narayanan Narasimhan
Department of Chemical and Material engineering, Auckland, New Zealand

Ms Nadia Norhakim
Department of Mechanical Engineering, Nanotechnology Research Laboratory, Center of Innovative Nanostructures and Nanodevices (COINN), Universiti Teknologi PETRONAS, Perak, Malaysia

Ms Lekshmi Omana
Department of Physics, St. Berchmans College, Changanassery, Kerala, India

Professor Chang-Soo Park
Functional Materials and Components R&D Group, Korea Institute of Industrial Technology, Republic of Korea

Ms Jihyun Park
Hydrogen Energy Technology Laboratory Korea Institute of Energy Technology (KENTECH), Jeollanamdo, Republic of Korea

Dr K P Pradhan
Department of Electronics and Communication Engineering, Indian Institute of Information Technology Design and Manufacturing, Vandalur, Chennai, India

Dr Premlal Balakrishna Pillai
Department of Chemical Engineering and Analytical Science (CEAS), Sackville Street, Manchester, UK

Professor Maria Raposo
Laboratory of Instrumentation, Biomedical Engineering and Radiation Physics (LIB Phys-UNL), Department of Physics, NOVA School of Science and Technology, NOVA University Lisbon, Caparica, Portugal

Dr I Rejeena
MSM College, Kayamkulam, Kerala, India

Professor Paulo Ribeiro
Laboratory of Instrumentation, Biomedical Engineering and Radiation Physics (LIB Phys-UNL), Department of Physics, NOVA School of Science and Technology, NOVA University Lisbon, 2829-516 Caparica, Portugal

Dr M Shuaib M Saheed
Department of Mechanical Engineering, Nanotechnology Research Laboratory, Center of Innovative Nanostructures and Nanodevices (COINN), Universiti Teknologi PETRONAS, Perak, Malaysia

Dr Neha Sardana
Department of Metallurgical and Materials Engineering of Indian Institute of Technology, Indian Institute of Technology Ropar, Ropar Main Campus, Rupnagar, Punjab, India

Mr Sagar Sardana
Department of Physics, Material Science Laboratory, Guru Nanak Dev University, Amritsar, India

Mr Rodrigo Schneider
Nanotechnology National Laboratory for Agriculture (LNNA), Embrapa Instrumentação, Centro, São Carlos, SP, Brazil
PPGQ, Department of Chemistry, Center for Exact Sciences and Technology, Federal University of São Carlos (UFSCar), São Carlos, SP, Brazil

Professor Susana Sério
Laboratory of Instrumentation, Biomedical Engineering and Radiation Physics (LIB Phys-UNL), Department of Physics, NOVA School of Science and Technology, NOVA University Lisbon, 2829-516 Caparica, Portugal

Dr Anshul Kumar Sharma
Department of Physics, Material Science Laboratory/Centre for Sustainable Habitat, Guru Nanak Dev University, Amritsar, India

Professor Uk Sim
Hydrogen Energy Technology Laboratory Korea Institute of Energy Technology (KENTECH), Jeollanamdo, Republic of Korea
and
Research Institute, NEEL Sciences, INC., Jeollanamdo, Republic of Korea
and
Center for Energy Storage System, Chonnam National University, Gwangju, Republic of Korea

Mr Gaurav Pal Singh
Department of Metallurgical and Materials Engineering of Indian Institute of Technology, Indian Institute of Technology Ropar, Ropar Main Campus, Rupnagar, Punjab, India

Dr Subramani Surendran
Hydrogen Energy Technology Laboratory Korea Institute of Energy Technology (KENTECH), Jeollanamdo, Republic of Korea

Dr Vinoy Thomas
Christian College, Chengannur, Kerala, India

Professor (Retired) N V Unnikrishnan
School of Pure and Applied Physics, Mahatma Gandhi University, Kottayam, Kerala, India

Dr Steffy Sara Varghese
Space and Planetary Science Centre, Khalifa University, Abu Dhabi, UAE

Dr Professor Gyu-Chul Yi
Department of Physics and Astronomy, Institute of Applied Physics, and Research Institute of Advanced Materials (RIAM), National University, Seoul, Republic of Korea

Ms Elsa Zachariah
Christian College, Chengannur, Kerala, India

Mr Paulo Zagalo
Laboratory of Instrumentation, Biomedical Engineering and Radiation Physics (LIB Phys-UNL), Department of Physics, NOVA School of Science and Technology, NOVA University Lisbon, Caparica, Portugal

IOP Publishing

Recent Advances in Graphene and Graphene-Based Technologies

Anoop Chandran, N V Unnikrishnan, M K Jayaraj, Reenu Elizabeth John and Justin George

Chapter 1

Graphene: an introduction

Justin George and Debes Bhattacharyya

Graphene is a 2D wonder material that has opened up a new avenue of research into 2D material since its discovery in 2004 by Sir A K Geim and Sir K S Novoselov. It is a 2D sheet of sp^2 hybridized carbon arranged in a hexagonal lattice, which makes it the strongest material on Earth with many peculiar optical, electrical and electronic properties. This chapter discusses the structure and properties of graphene and its characterization. The chapter begins by exploring the atomic and electronic structure of graphene, followed by its optical, mechanical and electronic properties, such as the quantum hall effect and the Klein paradox. In the last part of the chapter, characterization of graphene through various techniques, such as atomic force microscope (AFM), scanning electron microscope (SEM), and transmission electron microscope (TEM), are discussed. The chapter aims to provide the fundamental understanding of the properties of graphene and its characterization.

1.1 Introduction

Carbon, one of the Earth's most abundant and intriguing elements, appears in different allotropes (graphite, graphene, diamond, carbon nanotube, amorphous carbon, and many more), and has attracted research attention for centuries. The properties of these allotropes (mechanical, optical, electrical, and electronic) differ significantly depending on the arrangement of carbon atoms in the material. For example, sp^2 hybridized graphite appears as a soft, black, and excellent conductor of electricity, while sp^3 diamond is a hard, transparent, and bad conductor of electricity [1–3]. Graphite is a naturally occurring crystalline form of carbon with extreme stability at high temperatures. It is highly inert and shows high electrical conductivity. It has been used for several applications, such as in energy storage and electrical devices, lubricants and refractories [4]. A graphite crystal is a stack of several layers of sp^2 hybridized honeycomb structure atomic layers (graphene sheets) to form a 3D crystalline structure [5]. These 2D layers of carbon (graphene) can slide apart as there is only a weak Van der Waals interaction between the layers [6]. For example, a few graphene layers can stick to the paper surface while writing with a graphite pencil. However, separating a single graphene layer requires mechanical or chemical exfoliation methods [7, 8]. Other methods are also used to segregate a single or few layers of graphene sheets and their derivatives from a graphite block [9–12].

Although there were attempts to study graphene since 1859, research interest in this wonder material increased when Professor Sir Andre Geim and Professor Sir Kostya Novoselov of the University of Manchester isolated this wonder material in 2004 by mechanical exfoliation, a process of repeated peeling [1]. Theoretically, it was considered that the isolated 2D carbon structure or graphene sheet, would be thermodynamically unstable [13], and that a 2D atomic sheet of carbon could not be isolated without an epitaxial surface that provides an additional atomic bonding [5]. However, Geim and Novoselov were able to isolate single graphene layers with a 2D honeycomb structure (hexagonal lattice) from graphite through Scotch Tape exfoliation (figure 1.1) [14]. Experiments showed that graphene had superior electrical, thermal, mechanical, and optical properties compared to many other natural materials, making it a wonder material [9, 15, 16]. Carbon allotropes with a crystal structure similar to graphene include carbon nanotubes and graphite [17, 18]. In these sp^2 hybridized carbon materials, the aromatic ring or hexagonal lattice, is the building block. A 2D sheet of sp^2 aromatic carbon forms graphene, while a 0D spherical-wrapped graphene with introduced pentagon forms fullerene (C_{60}). Rolled graphene sheets form carbon nanotubes (1D), and stacked graphene sheets form graphite (3D) crystals [19]. As graphene exhibits superior mechanical and electrical

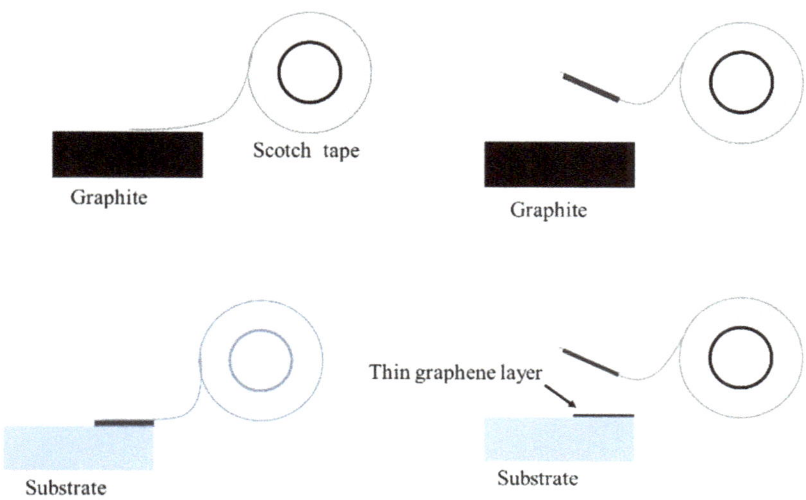

Figure 1.1. Scotch Tape method of graphene production. Cleaving a few layers of graphene using adhesive tape (top row). Pressing tape on the substrate (bottom row).

properties compared to other allotropes, it has a significant potential for numerous electrical and electronic applications. As a result, it has been extensively researched over the past decade and is one of the most studied materials. With rapid advancements occurring in graphene-based technologies every year, it is crucial to keep the scientific community informed of these developments. Therefore, this book aims to review the recent progress in research on graphene and graphene-based technologies. The introductory chapter provides an overview of the basic properties of graphene and its characterization techniques.

1.2 Atomic and electronic structure of graphene

The electronic configuration of an isolated carbon atom is $1s^22s^22p^2$. In graphene, the carbon atoms are arranged in a hexagonal lattice forming a honeycomb structure. Only the valence orbitals ($2s^22p^2$) get hybridized to form three sp^2 orbitals and one p_z orbital, figure 1.3(c). The sp^2 orbitals form σ bonds, while the p_z orbitals form the valence band (π band) containing one electron and the conduction band (π^* band) with no electron. In graphene, sp^2 hybridized carbon is arranged in a honeycomb crystal lattice meaning each carbon atom has sigma bonds with three other carbon atoms, and π electrons are delocalized [20]. In contrast, carbon atoms in diamond crystals form a sigma bond with four different carbon atoms (sp^3 hybridisation), hence no free electrons are there in the valence band [3, 21]. Studies have shown that the bond length in the sp^2 hybridized graphene is smaller than the bond length in the sp^3 hybridized diamond structure. This substantial overlap makes the sp^2 sigma bond stronger than the sp^3 hybridized bonds [22, 23]. Hence the carbon–carbon bonding in graphene is shorter and stronger than in diamond, making it a very stable and robust material. The sigma bonding of sp^2 hybridization orbitals is formed by the superposition of s, p_x and p_y orbitals of the carbon atom. The p_z orbital is unhybridised and bound in c-direction by weak Van der Waals force with neighbouring carbon atoms above and below to form π-bonds. These delocalized loosely bound π-electrons are involved in the electrical conduction in graphene [20].

In graphene, carbon atoms are arranged in a honeycomb lattice (hexagonal lattice) as shown in figure 1.2. The graphene structure can be considered as a

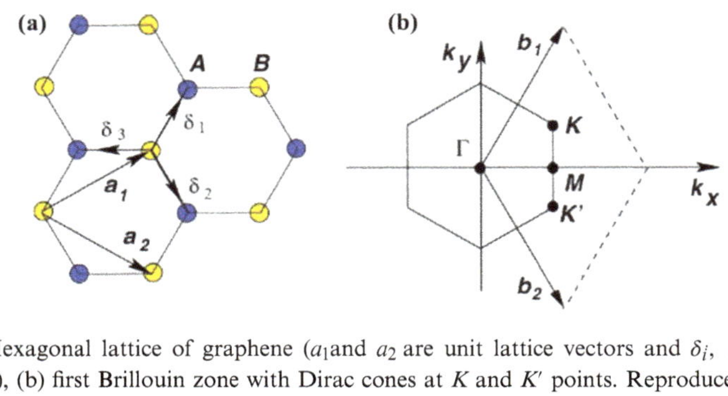

Figure 1.2. (a) Hexagonal lattice of graphene ($a_1$and a_2 are unit lattice vectors and δ_i, $i = 1, 2, 3$ nearest neighbour vector), (b) first Brillouin zone with Dirac cones at K and K' points. Reproduced with permission [24], copyright 2009, American Physical Society.

bipartite lattice of two triangular sublattices (A and B) with two atoms per unit cell, figure 1.2(a). The lattice vectors can be written as:

$$a_1 = 0.5\, a\, (\sqrt{3},\, 1) \tag{1.1}$$

$$a_2 = 0.5\, a\, (\sqrt{3},\, -1) \tag{1.2}$$

where $a = \sqrt{3}\, a_{C-C}$ and $a_{C-C} = 1 \cdot 42$ Å is the distance between neighbouring carbon atoms. The reciprocal lattice vectors can be written as:

$$b_1 = \frac{2\pi}{a}\left(\frac{1}{\sqrt{3}},\, 1\right) \tag{1.3}$$

$$b_2 = \frac{2\pi}{a}\left(\frac{1}{\sqrt{3}},\, -1\right) \tag{1.4}$$

with reciprocal lattice constant $\frac{4\pi}{\sqrt{3}\,a}$. In real space, the immediate three nearest neighbours vectors are

$$\delta_1 = \frac{a}{2}\left(1,\, \sqrt{3}\right) \tag{1.5}$$

$$\delta_2 = \frac{a}{2}\left(1,\, -\sqrt{3}\right) \tag{1.6}$$

$$\delta_3 = -a\,(1,\, 0) \tag{1.7}$$

The six next neighbour locations are

$$\delta'_1 = \pm a_1 \tag{1.8}$$

$$\delta'_2 = \pm a_2 \tag{1.9}$$

$$\delta'_3 = \pm(a_2 - a_1) \tag{1.10}$$

In the k-space (reciprocal space), the first Brillouin zone is a hexagon with two inequivalent k-points, K and K' (Dirac points), located at the corners of the graphene Brillouin zone. Their positions in momentum space are shown in figure 1.2(b)

$$K = \left(\frac{2\pi}{3a},\, \frac{2\pi}{3\sqrt{3a}}\right) \tag{1.11}$$

$$K' = \left(\frac{2\pi}{3a},\, -\frac{2\pi}{3\sqrt{3a}}\right) \tag{1.12}$$

Using tight-binding approximation and assuming the electrons from the carbon atom can only hop to the nearest neighbour Hamiltonian for graphene (here $\hbar = 1$) can be written as:

$$H = -t \sum_{(i,j)} \left(\hat{a}_i^\dagger \hat{b}_j + h.c \right) \tag{1.13}$$

where i and j denote the sublattice sites A and B, respectively. The operators are creation and annihilation operators of an electron at site A with position r_i, whereas $\hat{b}_j^\dagger$ and $\hat{b}_j$, are creation and annihilation operators for an electron at site B. The sum over the nearest neighbour is

$$\sum_{(i,j)} \left(\hat{a}_i^\dagger \hat{b}_j + \hat{b}_j^\dagger \hat{a}_l \right) = \sum_{i \in A} \sum_{\delta} \hat{a}_i^\dagger \hat{b}_{i+\delta} + \hat{b}_{i+\delta}^\dagger \hat{a}_i \tag{1.14}$$

where $\hat{b}_{i+\delta}$, is the annihilation operator of an electron at site B at position $r_i + \delta$. In the Fourier space, we can write

$$\hat{a}^\dagger{}_i = \frac{1}{\sqrt{N/2}} \sum_k e^{ik.r_i} \hat{a}_k^\dagger \tag{1.15}$$

where $N/2$ number of A sites. Substituting this to the tight-binding Hamiltonian, one can write

$$\hat{H} = -\frac{1}{\frac{N}{2}} \sum_{i \in A} \sum_{\delta, k, k'} [\mathrm{e}^{i(k-k').r_i} \, \mathrm{e}^{-ik'.\delta} \, \hat{a}_k^\dagger \hat{b}_{k'} + \mathrm{h.c.}] \tag{1.16}$$

$$H = -t \sum_{\delta, k} [\mathrm{e}^{-ik.\delta} \, \hat{a}_k^\dagger \hat{b}_k + \mathrm{h.c.}] \tag{1.17}$$

where $\sum_{i \in A} e^{i(k-k').r_i} = \frac{N}{2} \delta_{kk'}$.

The Hamiltonian can be written as

$$H = \sum_k \Psi^\dagger \, h[k] \Psi \tag{1.18}$$

where $\Psi = \begin{pmatrix} \hat{a}_k \\ \hat{b}_k \end{pmatrix}$ and $\Psi^\dagger = (\hat{a}_k^\dagger \; \hat{b}_k^\dagger)$

$h(k) = -t \begin{bmatrix} 0 & \Delta_k \\ \Delta_k^* & 0 \end{bmatrix}$, represents the Hamiltonian in matrix form. Where $\Delta_k = \sum_\delta e^{ik.\delta}$.

The energy eigenvalues are

$$E_\pm = \pm \, t \sqrt{\Delta_k \, \Delta_k^*} \tag{1.19}$$

where Δ_k can be computed as

$$\Delta_k = e^{ik.\delta_1} + e^{ik.\delta_2} + e^{ik.\delta_3} \tag{1.20}$$

$$\Delta_k = e^{ik.\delta_3}\,[1 + e^{ik.(\delta_1-\delta_3)} + e^{ik.(\delta_2-\delta_3)}] \tag{1.21}$$

Substituting δ_1, δ_2 and δ_3

$$\Delta_k = e^{-ik_xa}\left[1 + e^{i\frac{3k_xa}{2}}\left(e^{i\frac{\sqrt{3}k_ya}{2}} + e^{-i\frac{\sqrt{3}k_ya}{2}}\right)\right] \tag{1.22}$$

$$\Delta_k = e^{-ik_xa}\left[1 + 2e^{i\frac{3k_xa}{2}}\cos\frac{\sqrt{3}k_ya}{2}\right] \tag{1.23}$$

Similarly, Δ_k^* and substituting in equation $E_\pm = \pm t\sqrt{\Delta_k\Delta_k^*}$

$$E_\pm = \pm t\sqrt{1 + 4\cos\left(\frac{3k_x\,a}{2}\right)\cos\left(\frac{\sqrt{3}k_y\,a}{2}\right) + 4\cos^2\left(\frac{\sqrt{3}k_y\,a}{2}\right)} \tag{1.24}$$

In 1947, Wallace used tight-binding approximation to calculate the energy dispersion relation of a single layer of graphite, which is now known as graphene see (figure 1.3 (b)). Within the nearest neighbour approximation, energy bands in energy–momentum space can be obtained as in equation (1.24). This can also be written as

$$E_\pm(k) = \pm t\sqrt{3 + f(k)} \tag{1.25}$$

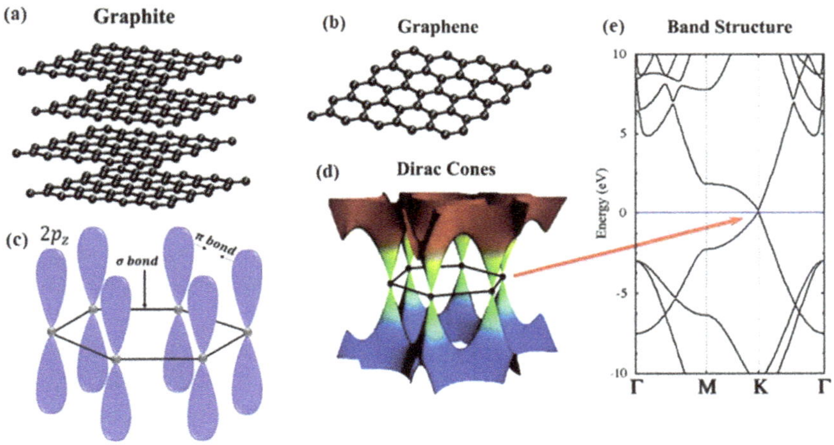

Figure 1.3. Graphene 2D hexagonal lattice. (a) Graphite structure stacked graphene layers. (b) Graphene—a single layer of graphite with a hexagonal structure. (c) Dirac cones in graphene. (d) Graphene band structure (Γ–M–K–Γ) and Fermi level shifting is depicted with a blue horizontal line. Reproduced from [25], with permission from Royal Society of Chemistry.

where

$$f(k) = 2\cos(\sqrt{3}k_y a) + 4\cos\left(\frac{3k_x a}{2}\right)\cos\left(\frac{\sqrt{3}k_y a}{2}\right) \quad (1.26)$$

The nearest neighbour hopping energy at t (~2.8 eV), and a represents the lattice constant. The energy band of 2D graphene, according to equation (1.25), is plotted in figure 1.3(d) as a function of the wave vector. The upper half of the curves is the π^* (anti-bonding band) and the lower half of the curves is the π (bonding band). These two bands degenerate (have the same energy) at K points in the reciprocal space, which is also the Fermi energy level of undoped graphene. Graphene has two atoms per unit cell (two electrons per unit cell) and the lower π band is completely filled while the upper π^* band is unfilled. We can observe, the Fermi energy corresponds to $E = 0$ at the K point. Near to the K points, $|k|a \ll 1$, Taylor's series expansion of energy dispersion in equation (1.25) reduces to a linear dispersion relation,

$$E(k) = \pm\frac{\sqrt{3}}{2}\gamma ak = \pm v_F \hbar k \quad (1.27)$$

where Fermi velocity $v_F = \dfrac{\sqrt{3}}{2}\dfrac{\gamma a}{\hbar} \approx 10^6$ m s^{-1}, here positive energy corresponds to the kinetic energy of electrons in the conduction band, and negative energy corresponds to the kinetic energy of electrons in the valence band. This linear dispersion with constant fermi velocity near the Dirac points in graphene is a characteristic of relativistic massless fermions [24].

The density of states of 2D material can be calculated using

$$D_{2D}(E) = \frac{1}{A}\sum_k \delta(E - E_k) = \frac{1}{A}\frac{A}{(2\pi)^2}2\int_\infty^0 \delta(E - E_k)2\pi k dk \quad (1.28)$$

where $E(k) = \hbar v_F k$, $dE = \hbar v_F dk$, $kdk = \dfrac{EdE}{(\hbar v_F)^2}$.

$$D_{2D}(E) = \frac{g}{\pi\hbar^2 v_F{}^2}\int_\infty^0 \delta(E - E_k)EdE = \frac{2\,|\,E\,|}{\pi\hbar^2 v_F{}^2} \quad (1.29)$$

where g is the degeneracy of the electron. Here $D(E)\alpha\,|\,E\,|$, i.e., the density of states is linear and vanishes at zero energy, which contrasts with conventional 2D materials, where the density of states is a constant.

As described, graphene shows a unique band structure, and it has a unique electronic structure with zero bandgap, near to the Fermi level ($E = 0$), the effective Hamiltonian for the highly symmetric points the K and K' can be written as $\hat{H}_K$ and $\hat{H}_{K'}$

$$\widehat{H}_K(k) = \hbar v_F(0\ k_x - ik_y\ k_x + ik_y\ 0) = \hbar v_F\,\vec{\sigma}.\vec{k} = -\widehat{H}^T{}_{K'}(k) \quad (1.30)$$

where $\vec{\sigma} = (\sigma_x, \sigma_y)$ is the 2D Pauli matrix, $\vec{k}$ is the massless quasiparticle (Dirac massless fermions) momentum and v_F, the Fermi energy of the charge carriers.

The wavefunction of these Dirac particles for K and K' points are $\Psi^{K}_{s,k} = e^{ik.r}(s\ e^{i\theta_k})$ and $\Psi^{K'}_{s,k} = e^{ik.r}(e^{i\theta_k}\ s)$, respectively. Where $s = 1$ for the upper band and $s = -1$ for the lower band, $\theta_k = \tan^{-1}\left(\frac{k_y}{k_x}\right)$. The upper and lower terms represent the amplitude (*pseudospin*) of finding a particle in either sublattice A or B. In the case of graphene, the pseudospin direction is related to the momentum of particles, which means wavefunction near the Dirac points is *chiral.* Hence, any backscattering (particle scattering from wave vector $\vec{k}$ and $-\vec{k}$) is prohibited. The chirality of particles in K and K' valleys, electron and hole bands are opposite. The energy bands meet at the Dirac point, where $k = 0$ and the charge carriers in graphene are called *massless Dirac fermions.* These transformations of electrons in graphene to massless relativistic particles lead to interesting effects such as the anomalous integer quantum Hall effect, the novel effect at the edge and the Klein paradox.

1.3 Properties of graphene

The nanoscale properties of graphene are fascinating due to the potential industrial applications and fundamental scientific research. Some important properties are discussed as follows.

1.3.1 Optical properties

The optical conductivity (G) of the 2D Dirac spectrum with conical dispersion is theoretically predicted to show a universal value of $G_0 = \frac{e^2}{4\hbar}$, provided the energy of a photon is much higher than the temperature and Fermi energy (E_F). As there is a universal conductivity for graphene, the optical properties can be expressed as a fundamental constant. The transmittance can be expressed as $T_{opt} = \left(1 + \frac{\pi\alpha}{2}\right)^{-2} \approx 1 - \pi\alpha \approx 0.977$ for normal incidence of light, where $\alpha = \frac{e^2}{\hbar c}$, the fine structure constant. The transmittance (T) and reflection (R) are observable quantities, which were observed by Nair *et al* and found to follow the universal conduction behaviours in graphene [26].

The light absorption in graphene can be calculated in an alternative way other than dynamic conductivity. If a light wave with electric field E and frequency ω is incident perpendicular to a unit area of the graphene sheet, the incident energy flux Φ_i can be calculated as

$$\Phi_i = \frac{c}{4\pi} \mid E \mid^2 \tag{1.31}$$

Considering the momentum conservation of the initial and final state, only excitation with energy $E = \hbar\omega$ leads to light absorption. If the number of absorptions per unit time is n the total absorbed energy can be written as $E_a = n\hbar\omega$. The number of absorptions per unit time can be calculated using the famous Fermi's golden rule as $n = \frac{2\pi}{\hbar} \mid M \mid^2 D$, where M is a matrix that represents the interaction of a photon with Dirac fermions and D is the density of states at energy $E = \frac{\hbar\omega}{2}$. The 2D fermion density of states can be calculated by substituting $\frac{\hbar\omega}{2}$ into 2D density equation

$$D\left(\frac{\hbar\omega}{2}\right) = \frac{2\,|\,E\,|}{\pi\hbar^2{\nu_F}^2} = \frac{\hbar\omega}{\pi\hbar^2 v_F^2} = \frac{\omega}{\pi\hbar{v_F}^2} \tag{1.32}$$

The interaction Hamiltonian (between light and Dirac fermion) is

$$\hat{H} = v_F\vec{\sigma}\cdot\vec{p} = v_F\vec{\sigma}.\left(\hat{p} - \frac{e}{c}\vec{A}\right) = \hat{H}_0 + \hat{H}_{\text{int}} \tag{1.33}$$

where the second term represents the interaction Hamiltonian

$$\hat{H}_{\text{int}} = -v_F\vec{\sigma}.\,\frac{e}{c}\vec{A} = v_F\vec{\sigma}.\,\frac{e}{i\omega}\vec{E} \tag{1.34}$$

where the magnetic vector potential is $\vec{A} = \frac{ic}{\omega}\vec{E}$ and σ is Pauli spin matrices. We get after averaging over all the states as:

$$|M|^2 = \left|\left\langle i\left|v_F\vec{\sigma}.\frac{e}{i\omega}\vec{E}\right|f\right\rangle\right|^2 = \frac{e^2{v_F}^2}{8\,\omega^2}\,|E|^2 \tag{1.35}$$

The absorbed energy flux can be calculated as:

$$\Phi_a = n\hbar\omega = \frac{2\pi}{\hbar}\frac{e^2{v_F}^2}{8\,\omega^2}\,|E|^2\,\frac{\omega}{\pi\hbar{v_F}^2}\hbar\omega = \frac{e^2}{4\hbar}\,|E|^2 \tag{1.36}$$

Hence the absorption can be calculated as: $\frac{\Phi_a}{\Phi_i} = \frac{\pi e^2}{\hbar c} = \pi\alpha$, where $\alpha = \frac{e^2}{\hbar c} \approx \frac{1}{137}$ the fine structure constant. The absorption does not dependent on material parameter v_F (Fermi velocity). Moreover, the dynamic conductivity is $G \equiv \frac{\Phi_a}{|E|^2} = \frac{e^2}{4\hbar}$.

In practice, graphene does not reflect light ($R \ll 1$), hence the graphene opacity $(1 - T) \approx \pi\alpha$. The absorption for bilayer graphene using Fermi's golden rule is $2\pi\alpha$. Although graphene is only one atom in thickness, the light absorption of graphene is $\pi\alpha = 2.3\,\%$ of the incident white light due to its peculiar electronic structure [26].

1.3.2 Mechanical properties

Graphene has exceptional mechanical properties, making it suitable for many applications, such as filler material in composites and specific electronic applications. The main reason for its impressive mechanical properties is the stability of sp^2 hybridized hexagonal lattice. Mechanical properties measured using AFM by

Lee *et al* concluded that graphene is the strongest material in the world [15]. The elastic properties were obtained from the force–displacement response of the graphene membrane. After theoretical and experimental evaluation, Young's modulus of graphene with an effective thickness of 0.335 nm is found to be $E = 1.0 \pm 0.1$ T Pa. Nanoindentation of the suspended multilayer graphene flakes shows a bending stiffness of 2×10^{-14} N m^{-1} to 2×10^{-11} N m^{-1} for 8–100 layers of graphene, respectively [27]. Static nanoindentation based on the AFM cantilever pressed on a double-clamped graphene sheet has shown Young's modulus of 0.5 TPa. The static nanoindentation mechanical property measurement requires suspension of the graphene sheet; however, this is hard when the graphene is deposited on a substrate. However, Raman spectroscopy has been employed to measure the mechanical properties of both freestanding and deposited graphene films at room temperature and elevated temperature by probing the bond vibration of the graphene through optical spectroscopy and retrieving the mechanical and structural properties.

Monolayer graphene with no defect is considered one of the strongest materials, and its intrinsic strength is 130 GPa. It is to be noted that the strength and stiffness of graphene are not diminished even with induced sp^3 type defects. However, unlike pristine graphene, breaking strength is reduced by 14% in a defective regime [28].

One of the crucial properties of graphene for engineering applications is its fracture toughness. Zhang *et al* investigated the fracture toughness of CVD synthesised graphene. It was tested using *in situ* micromechanical devices available in a scanning electron microscope [29]. A central crack was introduced using a focused ion beam (FIB), and the brittle fracture was observed when the load was applied. From the experiment, the fracture toughness (stress intensity factor) of the graphene was found to be 4.0 ± 0.6 MPa. Gang Seob Jung *et al* conducted a theoretical simulation of the fracture mechanics of polycrystalline graphene. The fracture toughness of polycrystalline graphene has been found to be 20%–30% higher than that of pristine graphene [30]. The increase in fracture toughness can be attributed to the large energy release rate, which can increase with smaller grain sizes.

1.3.3 Electronic properties

The atomic arrangements in the graphene form a honeycomb lattice structure in a plane. Each carbon atom forms three-sigma bonds to three other neighbouring carbon atoms and the p_z orbital in carbon atom is unhybridised, and these unhybridized orbitals form delocalized π electrons, which engage in electrical conduction. These pi-electrons can move through the lattice and make graphene a good electrical conductor. Hence, graphene exhibits fascinating properties such as semi-integer quantum Hall effect, superconductivity, bipolar electric field effect and many more at room temperature. At room temperature, the carrier mobility of graphene can even reach up to 15 000 cm^2 V^{-1} s^{-1}.

(i) **Klein paradox and chiral tunnelling**

According to quantum mechanics, the Klein paradox is observed when an incoming relativistic electron penetrates through a potential barrier (V_0) that is higher than the electron's rest energy, mc^2 (where m is the mass of the electron and c, the speed of light). In this case, the transmittance probability (T) is not significantly affected by the barrier height and approaches 100% for higher potential barriers; however, for non-relativistic tunnelling, the transmittance decays exponentially with an increase in potential barrier height. The Klein paradox can be explained by the barrier height V reaching a critical potential ($V_c = E + m$), the electron energy levels of the barrier regions are lifted by V, and the occupied lower energy states continuum extends and overlaps a positive energy spectrum outside the barrier. Thus, an incident electron impact on the potential barrier can knock out electrons from these states in the barrier and initiate an electron–positron pair production at the potential barrier. When resonance penetration occurs, a positron wave is created inside the barrier [31].

Electronically, graphene can be described as a 2D zero-gap semiconductor with an electronic spectrum, as shown in figure 1.4. The low-energy quasiparticles in graphene can be written in the form of Dirac-like Hamiltonian, $\widehat{H_0} = -i\hbar v_F \sigma \nabla$, where v_F is the Fermi velocity and $\sigma = (\sigma_x, \sigma_y)$, the Pauli matrices. This Dirac-like description of quasiparticles in graphene results from graphene's special crystal structure (two equivalent sublattices), and the quantum mechanical hopping of the particles between the two sublattices leads to two consine-like energy bands which intersect near to the Brillion zone edges to form a conical spectrum. Hence, the quasiparticles in graphene show a linear dispersion relation ($E = \hbar k v_F$). These quasiparticles behave as massless relativistic particles like photons with momentum k and a fermi velocity $v_F \approx \frac{c}{300}$.

Graphene has been demonstrated as an effective medium (vacuum) for relativistic quantum tunneling which is described by the Klein paradox.

In the case of graphene, the quasiparticles resemble the fermions in quantum electrodynamics; hence an experiment considering an infinite potential barrier can

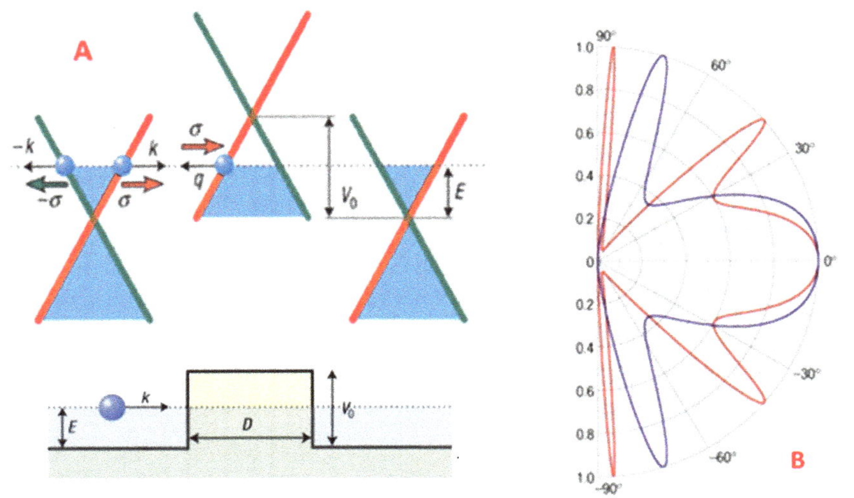

Figure 1.4. (A) Schematic diagram of the energy spectrum of electrons in single-layer graphene. (B) Tunneling probability of quasiparticle through a potential barrier as a function of incident angle for single-layer graphene. Reproduced with permission [32], copyright 2014, with permission from Elsevier.

be used. From the experiment, the angular-dependent transmission probability electron in graphene with a high barrier can be written as $T = \dfrac{\phi}{1 - \cos^2(q_x D)\sin^2\phi}$, where $q_x = \sqrt{(E - V)/(\hbar^2 v_F{}^2 - k_y{}^2)}$ is the wave vector in the boundary, and V is the height of the potential barrier. Under resonance condition $q_x D = N\pi$, where $N = 0,\ \pm1, \ldots$ $T = 1$, and the barrier become transparent for electrons, which is also valid when the electron approaches normal to the barrier [31].

(ii) **Ballistic transport and ambipolar electric field effect**
In graphene, the mean free path of electrons is 1 μm, which is much longer than in other materials, which means the scattering is much less; hence at room temperature these electrons can move faster than the electrons in many conductors. Electrons can get scattered by defects, impurities, thermal fluctuations and free-moving atoms. The ballistic transport in materials is the travel of charge carriers for long distances without impedance. In graphene, electrons experience zero mass at the Dirac points, making them faster and less scattered by defects in the crystal lattice. The conduction and valence band are symmetric, and electrons and holes as carriers can be tuned at high concentrations ($\approx 10^{13}$ cm^{-2}) with high mobility. Figure 1.5 shows the ambipolar electric field effect in graphene. The position of Fermi energy (E_F) in the conical low energy spectrum changes with gate voltage (V_g).

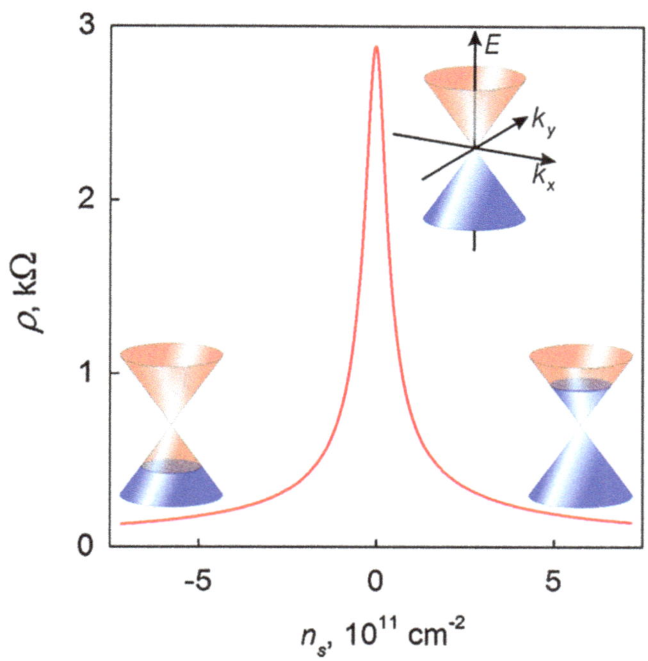

Figure 1.5. Ambipolar electric field effect in single-layer graphene. Reproduced with permission [33], copyright (2011) by the American Physical Society.

(iii) **Quantum Hall effect (QHE)**
The quantum Hall effect (QHE) is observed usually only in 2D metals at low temperatures and is unlikely to be observed at room temperature. However, graphene exhibits the QHE at room temperature due to the peculiar nature of its charge carriers, which behave as massless particles without scattering in ambient conditions. Although the charge carrier density is low near the Dirac points in graphene, it still exhibits some electrical conductivity of $4e^2/h$. Furthermore, the mobility of charge carriers remains independent of temperature between 10 K and 100 K [34]. While the QHE is observed in 2D materials at low temperatures and in a strong magnetic field, graphene exhibits the Hall effect at room temperature, figure 1.6. The QHE at a particular temperature can be attributed to the cyclotron gap $\hbar\omega_c$, which is a feature of graphene Dirac fermions. The Hall conductivity (σ_{xy}) at room temperature is an odd integer multiples of $\frac{2e^2}{h}$. When an external magnetic field is applied, the QHE exhibits an anomaly $\sigma_{xy} = \pm\frac{4e^2}{h}(n + \frac{1}{2})$ where 4 is the multiple factor due to double energy valley and double spin, and n indicates Landau energy [35].

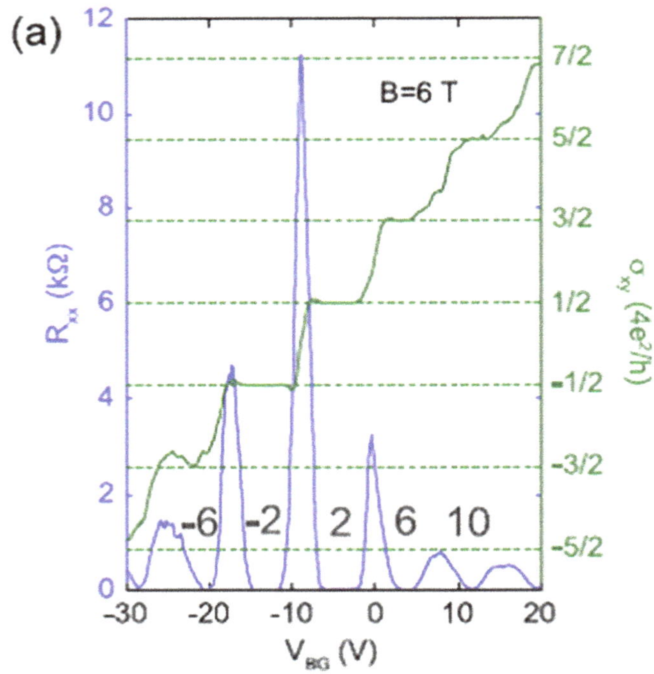

Figure 1.6. QHE of massless Dirac fermion. Reproduced from [35], copyright IOP Publishing Ltd. All rights reserved.

1.3.4 Ferromagnetism in graphene

Ferromagnetism in 2D material is both important and exciting, as such materials have the potential to serve as candidates for molecular-based magnets. Several theoretical studies have suggested that graphene with defects is an excellent candidate for the 2D ferromagnetic material. Furthermore, many researchers have reported the presence of ferromagnetism in highly oriented pyrolytic graphitic (HOPG) material [36, 37]. When HOPG was bombarded with protons, it showed ferromagnetism with strong ordering even at room temperature [38].

1.4 Pristine graphene

Single-sheet graphene and few-layer graphene can be prepared using several techniques. The first successful method for preparing single-layer graphene, also known as pristine graphene, involved 'peeling off' and epitaxial growth [14, 39]. More recently, Xiaolin Li *et al* developed graphite's exfoliation intercalation expansion to form a highly stable single-sheet graphene suspension in organic solvents [40]. These single graphene layer graphene sheets demonstrate excellent electrical conductivity at both room temperature and low temperatures. While electrically conductive graphene can be obtained through the chemical reduction of graphene oxide, its conductivity is significantly lower than that of pristine graphene due to the defects in the graphene oxide. Reduced graphene oxide is non-metallic in behaviour, whereas pristine graphene shows a nearly metallic behaviour [14].

1.5 Characterization of graphene

There are various methods for characterizing graphene to understand its size, number of layers, defects, and associated functional groups. Some of graphene's properties such as its electronic band structure depend on the number of layers and the quality of the layer deposited on the substrate. Therefore, it is crucial to properly characterize graphene before conducting any experiments to obtain reliable results.

1.5.1 Atomic force microscopy of graphene

AFM is a reliable technique for probing graphene and understanding the layer thickness and number of layers of graphene. This method could easily identify the layer thickness of 0.34 nm, and it comes within the detection limit of the instrument. The electronic and optical properties of graphene depend on the number of graphene layers [26, 41, 42]. In AFM, differential height measurement at the foldable edge can provide information about the layer thickness (~ 4Å) close to the monolayer thickness of graphene. Figure 1.7 shows the AFM imaging of graphene.

1.5.2 Raman spectroscopy of graphene

Raman spectroscopy is a powerful, non-destructive, and non-inversive technique that uses a laser to characterize the quality and defects of carbon material. The quality of graphene can be identified through this technique. Raman spectroscopy of

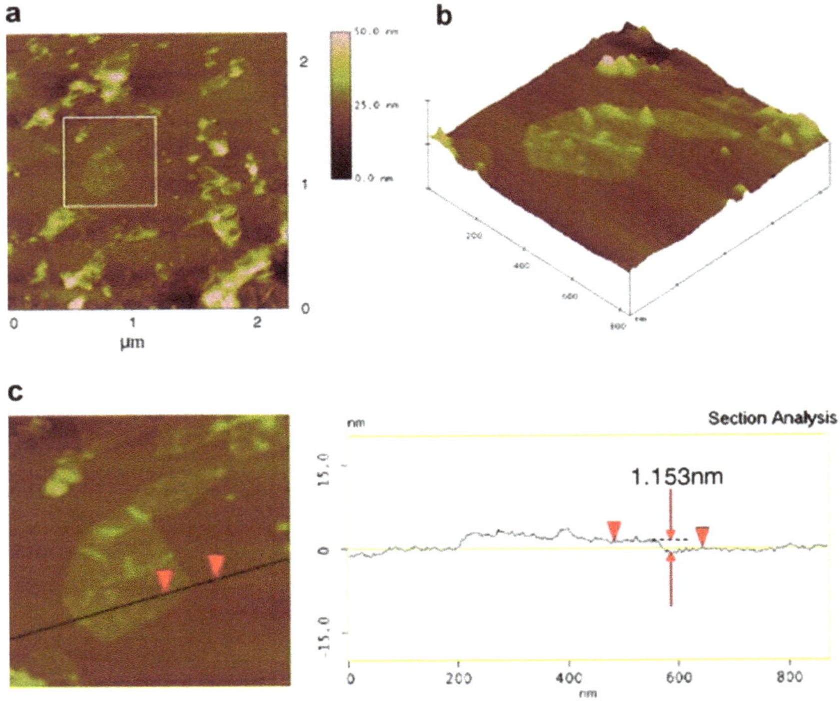

Figure 1.7. (a) AFM topography image of graphene; (b) 3D representation of the selected area in (a); and (c) line scan of the selected individual graphene. Reproduced from [43], copyright (2009), with permission from Elsevier.

graphene shows two distinct peaks called the G (~1580cm^{-1}) and D (~1350 cm^{-1}) bands, (figure 1.8). The G band is associated with the plain vibrational mode sp^2 hybridized carbon atoms. The D is associated with the radial breathing mode of the aromatic carbon. The D band is not present in the case of perfect crystalline defect-free graphene; however, it is dominant in defective graphene and functionalized graphene (graphene oxide).

The 2D band is the second order of the D band (overtone of the D band). The 2D band is a result of two phonon lattice vibration process; however, unlike the D band, it is not activated near the defects. Hence, the 2D band is always strong in graphene even when the D band is not present (figure 1.8), and this band does not represent defects. Additionally, the D and 2D band position depends on the laser excitation energy [44].

The D band and G band positions and shape of the peaks change with the number of graphene layers. The G band position is shifted to lower energy. Empirically, the band position and number of layers can be corrected using the following relation: $\omega_G = 1581.6 + 11/(1 + n^{1.6})$, where ω_G represents band position wavenumber and n the number of layers of graphene [45].

The ratio of intensities of the D and G bands, I_D/I_G, can be used to determine the level of defects in graphene [46].

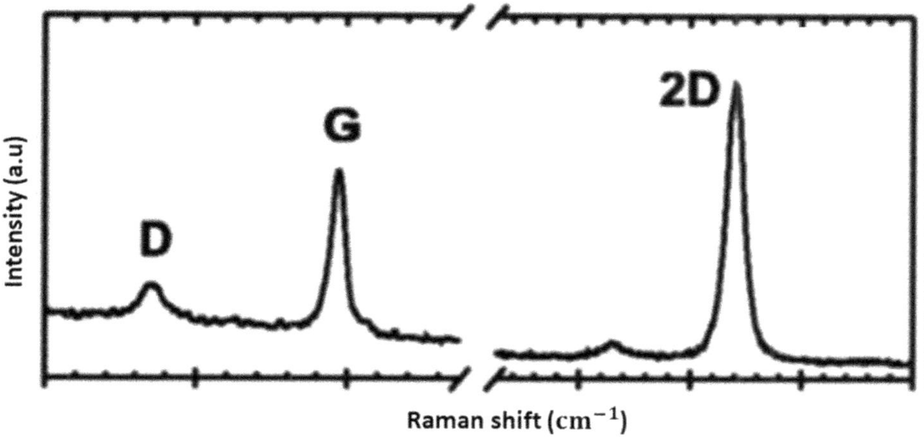

Figure 1.8. Raman spectrum of single-layer graphene. Reproduced from [35], copyright 2011, IOP Publishing Ltd. All rights reserved.

1.5.3 X-ray diffraction (XRD) of graphene

XRD is a characterization technique to determine the structure of a material. As graphene is a crystalline material, XRD is a valuable technique for determining the structure and measuring the lattice spacing between the atoms. In XRD, nano-crystalline and amorphous materials show a broad peak, while crystalline materials show a sharp peak. The XRD measurement of 2θ in the range of 5° and 70° shows (002) diffraction peak at $2\theta \sim 10°$ (graphene oxide) and $2\theta \sim 24°$ (reduced graphene oxide). For reduced graphene oxide (RGO), XRD analysis shows sharp diffraction peaks at 2θ ~24° from (002), indicating the *d* spacing between the graphene layers. In contrast, another peak from (100) at $2\theta \sim 43°$ indicates stacked graphene short-range orders. Using Bragg's equation, the average spacing between the graphene layers can be determined, (see figure 1.9(a))

Additionally, Scherrer's equation can be used to determine the average height of the stacking layer using the (002) reflection. The average diameter of the stacking layers can also be determined using Scherrer's equation and Warren's constant (1.84). Commercially available reduced graphene oxide (FL-RGOc) with 2–3 layers of nanostructures shows an average diameter of 8 nm, and height of 1 nm, while the graphene layer distance is 0.4 nm [47].

Figure 1.9(b) shows the XRD analysis of the pristine graphite conversion to graphene oxide. During oxidation using chemicals, the (200) peak (26.23°, *d* spacing 3.4 Å) disappears entirely after oxidation. The *d*-spacing of the graphene oxide is much larger than the single-layer graphene. This is because the presence of oxygen-containing functional groups attached to two sides of the graphene sheet leads to atomic scale surface roughness due to structural defects (sp^3 bonding) which were not presence in graphene sheets [48].

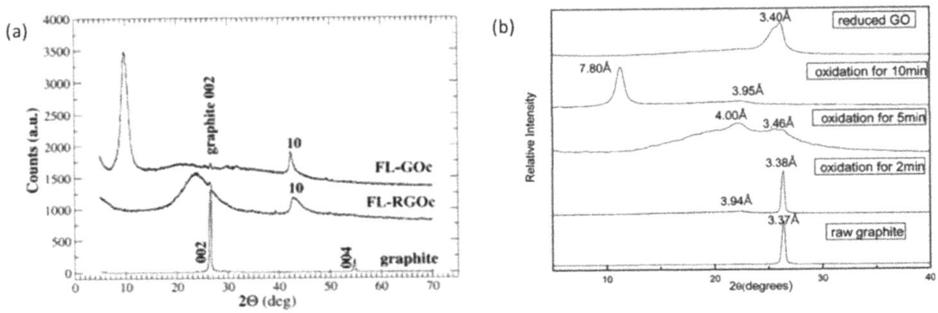

Figure 1.9. (a) XRD of FL-RGOc, FL-GOc and graphite. Reproduced from [47], copyright (2014), with permission from Elsevier. (b) XRD analysis of graphite, graphene oxide prepared after oxidation (2 min, 5 min and 10 min) and reduced graphene oxide. Reprinted with permission from [48]. Copyright (2009) American Chemical Society.

1.5.4 X-ray photoelectron spectroscopy (XPS) of graphene

X-ray photoelectron spectroscopy (XPS) is a versatile technique for surface elemental analysis, determination and quantification of hybridization. This technique can measure the percentage of functional groups associated with chemical compounds, especially with graphene and carbon materials. Stobinski *et al* characterized the commercially available few-layered reduced graphene (FL-RGOc) prepared through the modified Hummer method using XPS. FL-RGOc showed a carbon content (C) of ~ 73% and oxygen content (O) of ~26%. However, the C/O ratio of graphene depends on the preparation methods, purity of the sample and contamination on the surface. While graphite has a carbon content of ~96% and oxygen content of ~4%, FL-RGOp prepared showed a higher percentage of carbon content and lower oxygen content.

A detailed investigation of core level spectrum of C1s and O1s can provide an understanding of the percentage of functional groups associated with the graphene. Most common functional groups attached to graphene are hydroxyl (–OH), carbonyl (–COOH) and epoxy (–C=O) groups. The percentage of hybridization can also be determined from core level C1s spectrum. Generally, graphene shows a higher sp^2 hybridization compared to sp^3 hybridized carbon [47].

1.5.5 Fourier transform infrared analysis (FTIR) of graphene

Fourier transform infrared analysis (FTIR) technique is used to conduct a qualitative study on the functional groups associated with a material. This technique is used to identify the polymer molecules as it provides a fingerprint of such molecules. Graphene, graphene oxide, reduced graphene oxide, and graphite can be distinguished through this technique. This technique is used to investigate the presence of functional groups from the vibrational spectra obtained. Graphene oxide contains a high concentration of oxygen compared to reduced graphene oxide. However, graphite contains a minimal amount of oxygen. An FTIR peak at ~3400 cm^{-1} represents the hydroxyl group (-OH). Graphene oxide has a strong

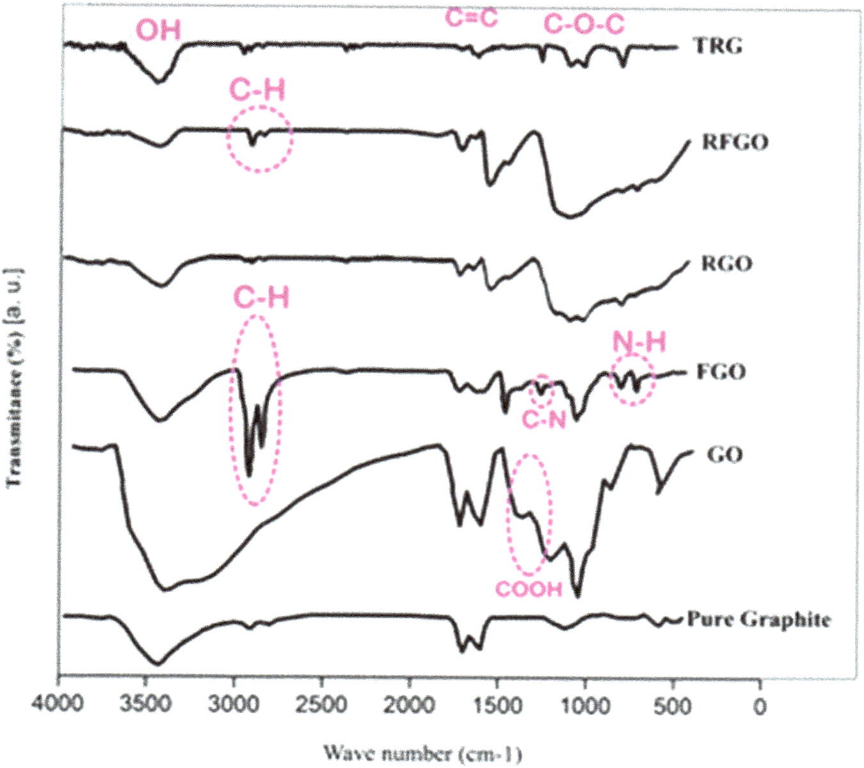

Figure 1.10. FTIR of graphene. Reproduced from [50] with permission from the Royal Society of Chemistry.

and broad peak at ~3400 cm^{-1} indicating a high hydroxyl group concentration; however, the peak intensity is low in the case of reduced graphene oxide and very low for graphite. Hence, FTIR can be used to determine the degree of oxidation in graphene and other functional groups such as carbonyl (–C=O), carboxyl group (–COOH) and epoxy group (C–O–C) associated with graphene and graphene oxide [49].

Figure 1.10 shows the thermally reduced graphene (TRG) with C–O–C functional group and –OH groups. Reduced functional graphene oxide (RFGO) and reduced graphene oxide (RGO) show similar functional groups. Moreover, functionalized graphene oxide (FGO) shows additional peaks of –NH and –CN and strong –CH peaks. Pure graphene shows only C=C peaks and small –OH concentrations of –OH functional groups. [50].

1.5.6 Electron microscopy of graphene (SEM, TEM, HRTEM)

Scanning electron microscopy (SEM) uses a high-energy focused electron beam that interacts with samples to reveal the surface morphology, chemical composition and elemental distribution of materials, making it a suitable technique for imaging microstructure of graphene. Wentian Gu *et al* investigated the microstructure of

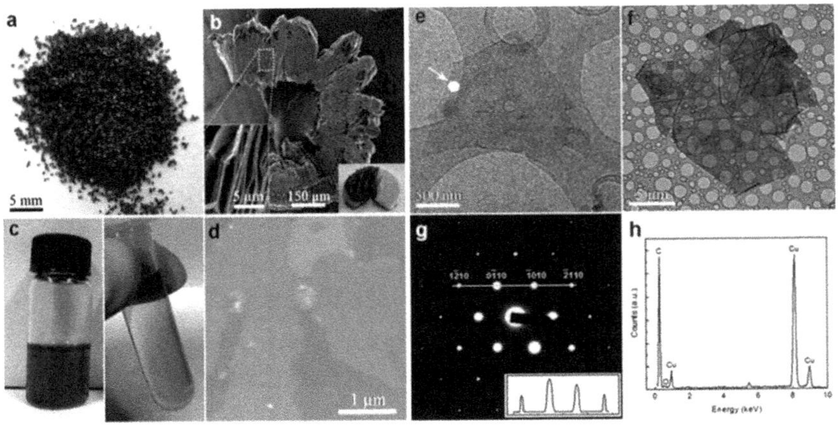

Figure 1.11. Graphene from liquid phase exfoliation of worm-like exfoliated graphite (WEG). (a) Photograph of WEG. (b) SEM image of WEG. (c) Uniform dispersions upon sonication (left) and centrifugation (right). (d) AFM image and (e) TEM image of the monolayer graphene sheets. (f) TEM image of a folded graphene sheet. (g) Electron diffraction from the white spot marked in (e). (h) EDX spectrum of the graphene sheet in (f). Reproduced from [51] with permission from the Royal Society of Chemistry.

high-quality single- and few-layer graphenes obtained through chemical treatment and thermal exfoliation of graphite. The SEM microanalysis exposed graphene morphology and layer delamination through the chemical process (figure 1.11) [51]. The intercalated graphene expanded to different layers since there is only weak Van der Waals interaction between the layers of graphene sheets.

Transmission electron microscopy (TEM) is another characterization technique used for graphene structure analysis. TEM studies can determine the number of graphene layers and the morphology of graphene. Studies using selected area electron diffraction (SAED) of the bright field image of graphene show that the graphene has a hexagonal lattice, figure 1.11. Moreover, the intensity ratio of planes and $(I_{[40]}/I_{\{2110\}})$ indicate whether the graphene is monolayer or multilayer. An intensity ratio greater than $(I_{\{1100\}}/I_{\{2110\}}) > 1$ indicates that the graphene is monolayer; while $(I_{\{1100\}}/I_{\{2110\}}) < 1$ indicates that the graphene is multilayer [51].

Energy dispersive x-ray (EDX) analysis is used to find the atomic percentage of different elements present in a substance. This method can be used to identify the contaminants on the graphene surface or the presence of small quantities of functional groups on graphene defects and edges [40]. For pure graphene, the only element present is carbon; however, the elements from the grid used for TEM studies may appear in the spectrum, but they can be subtracted from the spectrum to calculate the actual atomic percentage. Figure 1.12 shows the single-layer and multilayer graphene, where the darkeness increases with number of layers.

High-resolution transmission electron microscopy (HRTEM) and scanning tunnelling microscope (STM) can be used to image atoms in the lattice. Figure 1.13(c) shows the hexagonal carbon rings and lattice constant can be estimated through these techniques [51]. TEM STM studies can also provide information on the number and the type of defects in graphene layers. Through this method, the

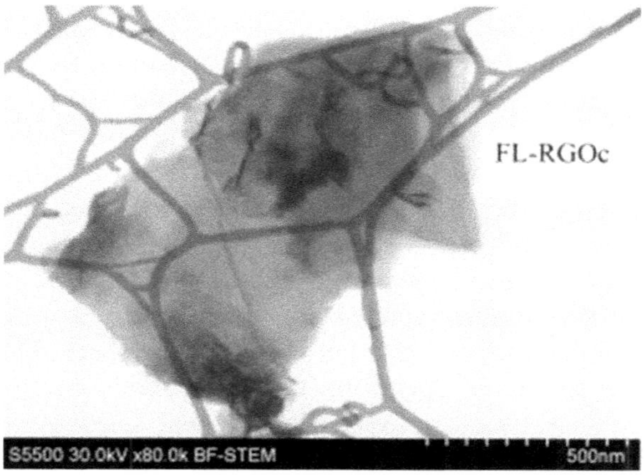

Figure 1.12. TEM analysis of FL-RGOc. Reproduced from [47], copyright (2014), with permission from Elsevier.

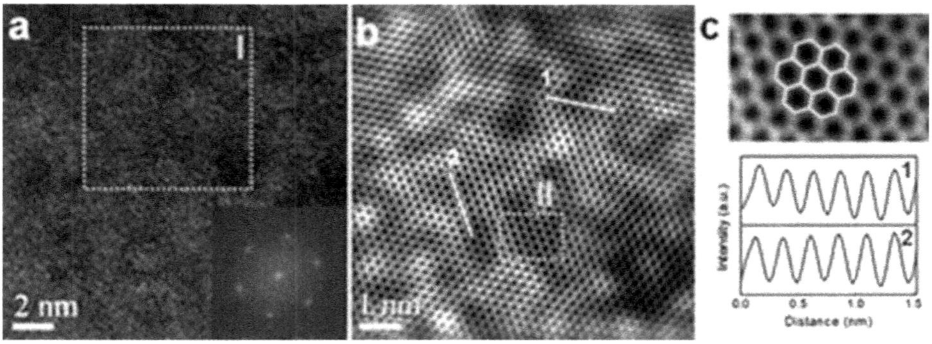

Figure 1.13. HRTEM image of (a) monolayer graphene and its FFT pattern from the selected area 1, (b) inverse FFT (region I) and (c) enlarged view of region II and line profile (1 and 2). Reproduced from [51] with permission from the Royal Society of Chemistry.

layer orientation of a few layers of graphene can be understood. In the TEM study, graphene shows a sheet-like structure with folding. It is transparent for a few graphene layers, but it can be darker where the layers are folded or the number of layers is more than a few [47].

1.6 Defects in graphene

There are two types of defects in graphene: point defect, which is zero-dimensional with vacancies or interstitial atoms, and one-dimensional defect that appears along a line. These defects are not stationary but can migrate along the graphene plane. Defect migration usually depends on the activation energy barrier, which is dependent on the defect type and increases exponentially with temperature. Some of the major types are defects in graphene are discussed below.

Stone–Wales defects appear in graphene when the graphene lattice is reconstructed to form a non-hexagonal ring. In this defect, there is no addition or removal of atoms, figure 1.14. Here four hexagons are converted into two pentagons and two heptagons through the rotation of one C–C bond by 90°.

Another type of defect is the single vacancy where one of the carbon atoms is missing from the lattice. This defect was observed in TEM experiments and led to the formation of five-membered and nine-membered distorted rings (figure 1.15).

A double vacancy is another vacancy created by the coalescence of two single vacancies or the removal of two neighbouring atoms. There are no dangling bonds present in the case of a double vacancy but there appear two pentagons and one octagon instead of four hexagons in a perfect graphene sheet. The graphene sheets are terminated with edges with edge atoms that are free or passivated with hydrogen

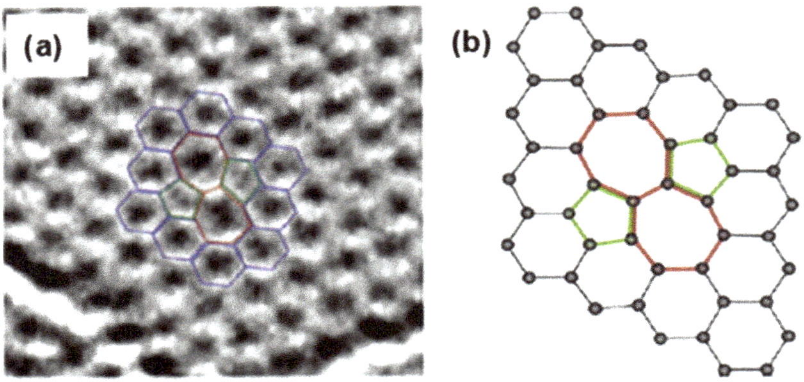

Figure 1.14. Stone–Wales defect formed by the 90° rotation of C–C bond. (a) Experimental TEM images of the defect, (b) the atomic arrangement obtained from density functional theory (DFT). Reproduced from [52], copyright (2015), with permission from Elsevier.

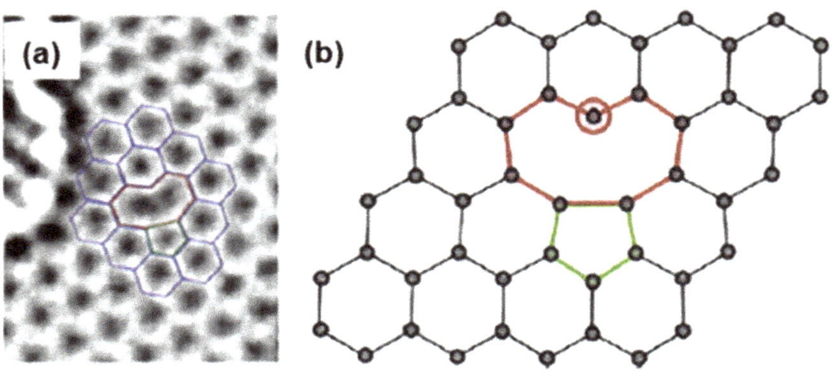

Figure 1.15. Single vacancy. (a) TEM experimental image and (b) atomic structure form DFT calculations. Reproduced from [52], copyright (2015), with permission from Elsevier.

atoms. The defects can appear at the edges of the graphene because of the changes in the local reconstruction.

1.7 Conclusions

Graphene is a wonder material with many unique mechanical, optical, and electronic properties. It is strictly a 2D material with peculiar electronic excitations described by Dirac fermions. It also exhibits a semimetal property with both zero density of states (semiconductor) and gapless conduction and valence bands (metallic nature). Also, the electron–electron interactions in graphene show a long mean free path. The graphene electronic states can be altered by doping pentavalent or trivalent impurity atoms. Moreover, graphene's physical and chemical properties can be manipulated by attaching functional groups to the graphene edges and defects. Many such graphene derivatives can be obtained by replacing carbon atoms and the suitable introduction of other atoms or functional groups. Due to its unusual properties, graphene can be used as a platform for many applications such as chemical sensors, strain sensors, and suitable material for electronic devices like a transistor. The abundance of precursor material (graphite) for graphene synthesis makes it a versatile, cheap material for many applications. The graphene and graphene-like (graphitic material) can literally be produced from any carbon source, making it attractive and sustainable for many industrial applications.

Through different characterisation techniques, graphene morphology, structure, functional groups and defects can be studied. The information about graphene's chemical and physical properties provides insight into graphene's behaviour and its usefulness in different applications.

References

[1] Speranza G and Laidani N 2004 Measurement of the relative abundance of sp^2 and sp^3 hybridised atoms in carbon based materials by XPS: a critical approach. Part I *Diam. Relat. Mater.* **13** 445–50

[2] Irifune T, Kurio A, Sakamoto S, Inoue T and Sumiya H 2003 Ultrahard polycrystalline diamond from graphite *Nature* **421** 599–600

[3] Falcao E H and Wudl F 2007 Carbon allotropes: beyond graphite and diamond *J. Chem. Technol. Biotechnol.* **82** 524–31

[4] Chehreh Chelgani S, Rudolph M, Kratzsch R, Sandmann D and Gutzmer J 2016 A review of graphite beneficiation techniques *Miner. Process. Extr. Metall. Rev.* **37** 58–68

[5] Geim A K and Novoselov K S 2009 The rise of graphene *Nanosci. Technol.* 11–19

[6] Golkarian A R and Jabbarzadeh M 2013 The density effect of Van der Waals forces on the elastic modules in graphite layers *Comput. Mater. Sci.* **74** 138–42

[7] Chen J, Duan M and Chen G 2012 Continuous mechanical exfoliation of graphene sheets via three-roll mill *J. Mater. Chem.* **22** 19625–28

[8] Zhang L, Li X, Huang Y, Ma Y, Wan X and Chen Y 2010 Controlled synthesis of few-layered graphene sheets on a large scale using chemical exfoliation *Carbon* **48** 2367–71

[9] Lee X J, Hiew B Y Z, Lai K C, Lee L Y, Gan S, Thangalazhy-Gopakumar S and Rigby S 2019 Review on graphene and its derivatives: synthesis methods and potential industrial implementation *J. Taiwan Inst. Chem. Eng.* **98** 163–80

[10] Chua C K and Pumera M 2014 Chemical reduction of graphene oxide: a synthetic chemistry viewpoint *Chem. Soc. Rev.* **43** 291–312

[11] Güler Ö, Güler S H, Selen V, Albayrak M G and Evin E 2016 Production of graphene layer by liquid-phase exfoliation with low sonication power and sonication time from synthesized expanded graphite *Fuller. Nanotub. Carbon Nanostructures* **24** 123–7

[12] Wu X, Liu Y, Yang H and Shi Z 2016 Large-scale synthesis of high-quality graphene sheets by an improved alternating current arc-discharge method *RSC Adv.* **6** 93119–24

[13] Geim A K 2011 Nobel lecture: random walk to graphene *Rev. Mod. Phys.* **83** 851

[14] Novoselov K S, Geim A K, Morozov S V, Jiang D E, Zhang Y, Dubonos S V, Grigorieva I V and Firsov A A 2004 Electric field effect in atomically thin carbon films *Science* **306** 666–9

[15] Lee C, Wei X, Kysar J W and Hone J 2008 Measurement of the elastic properties and intrinsic strength of monolayer graphene *Science* **321** 385–8

[16] Li X, Sun M, Shan C, Chen Q and Wei X 2018 Mechanical properties of 2D materials studied by *in situ* microscopy techniques *Adv. Mater. Interfaces* **5** 1701246

[17] Dresselhaus M S, Dresselhaus G and Jorio A 2004 Unusual properties and structure of carbon nanotubes *Annu. Rev. Mater. Res.* **34** 247–78

[18] Smalley R E and Yakobson B I 1998 The future of the fullerenes *Solid State Commun.* **107** 597–606

[19] Tiwari S K, Kumar V, Huczko A, Oraon R, Adhikari A D and Nayak G C 2016 Magical allotropes of carbon: prospects and applications *Crit. Rev. Solid State Mater. Sci.* **41** 257–317

[20] Chung D D L 2002 Review graphite *J. Mater. Sci.* **37** 1475–89

[21] Mezzi A and Kaciulis S 2010 Surface investigation of carbon films: from diamond to graphite *Surf. Interface Anal.* **42** 1082–4

[22] Shames A I, Mogilyansky D, Panich A M, Sergeev N A, Olszewski M, Boudou J P and Osipov V Y 2015 XRD, NMR, and EPR study of polycrystalline micro-and nano-diamonds prepared by a shock wave compression method *Phys. Status Solidi* A **212** 2400–9

[23] Li Z Z, Wang J T, Mizuseki H and Chen C 2018 Computational discovery of a new rhombohedral diamond phase *Phys. Rev.* B **98** 094107

[24] Neto A C, Guinea F, Peres N M, Novoselov K S and Geim A K 2009 The electronic properties of graphene *Rev. Mod. Phys.* **81** 109

[25] Miró P, Audiffred M and Heine T 2014 An atlas of two-dimensional materials *Chem. Soc. Rev.* **43** 6537–54

[26] Nair R R, Blake P, Grigorenko A N, Novoselov K S, Booth T J, Stauber T, Peres N M and Geim A K 2008 Fine structure constant defines visual transparency of graphene *Science* **320** 1308–8

[27] Ferralis N 2010 Probing mechanical properties of graphene with Raman spectroscopy *J. Mater. Sci.* **45** 5135–49

[28] Zandiatashbar A, Lee G H, An S J, Lee S, Mathew N, Terrones M, Hayashi T, Picu C R, Hone J and Koratkar N 2014 Effect of defects on the intrinsic strength and stiffness of graphene *Nat. Commun.* **5** 1–9

[29] Zhang P *et al* 2014 Fracture toughness of graphene *Nat. Commun.* **5** 1–7

[30] Jung G, Qin Z and Buehler M J 2015 Molecular mechanics of polycrystalline graphene with enhanced fracture toughness *Extreme Mech. Lett.* **2** 52–9

[31] Katsnelson M I, Novoselov K S and Geim A K 2006 Chiral tunnelling and the Klein paradox in graphene *Nat. Phys.* **2** 620–5

[32] Wang J, Ma F, Liang W and Sun M 2017 Electrical properties and applications of graphene, hexagonal boron nitride (h-BN), and graphene/h-BN heterostructures *Mater. Today Phys.* **2** 6–34

[33] Novoselov K S 2011 Nobel lecture: graphene: materials in the flatland *Rev. Mod. Phys.* **83** 837

[34] Novoselov K S, Geim A K, Morozov S V, Jiang D, Katsnelson M I, Grigorieva I, Dubonos S and Firsov A 2005 Two-dimensional gas of massless Dirac fermions in graphene *Nature* **438** 197–200

[35] Molitor F, Güttinger J, Stampfer C, Dröscher S, Jacobsen A, Ihn T and Ensslin K 2011 Electronic properties of graphene nanostructures *J. Phys.: Condens. Matter* **23** 243201

[36] Kopelevich Y and Esquinazi P 2006 Ferromagnetism and superconductivity in carbon-based systems arXiv preprint cond-mat/0609497

[37] Ávila M, Venosta L, Bajales N and Bercoff P 2015 Structural and magnetic changes induced by electron and ion irradiation on HOPG *Procedia Mater. Sci.* **9** 62–8

[38] Esquinazi P, Spemann D, Höhne R, Setzer A, Han K H and Butz T 2003 Induced magnetic ordering by proton irradiation in graphite *Phys. Rev. Lett.* **91** 227201

[39] Berger C *et al* 2006 Electronic confinement and coherence in patterned epitaxial graphene *Science* **312** 1191–6

[40] Li X, Zhang G, Bai X, Sun X, Wang X, Wang E and Dai H 2008 Highly conducting graphene sheets and Langmuir–Blodgett films *Nat. Nanotechnol.* **3** 538–42

[41] Falkovsky L A 2008 Optical properties of graphene *J. Phys.: Conf. Ser.* **129** 012004

[42] Nilsson J, Neto A C, Guinea F and Peres N M R 2008 Electronic properties of bilayer and multilayer graphene *Phys. Rev.* B **78** 045405

[43] Geng Y, Wang S J and Kim J K 2009 Preparation of graphite nanoplatelets and graphene sheets *J. Colloid Interface Sci.* **336** 592–8

[44] Malard L M, Pimenta M A, Dresselhaus G and Dresselhaus M S 2009 Raman spectroscopy in graphene *Phys. Rep.* **473** 51–87

[45] Wang H, Wang Y, Cao X, Feng M and Lan G 2009 Vibrational properties of graphene and graphene layers *J. Raman Spectrosc.* **40** 1791–6

[46] Jorio A, Ferreira E H M, Moutinho M V, Stavale F, Achete C A and Capaz R B 2010 Measuring disorder in graphene with the G and D bands *Phys. Status Solidi* B **247** 2980–2

[47] Stobinski L, Lesiak B, Malolepszy A, Mazurkiewicz M, Mierzwa B, Zemek J, Jiricek P and Bieloshapka I 2014 Graphene oxide and reduced graphene oxide studied by the XRD, TEM and electron spectroscopy methods *J. Electron. Spectrosc. Relat. Phenom.* **195** 145–54

[48] Shen J, Hu Y, Shi M, Lu X, Qin C, Li C and Ye M 2009 Fast and facile preparation of graphene oxide and reduced graphene oxide nanoplatelets *Chem. Mater.* **21** 3514–20

[49] Hidayah N M S, Liu W W, Lai C W, Noriman N Z, Khe C S, Hashim U and Lee H C 2017 Comparison on graphite, graphene oxide and reduced graphene oxide: synthesis and characterization *AIP Conf. Proc.* **1892** 150002

[50] Dehghanzad B, Aghjeh M K R, Rafeie O, Tavakoli A and Oskooie A J 2016 Synthesis and characterization of graphene and functionalized graphene via chemical and thermal treatment methods *RSC Adv.* **6** 3578–85

[51] Gu W *et al* 2009 Graphene sheets from worm-like exfoliated graphite *J. Mater. Chem.* **19** 3367–9

[52] Liu L, Qing M, Wang Y and Chen S 2015 Defects in graphene: generation, healing, and their effects on the properties of graphene: a review *J. Mater. Sci. Technol.* **31** 599–606

IOP Publishing

Recent Advances in Graphene and Graphene-Based Technologies

Anoop Chandran, N V Unnikrishnan, M K Jayaraj, Reenu Elizabeth John and Justin George

Chapter 2

Synthesis methods of graphene

Reenu Elizabeth John, N V Unnikrishnan, Steffy Sara Varghese, O Lekshmi and Anoop Chandran

Due to their remarkable physicochemical properties, graphene and its derivatives have gained significant attention from the research community since the discovery of the former in 2004. In this chapter, various top-down and bottom-up synthesis methods of graphene have been discussed with a special focus on the latest graphene bulk-production method—the flash Joule heating (FJH). The current difficulties associated with graphene bulk-production and how FJH can be a possible remedy for them in a not-too-distant future have also been discussed.

2.1 Introduction

The world we live in is jam-packed with a variety of materials from which many serve as the backbone of modern society. Among them, carbon and carbon-based materials play a key role in growing the world economy today as they have triggered distinctive development in materials science and engineering. The ability of carbon to exist in various allotropic forms enhances its potential for advanced technological applications. The unique catenation property of carbon atoms enables them to exhibit contradictory traits like softness as in graphite and hardness as in diamond which is rare in non-carbon materials. Out of the multifarious allotropic forms, the discovery of graphene in 2004 paved the path to a major leap in the field of material science, thereby stirring up advanced scientific research for sophisticated applications and endless possibilities.

According to International Organization for Standards (ISO), graphene is a single layer of carbon atoms, with each atom bound to three neighbors in a honeycomb structure. In other words, it is a two-dimensional (2D) sheet of covalently bonded carbon atoms in sp^3 hybridization which maintains a hexagonal or honeycomb structure. However, other terminologies also exist both academically and

doi:10.1088/978-0-7503-3999-5ch2

industrially. Some suggest that ten or fewer layers can still be considered graphene and ISO define such structures with 3 to 10 layers as ‘few-layer graphene’ [1]. Each monolayer of graphene sheets is held together by weak van der Waals force of attraction. Graphene is not only noted for its thinness, but also it is about 200 times stronger than steel and a good conductor of electricity compared to any other known materials at room temperature. The unsurpassed and excellent properties of graphene like high carrier mobility (~200 000 cm^2 V^{-1} s^{-1}), outstanding thermal conductivity (~5000 W m^{-1} K^{-1}), large surface area (2630 m^2 g^{-1}), exceptional transparency toward visible light (~97.7%), ambipolar electric field effect and room temperature quantum Hall effect, has sparked a graphene ‘gold rush’ in almost all fields of science and technology. These remarkable features of graphene can be utilized in electronic applications such as printed circuits, microchips, transistors, and flexible electronics [2–5]. Furthermore, the high strength and low thickness of graphene are increasingly being used in the research and development of bioelectric sensors used to monitor glucose levels and blood components like hemoglobin, cholesterol etc [6–9]. It is also being researched for its uses in gene and small molecular drug delivery, dental implants, therapeutic tools, and even anti-cancer therapy applications [10–14]. The immense scope of graphene-based technologies is obvious from its market studies, which suggest that the global market for graphene has accounted for USD 71.0 Million in 2020 and is expected to reach USD 785.2 Million by 2028, with a compound annual growth of 35% [15]. It seems the COVID-19 pandemic has somewhat boosted the graphene market because of the usage of graphene in the coating of facemasks, the sales of which skyrocketed during the pandemic period and still continue to be high. A market study published in 2021 analyzed the impact of COVID-19 on the graphene market. It predicts an estimated market value of $6.2 billion in 2021 and a projected $7.7 billion in 2026 [16]. However, it would be interesting to see if this trend would sustain in the global graphene market in the post-COVID years.

Even though naturally occurring graphite has been known for centuries, the first reported method for the production of graphene can be traced back to 1970 and high quality single-layer graphene was produced in 2004 by micromechanical cleavage (Scotch Tape method) of graphite [17]. The delay that led to the discovery of graphene is mainly attributed to its monolayer nature and fallacies related to its thermodynamic stability [18]. However, immediately after its discovery, the research on graphene, including the control of the graphene layers on substrates, functionalizing graphene and exploring the applications of graphene grew exponentially. Governments all over the world have spent billions for graphene-based research mainly in establishing research laboratories and aiding fundamental research. Also, there are large consortia such as the graphene flagship [19] which provides long-term funding of €1 billion to commercialize graphene products out in the market and an additional amount of €20 million to support sufficient research in graphene-based electronics, optoelectronics and sensors applications. Despite these, real commercialization of graphene is miles away from an overnight success story as it is still primarily hidden in research laboratories. Commercialization of graphene is heavily reliant on its quality, cost, reproducibility, processability and safety of the method adopted. Nevertheless, the key requirement would be the large-scale fabrication of high quality graphene [20, 21]. A review of the literature on graphene would clearly imply that in most of the reports, the quantity of graphene synthesized is only in the milligram scale or below, despite the good quality. Furthermore, critical quality issues might arise when taking graphene from the lab scale to industrial scale, addressing which is vital in the path towards commercialization of graphene. Hitherto, the mass production of graphene suffers a technological trade-off dilemma between quality and production rate. Setting-up more industrial laboratories which are commercially focused would aid overcoming this hurdle as their R&D would be more inclined towards obtaining a practical solution to this problem compared to laboratories in academic institutions.

Current popular industrial synthesis methods of graphene can be broadly classified into two categories: top-down methods and bottom-up methods [22]. Top-down method is generally a destructive technique in which a bulk material is broken down into fragments. Individual graphene layers are delaminated from pyrolytic graphite by overcoming van der Waals force of attraction between the layers. At the industrial level, only the top-down methods are used for the large-scale synthesis of graphene. The principal advantage of this approach is that it facilitates possibilities of scaling up, requires no substrate-transfer and is fairly a cost effective method. However, they have some limitations in that it is arduous to separate out the graphene layers efficiently and strenuous to prevent the restacking of graphene sheets after exfoliation [23]. The ineffectiveness to precisely configure graphene is also another setback of this scheme. Alternatively, bottom-up fabrication involves the building up of the required materials from smaller entities. In these methods, synthesis of graphene involves the implementation of carbon precursors as a building block to form graphene layers. The major benefit of this synthesis route is that it enables precise fabrication and fine tuning of graphene configuration, thereby offering the most tangible solution for the fabrication of next generation graphene devices. Most of these methods are used in industries for the development of continuous and large-scale graphene films which are used in applications such as solar cells and sensors [24–26]. Yet, their high cost, requirement of high temperatures and the difficulty of scaling up compared to the top-down approach remains a challenge for industries. Some popular top-down and bottom-up graphene synthesis methods are shown in figure 2.1. Figure 2.2 suggests that the research community has been promoting the molecular build up model for graphene synthesis. Conspicuously, the figure also suggests that among the bottom-up methods, chemical vapour deposition (CVD) is the most popular one, while exfoliation and chemical synthesis remain the favorite top-down methods for graphene production. The techno-economic viability of these synthesis routes is valued based on their proficiency to produce graphene with low cost, quality, suitability, and sustainability. In this chapter a thorough discussion on some popular top-down and bottom-up methods for graphene production is included. The methodology along with merits, demerits and future directions are presented in detail.

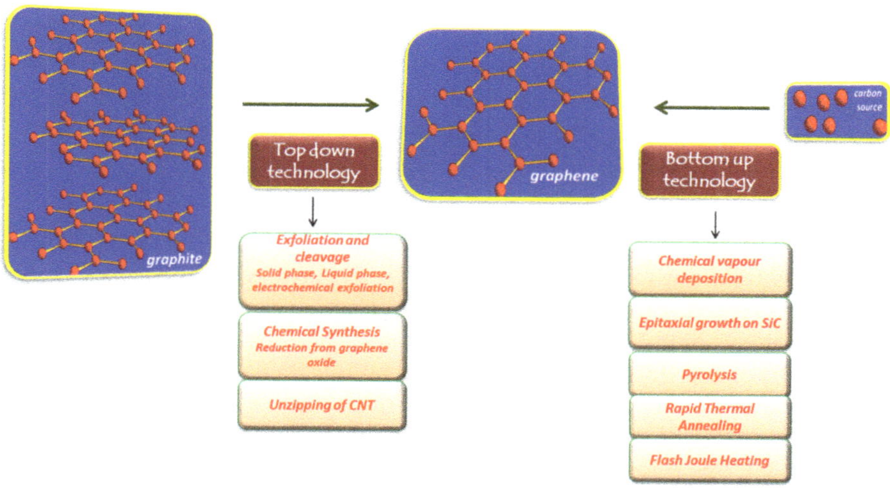

Figure 2.1. Illustration of the classification of various fabrication techniques developed for the production of graphene.

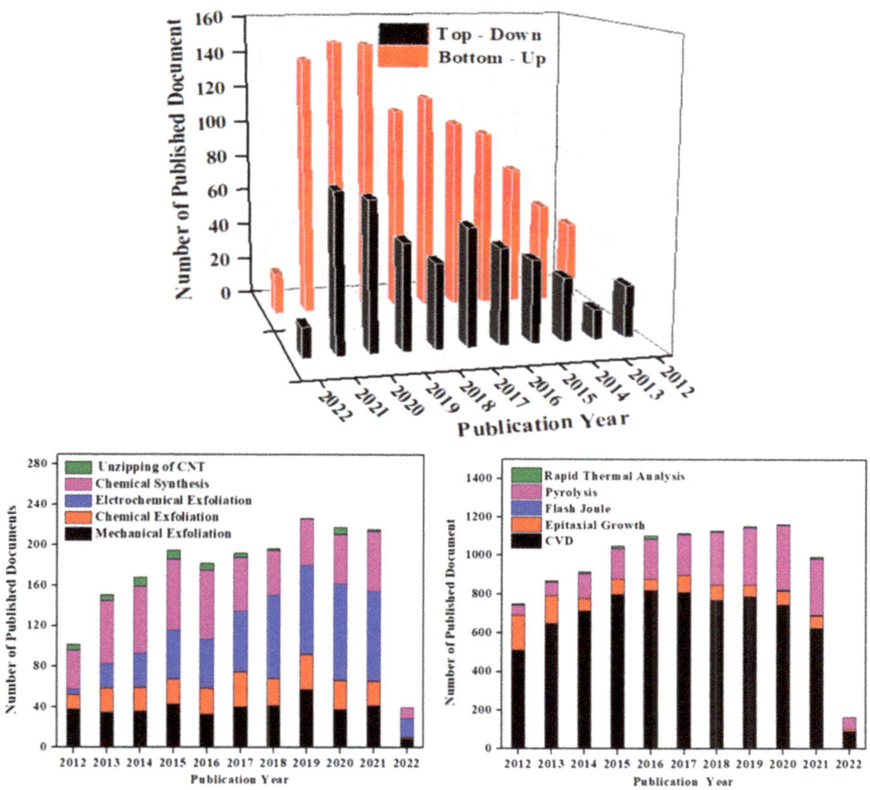

Figure 2.2. An overview of number of publications in graphene synthesis with reference to top-down and bottom-up categories for the past 10 years. Data collected in February 2022 from the Scopus database.

2.2 Top-down approach

Among the various top-down methods used for industrial and laboratory production of graphene, the most widely used are: (1) exfoliation and cleavage; (2) reduction from graphene oxide (GO); and (3) unzipping of carbon nanotubes (CNTs). They are discussed as follows.

2.2.1 Exfoliation and cleavage

Mechanical exfoliation marks the beginning in the history of graphene fabrication [27]. In 2004 Andrie Geim and Konstantin Novoselove through their 'groundbreaking experiments' were the first to propose the cleavage of pyrolytic graphite to isolate single-layer graphene. In this method, traditionally called 'Scotch Tape method', a Scotch Tape was used to repeatedly peel-off layers from graphite crystal to obtain graphene flakes. These peeled off layers were then deposited on silicon dioxide/silicon (SiO_2/Si) substrates. The graphene flakes acquired through this procedure were bestowed with versatile electrical, optical and thermal conductivity properties capable of opening myriad opportunities in sophisticated applications. On the flip side, even when graphene obtained through this technique is considered pristine and of high quality, it can only be implemented for fundamental research as a consequence of its moderate yield. Despite the thinnest flake produced being ~10 nm thick, this technique failed to yield single- or bilayer graphene. Considering the shortcomings of the Scotch Tape method, Huang *et al* improvised this method by adopting two additional steps prior to exfoliation: oxygen plasma cleaning of the substrate and an additional heat treatment [28]. As a result, they were successful in producing graphene flakes of thickness 1–4 layers. Beyond that, these simple steps were effective in augmenting the yield and the area of the flakes by more than 50 times compared to the established exfoliation methods. Inspired by the traditional Scotch Tape method, researchers developed new mechanically exfoliating techniques like micromechanical cleavage [29] and ball milling [30]. A micro-cleavaging technique was introduced by Jayasena *et al* where an ultra-sharp wedge shaped crystal diamond was used to cleave off a few layers of graphene flakes from graphite [31]. In this method, an understanding of the wedge speed and depth of insertion enabled precise tuning and controlling of the exfoliation technique for the production of flat or folded graphene layer. Likewise, ball milling is also a technique employed to yield few-layer graphene through a low cost and environmentally friendly conditions. Here a shear force is applied to laterally exfoliate graphite to graphite flakes. However, the vertical forces and collisions applied by the balls tend to deteriorate the graphite flakes, thereby depreciating the quality of graphene layers. To favour the shearing forces exfoliating agents can be used to offer multipoint interaction sites for the graphene flakes [32]. Melamine is considered as a good exfoliated agent since it has an aromatic nucleus that can efficiently interact with the graphene π-system and can form extended networks on its surface due to the presence of hydrogen bonds. Furthermore, melamine can also be easily washed away leaving behind pure graphene flakes [33]. In another work, Deepak *et al* introduced bovine serum albumin protein as an exfoliating agent during the ball milling procedure. The protein acted not only as an effective exfoliation agent, but it also prevented restacking of the graphene layers [34]. Although the research community has made tremendous progress in this field, mechanical exfoliation methods face a lot of issues demanding a systematic in-depth study. All the aforementioned methods fail to provide high yield and defect-free graphene flakes. Furthermore, exfoliation techniques are slow and imprecise, which limits their use for research studies rather than for commercial applications

Chemical exfoliation is another advancing method employed for fabrication of graphene. Here, alkali metals are intercalated into the graphite structure to increase the interplanar spacing, thereby facilitating easy peel-off of the layers. The intercalated products can be easily exfoliated using sonication process. This solution-based process can produce large-scale exfoliation at low ambient temperature, thus making it distinctive among other graphene fabrication processes. The inclusion of chemical contamination is one of the serious downsides of this process. Recently, Wu *et al* developed an environmentally friendly fast chemical exfoliation method to produce graphene with high quality, large size and fewer layer features [35]. The obtained graphene film exhibited excellent electrical conductivity of up to 2.03×10^5 S m^{-1} which is one of the highest ones reported so far for graphene. Mass production of graphene sheets was enabled using liquid phase exfoliation (LPE) where graphite dispersed in stabilizing liquids can be exfoliated using sonication [36]. In selecting a solvent, it is necessary to minimise the interfacial tension between the graphite and the liquid in order to curtail the aggregation of single-layer graphene. Microfluidizer method is an emerging LPE technology, which can exfoliate graphite successfully using high-shear mixer-driven fluid dynamics. Recently, Shang *et al* introduced a combination of microfluidizer and supercritical CO_2 to yield graphene nanosheets of less than three-layer thickness [37].

Among the different exfoliation methods, electrochemical exfoliation is an economical and eco-friendly fabrication technique for the production of high quality graphene sheets. It opens up the opportunity for bulk production of graphene from graphite. This exfoliation procedure employs an applied voltage (5–35 V) to drive the ionic species in the electrolyte to get intercalated between the layers of graphite electrode [38]. The intercalated ionic species weakens the van der Waals forces between the layers, thereby providing an easy path for layer separation. For example, for conventionally used H_2SO_4 electrolyte, the SO_4^{2-} ions of ionic radii 0.46 nm can easily be intercalated into graphite layers because of their ionic size, which is comparable to graphite *d* spacing (0.34 nm). Based on the type of potential employed, electrochemical synthesis can be categorized into two; anodic and cathodic exfoliation of graphite. In the anodic type the negative ions move to the graphite anode on applying positive voltage, whereas in cathodic type of exfoliation, positively charged species move toward the graphite cathode on the application of the negative voltages.

A major issue centred with these exfoliation techniques is the high electrolysis voltage used during the process, which leads to inadequate intercalation and subsequently low yield of few-layered graphene. In order to overcome these drawbacks researchers have upgraded the electrochemical exfoliation method by introducing modified electrolytes and additives [39]. Liu *et al* and his co-workers introduced ultrasound during the exfoliation technique in Na_2SO_4 solution. The ultrasound provided extra mechanical forces aiding effective exfoliation [40]. In another work, Chen *et al* demonstrated an electrochemical exfoliation of graphite powder by using melamine as an electrolyte additive. The inclusion of melamine increased the yield of graphene up to 25% compared to the exfoliation with any other additives. In yet another work, a novel non-electrified electrochemical exfoliating method was introduced to fabricate high quality graphene sheets [41]. In this report 1 M $LiPF_6$/propylene carbonate (PC) was used as the electrolyte, which effectively exfoliated the graphite electrode into sheets of graphene. The intercalation of Li^+–$(PC)_4$ into the layers led to the formation of Li||graphite micro-cells, which helped in the separation of graphene layers without the aid of any applied electrical voltage. The prepared graphene layers were of high quality with a high C/O ratio (27.74) and excellent electronic conductivity of 102.5 S cm^{-1}. An undesirable consequence of these synthesis methods is that the graphite pieces that have been prematurely exfoliated, cause the quality of graphene sheets to deteriorate. To overcome this hurdle, Wang *et al* designed a novel electrochemical exfoliation technique to synthesize graphene films in which the graphite was placed in a confined space, thereby preventing untimely falling of graphene sheets [42]. For this purpose the working electrode (graphite foil) was coated with paraffin and the bottom part was exposed to the electrolyte. Nickel was used as the counter electrode and a dc supply of 5 V was applied in NaOH electrolyte. The transparent conductive graphene films thus prepared exhibited excellent transparency and conductivity with the potential to replace the conventionally used indium tin oxide. Achee *et al* modified the above method by utilising compressed graphitic flakes inside a permeable container as the electrode [43]. The prime advantage of this technique is that a wide range of graphitic material can be used as the electrode, thereby replacing the conventionally required graphite powder for electrode purpose. Moreover, in this work the authors proposed a design method for graphene reactors promising graphene production with high yield and quantity. Figure 2.3 depicts pictorial representation of exfoliation of graphene by both intercalation and mechanical method. Recently, an ultrafast chemical free exfoliation methods was adopted by Islam *et al* [44]. In their work, instead of using intercalants or chemical, they utilised plasma spray exfoliation technique which resulted in a high production rate of graphene (48 g h^{-1}) with high single-layer selectivity (85%). The exfoliated graphene exhibited almost no basal defects and had no structural distortion. The obtained sheets can be used in applications like frictionless, transparent conductive coatings, mechanical reinforcements and energy storage devices.

While exfoliation technique is very popular in the graphene industry, it has many disadvantages. A major disadvantage of solvent based exfoliation is that depending on the surface tension of the solvent used, the ions usually get trapped in between the graphene sheets adversely affecting the purity of the sheets. Moreover, the solvents used are not harmful to the human body and to the environment. On the other hand, the mechanical method has no real control over the number of graphene layers.

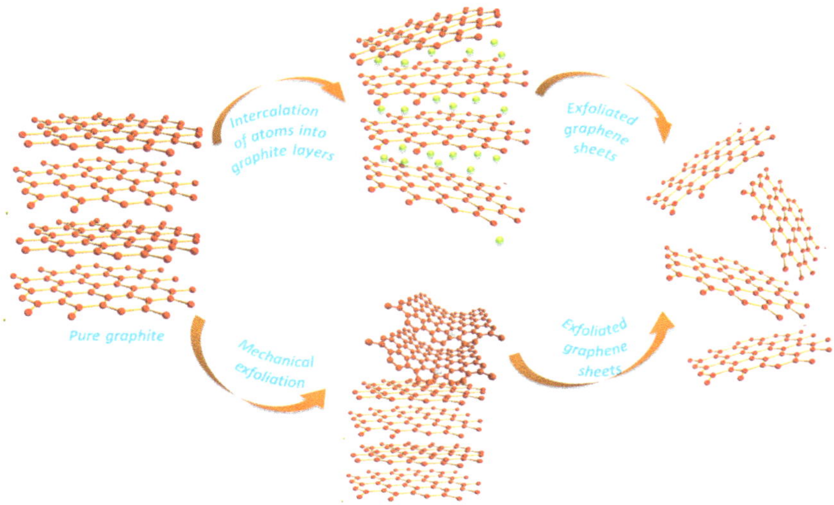

Figure 2.3. Separation of graphene layer from graphite using intercalation technique (upper half) and by mechanical cleavage (lower half).

Therefore, improvisations have to be adopted in exfoliation techniques so as to control the number of graphene layers and minimize impurity levels in the graphene.

2.2.2 Chemical synthesis: reduction from graphene oxide

While mechanical exfoliation works on weakening the van der Waals forces between the graphite layers to structure graphene layers, a chemical route can also be used based on the same principle. Large-scale production of graphene by the reduction of GO is a cost effective chemical synthesis method. Chemical reduction of graphite oxide using reducing agents is one of the best techniques to obtain a large quantity of graphene sheets (figure 2.4). Instead of pristine graphene sheets, the end product obtained by this method is reduced GO or modified graphene. Generally, the principle involved is the intercalation of ions within the graphitic layers leading to the expansion of layers with their simultaneous oxidation by strong oxidising agent like concentrated sulfuric acid or nitric acid. Similar to graphene, GO is also a potential material for various industrial and electronic applications. The unusual thermal, optical, and electrical properties of GO, make it suitable for applications like photovoltaic cell, capacitors and sensors. First and foremost, GO can be synthesised from the three well acclaimed methods—Brodie, Hummers and Staudenmaier methods [45]. In Brodie's oxidation method, graphite is oxidised using concentrated nitric acid (HNO_3) and potassium chlorate ($KClO_3$). The resulting solid end product contains carbon, hydrogen and oxygen. The grave drawback of this method is that it is time consuming, tedious and expels several toxic gases. The upgraded version of Brodie's method is the Staudenmaier method, where concentrated sulfuric acid (H_2SO_4) is used as an extra additive apart from HNO_3 and $KClO_3$. The presence of H_2SO_4 enhances the rate of oxidation, thereby

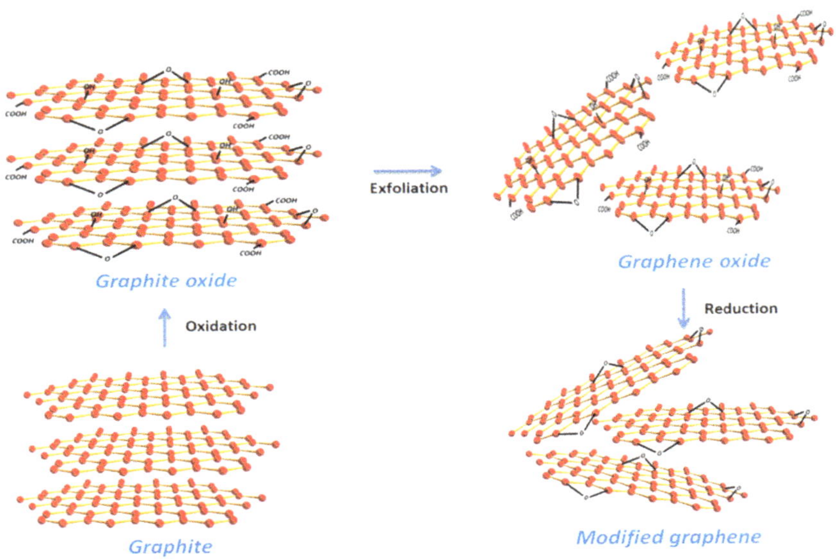

Figure 2.4. Portrayal of the formation of graphene from graphite powders.

expediting the formation of graphite oxide. Though this method produces high quality graphite oxide, it suffers similar detriments of the original Brodie method. Later, in 1958, Hummers developed an alternative technique to obtain graphite oxide, wherein graphite, potassium permanganate, sodium nitrate and concentrated sulfuric acid were used as the oxidizing mixture. Currently, Hummers method and its modified versions are widely accepted to achieve highly oxidized graphite oxide to serve as an efficient precursor for graphene [46]. The value of this method is that it is environmentally sound, since it emits no toxic gases, and it is viable to harvest graphene in large volume for commercial utilization.

By the proper elimination of oxygen functional group from GO structure, GO can be appropriately used as a precursor for graphene synthesis. The three main approaches performed to accomplish this task include thermal, electrochemical and chemical methods. Thermal reduction involves heating of GO to about 1000 °C which generates gaseous species that intercalate onto the GO structure [47]. This intercalation paves a path to an increase in the internal pressure, thereby aiding in exfoliation and eliminates the oxygen function group. Annealing temperature and annealing atmosphere are the important parameters in this method. Studies indicate that even though thermally reduced graphene shows remarkable conductivity, it suffers prominent structural defects which hinder its exploitation in electronic device applications. Also, this process is ineffective in fabricating GO thin film on low melting point substrates like polymer [48]. In the electrochemical method, the potential applied can efficiently remove the oxygen functional from the GO structure. The main advantage of this reduction method is that it does not emit any toxic gases and the whole technique is controllable and employs simple instrumentation. For a much more high quality yield of graphene, generally chemical methods which employ chemical reagents are commonly used. Here, the reduction of graphite oxide can be achieved by using several reducing agents like hydrazine hydrate, sodium borohydride, hydroiodic acid, hydroquinone, amines and sulfur-containing agents [49–51]. However, even if hydrazine hydrate is generally used as the reducing agent owing to its strong reducing property it can be replaced by green reducing agents like vitamin C, ascorbic acid, vancomycin, amino acids, tea leaves extract, reducing sugar, alcohols, hydrohalous acids, annona squamosa leaf extract, lysosomes etc [52–58].

Apart from these, several other methods like hydrothermal, electrochemical and photolysis-based processes, solvothermal, microwave techniques have been adopted to reduce GO to yield graphene sheets [59–63]. Recently, researchers have used camera flashes and laser scribes from optical drives to reduce GO films [64–66]. Although chemical reduction methods are widely employed for graphene fabrication, the quality of graphene obtained has always been inferior to pristine graphene. However, the remnant oxygen functional group and persisting defects in graphene can be used as a signal carrier in electrochemical sensing and transducers in devices. Each of the aforementioned chemical reduction methods has its own advantages and disadvantages, due to which their preference can vary depending on what can be afforded to compromise. Rigorous investigations and modification of these methods are necessary to develop graphene flakes from GO, with fewer defects and no oxygen functional group.

2.2.3 Unzipping of carbon nanotubes

Another important top-down method for graphene synthesis is the opening or unzipping of CNTs. Instead of the conventionally used graphite, this method employs CNT as the precursor for graphene preparation. The unzipping mechanism yields thin elongated strip graphene nanoribbon which has a high aspect ratio (at least 10) and a greater width (<50 nm) when compared to graphene. Figure 2.5 illustrates the different stages of the unzipping mechanism of a CNT into a graphene strip. During the slicing process, graphene transforms its electronic state from

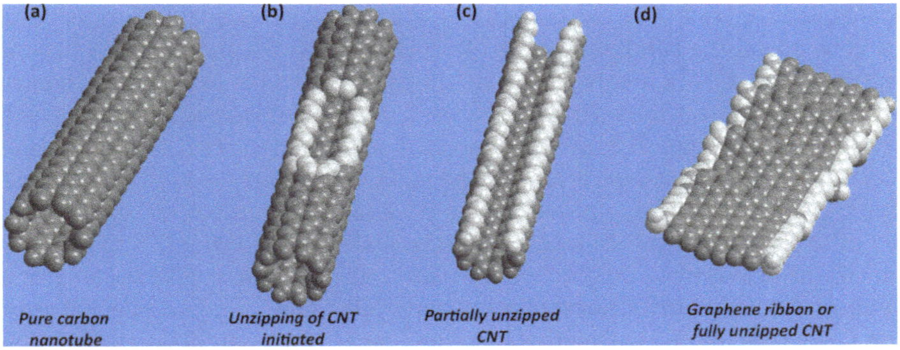

Figure 2.5. Schematic illustration of the unzipping of CNT to graphene. In the figure the atoms around the unzipped portion are faded to highlight the mechanism of slicing of CNT.

semimetal to a semiconductor [67]. Otherwise stated, graphene nanoribbons have a phenomenal capability of opening a bandgap in their electronic structure. In this regard, the electronic properties of thin strip graphene nanoribbons are presently under vigorous investigation [68]. The width of the forbidden gap depends on the width of the nanoribbon. Since the circumference of the nanotube defines the width of the nanoribbon, the usual protocol adopted is the longitudinal opening of CNTs. Depending upon whether the CNT is multiwalled or single-walled, the final end product will be multilayered graphene or single-layer graphene, respectively.

Generally, two methods of chemical–thermal processes are adopted to derive graphene layers from CNTs. The first one involves the oxidation of CNT in an acid medium followed by reduction to obtain graphene ribbons [69]. Kosynkin and his co-workers introduced a scalable method in which multiwalled CNTs were treated with sulfuric acid and potassium permanganate. In this solution-based oxidative technique, even though efficient unzipping of CNT occurred, oxygen functional groups were introduced into the structure. A reductive treatment in due course expelled the oxygen functional from the structure. This oxidation process is a major drawback of this method, since the structural defects that get integrated with the structure cannot be healed. To circumvent the deleterious effects of the oxidation process, the very same authors introduced the second method, which encompasses the preparation of graphene by unzipping of CNTs by a reaction with potassium [71]. However, uncontrollable damage of the framework during harsh unzipping reaction has remained a key challenge in this synthesis technique. To overcome this challenge, Lim and his co-workers formulated a new technique, where a dopant selectively initiated the unzipping of the CNT. The obtained graphene nanoribbons are large in size and suitable for high power supercapacitor application [72]. Highly conducting graphene nanoribbons, derived from CNT opening are highly desirable for many electronic device applications. Suresh *et al* introduced a facile method for the fabrication of highly conducting graphene nanoribbons through microwave treatment of acid processed CNT. The enhancement in the electrical conductivity observed in this case proves the efficiency of microwave treatment in simultaneous exfoliation and reduction of CNTs. In another work, photoluminescent graphene quantum dots with prominent blue emission were fabricated by the partial unzipping of multiwalled CNTs. The electrocatalytic studies of these quantum dots revealed them to be promising candidates for oxygen electrodes in fuel cells. Conventionally, the unzipping of CNTs involves a lot of strong oxidants or alkali metals which are environmentally hazardous. Recently, a facile and scalable green synthesis method was introduced, where the application of a constant current to MWCNTs in concentrated sulfuric acid generated graphene nanoribbons through an intercalated unzipping. This method is unique due to its ability to control the oxygen/carbon (O/C) ratio from 0.10 to 0.24 and can be the scaled up to 100 g h^{-1} m^{-1} [73].

In all the top-down approaches discussed so far, the synthesis of graphene depends on two main precursors: pyrolytic graphite and CNT. The selected graphite material can be of natural or synthetic origin. Even though synthetic graphite (purity level 99.9%) obtained by the graphitization of carbon source like coal produces high purity and good quality graphene, natural graphite is widely chosen as the precursor owing to its low cost and large availability. A purity level of merely 80%–98% in natural graphite with the presence of impurities like nickel and iron is impassable, which further impedes its commercialisation. Though the purification process can remove the weakly bonded impurities, they fail to wipe out the strongly intercalated impurities within the graphite layers. Concurrently, CNTs used for fabricating graphene nanoribbons cannot flee from the detrimental effects of impurities. As CNTs are grown on metal nanoparticle catalyst at high temperature, some of the metal atoms get trapped in the nanotube. In addition to that, certain carbonaceous materials are also incorporated during the fabrication process. Overall, irrespective of which top-down method is used, the purity of the source material is a primary factor deciding the quality of the graphene produced.

2.3 Bottom-up approach

Some of the most important bottom-up methods used for the synthesis of graphene are: (a) CVD, (b) pyrolysis, (c) rapid thermal annealing, (d) epitaxial growth, and (e) flash Joule heating. A detailed discussion on them can be seen below.

2.3.1 Chemical vapour deposition

CVD method is known to be a prominent and high throughput production method to fabricate uniform films of monolayer and bilayer graphene [74, 75]. In this method the decisive step involved is the controlled growth of graphene layers on a catalytic substrate like transition metals (Fe, Ru, Rh, Ni, Pd, Pt, Cu, Au etc) and alloys (Co–Ni, Au–Ni, stainless steel, Ni–Mo) [76–81]. Figure 2.6 represents a schematic illustration of the growth of graphene layer on a substrate by CVD. The CVD technique involves a range of deposition techniques viz., thermal processes, plasma-enhanced CVD (PECVD), photon-initiated CVD, and atomic layer deposition (ALD), though the first two are the most popular ones employed for fabrication of graphene. In thermal CVD, the substrate is placed in the furnace at high temperature (750 °C–1200 °C) and a mixture of hydrogen, argon and methane gases is passed through it. Once the chemical reaction is complete, a layer of

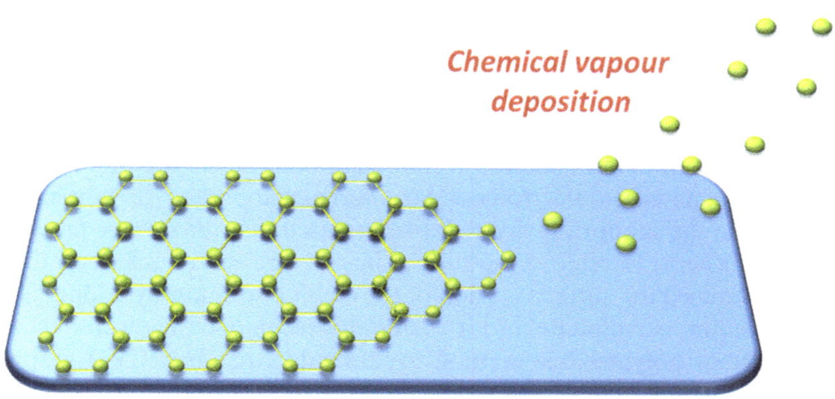

Figure 2.6. Schematic illustration of bottom-up graphene preparation by CVD.

graphene is deposited on the substrate. Jacobberger and his co-workers utilised CVD technique to fabricate self-aligned graphene ribbons on Ge (001) substrate with potential application in the field of integrated circuits [82]. Recently, defect-free graphene was fabricated on a Ti Catalytic layer through a low temperature CVD method. This is widely considered to be a breakthrough in the field of transparent thin film capacitors [83]. Through a different approach called advancing local control of precursor concentration, Vlassiouk *et al* were able to synthesize a foot-long monolayer single crystal of graphene films on polycrystalline substrates [84]. Unlike thermal CVD, PECVD involves the generation of plasma inside a vacuum chamber by the reacting gases, leading to the deposition of thin film on the substrate surface. In this technique, the synthesis process can be carried out at relatively low temperatures compared to the normal thermal CVD due to which the former is more industrially sustainable. Bilayer graphene films fabricated with a low temperature PECVD help to realize many of the potential uses of graphene in a variety of applications [85]. Site and alignment controlled narrow graphene devices can be directly fabricated from nickel nanobars by a plasma-assisted CVD method, and is expected to allow integration of graphene nanoribbons with silicon technology [86]. Recent advancement in the field of PECVD involves horizontal growth of mono-layer and multilayer graphene sheets, vertical growth of graphene nanostructures, deposition of graphene layers on nanostructured substrates, and growth of multiwall carbon nanotubes [87]. Another key feature of this CVD technique is that it can be used for the fabrication of doped and functionalized graphene. Zhai and his co-workers synthesized N-doped graphene films through a metal-free PECVD technique [88]. Large-area, wafer-scalable graphene field effect transistor, functionalized with 1-pyrenebutyric acid *N*-hydroxysuccinimide ester and conjugated with anti-CD63 antibodies was prepared, which could be an effective tool for early detection of cancer [89].

Another classification of CVD is based on the transfer technique of graphene films from a substrate to the target substrate. The different transfer techniques are: (1) transfer with the help of a support layer onto target substrates; (2) transfer without any support layer onto target substrates; and (3) direct growth of CVD graphene on target substrates without any post-growth transfer process. The use of supporting layer is one of the popular ways of obtaining graphene films. Usually, metals and polymers serve as supporting layers since they can be easily removed using etching solutions after the transfer of graphene films. However, it is worth mentioning that the current metal-assisted or polymer-assisted graphene transfer inhibits complete removal of the supporting layer and hence it cannot be used for ultraclean graphene-based electronics. An alternate approach to fabricating graphene is to avoid the supporting layer. It is recognised that the surface tension of the etching solution helps to remove the supporting substrate. Focussing on this point, Lin *et al* [90] reduced the surface tension of the etching solution by adding isopropyl alcohol ammonium persulphate into it. Using this technique they were successful in introducing a new polymer-free method to fabricate large-area CVD graphene on any substrate. An ideal way of growing high quality graphene is through direct growth of the material on the desired substrate. This directly grown graphene film on dielectric substrates is found to exhibit high electrical conductance and optical transmittance, thus making it suitable for transparent conductive electrodes [91]. Bilge *et al* elaborately studied the growth mechanism of directly grown graphene on germanium substrate through thermal CVD and PECVD [92]. The novelty of this PECVD method was that it successfully employed a growth temperature much lower than (by 200 °C) the melting point of germanium for graphene production. Based on the grain size and strain properties, a model was proposed for the growth of graphene on any substrate. High quality graphene was directly grown on Cu ink deposited on polyimide by PECVD method which has the potential to be used in flexible electronics and Internet of Things (IoT) [93].

2.3.2 Epitaxial growth on silicon carbide

The epitaxial growth of graphene by thermal decomposition of SiC is a widely approved technique to prepare high-quality graphene [94]. On annealing SiC substrate at high temperature and ultra-vacuum conditions, Si atoms will be desorbed from the surface leaving behind carbon atoms to form graphene films of thickness less than 50 μM (figure 2.7). The fabrication of uniform epitaxial graphene on C-face 6H-SiC substrates with a sputtered SiC film could be used for manufacturing analogue high frequency devices [95]. The latest advancement in this field is the fabrication of multilayer graphene on SiC using a high-power continuous laser. The obtained graphene/SiC samples were highly conductive and had a sheet resistance of as low as 0.43 Ω [96]. Even though these samples are widely used for high conducting electronic application, a major flaw in this method is that the underlying SiC substrate always tries to retard the carrier mobility of the graphene layers. Briggs and his co-workers introduced a novel approach to separate out graphene layers formed on SiC substrate by intercalation of atomic species between the layers, resulting in quasi free-standing graphene layers with improved carrier transport [97]. Hongyan and his co-workers employed Ga–Pd alloy as a catalyst to induce decomposition of S–C bonds from SiC substrate, which facilitated the

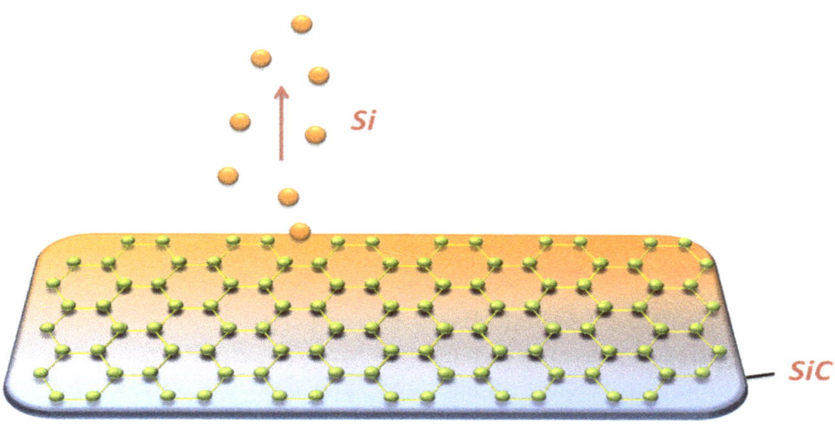

Figure 2.7. Diagrammatic representation of graphene synthesis by epitaxial growth on SiC.

deposition of epitaxial layers of graphene films on both the interface between Ga–Pd alloy and SiC substrate and the upper surface of the alloy [98]. This transfer-free strategy of graphene synthesis can also be employed to grow graphene on other dielectrics too, by simply placing the material face down on the surface of the alloy.

2.3.3 Pyrolysis

In addition to the aforementioned synthesis techniques, various forms of pyrolysis techniques are also used for graphene preparation. Pyrolysis involves the thermal decomposition of carbon-based materials in the absence of oxygen. Thermal pyrolysis technique involves the fabrication of graphene on metal surface by the pyrolysis of different carbon sources like ethanol or biomass [99, 100]. However, in this method the final yield of graphene is limited. In another work, ultra-low density three-dimensional (3D) macroassemblies of graphene sheets with high electrical conductivities and large internal surface areas were synthesised by pyrolysis of a gel mixture of resorcinol, formaldehyde and GO [101]. A latest research article reported the synthesis of foamed graphene flakes by a one-step pyrolysis of polyethylene glycol/MgO composites which can be applied in supercapacitors and lithium-ion batteries [102]. In a most recent study, ultrasonic pyrolysis was used to fabricate graphene with ultrahigh carrier mobility and remarkable optical transparency for applications in photodetectors [103].

Another variant of this technique, called spray pyrolysis, is used for the fabrication of graphene. A notable advantage of this method is that it allows thin and homogenous deposition of graphene films on a heated substrate through the deposition of aerogel droplets of nebulised graphene dispersion [104]. Another key benefit of this technique is that it does not require high temperature and vacuum conditions for the synthesis. Recently, a cost effective spray pyrolysis approach was proposed by AlShammari and his co-workers for the preparation of reduced graphene oxide (rGO) from GO dispersion [105]. Their study conveyed the reduction in optical bandgap of rGO from 4.2 to 3.27 eV, when the spray cycles were increased from 2 to 10. The enhanced photoelectric and photodetection properties observed for this rGO thin film could be utilized in advanced metal–semiconductor–metal (MSM) optoelectronic devices. In another work, a novel route was introduced to fabricate high quality graphene quantum dots through a low vacuum hydrogen-assisted pyrolysis of silicon carbide (SiC). The obtained quantum dotes are highly-ordered crystalline structure with very high purity [106, 107].

In recent years, biomass pyrolysis method has gained a lot of interest in the scientific world since it facilitates the production of graphene from organic waste and other renewable carbonous sources. Graphene synthesis from biomass leads to a zero waste scenario. The most popularly used biomass pyrolysis methods can be categorized as salt-based, template based and chemical blowing based [108]. The salt-based pyrolysis of waste materials enables the production of graphene on a large scale. This method is attractive in that the salts used can be recycled by water washing and crystallisation. The salts included can serve as either an inert solvent forbidding the shrinkage of graphene sheets or a graphitic catalyst. In an experiment performed by Ali *et al*, the pyrolysis of plastic waste along with molten NaCl led to the formation of graphene sheets with a very high conductivity of 1150 S m^{-1} and surface area of 522 m^2 g^{-1} [109]. The molten salt enhanced graphitisation and crystallinity of graphene product due to the high diffusion rate of oxygen ions in the solvent [110]. This single step plastic derived graphene synthesis method produces the highest quality graphene which has been reported so far. In another work, the same authors compared the effect of various molten salt on the synthesis of graphene [111]. They observed that the quality of graphene obtained with molten NaF, KF and NaCl was much greater than that obtained when using $ZnCl_2$ and $MgCl_2$ and KCl. They proposed that the wettability of carbon surface by molten salt plays a key role in enhancing the graphitisation degree and crystallinity of graphene. The graphene sheets obtained using KF exhibited better crystallinity owing to their greater ionic radii of K^+ (152 pm) than Na^+ ion. Although the reaction mechanism is different, chemical blowing is similar to the salt-based synthesis method. Here, the conventional strategy is to blow molten sugar using NH_4Cl as a blowing agent. The gases released during the process made the walls of the biomass precursor thinner and helped in efficient exfoliation. Several modifications were made on the above synthesis method so as to obtain high specific area and high specific capacitance. Jung *et al* included CO_2 and KOH along with NH_4Cl. The obtained porous graphene exhibited high specific area of 3657 m^2 g^{-1} and specific capacitance of 175 F g^{-1} [112]. The use of agaric solution along with NH_4Cl, leads to the production of graphene with a specific surface area of 2200 m^2 g^{-1} [113]. In some biomass pyrolysis techniques, layered structures of g-C_3N_4, clays, gelatine and zeolites can be used as a template for graphene synthesis. This method needs to be modified further to use biomass material as template material.

Roy and her co-workers developed an eco-friendly pyrolysis method, where non-conventional sources like tannic acid, alginic acid, and green tea were used for graphene synthesis [114]. The obtained graphene films exhibited a specific capacitance of 315 F g^{-1}, making them suitable for green supercapacitors. The most interesting property exhibited by the prepared sample was room temperature ferromagnetism, due to which they can be used for spintronic applications. High quality 3D graphene foams (3D GFs) were synthesized by the use of renewable precursors and waste heat [115]. Here, the pyrolysis of biomass generated a large amount of gases and waste heat. Utilising these by-products as the carbon precursors, resulted in 3D GFs with superior performance in energy storage applications. The authors also claimed that the incorporation of pyrolytic route in the conversion of biomass into graphene has much less impact on human health and the ecosystem when compared to the conventional CVDs. Yuxin *et al* successfully fabricated few-layer GO through a shear exfoliation of carbon containing energy crop miscanthus [116]. The pyrolysis temperature played a key role in converting the biomass to pyrolytic biochar. At 1200 °C the resultant biochar was highly aromatic with functional groups that facilitate efficient exfoliation to form six-layer GO. Today, more efforts are being made to use renewable sources as the starting material for the fabrication of graphene and graphene-based compounds.

2.3.4 Rapid thermal annealing

Rapid thermal annealing (RTA) technology helps to realize graphene synthesis on a wide range of substrates from different carbon precursors [117]. The main advantage of rapid thermal annealing compared to the classical furnace is that this method offers precise control of the annealing temperature as well as the cooling process. The basic steps involved in the synthesis of graphene using RTA are: (i) pre-treatment of substrate enabling defect-free growth of graphene films; (ii) deposition of carbon and metal source on the substrate; and (iii) growth of graphene films under rapid thermal annealing and cooling [118]. Usually, the preferred support substrate includes Si, SiC or graphite. The presence of metal catalysts like Ni, Cu and Co films provides a better route for the synthesis of graphene layers of low defect concentration. Bleue *et al* experimentally demonstrated that pulsed laser deposition of solid carbon source in the presence of Ni catalyst on Si and SiO_2 substrate leads to the formation of high quality graphene films [119]. Several researchers have exploited the RTA method to investigate the effect of different carbon precursors on the quality of graphene films obtained. Prekodravac and his co-workers concluded that the selection of amorphous carbon as the initial source has no prospects for graphene synthesis, whereas adopting graphene quantum dots and fullerol, even at very low concentrations, resulted in the formation of single-layer graphene films [120].

The main disadvantage associated with the RTA method is the inclusion of metal films as catalysts. A strong acid treatment is involved to etch out the metal film for transferring the graphene films. To overcome this difficulty, a capping agent like polymethyl methacrylate (PMMA) served as a good protector for graphene layers during the transfer process. Figure 2.8 shows different ways of removal of PMMA

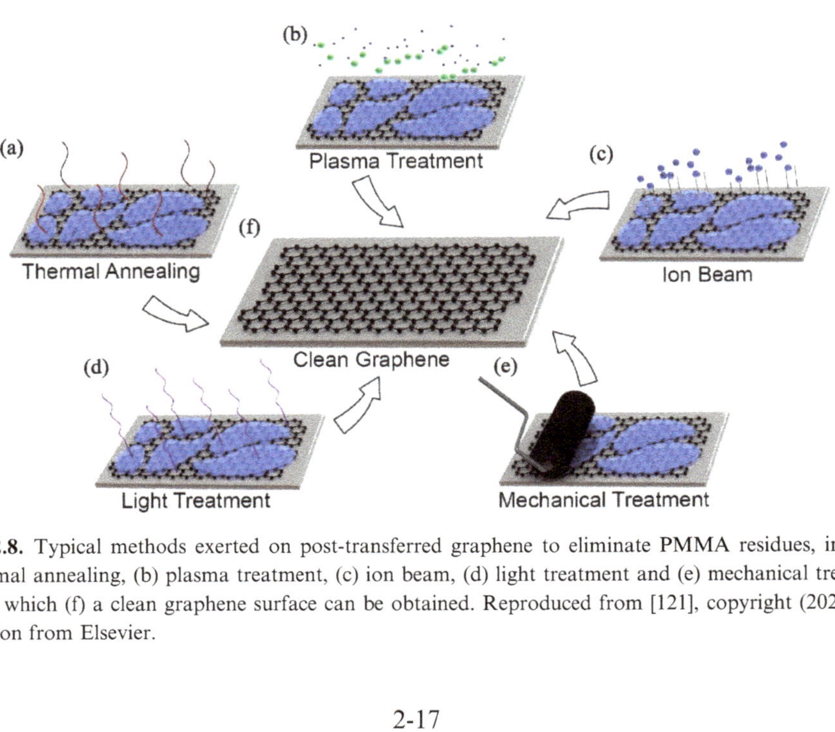

Figure 2.8. Typical methods exerted on post-transferred graphene to eliminate PMMA residues, including (a) thermal annealing, (b) plasma treatment, (c) ion beam, (d) light treatment and (e) mechanical treatment, through which (f) a clean graphene surface can be obtained. Reproduced from [121], copyright (2021), with permission from Elsevier.

for superclean graphene [121]. PMMA can also be used as a carbon source for graphene synthesis in SiO_2 or Cu foil substrate.

2.3.5 Flash Joule heating

Although graphene has been regarded as the 'wonder material' since its discovery in 2004, its wide-scale implementation in the areas in which it was predicted to bring about revolutionary changes, still remains far from reality [122]. Some experts believe that decades of research and technological innovations are required before graphene-based technologies could live up to the expectations surrounding them [123]. One of the major reasons for this belief has been the inability of popular bottom-up graphene synthesis methods like CVD, to produce graphene on a large scale. While top-down approaches like exfoliation method and advanced synthetic organic method are the most popular methods used in industries for bulk graphene production, what these methods do is only to separate physically or chemically the already fully formed graphene from graphite or CNT [124]. However, in applications like solar cells, sensors etc which require large and continuous graphene films, these methods cannot be of much help. This is where bottom-up methods like CVD are useful. Nevertheless, the bottom-up methods also have some limitations such as low yield in the case of CVD and highly defect ridden structure in the case of preparation in bulk solution [122]. It is obvious that at the industrial level, product prototyping requires hundreds of such high quality films with good reproducibility, which is tedious and expensive. In addition, for realising all the predicted revolutionary changes, kilograms of graphene sheets are to be produced every day, not necessarily as films on substrates, which is something the most popular bottom-up methods are not capable of. Thus the graphene industry is still continuing the quest for an efficient bottom-up synthesis method which can produce graphene in bulk quantities.

A promising breakthrough occurred when Luong *et al* in 2020 employed Flash Joule heating (FJH) to synthesis graphene structures in gram-scale from carbon sources such as carbon black, coal, rubber tyres, biochar etc [125]. Even though FJH has been reported to be used in the sintering process of ceramic materials for more than a decade, this was the first time it was used for the synthesis of high quality graphene. It is apparent that the inspiration to use FJH comes from the success of LASER induced graphene (LIG) which converts carbon rich materials such as polymers and paper to graphene foam through rapid localized heating. FJH too relies on this principle, except that instead of using LASER to produce high temperature, it operates on the principle of conversion of electric energy directly into thermal energy by means of a resistor. Luong *et al* used the setup shown in figure 2.9 to synthesize graphene, in which two electrodes can be seen placed inside a quartz tube, compressing the carbon source. The resistance could be minimized by controlling the compressing force to obtain ample flash reaction. Although the experiment was initially conducted at low pressure, it was clarified that FJH method can perform equally well at 1 atm. A capacitor bank charged by a DC supply with a maximum voltage of 400 V was used to produce electric discharge which expose the carbon source to temperatures in excess of 3000 K in about

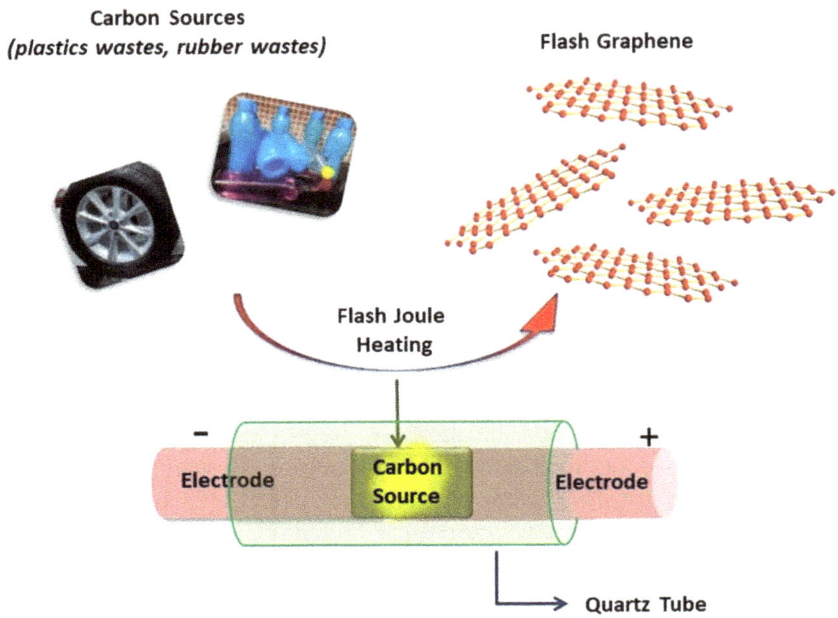

Figure 2.9. Schematic diagram of the sample setup of FJH system for conversion of various waste materials into flash graphene.

100 ms. This converted the carbon source into so-called flash graphene (FG) with turbostratic arrangement. The authors confirmed the high quality of the graphene thus produced using Raman spectroscopy. The oxygen content and defects in the samples varied with the starting material. The yield was the highest (80%–90%) for high carbon sources like carbon black and calcined coke with a 99% purity. Thus, the FJH method is now expected to remove many bottlenecks in the industrial production of graphene not only because of its high yield but also due to the fact that it does not use solvents or reactive gases.

The same group in another report, investigated the morphology and properties of FG produced from carbon black. In this work they mainly tried to comprehend the assembling mechanism of turbostratic flash graphene (tFG). The setup employed was the original one used in the previous work and the electrodes were charged to 120 V. The carbon black was heated to a temperature of ~3000 K in a few milliseconds, which converted it to tFG. The planar crystals of fFG thus formed were found to be aligned in the direction of the current. After 100 ms of heating the bulk production of tFG was confirmed with a small quantity of graphitized carbon. The graphitic particles were called wrinkled graphene as they have many bends and were about 3–8 layers thick. From atomistic simulation the authors concluded that the generic thermal annealing predominantly produces wrinkled graphene, which has minimum alignment of graphitic planes unlike tFG. They were also able to exfoliate tFG easily by applying shear without chemical treatments. Transmission electron microscope (TEM) images of tFG and wrinkled graphene are shown in

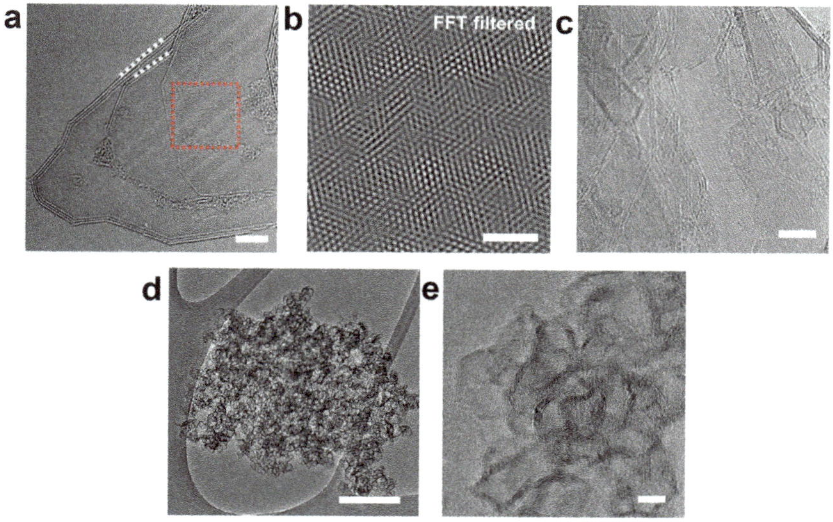

Figure 2.10. (a) TEM images of tFG showing rotational mismatch as striations (scale bar: 5 nm). (b) FFT image of a portion of the first image (region in red box) (scale bar: 2 nm). (c) TEM image of tFG showing more rotational mismatch (scale bar: 5 nm). (d) TEM image of wrinkled graphene (scale bar: 200 nm). (e) High-resolution TEM image of wrinkled graphene (scale bar: 10 nm). Reproduced with permission from [126], copyright 2020, American Chemical Society.

figure 2.10. Algozeeb and his co-workers in the same year reported the synthesis of flash graphene from plastic waste (PW) [127]. The authors presented this as an alternative approach to the physical and chemical recycling of plastic. They pointed out that flash synthesis avoids the requirements of human labour-intensive sorting of plastic materials associated with physical recycling and the huge energy consumption and poisoning of catalysts associated with chemical recycling method such as pyrolysis, due to which it is more advantageous at the industrial level. Both direct current (DC) FJH and alternating current (AC) FJH were used to synthesize FG and it was found that the AC-FJH method, which can be sustained for seconds, is more advantageous with PW. This is because this extra time helps in the release of sufficient volatiles and converts the PW into intermediate AC-flash graphene with high I_{2D}/I_G peak ratio from Raman spectra. This process helps to iron out the pyrolysis process. However, the AC-FG graphene thus produced has to be subjected to single DC-FJH pulse to produce high quality tFG. The authors primarily verified this by noting the enhancement in the I_{2D}/I_G peak ratio and a reduction in the D band intensity in Raman spectra. The evolution of Raman spectra of PW such as high density polyethylene (HDPE), poly(vinyl chloride) (PVC), polyethylene terephthalate (PET), low density polyethylene (LDPE), and polystyrene (PS), polypropylene (PP) and their mixtures can be seen in figure 2.11. In addition to FG, FJH also produces carbon oligomers, light hydrocarbons and hydrogen. Despite this, the authors project FJH as an environmentally beneficial method for the upcycling of PW. Yet, they admit that a full-lifecycle analysis is necessary to explore the full utility this method.

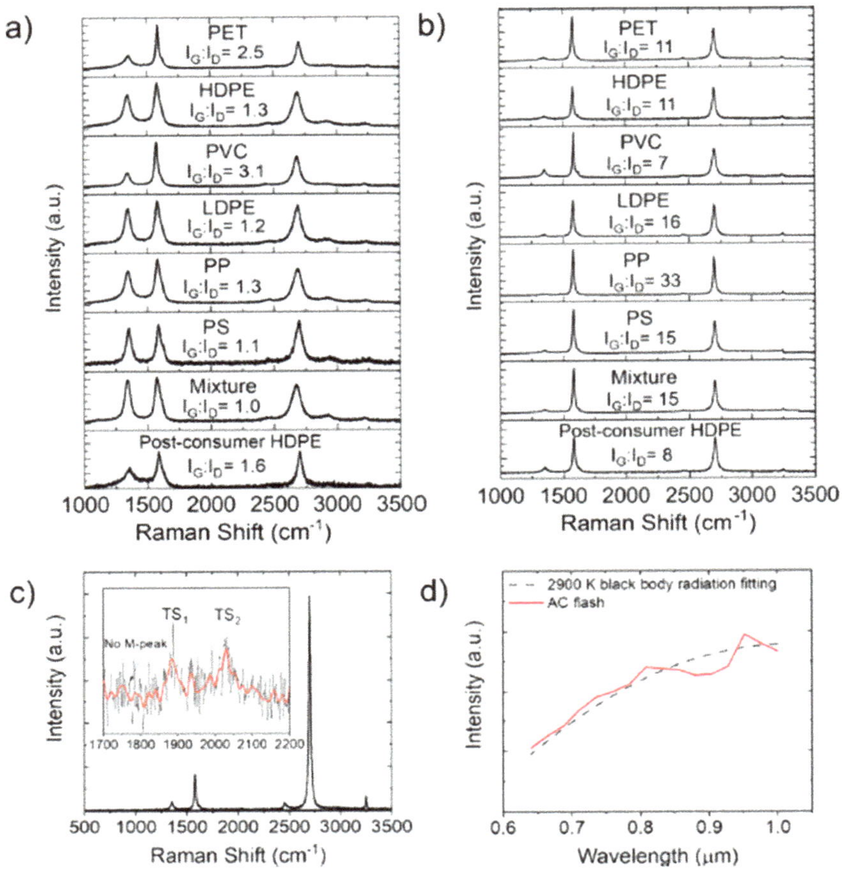

Figure 2.11. Raman spectra of (a) AC-FG and (b) ACDC-tFG. Plastic mixture in the samples are 40% HDPE, 20% PP, 20% PET, 10% LDPE, 8% PS, and 2% PVC. (c) Raman spectrum of tFG obtained for ACDC-tFG from PVC, with the turbostratic FG bands in the inset. (d) Temperature profile of the AC-FJH processes obtained using an IR spectrometer and blackbody radiation fitting. Reproduced with permission [127], copyright 2020, American Chemical Society.

Along with these developments, the very same group reported the production of tFG from rubber waste feedstocks [128]. By adjusting pulse voltage and time, they developed tFG with high I_{2D}/I_G peak ratio. From their studies they concluded that incorporation of tFG in Portland cement leads to an enhancement in the compressive strength of the composite, which could greatly curtail the energy use and greenhouse gas emissions that occur during the production of cement. In a most recent work, Chen *et al* used the FJH to develop seven different types of heteroatom doped graphene [129]. This included single-element-doped (boron, oxygen, nitrogen, phosphorus, sulfur), two-element co-doped (boron and nitrogen), and three-element co-doped (boron, nitrogen and sulfur) FG. These samples exhibited good tFG structure and were dispersible in water-pluronic solution. They also concluded that since heteroatom doping changes the electronic structure of graphene, its applicability as an electrochemical energy storage material would be substantially improved. Thus, FJH can produce both doped and undoped FG from almost any carbon source. Its advantage is the bulk production capability and good quality of graphene. However, although FJH is projected to be the next big thing in the field of graphene bulk production, it is yet to prove its mettle at the industrial level.

2.4 Challenges and the way ahead

Scientific advancements in the field of graphene synthesis have been rapidly progressing to bring forth the wonder material's superlative properties for high-end industrial applications. Industrialization requires the production of graphene with large area/quantity and high purity under stable and low-cost conditions. Currently, all the methods mentioned so far are immature or expensive, thereby preventing global industrialization and commercialization of this miracle material. The CVD currently remains the most preferable industrial method for the production of graphene films, whereas physical exfoliation remains the most popular one for the growth of bulk graphene. However, physical exfoliation, like other top-down methods is limited by the quality of the starting material and strong inter-sheet interactions. This highlights the necessity of developing a novel bottom-up synthesis method for bulk graphene production which ensures high quality as well as quantity. FJH seems to be the frontrunner for this position. Its ability to synthesize turbostratic graphene in gram-scale from almost any carbon feedstock is observed with great interest by the graphene research community all over the world. Also, this method has substantial environmental benefits in that it does not use solvents and utilizes comparatively less energy to produce graphene. However, as this method has only been in the picture since 2018, questions about its industrial opportunities and challenges pertain. The flash Joule synthesis of graphene is an intellectual property of Rice University and the method is currently licenced by the company Universal Matter Ltd [123]. The company was able to make their manufacturing prototype line operational in the first quarter of 2022. Therefore, if everything goes as planned, the world will witness the first industrial-scale bottom-up synthesis of graphene very soon. Yet, this method suffers certain drawbacks which need the immediate attention of researchers. Firstly, the optimization of reaction parameters such as heating rate, energy density etc for new feedstocks is done through a laborious trial and error method. It will take a considerable amount of time and laboratory level characterization to incrementally improve the properties and eventually obtain the desired product. Secondly, the flash Joule mechanism is extremely difficult to study since the formation of graphene occurs in just hundreds of milliseconds. Furthermore, most FJH reactors in use today do not provide precise control over current discharge profile due to which even a slight fluctuation in circuit to sample contact can sometimes make it difficult to map the relationship between structure and properties. Recent studies predict that machine learning could be used to overcome these hurdles as it has the potential to predict the level of crystallinity, extract fundamental information regarding structure and properties and drive process optimization. Certain machine learning models have already been employed to optimize parameters so as to improve the crystallinity of FG [130]. A continuation of these studies could help to better understand the FG synthesis process more deeply and enable further optimization of parameters leading to maximization of its potential for commercialization.

Along with FJH, other graphene synthesis methods are also incrementally improving themselves to meet the industrial requirements for graphene production. However, irrespective of the synthesis method used, the most worrisome issue which inhibits the successful development and applications of graphene related technologies is the poor quality of graphene available in the market. Therefore, industrial graphene production requires uncompromising quality control. To ensure this, the industry must utilize characterization techniques like Raman spectroscopy, AFM, TEM and so forth in a more comprehensive way than how they are used in academic research laboratories. This means that these techniques have to be used multiple times on multiple regions to ensure the purity and size uniformity of the 'black powder'. Another factor stalling graphene research is the indiscriminate financing by governments in the form of grants and other commercial funds. Investing for research labs and fundamental research will not always help to overcome fundamental obstacles for commercialization like the problems associated with the present industrial synthesis methods. Instead, funds must be allocated at the right time and to the right places. Consequently, funding focussed explicitly on the large-scale fabrication of high-quality graphene is of utmost importance as it might help to remove one of the greatest bottlenecks towards graphene commercialization. However, funding is not everything; strategic partnership between companies including start-ups and government-sponsored academia and R&Ds as well as cross-industry collaborations could help to eliminate many hurdles associated with graphene production and help in the tailored tuning of its properties according to the requirement. Apart from these steps, to improve the quality of industrial graphene, the future methods for graphene production must also consider going 'green'. Currently, many methods used for graphene synthesis do not fit under the label 'environment friendly' which is worrisome. Along with this, the focus must also be on developing graphene from waste materials including non-biodegradable ones like plastics and glass. These are some of the important considerations for the industrial level large-scale preparation of high quality graphene suitable for commercial applications. Having discussed this, it must be recognised that graphene research is still in its premature phase and the developmental work to transform ideas into realistic procedures and products takes time and money.

References

[1] Chen C H, Yang S W, Chuang M C, Woon W Y and Su C Y 2015 Towards the continuous production of high crystallinity graphene via electrochemical exfoliation with molecular *in situ* encapsulation *Nanoscale* **7** 15362–73

[2] Sang M, Shin J, Kim K and Yu K J 2019 Electronic and thermal properties of graphene and recent advances in graphene based electronics applications *Nanomaterials* **9** 374

[3] Sharma B K and Ahn J H 2013 Graphene based field effect transistors: efforts made towards flexible electronics *Solid-State Electron.* **89** 177–88
[4] Song D, Mahajan A, Secor E B, Hersam M C, Francis L F and Frisbie C D 2017 High-resolution transfer printing of graphene lines for fully printed, flexible electronics *ACS Nano* **11** 7431–9
[5] Liu Y, Zhou H, Cheng R, Yu W, Huang Y and Duan X 2014 Highly flexible electronics from scalable vertical thin film transistors *Nano Lett.* **14** 1413–8
[6] Siddique S, Iqbal M Z and Mukhtar H 2017 Cholesterol immobilization on chemical vapor deposition grown graphene nanosheets for biosensors and bioFETs with enhanced electrical performance *Sens. Actuators* B **253** 559–65
[7] Lawal A T 2018 Progress in utilisation of graphene for electrochemical biosensors *Biosens. Bioelectron.* **106** 149–78
[8] Liu Y, Dong X and Chen P 2012 Biological and chemical sensors based on graphene materials *Chem. Soc. Rev.* **41** 2283–307
[9] Liu J, Liu Z, Barrow C J and Yang W 2015 Molecularly engineered graphene surfaces for sensing applications: a review *Anal. Chim. Acta* **859** 1–19
[10] Gaur M, Misra C, Yadav A B, Swaroop S, Maolmhuaidh F Ó, Bechelany M and Barhoum A 2021 Biomedical applications of carbon nanomaterials: fullerenes, quantum dots, nanotubes, nanofibers, and graphene *Materials* **14** 5978
[11] Nizami M Z I, Takashiba S and Nishina Y 2020 Graphene oxide: a new direction in dentistry *Appl. Mater. Today* **19** 100576
[12] Feng L and Liu Z 2011 Graphene in biomedicine: opportunities and challenges *Nanomedicine* **6** 317–24
[13] Majumder S, Kumari S and Ghosh D 2021 Graphene based biopolymer nanocomposite applications in drug delivery *Graphene Based Biopolymer Nanocomposites* (Singapore: Springer) pp 287–309
[14] Pal K, Asthana N, Aljabali A A, Bhardwaj S K, Kralj S, Penkova A, Thomas S, Zaheer T and Gomes de Souza F 2021 A critical review on multifunctional smart materials 'nanographene' emerging avenue: nano-imaging and biosensor applications *Crit. Rev. Solid State Mater. Sci.* **47** 691–707
[15] Zionmarketresearch.com 2022 Graphene Market By Product (Mono-Layer and Bi-Layer Graphene, Few Layer Graphene, Graphene Oxide, Graphene Nano Platelets, and Others), By Application (Composites, Energy Storage, Electronics, and Others): Global Industry Perspective, Comprehensive Analysis and Forecast, 2020–2028 https://Zionmarketresearch.com
[16] GlobeNewswire R 2022 Global EMI Shielding Market with COVID-19 Impact, by Material (EMI Shielding Tapes and Laminates, Conductive Coatings and Paints, Metal Shielding, Conductive Polymers, EMI/EMC Filters), Method (Radiation, Conduction), Industry, and Region – Forecast to 2026
[17] Sumdani M G, Islam M R, Yahaya A N A and Safie S I 2021 Recent advances of the graphite exfoliation processes and structural modification of graphene: a review *J. Nanopart. Res.* **23** 1–35
[18] Meyer J C, Geim A K, Katsnelson M I, Novoselov K S, Booth T J and Roth S 2007 The structure of suspended graphene sheets *Nature* **446** 60–3
[19] Ferrari A C *et al* 2015 Science and technology roadmap for graphene, related two-dimensional crystals, and hybrid systems *Nanoscale* **7** 4598–810

[20] Zhang P, He P, Yu Q, Wang G, Huang T, Yang S, Liu Z, Xie X and Ding G 2022 Delocalized electrochemical exfoliation toward high-throughput fabrication of high-quality graphene *Chem. Eng. J.* **428** 131122
[21] Tian B, Li J, Samad A, Schwingenschlögl U, Lanza M and Zhang X 2022 Production of large-area nucleus-free single-crystal graphene-mesh metamaterials with zigzag edges *Adv. Mater.* **34** 2201253
[22] Adetayo A and Runsewe D 2019 Synthesis and fabrication of graphene and graphene oxide: a review *Open J. Compos. Mater.* **9** 207
[23] Olatomiwa A L, Adam T, Gopinath S C, Kolawole S Y, Olayinka O H and Hashim U 2022 Graphene synthesis, fabrication, characterization based on bottom-up and top-down approaches: an overview *J. Semicond.* **43** 061101
[24] Villarreal C C, Sandoval J I, Ramnani P, Terse-Thakoor T, Vi D and Mulchandani A 2022 Graphene compared to fluorine-doped tin oxide as transparent conductor in ZnO dye-sensitized solar cells *J. Environ. Chem. Eng.* **10** 107551
[25] Dai T, Xu H, Chen S and Zhang Z 2022 High performance Hall sensors built on chemical vapor deposition-grown bilayer graphene *ACS Omega* **7** 25644–9
[26] Ramaraj S G, Muruganathan M, Agbonlahor O G, Maki H, Onda Y, Hattori M and Mizuta H 2022 Carbon molecular sieve-functionalized graphene sensors for highly sensitive detection of ethanol *Carbon* **190** 359–65
[27] Yi M and Shen Z 2015 A review on mechanical exfoliation for the scalable production of graphene *J. Mater. Chem.* A **3** 11700–15
[28] Huang Y, Sutter E, Shi N N, Zheng J, Yang T, Englund D, Gao H J and Sutter P 2015 Reliable exfoliation of large-area high-quality flakes of graphene and other two-dimensional materials *ACS Nano* **9** 10612–20
[29] Jayasena B and Subbiah S 2011 A novel mechanical cleavage method for synthesizing few-layer graphenes *Nanoscale Res. Lett.* **6** 1–7
[30] Knieke C, Berger A, Voigt M, Taylor R N K, Röhrl J and Peukert W 2010 Scalable production of graphene sheets by mechanical delamination *Carbon* **48** 3196–204
[31] Jayasena B, Reddy C D and Subbiah S 2013 Separation, folding and shearing of graphene layers during wedge-based mechanical exfoliation *Nanotechnology* **24** 205301
[32] Backes C *et al* 2020 Production and processing of graphene and related materials *2D Mater.* **7** 022001
[33] Servant A, Leon V, Jasim D, Methven L, Limousin P, Fernandez-Pacheco E V, Prato M and Kostarelos K 2014 Graphene-based electroresponsive scaffolds as polymeric implants for on-demand drug delivery *Adv. Healthcare Mater.* **3** 1334–43
[34] Thomas D G, De-Alwis S, Gupta S, Pecharsky V K, Mendivelso-Perez D, Montazami R, Smith E A and Hashemi N N 2021 Protein-assisted scalable mechanochemical exfoliation of few-layer biocompatible graphene nanosheets *R. Soc. Open Sci.* **8** 200911
[35] Wu W, Liu M, Gu Y, Guo B, Ma H, Wang P, Wang X and Zhang R 2020 Fast chemical exfoliation of graphite to few-layer graphene with high quality and large size via a two-step microwave-assisted process *Chem. Eng. J.* **381** 122592
[36] Eredia M, Ciesielski A and Samorì P 2017 Graphene via molecule-assisted ultrasound-induced liquid-phase exfoliation: a supramolecular approach *Chemistry of Carbon Nanostructures* (Berlin: De Gruyter) ch 6 pp 173–93

[37] Shang T, Feng G, Li Q and Zheng Y 2017 Production of graphene nanosheets by supercritical CO_2 process coupled with micro-jet exfoliation *Fuller. Nanotub. Carbon Nanostructures* **25** 691–8
[38] Abdelkader A M, Cooper A J, Dryfe R A and Kinloch I A 2015 How to get between the sheets: a review of recent works on the electrochemical exfoliation of graphene materials from bulk graphite *Nanoscale* **7** 6944–56
[39] Parvez K, Li R, Puniredd S R, Hernandez Y, Hinkel F, Wang S, Feng X and Mullen K 2013 Electrochemically exfoliated graphene as solution-processable, highly conductive electrodes for organic electronics *ACS Nano* **7** 3598–606
[40] Liu J, Notarianni M, Will G, Tiong V T, Wang H and Motta N 2013 Electrochemically exfoliated graphene for electrode films: effect of graphene flake thickness on the sheet resistance and capacitive properties *Langmuir* **29** 13307–14
[41] Shi P C, Guo J P, Liang X, Cheng S, Zheng H, Wang Y, Chen C H and Xiang H F 2018 Large-scale production of high-quality graphene sheets by a non-electrified electrochemical exfoliation method *Carbon* **126** 507–13
[42] Wang H, Wei C, Zhu K, Zhang Y, Gong C, Guo J, Zhang J, Yu L and Zhang J 2017 Preparation of graphene sheets by electrochemical exfoliation of graphite in confined space and their application in transparent conductive films *ACS Appl. Mater. Interfaces* **9** 34456–66
[43] Achee T C, Sun W, Hope J T, Quitzau S G, Sweeney C B, Shah S A, Habib T and Green M J 2018 High-yield scalable graphene nanosheet production from compressed graphite using electrochemical exfoliation *Sci. Rep.* **8** 1–8
[44] Islam A, Mukherjee B, Pandey K K and Keshri A K 2021 Ultra-fast, chemical-free, mass production of high quality exfoliated graphene *ACS Nano* **15** 1775–84
[45] Yu H, Zhang B, Bulin C, Li R and Xing R 2016 High-efficient synthesis of graphene oxide based on improved hummers method *Sci. Rep.* **6** 1–7
[46] Yoo M J and Park H B 2019 Effect of hydrogen peroxide on properties of graphene oxide in Hummers method *Carbon* **141** 515–22
[47] Tian S, Sun J, Yang S, He P, Ding S, Ding G and Xie X 2015 Facile thermal annealing of graphite oxide in air for graphene with a higher C/O ratio *RSC Adv.* **5** 69854–60
[48] Pei S and Cheng H M 2012 The reduction of graphene oxide *Carbon* **50** 3210–28
[49] Shin H J *et al* 2009 Efficient reduction of graphite oxide by sodium borohydride and its effect on electrical conductance *Adv. Funct. Mater.* **19** 1987–92
[50] Chua C K and Pumera M 2016 The reduction of graphene oxide with hydrazine: elucidating its reductive capability based on a reaction-model approach *Chem. Commun.* **52** 72–5
[51] Kim M, Hwang H M, Park G H and Lee H 2017 Graphene-based composite electrodes for electrochemical energy storage devices: recent progress and challenges *Flat Chem.* **6** 48–76
[52] Choi Y J, Kim E, Han J, Kim J H and Gurunathan S 2016 A novel biomolecule-mediated reduction of graphene oxide: a multifunctional anti-cancer agent *Molecules* **21** 375
[53] Ismail Z 2019 Green reduction of graphene oxide by plant extracts: a short review *Ceram. Int.* **45** 23857–68
[54] Xu L Q, Liao Y B, Li N N, Li Y J, Zhang J Y, Wang Y B, Hu X F and Li C M 2018 Vancomycin-assisted green synthesis of reduced graphene oxide for antimicrobial applications *J. Colloid Interface Sci.* **514** 733–9

[55] Chandu B, Mosali V S S, Mullamuri B and Bollikolla H B 2017 A facile green reduction of graphene oxide using *Annona squamosa* leaf extract *Carbon Lett.* **21** 74–80
[56] Zhang J, Yang H, Shen G, Cheng P, Zhang J and Guo S 2010 Reduction of graphene oxide via L-ascorbic acid *Chem. Commun.* **46** 1112–4
[57] Fernández-Merino M J, Guardia L, Paredes J I, Villar-Rodil S, Solís-Fernández P, Martínez-Alonso A and Tascón J M D 2010 Vitamin C is an ideal substitute for hydrazine in the reduction of graphene oxide suspensions *J. Phys. Chem.* C **114** 6426–32
[58] Rattan S, Kumar S and Goswamy J K 2020 Graphene oxide reduction using green chemistry *Mater. Today Proc.* **26** 3327–31
[59] Marlinda A R, Yusoff N, Sagadevan S and Johan M R 2020 Recent developments in reduced graphene oxide nanocomposites for photoelectrochemical water-splitting applications *Int. J. Hydrog. Energy* **45** 11976–94
[60] Toh S Y, Loh K S, Kamarudin S K and Daud W R W 2014 Graphene production via electrochemical reduction of graphene oxide: synthesis and characterisation *Chem. Eng. J.* **251** 422–34
[61] Duan L, Zhang T, Song W, Jiang C, Hou Y, Zhao W, Chen W and Alvarez P J 2019 Photolysis of graphene oxide in the presence of nitrate: implications for graphene oxide integrity in water and wastewater treatment *Environ. Sci. Nano* **6** 136–45
[62] Xie X, Zhou Y and Huang K 2019 Advances in microwave-assisted production of reduced graphene oxide *Front. Chem.* **7** 355
[63] Hu K, Xie X, Szkopek T and Cerruti M 2016 Understanding hydrothermally reduced graphene oxide hydrogels: from reaction products to hydrogel properties *Chem. Mater.* **28** 1756–68
[64] Scardaci V, Fichera L, Fragalà M E, Tuccitto N, Marletta G and Compagnini G 2020 Reduction of graphene oxide by laser scribing in different atmospheres and application in humidity sensing *J. Nanomater.* **2020** 4946954
[65] Evlashin S *et al* 2016 Controllable laser reduction of graphene oxide films for photoelectronic applications *ACS Appl. Mater. Interfaces* **8** 28880–87
[66] Jakhar R, Yap J E and Joshi R 2020 Microwave reduction of graphene oxide *Carbon* **170** 277–93
[67] Chen Z, Lin Y M, Rooks M J and Avouris P 2007 Graphene nano-ribbon electronics *Physica* E **40** 228–32
[68] Jiao L, Wang X, Diankov G, Wang H and Dai H 2010 Facile synthesis of high-quality graphene nanoribbons *Nat. Nanotechnol.* **5** 321–5
[69] Jiao L, Zhang L, Wang X, Diankov G and Dai H 2009 Narrow graphene nanoribbons from carbon nanotubes *Nature* **458** 877–80
[70] Kosynkin D V, Higginbotham A L, Sinitskii A, Lomeda J R, Dimiev A, Price B K and Tour J M 2009 Longitudinal unzipping of carbon nanotubes to form graphene nanoribbons *Nature* **458** 872–6
[71] Kosynkin D V, Lu W, Sinitskii A, Pera G, Sun Z and Tour J M 2011 Highly conductive graphene nanoribbons by longitudinal splitting of carbon nanotubes using potassium vapor *ACS Nano* **5** 968–74
[72] Lim J *et al* 2016 Dopant-specific unzipping of carbon nanotubes for intact crystalline graphene nanostructures *Nat. Commun.* **7** 1–9

[73] Ko D, Choi J, Yan B, Hwang T, Jin X, Kim J M and Piao Y 2020 A facile and scalable approach to develop electrochemical unzipping of multi-walled carbon nanotubes to graphene nanoribbons *J. Mater. Chem.* A **8** 22045–53

[74] Hawaldar R, Merino P, Correia M R, Bdikin I, Grácio J, Méndez J, Martín-Gago J A and Singh M K 2012 Large-area high-throughput synthesis of monolayer graphene sheet by hot filament thermal chemical vapor deposition *Sci. Rep.* **2** 1–9

[75] Zhang Y I, Zhang L and Zhou C 2013 Review of chemical vapor deposition of graphene and related applications *Acc. Chem. Res.* **46** 2329–39

[76] An H, Lee W J and Jung J 2011 Graphene synthesis on Fe foil using thermal CVD *Curr. Appl. Phys.* **11** S81–5

[77] Huang M *et al* 2018 Highly oriented monolayer graphene grown on a Cu/Ni (111) alloy foil *ACS Nano* **12** 6117–27

[78] Starodub E, Maier S, Stass I, Bartelt N C, Feibelman P J, Salmeron M and McCarty K F 2009 Graphene growth by metal etching on Ru (0001) *Phys. Rev.* B **80** 235422

[79] Preobrajenski A B, Ng M L, Vinogradov A S and Mårtensson N 2008 Controlling graphene corrugation on lattice-mismatched substrates *Phys. Rev.* B **78** 073401

[80] Ye C *et al* 2021 Layer-by-layer stacked graphene nanocoatings by Marangoni self-assembly for corrosion protection of stainless steel *Chin. Chem. Lett.* **32** 501–5

[81] Jones A C and Hitchman M L (ed) 2009 *Chemical Vapour Deposition: Precursors, Processes and Applications* (Cambridge: Royal Society of Chemistry)

[82] Jacobberger R M *et al* 2015 Direct oriented growth of armchair graphene nanoribbons on germanium *Nat. Commun.* **6** 1–8

[83] Park B J, Choi J S, Eom J H, Ha H, Kim H Y, Lee S, Shin H and Yoon S G 2018 Defect-free graphene synthesized directly at 150 °C via chemical vapor deposition with no transfer *ACS Nano* **12** 2008–16

[84] Vlassiouk I V *et al* 2018 Evolutionary selection growth of two-dimensional materials on polycrystalline substrates *Nat. Mater.* **17** 318–22

[85] Pekdemir S, Onses M S and Hancer M 2017 Low temperature growth of graphene using inductively-coupled plasma chemical vapor deposition *Surf. Coat. Technol.* **309** 814–9

[86] Kato T and Hatakeyama R 2012 Site-and alignment-controlled growth of graphene nanoribbons from nickel nanobars *Nat. Nanotechnol.* **7** 651–6

[87] Yeh N C, Hsu C C, Bagley J and Tseng W S 2019 Single-step growth of graphene and graphene-based nanostructures by plasma-enhanced chemical vapor deposition *Nanotechnology* **30** 162001

[88] Zhai Z, Shen H, Chen J, Li X and Jiang Y 2019 Direct growth of nitrogen-doped graphene films on glass by plasma-assisted hot filament CVD for enhanced electricity generation *J. Mater. Chem.* A **7** 12038–49

[89] Tsang D K H, Lieberthal T J, Watts C, Dunlop I E, Ramadan S, Armando E and Klein N 2019 Chemically functionalised graphene FET biosensor for the label-free sensing of exosomes *Sci. Rep.* **9** 1–10

[90] Lin W H, Chen T H, Chang J K, Taur J I, Lo Y Y, Lee W L, Chang C S, Su W B and Wu C I 2014 A direct and polymer-free method for transferring graphene grown by chemical vapor deposition to any substrate *ACS Nano* **8** 1784–91

[91] Chen Y, Jiang K, Zang H, Ben J, Zhang S, Shi Z, Jia Y, Lü W, Li D and Sun X 2021 Growth of high-quality wafer-scale graphene on dielectric substrate for high-response ultraviolet photodetector *Carbon* **175** 155–63

[92] Bekdüz B, Kaya U, Langer M, Mertin W and Bacher G 2020 Direct growth of graphene on Ge (100) and Ge (110) via thermal and plasma enhanced CVD *Sci. Rep.* **10** 1–11

[93] Lu C H, Leu C M and Yeh N C 2021 Single-step direct growth of graphene on Cu ink toward flexible hybrid electronic applications by plasma-enhanced chemical vapor deposition *ACS Appl. Mater. Interfaces* **13** 6951–9

[94] Coroş M, Pogăcean F, Măgeruşan L, Socaci C and Pruneanu S 2019 A brief overview on synthesis and applications of graphene and graphene-based nanomaterials *Front. Mater. Sci.* **13** 23–32

[95] Mitsuhashi F, Okada M, Tateno Y, Nakabayashi T, Ueno M, Nagasawa H, Fukidome H and Suemitsu M 2017 Extremely uniform epitaxial growth of graphene from sputtered SiC films on SiC substrates *MRS Adv.* **2** 51–6

[96] Liu Z *et al* 2020 Laser-induced growth of large-area epitaxial graphene with low sheet resistance on 4H-SiC (0001) *Appl. Surf. Sci.* **514** 145938

[97] Briggs N, Gebeyehu Z M, Vera A, Zhao T, Wang K, Duran A D L F, Bersch B, Bowen T, Knappenberger K L and Robinson J A 2019 Epitaxial graphene/silicon carbide intercalation: a minireview on graphene modulation and unique 2D materials *Nanoscale* **11** 15440–47

[98] Zhu H, Shi Z, Zhang C, Gao B, Chen J, Ding J, Jin M, Wu T and Yu G 2022 Molten Ga–Pd alloy catalyzed interfacial growth of graphene on dielectric substrates *Appl. Surf. Sci.* **576** 151806

[99] Wu F C and Tseng R L 2006 Preparation of highly porous carbon from fir wood by KOH etching and CO_2 gasification for adsorption of dyes and phenols from water *J. Colloid Interface Sci.* **294** 21–30

[100] Li X, Xing W, Zhuo S, Zhou J, Li F, Qiao S Z and Lu G Q 2011 Preparation of capacitor's electrode from sunflower seed shell *Bioresour. Technol.* **102** 1118–23

[101] Worsley M A, Pauzauskie P J, Olson T Y, Biener J, Satcher J H Jr and Baumann T F 2010 Synthesis of graphene aerogel with high electrical conductivity *J. Am. Chem. Soc.* **132** 14067–69

[102] Yu L and Song R 2019 Foamed graphene flakes from the one-step pyrolysis of PEG/MgO composites for supercapacitors and lithium-ion batteries *New J. Chem.* **43** 17298–305

[103] Gao Z, Jin Z, Ji Q, Tang Y, Kong J, Zhang L and Li Y 2018 Crumpled graphene prepared by a simple ultrasonic pyrolysis method for fast photodetection *Carbon* **128** 117–24

[104] Yan Y, Nashath F Z, Chen S, Manickam S, Lim S S, Zhao H, Lester E, Wu T and Pang C H 2020 Synthesis of graphene: potential carbon precursors and approaches *Nanotechnol. Rev.* **9** 1284–314

[105] AlShammari A S, Halim M M, Yam F K, Chahrour K M, Raypah M E and Kaus N H M 2021 The effect of spray cycles on the morphological, structural, and optical properties of rGO thin film deposited using spray pyrolysis technique *Mater. Sci. Semicond. Process.* **127** 105655

[106] Zhang H, Ding F, Li H, Qu F, Meng H and Gu H 2019 Controlled synthesis of monolayer graphene with a high quality by pyrolysis of silicon carbide *Mater. Lett.* **244** 171–4

[107] Lee N E, Lee S Y, Lim H S, Yoo S H and Cho S O 2020 A novel route to high-quality graphene quantum dots by hydrogen-assisted pyrolysis of silicon carbide *Nanomaterials* **10** 277

[108] Kong X *et al* 2020 Synthesis of graphene-like carbon from biomass pyrolysis and its applications *Chem. Eng. J.* **399** 125808

[109] Kamali A R, Yang J and Sun Q 2019 Molten salt conversion of polyethylene terephthalate waste into graphene nanostructures with high surface area and ultra-high electrical conductivity *Appl. Surf. Sci.* **476** 539–51

[110] Kamali A R 2020 Production of advanced materials in molten salts *Green Production of Carbon Nanomaterials in Molten Salts and Applications* (Singapore: Springer) pp 5–18

[111] Kamali A R and Yang J 2020 Effect of molten salts on the structure, morphology and electrical conductivity of PET-derived carbon nanostructures *Polym. Degrad. Stab.* **177** 109184

[112] Jung S, Myung Y, Kim B N, Kim I G, You I K and Kim T 2018 Activated biomass-derived graphene-based carbons for supercapacitors with high energy and power density *Sci. Rep.* **8** 1–8

[113] Zhang H, Zhang Z, Luo J D, Qi X T, Yu J, Cai J X, Wei J C and Yang Z Y 2019 A chemical blowing strategy to fabricate biomass-derived carbon-aerogels with graphene-like nanosheet structures for high-performance supercapacitors *Chem. Sus. Chem.* **12** 2462–70

[114] Roy A, Kar S, Ghosal R, Naskar K and Bhowmick A K 2021 Facile synthesis and characterization of few-layer multifunctional graphene from sustainable precursors by controlled pyrolysis, understanding of the graphitization pathway, and its potential application in polymer nanocomposites *ACS Omega* **6** 1809–22

[115] Zhang S *et al* 2020 Sustainable production of value-added carbon nanomaterials from biomass pyrolysis *Nat. Sustain.* **3** 753–60

[116] Yan Y, Meng Y, Zhao H, Lester E, Wu T and Pang C H 2021 Miscanthus as a carbon precursor for graphene oxide: a possibility influenced by pyrolysis temperature *Bioresour. Technol.* **331** 124934

[117] Massoud H Z 1993 Rapid Thermal Growth and Processing of Dielectrics *Rapid Thermal Processing* ed R B Fair (New York: Elsevier) pp 45–77

[118] Prekodravac J R, Kepić D P, Colmenares J C, Giannakoudakis D A and Jovanović S P 2021 A comprehensive review on selected graphene synthesis methods: from electrochemical exfoliation through rapid thermal annealing towards biomass pyrolysis *J. Mater. Chem.* C **9** 6722–48

[119] Bleu Y, Bourquard F, Gartiser V, Loir A S, Caja-Munoz B, Avila J, Barnier V, Garrelie F and Donnet C 2019 Graphene synthesis on SiO_2 using pulsed laser deposition with bilayer predominance *Mater. Chem. Phys.* **238** 121905

[120] Prekodravac J R, Jovanović S P, Holclajtner-Antunović I D, Peruško D B, Pavlović V B, Tošić D D, Todorović-Marković B M and Marković Z M 2014 Monolayer graphene films through nickel catalyzed transformation of fullerol and graphene quantum dots: a Raman spectroscopy study *Phys. Scr.* **2014** 014030

[121] Zhuang B, Li S, Li S and Yin J 2021 Ways to eliminate PMMA residues on graphene superclean graphene *Carbon* **173** 609–36

[122] Barkan T 2019 Graphene: the hype versus commercial reality *Nat. Nanotechnol.* **14** 904–6

[123] Zurutuza A and Marinelli C 2014 Challenges and opportunities in graphene commercialization *Nat. Nanotechnol.* **9** 730–4

[124] Wyss K M, Luong D X and Tour J M 2022 Large-scale syntheses of 2D materials: flash joule heating and other methods *Adv. Mater.* **34** 2106970

[125] Luong D X *et al* 2020 Gram-scale bottom-up flash graphene synthesis *Nature* **577** 647–51

[126] Stanford M G *et al* 2020 Flash graphene morphologies *ACS Nano* **14** 13691–99

[127] Algozeeb W A, Savas P E, Luong D X, Chen W, Kittrell C, Bhat M, Shahsavari R and Tour J M 2020 Flash graphene from plastic waste *ACS Nano* **14** 15595–5604
[128] Advincula P A, Luong D X, Chen W, Raghuraman S, Shahsavari R and Tour J M 2021 Flash graphene from rubber waste *Carbon* **178** 649–56
[129] Chen W *et al* 2022 Heteroatom-doped flash graphene *ACS Nano* **16** 6646–56
[130] Beckham J L, Wyss K M, Xie Y, McHugh E A, Li J T, Advincula P A, Chen W, Lin J and Tour J M 2022 Machine learning guided synthesis of flash graphene *Adv. Mater.* **34** 2106506

Chapter 3

Forms of graphene I—graphene oxide and reduced graphene oxide

Murilo H M Facure, Rodrigo Schneider, Augusto D Alvarenga, Luiza A Mercante and Daniel S Correa

Since the discovery of graphene in 2004, 2D materials have attracted huge interest from the research community. Among several graphene-based materials, graphene oxide (GO) and reduced graphene oxide (rGO) stand out in this scenario, being widely investigated and applied in many different areas. GO is the oxidized form of graphene, in which oxygen-containing functional groups are attached to its surface. GO reduction is performed by removing its oxygen functionalities, giving rise to rGO, a material with structure and properties much closer to that of pristine graphene. An important advantage that enabled their extensive use is that they can be easily synthesized using simple and inexpensive solution-based methods with high yields. GO and rGO present many opposite characteristics, and the routes employed during the synthesis play a crucial role in their final properties. In this chapter, the basic aspects of GO and rGO are addressed and their main synthetic routes and functionalization methods are reviewed. Next, their physicochemical properties are discussed and some fundamental characterization techniques are explored. Finally, up-to-date challenges and prospects involving graphene-based materials are also discussed.

3.1 Introduction

Since the mid-20th century, many advances related to nanotechnology have been achieved. Among them, the discovery of graphene in 2004 is certainly one of the most notable [1–4]. Along with the great previous interest in carbon nanotubes (CNTs) [5], the development of 2D materials (which were previously not believed to be possible) with enhanced electrical and thermal conductivities, high

doi:10.1088/978-0-7503-3999-5ch3

mechanical strength, optical transparency, and good chemical stability, boosted even more the research on carbon-based materials [3, 6, 7]. Such features made graphene be considered the wonder material of the 21st century. Although some recent advances have been reported [8], the difficulty of producing high-quality graphene on a large scale, in addition to some limitations of use, such as in solutions, were challenges that needed to be overcome to use graphene in commercial applications [3, 7, 9].

Even so, the discovery of graphene has raised a huge interest in the scientific community about graphene-based materials, especially GO and rGO [9, 10]. GO is the oxidized form of graphene. In addition to the hexagonal carbon structure similar to that of graphene, GO also presents oxygen functionalities, such as hydroxyls, carbonyls, carboxylic acids, epoxy, and alkoxy groups, and its degree of oxidation can be evaluated by the C/O ratio [3, 9–11]. GO is generally obtained through the oxidation of graphite, which will initially give rise to graphite oxide. Such a process, responsible for anchoring the functional oxygen groups in the graphite layers, enables the dispersion of the material in a wide range of solvents. In this way, when graphite oxide is exfoliated, it gives rise to GO, i.e., the materials differ only structurally, in relation to the number of stacked layers, and not chemically [3]. In addition to the easy dispersion acquired, the oxidation of the graphite layers gives rise to many different characteristics compared to graphene, especially regarding the electrical and electrochemical properties [3, 10].

One strategy to synthesize a material with characteristics closer to those of graphene is to perform the chemical reduction of GO to obtain rGO [10, 12]. The reduction process can be carried out using different synthetic routes, such as chemical reactions, thermal and electrochemical processes [3, 10, 13]. The reduction is responsible for removing the oxygen functionalities from the GO layers, increasing the C/O ratio, and, consequently, increasing the graphitic characteristic of the material. However, such a process, regardless of the route used to perform the reduction, is not able to entirely remove the oxygen functionalities and can generate defects in the structure of the material [3, 12]. Thus, it is important to differentiate rGO from graphene, which is often called pristine graphene to emphasize the presence of a perfect structure. The formation of rGO can be observed by a change in the reaction medium color and the dispersion of the material since the oxygen groups are removed and the hydrophobicity is regained. The synthesis of rGO is an interesting alternative to achieve, in an easy and viable way, and with a high yield, a material with properties similar to those of pristine graphene with excellent performance in many applications [3, 10, 12].

The term 'graphene' as a reference to the single-atom-thick sheet of sp^2-bonded carbon atoms arranged in hexagonal rings was coined in 1986 by Boehm *et al* [14]. Although 'graphene' term has been overused since it should refer only to the pristine isolated monolayer [15], it has been demonstrated that the increase in the stacked-layer number can significantly change the electronic properties of the material [7]. Bilayer graphene still holds approximately the same electronic properties and is also considered a zero-gap semiconductor. However, from three layers or more, the valence and conduction bands start to overlap. The presence of more than 10 layers leads to graphite-like characteristics. In this way, the number of layers can be furnished by the appropriate nomenclature, i.e., bilayer (2 sheets) or few-layer (3–10 sheets) graphene [7, 15].

Interestingly, the first reports about the synthesis of graphite oxide and attempts to remove its oxygen functionalities are from much earlier than the discovery of graphene, dating back to 1859 [16] and 1963 [17], respectively [3]. However, the achievement of graphene increased the interest in similar materials and, consequently led to an enhancement in the efforts to exfoliate and reduce graphite oxide. In figure 3.1, it is possible to observe how the occurrence of the term 'graphene oxide' increased in published works after 2010, the year in which Geim and Novoselov were awarded the Nobel prize for the discovery of graphene [1]. Therefore, it is undeniable that the development of graphene represents a milestone from which many advances in nanotechnology have been achieved [4]. In addition to those involving graphene itself, there are also those involving similar materials, such as GO, rGO, graphene quantum dots (GQDs), and other 2D materials.

The GO structure has been the subject of discussion since the first synthesis of graphite oxide, which was further intensified after the discovery of graphene. The elucidation of the structure is essential to understand the materials' properties and obtain the best benefit for a given application [3]. The first structure model of graphite oxide was proposed by Hofmann and Holst in 1939 [18], whose model was

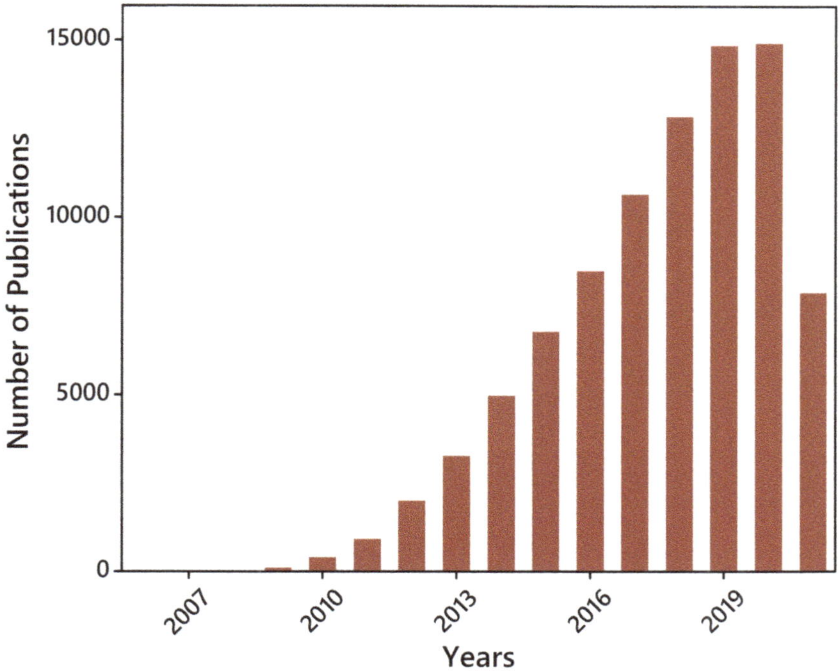

Figure 3.1. Number of scientific works published related to GO per year. Data obtained from Web of Science accessed in August 2021 using the keyword 'Graphene Oxide'.

based on 1,2-epoxides throughout the whole sp^2 graphitic structure. To add hydrogens, in 1947 [19] Ruess introduced hydroxyl and 1,3-epoxides groups to the structure model. After some advances obtained by other groups, the use of new techniques, such as nuclear magnetic resonance (NMR), enabled the achievement of more precise graphite oxide structure models [3, 10]. Lerf and Klinowski, at the end of the 1990s [20], proposed a graphite oxide structure with two different regions: the aromatic regions containing unoxidized benzene rings and the regions with aliphatic six-membered rings, as depicted in figure 3.2(a). The authors proposed that hydroxyl and epoxy groups were in the core of the sheets, while the carboxylic acids were at the edges. The Dékány model (figure 3.2(b)) proposed in 2006 introduced phenol groups in addition to hydroxyl, ether, ketone, and quinone groups [21]. The model also differentiates two regions: one with trans-linked cyclohexane chairs and the other with flat hexagons. Later, in 2009, Ajayan's Group proposed a new model [22] including the presence of five- and six-membered-ring lactols in the structure of graphite oxide, as shown in figure 3.2(c). Since GO is the exfoliated form of graphite oxide and the synthetic approach to produce GO can result in different structures, these models are conveniently used for GO. In this regard, the Lerf–Klinowski, Dékány, and Ajayan models are considered the most suitable to represent the GO structure and the most likely to be obtained [3, 22].

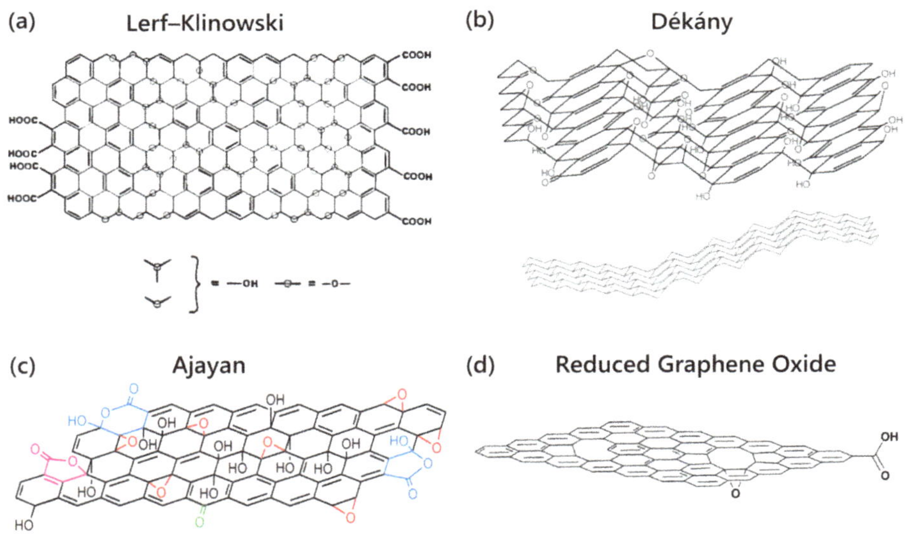

Figure 3.2. (a) Lerf–Klinowski model of the GO structure with hydroxyl and epoxy groups in the core and carboxylic acids at the edges. Reproduced with permission [20]. Copyright 1998 American Chemical Society. (b) Dékány model of the GO with the surface oxygen groups and the folded carbon skeleton. Reproduced with permission from [21]. Copyright 2006 American Chemical Society. (c) The GO structure proposed by Ajayan and co-authors introducing the lactol rings (blue). Reproduced with permission [22]. Copyright 2009 Springer Nature. (d) rGO model with a high degree of reduction with few remaining oxygen functionalities and defects in the carbon structure. Reproduced with permission [9]. Copyright 2010 Wiley.

The structure of rGO is not a topic of intense discussion, as is the case of GO. The route used for the reduction process is surely a determining factor in the final structure of rGO [10, 12, 23]. Since the presence of oxygen functionalities leads to a decrease in the electrical conductivity of the material, most of the processes usually attempt to remove most of these groups without causing defects in the layer's structure. In this regard, combining more than one reduction route seems to be the most appropriate alternative [13, 22]. However, a slightly inferior reduction degree can also be chosen to maintain a higher dispersibility of the rGO in a given solvent while gaining enhanced electrical properties. In this case, the strategy is therefore related to the parameters used in the synthetic reduction routes, such as the reducing agent, reduction time, temperature, among others [23–25]. Figure 3.2(d) shows the schematic illustration of a highly reduced GO structure, but still containing remaining functionalities and defects in the carbon skeleton.

This chapter aims to provide the most important concepts of GO and rGO, starting with the synthetic routes used to obtain such materials. Next, the physicochemical properties of GO and rGO, as well as their characterization techniques and potential applications are addressed. Finally, the potential benefits to be achieved by GO and rGO applications in the near future are also discussed.

3.2 Synthesis

3.2.1 GO synthesis

The first oxidation of graphite was reported in 1859 by Benjamin C Brodie in an attempt to measure its molecular weight [16]. The basis of Brodie's graphite oxidation method relies on the reaction of graphite using strong acidic conditions, potassium chlorate, and fuming nitric acid for up to four days at 60 °C. This method took many days and consisted of repetitions of the oxidation steps intercalated with water rinsing. After drying in a vacuum at 100 °C, a transparent dried product was obtained. Similar to Brodie's oxidation reaction, the Staudenmaier procedure, described in 1898, proposed the oxidation of graphite by a mixture of nitric and sulfuric acids [26]. Another noteworthy approach related to GO synthesis was made in the 1930s by Hofmann and collaborators [18, 27, 28], employing concentrated sulfuric acid, nitric acid, and potassium chlorate. Over time, the oxidation of graphite went through optimizations and improvements to decrease the reaction time, the amount of reagents used, and to increase the safety regarding the chemical reactions for the oxidation process [3]. In 1958, Hummers and Offeman introduced the use of potassium permanganate as oxidizing agent, giving rise to a safer synthesis and a graphite oxide with a higher degree of oxidation [29]. This method contrasts the previous ones because it uses only one acid as solvent (concentrated sulfuric acid), sodium nitrate, and sodium permanganate salts, which is treated with 3% oxygen peroxide to produce the reduced form of manganese, manganese sulfate. The synthesis lasts approximately 2 h and is schematically illustrated in figure 3.3(a) [29, 30].

The method employed to obtain GO severely impacts its physicochemical properties and morphology. Poh *et al* compared three popular methodologies of

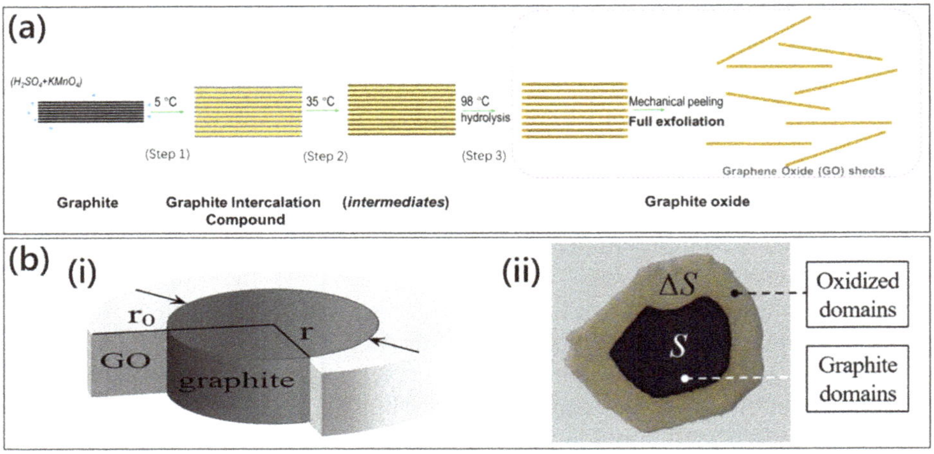

Figure 3.3. (a) Schematic diagram of Hummers' method for GO synthesis. Reproduced from [30]. Copyright 2019 with permisison from Elsevier. (b) (i) Edge-to-center oxidation progress direction and (ii) oxidized domains compared to graphite domains. ΔS and S are the oxidized domain and graphitic areas (μm^2), respectively. Reproduced from [32]. Copyright 2018 with permission from Elsevier.

GO production: Staudenmaier, Hofmann and Hummers [31]. The authors verified that the Hummers' method produced the largest distance between graphene layers. Also, the Hummers' method presented the highest degree of oxidation and a lower amount of N and S atoms in the GO sheets [31].

The mechanism of GO synthesis from graphite with the introduction of oxygen-containing moieties into graphite sheets, despite not yet completely described and understood, can be summarized by the following steps: (i) formation of oxidizing species, (ii) graphite intercalation compound formation, (iii) diffusion of oxidizing species between graphite sheets, and (iv) oxidation reaction [30, 32, 33]. Experimental [32] and theoretical studies [34] indicate that the mechanism of introducing oxygen moieties into graphite layers takes place initially at the edges and the defects of the sheets, as schematized in figure 3.3(b).

3.2.2 Modifications in GO synthesis

The Hummers' method has been widely used to obtain GO, as well as several other methods adapted from it. These new methods are so-called modified or improved Hummers' methods, in which some parameters are changed aiming to avoid the production of toxic gases (such as NO_2, N_2O_4, explosive ClO_2), reaching a less time consuming reaction, generating a graphite oxide with a higher degree of oxidation, and performing the doping of the oxidized material [3, 10, 35–37]. In this sense, in 2010, researchers developed a new methodology, called the Tour method, to produce improved GO (IGO) by performing the oxidation step with a mixture of sulfuric and phosphoric acids (9:1) and adding an equivalent amount of sodium permanganate salt with respect to the graphite mass [38]. The authors used x-ray photoelectron spectroscopy (XPS) to show that the oxidation level of IGO synthetized by the Tour method was higher than those obtained using the Hummers' process.

Features such as size [39, 40] and origin (natural or synthesized) [41] of the graphite employed as the carbon-based source in the GO synthesis will also affect the final properties of the product. For example, synthetic powder and natural flake graphite processed by the same Hummers' reaction conditions lead to the formation of GO with different oxidation degree levels and, therefore, different relative amounts of epoxy/hydroxyls, carbonyl, and carboxylates functionalities [42]. Lavin-Lopez and collaborators, aiming for less time consuming and lower-cost routes to obtain GO structures, investigated the influence of different synthetic parameters, including oxidation conditions, reaction time, and the number of washes [43]. The authors demonstrated that the reduction in reaction time, from 12 h to 3 h, presented no drastic effect on the structural disorder of the GO caused by the introduction of oxygen groups during the oxidation step. Additionally, it was also demonstrated that the use of 400 mL of H_2SO_4 and 15 g of graphite was more efficient than using 40 mL of H_3PO_4 and 360 mL of H_2SO_4 in the presence of only 3 g of graphite.

In the past years, GO synthesis methods have been extensively explored by researchers and, besides the traditional renowned oxidative methods (Hummer, Staudenmaier, Brodie, Tour), additional novel steps, such as mechanical milling [44], ultrasonic [45], hydrothermal/solvothermal [44], and microwave [46] processes have been employed to assist GO synthesis and modulate the final material's properties. In brief, these procedures contribute to enhancing layer intercalation, weakening van der Waals interactions, enhancing material dispersion, and avoiding restacking of the layers [10, 30].

3.2.3 rGO synthesis

rGO is obtained by the removal of the oxygen functionalities anchored on the GO surface, which partially restores of the π-conjugation system of the honeycomb carbon structure found in pristine graphene. As previously stated, the reduction of GO does not completely remove the oxygen functionalities and can generate structural defects in the carbon skeleton [3, 12, 47]. Kuila and co-workers reviewed the most commonly employed methodologies for GO reduction, including chemical, thermal, electrochemical, microwave, hydrothermal, and photo-reduction strategies, [48]. The analyses were made based on the modulation of rGO properties. Moreover, the authors also discussed in detail the energy costs, time consumption, and level of reduction achieved by each reduction method. Some interesting conclusions could be drawn from their studies, such as chemical modifications can be performed with toxic or eco-friendly reducing agents, the electrochemical method allowed the control of the reduction degree, the thermal method was not economical on energy consumption, and the microwave synthesis showed to be a fast reduction method [48].

Different traditional chemical reducing reactants can be employed to convert GO to rGO by chemical reaction, including sodium borohydride [49], hydrazine [50], hydrazine monohydrate [51], hydroxylamine hydrochloride [52], and hydroquinone [53]. In this sense, Lesiak and collaborators conducted the GO reduction with different reducing agents, including sodium borohydride ($NaBH_4$), hydrazine (N_2H_4), sodium hydroxide (NaOH), formaldehyde (CH_2O), and L-ascorbic acid ($C_6H_8O_6$) [43]. X-ray diffraction (XRD) and XPS analyses demonstrated that using $NaBH_4$ as the reduction agent yielded less-stacked rGO (with an interlayer spacing of 0.549 nm), while N_2H_4 yielded rGO with lower content of oxygen functionalities [43]. Recently, a novel eco-friendly GO reduction method using bio-reductants has gained popularity [54–56]. In this direction, Fernández-Merino *et al* reported the use of ascorbic acid as an ideal substitute for hydrazine in the GO reduction [57]. The authors compared many reducing agents, including sodium borohydride, pyrogallol, and ascorbic acid, and showed that only the latter presented highly reduced suspensions comparable to those provided using hydrazine. Additionally, they showed that ascorbic acid could also be used in common organic solvents, such as *N,N*-dimethylformamide and *N*-methyl-2-pyrrolidone. In another work, Abulizi and co-authors demonstrated that ultrasound-assisted GO reduction accelerated the rGO formation compared to the conventional mechanical mixing method, using ascorbic acid as the reducing agent [25].

Thermal reduction methods for rGO synthesis eliminate oxygen-containing groups by annealing (or heat treatment). In general, this reduction is achieved at high temperatures under different atmospheres (Ar, Ar/H_2, N_2) [48, 58, 59]. The reduction by thermal approach possibly starts with the remotion of intercalated water (≈140 °C–180 °C), followed by carboxyl evolution (≈180 °C–600 °C), and then the remotion of the hydroxyl groups (≈1000 °C) [60, 61]. Gao *et al* investigated the kinetics and thermodynamics of the deoxygenation of GO sheets and demonstrated that thermally rGO might differ from chemically rGO through the full/partial removal of different oxygen functionalities [62]. Additionally, microwave [63, 64] and hydrothermal/solvothermal [44, 50] methods can also be applied to GO reduction and are usually conducted at temperatures lower than ≈250 °C.

The electrochemical reduction is a rapid and effective reduction method in which the current applied against the GO allows the modulation of the rGO charge transport [65], reduction degree [66], and avoids the generation of impurities [47, 48, 67]. Regarding the sample preparation, electrochemically reduced GO can be prepared in a single step, when the GO is directly reduced from electrolyte dispersion, or in two steps, when the GO is firstly deposited onto the electrode substrate and then submitted to the electrochemical reduction [47].

Other important reducing methods include photo-reduction [68, 69] and laser irradiation-assisted methods [70–73]. The latter is a fast and easy operational procedure and chemical reactants-free. However, the reaction is confined to the specific site of GO that is laser irradiated. In this way, laser-irradiated GO substrates are promising platforms for achieving highly accurate patterns with a controlled reduction degree. This is highly suitable, for instance, to fabricate electrodes for electronics and sensor developments [72, 74, 75].

3.3 Functionalization of GO and rGO

Pre- and post-modification of carbon-based structures are commonly used strategies to boost their properties in order to achieve specifications for an intended application. For instance, preserving sp^2 carbon hybridization nature, and consequently, π-orbitals that guarantee an improved electron transport characteristic can be easily achieved by non-covalent modifications of rGO. The non-covalent functionalization occurs basically through π–π interactions, van der Waals forces, *in situ* growth/coating of other nanomaterials, ionic interactions, and hydrogen bonds with oxygen-functional moieties, which lead to substantial improvements in electrical and electrochemical properties [76, 77].

10,15,20-Tetrakis(4-trimethylammoniophenyl)porphyrin tetra(*p*-oluenesulfona-te) (TPP-ammonium) is a π-bond rich porphyrin with recognized electron donor character. Multiple-bilayered GO/TPP-ammonium films were fabricated due to electrostatic and π–π interactions between porphyrin molecules and GO sheets [78]. The GO was then vapor reduced to rGO. The authors indicated that π–π interaction between rGO/TPP-ammonium was stronger than in GO/TPP-ammonium due to the enhanced π character of the rGO. A complete quenching of TPP-ammonium photoluminescence was observed in both composites. Several reports regarding the fabrication of nanodevices employing the non-covalent interactions using rGO and/or GO to obtain nanocomposites demonstrated the versatility of such modification. For instance, this approach can be used for the obtention of rGO-based composites for different sensing applications, including rGO-poly(*p*-phenylenediamine) for dopamine detection [79], polyethyleneimine-rGO for gallic acid detection [80], DNA and GO aptasensor (GO/Apt) for glycated human serum albumin (GHSA) detection [81], and MoS_2:GO composite for antibiotics detection [82].

In contrast, the covalent modification enhances the sp^3 character, affecting the π-orbital overlap. Covalent functionalization in rGO and GO sheets can be performed in edge and basal plane sites. Edge reactions generally occur through the activation of carboxyl groups, while basal covalent reactions take place mainly through ring-opening mechanism [83]. Among the several chemical reactions available, cycloaddition [84], grafting [85], radical polymerization [86], photo-chemical [87], grafting [85], Diels–Alder click reaction [88], amination [89], are examples of strategies used to obtain covalently functionalized rGO and GO materials.

3.4 Physical and chemical properties

3.4.1 Structure

As previously stated, GO and rGO can be described as typical monolayer graphene derivatives functionalized with different oxygen-functional groups, consisting of carbon, hydrogen, and oxygen [90]. However, these elements do not present fixed stoichiometric ratio due to various factors, including different aspects of graphite precursor, oxidation conditions (concentration, temperature, pH, etc), and reduction degree, being largely dependent upon the particular synthetic route employed [91, 92].

Furthermore, the oxygen group is not from a single kind but a collection of oxygen-containing groups, including hydroxyl and epoxy, generally located at the basal plane, and carbonyl, carboxyl (among others) located at the edges of the sheets [90]. Moreover, the topological defects, such as holes and disordered regions (areas of high oxidation), are also present and scattered across the basal plane, and the density of defects is also dependent on the oxidation level [91, 93]. As a result, the exact structure of graphene derivatives is still largely unknown, and GO/rGO may be described as a family of materials instead of a material with a specific composition [94]. Consequently, their structures must be properly described in terms of C/O ratio, surface area, lattice defects, particle size, spectroscopic features, among other aspects [91, 93, 95]. In this scenario, a broad collection of complementary analytic tools has been employed to investigate GO and rGO chemical composition and structural features [93, 96], as will be discussed in the next section.

3.4.2 Dispersibility

The essential criteria for a dispersion to be useful is that the graphene-based material must disperse in an appropriate solvent, at a useful concentration, and remain dispersed over a reasonable period of time [97]. In summary, the dispersibility follows the order: graphene < rGO < GO. The presence of the oxygen functionalities alters the non-covalent interactions between the sheets, endowing GO with excellent dispersibility and stability in water and other organic solvents, as illustrated in figure 3.4(a) [90, 98]. This is important for processing, allowing the use of solution processing techniques, including spraying, drop-casting, and spin-coating, and for further derivatization [93, 95]. The affinity of GO sheets towards solvent molecules has been assessed by physical parameters, such as dielectric constant, dipole moment, surface tension, and Hildebrand solubility parameter [90, 98, 99].

On the other hand, due to the partial removal of oxygen-containing groups, rGO is more hydrophobic than oxygen-enriched GO and much closer to the aromatic character of graphene [95]. As a consequence, in the absence of dispersing agents, rGO tends to agglomerate due to strong intermolecular (especially π–π and van der Waals) interactions, thus forming aggregates or restacking to form graphite in aqueous/polar medium [100]. For instance, Konios and co-workers evaluated the effect of the reduction process on GO dispersibility and stability in eighteen solvents [99]. The authors observed that GO presented long-term stability in polar solvents, such as *N*-methyl-2-pyrrolidone, ethylene glycol, and water. After the reduction process, rGO presented greater interaction with non-polar solvents, presenting high stability in *o*-dichlorobenzene and 1-chloronaphthalene. In another work, Konkena and Vasudevan evaluated the aqueous dispersibility of GO and rGO at different pHs [100]. GO sheets formed stable dispersions in the pH range of 4–11.5, with the lowest zeta potential of −54.3 mV at pH 10.3, while for rGO sheets, stable dispersions were observed in more basic media (8–11.5). Their results suggest that the stability of GO and rGO in water is a consequence of the presence of negative charges on the sheets due to the ionization of the different oxygen functionalities. They also revealed, by infrared spectroscopy studies, that in both GO and rGO, the negative charges arise

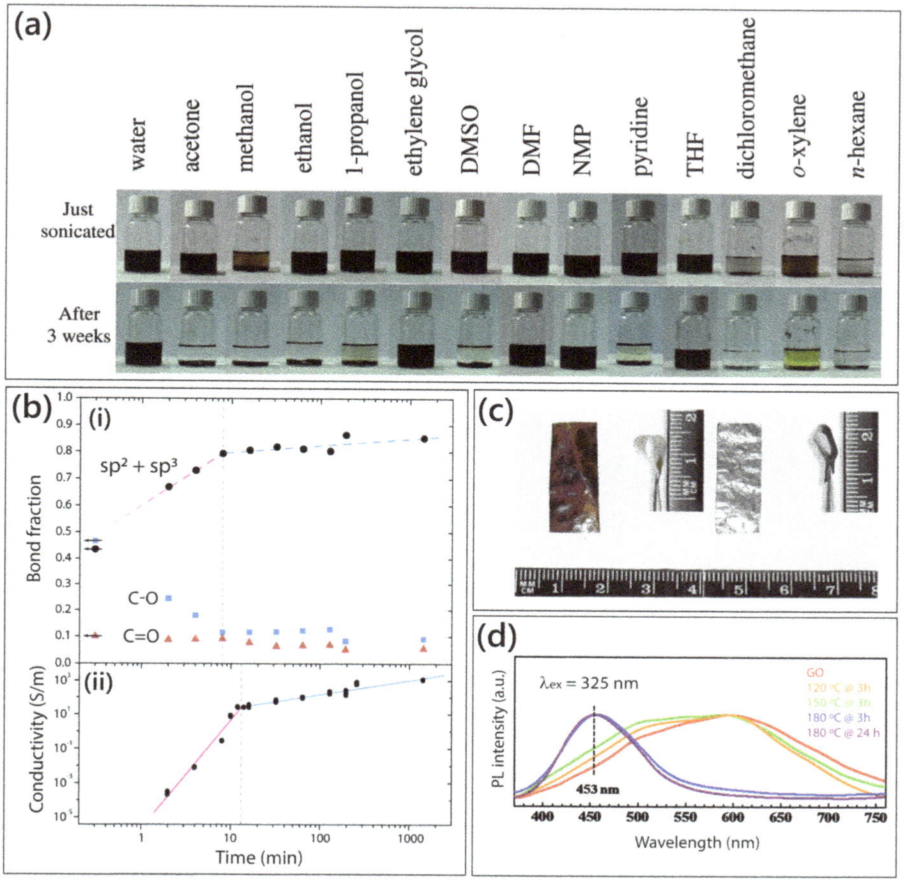

Figure 3.4. (a) Digital photographs of GO dispersions in water and 13 organic solvents immediately after sonication (top) and after three weeks (bottom). Reprinted with permission [98]. Copyright 2008 American Chemical Society. (b) (i) Evolution of (Csp^2 + Csp^3), C–O, and C=O functional groups as a function of the reduction reaction time and (ii) conductivity as a function of the reduction reaction time. Reprinted from [101], with permission from The Royal Society of Chemistry. (c) Optical photographs of GO (left) and rGO (right). After 1 h reduction in HI at 100 °C, the rGO film showed a shining metallic luster, due to the increase of electrical conductivity and reflectivity of visible light. Reprinted from [102], copyright (2010) with permission from Elsevier. (d) Normalized photoluminescence spectra of the GO and rGOs reduced at different temperatures for different periods. Reproduced [103] under the terms of the Creative Commons Attribution 3.0 unported license. Copyright 2014 *Nature*.

predominantly from the ionization of the carboxylic groups. However, the presence of phenolic and hydroxyl groups near the carboxylic groups lowers the pKa value of GO by stabilizing the carboxylate ion, which leads to superior water dispersibility. Therefore, according to the authors, the superior water dispersibility of GO is not only due to the larger amount of ionizable oxygen groups, but also due to the chemical nature of these functionalities.

3.4.3 Electrical conductivity

Pristine graphene is known as a zero-gap semiconductor and its electrical conductivity is associated with the long-range ordering of graphitic lattice [12, 104]. The introduction of functional groups breaks the conjugated structure and localizes π-electrons, decreasing the carrier mobility and carrier concentration [12]. Therefore, the electronic properties of the graphene derivatives are highly dependent on the degree of structural disorder and the electrical conductivity can be modulated by appropriately fine-tuning the oxidation/reduction parameters [12, 93, 105]. In this scenario, Guex *et al* investigated the electrical properties of GO as a function of reduction time by monitoring the increasing conductivity with a decreasing amount of C–O bonds [101]. XPS analysis showed that during the first 8–12 min of reaction, the increase in conductivity was mainly due to the reduction of hydroxyl and epoxy groups. Moreover, the variation of the content of C–O bond functionalities with the sum of Csp^2 and Csp^3 (figure 3.4(b-i)) revealed that the reduction of epoxy or hydroxyl groups had a greater influence on the re-establishment of the conductive network of rGO, as shown in figure 3.4(b-ii). Their reduction protocol allowed the achievement of high conductivity values (1500 S m^{-1}). It is important to mention that, although the reduction of GO can increase its conductivity, there is a consensus that the inferior electrical property of rGO, in comparison with pristine graphene sheet, originates from the presence of residual functional groups remaining after the reduction process and defects on the carbon structure. In addition to the surface chemistry (carbon and oxygen-functional groups), the surface topography (roughness) and physical parameters (film thickness) also have an influence on the electrical conductivity of GO and rGO, as demonstrated by Mohan and co-workers [106]. The authors observed that for both GO and rGO, films with smoother surfaces and lower thickness showed higher electrical conductivity.

3.4.4 Electrochemical properties

The high surface-to-volume ratio of GO and rGO, in conjunction with their reactive surface functional groups, make them very attractive for electrochemical applications, such as (bio)sensing, electrocatalysis, energy conversion, and field-effect transistors [93, 107, 108]. Furthermore, modulating the electrical conductivity allows the tunability of their electrochemical properties [91, 93, 108, 109]. Electrochemical impedance spectroscopy (EIS) has been commonly employed to investigate the electrochemical properties of graphene-based materials [110]. For instance, Casero *et al* reported the use of EIS to distinguish between the two forms of graphene: GO and rGO [111]. They observed that the GO-based electrode showed a finite-length diffusion, due to the covering of the GO surface by oxygen-containing functional groups. In addition, the resistance of charge transfer was found to be 70 and 21 Ω cm^2 for GO and rGO, respectively, showing that EIS is a reliable method for determining whether the graphene derivative is in its oxidized or reduced form. Lei and co-workers employed cyclic voltammetry and galvanostatic charge–discharge techniques to evaluate the electrochemical properties of

GO after reduction with urea [112]. The experimental results showed that the porous and partially repaired graphitic structure favored ion and electron transports, and the residual oxygen-functional groups enhanced the capacitance through additional Faradic redox processes. The outstanding electrochemical properties of GO and rGO have also been explored to develop electrochemical sensors and electroanalysis [107, 113–115].

3.4.5 Optical properties

The transition between insulating GO and conducting graphene state is accompanied by changes in optical properties [116]. For example, the reduction process will improve the reflection of the incident light, giving rGO film a metallic aspect (figure 3.4(c)), as a consequence of the increase in the charge carrier concentration and mobility [102]. In addition, the reduction of the yellow–brown GO colloidal suspension usually results in a black precipitate, which is probably a result of the hydrophobicity increase [12].

In contrast to pristine graphene, as a result of the heterogeneous electronic and atomic structures, GO and rGO present intrinsic fluorescence in the near-infrared, visible and ultraviolet regions [117]. The fluorescence originates from the recombination of electron–hole pairs, localized within the sp^2 domain embedded within a sp^3 matrix. The sp^2 domain regulates the local energy gap and consequently the emitted fluorescence wavelength [105]. Therefore, the maximum emission wavelength can be modulated by their lateral size and the amount and type of oxygen-containing groups [105, 117]. For instance, larger sp^2 domains (>2 nm) have smaller gaps, which may account for red-to-NIR emission [105]. Chuang *et al* evaluated the influence of the temperature during the thermal reduction on the photoluminescence properties of GO [103], as illustrated in figure 3.4(d). The PL spectra of GO and weakly reduced rGO (120 °C@3 h and 150 °C@3 h) showed asymmetric broad bands, ranging from 400 to 800 nm. On the other hand, strongly reduced rGOs (180 °C@3 h and 180 °C@24 h) mostly exhibited one band centered at around 453 nm. The original GO consisted of a significant amount of disorder-induced defect states within the π–π^* gap and exhibited a broad PL band centered at longer wavelengths. After the reduction process, the number of disorder-induced states decreased and an increased number of small sp^2 domains in the rGO resulted in blue luminescence.

GO-based materials also possess nonlinear optical (NLO) properties, which are highly beneficial for optoelectronic applications [93]. GO presents two NLO regimes (sp^2 and sp^3 domains): two-photon absorption dominates the nonlinear absorption for picosecond pulses in the sp^2 domain, while for nanosecond pulses, excited-state absorption influences the nonlinear response in the sp^3 domains. In this way, the NLO response can be modulated by controlling the degree and location of oxidation on the GO, as demonstrated by Wang and co-workers [118]. Upon excitation by a laser delivering 4 ns pulses at 532 nm, rGO with lower sp^3 domain size displayed larger NLO properties than GO and rGO with relatively few sp^2 hybrid carbon configuration ratios.

3.4.6 Magnetic properties

The magnetic properties of graphene derivatives are of great interest due to their potential for spintronics [119]. Due to the delocalized π-bonding network, pristine graphene has no unpaired electrons and, consequently, is a non-magnetic material [120]. However, over the years, it has been proved that cutting graphene sheets to create zig-zag edges and introducing structural defects by functionalization or chemical doping is responsible for the uncommon magnetic properties of graphene-based materials [119, 121]. Therefore, it can be expected that GO in different oxidation degrees exhibits different magnetic responses. For example, Bagani *et al* investigated the magnetic properties of GO upon chemical reduction [122]. Magnetic hysteresis curves revealed that the paramagnetic contribution of the magnetic moment was larger in GO compared to rGO. They claimed that the presence of paramagnetic moments was related to the topological defects and the removal of the OH-containing groups and, therefore, the topological defects lead to a lower paramagnetic moment in rGO compared to GO.

3.4.7 Mechanical properties

It is well-known that graphene exhibits fascinating mechanical properties with Young's modulus of approximately 1 TPa and intrinsic strength of around 130 GPa, which are promising for varied applications [104, 123]. The graphene functionalization generally reduces its mechanical robustness due to the insertion of structural defects [124]. Liu and colleagues, for instance, demonstrated that the oxidation of graphene has led to a decrease of Young's modulus [123]. Using first-principles calculations, they investigated the effects of crystallinity, coverage, and type of functional groups on the mechanical properties of GO. They found that mechanical properties of GO decrease with the increase of the amount of hydroxyl and epoxy groups. Recently, the mechanical properties of GO samples with different concentrations of epoxide and hydroxyl groups were studied through molecular dynamics simulation [125]. The obtained results revealed that increasing the epoxide coverage led to a deterioration in the mechanical properties of GO, which was associated with the increase of GO structural defects. Furthermore, the mechanical properties of GO/rGO sheets can be tuned by preparing composite materials [123]. For instance, Yang *et al* investigated the mechanical properties of rGO/poly(2-acrylamido-2-methylpropanesulfonic acid-co-acrylamide) (rGO/poly (AMPS-co-AAm)) nanocomposite hydrogels. They observed that the hydrogen-bond interactions between the residual oxygen groups of rGO and the polymer chains contributed to the enhancement of mechanical properties on both tensile strength and compressive strength [126].

3.5 Characterization

The differences of properties between GO and rGO can be evaluated using physicochemical characterization techniques [3, 10], as illustrated in figure 3.5. The most employed physicochemical characterization techniques for these materials

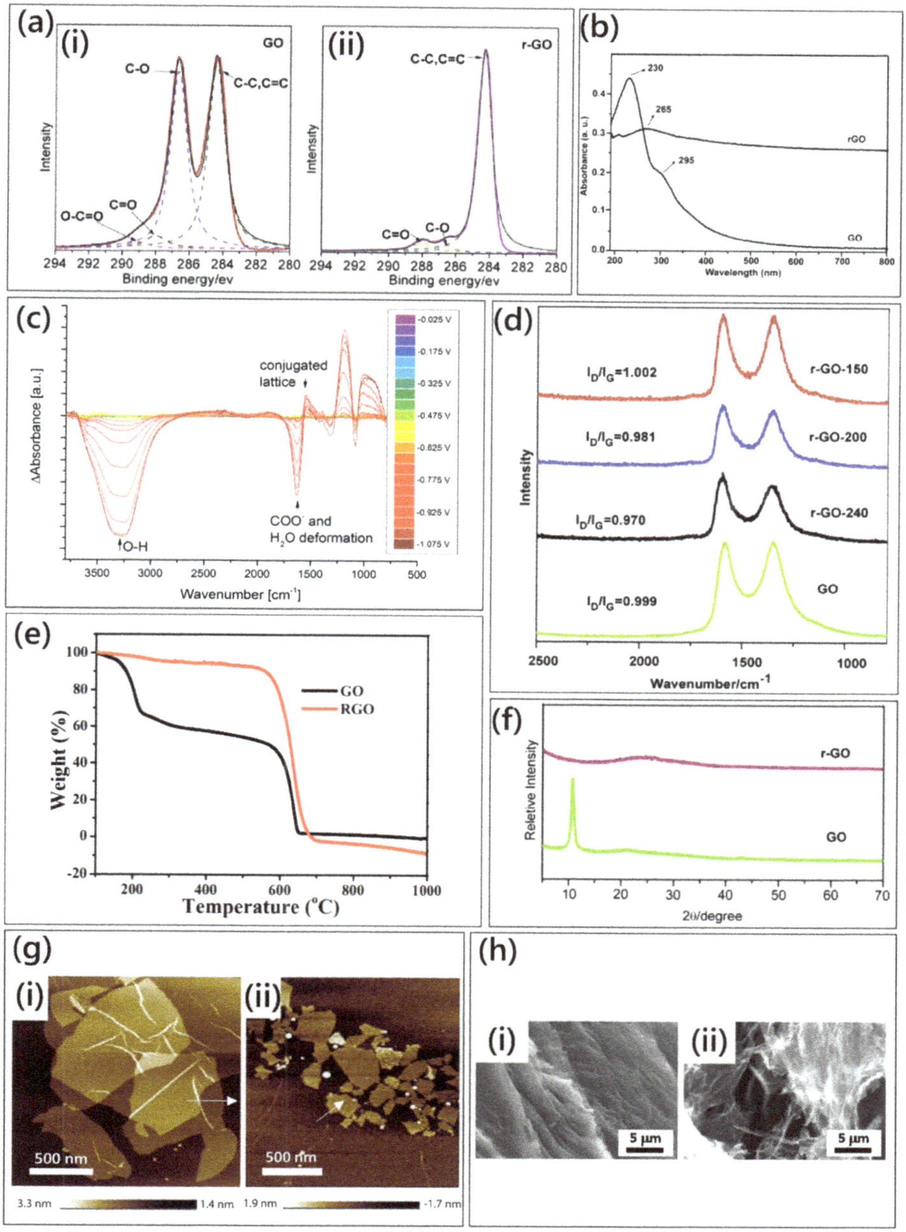

Figure 3.5. (a) C1s XPS spectra of (i) GO and (ii) rGO showing the peaks intensity reduction related to the oxygen-functional groups after chemical reduction. Reproduced from [135], copyright (2013) with permission from Elsevier. (b) UV–vis spectra of GO and rGO showing the disappearance of the band at approximately 300 nm (n–π* transitions) and the red-shift of the peak in 230–265 nm (π–π* transitions) after chemical reduction of the GO. Reproduced from [143] with permission from The Royal Society of Chemistry. (c) FTIR spectra of the progression of the GO chemical reduction in H_2O. Reproduced from [139], copyright (2017) with permission from Elsevier. (d) Raman spectra and I_D/I_G values of GO (green) and rGO at 150 °C (red), 200 °C

(blue), and 240 °C (black). The decrease in the I_D/I_G values indicates a higher reduction degree. Reproduced from [135], copyright (2013) with permission from Elsevier. (e) Typical TGA curves of GO and rGO. The weight loss between 200 °C and 600 °C for the GO curve is related to the evolution of the oxygen-functional groups. Reproduced from [144] with permission from The Royal Society of Chemistry. (f) XRD patterns of GO and rGO showing peak displacement upon GO reduction. Reproduced from [135], copyright (2013) with permission from Elsevier. (g) AFM images of (i) GO and (ii) rGO. Reproduced from [24], copyright (2018) with permission from Elsevier. (h) SEM images of (i) GO and (ii) rGO. rGO presents a more wrinkled surface due to the lack of oxygen functionalities. Reproduced from [24], copyright (2018) with permission from Elsevier.

are XPS, Fourier-transform infrared (FTIR), Raman, and UV–vis spectroscopies, XRD, zeta potential, among others, while their morphologies are generally investigated by atomic force microscopy (AFM), transmission electron microscopy (TEM), and scanning electron microscopy (SEM) [10, 127].

As previously mentioned, the oxygen functionalities present in GO are mainly in the form of hydroxyl and epoxy groups at the carbon core, while the carboxylic and carbonyl groups are predominantly distributed at the edges [90, 128]. Thus, larger sheets of GO tend to have a higher level of sp^2 hybridization, and epoxy and hydroxyl groups, while smaller sheets tend to have higher levels of carbonyl and carboxylic features. In this regard, XPS technique can be used to estimate the presence and quantity of these oxygen functionalities. XPS can also be used to evaluate the oxidation or reduction degree of GO and rGO, respectively. Such information can be obtained by the sp^2/sp^3 carbon ratio or the atomic ratio of carbon to oxygen (C/O) [129–131]. As illustrated in C1s XPS spectra presented in figure 3.5(a), the peaks related to the oxygenated groups decrease in intensity after the chemical reduction of GO.

Another way to monitor the GO chemical reduction is through UV–vis spectroscopy. The GO spectrum presents a characteristic peak near 230 nm ascribed to the π–π^* transitions related to the C=C of the aromatic rings and a shoulder at ca. 300 nm related to the n–π^* transitions of the C=O bonds [127]. As the GO is reduced, the peak at 230 nm redshifts to a region between 260 and 290 nm, and the band at 300 nm decreases or even disappears, as illustrated in figure 3.5(b) [127, 132, 133].

The presence of oxygen functionalities can also be observed using FTIR spectroscopy. The typical peak at c.a. 3412 cm^{-1} is representative of the hydroxyl groups, the peak related to the C=O carbonyl stretching is located at c.a. 1723 cm^{-1}, while the peaks of the C–O epoxy groups stretching are located at c.a. 1178 and 1073 cm^{-1}. The peak referring to the chain of alkenes of the aromatic rings stretching is found at c.a. 1620 cm^{-1}. When the GO is chemically reduced, all these representative peaks decrease in intensity (figure 3.5(c)), and the peak of the alkenes tends to blueshift from 1620 to 1630 cm^{-1} [134].

The monitoring of the extent of the GO reduction can also be followed by Raman spectroscopy, in which the G band (E_{2g} symmetry) at ~1580 cm^{-1} is related to sp^2 bonds, while the D band (A_{1g} symmetry) related to sp^3 bonds at ~1340 cm^{-1}, indicates the presence of oxygenated functional groups. In this way, monitoring GO reduction is usually assessed by the ratio between the intensities of the D and the G peaks (I_D/I_G). The higher its value, the more defects (oxygen functionalities) are present in the graphene sheet. Thus, the reduction of GO leads to a decrease in the I_D/I_G ratio, as shown in figure 3.5(d) [127, 128, 135]. However, the difference in proportions can be subtle. It is also possible to identify the reduction of GO in the 2D band at ~2700 cm^{-1} [64, 136]. In addition, recent works have reported a broad peak in the range of 1150–1200 cm^{-1} (D*) and a small peak located between 1500 and 1550 cm^{-1} (D″), which exhibit a dependence on the oxygen quantity level [136].

The crystalline degree and the presence of oxygen groups affect the thermal stability of GO and rGO. Thus, thermogravimetric analysis (TGA) can be used to evaluate such differences between GO and rGO. Specifically, due to the oxygen functionalities, the decomposition of GO starts at temperatures lower than that of rGO [134]. Therefore, the weight loss can also be used to evaluate the degree of oxidation/reduction of the material. The weight loss at temperatures lower than 100 °C can be attributed to the water adsorbed in the materials. Then, with temperatures between 100 °C and 360 °C, the unstable oxygen-containing functional groups in GO start to decompose. In temperatures superior to 360 °C, the weight loss is attributed to the combustion of the carbon skeleton [137]. In this way, highly reduced GO only exhibits this final stage in the TGA curves [138]. The TGA curves of GO and rGO are shown in figure 3.5(e).

XRD analyses have been used to analyze the number and distance between the layers of GO and rGO. The graphite has a strong and narrow peak at $2\theta = 26.6°$ with a spacing between layers of about 3.35 Å. The oxidation and exfoliation of graphite produce GO for which a diffractogram shows a broad peak at $2\theta \approx 10°$, due to the deterioration of the crystalline structure. The spacing between GO sheets is close to 9.30 Å due to the oxygen functionalities that prevent the compaction of the sheets. Upon reduction, this peak moves to higher 2θ values (up to 25°), indicating the removal of these oxygen groups and a consequent decrease in the space between rGO layers, reaching a value close to those of graphite (figure 3.5(f)). Like the GO peak, this peak is not perfectly defined due to the lack of order and residual oxygen groups [24].

As previously discussed, the oxygen functionalities or the lack of them also reflect in the material's surface charge and, consequently, in their dispersion. The zeta potential of GO is −28 mV at pH 2 and reaches the lowest value of −54.3 mV at pH 10.3 [100, 139]. The rGO, on the other hand, has a surface charge of −8 mV at pH 2, maintaining a value of about −25 mV between pH 4 and 8, and reaching the lowest value of −35 mV at pH 11 [140].

Regarding morphological characterizations, AFM, SEM, and TEM techniques are the most employed ones. AFM images (as the one in figure 3.5(g)) allow the determination of 2D material sizes and evaluation of the presence of single or multiple layers [141]. A field emission scanning electron microscope (FESEM) allows one to see the shape and size of the sheets. The rGO sheets have a higher level of roughness (compared to GO) due to the lack of oxygen functionalities, once these can hinder the formation of wrinkles, which is the case of GO [134]. High-resolution transmission electron microscopy (HR-TEM) allows the visualization of the ordering and the presence of defects in the 2D sheets of GO and rGO. This ordering reflects in the electrical properties, as well as the presence of more intense oxygen clusters in the GO that gives it an insulating characteristic observed in cyclic voltammetry [64, 142].

3.6 Applications

Due to the distinct properties of GO and rGO, the level of reduction/oxidation can be modulated to attain customized performances for a given application. For instance, for pollutant adsorption applications, the level of oxidation or reduction will depend on the target pollutant. Pollutants with a strong cationic character without the presence of aromatic rings will have a greater chance of being adsorbed on platforms containing GO [145]. Pollutants with aromatic chains, on the other hand, will have better interaction with rGO through π–π interactions [146]. For electrochemical sensors, the oxygenation level needs to be modulated because a high oxygenation level impairs the electronic conductivity. Thus, the restoration of sp^2 connections in rGO enables application in sensors [142], photovoltaic cells [147], supercapacitors [148], among others.

Due to the presence of the oxygen-containing functional groups, the choice to use GO can be related to applications in which a good dispersibility of the material is important [99], in applications where the functionalities act as actives sites for target molecules aiming at adsorption or detection [149, 150], or to combine materials in nanocomposites fabrication [82, 151], for example. On the other hand, the use of rGO is generally related to applications in which good electrical or thermal conductivities are desired [115, 152] or in applications in which the π–π interaction is fundamental [146]. Applications of graphene-based materials will be further discussed and illustrated in the next chapters of this book.

3.7 Conclusions and perspectives

In this book chapter, a theoretical background and a brief scientific history involving GO and rGO have been explored, as well as their most important production processes, their main characteristics, and the characterization techniques usually employed to characterize these materials.

The tremendous interest in graphene-based materials has led to many advances regarding GO and rGO. While pristine graphene still has limitations to be used in some laboratory applications and to be produced at large scales for commercial use, the development of methods for the synthesis of GO and rGO has experienced considerable advances recently. The accessibility of production and excellent performance of these materials in many areas have led to a significant increase in research and investigation. For all the above-mentioned reasons, the interest and applications of GO and rGO must remain very intense for the next years. Progress related to their physical–chemical and structural characteristics appears to be well advanced and some specific progress can still be achieved. Regarding their fabrication, despite the numerous milestones, advances can still be made, and certainly, in the relatively near future, such materials will tend to gain even more space in commercial applications, as many companies worldwide have already begun working with them. Even with the arrival of a commercially viable production process for graphene, GO and rGO will certainly enjoy a prominent place and will be widely used in several areas as their production costs are expected to drop more rapidly in the coming years.

Acknowledgments

The authors acknowledge the financial support from Fundação de Amparo à Pesquisa do Estado de São Paulo (FAPESP) (2018/22214-6, 2018/18468-2, 2017/10582-8), Conselho Nacional de Desenvolvimento Científico e Tecnológico (CNPq) (CNPq/445982/2020-9, 381405/2021-4), MCTI-SisNano (CNPq/402287/2013-4), and Rede Agronano-EMBRAPA from Brazil.

References

[1] Novoselov K S, Geim A K, Morozov S V, Jiang D E, Zhang Y, Dubonos S V, Grigorieva I V and Firsov A A 2004 Electric field effect in atomically thin carbon films *Science* **306** 666–9

[2] Novoselov K S, Jiang D, Schedin F, Booth T J, Khotkevich V V, Morozov S V and Geim A K 2005 Two-dimensional atomic crystals *Proc. Natl Acad. Sci. USA* **102** 10451–3

[3] Chua C K and Pumera M 2014 Chemical reduction of graphene oxide: a synthetic chemistry viewpoint *Chem. Soc. Rev.* **43** 291–312

[4] Ren W and Cheng H M 2014 The global growth of graphene *Nat. Nanotechnol.* **9** 726–30

[5] Iijima S 1991 Helical microtubules of graphitic carbon *Nature* **354** 56–8

[6] Novoselov K S, Colombo L, Gellert P R, Schwab M G and Kim K 2012 A roadmap for graphene *Nature* **490** 192–200

[7] Geim A K and Novoselov K S 2010 The rise of graphene *Nanoscience and Technology: A Collection of Reviews from Nature Journals* 11–9

[8] Islam A, Mukherjee B, Pandey K K and Keshri A K 2021 Ultra-fast, chemical-free, mass production of high quality exfoliated graphene *ACS Nano* **15** 1775–84

[9] Compton O C and Nguyen S T 2010 Graphene oxide, highly reduced graphene oxide, and graphene: versatile building blocks for carbon-based materials *Small* **6** 711–23

[10] Mao S, Pu H and Chen J 2012 Graphene oxide and its reduction: modeling and experimental progress *RSC Adv.* **2** 2643–62

[11] Smith A T, LaChance A M, Zeng S, Liu B and Sun L 2019 Synthesis, properties, and applications of graphene oxide/reduced graphene oxide and their nanocomposites *Nano Mater. Sci.* **1** 31–47

[12] Pei S and Cheng H M 2012 The reduction of graphene oxide *Carbon* **50** 3210–28

[13] Ju H M, Huh S H, Choi S H and Lee H L 2010 Structures of thermally and chemically reduced graphene *Mater. Lett.* **64** 357–60

[14] Boehm H P, Setton R and Stumpp E 1986 Nomenclature and terminology of graphite intercalation compounds *Carbon* **24** 241–5

[15] Bianco A *et al* 2013 All in the graphene family—a recommended nomenclature for two-dimensional carbon materials *Carbon* **65** 1–6

[16] Brodie B C 1859 XIII. On the atomic weight of graphite *Philos. Trans. R. Soc.* **149** 249–59

[17] Brauer G (ed) 2012 *Handbook of Preparative Inorganic Chemistry V2* 2 (Amsterdam: Elsevier)

[18] Hofmann U and Holst R 1939 Über die Säurenatur uuund die Methylierung von Graphitoxyd *Ber. Dtsch. Chem. Ges. A and* B **72** 754–71

[19] Ruess G 1947 Über das graphitoxyhydroxyd (graphitoxyd) *Monatsh.efte für Chem. Verw. Teile Anderer Wiss.* **76** 381–417

[20] Lerf A, He H, Forster M and Klinowski J 1998 Structure of graphite oxide revisited *J. Phys. Chem.* B **102** 4477–82

[21] Szabó T, Berkesi O, Forgó P, Josepovits K, Sanakis Y, Petridis D and Dékány I 2006 Evolution of surface functional groups in a series of progressively oxidized graphite oxides *Chem. Mater.* **18** 2740–9

[22] Gao W, Alemany L B, Ci L and Ajayan P M 2009 New insights into the structure and reduction of graphite oxide *Nat. Chem.* **1** 403–8

[23] Lavin-Lopez M D P, Paton-Carrero A, Sanchez-Silva L, Valverde J L and Romero A 2017 Influence of the reduction strategy in the synthesis of reduced graphene oxide *Adv. Powder Technol.* **28** 3195–203

[24] De Silva K K H, Huang H H and Yoshimura M 2018 Progress of reduction of graphene oxide by ascorbic acid *Appl. Surf. Sci.* **447** 338–46

[25] Abulizi A, Okitsu K and Zhu J J 2014 Ultrasound assisted reduction of graphene oxide to graphene in L-ascorbic acid aqueous solutions: kinetics and effects of various factors on the rate of graphene formation *Ultrason. Sonochem.* **21** 1174–81

[26] Staudenmaier L 1899 Verfahren zur darstellung der graphitsäure *Ber. Dtsch. Chem. Ges.* **32** 1394–9

[27] Hofmann U and Frenzel A 1930 Quellung von Graphit uuund die Bildung von Graphitsäure *Ber. Dtsch. Chem. Ges. A and* B **63** 1248–62

[28] Hofmann U and König E 1937 Untersuchungen über graphitoxyd *Z. Anorg. Allg. Chem.* **234** 311–36

[29] Hummers W S and Offeman R E 1958 Preparation of graphitic oxide *J. Am. Chem. Soc.* **80** 1339

[30] Sun L 2019 Structure and synthesis of graphene oxide *Chin. J. Chem. Eng.* **27** 2251–60

[31] Poh H L, Šaněk F, Ambrosi A, Zhao G, Sofer Z and Pumera M 2012 Graphenes prepared by Staudenmaier, Hofmann and Hummers methods with consequent thermal exfoliation exhibit very different electrochemical properties *Nanoscale* **4** 3515–22

[32] Li C, Shi Y, Chen X, He D, Shen L and Bao N 2018 Controlled synthesis of graphite oxide: formation process, oxidation kinetics, and optimized conditions *Chem. Eng. Sci.* **176** 319–28

[33] Dimiev A M and Tour J M 2014 Mechanism of graphene oxide formation *ACS Nano* **8** 3060–8

[34] Shao G, Lu Y, Wu F, Yang C, Zeng F and Wu Q 2012 Graphene oxide: the mechanisms of oxidation and exfoliation *J. Mater. Sci.* **47** 4400–9

[35] Singh R K, Kumar R and Singh D P 2016 Graphene oxide: strategies for synthesis, reduction and frontier applications *RSC Adv.* **6** 64993–5011

[36] Chen J, Yao B, Li C and Shi G 2013 An improved Hummers method for eco-friendly synthesis of graphene oxide *Carbon* **64** 225–9

[37] Ikram R, Jan B M and Ahmad W 2020 An overview of industrial scalable production of graphene oxide and analytical approaches for synthesis and characterization *J. Mater. Res. Technol.* **9** 11587–610

[38] Marcano D C, Kosynkin D V, Berlin J M, Sinitskii A, Sun Z, Slesarev A, Alemany L B, Lu W and Tour J M 2010 Improved synthesis of graphene oxide *ACS Nano* **4** 4806–14

[39] Hou Y, Lv S, Liu L and Liu X 2020 High-quality preparation of graphene oxide via the Hummers' method: understanding the roles of the intercalator, oxidant, and graphite particle size *Ceram. Int.* **46** 2392–402

[40] Yu H, Xie K, Hu J, Shen C, Wang J G and Wei B 2016 The importance of raw graphite size to the capacitive properties of graphene oxide *RSC Adv.* **6** 17023–8

[41] Mahmoudi E, Ang W L, Ng C Y, Ng L Y, Mohammad A W and Benamor A 2019 Distinguishing characteristics and usability of graphene oxide based on different sources of graphite feedstock *J. Colloid Interface Sci.* **542** 429–40

[42] Kim J H, Shim G H, Vo T T N, Kweon B, Kim K M and Ahn H S 2021 Building with graphene oxide: effect of graphite nature and oxidation methods on the graphene assembly *RSC Adv.* **11** 3645–54

[43] Lesiak B, Trykowski G, Tóth J, Biniak S, Kövér L, Rangam N, Stobinski L and Malolepszy A 2021 Chemical and structural properties of reduced graphene oxide—dependence on the reducing agent *J. Mater. Sci.* **56** 3738–54

[44] Caicedo F M C, López E V, Agarwal A, Drozd V, Durygin A, Hernandez A F and Wang C 2020 Synthesis of graphene oxide from graphite by ball milling *Diam. Relat. Mater.* **109** 108064

[45] Muthoosamy K and Manickam S 2017 State of the art and recent advances in the ultrasound-assisted synthesis, exfoliation and functionalization of graphene derivatives *Ultrason. Sonochem.* **39** 478–93

[46] Chang W T, Chao Y H, Li C W, Lin K L, Wang J J, Kumar S R and Lue S J 2019 Graphene oxide synthesis using microwave-assisted vs. modified Hummer's methods: efficient fillers for improved ionic conductivity and suppressed methanol permeability in alkaline methanol fuel cell electrolytes *J. Power Sources* **414** 86–95

[47] Toh S Y, Loh K S, Kamarudin S K and Daud W R W 2014 Graphene production via electrochemical reduction of graphene oxide: synthesis and characterisation *Chem. Eng. J.* **251** 422–34

[48] Kuila T, Mishra A K, Khanra P, Kim N H and Lee J H 2013 Recent advances in the efficient reduction of graphene oxide and its application as energy storage electrode materials *Nanoscale* **5** 52–71

[49] Hu J, Kong G, Zhu Y and Che C 2021 Ultrafast room-temperature reduction of graphene oxide by sodium borohydride, sodium molybdate and hydrochloric acid *Chin. Chem. Lett.* **32** 543–7

[50] Olumurewa K O, Olofinjana B, Fasakin O, Akhigbe G E, Eleruja M A and Ajayi E O B 2020 Effect of hydrothermal and chemical treatment on the optical and electrical properties of reduced graphene oxide deposited on ITO glass *Mater. Res. Express* **7** 105606

[51] Malas A, Bharati A, Verkinderen O, Goderis B, Moldenaers P and Cardinaels R 2017 Effect of the GO reduction method on the dielectric properties, electrical conductivity and crystalline behavior of PEO/rGO nanocomposites *Polymers* **9** 613

[52] Mao S, Yu K, Cui S, Bo Z, Lu G and Chen J 2011 A new reducing agent to prepare single-layer, high-quality reduced graphene oxide for device applications *Nanoscale* **3** 2849–53

[53] Valles-Garcia C, Montero-Lanzuela E, Navalon S, Alvaro M, Dhakshinamoorthy A and Garcia H 2020 Tuning the active sites in reduced graphene oxide by hydroquinone functionalization for the aerobic oxidations of thiophenol and indane *Mol. Catal.* **493** 111093

[54] Mahmoud A E D 2020 Eco-friendly reduction of graphene oxide via agricultural byproducts or aquatic macrophytes *Mater. Chem. Phys.* **253** 123336
[55] Khojasteh H, Safajou H, Mortazavi-Derazkola S, Salavati-Niasari M, Heydaryan K and Yazdani M 2019 Economic procedure for facile and eco-friendly reduction of graphene oxide by plant extracts; a comparison and property investigation *J. Clean. Prod.* **229** 1139–47
[56] Mindivan F and Göktaş M 2020 Rosehip-extract-assisted green synthesis and characterization of reduced graphene oxide *ChemistrySelect* **5** 8980–5
[57] Fernández-Merino M J, Guardia L, Paredes J I, Villar-Rodil S, Solís-Fernández P, Martínez-Alonso A and Tascón J M D 2010 Vitamin C is an ideal substitute for hydrazine in the reduction of graphene oxide suspensions *J. Phys. Chem.* C **114** 6426–32
[58] Jao W C, Lin C H, Hsieh J Y, Yeh Y H, Liu C Y and Yang M C 2010 Effect of immobilization of polysaccharides on the biocompatibility of poly (butyleneadipate-co-terephthalate) films *Polym. Adv. Technol.* **21** 543–53
[59] Tu N D K, Choi J, Park C R and Kim H 2015 Remarkable conversion between n-and p-type reduced graphene oxide on varying the thermal annealing temperature *Chem. Mater.* **27** 7362–9
[60] Huh S H 2011 Thermal reduction of graphene oxide *Physics and Applications of Graphene – Experiments* ed S Mikhailov (Rijeka: InTech Open), pp 73–90
[61] Agarwal V and Zetterlund P B 2021 Strategies for reduction of graphene oxide–A comprehensive review *Chem. Eng. J.* **405** 127018
[62] Gao X, Jang J and Nagase S 2010 Hydrazine and thermal reduction of graphene oxide: reaction mechanisms, product structures, and reaction design *J. Phys. Chem.* C **114** 832–42
[63] Rosli N H A, Lau K S, Winie T, Chin S X and Chia C H 2021 Microwave-assisted reduction of graphene oxide for an electrochemical supercapacitor: structural and capacitance behavior *Mater. Chem. Phys.* **262** 124274
[64] Voiry D, Yang J, Kupferberg J, Fullon R, Lee C, Jeong H Y, Shin H S and Chhowalla M 2016 High-quality graphene via microwave reduction of solution-exfoliated graphene oxide *Science* **353** 1413–6
[65] Faucett A C, Flournoy J N, Mehta J S and Mativetsky J M 2017 Evolution, structure, and electrical performance of voltage-reduced graphene oxide *FlatChem* **1** 42–51
[66] AL-Gahouari T, Sayyad P, Bodkhe G, Ingle N, Mahadik M, Shirsat S and Shirsat M 2021 Controlling reduction degree of graphene oxide-based electrode for improving the sensing performance toward heavy metal ions *Appl. Phys.* A **127** 1–16
[67] Feng J, Ye Y, Xiao M, Wu G and Ke Y 2020 Synthetic routes of the reduced graphene oxide *Chem. Pap.* **74** 3767–83
[68] Yin R, Shen P and Lu Z 2019 A green approach for the reduction of graphene oxide by the ultraviolet/sulfite process *J. Colloid Interface Sci.* **550** 110–6
[69] Renteria J A Q, Ruiz-Garcia C, Sauvage T, Chazaro-Ruiz L F, Rangel-Mendez J R and Ania C O 2020 Photochemical and electrochemical reduction of graphene oxide thin films: tuning the nature of surface defects *Phys. Chem. Chem. Phys.* **22** 20732–43
[70] Qahtan T F, Cevik E, Gondal M A, Bozkurt A, Akhtar S and Hassan M 2021 Synthesis of manganese (IV) oxide at activated carbon on reduced graphene oxide sheets via laser irradiation technique for organic binder-free electrodes in flexible supercapacitors *Ceram. Int.* **47** 7416–24

[71] Tembei S A, El-Bab A M F, Hessein A and Abd El-Moneim A 2020 Ultrasonic doping and photo-reduction of graphene oxide films for flexible and high-performance electrothermal heaters *Flat Chem.* **24** 100199
[72] Kavinkumar T, Shobin L R and Manivannan S 2019 Effect of laser irradiation on electrical and gas sensing properties of reduced graphene oxide-graphene oxide heterostructure films *J. Alloys Compd.* **784** 301–12
[73] Paula K T, Santos M V, Facure M H, Andrade M B, Araujo F L, Correa D S, Ribeiro S J and Mendonca C R 2020 Laser patterning and induced reduction of graphene oxide functionalized silk fibroin *Opt. Mater.* **99** 109540
[74] Li Q, Ding Y, Yang L, Li L and Wang Y 2021 Periodic nanopatterning and reduction of graphene oxide by femtosecond laser to construct high-performance micro-supercapacitors *Carbon* **172** 144–53
[75] Liu Y Q, Zhang Y L, Liu Y, Jiang H B, Han D D, Han B, Feng J and Sun H B 2016 Surface and interface engineering of graphene oxide films by controllable photoreduction *Chem. Rec.* **16** 1244–55
[76] Facure M H, Schneider R, Mercante L A and Correa D S 2020 A review on graphene quantum dots and their nanocomposites: from laboratory synthesis towards agricultural and environmental applications *Environ. Sci. Nano* **7** 3710–34
[77] Georgakilas V, Tiwari J N, Kemp K C, Perman J A, Bourlinos A B, Kim K S and Zboril R 2016 Noncovalent functionalization of graphene and graphene oxide for energy materials, biosensing, catalytic, and biomedical applications *Chem. Rev.* **116** 5464–519
[78] Sun J, Meng D, Jiang S, Wu G, Yan S, Geng J and Huang Y 2012 Multiple-bilayered RGO–porphyrin films: from preparation to application in photoelectrochemical cells *J. Mater. Chem.* **22** 18879–86
[79] Liu S, Yu B and Zhang T 2013 Preparation of crumpled reduced graphene oxide–poly (p-phenylenediamine) hybrids for the detection of dopamine *J. Mater. Chem.* A **1** 13314–20
[80] Luo J H, Li B L, Li N B and Luo H Q 2013 Sensitive detection of gallic acid based on polyethyleneimine-functionalized graphene modified glassy carbon electrode *Sens. Actuators* B **186** 84–9
[81] Waiwinya W, Putnin T, Pimalai D, Chawjiraphan W, Sathirapongsasuti N and Japrung D 2021 Immobilization-free electrochemical sensor coupled with a graphene-oxide-based aptasensor for glycated albumin detection *Biosensors* **11** 85
[82] Facure M H M, Schneider R, Dos Santos D M and Correa D S 2020 Impedimetric electronic tongue based on molybdenum disulfide and graphene oxide for monitoring antibiotics in liquid media *Talanta* **217** 121039
[83] Park M, Kim N, Lee J, Gu M and Kim B S 2021 Versatile graphene oxide nanosheets via covalent functionalization and their applications *Mater. Chem. Front.* **5** 4424–44
[84] Sviridova E, Li M, Barras A, Addad A, Yusubov M S, Zhdankin V V, Yoshimura A, Szunerits S, Postnikov P S and Boukherroub R 2021 Aryne cycloaddition reaction as a facile and mild modification method for design of electrode materials for high-performance symmetric supercapacitor *Electrochim. Acta* **369** 137667
[85] Khan R and Nishina Y 2020 Grafting conductive polymers on graphene oxide through cross-linker: a stepwise approach *J. Mater. Chem.* A **8** 13718–24
[86] Li Y, Jian Z, Lang M, Zhang C and Huang X 2016 Covalently functionalized graphene by radical polymers for graphene-based high-performance cathode materials *ACS Appl. Mater. Interfaces* **8** 17352–9

[87] Yang X, Chen F, Kim M A, Liu H, Wolf L M and Yan M 2021 On the reactivity enhancement of graphene by metallic substrates towards aryl nitrene cycloadditions *Chem. Eur. J.* **27** 7887–96
[88] Ramesh K, Siboro S A, Kim D W and Lim K T 2020 Ultrasound-accelerated covalent-functionalization of reduced graphene oxide with imidazolium-based poly (ionic liquid) s by Diels–Alder click reaction for supercapacitors *React. Funct. Polym.* **152** 104605
[89] Li R, Yang Y, Wu D, Li K, Qin Y, Tao Y and Kong Y 2019 Covalent functionalization of reduced graphene oxide aerogels with polyaniline for high performance supercapacitors *Chem. Commun.* **55** 1738–41
[90] Du W, Wu H, Chen H, Xu G and Li C 2020 Graphene oxide in aqueous and nonaqueous media: dispersion behaviour and solution chemistry *Carbon* **158** 568–79
[91] Dreyer D R, Todd A D and Bielawski C W 2014 Harnessing the chemistry of graphene oxide *Chem. Soc. Rev.* **43** 5288–301
[92] Erickson K, Erni R, Lee Z, Alem N, Gannett W and Zettl A 2010 Determination of the local chemical structure of graphene oxide and reduced graphene oxide *Adv. Mater.* **22** 4467–72
[93] Chen D, Feng H and Li J 2012 Graphene oxide: preparation, functionalization, and electrochemical applications *Chem. Rev.* **112** 6027–53
[94] Wang X and Shi G 2015 An introduction to the chemistry of graphene *Phys. Chem. Chem. Phys.* **17** 28484–504
[95] Huang H *et al* 2020 The chemistry and promising applications of graphene and porous graphene materials *Adv. Funct. Mater.* **30** 1909035
[96] Stobinski L, Lesiak B, Malolepszy A, Mazurkiewicz M, Mierzwa B, Zemek J, Jiricek P and Bieloshapka I 2014 Graphene oxide and reduced graphene oxide studied by the XRD, TEM and electron spectroscopy methods *J. Electron. Spectrosc. Relat. Phenom.* **195** 145–54
[97] Johnson D W, Dobson B P and Coleman K S 2015 A manufacturing perspective on graphene dispersions *Curr. Opin. Colloid Interface Sci.* **20** 367–82
[98] Paredes J I, Villar-Rodil S, Martínez-Alonso A and Tascon J M D 2008 Graphene oxide dispersions in organic solvents *Langmuir* **24** 10560–4
[99] Konios D, Stylianakis M M, Stratakis E and Kymakis E 2014 Dispersion behaviour of graphene oxide and reduced graphene oxide *J. Colloid Interface Sci.* **430** 108–12
[100] Konkena B and Vasudevan S 2012 Understanding aqueous dispersibility of graphene oxide and reduced graphene oxide through pK_a measurements *J. Phys. Chem. Lett.* **3** 867–72
[101] Guex L G, Sacchi B, Peuvot K F, Andersson R L, Pourrahimi A M, Ström V, Farris S and Olsson R T 2017 Experimental review: chemical reduction of graphene oxide (GO) to reduced graphene oxide (rGO) by aqueous chemistry *Nanoscale* **9** 9562–71
[102] Pei S, Zhao J, Du J, Ren W and Cheng H M 2010 Direct reduction of graphene oxide films into highly conductive and flexible graphene films by hydrohalic acids *Carbon* **48** 4466–74
[103] Chuang C H *et al* 2014 The effect of thermal reduction on the photoluminescence and electronic structures of graphene oxides *Sci. Rep.* **4** 1–7
[104] Dasari B L, Nouri J M, Brabazon D and Naher S 2017 Graphene and derivatives—synthesis techniques, properties and their energy applications *Energy* **140** 766–78
[105] Loh K P, Bao Q, Eda G and Chhowalla M 2010 Graphene oxide as a chemically tunable platform for optical applications *Nat. Chem.* **2** 1015–24

[106] Mohan V B, Jayaraman K, Stamm M and Bhattacharyya D 2016 Physical and chemical mechanisms affecting electrical conductivity in reduced graphene oxide films *Thin Solid Films* **616** 172–82
[107] Pavinatto A, Mercante L A, Facure M H, Pena R B, Sanfelice R C, Mattoso L H and Correa D S 2018 Ultrasensitive biosensor based on polyvinylpyrrolidone/chitosan/reduced graphene oxide electrospun nanofibers for 17α-ethinylestradiol electrochemical detection *Appl. Surf. Sci.* **458** 431–7
[108] Chen D, Tang L and Li J 2010 Graphene-based materials in electrochemistry *Chem. Soc. Rev.* **39** 3157–80
[109] Martin A and Escarpa A 2014 Graphene: the cutting–edge interaction between chemistry and electrochemistry *TrAC Trends Anal. Chem.* **56** 13–26
[110] Bonanni A, Loo A H and Pumera M 2012 Graphene for impedimetric biosensing *TrAC Trends Anal. Chem.* **37** 12–21
[111] Casero E, Parra-Alfambra A M, Petit-Domínguez M D, Pariente F, Lorenzo E and Alonso C 2012 Differentiation between graphene oxide and reduced graphene by electrochemical impedance spectroscopy (EIS) *Electrochem. Commun.* **20** 63–6
[112] Yang T, Lin H, Zheng X, Loh K P and Jia B 2017 Tailoring pores in graphene-based materials: from generation to applications *J. Mater. Chem.* A **5** 16537–58
[113] Zhu C, Du D and Lin Y 2017 Graphene-like 2D nanomaterial-based biointerfaces for biosensing applications *Biosens. Bioelectron.* **89** 43–55
[114] Migliorini F L, Sanfelice R C, Mercante L A, Facure M H and Correa D S 2019 Electrochemical sensor based on polyamide 6/polypyrrole electrospun nanofibers coated with reduced graphene oxide for malathion pesticide detection *Mater. Res. Express* **7** 015601
[115] Mercante L A, Facure M H, Sanfelice R C, Migliorini F L, Mattoso L H and Correa D S 2017 One-pot preparation of PEDOT: PSS-reduced graphene decorated with Au nanoparticles for enzymatic electrochemical sensing of H_2O_2 *Appl. Surf. Sci.* **407** 162–70
[116] Schöche S, Hong N, Khorasaninejad M, Ambrosio A, Orabona E, Maddalena P and Capasso F 2017 Optical properties of graphene oxide and reduced graphene oxide determined by spectroscopic ellipsometry *Appl. Surf. Sci.* **421** 778–82
[117] Morales-Narváez E, Baptista-Pires L, Zamora-Gálvez A and Merkoçi A 2017 Graphene-based biosensors: going simple *Adv. Mater.* **29** 1604905
[118] Wang S, Dong Y, He C, Gao Y, Jia N, Chen Z and Song W 2017 The role of sp^2/sp^3 hybrid carbon regulation in the nonlinear optical properties of graphene oxide materials *RSC Adv.* **7** 53643–52
[119] Tang T, Liu F, Liu Y, Li X, Xu Q, Feng Q, Tang N and Du Y 2014 Identifying the magnetic properties of graphene oxide *Appl. Phys. Lett.* **104** 123104
[120] Si N, Zhang F, Jiang W and Zhang Y L 2018 Magnetic and thermodynamics properties graphene monolayer with defects *Physica* A **510** 641–8
[121] Sarkar S K, Raul K K, Pradhan S S, Basu S and Nayak A 2014 Magnetic properties of graphite oxide and reduced graphene oxide *Physica* E **64** 78–82
[122] Bagani K, Bhattacharya A, Kaur J, Rai Chowdhury A, Ghosh B, Sardar M and Banerjee S 2014 Anomalous behaviour of magnetic coercivity in graphene oxide and reduced graphene oxide *J. Appl. Phys.* **115** 023902
[123] Liu L, Zhang J, Zhao J and Liu F 2012 Mechanical properties of graphene oxides *Nanoscale* **4** 5910–6

[124] Wei W and Qu X 2012 Extraordinary physical properties of functionalized graphene *Small* **8** 2138–51
[125] Tavakol M, Montazeri A, Aboutalebi S H and Asgari R 2020 Mechanical properties of graphene oxide: the impact of functional groups *Appl. Surf. Sci.* **525** 146554
[126] Yang C, Liu Z, Chen C, Shi K, Zhang L, Ju X J, Wang W, Xie R and Chu L Y 2017 Reduced graphene oxide-containing smart hydrogels with excellent electro-response and mechanical properties for soft actuators *ACS Appl. Mater. Interfaces* **9** 15758–67
[127] Trusovas R, Račiukaitis G, Niaura G, Barkauskas J, Valušis G and Pauliukaite R 2016 Recent advances in laser utilization in the chemical modification of graphene oxide and its applications *Adv. Opt. Mater.* **4** 37–65
[128] Hernández C N, García M B G, Santos D H, Heras M A, Colina A and Fanjul-Bolado P 2016 Aqueous UV–Vis spectroelectrochemical study of the voltammetric reduction of graphene oxide on screen-printed carbon electrodes *Electrochem. Commun.* **64** 65–8
[129] Lee D W and Seo J W 2011 sp^2/sp^3 Carbon ratio in graphite oxide with different preparation times *J. Phys. Chem.* C **115** 2705–8
[130] Al-Gaashani R, Najjar A, Zakaria Y, Mansour S and Atieh M A 2019 XPS and structural studies of high quality graphene oxide and reduced graphene oxide prepared by different chemical oxidation methods *Ceram. Int.* **45** 14439–48
[131] Krishnamoorthy K, Veerapandian M, Yun K and Kim S J 2013 The chemical and structural analysis of graphene oxide with different degrees of oxidation *Carbon* **53** 38–49
[132] Kwan Y C G, Ng G M and Huan C H A 2015 Identification of functional groups and determination of carboxyl formation temperature in graphene oxide using the XPS O 1s spectrum *Thin Solid Films* **590** 40–8
[133] Bhattachraya S, Maiti R, Das A C, Saha S, Mondal S, Ray S K, Bhaktha S N B and Datta P K 2016 Efficient control of ultrafast optical nonlinearity of reduced graphene oxide by infrared reduction *J. Appl. Phys.* **120** 013101
[134] Sharma N, Sharma V, Jain Y, Kumari M, Gupta R, Sharma S K and Sachdev K 2017 Synthesis and characterization of graphene oxide (GO) and reduced graphene oxide (rGO) for gas sensing application *Macromol. Symp.* **376** 1700006
[135] She X, Liu T, Wu N, Xu X, Li J, Yang D and Frost R 2013 Spectrum analysis of the reduction degree of two-step reduced graphene oxide (GO) and the polymer/r-GO composites *Mater. Chem. Phys.* **143** 240–6
[136] Claramunt S, Varea A, Lopez-Diaz D, Velázquez M M, Cornet A and Cirera A 2015 The importance of interbands on the interpretation of the Raman spectrum of graphene oxide *J. Phys. Chem.* C **119** 10123–9
[137] Farivar F, Yap L, Karunagaran R U and Losic D 2021 Thermogravimetric analysis (TGA) of graphene materials: effect of particle size of graphene, graphene oxide and graphite on thermal parameters *C* **7** 41
[138] Alam S N, Sharma N and Kumar L 2017 Synthesis of graphene oxide (GO) by modified hummers method and its thermal reduction to obtain reduced graphene oxide (rGO) *Graphene* **6** 1–18
[139] Pfaffeneder-Kmen M, Casas I F, Naghilou A, Trettenhahn G and Kautek W 2017 A multivariate curve resolution evaluation of an *in situ* ATR-FTIR spectroscopy investigation of the electrochemical reduction of graphene oxide *Electrochim. Acta* **255** 160–7

[140] Kang T, Choi H, Joo S W, Lee S Y, Yoon K A and Lee K 2014 Peptide nucleic acid-mediated aggregation of reduced graphene oxides and label-free detection of DNA mutation *J. Phys. Chem.* B **118** 6297–301

[141] Rathour R K S and Bhattacharya J 2018 A green approach for single-pot synthesis of graphene oxide and its composite with Mn_3O_4 *Appl. Surf. Sci.* **437** 41–50

[142] Imran H, Manikandan P N and Dharuman V 2018 Graphene oxide supported liposomes for efficient label free electrochemical DNA biosensing *Sens. Actuators* B **260** 841–51

[143] Yang S, Yue W, Huang D, Chen C, Lin H and Yang X 2012 A facile green strategy for rapid reduction of graphene oxide by metallic zinc *RSC Adv.* **2** 8827–32

[144] Huang L, Zhu P, Li G, Lu D D, Sun R and Wong C 2014 Core–shell SiO_2@ RGO hybrids for epoxy composites with low percolation threshold and enhanced thermo-mechanical properties *J. Mater. Chem.* A **2** 18246–55

[145] Jiao X, Zhang L, Qiu Y and Guan J 2017 Comparison of the adsorption of cationic blue onto graphene oxides prepared from natural graphites with different graphitization degrees *Colloids Surf.*, A **529** 292–301

[146] Mercante L A, Facure M H, Locilento D A, Sanfelice R C, Migliorini F L, Mattoso L H and Correa D S 2017 Solution blow spun PMMA nanofibers wrapped with reduced graphene oxide as an efficient dye adsorbent *New J. Chem.* **41** 9087–94

[147] Kumar S, Singh R, Mahajan A, Bedi R K, Saxena V and Aswal D K 2018 Optimized reduction of graphite oxide for highly exfoliated silver nanoparticles anchored graphene sheets for dye sensitized solar cell applications *Electrochim. Acta* **265** 131–9

[148] Jha P K, Kashyap V, Gupta K, Kumar V, Debnath A K, Roy D, Rana S, Kurungot S and Ballav N 2019 In-situ generated Mn_3O_4-reduced graphene oxide nanocomposite for oxygen reduction reaction and isolated reduced graphene oxide for supercapacitor applications *Carbon* **154** 285–91

[149] Shetti N P, Malode S J, Malladi R S, Nargund S L, Shukla S S and Aminabhavi T M 2019 Electrochemical detection and degradation of textile dye Congo red at graphene oxide modified electrode *Microchem. J.* **146** 387–92

[150] Lu Z, Zhang L, Deng Y, Li S and He N 2012 Graphene oxide for rapid microRNA detection *Nanoscale* **4** 5840–2

[151] Li G L, Liu G, Li M, Wan D, Neoh K G and Kang E T 2010 Organo- and water-dispersible graphene oxide–polymer nanosheets for organic electronic memory and gold nanocomposites *J. Phys. Chem.* C **114** 12742–8

[152] Robinson J T, Perkins F K, Snow E S, Wei Z and Sheehan P E 2008 Reduced graphene oxide molecular sensors *Nano Lett.* **8** 3137–40

Chapter 4

Forms of graphene II: graphene quantum dots: properties, preparation and applications

Elsa Susan Zachariah, I Rejeena and Vinoy Thomas

Graphene quantum dots (GQDs), a form of graphene, have drawn research interest in various fields due to their interesting physico-chemical properties caused by a combination of graphene properties, significant quantum confinement and edge effects. This relatively new class of carbon material exhibits unique and intriguing properties with novel applications in varied fields such as energy storage, biomedical and environmental applications. This chapter presents the current technical progress on the properties, preparation and application of GQDs reported in literature. The structural, optical and electronic properties of this potent material are described in detail. Various top-down and bottom-up synthesis techniques and their implementation in various fields for novel applications are discussed. Finally, a brief outlook specifying the future prospects for further development in the field is given.

4.1 Introduction

GQDs are zero-dimensional graphene structures having single or few layers of graphene sheets and lateral dimensions less than 10 nm. This interesting material exhibits unique and outstanding properties compared to its two-dimensional counterpart graphene due the quantum confinement and increased surface area-to-volume ratio effects arising from size reduction. For example, the alterations in the electron distribution and opening up of bandgap in the material arises as a result of the reduction of graphene into GQDs, which in turn causes the latter to fluoresce [1]. Moreover, GQDs exhibit certain other important properties such as non-toxicity, widely tunable photoluminescence (PL), high chemical- and photo-stabilities and biocompatibility [2]. Most of these characteristic properties render GQDs overwhelming superiority over conventional quantum dots in several applications like bioimaging, optical sensing, photovoltaics, and energy storage and conversion.

doi:10.1088/978-0-7503-3999-5ch4

4.2 Properties of graphene quantum dots

4.2.1 Structural properties

Most of the reported GQDs are quasi spherical crystals consisting of 1–3 sheets of graphene layers and lateral dimensions in the range of 2–20 nm. The interplanar spacing of the quantum dots varies from 0.22 nm to 0.28 nm conforming to the graphitic (100) plane. The structure of GQDs is highly dependent on the synthesis procedure and conditions; however, they typically possess a honeycomb lattice structure with each carbon atom covalently linked to three other carbon atoms giving rise to a sp^2 hybridised network of carbon atoms with lateral dimensions less than 10 nm resulting in delocalized π orbital electrons [3]. In addition to the sp^2 carbon core, GQDs also possess oxygen-containing functional groups the amount and type of which depend on the synthesis method and experimental conditions. These functional groups exist either on the surface or on the edges of the GQDs and are responsible for the solubility and dispersibility of the GQDs in water and other solvents [4]. Figure 4.1 depicts the general structure, transmission electron microscope (TEM) and atomic force miscroscope (AFM) images of typical GQDs.

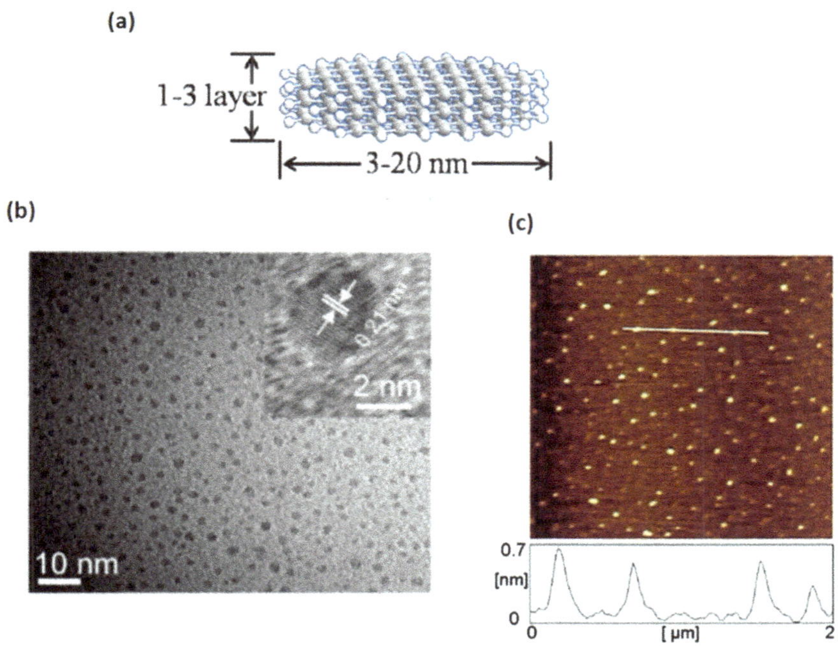

Figure 4.1. (a) A depiction of structure of GQDs, reproduced from [5] with permission from the Royal Society of Chemistry. (b) TEM image, reproduced with permission from [6] John Wiley & Sons, Wiley-VCH, Weinheim and (c) AFM image of GQD, reproduced from [7] with permission from the Royal Society of Chemistry.

4.2.2 Optical properties

The optical absorption of GQDs synthesized through various techniques is consistent in the ultraviolet region along with an extension into the visible region. A prominent peak in the absorption spectra in the wavelength range of 230 nm to 280 nm is attributed to the π–π^* transition of the sp^2 carbon core atoms, while another peak at 270 nm to 390 nm is present for most GQDs mainly contributed by the n–π^* transition of the C=O bond depicted in figure 4.2 [8]. The position of these absorption bands depends more on the different functional groups and surface passivation than on the size or shape of GQDs [9].

PL is the most intriguing property of GQDs which is absent in 2D graphene, rendering it versatile application in various fields. Even though the photoluminescent efficiency of GQDs is lower than traditional semiconductor quantum dots, their better photostability against photobleaching and blinking along with non-toxicity make them a superior alternative. PL of GQDs is mainly credited to the bandgap introduction which arises due to the size reduction. However, some interesting PL behaviors like excitation dependent emission and up-conversion luminescence properties have made PL mechanism of GQDs a subject of debate in the scientific community. The general consensus is that the PL of GQDs originates mainly from two different categories of emission which are intrinsic state emission and defect state emission. The former is size dependent and is affected by the edge effects and electron–hole recombination while the latter is the result of the defects introduced into the carbon core due to functionalization of the dot surface [11]. Transitions among the bonding and antibonding molecular orbitals of the graphene core and the amino, hydroxyl or carboxylate functional groups on the surface are a probable explanation for the PL mechanism in GQDs [12]. The photoluminescent properties

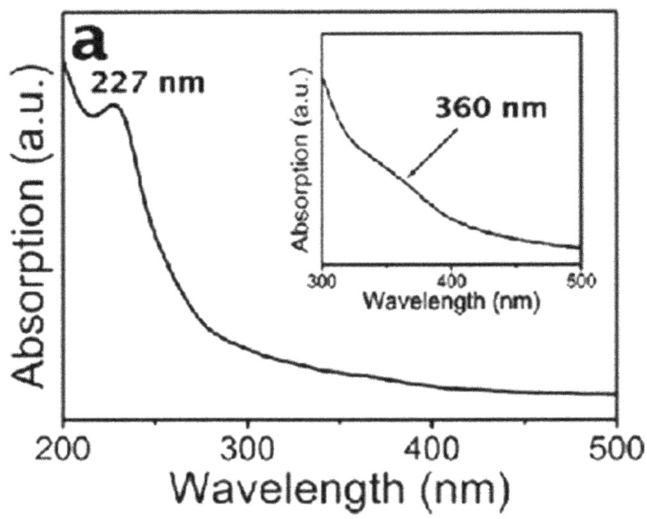

Figure 4.2. UV–visible absorption spectra of GQDs, reproduced from [10] with permission from the Royal Society of Chemistry.

of GQDs depend not only on the synthesis procedure but also on the synthesis conditions like precursor, reaction time, temperature etc. However, almost all the GQDs fluoresce in the blue–green region of the spectrum. The smaller dielectric constant and weak spin–orbit coupling of GQDs compared to other inorganic semiconductor QDs with comparable size leads to stronger interaction among the carriers and well-defined spin multiplicity which in turn leads to a larger energy band fluorescing in the blue–green region [13].

The underlying mechanism of fluorescence in GQDs is influenced by a variety of factors from physical parameters like lateral height, size, zigzag/armchair edge effects, defects to the chemical composition such as functional groups, oxidative state among others. Understanding the mechanism behind the PL of this potent material involves the design of experiments which can finely tune the above attributes. Many of the earlier reported works showcase the size dependent PL properties. For instance, Peng *et al* synthesised GQDs with different size ranges through chemical exfoliation by varying the reaction parameters. These dots of sizes in the ranges 1–4 nm, 4–8 nm and 7–11 nm exhibit different optical properties fluorescing in the blue, green and yellow regions, respectively, demonstrating the fine tuning of PL and bandgap with particle size [14]. Sarkar *et al* conducted some density functional theory studies to reveal the size dependent bandgap and tunable PL of GQDs synthesised through sonochemical route, which is in accordance with the quantum confinement effect. Richly photoluminescent GQDs with excitation independent emission, up-conversion luminescence and delayed fluorescence were observed. The size dependence was found to be more prominent in functional GQDs compared to those without functionalization [15].

The contributions of other structural and physical properties of GQDs to PL like edge state, functional groups, functionalization, defects etc have also been studied by many groups. Pan *et al* proposed that the blue luminescence of GQDs originates from the zigzag sites with a triplet ground state represented by $\sigma^1\pi^1$, the multiplicity which is dependent on the difference in energy of the σ and π orbitals. The peaks at 320 nm and 257 nm in PL excitation (PLE) spectra can be assigned to the transitions from the highest occupied molecular orbitals (HOMO) to the lowest unoccupied molecular orbital (LUMO) and the bright blue emission can be attributed to the decay of excited electrons from LUMO to HOMO [16]. Tuning the PL behaviour of GQDs through surface chemistry was demonstrated by Zhu *et al*, in which green emitting GQDs were tuned to fluoresce in the blue region with modification with methylamine (m-GQD) as well as reduction (rGQD). The modification process passivates the surface epoxy and carboxylate groups and the resultant groups suppress the non-radiative recombination of election–hole pairs making intrinsic state emission (blue) the dominant PL behaviour. Similarly, reduction process changes all epoxy, carbonyl and amide groups into OH groups, which further mask the non-radiative recombination leading to dominance in the intrinsic state blue emission [17]. PL mechanism of GQDs with tunable properties emitting in yellow, blue and red regions through coating with polyethyleneimine (PEI) of different molecular weights was studied by Gao *et al*. The uncoated GQDs emit in the yellow region, PEI1800 coated GQDs having a single GQD core emit in the blue region and PEI600 coated GQDs having multiple GQD in a large PEI cage emit in the red region. In the case of blue emitting GQDs, formation of amides on the surface of GQDs through passivation with PEI decreases the conjugation and increases the bandgap leading to a blue shift, while in the case of red emitting GQD amidation increases the conjugation area due to the interaction of GQDs with PEI cage leading to a red shift in emission [18]. The effect of size as well as surface oxidation degree on the PL properties of GQDs were studied by Qi *et al*, in which the red shift of PL emission was observed for increasing size as well as increasing surface oxidation degree. The PL phenomenon was dominated by the conjugated structure since GQDs possessed minimum surface defects [19]. Wang *et al* studied in detail the contributions of aromatic core, functional groups and defects on the PL properties of oxygenated GQDs. Particularly, functional groups were found to execute a double role towards the overall emission contributing to the π*–n and σ*–n molecular type PL which is not spectrally affected by particle size and excitation energy. The functional groups also bring about structural distortions in the core which results in emissions due to transitions from these mid gap states. The PL mechanism proposed is illustrated in figure 4.3 [20]. Even though numerous studies have been undertaken for the study of PL mechanism in GQDs, a clarification on the matter is not yet achieved. It is thus empirical to conduct more experimental and theoretical studies on the subject to elucidate the full potential of this potent material.

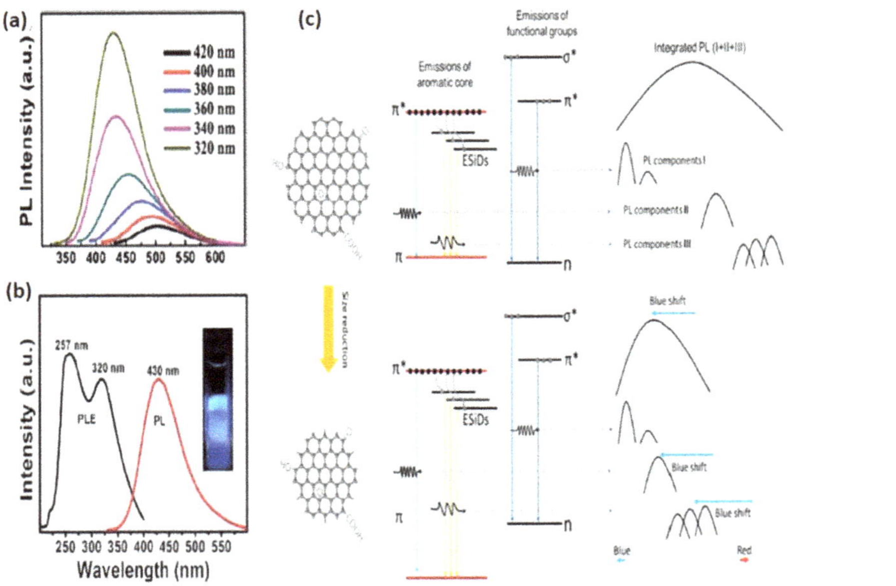

Figure 4.3. (a) Excitation dependent PL spectra of GQD. (b) PL excitation spectra and photoluminescent spectra of GQD excited at 257 nm, reproduced with permission from [16] John Wiley & Sons, Wiley-VCH, Weinheim. (c) PL mechanism of GQDs from aromatic core and functional groups, reproduced from [20] with permission from the Royal Society of Chemistry.

4.2.3 Electronic properties

The electronic properties of GQDs are highly dependent on the lateral size, types of edge and corner geometry of the material. Variation of the electronic structure with edge lattice symmetry was experimentally shown by Ritter *et al*, in which metallic properties of GQDs with predominantly zigzag edges were demonstrated [21]. Continuous tuning of the energy band of the GQDs with variation in size, shape, edge character and number of layers was reported by Ozfidan *et al.* In this study, for weak interactions where the electron–electron interactions are screened by the environment, the model of Dirac fermions was used to describe the electronic properties, in which the electron–electron interactions renormalise the Fermi bandgap by self-energy and vertex corrections [22]. The electronic properties of GQDs with different functionalities were studied theoretically and determined to be a result of collaboration and competition between frontier orbital hybridisation of the functional group and GQD and the charge transfer. It was found that functional groups containing –C=O were more favourable for tuning energy gap [23]. Electron–electron interaction and electron–hole interaction are reported to significantly influence electronic structure due to quantum confinement and screening effects [1]. Even though both the bandgap and exciton binding energy decreased with the size of GQDs, GQDs with armchair edges exhibited higher bandgap and exciton energy compared to the ones with zigzag edges, though they were larger in size. Moreover, the larger spin singlet–triplet splitting of GQDs compared to higher dimensional carbonaceous materials reduced the non-radiative transitions from triplet to singlet state leading to higher fluorescence quantum yield [1]. These properties of GQDs arising from quantum confinement effects and screening effects due to lower dimensionality provide them with potential applications for optoelectronic devices.

4.3 Synthesis

The well-established synthetic strategies of GQDs can be categorised into top-down and bottom-up strategies. The former requires the breaking of bulk carbonaceous materials such as graphite, carbon black etc into tiny fragments to form GQDs through physical or chemical processes. The synthesis methods that come under this category include hydrothermal/solvothermal processes, microwave/ultrasound exfoliation, oxidative cutting and electrochemical oxidation. On the other hand, the bottom-up technique involves nano-sized GQDs built gradually from small precursor molecules. Carbonisation/pyrolysis and step-wise organic synthesis comes under this bottom-up strategy. The top-down and bottom synthetic techniques are depicted in figure 4.4. The following section elaborates the synthesis procedures followed by various published works, which come under the broad categories of top-down and bottom-up techniques.

4.3.1 Top-down strategy

4.3.1.1 Hydrothermal process

The hydrothermal method for the synthesis of blue emitting GQDs reduced from preoxidised graphene sheets was first demonstrated by Pan *et al.* A suspension of

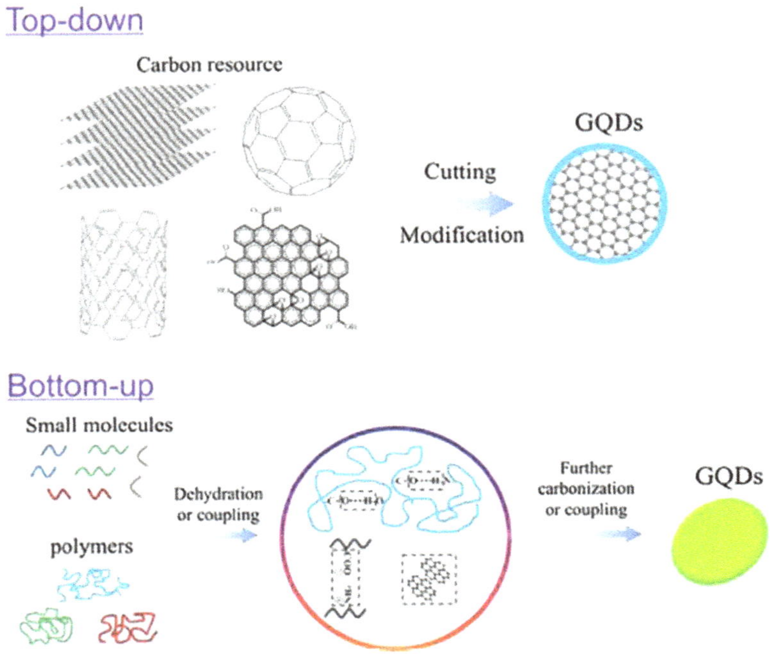

Figure 4.4. Top-down and bottom-up synthesis techniques, reproduced from [13], copyright (2016) with permission from Elsevier.

purified oxidized graphene sheets was heated in a Teflon-lined autoclave and the obtained solution was filtered and purified to get strongly fluorescent GQDs. Hydrothermal deoxidation leads to the breakage of ultrafine pieces of oxidized graphene sheets surrounded by epoxy chains through the elimination of connecting oxygen atoms resulting in the formation of GQDs [16]. Shen and coworkers developed a facile hydrothermal technique for the synthesis of GQDs passivated with polyethylene glycol. Graphene oxide (GO) treated with nitric acid solution was neutralised, broken down into tiny fragments by an ultrasonic cell crusher and filtered to remove large fragments of GO. Polyethylene glycol was mixed with the prepared GO solution, and heated in a Teflon-lined autoclave. The resulting solution after cooling was dialysed for a week to obtain PEG surface-functionalized GQDs with strong blue fluorescence [24]. Amino-functionalised GQDs with bright fluorescence and wide photoluminescent tunability for the same excitation were synthesised by Tetsuka *et al.* A modified Hummers method was followed to prepare oxidised graphene sheets (OGS) from synthetic graphite powder. For extracting amino-functionalised GQDs, the as-prepared homogenous OGS solution was mixed with ammonia solution and water and heated in a Teflon-lined autoclave. The resulting supernatant is filtered out and heated to remove ammonia and further filtered and dialysed to obtain the amino-functionalised GQD [25]. Hydrothermal synthesis of GQDs by means of graphene as the precursor material was described by Far'ain *et al.* Graphene was diluted with *n*-butyl acetate and sonicated hydrothermally heated for various times. The naturally cooled black solution thus obtained was filtered and dialysed to obtain GQDs [26].

4.3.1.2 Solvothermal process

Solvothermal technique is another extensively used synthesis procedure of GQDs. Tian *et al* presented a facile solvothermal synthesis of GQD from expanded graphite and hydrogen peroxide with dimethyl formamide (DMF) as the solvent. In a typical reaction, expanded graphite prepared by thermal expansion method was mixed with DMF through ultrasonication and hydrogen peroxide was added to the solution and stirred to get a homogenous solution. The solution obtained after heating the solution in an autoclave was vacuum filtered and solvent evaporated. Pure GQDs with strong blue fluorescence can be obtained by passing the residue redissolved in deionized water through filter membrane. This solvothermal process which is facile, ecofriendly, less time-consuming and does not require tedious dialysis process or high-cost instruments is illustrated in figure 4.5 [27]. In another report, Zhang *et al* prepared GQDs from GO through a solvothermal route. A thoroughly dispersed GO solution in dimethyl formamide solvent was heated in a Teflon-lined autoclave. The heated solution was thereafter vacuum filtered and rotated for evaporation to remove DMF and obtain GQDs, which can then be dissolved in solvents to obtain a suspension of GQDs [28]. Liu *et al* presented a facile solvothermal synthesis route which used graphite as the precursor. Sulfuric acid was gradually added into a mixture of graphite powder and potassium permanganate followed by the

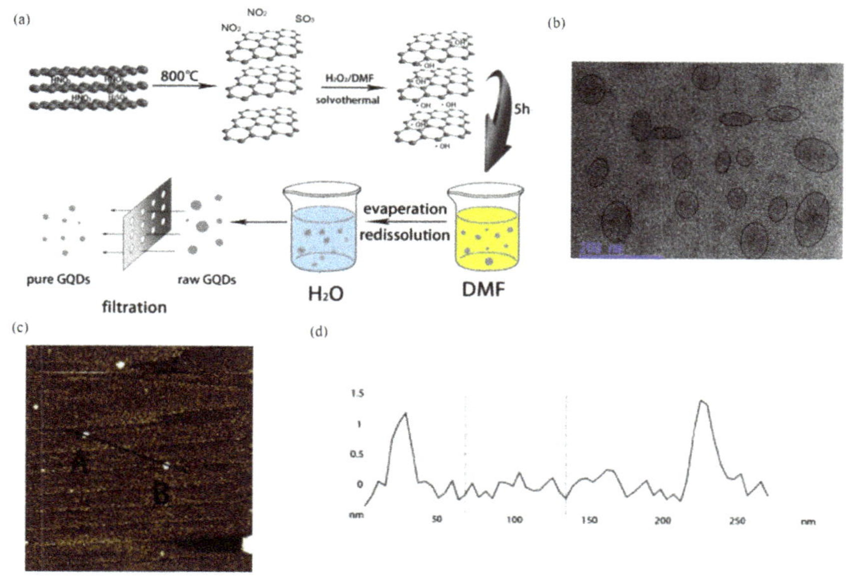

Figure 4.5. (a) Schematic representation of solvothermal synthesis of GQD from expanded graphite. (b) TEM image and (c) AFM image with (d) height profile of the prepared GQD with average size of 35 nm with an average thickness of 1–1.5 nm (3–4 graphene layers), reproduced from [27], copyright (2016) with permission from Elsevier.

introduction of nitric acid. The desired product of GQD solution was obtained by heating an autoclave containing the above solution in an oven. This study revealed that ratio of potassium permanganate and graphite determined the size of the GQDs while either GO or GQDs could be obtained by adjusting the temperature [29].

4.3.1.3 Ultrasound exfoliation

Ultrasound waves, which can produce alternating high and low frequency waves in a medium, have the capability to dissociate graphene into GQDs. An ultrasonic exfoliation technique for the synthesis of GQDs was presented by Hassan *et al.* Ultrasonication of few- layered graphene sheets in ethanol–H_2O and the subsequent filtration and purification of the suspension yielded GQDs with blue emission [30]. Ultrasonication-assisted liquid phase exfoliation for the synthesis of defect-selective GQDs was presented by Lu *et al* in which high defect GQDs and low defect GQDs were attained through the ultrasound exfoliation of acetylene black and nano-graphite, respectively [31]. In a similar work GQD prepared through the ultrasound exfoliation of graphite powder in a mixture of ethyl acetoacetate and NaOH was demonstrated by Sarkar *et al.* The dark brown mixture after cooling down was centrifuged to remove unexfoliated graphite and the larger GQDs were removed by filtration with a syringe filter. Further, dialysis of the solution yielded GQD solution which was dried using a rotary evaporation system to obtain GQD powder [15].

4.3.1.4 Oxidative cutting

Oxidative cutting or cleavage of large carbonaceous materials by strong oxidative agents is a commonly used technique to synthesize GQDs. Dong *et al* synthesized GQDs from carbon black by refluxing with concentrated nitric acid. The supernatant obtained after centrifugation was dried yielding single-layered GQD while the sediment was washed, dried and filtered after pH adjustments to yield multi-layered GQDs [32]. Zhu *et al* adopted a simple technique void of byproducts through the oxidation of GO by hydroxyl radicals obtained by the decomposition of H_2O_2 in the presence of tungsten oxide ($W_{18}O_{49}$) nanowires catalyst in a Teflon-lined autoclave. Centrifugation was used to remove the catalyst from the obtained aqueous GQD solution [33]. The synthesis of GQDs from GO through oxidative cleavage by acid free oxidant, oxone (potassium monopersulfate) through photo and sono irradiation was shown by Shin *et al* as illustrated in figure 4.6 [34]. Liu *et al* synthesised GQDs from a cheap material source, coal tar pitch through oxidation by hydrogen peroxide under mild conditions. Coal tar pitch was suspended in hydrogen peroxide through sonication and refluxed. The solution was cooled down and the filtrate was vacuum freeze dried to obtain highly fluorescent GQDs [35].

4.3.1.5 Electrochemical oxidation

Electrochemical exfoliation of large carbonaceous materials is a proven technique for the synthesis of GQDs with uniform size and high yield. Green luminescent GQDs prepared through an electrochemical method for application in photovoltaics were reported by Li *et al.* A cyclic voltammetry scan rate of $0.5\ V\ s^{-1}$ was utilized for the preparation of GQD in which Pt and Ag/AgCl act as the counter and reference

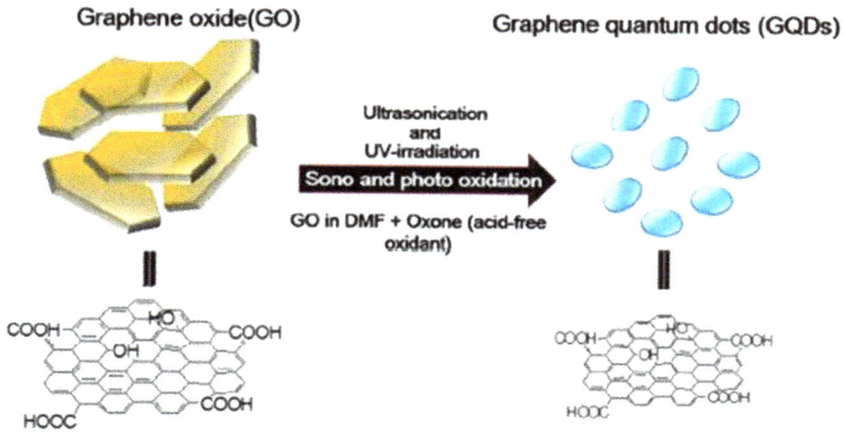

Figure 4.6. Illustration of GQD synthesis from GO through the oxidation using oxone, reproduced from [34] with permission from the Royal Society of Chemistry.

electrodes while a graphene film acts as the working electrode [36]. Shinde *et al* demonstrated the electrochemical preparation of highly luminescent GQDs from multiwalled carbon nanotubes (MWCNTs). GQDs were cut out of MWCNT through a two-step electrochemical oxidation process. The breakage of sp^2 carbon atoms is initiated in the initial step by the application of potential of 1 V versus Pt QRE (quasi-reference electrode) to the working electrode which is coated with MWCNTs in electrolyte propylene $LiClO_4$. The final step includes the application of potential −1 V versus QRE upon which variations with respect to time of oxidation due to the intercalation of Li^+/propylene carbonate complexes results in the exfoliation of MWCNTs into GQDs [37]. A facile method to electrochemically exfoliate three-dimensional graphene was reported by Anatharamayanan *et al*, in which 1-butyl 3-methylimidazolium hexafluorophosphate ($BMIMPF_6$) in acetonitrile was used as the electrolyte, Pt wire as the counter electrode, Ag/AgCl as reference electrode and 3D graphene as the working electrode. This approach yielded highly crystalline GQDs with uniform distribution in dimensions [38]. Red fluorescent GQDs, which showed great potential as labels in cellular imaging, were synthesised by Tan *et al* through electrochemical exfoliation of graphite in $K_2S_2O_8$ solution. The electrolysis of graphite rods was performed with a current intensity in the range of 80–200 mA cm^{-2} (potential: 5 V) in 7 ml 0.01 M $K_2S_2O_8$ aqueous solution to yield pure GQD solution [39]. Phosphorous-doped GQDs were synthesised by Li *et al* through an electrochemical approach where sodium phytate was used as the electrolyte at the same time acting as the phosphorous source. Scanning voltage set at 5 V utilizing high-quality graphite rod and Pt as the working electrode and reference electrode, respectively, led to phosphorous-doped GQDs with excellent ability to scavenge free radicals [40].

4.3.1.6 Microwave-assisted exfoliation

Microwave-assisted technique is being increasingly used for the synthesis of GQDs because of their advantages such as homogenous heating and shorter reaction time.

For instance, Li *et al* reported a microwave-assisted synthesis technique for green luminescent GQDs through the cleaving of GO under acidic conditions. The cleavage of GO and reduction occurred simultaneously under microwave irradiation without the need for further reducing agents. However, further reduction with $NaBH_4$ resulted in bright blue luminescent GQDs [41]. A microwave-assisted hydrothermal approach adopted by Luo *et al* for the synthesis of white fluorescent GQDs from graphite exhibited application in white light emitting diodes. Microwave treatment of nanographite dispersed in a mixture of nitric acid and sulphuric acid after filtration and purification yielded yellow–green fluorescent GQDs. White light emitting GQD was obtained through the further microwave irradiation of this GQD solution [42]. Zhang *et al* presented a method for the synthesis of blue light emitting GQDs, which were used as fluorescent probes for detecting iron through the pyrolysis of aspartic acid and NH_4HCO_3 mixture by microwave irradiation [43]. Oxidation of cheap deoiled asphalt in acidic solution though microwave irradiation was used by Zhao *et al* for the yield of high-quality GQDs. These GQDs exhibited amphipathic properties providing use as surfactants in asphalt emulsions [44].

4.3.2 Bottom-up strategy

4.3.2.1 Pyrolysis/carbonisation

Carbonisation or pyrolysis of small carbon-based molecules is a simple and straightforward approach for the synthesis of GQDs, which has been extensively researched in recent times. For instance, controlled carbonisation of citric acid was used by Alizadeh *et al* for the synthesis of GQDs which exhibited exceptional sensitivity to environmental humidity [45] while the yield of ultra-small monolayer GQDs by the pyrolysis of trisodium citrate was demonstrated by Hong *et al* [46]. Fabrication of N-doped GQDs through carbonisation of citric acid and urea was reported by Gu *et al* utilizing IR heaters that were capable of transferring energy to a body through electromagnetic radiation [47]. A simple synthesis of GQDs that involved pyrolysis of silicon carbide under vacuum while introducing hydrogen etching gas with no harmful chemicals was demonstrated by Lee *et al.* Few-layer GQDs with high crystallinity, low defects and high purity were successfully fabricated [48].

4.3.2.2 Step-wise organic synthesis

Step-wise organic synthesis is an effective solution chemistry method for the fabrication of GQDs. For instance, Yan *et al* utilized 3-iodo 4-bromo aniline as a precursor for the synthesis of GQDs. In this step-wise synthesis process, oxidative condensation of the precursor molecules form fused graphene moieties, which resulted in the formation of GQDs [49]. In another work, Li *et al* took on a two-step solution chemistry approach for the synthesis of N-doped GQDs. The first step involved the formation of GQD precursor through the Suzuki coupling of nitrogen containing intermediates formed using benzene derivatives while the second step comprised the formation of GQD through the reaction of the precursor with iron

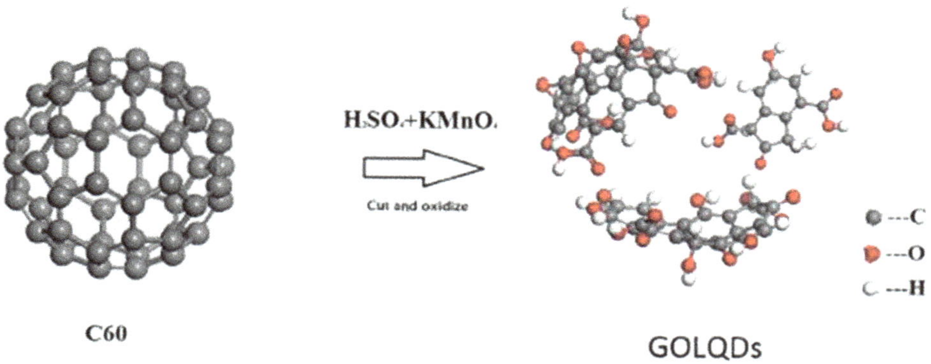

Figure 4.7. Rapturing of C_{60} molecule into GO-like GQDs, reproduced with permission from [52], John Wiley & Sons, Wiley-VCH, Weinheim.

(III) chloride in an inert atmosphere. GQDs with excellent activity towards electrocatalytic oxygen reduction reaction (ORR) were obtained [50].

4.3.2.3 Cage opening

Cage opening of fullerene to produce GQDs has also been adopted. For example, the breakage of C_{60} in the presence of ruthenium which led to the formation of GQDs was reported by Lu *et al.* The thermal annealing above 650 K completely dissociated C_{60} molecule to form GQDs, the shape which depended on the annealing temperature [51]. In a similar work, Chen *et al* oxidised C_{60} to form GQDs through a modified Hummers method, shown in figure 4.7, which resulted in a good yield of GO-like QDs, with a diameter from 0.6 to 2.2 nm, thickness of 1.2 nm and displaying higher water solubility [52].

4.4 Applications

4.4.1 Energy related applications

4.4.1.1 Supercapacitor

The ever-increasing demand for flexible and portable energy storage devices has given supercapacitors or electrochemical capacitors much attention owing to fast charge–discharge rate, elevated charge density and extended life cycle. The possibility of using GQDs as a potential material for supercapacitors was identified by Liu *et al.* GQD-based supercapacitors were fabricated and their electrochemical properties studied, to which end a simple electrodeposition route was followed to fabricate a GQDs//GQDs symmetric micro- supercapacitor with a higher rate capability up to 1000 V s^{-1}, good power response and cyclability while a GQDs// MnO_2 asymmetric supercapacitor fabricated using MnO_2 nanoneedles and GQD as the positive and negative electrodes, respectively, showed almost two times energy density and specific capacitance compared to the symmetric supercapacitor. Designing of GQD-based material for electrodes in supercapacitors with outstanding properties was thus given a new light [53]. Edge enriched activated GQDs were found to possess two times electrocapacitance compared to non-activated GQDs for application as electrodes in supercapacitors. The presence of more edge sites that have increased ability to accumulate charges leads to an enhanced accessibilty of electrolytes through free and bound edges which in turn enhances the electrocapacitance [30]. The remarkable properties of GQDs as electrodes for supercapacitor applications were further proved through electrochemical studies of uniform size GQDs synthesized through strong acid cutting of GO. Superior electrochemical double layer capacitance (EDLC) properties were exhibited by the GQDs with a specific capacitance of 307.6 F g^{-1}, greater energy density and excellent cyclability. The superior supercapacitance of GQDs compared to graphene is shown in figure 4.8, in which defects and edge effects overcome the restacking of graphene sheets providing higher conductivity and electrochemical activity between the electrolyte and electrode materials, thus proving their potential application as electrode materials in supercapacitor applications [54].

Composites of GQDs combined with other well-defined carbon-based supercapacitor electrodes are reported to enhance the device performance. For instance, GQD-deposited three-dimensional graphene (3DG) electrodes exhibit a 90% enhancement in capacitance in comparison to electrodes made from 3DG alone [55]. A composite of GQDs and carbon cloth were used for the fabrication of a flexible EDLC electrode wherein the latter acts as a conductive, flexible substrate for the former active material. A capacitance of 70.7 mF cm^{-2} at a scan rate of 50 mV s^{-1} and 77.5 mF cm^{-2} with a discharge 0.2 mA cm^{-2} were exhibited by the composite electrode and the full cell supercapacitor made with the composite material, respectively, paving its way to be used in flexible electronic devices [56]. The electrochemical performance of activated carbon which is commonly used as electrodes for supercapacitors was seen to improve on embedding with highly crystalline GQD. The microporous carbon exhibited a capacitance of 388 F g^{-1} much higher than that of most reported activated carbon which may be due to the better charge transfer kinetics and ion migration due to the conductive networks formed inside activated carbon on being embedded with GQDs [57]. GQDs were also combined with non-carbon materials aimed at the development of electrodes for

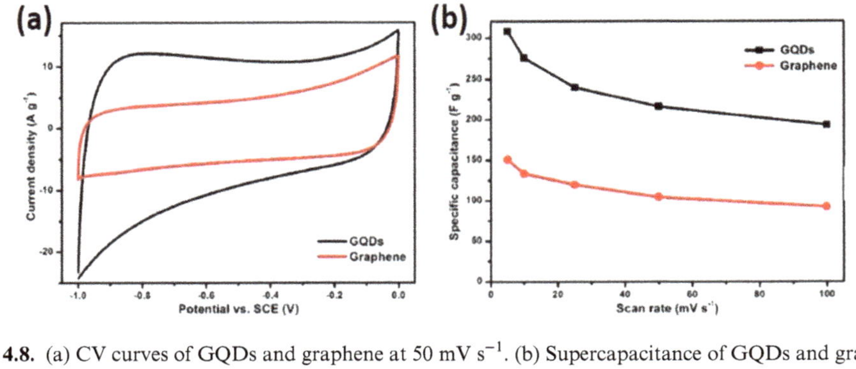

Figure 4.8. (a) CV curves of GQDs and graphene at 50 mV s^{-1}. (b) Supercapacitance of GQDs and graphene, reproduced with permission [54], copyright (2018) American Chemical Society.

supercapacitors. For instance, introducing GQDs induced a change in morphology of $CuCo_2S_4$ nanocomposite forming a unique nanostructure with a rough surface. An enhancement in the electrochemical performance of the GQD/$CuCo_2S_4$ nanocomposite compared to $CuCo_2S_4$ nanosheets may be attributed to the presence of abundant electroactive sites and shorter charge transfer paths owing to the larger surface area of the composite [58].

4.4.1.2 Solar cells

Increasing the efficiency of dye-sensitised solar cells (DSSCs) and organic or polymer solar cells (PSCs) with the introduction of GQDs is a field of intense research since it is an environment-friendly and cost-effective technique. GQDs are proven to be efficient materials as anode interlayer (AIL) and cathode interlayer (CIL) of solar cells to improve efficiency and stability. Tetramethylammonium (TMA) functionalised GQDs were used as CIL for polymer solar cells with good device performance. The peripheral methylammonium groups can form interfacial dipoles with metal cathode, which decreases the metal cathode work function. π-conjugated graphene structure of the GQD core rendered high conductivity and excellent film-forming properties. A power conversion efficiency as high as 8.8% was achieved for the fabricated polymer solar cell with a conventional configuration using TMA functionalised GQDs as CIL and PTB7-Th: PC71BM as the active layer [59]. GQDs were also demonstrated as a hole extraction layer for highly efficient polymer solar cells. The small-sized GQD rich in –COOH groups increases the work function leading to the formation of ohmic contacts with active layer polymers. This can increase the hole extraction property and reduce the electron–hole recombination at the anode, thereby improving the efficiency. A high-power conversion efficiency of 7.9% was accomplished through the PSC employing PTB7: PC71BM as the active layer and few-layer GQDs as the hole extraction layer [60]. Another factor that broadens the application potential of GQDs in photovoltaics is the fact that their properties can be tuned with proper functionalization. Tuning the work function of GQD-modified electrodes with alkali metal ions utilized as cathode interlayer in polymer solar cells was reported. The results indicated that the incorporation of an alkali metal cation could successfully tune the work function of the corresponding electrodes. The work functions of GQD with various alkali metals are shown in figure 4.9(a). The ITO electrodes modified with alkali metal-incorporated GQDs possess a lower work function, which also decreases with increasing atomic number of the alkali metal. This proves that these modified GQDs are suitable materials as CIL for polymer solar cells with efficiency comparable to those of conventional CIL materials [61]. Work function tuning of GQDs with quantum size for use in high-performance cathode interlayer materials was demonstrated by Wang *et al.* In this work, amino-functionalised GQDs efficiently reduced the work function of the electrode, which was tuned with the quantum size of the amino-functional GQD. The density of the amino-functional groups increases with decreasing size of GQD which results in stronger interfacial dipole moments and hence decreased work function. CIL electrode with medium-sized GQDs exhibited

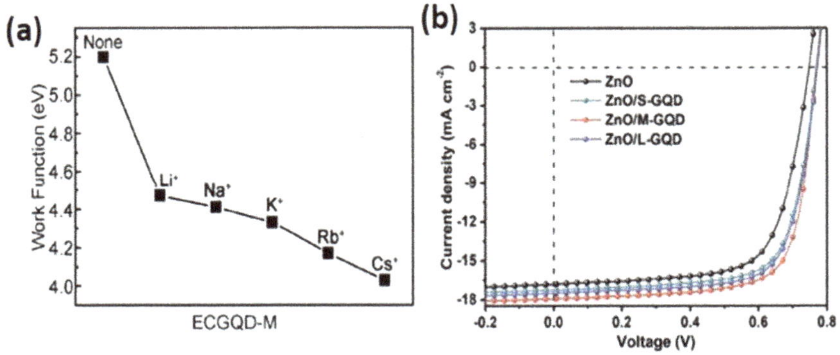

Figure 4.9. (a) The work function of GQD with varied alkali metal ions [58], reproduced from [61] with permission from the Royal Society of Chemistry. (b) Current density curves of optimised solar cell with and without amino-functionalised GQD under illumination, reproduced with permission from [62], John Wiley & Sons, Wiley-VCH, Weinheim.

the optimal balance between interfacial modification and conductivity leading to high power conversion efficiency of 12.81% with J71:ITIC as the active layer. Current density curves of optimized solar cells with and without amino-functionalised GQDs under illumination is shown in figure 4.9(b) [62]. GQDs synthesized through a facile microwave method was successfully utilized as hole transport layer in organic solar cells exhibiting an increase in efficiency up to 44% compared to the pristine solar cells without the incorporation of GQDs. It was found that GQDs function as an intermediate buffer layer decreasing the potential between active layers and electrodes in the cells incorporated with GQD in comparison to pristine solar cells correspondingly increasing the efficiency [63].

4.4.1.3 Fuel cells

The escalating demand of energy, diminishing conventional fuels and pollution associated with the production of conventional sources of energy has led to considerable attention in the field of fuel cells. GQDs are attracting much research interest as an alternative to Pt electrocatalysts in fuel cells due to their unique properties. A platinum free synergetic electrocatalyst composed of iron phthalocyanine functionalised GQD for enhanced oxygen reduction reaction was reported. Enhanced electrochemical activity was exhibited by these catalysts with a direct four-electron pathway in alkaline medium. A greater tolerance towards methanol crossover and carbon monoxide poisoning effect were also exhibited by this material, which promises the alternative to Pt in electrocatalysts for fuel cells [64]. Colak *et al* demonstrated a simple synthesis of cost-effective catalyst for working electrode in direct methanol fuel cells, with polyoxometalate (NaPWO) functionalised GQDs, several mono- and bimetallic nanoparticles. The electrochemical measurement of the prepared composites shows catalytic activity of the composite based on bimetallic nanoparticles towards methanol as fuel [65]. An efficient electrocatalyst with high electro-oxidative property and superior carbon monoxide

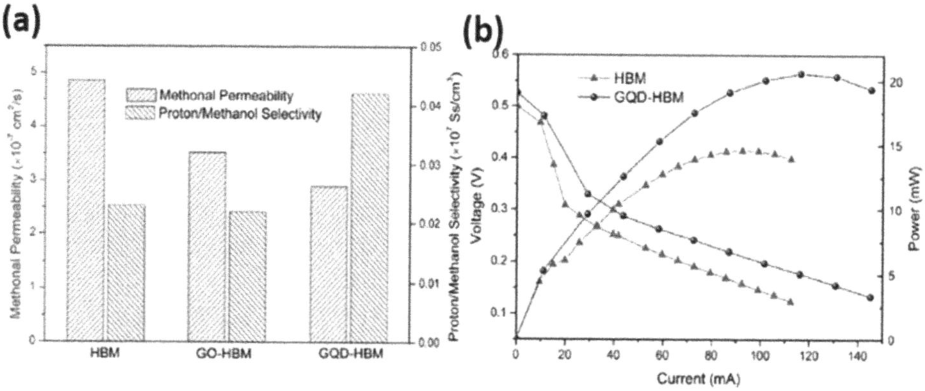

Figure 4.10. (a) Methanol permeability and proton/methanol selectivity of pristine HBM, GO-HBM and GQD-HBM membranes. (b) DMFC performance of pristine HBM and GQD-HBM membranes, reproduced from [67], copyright (2020) with permission from Elsevier.

tolerance based on a composite of GQDs and functionalised multiwalled carbon nanotubes was reported. The composites displayed exceptional electrochemical activity against methanol attributed to the high effective surface area and synergetic effects between composite components thus proving their application in direct methanol fuel cells [66]. In another work, GQD and GO were incorporated into a hyperbranched macromolecule (HBM) polyamide proton exchange membrane to enhance the proton/methanol selectivity of the membrane to be used in direct methanol fuel cells. The functional groups of GQD and GO interact with the $–SO_3H$ groups on HBM through hydrogen bond and participate in the proton conduction. The permeation of methanol molecules can be effectively prevented by the GQD and GO. It was observed that due to the smaller size of GQD compared to GO, GQD-HBM membrane could effectively avoid the large-scale phase separation. The GQD-HBM membrane exhibits a proton/methanol selectivity almost twice that of pristine HBM membrane, with good stability, as depicted in figure 4.10 [67].

4.4.1.4 Water splitting for hydrogen production

Finding alternative fuels for sustainable development remains a challenge in the modern world. Hydrogen fuel has proven to be a renewable and environment-friendly energy source. Water and sunlight being two of the world's most abundantly available resources, use of solar power for the production of H_2 to be used as a fuel from water has been a hot topic of research in recent times. GQDs are generally combined with TiO_2 photocatalysts to enhance their photocatalytic properties for hydrogen evolution through water splitting. Yu *et al* synthesized GQD composites with TiO_2 nanotube arrays (TiO_2–NTA) and CdS modified TiO_2 nanotube arrays (CdS/TiO_2–NTA). The hydrogen generation ability of the GQD-modified catalysts were found to be elevated, which is ascribed to the enhancement in charge separation and transfer in the presence of GQD. Moreover, the TiO_2–NTA shows better morphological stability after the internalisation of GQDs [68]. In another work, hydrothermal method was followed for coupling GQDs with TiO_2 photocatalysts using 1,3,6-trinitropyrene (TNP) as the precursor for GQDs. TNP molecules undergo molecular fusion to form GQDs and also decorate onto TiO_2 nanoparticles leading to effective coupling between these molecules. This coupling of GQD and TiO_2 lead to an extension of the optical absorption into the visible region and enhances the charge carrier separation and transportation. The hydrogen evolution rate of TiO_2/GQDs was almost seven times that of TiO_2 alone [69]. The excellent electron transfer and photosensitizing properties of GQDs for photocatalytic water splitting for hydrogen generation were demonstrated by Hao *et al.* In this work, a GQD/TiO_2 system exhibited photocatalytic hydrogen evolution from water under UV–visible irradiation. In addition to being efficient electron reservoirs, GQDs act as electron transfer reagents as well as photosensitizing agents to sensitize TiO_2 leading to the production of H_2 as illustrated in figure 4.11 [70]. In yet another work, GQDs were coupled with {001} faceted anatase TiO_2 for efficient photocatalytic hydrogen evolution. The internal structure of TiO_2 nanosheets were not altered by the incorporation of GQD. The increased charge separation efficiency in the composite leads to a higher photocatalytic efficiency for hydrogen evolution by the composite [71]. Improved solar energy utilization for hydrogen evolution by coupling GQDs with semiconductor photocatalysts was demonstrated by Xie *et al.* In their work S, N co-doped GQD/P-25 TiO_2 nanocomposite exhibited better photocatalytic efficiency for hydrogen production than P-25 TiO_2 alone. The enhancement in the efficiency of composite can be attributed to the broadening of the visible light absorption of P-25 TiO_2 when decorated with GQDs [72]. In a more

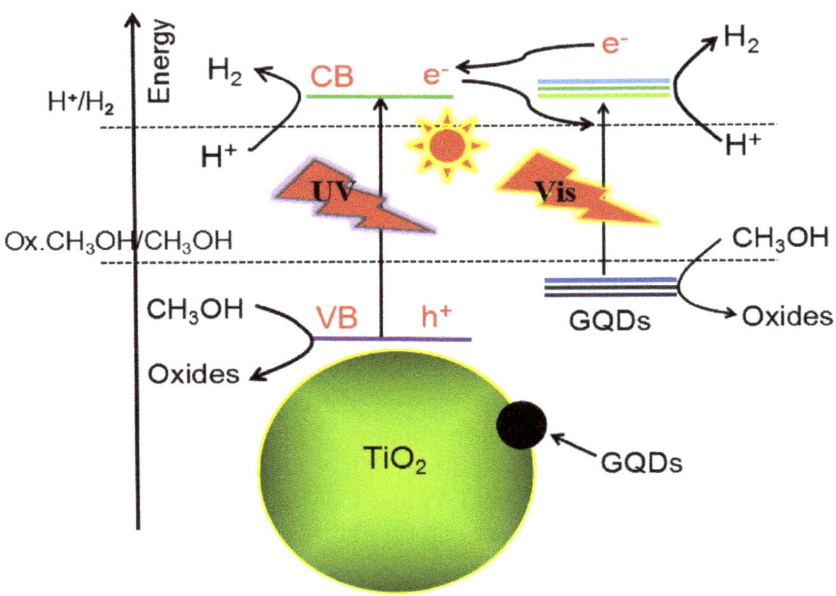

Figure 4.11. Photocatalytic H_2 evolution mechanism over GQDs/TiO_2 under UV–visible light irradiation, reproduced from [70], copyright (2016) with permission from Elsevier.

recent work, the effect of varying morphologies of TiO_2 in TiO_2/GQD nanocomposites on photocatalytic hydrogen evolution was studied. The TiO_2/GQD composite with the bi-phasic P-25 TiO_2 was the most efficient for photocatalytic hydrogen evolution. The higher efficiency of this composite was attributed to the bi-phasic TiO_2 as well the role of GQDs as photosensitiser as well as a co-catalyst [73].

GQDs have also been integrated with well-known electrocatalysts for overall water splitting. For instance, Lv *et al* developed a composite which is a high-performance, inexpensive and earth abundant substitute for the noble metal-based conventional electrocatalysts for water splitting, by coupling nitrogen-doped GQDs (NGQD) with Ni_3S_2 nanosheets on a Ni foam (Ni_3S_2-NGQD/NF). The electrocatalytic measurements of the composite show higher negative onset potential and much higher catalytic current density for the Ni_3S_2-NGQD/NF composite than for Ni_3S_2/NF or NF and display a much higher durability and negligible degradation [74]. The fabrication of a transition metal GQD, that is Co/GQD composite as an electrocatalyst for the otherwise sluggish oxygen evolution reaction(OER) in electrocatalytic water splitting was reported. A higher current density at a given voltage and lower overpotential was exhibited by these composites compared to the benchmark electrocatalysts, such as Pt/C and Co nanoparticles in addition to giving stable responses. This nanocomposite reveals an exciting pathway of transition metal/GQD-based electrocatalysts for clean energy [75]. A bifunctional electrocatalyst for water electrolysis was put forward by Li *et al* in the form of GQD loaded and molybdenum-doped Ni_3S_2 nanosheets. Negligible hydrogen evolution reaction (HER) and OER activity was exhibited by pure molybdenum loaded Ni_3S_2 sheets with no GQD loading while the same composite with proper GQD loading shower superior electrocatalytic activity towards HER and OER with overpotentials of 68 mV at 10 mA cm^{-2} and 326 mV at 20 mA cm^{-2}, respectively. Moreover, an excellent potential of 1.58 V towards overall water splitting with good stability was exhibited by the composite which may be attributed to the modulation of electrical structure and increase of active site density [76]. A carbon-based composite of boron-doped GQDs anchored on graphene hydrogels synthesized through a facile hydrothermal technique was demonstrated. The composite functions as an electrocatalyst towards OER and HER as well as ORR with the water electrolysis cell using the composite electrode and displaying a current density of 10 mA cm^{-2} with a stability of up to 70 h [77].

4.4.2 Biomedical applications

4.4.2.1 Bioimaging

GQDs are emerging as excellent tools in biomedical applications owing to their non-toxicity, biocompatibility, water solubility, and ease of preparation. The application of GQD as a safe fluorescent probe for cell imaging was demonstrated by Wang *et al.* Electron beam irradiation method was employed to synthesise GQDs to image MCF7 cells as a model. The cytoplasm of MCF7 cells could be brightly imaged with GQDs compared to bright field images. The high quantum yield of GQD made cell imaging possible at concentrations well below the cytotoxic limit of GQD, which proved that GQD is a safe and efficient fluorescent probe for cell imaging [78].

Stable and bright blue fluorescent sulfur-doped GQDs (S-GQDs) synthesized through a hydrothermal approach were demonstrated to be an excellent cell imaging material by Jin *et al*, wherein cell viability tests confirmed the low cytotoxicity of the SGQD. Live HeLa cells were imaged after incubation for sufficient time with SGQD using laser scanning confocal microscope under 365 nm excitation, as shown in figure 4.12. The imaging confirmed the penetration of S-GQDs into the cytoplasm but not to the nuclei, which confirmed that the SGQD will not lead to genetic disruption and toxicity [79]. An environment-friendly route for the preparation of GQDs with tunable fluorescence was used for multicolour imaging of fibroblast cells [80]. The great bioimaging possibility of nitrogen-doped GQDs with low cytotoxicity and high biocompatibility was demonstrated by incubation with three different cell lines HeLa, HepG2 and LO2. The cell lines incubated with NGQD exhibited bright blue fluorescence under an excitation of 405 nm. The fluorescence was mainly concentrated in the cytoplasm and not in the nuclei, which is an indication that GQDs entered the cytoplasm without causing any genetic damage. This underlines the broad application potential of GQD in the field of bioimaging [81].

4.4.2.2 Biosensing

GQD-based biosensors that specifically and sensitively detect target biomolecules have made a leap in the research field due to the extraordinary characteristics of GQDs. A GQD-based cholesterol sensor was demonstrated by Sun *et al*, where hydrothermally synthesised nitrogen-doped GQDs were combined with chromium picolinate (CrPic) to detect cholesterol by analysing the fluorescence quenching through photoinduced energy transfer (PET). CrPic also acted as a potential cholesterol receptor leading to the formation of a favorable complex with NGQD/CrPic, which in turn enhanced the fluorescence. This cholesterol sensor could selectively determine cholesterol with a detection limit of 0.4 μM [82]. A simple and sensitive approach for detecting glucose through a specific enzymatic reaction and fluorescence quenching of GQDs was presented in which the enzyme

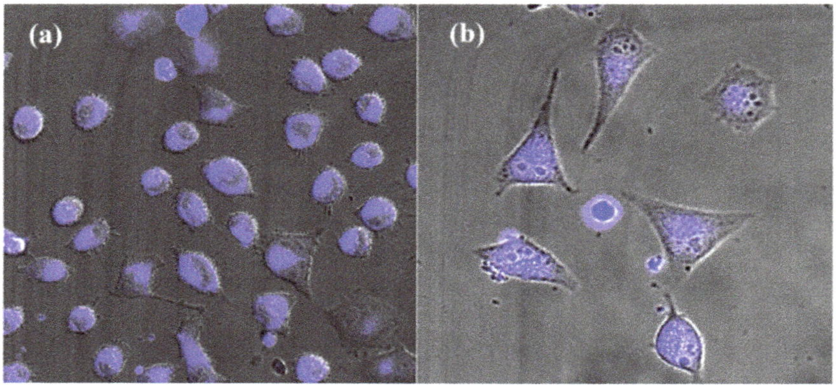

Figure 4.12. (a) Fluorescent images of HeLa cells incubated with SGQD. (b) Enlarged fluorescent image of HeLa cells incubated with SGQD, reproduced with permission [79] Copyright 2018 Elsevier B.V.

glucose oxidase (GO_x) oxidized glucose to form hydrogen peroxide through catalysis by horseradish peroxidase (HRP), which in turn oxidized phenol into quinone compounds, thereby quenching the fluorescence of GQD. The analysis results showed that the quenching degree was proportional to concentration of glucose and hence this system can be effectively used for detecting glucose [83]. Sensing of circulating tumor cells with the help of GQDs-functionalised with an aptamer (epithelial cell adhesion molecule (EpCAM) receptor) and a magnetic material Fe_3O_4 was presented. MoS_2 nanosheets acting as fluorescent quencher and aptamer@ Fe_3O_4@GQD complex were assembled together for constructing a 'turn-on' biosensing magnetic nanocomposite. The system exhibiting low cytotoxicity was capable of identifying and labelling circulating tumor cells within 15 min with a high detection limit up to ten tumor cells in whole blood [84]. In another work, detecting *Escherichia coli* bacteria using carboxylated GQDs conjugated with antibacterial peptide cecropin P1 that attaches the *E. coli* to the GQD. This attachment of *E. coli* bacteria to the GQD affected the fluorescence-related characteristics of GQDs, such as bandgap width and thus enabled their detection in real time [85].

4.4.3 Environmental applications

4.4.3.1 Removal of organic dyes from waste water

Large-scale production of synthetic dyes, their high toxicity and slow degradation rate have led to increased research on ways to degrade such organic pollutants, out of which photocatalytic degradation of dyes from waste water is receiving much attention due to its effectiveness and ease of use. GQDs have been used as highly efficient photocatalytic agents for the degradation of organic dyes. The photocatalytic properties of GQDs prepared by pyrolyzing citric acid were demonstrated by the degradation of new Fuchsin dye, the kinetic rate constants and decolourisation efficiency of which were found to be independent of the dye concentration [86]. GQDs as efficient and visible-light photocatalysts via energy transfer mechanism were shown, in which the light energy converted the GQD into singlet oxygen, thereby degrading Rhodamine B dye [87]. In a more recent work, proven synthesis methods and different precursors were successfully used for the preparation of three different types of GQDs. Their photocatalytic performance for the degradation of Rhodamine B under identical experimental conditions indicated that the composition of the GQDs affects the photocatalytic activity. The influence of the intrinsic properties of GQDs like light absorptivity and surface charge on the degradation of organic dye Rhodamine B highlighted the importance of tailoring such intrinsic properties for enhancing the photocatalytic activity [88]. The study of the photocatalytic activity of polymer modified GQDs such as PEI and PEG for the photocatalytic degradation reveals that their performance rate decreased in the order GQD, PEI-GQD and PEG-GQD. This shows that the degradation rate can be qualitatively manipulated by the use of different polymers [89].

GQDs are also being extensively used along with other established photocatalysts such as TiO_2 for their performance enhancement in the degradation of dyes. For example, Qu *et al* synthesised N and S, N-doped GQDs by the hydrothermal treatment of citric acid and urea/thiourea for the fabrication of GQD/TiO_2 nanocomposite, which exhibited excellent photocatalytic degradation efficiency towards the visible light induced degradation of Rhodamine B. Co-doped GQD/TiO_2 nanocomposite displayed a higher degradation rate than single-doped GQD/TiO_2 nanocomposite and TiO_2 alone [90]. GQDs deposited on anatase TiO_2 nanoparticles show higher efficiency for the visible light induced degradation of methyl compared to TiO_2 alone [91]. A hybrid structure of GQDs impregnated into TiO_2 3D nanotube array showed enhanced efficiency for the photocatalytic degradation of methylene blue under UV illumination. The greater surface area of GQDs and narrow bandgap of TiO_2 in hybrid structure allowed greater absorption of light and efficient charge transfer from GQD to TiO_2 leading to enhanced photocatalytic decomposition of the dye [91]. The bandgap tuning of GQD/TiO_2 nanocomposites by controlling the size of the GQD was reported. The bandgap of the composite increases with increasing size of GQD extending the absorption of the composite into the visible region. This along with efficient charge separation in the GQD/TiO_2 nanocomposite leads to the efficient degradation of methylene blue [93]. Different sized GQDs synthesised using a wet chemical method at different temperatures were coupled with TiO_2 forming stable GQD/TiO_2 nanocomposites. The extension of optical absorption into the visible region decreases the rate of charge recombination leading to enhancement in photocatalytic efficiency [94]. GQDs and nitrogen-doped GQDs synthesized by hydrothermal method were combined with TiO_2 forming the corresponding nanocomposite out of which NGQD/TiO_2 nanocomposite showed better photocatalytic efficiency towards the degradation of Rhodamine B under visible light illumination due to the enhanced visible light absorption and photogenerated charge carrier separation and transfer [95]. A ternary nanocomposite of N, S co-doped GQD-rGO-TiO_2 nanotube prepared by alkaline hydrothermal reaction and physical stirring exhibited higher photodegradation efficiency towards the degradation of methylene blue under visible light illumination compared to those of rGO–TiO_2 and pure TiO_2 alone. The extension of photo-absorption region into the visible range along with enhanced charge carrier separation and transportation was responsible for the higher photocatalytic activity of GQD-based nanocomposites than TiO_2 alone [96].

GQDs are combined with several photocatalysts other than TiO_2 for the successful degradation of pollutants from water. For example, nanocomposites of zinc porphyrin-functionalised GQDs (GQDs/ZnPor) exhibited enhancement in the photocatalytic activity towards methylene blue degradation compared to ZnPor alone which was ascribed to the increased contact area and change in surface of ZnPor due to the presence of GQDs [97]. Even though the unique photocatalytic activity of ZnO has already been established, fast carrier recombination and photocorrosion inhibit the performance of these photocatalysts. Combining electron acceptor materials and surface modification are ways to overcome these challenges. For instance, a simple spin coating and annealing procedure was followed by Li *et al* for the synthesis of GQD-functionalised ZnO/Cu composites, which exhibited higher photocatalytic degradation of Rhodamine B under UV irradiation compared to ZnO. The combined effects of Cu nanoparticles and GQDs lead to better light absorption, charge separation and transfer and reduce photocorrosion [98]. Wang *et al* fabricated a ternary photocatalyst combining ZnO, Ag and GQDs with photocatalytic activity superior to pure ZnO or binary photocatalysts as Ag–ZnO and GQD–ZnO towards visible light-driven degradation of Rhodamine B [99]. In another work S, N co-doped GQD integrated with leaf templated ZnO (L-ZnO/N, SGQD) nanocomposites showed elevated photocatalytic activity due to the combined effects of higher effective charge separation in the composites and the enhanced surface reaction, which result in effective absorption of visible light [100]. Hydroxyl GQD-modified mesoporous graphitic carbon nitride (OH-GQD/mpg-C_3N_4) composites exhibited higher photocatalytic activity than pure g-C_3N_4 towards the degradation of Rhodamine B and colourless tetracyclin hydrochloride under visible light irradiation. The intimate contact between GQD on the surface of mpg-C_3N_4 was attributed to the enhanced photocatalytic activity [101]. Also, S, N co-doped GQDs coupled with g-C_3N_4, forming g-C_3N_4/S, NGQD composites with elevated photocatalytic performance towards Rhodamine B degradation under visible light irradiation compared to pure g-C_3N_4 alone. The enhancement was attributed to the efficient separation of photogenerated charge carriers in the composite [102]. In a more recent work, composites of GQDs-functionalised Sb_2WO_6 (GQD/Sb_2WO_6) nanocomposite were prepared by electrostatic self-assembly and used as photocatalyst for UV and visible light induced methyl orange degradation. The composite displayed higher degradation efficiency than Sb_2WO_6 alone and exhibited high recyclability and stability with potential application for decomposition of other organic contaminants in water [103] (figure 4.13).

4.4.3.2 Reduction of CO_2

Carbon dioxide, a greenhouse gas is one of the primary causes of global warming. Converting CO_2 to valuable hydrocarbons or other chemicals through photocatalytic reduction using either sunlight or artificial sunlight presents the best resolution to global warming and energy scarcity problems [104]. Functionalised GQDs for solar fuel production through photocatalytic CO_2 reduction were first reported by Yadav *et al.* Visible-light-driven GQD-based system as an integrated photocatalytic/bio catalytic system was developed for the efficient CO_2 reduction to formic acid. The functionalised GQDs possess advanced light harvesting capabilities which leads to higher conversion of CO_2 to formic acid by the integrated photocatalyst/bio catalyst system. The formic acid production exhibits a 1.7-fold enhancement compared to the conventional counterparts [105]. A photocatalyst system composing of TiO_2 nanotube arrays sensitized with electrodeposited GQDs (graphene-2,4,6-trinitrotoluene (G-TNT) films) display enhanced photocatalytic conversion of CO_2 to CH_4 compared to TNT alone. Photogenerated electrons react with CO_2 molecules adsorbed on the photocatalysts surface thereby generating CH_4. The efficient charge carrier extraction and transportation is responsible for the enhanced performance of GQD sensitized TNT [106]. GQDs deposited on the surface of MIL-101(Fe) displayed higher photocatalytic efficiency towards reduction

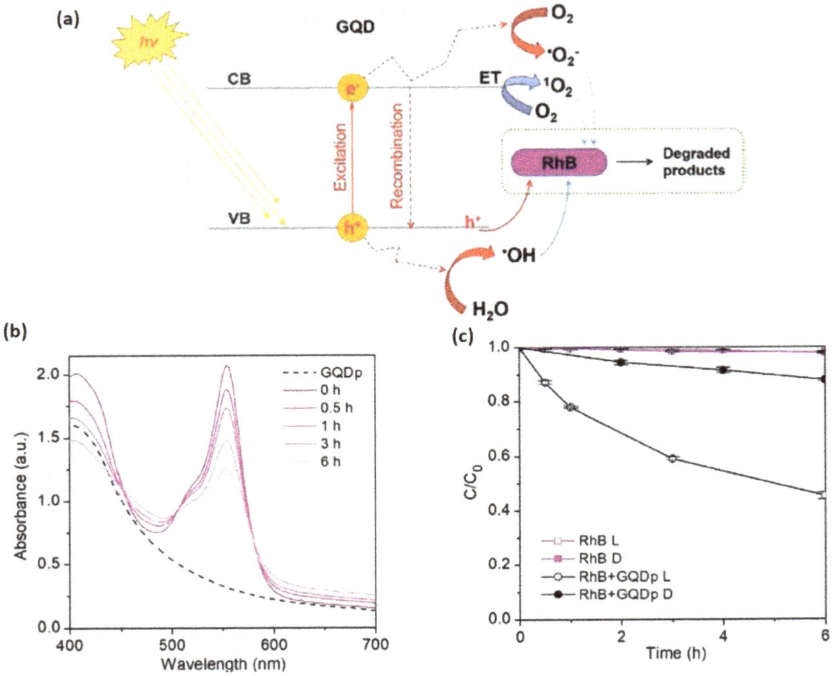

Figure 4.13. (a) Proposed photocatalytic degradation mechanism of Rhodamine B with GQD. (b) Photodegradation of Rhodamine B with GQD. (c) Relative Rhodamine B concentrations with and without GQD under CFL irradiation (L) and darkness (D), reproduced with permission from [88], John Wiley & Sons, Wiley-VCH, Weinheim.

of CO_2 compared to pristine MIL-101(Fe) under visible light irradiation. The content of GQDs in the photocatalyst was also found to have an effect on the reduction efficiency. The excellent optical properties of GQDs in the visible region as well as the particular combination of GQD and MIL-101(Fe) which reduce the recombination of photogenerated electrons and holes lead to enhanced photocatalytic activity of the composite [107]. In yet another work, GQDs mounted on a metal organic framework Zn-Bim-His displayed significant enhancements in photoactivity and selectivity for the reduction of CO_2 which can be attributed to the higher charge carrier separation efficiency. Moreover, the composite was found to be photostable for several cycles of the process [108]. These works provide a basis for design and synthesis of efficient GQD-based photocatalysts for the efficient conversion of CO_2 to fuels in the future.

4.4.3.3 Reduction of toxic elements

Excessive quantities of toxic elements in the atmosphere can be harmful to environmental well-being and health of human beings and animals. Photocatalytic oxidation or reduction of such toxic elements into harmless compounds is a way to detoxify the environment. For instance, Ag@GQD composites exhibited advanced

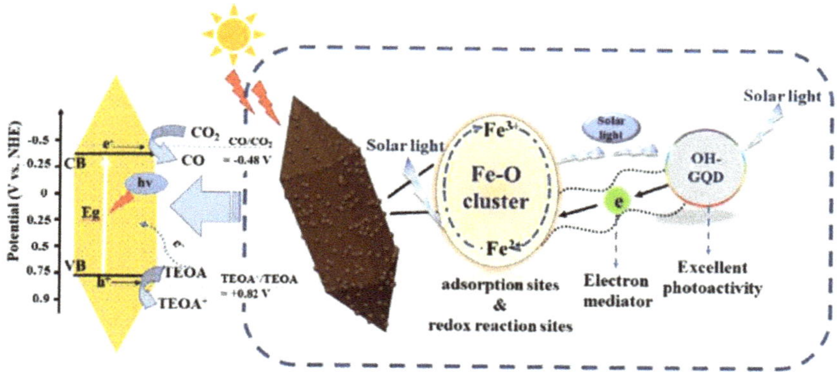

Figure 4.14. Schematic representing the reduction of CO_2 over G/M101 nanocomposite under visible light irradiation, reproduced with permission from [107], John Wiley & Sons, Wiley-VCH, Weinheim.

photothermal effect due to surface plasmon resonance-excited local heating of Ag NPs, which leads to enhanced electron transfer in the system and, thereby favours the selective reduction of 4-nitrophenol [109]. Photocatalytic removal of nitric oxide was successfully achieved by Xie *et al* using a tin oxide/GQD composite (SnO_2/GQD) under visible light illumination. The light harvesting and charge separation advanced without effecting the morphology and structure of the system. The increase in number of active species greatly enhanced the removal of NO compared to pure SnO_2 [110]. Liu *et al* successfully revealed the photoinduced charge carrier dynamics of nitrogen-doped $Bi_2O_2CO_3$/GQD composite (N-BOC/GQD) with enhanced photocatalytic performance towards photooxidation of NO. Electrons generated through irradiation of visible light move to BOC leaving holes in the valence band of GQD. UV irradiation results in a z scheme recombination of holes in the valence band of GQD and electron in the defect sites of N-BOC, leading to oxidation of NO [111] (figure 4.14).

4.5 Conclusions

This chapter presents a systematic review of the recent developments in GQD research. As evident from the plethora of published works in this field, GQD-based systems have proven their use in many applications and devices. However, the full advantage of these materials can be exploited only by filling up the research gaps in the field. It is imperative for successful and full-scale scientific use of the material that more experimental and theoretical studies are undertaken to get a clear picture of the mechanisms and effects of factors such as synthesis conditions and precursors on the PL property. Upscaling the production of GQDs with high yield and purity for commercial application through facile strategies is also a challenge. Taking into consideration the ongoing developments in the field, GQD-based systems with more functionalities are bound to emerge, which can replace conventional materials in the field of energy and environment.

References

[1] Li Y, Shu H, Wang S and Wang J 2015 Electronic and optical properties of graphene quantum dots: the role of many-body effects *J. Phys. Chem.* C **119** 4983–9

[2] Manikandan A, Chen Y Z, Shen C C, Sher C W, Kuo H C and Chueh Y L 2019 A critical review on two-dimensional quantum dots (2D QDs): from synthesis toward applications in energy and optoelectronics *Prog. Quantum Electron.* **68** 100226

[3] Facure M H, Schneider R, Mercante L A and Correa D S 2020 A review on graphene quantum dots and their nanocomposites: from laboratory synthesis towards agricultural and environmental applications *Environ. Sci. Nano* **7** 3710–34

[4] Chen J 2019 Graphene quantum dots and their applications in bioimaging and catalysis *Doctoral Thesis* Singapore Nanyang Technological University

[5] Shen J, Zhu Y, Yang X and Li C 2012 Graphene quantum dots: emergent nanolights for bioimaging, sensors, catalysis and photovoltaic devices *Chem. Commun.* **48** 3686–99

[6] Zhu J, Li X, Zhang Y, Wang J and Wei B 2018 Graphene-enhanced nanomaterials for wall painting protection *Adv. Funct. Mater.* **28** 1803872

[7] Zhang L, Xu H, Ding Z, Hu J, Liu J and Liu Y 2018 Amino N-oxide functionalized graphene quantum dots as a cathode interlayer for inverted polymer solar cells *J. Mater. Chem.* C **6** 5684–9

[8] Tajik S, Dourandish Z, Zhang K, Beitollahi H, Van Le Q, Jang H W and Shokouhimehr M 2020 Carbon and graphene quantum dots: a review on syntheses, characterization, biological and sensing applications for neurotransmitter determination *RSC Adv.* **10** 15406–29

[9] Shin Y, Park J, Hyun D, Yang J, Lee J H, Kim J H and Lee H 2015 Acid-free and oxone oxidant-assisted solvothermal synthesis of graphene quantum dots using various natural carbon materials as resources *Nanoscale* **7** 5633–7

[10] Zhang M *et al* 2012 Facile synthesis of water-soluble, highly fluorescent graphene quantum dots as a robust biological label for stem cells *J. Mater. Chem.* **22** 7461–7

[11] Tian R, Zhong S, Wu J, Jiang W and Wang T 2016 Facile hydrothermal method to prepare graphene quantum dots from graphene oxide with different photoluminescences *RSC Adv.* **6** 40422–6

[12] Lai S, Jin Y, Shi L, Zhou R, Zhou Y and An D 2020 Mechanisms behind excitation-and concentration-dependent multicolor photoluminescence in graphene quantum dots *Nanoscale* **12** 591–601

[13] Zhu S, Song Y, Wang J, Wan H, Zhang Y, Ning Y and Yang B 2017 Photoluminescence mechanism in graphene quantum dots: quantum confinement effect and surface/edge state *Nano Today* **13** 10–4

[14] Peng J *et al* 2012 Graphene quantum dots derived from carbon fibers *Nano Lett.* **12** 844–9

[15] Sarkar S, Gandla D, Venkatesh Y, Bangal P R, Ghosh S, Yang Y and Misra S 2016 Graphene quantum dots from graphite by liquid exfoliation showing excitation-independent emission, fluorescence upconversion and delayed fluorescence *Phys. Chem. Chem. Phys.* **18** 21278–87

[16] Pan D, Zhang J, Li Z and Wu M 2010 Hydrothermal route for cutting graphene sheets into blue-luminescent graphene quantum dots *Adv. Mater.* **22** 734–8

[17] Zhu S *et al* 2012 Surface chemistry routes to modulate the photoluminescence of graphene quantum dots: from fluorescence mechanism to up-conversion bioimaging applications *Adv. Funct. Mater.* **22** 4732–40

[18] Gao T, Wang X, Yang L Y, He H, Ba X X, Zhao J, Jiang F L and Liu Y 2017 Red, yellow, and blue luminescence by graphene quantum dots: syntheses, mechanism, and cellular imaging *ACS Appl. Mater. Interfaces* **9** 24846–56

[19] Qi B P, Zhang X, Shang B B, Xiang D and Zhang S 2018 Solvothermal tuning of photoluminescent graphene quantum dots: from preparation to photoluminescence mechanism *J. Nanopart. Res.* **20** 1–9

[20] Wang S, Cole I S, Zhao D and Li Q 2016 The dual roles of functional groups in the photoluminescence of graphene quantum dots *Nanoscale* **8** 7449–58

[21] Ritter K A and Lyding J W 2009 The influence of edge structure on the electronic properties of graphene quantum dots and nanoribbons *Nat. Mater.* **8** 235–42

[22] Ozfidan I, Korkusinski M and Hawrylak P 2016 Electronic properties and electron–electron interactions in graphene quantum dots *Phys. Status Solidi – Rapid Res. Lett.* **10** 13–23

[23] Li Y, Shu H, Niu X and Wang J 2015 Electronic and optical properties of edge-functionalized graphene quantum dots and the underlying mechanism *J. Phys. Chem.* C **119** 24950–7

[24] Shen J, Zhu Y, Yang X, Zong J, Zhang J and Li C 2012 One-pot hydrothermal synthesis of graphene quantum dots surface-passivated by polyethylene glycol and their photoelectric conversion under near-infrared light *New J. Chem.* **36** 97–101

[25] Tetsuka H, Asahi R, Nagoya A, Okamoto K, Tajima I, Ohta R and Okamoto A 2012 Optically tunable amino-functionalized graphene quantum dots *Adv. Mater.* **24** 5333–8

[26] Far'ain N, Salleh M M, Ashraf M, Abd Rahman M Y and Umar A A 2017 Synthesis of blue-luminescence graphene quantum dots using hydrothermal method *Diffus. Defect Data* B **268** 259–63

[27] Tian R, Zhong S, Wu J, Jiang W, Shen Y and Wang T 2016 Solvothermal method to prepare graphene quantum dots by hydrogen peroxide *Opt. Mater.* **60** 204–8

[28] Zhang J, Ma Y Q, Li N, Zhu J L, Zhang T, Zhang W and Liu B 2016 Preparation of graphene quantum dots and their application in cell imaging *J. Nanomater.* **2016** 1–9

[29] Liu B *et al* 2017 From graphite to graphene oxide and graphene oxide quantum dots *Small* **13** 1601001

[30] Hassan M, Haque E, Reddy K R, Minett A I, Chen J and Gomes V G 2014 Edge-enriched graphene quantum dots for enhanced photo-luminescence and supercapacitance *Nanoscale* **6** 11988–94

[31] Lu L, Zhu Y, Shi C and Pei Y T 2016 Large-scale synthesis of defect-selective graphene quantum dots by ultrasonic-assisted liquid-phase exfoliation *Carbon* **109** 373–83

[32] Dong Y, Chen C, Zheng X, Gao L, Cui Z, Yang H, Guo C, Chi Y and Li C M 2012 One-step and high yield simultaneous preparation of single-and multi-layer graphene quantum dots from CX-72 carbon black *J. Mater. Chem.* **22** 8764–6

[33] Zhu C *et al* 2015 A new mild, clean and highly efficient method for the preparation of graphene quantum dots without by-products *J. Mater. Chem.* B **3** 6871–6

[34] Shin Y, Park J, Hyun D, Yang J and Lee H 2015 Generation of graphene quantum dots by the oxidative cleavage of graphene oxide using the oxone oxidant *New J. Chem.* **39** 2425–8

[35] Liu Q, Zhang J, He H, Huang G, Xing B, Jia J and Zhang C 2018 Green preparation of high yield fluorescent graphene quantum dots from coal-tar-pitch by mild oxidation *Nanomaterials* **8** 844

[36] Li Y, Hu Y, Zhao Y, Shi G, Deng L, Hou Y and Qu L 2011 An electrochemical avenue to green-luminescent graphene quantum dots as potential electron-acceptors for photovoltaics *Adv. Mater.* **23** 776–80
[37] Shinde D B and Pillai V K 2012 Electrochemical preparation of luminescent graphene quantum dots from multiwalled carbon nanotubes *Chem. Eur. J.* **18** 12522–8
[38] Ananthanarayanan A, Wang X, Routh P, Sana B, Lim S, Kim D H, Lim K H, Li J and Chen P 2014 Facile synthesis of graphene quantum dots from 3D graphene and their application for Fe^{3+} sensing *Adv. Funct. Mater.* **24** 3021–6
[39] Tan X, Li Y, Li X, Zhou S, Fan L and Yang S 2015 Electrochemical synthesis of small-sized red fluorescent graphene quantum dots as a bioimaging platform *Chem. Commun.* **51** 2544–6
[40] Li Y, Li S, Wang Y, Wang J, Liu H, Liu X, Wang L, Liu X, Xue W and Ma N 2017 Electrochemical synthesis of phosphorus-doped graphene quantum dots for free radical scavenging *Phys. Chem. Chem. Phys.* **19** 11631–8
[41] Li L L, Ji J, Fei R, Wang C Z, Lu Q, Zhang J R, Jiang L P and Zhu J J 2012 A facile microwave avenue to electrochemiluminescent two-color graphene quantum dots *Adv. Funct. Mater.* **22** 2971–9
[42] Luo Z, Qi G, Chen K, Zou M, Yuwen L, Zhang X, Huang W and Wang L 2016 Microwave-assisted preparation of white fluorescent graphene quantum dots as a novel phosphor for enhanced white-light-emitting diodes *Adv. Funct. Mater.* **26** 2739–44
[43] Zhang C, Cui Y, Song L, Liu X and Hu Z 2016 Microwave assisted one-pot synthesis of graphene quantum dots as highly sensitive fluorescent probes for detection of iron ions and pH value *Talanta* **150** 54–60
[44] Zhao P, Li C and Yang M 2017 Microwave-assisted one-pot conversion from deoiled asphalt to green fluorescent graphene quantum dots and their interfacial properties *J. Dispersion Sci. Technol.* **38** 769–74
[45] Alizadeh T and Shokri M 2016 A new humidity sensor based upon graphene quantum dots prepared via carbonization of citric acid *Sens. Actuators* B **222** 728–34
[46] Hong G L, Zhao H L, Deng H H, Yang H J, Peng H P, Liu Y H and Chen W 2018 Fabrication of ultra-small monolayer graphene quantum dots by pyrolysis of trisodium citrate for fluorescent cell imaging *Int. J. Nanomed.* **13** 4807–15
[47] Gu S, Hsieh C T, Yuan C Y, Gandomi Y A, Chang J K, Fu C C, Yang J W and Juang R S 2020 Fluorescence of functionalized graphene quantum dots prepared from infrared-assisted pyrolysis of citric acid and urea *J. Lumin.* **217** 116774
[48] Lee N E, Lee S Y, Lim H S, Yoo S H and Cho S O 2020 A novel route to high-quality graphene quantum dots by hydrogen-assisted pyrolysis of silicon carbide *Nanomaterials* **10** 277
[49] Yan X, Cui X and Li L S 2010 Synthesis of large, stable colloidal graphene quantum dots with tunable size *J. Am. Chem. Soc.* **132** 5944–5
[50] Li Q, Zhang S, Dai L and Li L S 2012 Nitrogen-doped colloidal graphene quantum dots and their size-dependent electrocatalytic activity for the oxygen reduction reaction *J, Am. Chem. Soc.* **134** 18932–5
[51] Lu J, Yeo P S E, Gan C K, Wu P and Loh K P 2011 Transforming C60 molecules into graphene quantum dots *Nat. Nanotechnol.* **6** 247–52
[52] Chen G *et al* 2015 Rupturing C60 molecules into graphene-oxide-like quantum dots: structure, photoluminescence, and catalytic application *Small* **11** 5296–304

[53] Liu W W, Feng Y Q, Yan X B, Chen J T and Xue Q J 2013 Superior micro-supercapacitors based on graphene quantum dots *Adv. Funct. Mater.* **23** 4111–22
[54] Zhang S, Sui L, Dong H, He W, Dong L and Yu L 2018 High-performance supercapacitor of graphene quantum dots with uniform sizes *ACS Appl. Mater. Interfaces* **10** 12983–91
[55] Chen Q, Hu Y, Hu C, Cheng H, Zhang Z, Shao H and Qu L 2014 Graphene quantum dots–three-dimensional graphene composites for high-performance supercapacitors *Phys. Chem. Chem. Phys.* **16** 19307–13
[56] Tjandra R, Liu W, Zhang M and Yu A 2019 All-carbon flexible supercapacitors based on electrophoretic deposition of graphene quantum dots on carbon cloth *J. Power Sources* **438** 227009
[57] Qing Y *et al* 2019 Boosting the supercapacitor performance of activated carbon by constructing overall conductive networks using graphene quantum dots *J. Mater. Chem.* A **7** 6021–7
[58] Huang Y, Lin L, Shi T, Cheng S, Zhong Y, Chen C and Tang Z 2019 Graphene quantum dots-induced morphological changes in $CuCo_2S_4$ nanocomposites for supercapacitor electrodes with enhanced performance *Appl. Surf. Sci.* **463** 498–503
[59] Ding Z, Miao Z, Xie Z and Liu J 2016 Functionalized graphene quantum dots as a novel cathode interlayer of polymer solar cells *J. Mater. Chem.* A **4** 2413–8
[60] Ding Z, Hao Z, Meng B, Xie Z, Liu J and Dai L 2015 Few-layered graphene quantum dots as efficient hole-extraction layer for high-performance polymer solar cells *Nano Energy* **15** 186–92
[61] Zhang L, Ding Z C, Tong T and Liu J 2017 Tuning the work functions of graphene quantum dot-modified electrodes for polymer solar cell applications *Nanoscale* **9** 3524–9
[62] Wang S, Li Z, Xu X, Zhang G, Li Y and Peng Q 2019 Amino-functionalized graphene quantum dots as cathode interlayer for efficient organic solar cells: quantum dot size on interfacial modification ability and photovoltaic performance *Adv. Mater. Interfaces* **6** 1801480
[63] Hoang T T, Pham H P and Tran Q T 2020 A facile microwave-assisted hydrothermal synthesis of graphene quantum dots for organic solar cell efficiency improvement *J. Nanomater.* **2020** 1–8
[64] Koh K H, Noh S H, Kim T H, Lee W J, Yi S C and Han T H 2017 A graphene quantum dot/phthalocyanine conjugate: a synergistic catalyst for the oxygen reduction reaction *RSC Adv.* **7** 26113–19
[65] Çolak A T, Eren T, Yola M L, Beşli E, Şahin O and Atar N 2016 3D polyoxometalate-functionalized graphene quantum dots with mono-metallic and bi-metallic nanoparticles for application in direct methanol fuel cells *J. Electrochem. Soc.* **163** F1237
[66] Gizem Güneştekin B, Medetalibeyoglu H, Atar N and Lütfi Yola M 2020 Efficient direct-methanol fuel cell based on graphene quantum dots/multi-walled carbon nanotubes composite *Electroanalysis* **32** 1977–82
[67] Xu G, Wu Z, Xie Z, Wei Z, Li J, Qu K, Li Y and Cai W 2021 Graphene quantum dot reinforced hyperbranched polyamide proton exchange membrane for direct methanol fuel cell *Int. J. Hydrog. Energy* **46** 9782–9
[68] Yu Y, Ren J and Meng M 2013 Photocatalytic hydrogen evolution on graphene quantum dots anchored TiO_2 nanotubes-array *Int. J. Hydrog. Energy* **38** 12266–72

[69] Min S, Hou J, Lei Y, Ma X and Lu G 2017 Facile one-step hydrothermal synthesis toward strongly coupled TiO_2/graphene quantum dots photocatalysts for efficient hydrogen evolution *Appl. Surf. Sci.* **396** 1375–82

[70] Hao X, Jin Z, Xu J, Min S and Lu G 2016 Functionalization of TiO_2 with graphene quantum dots for efficient photocatalytic hydrogen evolution *Superlattices Microstruct.* **94** 237–44

[71] Yu S, Zhong Y Q, Yu B Q, Cai S Y, Wu L Z and Zhou Y 2016 Graphene quantum dots to enhance the photocatalytic hydrogen evolution efficiency of anatase TiO_2 with exposed {001} facet *Phys. Chem. Chem. Phys.* **18** 20338–44

[72] Xie H, Hou C, Wang H, Zhang Q and Li Y 2017 S, N co-doped graphene quantum dot/ TiO_2 composites for efficient photocatalytic hydrogen generation *Nanoscale Res. Lett.* **12** 1–8

[73] Raghavan A, Sarkar S, Nagappagari L R, Bojja S, MuthukondaVenkatakrishnan S and Ghosh S 2020 Decoration of graphene quantum dots on TiO_2 nanostructures: photosensitizer and cocatalyst role for enhanced hydrogen generation *Ind. Eng. Chem. Res.* **59** 13060–8

[74] Lv J J, Zhao J, Fang H, Jiang L P, Li L L, Ma J and Zhu J J 2017 Incorporating nitrogen-doped graphene quantum dots and Ni_3S_2 nanosheets: a synergistic electrocatalyst with highly enhanced activity for overall water splitting *Small* **13** 1700264

[75] Cirone J, Ahmed S R, Wood P C and Chen A 2019 Green synthesis and electrochemical study of cobalt/graphene quantum dots for efficient water splitting *J. Phys. Chem.* C **123** 9183–91

[76] Li J, Zhang X, Zhang Z, Li Z, Gao M, Wei H and Chu H 2019 Graphene-Quantum-Dots-induced facile growth of porous molybdenum doped Ni_3S_2 nanoflakes as efficient bifunctional electrocatalyst for overall water splitting *Electrochim. Acta* **304** 487–94

[77] Tam T V, Kang S G, Kim M H, Lee S G, Hur S H, Chung J S and Choi W M 2019 Novel graphene hydrogel/B-doped graphene quantum dots composites as trifunctional electrocatalysts for Zn–air batteries and overall water splitting *Adv. Energy Mater.* **9** 1900945

[78] Wang L, Li W, Wu B, Li Z, Pan D and Wu M 2017 Room-temperature synthesis of graphene quantum dots via electron-beam irradiation and their application in cell imaging *Chem. Eng. J.* **309** 374–80

[79] Jin K, Gao H, Lai L, Pang Y, Zheng S, Niu Y and Li X 2018 Preparation of highly fluorescent sulfur doped graphene quantum dots for live cell imaging *J. Lumin.* **197** 147–52

[80] Wang Z, Chen D, Gu B, Gao B, Wang T, Guo Q and Wang G 2020 Biomass-derived nitrogen doped graphene quantum dots with color-tunable emission for sensing, fluorescence ink and multicolor cell imaging *Spectrochim. Acta* A **227** 117671

[81] Su J, Zhang X, Tong X, Wang X, Yang P, Yao F, Guo R and Yuan C 2020 Preparation of graphene quantum dots with high quantum yield by a facile one-step method and applications for cell imaging *Mater. Lett.* **271** 127806

[82] Sun L, Li S, Ding W, Yao Y, Yang X and Yao C 2017 Fluorescence detection of cholesterol using a nitrogen-doped graphene quantum dot/chromium picolinate complex--based sensor *J. Mater. Chem.* B **5** 9006–14

[83] Wang B, Shen J, Huang Y, Liu Z and Zhuang H 2018 Graphene quantum dots and enzyme-coupled biosensor for highly sensitive determination of hydrogen peroxide and glucose *Int. J. Mol. Sci.* **19** 1696

[84] Cui F, Ji J, Sun J, Wang J, Wang H, Zhang Y, Ding H, Lu Y, Xu D and Sun X 2019 A novel magnetic fluorescent biosensor based on graphene quantum dots for rapid, efficient, and sensitive separation and detection of circulating tumor cells *Anal. Bioanal.Chem.* **411** 985–95

[85] Bruce J A and Clapper J C 2020 Conjugation of carboxylated graphene quantum dots with cecropin P1 for bacterial biosensing applications *ACS Omega* **5** 26583–91

[86] Roushani M, Mavaei M and Rajabi H R 2015 Graphene quantum dots as novel and green nano-materials for the visible-light-driven photocatalytic degradation of cationic dye *J. Mol. Catal. A: Chem.* **409** 102–9

[87] Ge J, Lan M, Liu W, Jia Q, Guo L, Zhou B, Meng X, Niu G and Wang P 2016 Graphene quantum dots as efficient, metal-free, visible-light-active photocatalysts *Sci. China Mater.* **59** 12–9

[88] Ibarbia A, Grande H J and Ruiz V 2020 On the factors behind the photocatalytic activity of graphene quantum dots for organic dye degradation *Part. Part. Syst. Charact.* **37** 2000061

[89] Fan J, Li D and Wang X 2016 Effect of modified graphene quantum dots on photocatalytic degradation property *Diam. Relat. Mater.* **69** 81–5

[90] Qu D, Zheng M, Du P, Zhou Y, Zhang L, Li D, Tan H, Zhao Z, Xie Z and Sun Z 2013 Highly luminescent S, N co-doped graphene quantum dots with broad visible absorption bands for visible light photocatalysts *Nanoscale* **5** 12272–7

[91] Pan D, Jiao J, Li Z, Guo Y, Feng C, Liu Y, Wang L and Wu M 2015 Efficient separation of electron–hole pairs in graphene quantum dots by TiO_2 heterojunctions for dye degradation *ACS Sustain. Chem. Eng.* **3** 2405–13

[92] Gupta B K, Kedawat G, Agrawal Y, Kumar P, Dwivedi J and Dhawan S K 2015 A novel strategy to enhance ultraviolet light driven photocatalysis from graphene quantum dots infilled TiO_2 nanotube arrays *RSC Adv.* **5** 10623–31

[93] Wang S, Cole I S and Li Q 2016 Quantum-confined bandgap narrowing of TiO_2 nanoparticles by graphene quantum dots for visible-light-driven applications *Chem. Commun.* **52** 9208–11

[94] Chinnusamy S, Kaur R, Bokare A and Erogbogbo F 2018 Incorporation of graphene quantum dots to enhance photocatalytic properties of anatase TiO_2 *MRS Commun.* **8** 137–44

[95] Sun X, Li H J, Ou N, Lyu B, Gui B, Tian S, Qian D, Wang X and Yang J 2019 Visible-light driven TiO_2 photocatalyst coated with graphene quantum dots of tunable nitrogen doping *Molecules* **24** 344

[96] Tian H, Shen K, Hu X, Qiao L and Zheng W 2017 N, S co-doped graphene quantum dots-graphene-TiO_2 nanotubes composite with enhanced photocatalytic activity *J. Alloys Compd.* **691** 369–77

[97] Lu Q, Zhang Y and Liu S 2015 Graphene quantum dots enhanced photocatalytic activity of zinc porphyrin toward the degradation of methylene blue under visible-light irradiation *J. Mater. Chem.* A **3** 8552–8

[98] Li Y, Wang L, Ge J, Wang J, Li Q, Wan W, Zhang B, Liu X and Xue W 2016 Graphene quantum dots modified ZnO + Cu heterostructure photocatalysts with enhanced photocatalytic performance *RSC Adv.* **6** 106508–15

[99] Wang J, Li Y, Ge J, Zhang B P and Wan W 2015 Improving photocatalytic performance of ZnO via synergistic effects of Ag nanoparticles and graphene quantum dots *Phys. Chem. Chem. Phys.* **17** 18645–52

[100] Cai A, Wang X, Qi Y and Ma Z 2017 Hierarchical ZnO/S, N: GQD composites: biotemplated synthesis and enhanced visible-light-driven photocatalytic activity *Appl. Surf. Sci.* **391** 484–90
[101] Liu J, Xu H, Xu Y, Song Y, Lian J, Zhao Y, Wang L, Huang L, Ji H and Li H 2017 Graphene quantum dots modified mesoporous graphite carbon nitride with significant enhancement of photocatalytic activity *Appl. Catal.* B **207** 429–37
[102] Cai A, Wang Q, Chang Y and Wang X 2017 Graphitic carbon nitride decorated with S, N co-doped graphene quantum dots for enhanced visible-light-driven photocatalysis *J. Alloys Compd.* **692** 183–9
[103] Li S, Chen Z, Wu D, Qin Y and Kong Y 2020 Graphene quantum dots enhanced photocatalytic activity of Sb_2WO_6 under ultraviolet-and visible-light irradiation *Bull. Korean Chem. Soc.* **41** 552–7
[104] Jeffrey C S 2010 Photocatalytic reduction of carbon dioxide (CO_2) *Developments and Innovation in Carbon Dioxide (CO_2) Capture and Storage Technology* (Cambridge: Woodhead Publishing), pp 463–501
[105] Yadav D, Yadav R K, Kumar A, Park N J and Baeg J O 2016 Functionalized graphene quantum dots as efficient visible-light photocatalysts for selective solar fuel production from CO_2 *Chem. Cat. Chem.* **8** 3389–93
[106] Zubair M, Kim H, Razzaq A, Grimes C A and In S I 2018 Solar spectrum photocatalytic conversion of CO_2 to CH_4 utilizing TiO_2 nanotube arrays embedded with graphene quantum dots *J. CO2 Util.* **26** 70–9
[107] Liu N, Tang M, Wu J, Tang L, Huang W, Li Q, Lei J, Zhang X and Wang L 2020 Boosting visible-light photocatalytic performance for CO_2 reduction via hydroxylated graphene quantum dots sensitized MIL-101 (Fe) *Adv. Mater. Interfaces* **7** 2000468
[108] Wei D, Tang W, Gan Y and Xu X 2020 Graphene quantum dot-sensitized Zn-MOFs for efficient visible-light-driven carbon dioxide reduction *Catal. Sci. Technol.* **10** 5666–76
[109] Wu C, Yuan Y, He Q and Song R 2016 Two-step synthesis of Ag@ GQD hybrid with enhanced photothermal effect and catalytic performance *Nanotechnology* **27** 48LT02
[110] Xie Y, Yu S, Zhong Y, Zhang Q and Zhou Y 2018 SnO_2/graphene quantum dots composited photocatalyst for efficient nitric oxide oxidation under visible light *Appl. Surf. Sci.* **448** 655–61
[111] Liu Y *et al* 2020 Defect state assisted Z-scheme charge recombination in $Bi_2O_2CO_3$/graphene quantum dot composites for photocatalytic oxidation of NO *ACS Appl. Nano Mater.* **3** 772–81

Chapter 5

Forms of graphene III: graphene nano-ribbons: preparation, assessments, and shock absorption applications

Prashant S Alegaonkar

In the strategic sector, effects of impulse pressure, in the form of shock waves, generated by detonation of explosives are of major interest. Interactions of transonic shock waves have been studied with graphene nano-ribbons (GNRs) having typical characteristics like Mach number >1.1, and built-in peak pressure >1.5 GPa. GNR specimens were synthesized by pyrolysis of biomass, carried out at rapid thermal chemical vapor deposition system. Starting material was chemically manipulated for purification to obtain GNRs and characterized using optical, electron, force microscopy, Raman, as well as Brunauer–Emmett–Teller (BET) and Barrett–Joyner–Halenda (BJH) techniques. Further, high-strain-rate measurements were carried out on GNR using Kolsky bar technique to collect temporal data of variations in stress, strain, and strain rates. A large number of involved dynamic mechanical parameters were extracted using Lagrange–Rankine–Hugoniot model. In fractographic analysis, statistics on areal flake dimensions, layer density, defect density, wrinkle, slip characteristics, etc, have been obtained. Incident shock energy was found to be dampened >60% of absorption loss with >10% transmittance. Details are presented.

5.1 Introduction and survey: blast and shock

In the defence sector, there is a constant demand for materials capable of enhancing energy dissipation rather than the conventional way of energy. Shock waves, typically, emerging from an explosion is the best example. By and large, explosion is typically an unpredictable phenomenon that encompasses the flow of energy; carried away by the components of the medium as compression and rarefaction.

Thus, an explosion is a tremendous acoustic (sound) wave that has been transfigured into a fragmented and stepper wave front of impulse pressure that has destructive repercussions on the surrounding medium. Blast can occur either naturally or synthetically. The synthetic explosions are further divided into two categories: accidental and intentional. Normally, accidental explosions are unplanned. They originate due to mishandling of the hazardous substances. There are number of incidences we read in news or hearsay such as burst of an LPG cooking tank in a kitchen, detonation of stored propellants, collision of gasoline wagons, ignition of stored firecrackers, explosion of dispersed hazardous dust, and blast because of loss of coolant in a nuclear reactor. These are all typical illustrations of explosions classified under various categories like confined, non-confined space, boiling liquid expansion vapour (BLEV), condensed phase, atmospheric dispersion, and nuclear explosion. The natural explosions that we occasionally witness are heavy lightning, volcanic eruption, impact of asteroid/comet, stellar bang, etc.

Historically, after the incidence of world war, dynamic deformation in materials was extensively investigated in a number of scientific disciplines and is the focus of intense research and development [1]. A dynamic process that eventually results in deformation is the application of time-varying external force to a body [2]. However, when high-strain-rate (HSR) deformation is imparted as a result of extremely temporal variations, the medium encounters a mechanical disturbance in the shape of stress. These deformation waves, which can be generically categorised as elastic, plastic, and shock waves, move through the material at specific speeds and amplitudes. Their classification is based on the medium's characteristics and the amount of dynamic force [3]. In elastic mode, the propagation of waves depends upon number of structural aspects of the material medium such as free surfaces (termed as Rayleigh waves), interfaces (tagged as Stoneley), microstructure (also called flexural), porosity (or shear), layers (Love), flow direction (longitudinal), etc. However, under the extreme strain rate conditions, amplitude of elastic wave exceeds the limit of elasticity of the medium and gets transform into a plastic wave. A plastic wave's wave front typically has a relatively narrow pulse width $\Delta\lambda$ of less than 1 μm and is compressed into a steep, spiky structure. The incoming wave has a peak pressure of greater than 50 GPa with a very strong amplitude. A shock wave of this enormity can transport through the material without modifying its macroscopic dimensions, and the material experiences compressive stress similar to hydrostatic compression. Such a plastic wave is, characteristically, termed as a *shock wave* [4]. Shock waves are lethal in nature, and depending on their interaction with materials, they can cause shear instabilities, dynamic fracture, blast, fragmentation, perforation, and even explosion. When they come into contact with interfaces and free surfaces, they have the ability to cause cracks, fractures, spalling, and other impacts. Notably, there are constructive aspects also to emergent shock waves. The propagation of shock waves through material has a wide range of significant applications [5].

In shock synthesis, a controlled consolidation of shock deposits intense energy between 0.1 and 2.0 GeV m^{-3} resulting phase transformations. Such effects cause chemical changes in a specific class of materials, such as the conversion of graphite to diamond. For making tools in high-speed machining shock treatment is used for bonding materials [6]. Further, HSR ranging from 0.1 to 3000 s^{-1} are of utmost importance by means of impact conditioning in mining, construction industries, rapid crack propagation studies, seismology, etc. Moreover, in the military sector, effect of impact for detonation of explosives is frequently encountered. Thus, investigating dynamic deformation becomes an important issue for shock dampening application. This necessitates a detailed understanding of both (a) the dynamics of material reaction and (b) the mechanics of HSR deformation. Even though explosions are unpredictable events, they can be simulated in a lab setting using a variety of approaches, including shock wave tubes [7], standoff or contact explosions [8], and the split Hopkinson pressure bar (SHPB) method [9]. In this, SHPB is a safe, time- and cost-efficient measuring technique to replicate shock wave conditions that help us better comprehend explosion-like scenarios off the field [10].

The mechanical characteristics of aluminium alloys subjected to SHPB at low and high temperatures have received little attention in recent years [9, 10]. In nano-systems, single, multiple-walled, and polymer-dispersed carbon nanotubes have been thoroughly investigated using experimental tensile, force–displacement curves, compressive data from structural mechanics, dynamic deformation, and molecular dynamic simulations [9, 11–15]. Most of the reports are based on quasi-static measurements rather than HSR and, mainly, focused onto quantification of reinforced parameters of the nano-composites [16, 17]. Deformation mechanics in graphene has received little attention [18, 19].

The scope of the book chapter is to present shock mitigation properties of GNRs). We mention specifically the interactions of shock waves with GNRs, address the deformation in morphology and microstructure under dynamic pressure, and examine the causes of intrinsic disorder and flaws that could lead to microscopic failures like slip and twinning, among other things. In experiments, after pyrolyzing bio-precursor by rapid thermal chemical vapor deposition, GNRs were subjected to SHPB to extract dynamic parameters like stress, strain, and strain rates. To uncover the link between structure and property, the GNRs were further examined employing optical spectroscopy techniques like Raman spectroscopy, Fourier transform infrared (FTIR), UV–visible spectroscopy and microstructure images captured using field emission scanning electron microscopy (FESEM). To extract damage feature information in GNR, signal processing of transmitted and reflected shock wave signals was performed. Shock wave/GNR interactions resulted in changes in microstructures that were essential for shock damping, physicochemical changes, plane gliding, phonon drag flow hardening and transient phase transformation. Specific details are furnished.

GNR was grown using synthesis protocols from [9, 18], and impulse pressure SHPB measurements were performed using the procedures described in [9]. GNRs subjected to SHPB were named as HSR GNRs. Figure 5.1 depicts a shock loading scheme with incident bar, GNRs, and transmission bar in an SHPB configuration. Both GNRs and HSR GNRs were compared using a variety of optical and structural characterization tools like Raman spectroscopy, with specification of wavelength equal to 532 nm and scan range given by 200–3000 cm^{-1}, confocal imaging, FESEM at 5 kV beam potential, and dark field transmission electron microscopy (HRTEM) at 200 kV at various magnifications.

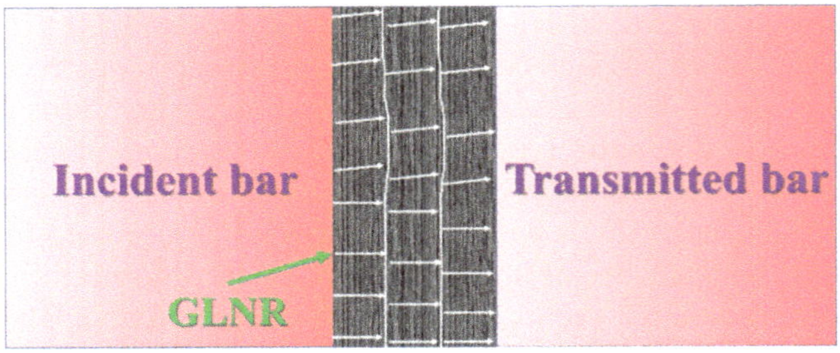

Figure 5.1. Shock loading scheme displaying impact of an incident pressure bar onto GNR specimen. Reproduced [18] under the terms of the Creative Common Attribution 4.0 International. Copyright 2020 MDPI.

5.2 Dynamic mechanical analysis

In a quasi-static range, it can be seen that whenever grade of stress surpasses elastic limit, in dynamic range, plastic deformation gets generated in a GNR [18]. Here the elastic and plastic wave component proliferates through the ribbon in the form of impulse [18]. This results in development of hydrostatic stress that exceeds by several orders of magnitude to dynamic flow stress. However, to a first approximation, our assumption is GNRs offers no impedance to the built-up shear. This assumption enables calculating the number of shock parameters attained using Lagrange–Rankine–Hugoniot (LRH) formulism [20–22]. The stress disturbance progress could be characterized using an equation:

$$\frac{\partial^2 u}{\partial t^2} = \frac{d\sigma/d\epsilon}{\rho_0}\left(\frac{\partial^2 u}{\partial x^2}\right) \tag{5.1}$$

at constant strain, ε, with

$$V_0 = \left(\frac{d\sigma/d\epsilon}{\rho_0}\right)^{1/2} \tag{5.2}$$

and $\frac{d\sigma}{d\epsilon} = E$ (elastic energy limit). In this, u is velocity of volume strain, ρ_0 is GNR density (~0.02 kg m^{-3}), and E is elastic energy in GPa. The interface conditions are $u = V_0 t$ at $x_i = 0$ and any $t > 0$; $u = 0$ at $x_f = \infty$. The variations for position, x and time, t is given by $\frac{x}{t} = \frac{\dot{\sigma}}{\rho_0}$. The variations in stress rate, $\dot{\sigma}$, could be elucidated by analyzing constitutive stress represented by σ—strain given by ε curve. Broadly speaking, in a dynamic region, the structural strength of a GNR are dependent on parameters like strain, ε, and the first dispersion derivative of strain, i.e., the strain rate, $\frac{d\varepsilon}{dt}$ (denoted by $\dot{\varepsilon}$). Figure 5.2(a) shows constitutive σ–ε curve for HSR GNR fitted using a bilinear function at the elastic regime i.e. first stage. The fitted power function is of a type

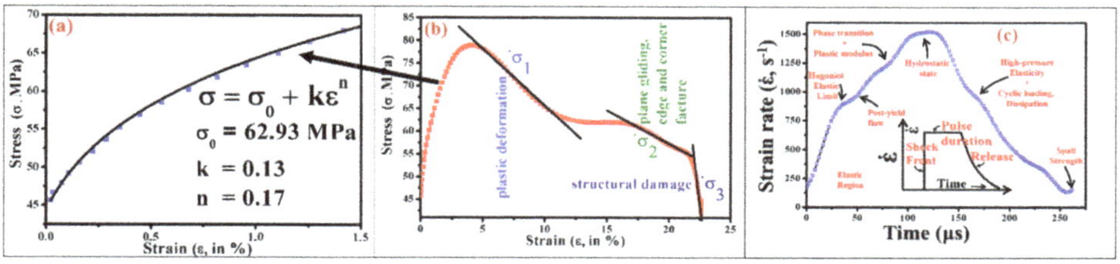

Figure 5.2. (a) Zoomed portion of σ versus ε curve (arrow represents the elastic region), (b) accompanying plastic regime displaying variations in $\dot{\sigma}_i$, The slope lines indicates the three stages indicating deformation in HSR GNR, and (c) recorded variations in $\dot{\varepsilon}$, with time. Reproduced [18] under the terms of the Creative Common Attribution 4.0 International. Copyright 2020 MDPI.

$$\sigma = \sigma_0 + k\,\epsilon^n \tag{5.3}$$

Here, k is a pre-exponential factor which is estimated to be 0.13 for HSR GNR, and the work-hardening coefficient to be $n \sim 0.17$. The incident shock velocity, V_S, is calculated to be ~3.5×10^3 m s^{-1}. It can be observed that the HSR GNR strongly drifts away from its elastic behavior with huge dispersion in V_0, as predicted by the LRH formulations. The elastic-to- plastic phase transformation characterized by transient time parameter t_c, is given by $t_c = \dfrac{\partial\sigma/\partial\dot{\varepsilon}}{\partial\sigma/\partial\varepsilon}$ and estimated to be $\approx 10^{-5}$ s^{-1}.

A progressive decline in figure 5.2(b), which represents the plastic regime, indicates that there has been stress propagation in the GNR matrix as a result of the strain flow hardening. The process is reported to be multi-stage, noting, typically, three stages. The $\dot{\sigma}_1$ is attributed to interparticle plastic deformation that emerged due to strain impediment offered by GNRs, $\dot{\sigma}_2$ is assigned to the conjugated plane gliding in a ribbon stack. This is shown very clear in the subsequent Raman and microscope analysis. However, the cause of $\dot{\sigma}_3$ is damage and spall to terminally rupture the ribbon structure. As shown graphically in the inset of figure 5.2(c), a discontinuous shear wave front strictly follows an ideal shock wave, which has a plateau and a steady decay of the hold-up pressure. However, realistic shock waves imprinted on a material exhibit clearly distinguishable characteristics that rely on the medium and pressure. Under elastic conditions, when the Hugoniot elastic limit (HEL) is around 8 GPa, the increase in volumetric stress rate is substantially higher in the case of HSR GNR. The flow strength in HSR GNR seems to be dependent upon the pressure employed and observed to be independent (Hugoniot) of change of pressure. This might have caused a material softening or hardening in GNR. In figure 5.2(c) elastic and plastic phases are observed to be distinctly separated, however, elastic fraction within HEL is seen to be propagated with greater velocity compared to that of plastic wave [18]. Beyond the HEL, a growth in pressure is seen to be steadily increased till the approach of the top plateau stage (figure 5.2(c)). Such behaviour results from the continuous nature of the plastic wave, mechanical characteristics of GNR control the rise in pressure. Due to a sizeable degree of elastic and plastic wave front separation, it should be noted that the detected phase variation is not as noticeable in the wave profile [18]. The plateau zone, observed at the top, is shock holding state. The pressure-unloading region in HSR GNRs suggested identical trend in which elastic deformation gives rise to plastic deformation as seen for shock loading prior to attaining elastic limit. A short tail feature represents the fracture of GNR structure [18].

5.3 Fractographic analysis

GNRs and HSR GNRs were both inspected and investigated using optical and electron microscopic technique. Optical microscopic analysis was carried out with a ×100 magnification. Micrographic images are presented in figure 5.3.

In this study, optical imaging was used to attain insight into the nature, form, and magnitude of GNRs before and after the shock impact. To obtain better imaging, suspended GNR and HSR GNR solutions were drop-casted onto two separate silicon wafers and dried. To gather data on ribbons, imaging was done in a number of locations using a raster scan-line technique.

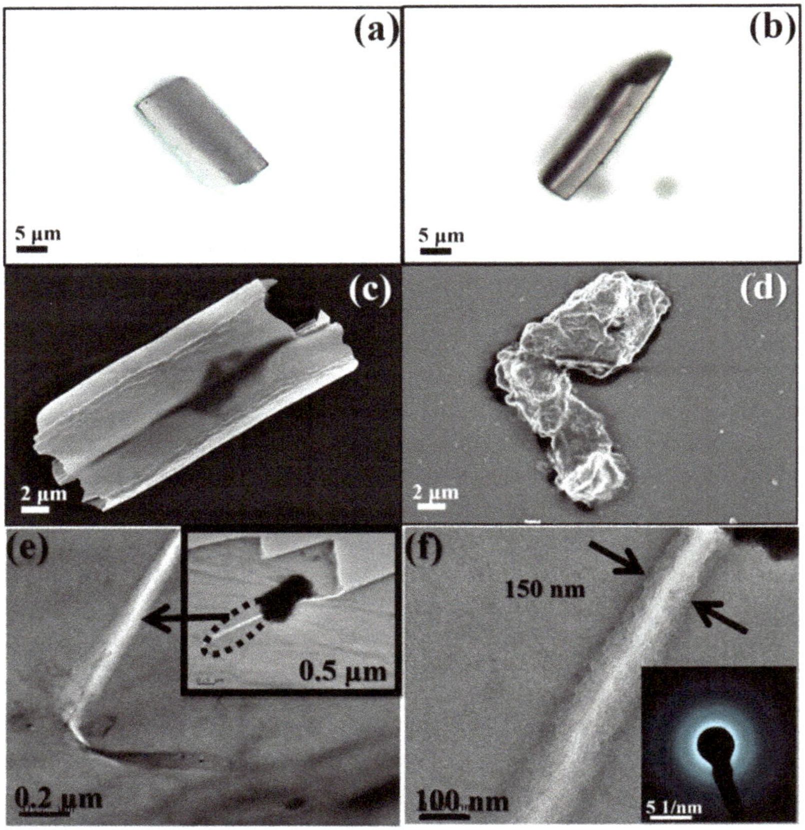

Figure 5.3. Optical micrographs for (a) GNRs and (b) HSR GNRs. (c and e) FESEM images of GNRs and (d and f) HRTEM images of HSR GNRs. Reproduced [18] under the terms of the Creative Common Attribution 4.0 International. Copyright 2020 MDPI.

Figure 5.3(a) displays typical recorded features of GNR and figure 5.3(b) shows HSR GNRs. To attain the microstructure features of ribbons the image processing was carried out with optimum sharpness, contrast and brightness. In GNRs, sharp edges are seen in the shape of a rectangle. Even smaller size ribbons had comparable structural characteristics (not provided herein). The ribbons are observed to be thick and encompassed with several conjugated graphene layers bunched together. Figure 5.3(b) depicts an HSR GNR dissimilar in its morphology when compared to a GNR. It appears that size and morphology are altered following the impact, along with some slight extensions and damage in the conjugated layers. Figures 5.3(c) and (d) depict the FESEM images of GNR and shock fractured HSR GNR. SEM micrographs resemble optical imaging with respect to change in thickness, sharp edges, and loss of rectangular shape for the shock treated ribbon geometry. HRTEM pictures, specifically for HSR GNRs are recorded in figures 5.3(e) and (f). It is clear from FESEM that each ribbon's surface served as a region of concentrated stress. The somewhat observed warps on the GNR surface are a sign of stress wave front flow within the GNR. The surplus flow stress has triggered breakdown and puncture in the graphene ribbon structure. Shock has affected the local structure by encountering dislocation, slip, and disorder, as discussed in TEM images. Broadly speaking, the SEM imaging revealed a change in surface of morphology of the ribbon to a certain extent to correlate the effect of impact.

In figure 5.3(e), HSR GNRs exhibited step-like features. The arrow in the inset displays the rise of a microcrack; initiated from the steps. The tip of the crack is positioned at the intersection of two orthogonal planes cut forming the step. The crack opening track is seen to be almost straight and narrowed down near the termination. In figure 5.3(e) such features are prominently seen as a spherical curve at the end of the crack following a warp-like exfoliation. In figure 5.3(f), inset, the blurred rings in selected area electron diffraction pattern depict an inferior grade of crystallinity in GNRs. By and large, the degree of crystallinity seems to be unchanged, even after the impact.

5.4 Raman spectroscopic studies

Raman spectroscopy plays a key role in quantifying the content of sp^2 and sp^3 carbon systems. Their proportion in sp^2 and sp^3 introduces new physical properties in the nanocarbons. By this technique, number of molecular parameters like dynamic force constant, k_q, crystalline, L_a, disordered length, L_D, areal defect density, n_D, electron–phonon coupling (EPC, Γ), and Fermi velocity, V_F can be extracted [18]. In figure 5.4, Raman spectra is displayed for (a) GNR and (b) HSR GNR at a wavelength of 532 nm laser excitation. Figures 5.4(c) and (d) depict the confocal Raman images. For GNR, D and G peaks are prominently visible, at ~1333 and 1593 cm^{-1}, respectively. From these values, k_q are estimated, according to a relationship: $\omega_{q=}\sqrt{k_q/\mu}$, where k_q is dynamical force constant, and μ is referred to as the reduced atomic mass of carbon. For GNR, the values assessed are 460 and 650 N m^{-1}, respectively, @ D (–C–C–) and G (–C=C–) with corresponding bond

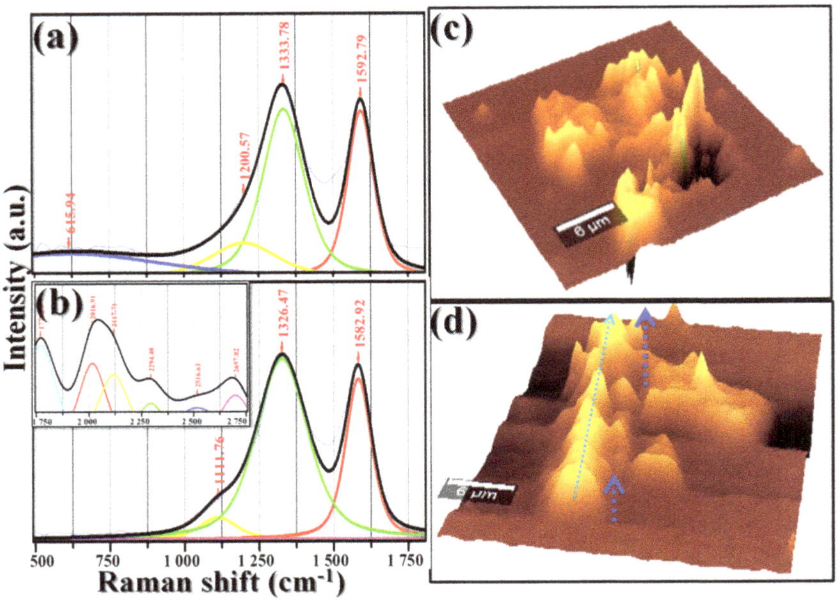

Figure 5.4. Raman spectra of (a) GNR and (b) HSR GNR with inset showing the inset 2D peaks, (c), and (d) confocal Raman mapping with scan area (40 × 40) μm^2. Reproduced [18] under the terms of the Creative Common Attribution 4.0 International. Copyright 2020 MDPI.

lengths 1.60 and 1.20 Å, respectively. According to post-impact study, the force constant is lowered to 400 N m^{-1} at D sites and to 600 N m^{-1} at G sites when coupled with bond deformations of 2.85 and 3.50. One can see that, at in-plane i.e., D-site, the elongation is less than double, whereas, at G-site it's nearly three times [18]. In general, plastic deformation in GNR cannot be described on the basis of specific theory of deformation, however, greater challenges emerge due to numerous channels of deformation like dislocation, glide, phase transformations, etc [18]. Raman analysis is capable of quantifying data associated with dislocation dynamics as a mediator of plastic deformation. Such induced dislocation evolves as a result of τ with a component of force given by $F = \tau b$; where, b is the offset (Burger) vector [18]. The dislocation movement is resisted by competitive frictional forces. At sp^3 site, such a force movement F_D, has a value of 0.3 nN, whereas, for sp^2, F_G has a value of 0.67 nN, revealing a 33% hike in-plane when related to the out-of-plane component [18].

The Raman features recorded are similar to the ones reported in our earlier work of graphene-like nanoflakes [9]. The value of I_D/I_G for GNRs is estimated to be 1.55 and after the impact it is reduced by 0.44. The relation between I_D/I_G and crystalline length, L_a is given by:

$$L_a(\text{nm}) = (2.4 \times 10^{-10})\, \lambda_L^4 \frac{I_D}{I_G} \tag{5.4}$$

The wavelength of laser excitation (533 nm) is given by λ_L. The estimated value of L_a is 17 nm for GNR and for HSR GNR a value of 7 nm; ascribed to dropping of sp^2 content with growth in sp^3 zone upon impact. Moreover, I_D/I_G is associated with inter-defect distance, L_D, by a modified Tuinstra–Koenig relationship:

$$\frac{I_D}{I_G} = \frac{C(\lambda_L)}{L_D^2} \tag{5.5}$$

where $C(\lambda)$ is the phenomenological parameter with an approximate value of 102 nm^2 at wavelength 532 nm. L_D, can be expressed in terms of defect density n_D, $(cm^{-2}) = \frac{10^{-14}}{L_D^2}$. For GNR, $L_D = 15$ nm and $n_D = 10^{-13}$ cm^{-2}. In HSR GNR, these values are 10 nm for L_D and 10^{-12} cm^{-2} for n_D. Notably, the values derived for L_a, L_D, and n_D have a connection with shear strain, γ. Under dynamic pressure, the chiral dislocation movement in GNR may result in an elastic shear strain γ_E. The relation $\gamma_E = n_D \times b \times L_D$; which yields to a value $>10^{-9}$. In the plastic regime, γ_P is greater than 10^{-9} for GNR. The speed with which disorder is moved (ν) is given by $\dot{\gamma} = n_D \times b \times \nu$, and ν is greater than 10 km s^{-1}.

Furthermore, the linewidth analysis of D peak in nanocarbon system has a close relationship with electron–phonon-coupling (Γ) coupled with phonons that have acoustic longitudinal and transverse modes, which entails inter-phonon interaction augmented by additional elementary excitations. However, one can ignore elementary anharmonic excitations. For GNR the value is 165 and 205 cm^{-1}. On the other hand, for HSR GNR with estimated Γ it is 5 and 35 N m^{-1}. Γ has a relationship with Fermi velocity, V_F, as

$$\Gamma = \frac{\sqrt{3}}{4}\frac{a_0^2}{\mu}\frac{EPC\Gamma^2}{\nu_F^2} \tag{5.6}$$

where μ represents the reduced mass of carbon molecule, $EPC\Gamma^2$ is labelled as hypothetical Γ-phonon-branching parameter with a value of 47 eV^2 Å^{-2} and a_0 has a value of 3.14 Å. The value of V_F for GNR and HSR GNR is, respectively, 10 and 5×10^5 ms^{-1} [23–25].

Curiously, for HSR GNR, a prominent 2D band is seen over 2000–2750 cm^{-1} which is in contrast to GNR, indicating longitudinal shear effect. Confocal mapping is a procedure successfully used for investigating composition [23]. We implemented a comparable methodology to assess our post-impacted ribbons. For the final image of a particular zone, the pixels are the Raman spectral features gathered from each particular spatial coordinate inside the chosen scan region. Raman spectra were recorded over 1000–1800 cm^{-1}. However, spectral variations in D and G peak over 1200–1600 cm^{-1} are used to map the modifications in the post-impacted GNR. Figures 5.4(c) and (d) reveal the topology of sp^3 and sp^2 fractions. The color contrast between emerged sp^3 and sp^2 zone is clearly visible in both the mapped images. The images were taken scanning areas 50×50 μm^2. The sp^3 zone illustrated in figure 5.4(c) is observed to be distributed homogeneously within the sp^2 network. In areas where the sp^3 disorder is remarkably high it appears to be noticeably a dark zone. However, in the planer region, no such distinctiveness is noted, as seen in figure 5.4(d). We have jotted down two notable observations: first, the sp^3 phase has a significant systematic pattern and, second, the sp^3 zone warps within sp^2 phase. Perhaps, imprinting of shock propagated in ribbon consequently strained the sp^3 zone (horizontal arrow). Relative strength is represented by vertical (blue) arrows, and it appears to be stronger near GNR boundary layers.

5.5 Signal processing investigations: pressure impulse interaction with GNR

Signal processing has added one more dimension to analysis stresses emerging within GNR. For this purpose, incident, transmitted, and reflected impulse signals form SHPB were used for signal processing. In SHPB, GNR specimens are placed between pressure bars and sandwiched between them. A striker bar was launched toward the incident bar by a gas pistol. When the stress wave strikes on the incident bar elastic compression arises which propagates through this input bar. Due to GNR being placed onto the input bar at the interface, part of the impact wave is reflected and part is transferred through the specimen to the input bar. Thereafter, the reflected wave rebounds as a tensile wave. With semiconductor strain gauge sensors linked to the incident bar, voltage drop signals that were used to quantify the resulting time-dependent stresses could be determined. The same sensor that is mounted to this bar may be capable of picking up the reflected signals. In a similar fashion, the output bar gathers the transmitted signal. The signals received are shown in figure 5.5(a). In order to understand more about the way strain energy propagates in GNR, the collected signals have been further analysed. For this, signal processing investigations employing the short-time Fourier transform (STFT)

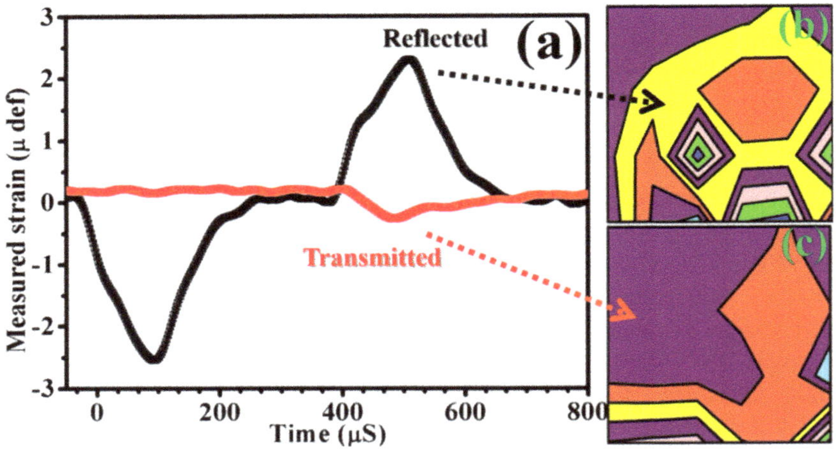

Figure 5.5. (a) Incident, reflected, and transmitted shock wave signals obtained from strain gauge sensors for GNR, (b) and (c) strain contours of STFT collected showing dispersion of shock energy for reflected and transmitted impact. Reproduced [18] under the terms of the Creative Common Attribution 4.0 International. Copyright 2020 MDPI.

method were carried out to look into the flow of strain energy in the time and frequency domain. The reflected and transmitted strain in the GNR are shown in figures 5.5(b) and (c), respectively.

The STFT functional is represented by the equation,

$$x\,([k],\ \omega') = X\,(n,\ \omega') = \sum_{n=\infty}^{n=0} x[k]\ \cdot \omega'\,[k - n]\mathrm{e}^{-j\omega' k} \tag{5.7}$$

Here, a time sequence mode of operation was performed. The discreet time locality is labelled as n, $x\ [k]$ gives data array variable of time, ω' is continuous signal frequency. The STFT was computed employing the succeeding process: (a) a definite number of signal data points (N) equal to window size were taken out from input shock signal; (b) window of selected type was employed to multiply the extracted data, point-by-point; and finally (c) zeros were added on either side of the window and (d) FFT were computed. Furthermore, the succeeding steps were executed to calculate STFT with the following settings: sampling interval as 1, both FFT length and window length as 256, overlap as 128, and rectangle as the type of window. The window function is given by

$$\omega'\,[k] = \ 1, \quad \text{if } 0 \leqslant k \leqslant N - 1; = 0, \quad \text{otherwise} \tag{5.8}$$

The output results are in the form of complex valued data matrix ($P \times Q$) with the following conditions:

$$P = \frac{F_\mathrm{w}}{2} + \ 1, \ \text{if input signal data real; else;} \ = F_\mathrm{w} \ \text{complex} \tag{5.9}$$

and

$$Q = f\left(\frac{D_\mathrm{s} - \ W_\mathrm{s}}{W_\mathrm{s} - \ O_\mathrm{s}}\right) + \ 1 \tag{5.10}$$

Here, F_w is the transform width, D_s is the size of the data, W_s is the width of the window, O_s is the overlap function, and f is the floor function. The discrete Fourier transform is applied to the signal's segments in order to slide the window over the full signal and obtain the transform coefficients. The spectral content of the signal can be attained using $|\ X(n,\ \omega')|^2$. Recurrence of the shock waves is represented by temporal variations of spectrograph illustrations. Blocks clearly show how power is dispersed or distributed within GNRs. The size of contour indicates quantity of energy lost, and each colored region symbolizes the energy dissipation.

Impulse pressure can be imaged as a kind of excitation perturbation, in the form of a δ-function. In the GNR, the impulse pressure employs negligibly small shear compared to the compressive hydrostatic component and induces disturbance, moving isentropically, given by an approximate relation: $\sqrt{\frac{d\sigma}{d\varepsilon}/\rho}$ is equivalent to $\sqrt{dP/dV}$, (where, ρ is the density of the GNRs, while P and V are the change in the pressure and volume, respectively). As it spreads over the GNRs, the mechanical disturbance that is being subjected to changes with time and gets stronger at its peak, which causes disorder activation [24]. For dynamic flow strength in nano-ribbons, inherent defect is a crucial factor. A discrepancy between the GNRs' material flow and shock wave front's speed is observed in terms of shock travel. This causes the GNR to distort past its elastic limit, which then causes plastic deformation and eventually fracture (spall region). The separation of conjugated layers would result from the plastic deformation's accumulation of residual internal tensions. The critical stress limit is determined by the ratio of the in-plane to out-of-plane stress as the slip dislocation begins to move in response to the applied shear stress [26, 27]. The slip ratio is given by

$$\frac{\sigma_{\parallel}}{\sigma_{\perp}} = \frac{4\pi}{(\sigma - 1)} \times \{5.8 + \log(\sigma - 1)\} \times e^{\frac{4\pi}{(\sigma-1)}} \tag{5.11}$$

The out-of-plane and in-plane stress components are given by $\sigma_{\perp}$ and $\sigma_{\parallel}$, respectively. For GNRs and HSR GNRs, values of σ are 0.07 and 0.1, respectively. The slip ratio is given by $\frac{\sigma_{\parallel}}{\sigma_{\perp}} \approx 10^{-4}$ for GNRs, which is as low as 10^{-6} after impact. As illustrated schematically in figure 5.6, a Mach umber of 1.0 for the incident transonic impulse nearly damps more than 80% of its shock energy within GNR, transmits a negligible fraction (5%), and the remaining shock energy is vulnerable to spall/fracture (table 5.1).

Before concluding, it should be noted that in some circumstances, a material's capacity to store energy within its structure becomes less significant than its capacity to release it internally. It is crucial when designing and constructing armour blocks,

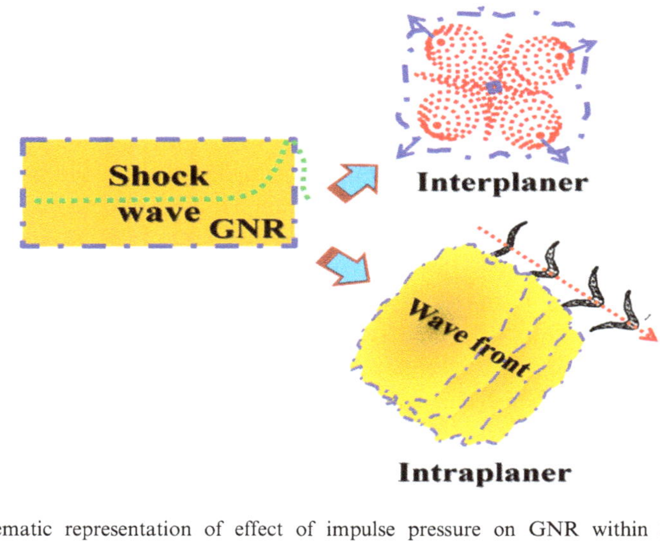

Figure 5.6. Schematic representation of effect of impulse pressure on GNR within layer conjugation. Reproduced [18] under the terms of the Creative Common Attribution 4.0 International. Copyright 2020 MDPI.

Table 5.1. Estimated dynamic mechanical parameters for GNR from Hugoniot formalism.

Sr No.	Parameter (SI Unit)	GNR	HSR GNR
1	Young's modulus	>60 MPa	—
2	Elastoplastic time constant, t_c	10^{-4} s^{-1} (metal-like)	—
3	Hugoniot elastic limit (GPa)	—	8 (ceramic-like)
4	Force constant, (–C–C–) (N m^{-1})	460	400
5	Force constant, (–C=C–) (N m^{-1})	650	600
6	Bond length (Å)	1.60	2.85
7	Bond deformation (Å)	1.20	3.50
8	Disorder velocity (km s^{-1})	—	13.00

bulletproof armour, bunkers, and many other structures where shock, acceleration, and jerk need to be precisely and efficiently suppressed.

The studies undertaken here areuseful in gaining a fundamental understanding of the shock wave interaction with GNR, which offers a crucial hint for creating body armour out of nanocarbons.

5.6 Conclusions

In conclusion, we have presented shock mitigation properties of GNRs that were fabricated using pyrolysis process carried out in a rapid thermal chemical vapor deposition set up. Prepared GNRs were subjected to an 1.5 GPa magnitude impulse pressure, with a speed greater than 1 Mach and microsecond duration. Such critical conditions were created using split Hopkinson pressure bar technique. Data for temporal variations in strain, stress, and strain rate was analyzed. Several dynamic mechanical factors, including constitutive (ε–σ) conjugate, Peierls–Nabarro stress, slip movement rate, work-hardening factor elastoplastic, shock dissipation constant transformation rate were extracted out. Curve σ–ε exhibited parabolic work hardening, that attained a value approximately 0.17. The elastoplastic phase transformation was obtained to a value of 10^{-5} s^{-1}. In the plastic region, the strain rate was seen to be varied, typically, in three stages, i.e., corner/edge fracture, separation of conjugated ribbons and layer exfoliation towards the spall stage. In elastic regime, the rise in volumetric stress ultimately tends to Hugoniot elastic limit at 8 GPa. GNRs were observed to depart significantly from the plastic behavior resembled closely to an ideal Rankian–Hugoniot model. It was observed that the shear waves had a threefold inline when compared to the out-of-plane impact within ribbons. Information on slip dynamics was obtained from Raman analysis by analyzing shear stresses that revealed a compressive force movement, F_D of approximate value 0.3 nN for sp^3 content and approximate value of 0.9 nN for sp^2 content location with a speed of disorder >10 km s^{-1}. Raman confocal imaging showed significant redistribution of sp^3 content with emergence of geometrically symmetric strain and an encroachment into the sp^2 phases. By and large, incident impulse velocity was decreased by an amount of 80% showing an effective shock mitigation character of GNR. The results clearly depict that the nanoribbon material had a complex hydrodynamic response that made it ideal for use as a shock absorber in the strategic field.

References

[1] Wong D M and Lee S Y 1963 Overpressure and dynamic-pressure impulse resulting from high-intensity explosion *J. Acoust. Soc. Am.* **35** 798

[2] Sun W B, Jiang Y and He W Z 2011 An overview on the blast loading and blast effects on the RC structures *Appl. Mech. Mater.* **94** 77–80

[3] Sencer B H, Maloy S A and Gray G T 2005 The influence of explosive-driven shock restraining at 35 GPa and of high deformation on the structure/property behavior of 316 L austenitic stainless steel *Metall. Mater. Trans.* A **36** 1825–31

[4] Lee O S, Lee J W and Kim S H 2005 Dynamic deformation behavior of rubber (NR/NBR) under high strain rate compressive loading *Key Eng. Mater.* **297** 172–7

[5] Johansson D and Ouchterlony F 2013 Shock wave interactions in rock blasting: the use of short delays to improve fragmentation in model-scale *Rock Mech. Rock Eng.* **46** 1–18

[6] Hui D and Dutta P K 2011 A new concept of shock mitigation by impedance-graded materials *Composites* B **42** 2181–4

[7] Briassulis G, Agui J H, Andreopoulos J and Watkins C B 1996 A shock tube research facility for high-resolution measurements of compressible turbulence *Exp. Therm. Fluid Sci.* **13** 430–46

[8] Sandhu I S, Sharma A, Singh M K, Kumari R, Alegaonkar P S and Saroha D R 2017 Study of blast wave pressure modification through rubber foam *Procedia Eng.* **173** 570–6

[9] Chinke S L, Sandhu I S, Saroha D R and Alegaonkar P S 2018 Graphene-like nanoflakes for shock absorption applications *ACS Appl. Nano Mater.* **1** 6027–37

[10] Chen W W and Song B 2010 *Split Hopkinson (Kolsky) Bar: Design, Testing and Applications* (Berlin: Springer Science & Business Media)

[11] Pandya K S and Naik N K 2015 Energy absorption capability of carbon nanotubes dispersed in resins under compressive high strain rate loading *Composites* B **72** 40–4

[12] Kai Y, Garen W and Teubner U 2018 Experimental investigations on microshock waves and contact surfaces *Phys. Rev. Lett.* **120** 064501

[13] Yu M F, Lourie O, Dyer M J, Moloni K, Kelly T F and Ruoff R S 2000 Strength and breaking mechanism of multiwalled carbon nanotubes under tensile load *Science* **287** 637–40

[14] Gojny F H, Wichmann M H, Fiedler B and Schulte K 2005 Influence of different carbon nanotubes on the mechanical properties of epoxy matrix composites—a comparative study *Compos. Sci. Technol.* **65** 2300–13

[15] Natsuki T and Endo M 2004 Stress simulation of carbon nanotubes in tension and compression *Carbon* **42** 2147–51

[16] Lee J H, Loya P E, Lou J and Thomas E L 2014 Dynamic mechanical behavior of multilayer graphene via supersonic projectile penetration *Science* **346** 1092–6

[17] Chinke S L, Sandhu I S, Bhave T M and Alegaonkar P S 2020 Surface interactions of transonic shock waves with graphene-like nanoribbons *Surfaces* **3** 505–15

[18] Kumar A, Patil S, Joshi A, Bhoraskar V, Datar S and Alegaonkar P 2013 Mixed phase, sp^2–sp^3 bonded, and disordered few layer graphene-like nanocarbon: synthesis and characterizations *Appl. Surf. Sci.* **271** 86–92

[19] Dresselhaus M S and Dresselhaus G 2002 Intercalation compounds of graphite *Adv. Phys.* **51** 1–186
[20] Kuhn H and Medlin D 2000 *Mechanical Testing and Evaluation* (Materials Park, OH: ASM International)
[21] Nemat-Nasser S (ed) 2014 *Mechanics Today: Pergamon Mechanics Today Series* (Amsterdam: Elsevier)
[22] Kumar A, Chouhan D K, Alegaonkar P S and Patro T U 2016 Graphene-like nanocarbon: an effective nanofiller for improving the mechanical and thermal properties of polymer at low weight fractions *Compos. Sci. Technol.* **127** 79–87
[23] Lee C, Wei X, Kysar J W and Hone J 2008 Measurement of the elastic properties and intrinsic strength of monolayer graphene *Science* **321** 385–8
[24] Casiraghi C 2012 Raman spectroscopy of graphene *Spectroscopic Properties of Inorganic and Organometallic Compounds: Techniques, Materials and Applications* vol 43 (Cambridge: Royal Society of Chemistry) pp 29–56
[25] Shepelev I A, Chetverikov A P, Dmitriev S V and Korznikova E A 2020 Shock waves in graphene and boron nitride *Comput. Mater. Sci.* **177** 109549
[26] Smotlacha J and Pincak R 2018 Electronic properties of phosphorene and graphene nanoribbons with edge vacancies in magnetic field *Phys. Lett.* A **382** 846–54
[27] Pincak R, Smotlacha J and Osipov V A 2015 Electronic states of zigzag graphene nanoribbons with edges reconstructed with topological defects *Physica* B **475** 61–5

Chapter 6

Forms of graphene IV—functionalized graphene

Neena D, Sesha Manuguri and Badri Narayanan Narasimhan

The functionalization of graphene/graphene oxide (GO) by different strategies has drawn significant research attention, owing to its ability to influence the interlayer spacing between graphene/GO sheets. The controlled manipulation of GO interlayer spacing leads to elevated cyclic stability and capacitance/power conversion efficiency of energy storage and conversion devices. Furthermore, functionalized graphene has been used in biomaterials and tissue engineering applications. Here, we present a brief introduction of functionalized graphene followed by the energy storage/conversion applications, biomedical applications/drug delivery and a summary of the chapter.

6.1 Brief introduction of functionalized graphene

Since its discovery in 2004, graphene has been hailed as a wonder material with the potential to revolutionize the technological world [1]. The unique 2D honeycomb structure confers graphene with outstanding electrical [2] and thermal conductivity [3], and with excellent chemical stability. Traditionally, graphene has been synthesized using mechanical and chemical exfoliation methods [4] or chemical vapor deposition-based procedures [5–7]. However, such fabrication methods render graphene difficult to process by solvent-assisted techniques that form the basis for functional device fabrication. Therefore, it is highly desirable to achieve suitable chemical functionalities that enable graphene to be utilized for practical applications.

The approach to functionalization is mostly dictated by the specific application and can be classified into two categories, i.e., covalent, and non-covalent. Covalent strategies mostly are linked to the chemical procedures that are used to prepare graphene. For example, pure graphene isolated from graphite that is dispersed in organic solvents like pyridine, chlorobenzene etc, results in functional groups making graphene suitable for further chemical functionalization [8, 9]. Furthermore, chemical

doi:10.1088/978-0-7503-3999-5ch6

modification of graphene invariably leads to disturbance in its honeycomb structure, resulting in modulation of its electronic properties [10–12]. On the other hand, non-covalent functionalization of graphene with organic compounds is performed by utilizing the π–π interactions [13, 14]. Compared to covalent modification, non-covalent procedures result in functional groups without disturbing the honeycomb structure.

In this chapter we focus upon the functionalization strategies specific to usage of graphene in energy storage and conversion applications. For example, the electronic properties of graphene are commonly enhanced by nitrogen (N) and boron (B) doping [15]. The process includes the reduction of graphene using nitrogen rich sources like ammonia or urea, and boron rich sources like boron hydride-tetrahydrofuran [16]. Such a process results in substitution of carbon atoms with N or B atom, resulting in modulation of conducting properties of graphene [17, 18]

6.2 Energy applications of functionalized graphene

Worldwide energy requirement is intensifying day by day and it is projected that energy requirement will reach 27.6 TW by 2050 [19–21]. Our conventional source of energy, fossil fuels, has caused global warming and environmental crisis. Due to the existing situation, renewable energy systems, namely solar cells, batteries, supercapacitors, carbon conversion and hydrogen production by water splitting, have become intense topics of research [22]. Consequently, it is essential to fabricate eco-friendly, cost-effective and high performing energy storage/conversion devices. Owing to its eco-friendliness, abundance and stability, carbon has been considered as a potential candidate for electrochemical devices. Amongst the carbon's allotropes, graphene has become the right choice for renewable energy applications because of its outstanding electrical conductivity, excellent chemical stability, good thermal conductivity, high specific surface area and easy functionalization [23]. Herein, we emphasize the discussion on various forms of graphene and its derivatives for energy applications.

6.2.1 Energy storage applications

6.2.1.1 Solar cells (DSSCs, PSCs)

Photovoltaic technology has experienced exponential growth since solar cells could be fabricated to harvest the solar energy with high power conversion efficiency [24]. One of the main aspects of the rapidly growing photovoltaic technology is the fabrication of countless materials that fulfil the device demands, together with abundant accessibility of materials, and cost-efficient fabrication process. Commonly, these types of devices are categorised into three groups, where silicon based solar cells are the first type of devices have dominated the solar industry for many years [25]. The second group of solar cells is based on semiconductors which are also known as thin-film solar cells [26, 27]. Further, another highly studied form of solar cells, including dye-sensitized solar cells (DSSCs) [28, 29], polymer solar cells [30, 31], and perovskite solar cells (PSCs) [32, 33], ise the third generation of solar cells. Among the third-generation devices, DSSCs and PSCs have gained considerable attention due to their fast-growing performance.

The dye-sensitized solar cell (DSSC) has become a promising contender as a renewable energy source, owing to low-cost, simple fabrication, flexibility, and high-power conversation rate as compared to silicone-based solar cells [34–36]. A typical DSSC comprises a counter electrode (CE) (a catalyst), photoanode (a dye-sensitized semiconducting metal oxide) and an electrolyte (redox couple) [37]. The main component of the device is the CE, which assembles the electrons from the external circuit and catalyses the charge reduction arbitrator [38]. Generally, platinum (Pt) has been employed as the conventional CE because of its good stability, superior electrocatalytic performance, low resistance of charge transfer and excellent conductivity [39]. However, its scarcity and high cost, hinders the mass production of dye-sensitized solar cells devices [40]. Thus, it is essential to develop an alternative for Pt CE of DSSC. Due to their abundance, cost-effectiveness and physiochemical characteristics, carbon-based materials have emerged as a promising alternative CE for DSSC. Most recently, graphene-based materials have been utilized as CE and demonstrated a power conversion efficiency (PCE) comparable with standard CE (Pt) of the device [21].

In one study, Yu *et al* [41] fabricated three-dimensional (3D) N-doped porous reduced graphene oxide (rGO) frameworks (NHGF) and used them as an efficient CE for dye-sensitized based devices. The synthesis process of a GO sheet to NHGF paper is presented in figure 6.1. The hierarchical porosity structure was incorporated to constitute a 3D nitrogen-doped rGO sheet. These hierarchical porosities delivered more active sites without hindering the electrical conductivity and interconnectivity between rGO sheets prevented the restacking process. Further, NHGF paper became strong enough to be a standalone CE for the cell. The NHGF CE-based cells provided higher efficiency (5.56%) than the pristine Pt based devices. The enhanced electrocatalytic activity achieved with NHGF suggests the presence of superior charge transport pathways and a higher density of active sites.

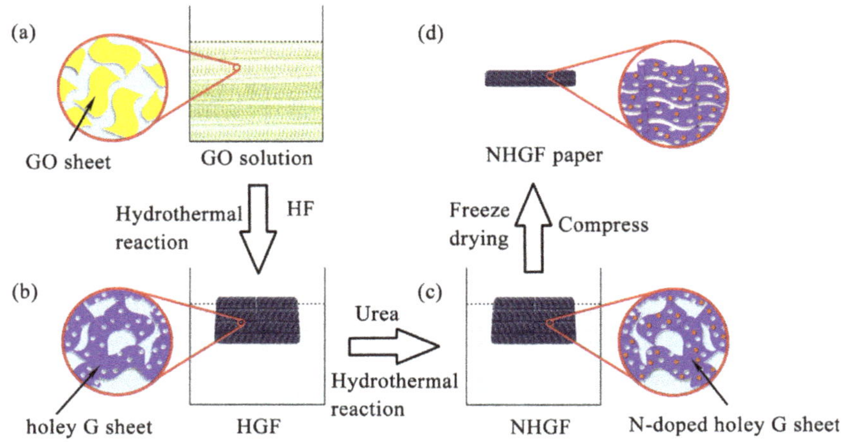

Figure 6.1. Schematic diagram for synthesis process of NHGF and NHGF paper. Reproduced (adapted) fromn [41], copyright 2016, with permission from Elsevier.

Similarly, Kannan *et al* [42] employed nitrogen and sulfur co-doped graphene (NS-G) as CE for DSSCs. The NG, and SG/NS-G were synthesised by a simple hydrothermal method using urea and thiourea, respectively. The cell with the NS-G CE showed a high conversion efficiency (7.42%), which was close to that of the Pt CE (7.56%) and higher than the bare NG and SG CEs. The improved electro-catalytic pattern of NS-G was attributed to the synergistic effect potentially aroused from the electronegativity difference between N and C, together with the distortion in the shape and structure of the composite initiated by the bulkier atomic size of S.

In another study, Roh *et al* [43] fabricated rGO on fluorine-doped tin oxide (FTO) substrates as transparent conducting electrodes in DSSCs. The bonding between rGO sheet and FTO surface was strong, which minimized the charge recombination rate at the TiO_2–FTO interface, resulting in favourable charge transfer process and eventually high PCEs of 8.44% in comparison with bare FTO and TiO_2 film.

Another research by Dissanayake *et al* [44] illustrated a composite consisting of tin oxide (SnO_2), rGO and polyaniline (PANI). They used rGO/SnO_2/PANI composite as an alternative to Pt CE for DSSCs. The newly prepared CEs exhibited 7.92% of PCE under 100 mW per cm^2 light illumination. The PCE increased to 8.68% for the present composite by using $TiCl_4$ treated TiO_2 photoanode, which correlates to an impressive 94% of value of 9.22% acquired from $TiCl_4$ treated photoanode and Pt CE-based DSSCs under the same light irradiation (figure 6.2). The outcomes from the mentioned composite indicated the beneficial use of this novel material for the fabrication of low-cost Pt-free CE-based DSSCs. Zheng [45] also introduced graphene into poly (acrylic acid) (PAA)/PEG, and designed stable DSSCs with 9.10% of PCE. The graphene-based devices generally conquered the pristine liquid and polymer electrolyte reference devices. This may be ascribed to the shorter charge transfer length and the presence of conducting and stable graphene terminals within the composite polymer and liquid electrolytes.

Also, perovskite solar cells (PSCs) have been considered as a capable applicant for thin-film photovoltaic technology, owing to solution processibility of materials, high absorption rate of light and long charge diffusion dimension [46]. Generally, the architecture of PSCs can be classified as direct/inverted (n–i–p/p–i–n) including either planar or mesoscopic structure [47, 48]. One can see in figure 6.3, the perovskite material is sandwiched between the hole transport layer (HTL) and electron transport layer (ETL). For improving the PSC performance including its cost-effectiveness, stability and power conversion efficiency, graphene-based material has been utilized for either of layers (HTL/ETL) or electrode [49, 50]

In one investigation, Chandrasekhar *et al* [51] adopted graphene/ZnO (G/ZnO) nanocomposite as electron transfer layer (ETL) in a PSC device. The optimum concentration (0.75 wt%) of graphene in G/ZnO nanocomposites films provided the improved cell performance. The cell conversion efficiency dramatically increased from 7.01% to 10.34% as compared to the pristine ZnO film. This improvement in photovoltaic performance may be attributed due to the refinement of surface morphology and crystallinity of the thin film.

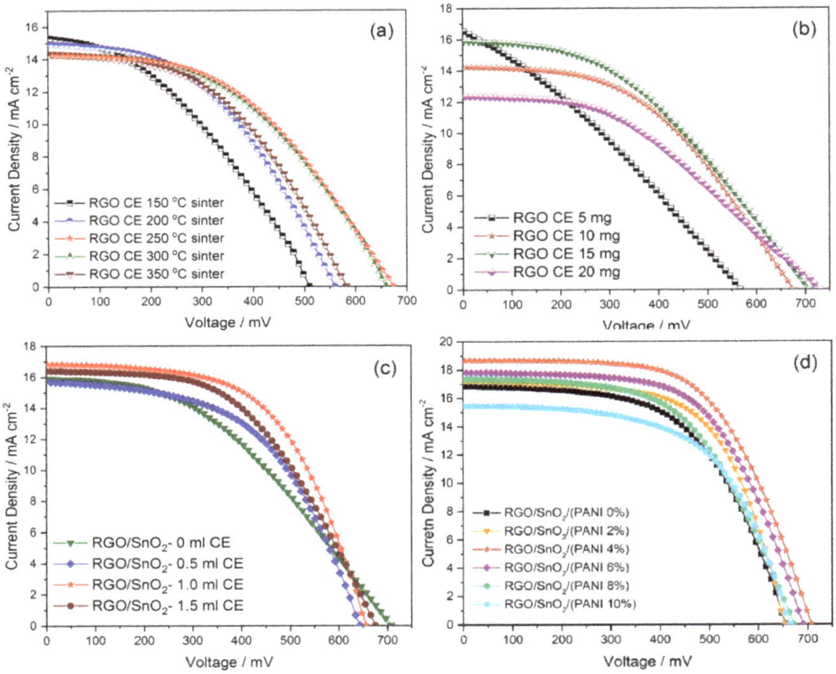

Figure 6.2. Current density–voltage characteristics of DSSCs with (a) pristine rGO CEs sintered at different temperatures, (b) pristine rGO CEs with different amounts of rGO used, (c) rGO/SnO_2 CEs with different SnO_2 amounts, and (d) rGO/SnO_2/PANI composite CEs with rGO:PANI wt% ratio. Reproduced (adapted) from [44], copyright 2021, with permission from Elsevier.

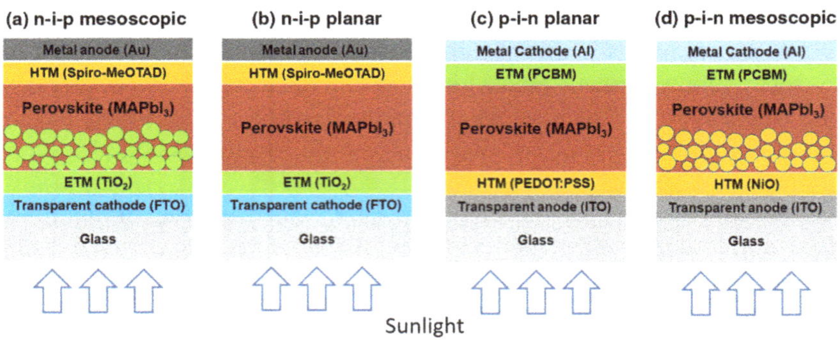

Figure 6.3. Illustration of layered structure of perovskite solar cells (a) n–i–p mesoscopic, (b) n–i–p planer, (c) p–i–n planer, and (d) p–i–n mesoscopic. Reproduced with permission from [47], copyright 2016, SPIE.

Furthermore, Mariani *et al* [52] fabricated a paintable carbon perovskite solar cell (C-PSC) by depositing a low-temperature-processed graphene-based carbon paste atop as a CE. The proposed small-area (0.09 cm^2) mesoscopic C-PSCs provided PCE of 15.81% along with an enhanced thermal stability as compared to Au CE-based reference devices. However, the PCEs rate was dropped to 14.06%, when the

CE was used for large-area (1 cm^2) mesoscopic devices for the same cell. It was also found that the low-temperature-processed planar n–i–p devices for graphene-based CE showed 13.85% of PCE for the given cell. Additionally, they reported record high PCE for C-PSCs with miniwafer-like area (aperture area = 4.00 cm^2, substrate area = 6.76 cm^2), confirming the metallization compatibility of the cell.

6.2.1.2 Supercapacitors

Among energy storage devices, supercapacitors have caught the attention of researchers due to their high-power density, good cycle stability and fast charging–discharging rate [53]. In past years, they were paired with batteries to acquire supplementary power which is needed in many fields [54]. Nevertheless, they were unable to function as separate objects in these fields due to their lesser energy density compared to batteries. To overcome these issues, researchers are trying to expand the power and energy density of supercapacitors by designing innovative electrode materials and electrolytes. Based on their storage mechanism, supercapacitors are divided into two categories, namely pseudocapacitors and electrochemical double-layer capacitors (EDLCs). EDLCs are able to perform as power sources, with 5–15 kW kg^{-1} of the power density range. It is found that pseudocapacitors chemically accumulate the charge through surface redox reaction, whereas double-layer capacitors physically gather charges through reversible ion adsorption at the interface of electrolyte and electrode. The graphene modified electrode materials have drawn a lot of attention because of their faster surface redox, electrochemically active sites, and appropriate diffusion pathway [54]. In one approach, Jeong *et al* [55] produced nitrogen-doped graphene materials by plasma method and used them in ultracapacitors. This simple doping process improved the capacitances (~280 F g^{-1} electrode) four-fold as compared to the pristine graphene-based material. In addition, the device also showed compatibility with flexible substrates, excellent life-cycle and high-power capability. According to characterization techniques, it was seen that the improved specific capacitance was mainly attributed to the certain nitrogen configuration at basal plane and edge, resulting in modification of the graphene structure. Similar results were reported by Elessawy *et al* [56]. They presented a green approach to synthesis of N-doped graphene with 3D porous architecture. They used polyethyleneterephthalate (PET) bottles and urea by using green synthesis through a novel single-step method. The influence of calcination temerature and urea doses on construction of N-doped graphene was examined. From impedence spectroscopy and cyclic votammetry studies, it was revealed that N content, which impacts the morophology and structure geometry of material enhances the ion diffusion and charge proliferation. The synthesised samples used as electrode material demonstrated a notable improvement in specific capacitance (~405 F g^{-1} at 1 A g^{-1}), power density (558.5 W kg^{-1}) and energy density of 68.1 W h kg^{-1}.

Furthermore, these materials displayed high charge/discharge duration and good cyclic stability until 5000 cycles at 4 A g^{-1} (figure 6.4). Eventually, these environmentally friendly and cost-effective materials will be considered as promosing candidates for fabricating engery storage devices. Similar to nitrogen dopant, boron can be coupled with graphene as it holds close proximity and similar

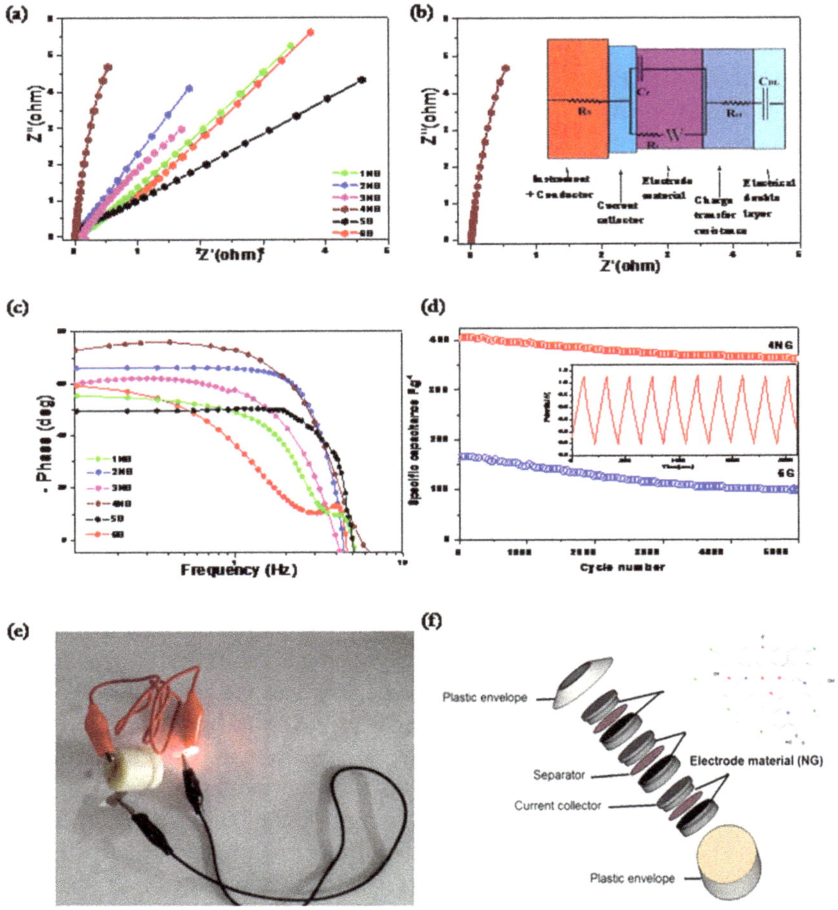

Figure 6.4. (a) Nyquist curves for all N-doped graphene samples as electrodes in 6 M KOH. (b) Nyquist curve of 4 N-graphene sample with equivalent circuit. (c) Bode plots for graphene-based supercapacitors devices. (d) The cycling stability of 4 N-graphene best supercapacitors and sample 6 G, during 5000 charge/discharge cycles, measured at a current density 4 A g^{-1} within the potential range from 0 to 1.1 V, with the first 10 charge/discharge cycles for sample 4NG. (e) Photograph and (f) schematic illustration of a stacking three-unit cell used to power a red LED. Reproduced [56] under the terms of the Creative Commons Attribution 4.0 International, Copyright 2019, Nature.

size to carbon in the periodic table of chemical elements. Doping can modify the electronic structure by changing the bandgap of material, resulting in improved conductivity [57].

Han *et al* [16] first modified graphene nanoplatelets chemically to achieve high surface area and then incorporated boron atom on the rGO surface via a solution process. The as-synthesised nanoplatelets displayed higher cycle stability and specific capacitance of 200 F g^{-1} than typical carbon-based supercapacitors. The Fried-Ice based cost-efficient approach was used to develop the porous B-doped graphene framework by Zuo *et al* [58]. When employed as an electrode, the prepared framework showed excellent supercapacitor performance including gravimetric capacitance up to 281 F g^{-1}, owing to high surface area of 662 m^2 g^{-1}. Furthermore, Li *et al* [59] synthesised boron-doped graphene aerogels (B-GA) through a simple hydrothermal route. These had fast ion diffusion as compared to pristine graphene aerogels, owing to some defects and high surface area. The B-GA had outstanding cycling stability with 308.3 F g^{-1} at 1 A g^{-1} of specific capacitance, which was higher than that of the pure counterparts.

In another study, Zou *et al* [60] developed boron and nitrogen co-modified holey graphene aerogels (BN-HGA) by a combined freeze-drying assembly and hydrothermal route (figure 6.5). They designed flexible supercapacitors (FCSCs) with outstanding performances (figures 6.5(a)–(c)). It is initiated in another carbon-based material that B and/or N can upsurge the charge transfer process, leading to improved electrochemical activities of carbon-based supercapacitors [61] BN-HGAs based FCSCs demonstrated an improved energy of 0.038 mWh cm^{-2}, good cycling stability, power of 22.8 mW cm^{-2}, and high specific capacitance of 456 F g^{-1} at 1 A g^{-1}, an outstanding rate performance (80% retention at 20 mA cm^{-2}) as compared to B-GAs, nitrogen-doped (N-GAs) and bare graphene (GAs) [62]. Additionally, the flexible supercapacitors exhibit very slight capacity loss as bent to arbitrary angles and follow the rudimentary arrangement of parallel/series of associations of capacitors. Figure 6.5(c) shows the ability of FCSCs for wearable electronics by linking two supercapacitors in series which light up a blue LED. Further, the microporous architecture of a 3D interconnected network was exhibited by as-synthesized BN-HGA, as shown in figure 6.5(d). The 3D network was advantageous for electron and ion transport in bulk electrode. After B and N doping, there is a possible surface polarization and aerogels devices can provide electrodes with minimized thickness and desirable sizes.

rGO was alo prepared via thermal treatment of GO platelets suspension at 150 °C–200 °C (Zhu *et al* [63]). The rGO exhibited high conductivity of 5230 S m^{-1}. The GO,

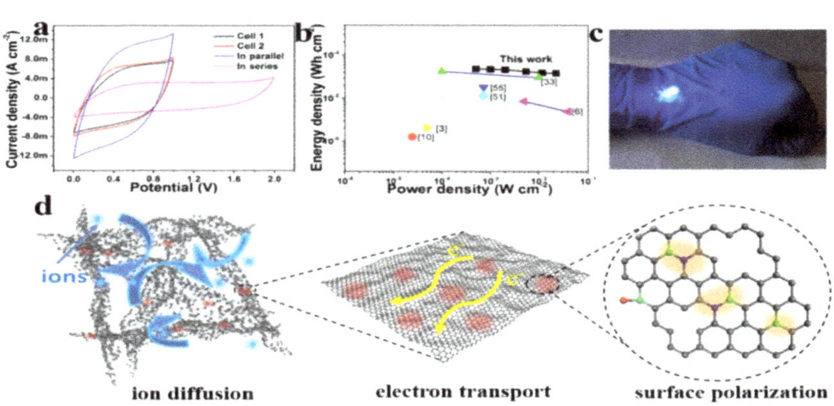

Figure 6.5. (a) CVs for two FCSCs in series or in parallel connection. (b) Ragone plot for the FSCs with areal energy/power densities and comparisons with other literature data. (c) Photograph that demonstrates a blue LED lit by two FSCs connected in series and twisted on the wrist. (d) Schematic illustration of capacitance performance of BN-HGA. Reproduced (adapted) from [60], copyright 2020, with permission from Elsevier.

which was treated at 200 °C, provided a specific capacitance of 122 F g^{-1} at 5 mA. Similarly, Wang *et al* [64] infused 30% of graphene sheets RuO_2 surface, which exhibited a high capacitance performance. The as-synthesised material showed a specific capacitance of 370 F g^{-1} at a scan speed rate of 2 mV s^{-1} and 280 F g^{-1} at 40 mV s^{-1} for the voltage range of 0.2–1.0 V in 1 M KOH electrolyte solution. Thus the functionalized graphene and GO are promising candidates for energy storage applications and can efficiently boost the performance of the various type of solar cells and supercapacitors.

6.2.2 Energy conversion applications

6.2.2.1 Hydrogen generation

Hydrogen has become a potential candidate for clean energy [65–68]. It attains a number of advantages such as environmental friendliness and high energy density, which is a necessity for new energy. In recent years, several strategies have been employed to overcome the energy problems. Although it remains a big challenge to solve the energy crisis, photocatalytic hydrogen evolution technology (i.e. photocatalysis process [69–74]), which is also known as water splitting, has become a potential solution to the energy problems. The elementary requirement for hydrogen production is that the reduction potential of H^+ to H_2 must be more positive than the lowermost of the conduction level, 0 V versus normal hydrogen electrode (NHE) (equation (6.1)) [75–77]. In particular, the valance band of the system should be more positive than the oxidation potential of water (equation (6.2)), 1.23 V versus NHE. The reaction process for water splitting is described here (see equations (6.1) and (6.2))

$$2H^+ + 2e^-(\text{electrons}) \rightarrow H_2 \; E^0\text{redox} = 0 \text{ V} \tag{6.1}$$

$$2H_2O + 4h^+(\text{holes}) \rightarrow 4H^2 + O_2 \; E^0\text{redox} = 1.23 \text{ V} \tag{6.2}$$

In the water-splitting process, a reaction only can occur if the theoretical bandgap of compound/semiconductor [78, 79] is more than 1.23 eV. Due to some thermal losses, the recommended bandgap [80] is higher than 1.7 eV to produce the hydrogen from water molecules. In a detailed procedure, the photons which contain higher energy than the bandgap of semiconductor are absorbed by the semiconductor. Afterwards, the excited electrons in the valance level move to the conduction level, providing the separation of photogenerated charge carriers [76]. Finally, in the conduction area, H^+ reduces to H_2 by a reductive electron. The key role of a co-catalyst is to reduce activation energy of the reaction and deliver redox reaction site [81, 82]. As an ideal co-catalyst, functionalized graphene facilitates the migration of photogenerated electron–hole pairs, resulting in hydrogen evolution.

In one study, Neena *et al* [83] designed N-doped reduced graphene oxide (N-rGO) and zinc oxide (ZnO) nanorods, which were prepared via solvothermal route at 180° centigrade. The prepared nanocomposite exhibited better photocatalytic performance as compared to the bare ZnO under visible light illumination. The strong coupling between ZnO and N-rGO increases the interfacial contact of the composite, resulting in improved photo-induced electron–hole pair separation and transfer process. The hydrogen generation rate and durability of nanocomposite was also found to be much better than that of pristine ZnO (figure 6.6). Additionally, from outcomes it was concluded that N-rGO was a superior supporting matrix for incorporating noble metals co-catalysts and semiconductors for improved water splitting. These findings can provide a considerable technique towards the facile synthetization of N-rGO nanocomposite along with crucial interfacial interaction between the moieties of the compound. Similarly, Cao *et al* [71] graphene was functionalized by thiourea (TU) and accumulated on the TiO_2 surface to design TiO_2/rGO-TU. The composites were prepared through a facile nucleophilic substitution reaction between the carboxyl (–COOH) and thiol (–SH) of thiourea of graphene. The TiO_2/rGO-TU exhibited the maximum photocatalytic activity rate of 241.83 μmol h^{-1} g^{-1}, which was 2.33- and 6.60-times fold of TiO_2 and TiO_2/rGO, respectively. This photocatalytic performance of TiO_2/rGO-TU composite might be attributed due to the synergistic effect of thiourea and graphene, specifically, thiourea functions as efficient H^+ absorbed active sites and graphene facilitates the photogenerated charge separation process to encourage the hydrogen evolution (figure 6.7). This finding offers new insights on designing the potential graphene-based photocatalysts in a sustainable energy field.

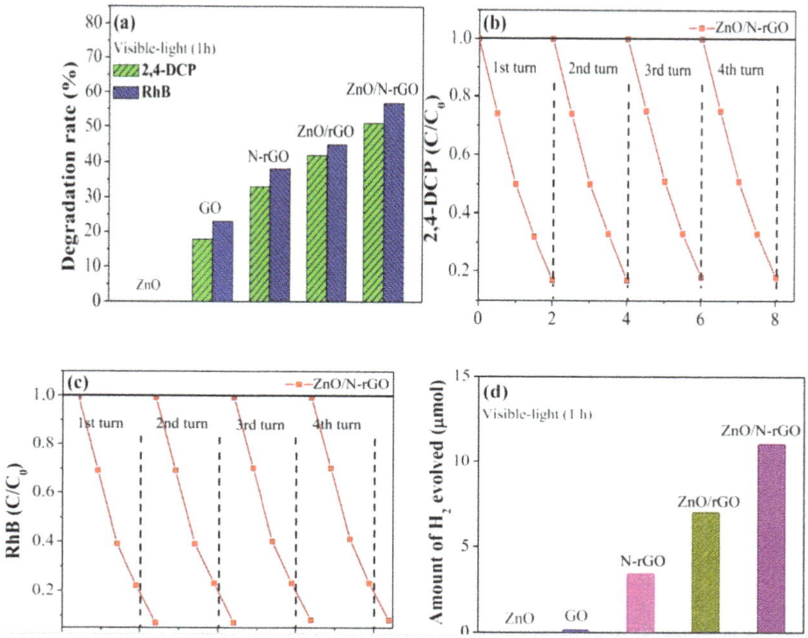

Figure 6.6. Visible-light activity of ZnO, GO, N-rGO, ZnO/rGO and ZnO/N-rGO samples, (a) for 2,4-dichlorophenol and RhB degradation, (b) stability and recyclability tests of 2,4-dichlorophenol degradation and (c) of RhB over ZnO/N-rGO, (d) amount of H_2 evolved over ZnO, GO, N-rGO, ZnO/rGO and ZnO/N-rGO. Reproduced (adapted) from [83], copyright 2019, with permission from Elsevier.

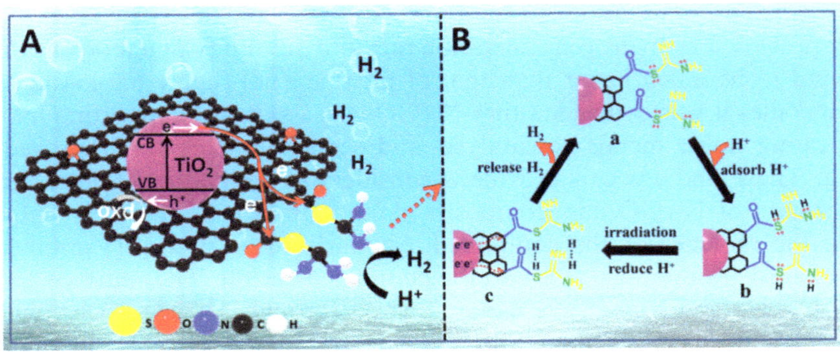

Figure 6.7. Schematic illustration for the photocatalytic H_2-generation mechanism of TiO_2/rGO-TU photocatalyst. Reproduced (adapted) from [84], copyright 2021, with permission from Elsevier.

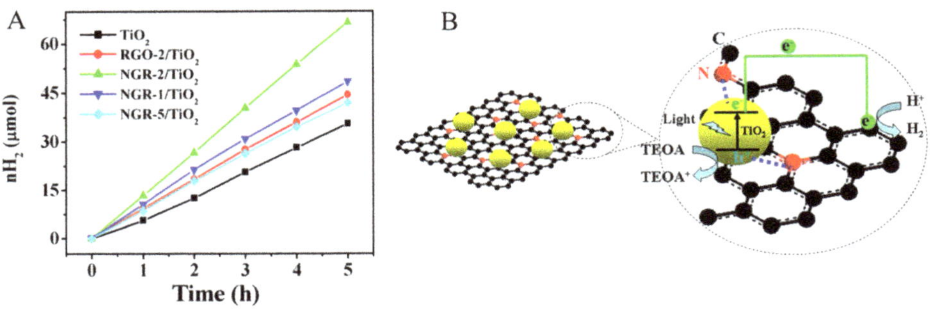

Figure 6.8. (a) Photocatalytic hydrogen production in different samples. (b) Schematic of the photocatalytic hydrogen production induced by the strong coupling between TiO_2 and NG. Reproduced (adapted) with permission from [85], copyright 2014, American Chemical Society.

The significance of the interface between semiconductor and functionalized graphene was highlighted by Mou *et al* [85]. They synthesised TiO_2-coupled N-doped graphene (NG) composite via solvothermal method. The nitrogen atoms were supposed to deliver favourable anchor and nucleation sites for TiO_2 to attach with NG sheets, serving as an interfacial contact between TiO_2 and NG sheet. Furthermore, this close contact suppressed the charge recombination rate and improved the separation and transfer process of electron–hole pairs. Due to the regaining of the sp^2 graphite phase and decrease of defects, NG possesses better electrical conductivity than rGO, leading to faster movement and separation of charges than rGO. Consequently, TiO_2/NG composites established higher hydrogen evolution rate (figures 6.8(a) and (b)) in comparison with its counterparts [85].

6.2.2.2 Carbon dioxide reduction

The world is facing a huge problem of pollution from burning of fossil fuels, resulting in the emission of greenhouse gases such as carbon dioxide. Although it is comprehensible that burning of fossils is the foremost energy source, we are causing a hazard to energy security by excessive diminution of them. Renewable energy sources have been considered a potential solution to resolve these issues. Consequently, there is an urgency to develop a capable approach which can utilize solar energy by transforming it into chemical energy or bonds [86–88]. This mode of utilization/transformation of solar energy, imitates the process of artificial photosynthesis. In artificial photosynthesis, photocatalytic carbon dioxide reduction (PCCR) is the main area. The photocatalysts are used to reduce carbon dioxide into renewable chemicals (e.g. methane, methanol, formic acid etc) [89]. The reduction process is carried out in the presence of some reducing agent such as hydrogen, water etc, under the irradiation of sunlight. Hence, we can receive a double benefit from PCCR, including minimization of environmental pollution and protection of the fossil fuels [90]. To increase the photocatalytic activities for CO_2 reduction, introduction of graphene in conventional photocatalysts has been a widely accepted approach amongst many research groups [90]. Graphene and its derivates have potential to enhance the surface area, eventually improving the light absorption on the surface of photocatalysts and reducing the recombination rate of photo-induced charges, which leads to higher photocatalytic performance [91, 92].

In one study, Wang *et al* [93] prepared ZnO/graphene nanocomposite through *in situ* chemical vapour deposition method. The photocatalytic performance of the composite was higher than that of pristine ZnO, attributed to fast facilitation of photogenerated charge carriers. This may be attributed to that of the intimate interfacial contact between the co-catalysts and semiconductor photocatalysts. Meanwhile, the presence of Schottky junction between the composites also played a crucial role in CO_2 reduction. Furthermore, π–π conjugation interaction between graphene and carbon dioxide molecules, in addition to photothermal effect of graphene, also subsidized to activity improvement. The possible pathway for photocatalytic performance for CO_2 activity is as shown in figure 6.9.

The photocatalytic CO_2 reduction (PCR) reaction mechanism of the mentioned composite is presented in figure 6.10. The free electrons mitigated to graphene from zinc oxide (figure 6.10(a)). Simultaneously, ZnO established an upward bending of

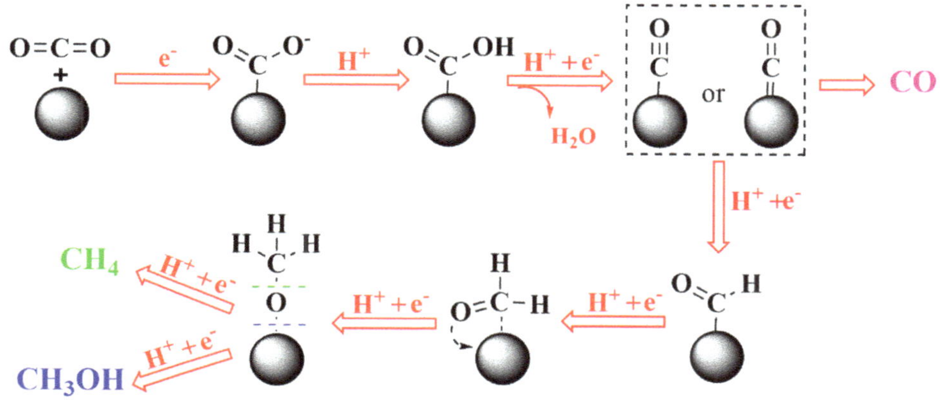

Figure 6.9. Possible CO_2 reduction pathways on the composite surface. Reproduced (adapted) from [93], copyright 2021, with permission from Elsevier.

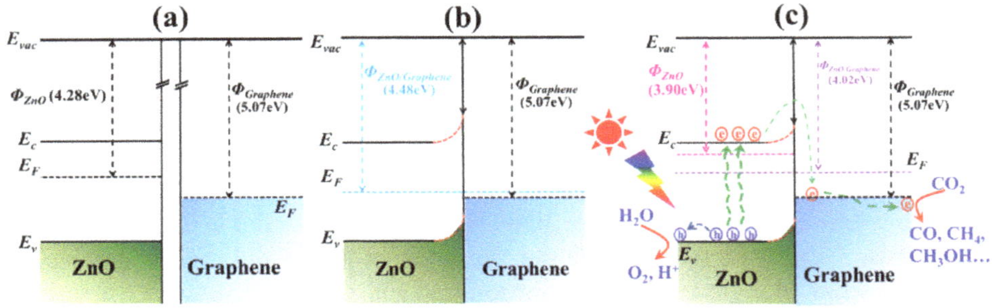

Figure 6.10. Schematic illustration of the PCR mechanism of ZnO/graphene. ϕ: work function; E_F: Fermi level; E_{vac}: vacuum energy; E_c: conduction band; E_V: valence band. (a) Before contact, (b) after contact and (c) under light irradiation. Reproduced (adapted) from [93], copyright 2021, with permission from Elsevier.

interface band due to the loss of electrons. In the meantime, graphene acted as electron sink and co-catalyst and received the electrons from zinc oxide. Subsequently, a constructed Schottky junction at the interface bent the ZnO band (figure 6.10(b)) [94]. The fermi level of ZnO shifted under the illumination of light, owing to electron's movement from valance band (VB) to conduction band (CB). However, light illumination had no influence on graphene Fermi level. Subsequently, electrons of ZnO received sufficient force to defeat the barrier and passage to graphene rapidly. Concurrently, water molecules reacted with photo-induced holes of ZnO. Additionally, graphene trapped the electrons and suppressed the recombination rate of charges. These photo-induced electrons on graphene participated in CO_2 reduction and improved the photocatalytic performance [93]

Similarly, the enhancement in the photoreduction performance for CO_2 was seen in a study by Yang *et al* [95]. They synthesised ZnO–rGO composite via one-step hydrothermal method and demonstrated five-fold performance for CO_2 reduction in comparison to pristine ZnO photocatalysts. The primary reason for this performance may be the photoexcited charges which drain to GO sheets and further react with CO_2 and complete the photoreduction reaction under light illumination. It is worth noting that excessive graphene can reduce the photocatalytic performance possibly because of the masking created by graphene on the ZnO surface which leads to high recombination rate of the charges. In another study, Lin *et al* [96] coupled N-rGO with TiO_2 at different nitrogen compositions through a wet route taking urea as nitrogen precursor. It was reported that N-rGO/TiO_2 exhibited higher carbon monoxide yield than RGO/TiO_2 and pure TiO_2 photocatalysts. The optimum nitrogen concentration was 3.6 wt.% in the composite and it was favourable to facilitate the charge separation and transfer process, thus proving improved CO_2 photoreduction. Furthermore, it was also found that N-rGO/TiO_2 was more stable/active than rGO/TiO_2. Wang *et al* [97] synthesized graphene-based tungsten-trioxide (WO_3) nanobelts composite by facile hydrothermal method. It was found that the WO_3 shifted to visible region with graphene doping. The visible light activation of composite may be associated with the uplifting of the conduction band which makes WO_3 suitable for CO_2/CH_4 redox reaction. Besides, the graphene intervened in the dispersion and size of WO_3 nanobelts. Furthermore, Xu *et al* [98] reported CO_2 reduction over graphene-based ternary composites under full spectrum sunlight. This ternary composite demonstrated almost double CO_2 reduction rate over its counterparts. In this study, GO nanosheets facilitate the charge separation and act as electron acceptor, eventually offering extra active sites for redox reaction.

Hence, functionalized graphene not only holds the intrinsic characteristics of graphene but also presents some new collective physicochemical characteristics such as large specific surface area, high porosity, low density, excellent electrochemical performance and so on. Due to its unique structures and characteristics, functionalized graphene is gaining extensive attention from researchers for a wide range of applications.

6.3 Biomedical applications of functionalized graphene

The biomedical applications of functionalized graphene can be classified into tissue engineering and drug delivery applications. Tissue engineering is an interdisciplinary field that applies principles from engineering and life sciences to develop biological alternatives for restoring, improving or maintaining tissue functions [99]. In recent years, one of the important goals of the tissue engineering field has been to develop scaffolds of tunable properties [100, 101].

6.3.1 Cytotoxicity of graphene-based materials

Materials used for cell culture and drug delivery applications should not be toxic to cells. Unfunctionalized graphene is hydrophobic in nature which causes the lipids of the cells to interact, leading to high toxicity in cells [102]. GO (functionalized graphene), on the other hand, has also been shown to be toxic to cells at a concentration greater than 50 $\mu g\ ml^{-1}$, as reported by Wang *et al* [103]. However, the toxicity also depends on the cell lines and dosage. For instance, adenocarcinomic alveolar epithelial A549 cell line exhibited no obvious toxicity [104] which is in contrast to Wang *et al*'s findings [103]. Some of the reasons argued for cytocompatibility of GO are its degree of functionalization, two-dimensional structure and its uptake by different cell lines [104]. Another reason could be the protein coating of GO by the proteins present in the cell culture media which may improve the cytocompatibility. In a study by Mu *et al* [105], cellular uptake of protein coated GO nanosheets with small (average disk diameter—0.42 μm) and large sizes (average disk diameter—0.86 μm) were investigated. It was found that protein coated graphene nanosheets with smaller sizes are internalized by clathrin-mediated endocytosis, whereas larger nanosheets are internalized by phagocytosis. Thus, the update of GO nanosheets by cells is size dependent. Furthermore, the cellular uptake did not result in toxicity up to concentrations of 100 $\mu g\ ml^{-1}$ which may be a result of the protein coating. Toxicity to bacterial cells exhibited by some graphene-based materials is due to the damaging of cell membrane through physical interactions with graphene materials exhibiting sharp edges [106, 107].

GO-based materials exhibit varying cytotoxic effects to various cell lines depending on dosage, size, physicochemical properties and preparation methods [108, 109].

One of the reasons for the toxicity is attributed to the oxidative stress and apoptosis. A study by Zhang *et al* [110] tested the cytotoxicity of GO with different levels of oxidation by treating the graphene with different levels of potassium permanganate (oxidizing agent). The cell experiments using mouse embryo fibroblasts revealed that GO toxicity is proportional to the oxidation degree where the toxicity is increased as the oxidation degree is decreased. The oxidative stresses in the cells were further analyzed by monitoring the reduced oxygen species (ROS). GO with low oxidation degree enabled a dramatic increase in ROS levels of the fibroblasts. The authors discuss that low oxidation degree of GO possesses more unpaired electrons, which exerts indirect oxidative damage on fibroblasts by production of hydroxyl radicals from H_2O_2 efficiently inducing higher oxidative stress on cells than GO with high oxidation degree.

One of the most commonly used approaches to improving biocompatibility of GO is by PEGylation. PEG functionalized GO showed no evidence of toxicity even at concentrations as high as 500 mg l^{-1} [111]. GO functionalization using PEG can be achieved by covalent and non-covalent methods. Non-covalent methods rely on the electrostatic and hydrogen bonding interactions. However, the covalent modification of GO with PEG resulted in insoluble precipitates and reduced electrical properties [112]. Hence, in applications where electrical properties are to be preserved, non-covalent modification is suggested (figure 6.11)

6.3.2 Scaffolds for tissue engineering

In tissue engineering, two approaches are adopted for restoring/healing of injured parts or for maintenance of diseased parts. In the first approach, biocompatible scaffolds are used to induce the growth of body parts which can potentially replace the diseased parts. For achieving this, stem cells were subjected to the extracellular environmental cues which result in directing the lineage towards desired cell types which in turn could be used to grow organs. For instance, a pioneering work by Engler *et al* [113] showed that mechanical cues resembling the extracellular matrix of a

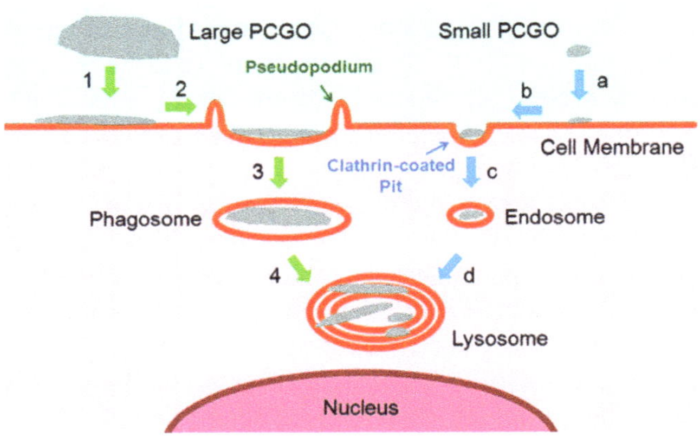

Figure 6.11. Mechanism of cellular uptake of protein coated GO (PCGO) nanosheets by cells. 1–4 and a–d represents the uptake of large and smaller sized PCGO. Reproduced (adapted) with permission from [105], copyright 2012, American Chemical Society.

particular cell type can direct the lineage of stem cells towards that particular cell type. For instance, stiffness of 1 kPa resembling the mechanical cue of brain directs the stem cell lineage towards neuronal cells.

In the second approach, scaffolds are used to deliver drugs or growth factors which can help in maintaining or regenerating diseased parts. Graphene-based materials offer excellent mechanical and electronic properties, which enables the production of scaffolds with tunable functionalities for cell culture. Some attractive features of functionalized graphene materials are their easy dispersibility and processability and specific interactions with various biomolecules that are important for tissue engineering applications [113].

In this section, an overview of graphene-based scaffolds for neural tissue engineering and growth factor delivery applications are discussed.

6.3.3 Scaffolds for neural tissue engineering

The nervous system consists of central nervous system (CNS) and peripheral nervous system (PNS). The CNS consists of the brain and the spinal cord while the PNS comprises sensory and motor neurons. When compared to the PNS, the CNS has less regeneration capacity [114]. Spinal cord injury is associated with the nerve damage resulting in the loss of motor and sensory functions in the CNS. Depending on the level of damage, the aftermath of the injury can result in serious conditions such as paralysis. In the United States alone, approximately 1.2 million patients were affected by spinal cord injury due to accidents with 17 000 new cases emerging every year [115]. After a spinal cord injury, a cavity forms at the injury site which will be surrounded by a dense glial scar. A complex inhibitory environment results comprising reactive astrocytes, glycosaminoglycans and other inhibitory molecules preventing neurons from reaching the injury site and thus disrupting regeneration [116]. In terms of scaffold design for spinal cord injury, several characteristics are required. The scaffolds should result in lesser glial scar formation enabling the neurons to reach the injury site and should serve as conduits to guide the neuronal cells to the injury site. Moreover, the scaffolds should support cell adhesion, biocompatibility, biodegradability, porosity and mechanical strength [117]. An attractive feature of graphene is its electrical conductivity, which can support neuronal stimulation that can aid in regeneration of lost functionalities.

Using of stem cells to direct the lineage towards neuronal cells is a common methodology used in neural tissue engineering. Park *et al* [118] first explored graphene as a conductive scaffold and demonstrated that neural stem cells differentiated into neurons. For attachments of cells, non-covalent functionalization of graphene with laminin was performed prior to cell seeding. Furthermore, electrical stimulation of cells was possible by using graphene as an electrode which recorded neural activity for the differentiated cells. In neural stem cell engineering, porosity and 3D structures are important as they mimic the *in vivo* environment. In a pioneering work by Li *et al* [119], 3D graphene scaffolds were produced using chemical vapour deposition using nickel foam as a template. The produced 3D scaffolds were then coated with laminin for cell culture. The 3D structure enabled excellent biocompatibility and supported neural stem cells differentiation into neurons and astrocytes. In addition, electrical stimulation was possible on the cultured cells. Severe CNS injuries result in damage of neurons and disrupted neuronal circuits. The majority of the neural stem cells transplanted to injured CNS sites undergo differentiation to astrocytes [120]. This strong bias towards astrocyte differentiation is one major problem encountered in regenerative therapy. For CNS regeneration, differentiation of neural stem cells to oligodendrocytes and neurons is highly advantageous. A study by Shah *et al* [121] addressed this problem by growing neural stem cells on graphene coated electrospun polycaprolactone (PCL) scaffolds. The surface of the scaffolds was functionalized with laminin for growing cells. The scaffolds promoted the differentiation of neural stem cells to oligodendrocytes, which were confirmed by the presence of oligodendrocyte markers. By tuning the concentration of GO, it was possible to control the differentiation of neural stem cells to mature oligodendrocytes. Another common technique to fabricate 3D scaffolds is by electrospinning of polymers. Electrospinning offers several advantages including its simplicity, alignment of fibers, control of fiber diameter, pore size and large surface area of scaffolds [102]. Blending GO into electrospinning polymer solution is one way to fabricate 3D nanofibrous scaffolds [122, 123]. However, this method can weaken the electrical properties and cytotoxicity of the scaffolds depending on the concentration. While graphene-based scaffolds support neuronal growth, it is still challenging to control the alignment of differentiated neurons, which is important for making interconnected neuronal networks. Qing *et al* [124] addressed this problem by casting electrospun silk fibroin on rGO paper. The produced scaffolds enabled neuronal differentiation of neuroblastoma cells and the cell alignment was found to be proportional to the fiber density and orientation.

Fibrous scaffolds can mimic the native extracellular matrix architecture. Magaz *et al* [123] produced micro/nanofibrous scaffolds made of fibers of silk fibroin and rGO by electrospinning. To elucidate the influence of conductivity, neuronal NG108-15 cells were cultured on scaffolds produced with GO and rGO. Both the scaffolds supported neurite growth. However, scaffolds made with rGO exhibited enhanced metabolic activity and higher cell proliferation when compared to scaffolds made with GO. While the exact mechanism responsible for varying cell responses on the scaffolds is difficult to interpret, it is speculated that the enhanced electrical conductivity of rGO is an important contributing factor. Andre *et al* [125] made 3D fibrous porous scaffolds that can support interconnected neuronal circuits formation. To achieve this, PCL, gelatin and GO nanofibers dispersed in 2,2,2-trifluoroethanol were blended and electrospun to produce 3D hydrogel-like scaffolds followed by freeze-drying. The fibrous porous architecture and the presence of GO synergistically enhanced the formation of neuronal networks.

While graphene-based scaffolds in the form of thin films and porous foams were used, *in vivo* cells reside inside extracellular matrix (ECM) consisting of large amounts of water and various fibrous proteins. To recapitulate the cellular niche, synthetic mimics are presented in the form of hydrogel scaffolds to cells. Hydrogels are water swollen cross-linked polymer networks wherein the water content and mechanical properties can be tunable depending on the cross-linking ratio.

Incorporation of conductive graphene-based materials can facilitate biological signal recording and electrical stimulation of living cells in the native cellular microenvironment. Huang *et al* produced composite hydrogels from polyurethane and graphene oxide using 3D printing methodology [126]. The composite hydrogels supported good cell viability and enabled neural differentiation of the stem cells. When incorporating GO into hydrogels, one of the main issues encountered is the solubility of GO. A strategy adopted by Cha *et al* [127] used methacrylated GO using 3-(trimethoxysilyl) propyl methacrylate (TMSPMA). The methacrylated GO was then dispersed into gelatin methacryloyl (GelMA) followed by UV cross-linking to yield GelMA/GO composite hydrogels. The GelMA/GO composite hydrogels exhibited good biocompatibility and fibroblasts cultured on the gels showed excellent cell viability. Interestingly, the covalently conjugated methacrylated GO in the hydrogels improved the toughness of the hydrogels. Thus, the composite gels can also be used to tune the mechanical properties in addition to the inherent electrical properties and has potential in studies that explore cells' response to mechanical properties. The solubility issue is more pronounced when using rGO as the conductive element in hydrogels as rGO is inherently hydrophobic. A study by Park *et al* [128] produced poly(acrylamide) (PAAm)/rGO hydrogels by first producing hydrogel comprising of GO and PAAm by polymerization. In the next step, the GO trapped inside the gels is reduced *in situ* by ascorbic acid to yield PAAm/rGO hydrogels. A recent research work utilized dopamine as a reducing agent in chitosan hydrogels incorporated with GO [129].

In conductive hydrogels produced from graphene-based materials, there is a trade-off between conductivity and mechanical properties [130]. For instance, increasing the conductive component enhances conductivity but also results in brittleness of the gels. In a recent work, Liu *et al* [131] developed positively charged hydrogels made of oligo(poly(ethylene glycol) fumarate) (OPF)–2-(methacryloyloxy)ethyltrimethylammonium chloride (MTAC) with two conductive components, i.e., functionalized GO acrylate and carbon nanotube poly(ethylene glycol) acrylate (CNTpega). The functionalized GO in the gels was reduced *in situ* using ascorbic acid in the second step. The obtained hydrogels exhibited compressive modulus of about 850 kPa and achieved high electrical conductivity of 5.75×10^{-3} S m^{-1}. The high conductivity was attributed to the synergistic effect of MTAC, functionalized rGO and CNTpega. Furthermore, these hydrogels supported good biocompatibility for PC12 cells and allowed neuronal differentiation of PC12 cells when supplemented with nerve growth factor (NGF).

3D printing is a promising strategy adopted in producing hydrogel scaffolds of desirable shapes for tissue engineering. Producing conductive scaffolds that are printable can lead to scaffolds of tunable geometries and mechanical properties. A recent work by Mendes *et al* [132] produced GelMA/GO composite hydrogels by mixing GelMA and GO, which resulted in soft conductive hydrogels due to the physical interactions between the two components. For 3D printing, the GelMA and GO solutions were mixed and printed using an extrusion-based 3D printer. After printing, the structures were exposed to UV for cross-linking which resulted in stable conductive scaffolds (figure 6.12).

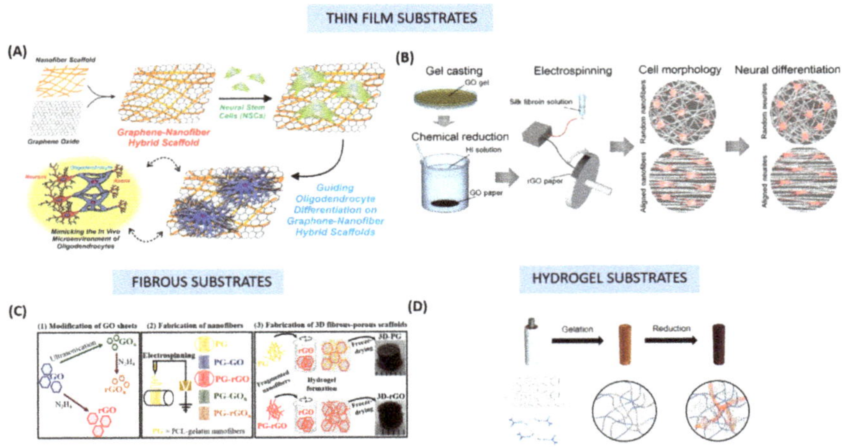

Figure 6.12. Various graphene-based scaffolds used in neural tissue engineering. (A) GO coated PCL scaffolds produced by electrospinning. Reproduced with permission from [121], John Wiley & Sons, copyright (2014) WILEY-VCH Verlag GmbH & Co. KGaA, Weinheim. (B) Electrospun silk fibroin deposited on GO paper. Reproduced (adapted) with permission from [124], Copyright 2018, American Chemical Society. (C) Production of fibrous porous scaffolds 3D scaffolds. Reproduced (adapted) with permission from [125], Copyright 2020, American Chemical Society. (D) Production of conductive rGO/GelMA hydrogel conduits by polymerization and its subsequent mild chemical reduction to rGO/GelMA composite hydrogel [34].

6.4 Graphene-based materials for growth factor proteins delivery

Growth factors are soluble proteins secreted by cells that bind to specific trans-membrane receptors on target cells [133]. Growth factors regulate various cellular processes such as proliferation, migration and differentiation [134] and also have roles to play in diseases such as cancer [135]. Controlled release of growth factors in combination with engineered scaffolds is an actively pursued approach in stem cell engineering.

Due to the surface functional groups available on GO, both immobilization strategies relying on non-covalent interactions such as π–π stacking and covalent linking of biomolecules or drugs are possible for GO [136]. Yoon *et al* [137] made GO sheets that can be loaded with transforming growth factor (TGF-β3) for chondrogenic differentiation of stem cells. TGF-β is a growth factor protein that plays an important role in stimulating mesenchymal stem cells' chondrogenesis and also in enhancing cartilage ECM production [138]. TGF-β3 and GO strongly interact and were mediated by the hydrophobic interactions (between hydrophobic cores of TGF-β3 and GO) and electrostatic interactions. Moreover, the large surface area of GO facilitates higher loading of TGF-β3. The release profile of TGF-β3 showed negligible release over 7 days and the native structure of TGF-β3 was preserved as confirmed by circular dichroism experiments. Thus, the interactions between GO and TGF-β3 were quite stable and can be used to deliver growth factor proteins. In a recent work by Zhou *et al* [138], TGF-β3 bound to GO was delivered to human mesenchymal cells in a 3D hydrogel matrix. With 14 days of cell culture, chondrogenic gene markers SOX9, COL2A1 and ACAN were expressed, and the differentiation was mediated by Smad2 phosphorylation. While this delivery method offers control over chondrogenesis by adjusting the amount of GO loaded TGF-β3, the limit is set up by the toxicity limit of GO to cells. In a Zhou *et al* [138] study, GO concentration up to 600 μg ml^{-1} was nontoxic to mesenchymal stem cells. A sustained and long-term release of growth factors will be more efficient for the biological control of cellular processes. Shen *et al* [139] incorporated GO inside a poly-D,L-lactic acid/polyethylene glycol (PDLLA) hydrogel and explored its applicability for sustained release of TGF-β3. The release of TGF-β3 from the hydrogels was measured in Hank's balanced salt solution for a period of 28 days. The controls without GO showed a burst release profile while the GO incorporated gels showed sustained release. In addition, the GO incorporated gels exhibited higher chondrogenic gene expression and ECM production confirmed by the levels of aggrecan, Collagen-II and SOX9.

Bone-morphogenetic proteins (BMPs) are a family of bone inducing growth factors that are used for bone regeneration. High dosage of BMPs is required for clinical applications. An optimal localized delivery of BMPs is essential for reducing the dosage. By using GO flakes loaded with BMP-2 in a fibrin hydrogel, La *et al* [140] found that the BMP-2 dosage required for inducing bone regeneration was reduced to half. Fu *et al* [141] explored the osteogenic differentiation of pre-osteoblasts by BMP-2 delivery using microcarriers made of GO incorporated poly (lactide-co-glycolide)/hydroxyapatite (PLGA/HA). Similar to La *et al*'s [140] findings, the introduction of GO considerably reduced the BMP-2 consumption for osteoinductive effect. Both the presence of negatively charged COO– groups of GO that interact with positively charged BMP-2 and the presence of π electrons of GO were speculated to be the reasons for high binding of BMP-2 on GO. A recent computational study outlined that GO has the highest affinity for BMP-2 adsorption when compared to rGO due to its polar nature [142]. For future and rational design of BMP-2 delivery systems, interactions with different forms of graphene should be taken into consideration.

In addition to being an efficient loading carrier of proteins, GO also has roles in stabilizing proteins. Proteins often undergo enzymatic cleavage in biological environments and has to be protected from damage before delivery to specific targets. In a study by Emadi *et al* [143], the protective effects of GO and chitosan modified GO on bovine serum albumin (BSA) against trypsin were studied. The mechanism of protective effect of GO and chitosan modified GO was argued to be a result of steric hindrance of GO and the reducing effect of BSA. The Fourier transform infrared spectroscopy (FTIR) revealed that loading of BSA on GO removed the oxygenated groups of GO which resulted in forming rGO aggregates. The aggregates in turn wrapped around trypsin and formed a barrier between BSA and trypsin which resulted in the increased digestion time. Furthermore, collagenase loaded GO and chitosan modified GO showed enzymatic activity similar to that of free collagenase, thus they can function as an efficient protein carrier.

The strong binding of growth factors to GO can also be coupled with dynamic triggers to deliver growth factors on demand. For instance, Zhang *et al* [144] coated iron oxide nanoparticles with GO and incubated the nanoparticles with bone-morphogenetic-protein-2 (BMP-2) growth factor that promotes osteogenesis. The loaded nanoparticle system was then incorporated into dental pulp stem cells. Such a system offers manipulation of cells to form sheets by using an external magnetic field. The authors further extended this concept to construct the complex osteochondral interface (cartilage to bone interface) which is quite challenging to accomplish using scaffold-based methodologies.

Electric field can also be used as a trigger for releasing growth factors by utilizing the conductivity of GO. In severe peripheral nerve injuries, nerve growth factors (NGFs) aid in the recovery but the effective administration of NGFs remains a challenge. A composite made of bombyx mori silk worm fibroin and rGO was recently explored for its potential as an NGF-β delivery system in response to electrical stimulation [145]. NGF-β is loaded electrostatically on to the rGO by applying a positive potential of 0.6 V for 30 min. The release profile was monitored by using an enzyme linked immunosorbent assay (ELISA). The application of reducing potential triggered the release of NGF-β and the amount released was higher (5–8 times) when compared to the passive release from the films at each time point. The growth factor release profile in the films can be further tuned by the frequency and duration of stimulation. This electrical stimulation of GO-based systems can be adopted to other scenarios such as wound healing, and stem cell engineering.

One interesting characteristic of graphene-based materials is their high absorption in the near infrared (NIR) region which can be used in photothermal therapy (generating heat from NIR radiation) for killing cancer cells [146, 147]. Thus, NIR can also be used as an external trigger for creating local changes in temperature which in turn can be used for releasing growth factors. Zhao *et al* [148] designed a composite microcarrier delivery system composed of GO, GelMA and poly(*N*-isopropyl acrylamide) (PNIPAM). PNIPAM is a thermoresponsive polymer that undergoes a volume phase transition at a lower critical solution temperature (LCST) of around 32 °C. This means beyond LCST, the hydrophobicity of PNIPAM increases and result in volume shrinkage [149] and can be used for releasing growth factors. GelMA in the composite acts a biocompatible coating and GO acts as the localized heat generator in response to NIR radiation. As a proof of concept, the composite microcarrier loaded with vascular endothelial growth factor (VEGF) which promotes migration and proliferation of endothelial cells to form blood vessels was implanted in a rat animal model. After 4 weeks of implantation, the NIR irradiated rats exhibited enhanced blood vessels formation. Thus, NIR irradiation can also be used as a trigger for release of growth factors in graphene-based materials.

6.5 Conclusions

In summary, the functional properties of graphene are largely influenced by the presence of the functional groups. In particular, in the case of usage of graphene for electronic applications the doping procedures play a crucial role. Therefore, the fabrication procedures for doping and understanding of the conductance mechanisms will result in development of high performing energy storage/conversion devices.

The usage of graphene towards biomedical applications is extremely sensitive to the functionalization features of graphene. A primary and important requisite is rendering graphene safe for such applications. Graphene-based materials with their excellent mechanical and electronic properties can revolutionize the field of tissue engineering. Particularly, the combination of graphene with polymers can lead to electrically responsive materials that can tune cell responses.

References

[1] Novoselov K S, Geim A K, Morozov S V, Jiang D E, Zhang Y, Dubonos S V, Grigorieva I V and Firsov A A 2004 Electric field effect in atomically thin carbon films *Science* **306** 666–9

[2] Khrapach I, Withers F, Bointon T H, Polyushkin D K, Barnes W L, Russo S and Craciun M F 2012 Novel highly conductive and transparent graphene-based conductors *Adv. Mater.* **24** 2844–9

[3] Balandin A A, Ghosh S, Bao W, Calizo I, Teweldebrhan D, Miao F and Lau C N 2008 Superior thermal conductivity of single-layer graphene *Nano Lett.* **8** 902–7

[4] Lotya M *et al* 2009 Liquid phase production of graphene by exfoliation of graphite in surfactant/water solutions *J. Am. Chem. Soc.* **131** 3611–20

[5] Kim K S, Zhao Y, Jang H, Lee S Y, Kim J M, Kim K S, Ahn J H, Kim P, Choi J Y and Hong B H 2009 Large-scale pattern growth of graphene films for stretchable transparent electrodes *Nature* **457** 706–10

[6] Li X *et al* 2009 Large-area synthesis of high-quality and uniform graphene films on copper foils *Science* **324** 1312–4

[7] Bae S *et al* 2010 Roll-to-roll production of 30-inch graphene films for transparent electrodes *Nat. Nanotechnol.* **5** 574–8

[8] Hamilton C E, Lomeda J R, Sun Z, Tour J M and Barron A R 2009 High-yield organic dispersions of unfunctionalized graphene *Nano Lett.* **9** 3460–2

[9] Bourlinos A B, Georgakilas V, Zboril R, Steriotis T A and Stubos A K 2009 Liquid-phase exfoliation of graphite towards solubilized graphenes *Small* **5** 1841–5

[10] Carey M, Hinton Z, Natu V, Pai R, Sokol M, Alvarez N J, Kalra V and Barsoum M W 2020 Dispersion and stabilization of alkylated 2D MXene in nonpolar solvents and their pseudocapacitive behavior *Cell Rep. Phys. Sci.* **1** 100042

[11] O'Neill A K U, Nirmalraj P N, Boland J and Coleman J N 2011 *J. Phys. Chem.* C **115** 5422

[12] Shih C J, Lin S, Strano M S and Blankschtein D 2010 Understanding the stabilization of liquid-phase-exfoliated graphene in polar solvents: molecular dynamics simulations and kinetic theory of colloid aggregation *J. Am. Chem. Soc.* **132** 14638–48

[13] Mann J A, Rodríguez-López J, Abruña H D and Dichtel W R 2011 Multivalent binding motifs for the noncovalent functionalization of graphene *J. Am. Chem. Soc.* **133** 17614–17

[14] Mann J A and Dichtel W R 2013 Noncovalent functionalization of graphene by molecular and polymeric adsorbates *J. Phys. Chem. Lett.* **4** 2649–57

[15] Panchakarla L S, Subrahmanyam K S, Saha S K, Govindaraj A, Krishnamurthy H R, Waghmare U V and Rao C N R 2009 Synthesis, structure, and properties of boron-and nitrogen-doped graphene *Adv. Mater.* **21** 4726–30

[16] Han J, Zhang L L, Lee S, Oh J, Lee K S, Potts J R, Ji J, Zhao X, Ruoff R S and Park S 2013 Generation of B-doped graphene nanoplatelets using a solution process and their supercapacitor applications *ACS Nano* **7** 19–26

[17] Martins T B, Miwa R, Da Silva A J and Fazzio A J R A 2007 Electronic and transport properties of boron-doped graphene nanoribbons *Phys. Rev. Lett.* **98** 196803

[18] Wang H B, Maiyalagan T and Wang X 2012 Review on recent progress in nitrogen-doped graphene: synthesis, characterization, and its potential applications *ACS Catal.* **2** 781–94

[19] Tale B, Nemade K R and Tekade P V 2021 Graphene based nano-composites for efficient energy conversion and storage in solar cells and supercapacitors: a review *Polym.-Plast. Technol. Mater.* **60** 784–97

[20] Theerthagiri J, Madhavan J, Lee S J, Choi M Y, Ashokkumar M and Pollet B G 2020 Sonoelectrochemistry for energy and environmental applications *Ultrason. Sonochem.* **63** 104960

[21] Lee S J, Theerthagiri J, Nithyadharseni P, Arunachalam P, Balaji D, Kumar A M, Madhavan J, Mittal V and Choi M Y 2021 Heteroatom-doped graphene-based materials for sustainable energy applications: a review *Renew. Sustain. Energy Rev.* **143** 110849

[22] Choi H J, Jung S M, Seo J M, Chang D W, Dai L and Baek J B 2012 Graphene for energy conversion and storage in fuel cells and supercapacitors *Nano Energy* **1** 534–51

[23] Theerthagiri J, Senthil R A, Senthilkumar B, Polu A R, Madhavan J and Ashokkumar M 2017 Recent advances in MoS_2 nanostructured materials for energy and environmental applications—a review *J. Solid State Chem.* **252** 43–71

[24] Safie N E, Azam M A, Aziz M F and Ismail M 2021 Recent progress of graphene-based materials for efficient charge transfer and device performance stability in perovskite solar cells *Int. J. Energy Res.* **45** 1347–74

[25] Zhang Z, Wu J, Wang L, Liu F, Jia P, Dai L, Lu Y and Bian T 2020 The analysis on simulation and invalidation of hot-spot temperature distribution in micro-defective crystalline silicon solar cells *Renew. Energy* **147** 2218–28

[26] Fthenakis V, Athias C, Blumenthal A, Kulur A, Magliozzo J and Ng D 2020 Sustainability evaluation of CdTe PV: an update *Renew. Sustain. Energy Rev.* **123** 109776

[27] Basnet R, Weigand W, Zhengshan J Y, Sun C, Phang S P, Sio H C, Rougieux F E, Holman Z C and Macdonald D 2020 Impact of pre-fabrication treatments on n-type UMG wafers for 21% efficient silicon heterojunction solar cells *Sol. Energy Mater. Sol. Cells* **205** 110287

[28] Freitag M and Boschloo G 2017 The revival of dye-sensitized solar cells *Curr. Opin. Electrochem.* **2** 111–9

[29] Lee W J, Ramasamy E, Lee D Y and Song J S 2008 Performance variation of carbon counter electrode based dye-sensitized solar cell *Sol. Energy Mater. Sol. Cells* **92** 814–8

[30] Yin H, Yan C, Hu H, Ho J K W, Zhan X, Li G and So S K 2020 Recent progress of all-polymer solar cells—from chemical structure and device physics to photovoltaic performance *Mater. Sci. Eng. R: Rep.* **140** 100542

[31] Kovrov A, Helgesen M, Boeffel C, Kröpke S and Søndergaard R R 2020 Novel acrylic monomers for organic photovoltaics encapsulation *Sol. Energy Mater. Sol. Cells* **204** 110210

[32] Li Y, Hu H, Chen B, Salim T, Lam Y M, Yuan N and Ding J 2017 Solution-processed perovskite-kesterite reflective tandem solar cells *Sol. Energy* **155** 35–8

[33] Lin Y *et al* 2019 Unveiling the operation mechanism of layered perovskite solar cells *Nat. Commun.* **10** 1–11

[34] Murugadoss V, Panneerselvam P, Yan C, Guo Z and Angaiah S 2019 A simple one-step hydrothermal synthesis of cobaltnickel selenide/graphene nanohybrid as an advanced platinum free counter electrode for dye sensitized solar cell *Electrochim. Acta* **312** 157–67
[35] Theerthagiri J, Senthil R A, Buraidah M H, Madhavan J and Arof A K 2016 Synthesis of α-Mo_2C by carburization of α-MoO_3 nanowires and its electrocatalytic activity towards tri-iodide reduction for dye-sensitized solar cells *J. Mater. Sci. Technol.* **32** 1339–44
[36] Theerthagiri J, Senthil A R, Madhavan J and Maiyalagan T 2015 Recent progress in non-platinum counter electrode materials for dye-sensitized solar cells *Chem. Electro. Chem.* **2** 928–45
[37] Duan Y, Tang Q, Liu J, He B and Yu L 2014 Transparent metal selenide alloy counter electrodes for high-efficiency bifacial dye-sensitized solar cells *Angew. Chem. Int. Ed.* **53** 14569–74
[38] O'regan B and Grätzel M 1991 A low-cost, high-efficiency solar cell based on dye-sensitized colloidal TiO_2 films *Nature* **353** 737–40
[39] Saranya K, Sivasankar N and Subramania A 2014 Microwave-assisted exfoliation method to develop platinum-decorated graphene nanosheets as a low cost counter electrode for dye-sensitized solar cells *RSC Adv.* **4** 36226–33
[40] Saranya K, Rameez M and Subramania A 2015 Developments in conducting polymer based counter electrodes for dye-sensitized solar cells—an overview *Eur. Polym. J.* **66** 207–27
[41] Yu M, Zhang J, Li S, Meng Y and Liu J 2016 Three-dimensional nitrogen doped holey reduced graphene oxide framework as metal-free counter electrodes for high performance dye-sensitized solar cells *J. Power Sources* **308** 44–51
[42] Kannan A G, Zhao J, Jo S G, Kang Y S and Kim D W 2014 Nitrogen and sulfur co-doped graphene counter electrodes with synergistically enhanced performance for dye-sensitized solar cells *J. Mater. Chem.* A **2** 12232–9
[43] Roh K M, Jo E H, Chang H, Han T H and Jang H D 2015 High performance dye-sensitized solar cells using graphene modified fluorine-doped tin oxide glass by Langmuir–Blodgett technique *J. Solid State Chem.* **224** 71–5
[44] Dissanayake M A K L, Kumari J M K W, Senadeera G K R and Anwar H 2021 Low cost, platinum free counter electrode with reduced graphene oxide and polyaniline embedded SnO_2 for efficient dye sensitized solar cells *Sol. Energy* **230** 151–65
[45] Zheng J 2017 Graphene tailored polymer gel electrolytes for 9.1%-efficiency quasi-solid-state dye-sensitized solar cells *J. Power Sources* **348** 239–45
[46] Wu T *et al* 2021 The main progress of perovskite solar cells in 2020–2021 *Nano-Micro Lett.* **13** 1–18
[47] Song Z, Watthage S C, Phillips A B and Heben M J 2016 Pathways toward high-performance perovskite solar cells: review of recent advances in organo-metal halide perovskites for photovoltaic applications *J. Photonics Energy* **6** 022001
[48] Hussain I, Tran H P, Jaksik J, Moore J, Islam N and Uddin M J 2018 Functional materials, device architecture, and flexibility of perovskite solar cell *Emergent Mater.* **1** 133–54
[49] Simari C, Stallworth P, Peng J, Coppola L, Greenbaum S and Nicotera I 2019 Graphene oxide and sulfonated-derivative: proton transport properties and electrochemical behavior of Nafion-based nanocomposites *Electrochim. Acta* **297** 240–9
[50] Nouri E, Mohammadi M R and Lianos P 2018 Improving the stability of inverted perovskite solar cells under ambient conditions with graphene-based inorganic charge transporting layers *Carbon* **126** 208–14

[51] Chandrasekhar P S and Komarala V K 2017 Graphene/ZnO nanocomposite as an electron transport layer for perovskite solar cells; the effect of graphene concentration on photovoltaic performance *RSC Adv.* **7** 28610–15
[52] Mariani P, Najafi L, Bianca G, Zappia M I, Gabatel L, Agresti A, Pescetelli S, Di Carlo A, Bellani S and Bonaccorso F 2021 Low-temperature graphene-based paste for large-area carbon perovskite solar cells *ACS Appl. Mater. Interfaces* **13** 22368–80
[53] Simon P and Gogotsi Y 2008 Materials for electrochemical capacitors *Nat. Mater.* **7** 845–54
[54] Yu Z, Tetard L, Zhai L and Thomas J 2015 Supercapacitor electrode materials: nanostructures from 0 to 3 dimensions *Energy Environ. Sci.* **8** 702–30
[55] Jeong H M, Lee J W, Shin W H, Choi Y J, Shin H J, Kang J K and Choi J W 2011 Nitrogen-doped graphene for high-performance ultracapacitors and the importance of nitrogen-doped sites at basal planes *Nano Lett.* **11** 2472–7
[56] Elessawy N A, Nady E, Wazeer W and Kashyout A B 2019 Development of high-performance supercapacitor based on a novel controllable green synthesis for 3D nitrogen doped graphene *Sci. Rep.* **9** 1–10
[57] Liu Y *et al* 2011 Nitrogen-doped single-walled carbon nanotubes grown on substrates: evidence for framework doping and their enhanced properties *Adv. Funct. Mater.* **21** 986–92
[58] Zuo Z, Jiang Z and Manthiram A 2013 Porous B-doped graphene inspired by Fried-Ice for supercapacitors and metal-free catalysts *J. Mater. Chem.* A **1** 13476–83
[59] Li J, Li X, Xiong D, Wang L and Li D 2019 Enhanced capacitance of boron-doped graphene aerogels for aqueous symmetric supercapacitors *Appl. Surf. Sci.* **475** 285–93
[60] Zou X, Wu D, Mu Y, Xing L, Zhang W, Gao Z, Xu F and Jiang K 2020 Boron and nitrogen Co-doped holey graphene aerogels with rich B–N motifs for flexible supercapacitors *Carbon* **159** 94–101
[61] Wang S, Iyyamperumal E, Roy A, Xue Y, Yu D and Dai L 2011 Vertically aligned BCN nanotubes as efficient metal-free electrocatalysts for the oxygen reduction reaction: a synergetic effect by co-doping with boron and nitrogen *Angew. Chem. Int. Ed.* **50** 11756–60
[62] Wang D W, Li F, Zhao J, Ren W, Chen Z G, Tan J, Wu Z S, Gentle I, Lu G Q and Cheng H M 2009 Fabrication of graphene/polyaniline composite paper via *in situ* anodic electropolymerization for high-performance flexible electrode *ACS Nano* **3** 1745–52
[63] Zhu Y, Stoller M D, Cai W, Velamakanni A, Piner R D, Chen D and Ruoff R S 2010 Exfoliation of graphite oxide in propylene carbonate and thermal reduction of the resulting graphene oxide platelets *ACS Nano* **4** 1227–33
[64] Wang H, Liang Y, Mirfakhrai T, Chen Z, Casalongue H S and Dai H 2011 Advanced asymmetrical supercapacitors based on graphene hybrid materials *Nano Res.* **4** 729–36
[65] Shen R, Xie J, Xiang Q, Chen X, Jiang J and Li X 2019 Ni-based photocatalytic H_2-production cocatalysts2 *Chin. J. Catal.* **40** 240–88
[66] Lu D, Yang M, Wang H, Kumar K K, Wu P and Neena D 2017 In situ hydrothermal synthesis of Y-TiO_2/graphene heterojunctions with improved visible-light-driven photocatalytic properties *Ceram. Int.* **43** 16753–62
[67] Kondamareddy K K, Bin H, Lu D, Kumar P, Dwivedi R K, Pelenovich V O, Zhao X Z, Gao W and Fu D 2018 Enhanced visible light photodegradation activity of RhB/MB from aqueous solution using nanosized novel Fe–Cd co-modified ZnO *Sci. Rep.* **8** 1–12
[68] Neena D, Humayun M, Zuo W, Liu C S, Gao W and Fu D J 2020 Hierarchical hetero-architectures of *in situ* g-C_3N_4-coupled Fe-doped ZnO micro-flowers with enhanced visible-light photocatalytic activities *Appl. Surf. Sci.* **506** 145017

[69] Zhang Y *et al* 2021 Controllable preparation and efficient visible-light-driven photocatalytic removal of Cr(VI) using optimized $Cd_{0.5}Zn_{0.5}S$ nanoparticles decorated H-titanate nanotubes *Adv. Powder Technol.* **32** 3788–800
[70] Zeng Y *et al* 2021 Interfacial optimization of Z-scheme Ag_3PO_4/MoS_2 nanoflower sphere heterojunction toward synergistic enhancement of visible-light-driven photocatalytic oxygen evolution and degradation of organic pollutant *J. Alloys Compd.* **888** 161583
[71] Lu D, Wang H, Shen Q, Kondamareddy K K and Neena D 2017 Highly efficient visible-light-induced photoactivity of magnetically retrievable Fe_3O_4@ SiO_2@ Bi_2WO_6@ g-C_3N_4 hierarchical microspheres for the degradation of organic pollutant and production of hydrogen *J. Phys. Chem. Solids* **106** 76–81
[72] Chen J L, Devi N, Li N, Fu D J and Ke X W 2018 Synthesis of Pr-doped ZnO nanoparticles: their structural, optical, and photocatalytic properties *Chin. Phys.* B **27** 086102
[73] Lu D, Zhang X, Wang S, Peng W, Wei M, Neena D, Fan H and Hao H 2019 Enhanced photocatalytic removal of Cr(VI) over 0D/2D anatase/graphene and its synergism with organic pollutants under visible light irradiation *Appl. Surf. Sci.* **470** 368–75
[74] Lu D, Yang M, Kumar K K and Wu P 2018 Investigation of structure and photocatalytic degradation of organic pollutants for protonated anatase/titanate nanosheets during thermal treatment *ACS Sustain. Chem. Eng.* **6** 4801–8
[75] Neena D, Humayun M, Bhattacharyya D and Fu D 2020 Hierarchical Sr–ZnO/g-C_3N_4 heterojunction with enhanced photocatalytic activities *J. Photochem. Photobiol.,* A **396** 112515
[76] Bie C, Yu H, Cheng B, Ho W, Fan J and Yu J 2021 Design, fabrication, and mechanism of nitrogen-doped graphene-based photocatalyst *Adv. Mater.* **33** 2003521
[77] Wu Q *et al* 2022 Interfacial optimization of CeO_2 nanoparticles loaded two-dimensional graphite carbon nitride toward synergistic enhancement of visible-light-driven photoelectric and photocatalytic hydrogen evolution *Int. J. Hydrog. Energy* **47** 2313–26
[78] Neena D, Shah A H, Deshmukh K, Ahmad H, Fu D J, Kondamareddy K K, Kumar P, Dwivedi R K and Sing V 2016 Influence of (Co–Mn) co-doping on the microstructures, optical properties of sol-gel derived ZnO nanoparticles *Eur. Phys. J.* D **70** 1–7
[79] Shah A H, Ahamed M B, Neena D, Mohmed F and Iqbal A 2014 Investigations of optical, structural and antibacterial properties of Al–Cr dual-doped ZnO nanostructures *J. Alloys Compd.* **606** 164–70
[80] Neena D, Humayun M, Lu D, Mohan V B, Fu D and Gao W 2020 Cadmium and iron decorated ZnO nanoparticles: effects on physical, electrochemical and antibacterial performance *Ceram. Int.* **46** 5278–88
[81] Xiang Q, Yu J and Jaroniec M 2012 Synergetic effect of MoS_2 and graphene as cocatalysts for enhanced photocatalytic H_2 production activity of TiO_2 nanoparticles *J. Am. Chem. Soc.* **134** 6575–8
[82] Zhang Y *et al* 2022 Enhanced visible light-driven photocatalytic hydrogen evolution and stability for noble metal-free $MoS_2/Zn_{0.5}Cd_{0.5}S$ heterostructures with W/Z phase junctions *Appl. Surf. Sci.* **586** 152770
[83] Neena D, Kondamareddy K K, Humayun M, Mohan V B, Lu D, Fu D and Gao W 2019 Fabrication of ZnO/N-rGO composite as highly efficient visible-light photocatalyst for 2,4-DCP degradation and H_2 evolution *Appl. Surf. Sci.* **488** 611–9

[84] Cao Y, Wang P, Fan J and Yu H 2021 Covalently functionalized graphene by thiourea for enhancing H_2-evolution performance of TiO_2 photocatalyst *Ceram. Int.* **47** 654–61

[85] Mou Z, Wu Y, Sun J, Yang P, Du Y and Lu C 2014 TiO_2 nanoparticles-functionalized N-doped graphene with superior interfacial contact and enhanced charge separation for photocatalytic hydrogen generation *ACS Appl. Mater. Interfaces* **6** 13798–806

[86] Ali S, Razzaq A and In S I 2019 Development of graphene based photocatalysts for CO_2 reduction to C1 chemicals: a brief overview *Catal. Today* **335** 39–54

[87] Das S and Daud W W 2014 A review on advances in photocatalysts towards CO_2 conversion *RSC Adv.* **4** 20856–93

[88] Nonhebel S 2005 Renewable energy and food supply: will there be enough land? *Renew. Sustain. Energy Rev.* **9** 191–201

[89] Low J, Yu J and Ho W 2015 Graphene-based photocatalysts for CO_2 reduction to solar fuel *J. Phys. Chem. Lett.* **6** 4244–51

[90] Yuan L and Xu Y J 2015 Photocatalytic conversion of CO_2 into value-added and renewable fuels *Appl. Surf. Sci.* **342** 154–67

[91] Tahir M and Amin N S 2013 Recycling of carbon dioxide to renewable fuels by photocatalysis: prospects and challenges *Renew. Sustain. Energy Rev.* **25** 560–79

[92] Low J, Cheng B and Yu J 2017 Surface modification and enhanced photocatalytic CO_2 reduction performance of TiO_2: a review *Appl. Surf. Sci.* **392** 658–86

[93] Wang L, Tan H, Zhang L, Cheng B and Yu J 2021 In-situ growth of few-layer graphene on ZnO with intimate interfacial contact for enhanced photocatalytic CO_2 reduction activity *Chem. Eng. J.* **411** 128501

[94] She X, Xu H, Yu Y, Li L, Zhu X, Mo Z, Song Y, Wu J, Yuan S and Li H 2019 Accelerating photogenerated charge kinetics via the synergetic utilization of 2D semi-conducting structural advantages and noble-metal-free schottky junction effect *Small* **15** 1804613

[95] Zhang L, Li N, Jiu H, Qi G and Huang Y 2015 ZnO-reduced graphene oxide nano-composites as efficient photocatalysts for photocatalytic reduction of CO_2 *Ceram. Int.* **41** 6256–62

[96] Lin L Y, Nie Y, Kavadiya S, Soundappan T and Biswas P 2017 N-doped reduced graphene oxide promoted nano TiO_2 as a bifunctional adsorbent/photocatalyst for CO_2 photo-reduction: effect of N species *Chem. Eng. J.* **316** 449–60

[97] Wang X J, Tian X, Sun Y J, Zhu J Y, Li F T, Mu H Y and Zhao J 2018 Enhanced Schottky effect of a 2D–2D CoP/gC_3N_4 interface for boosting photocatalytic H_2 evolution *Nanoscale* **10** 12315–21

[98] Xu D, Cheng B, Wang W, Jiang C and Yu J 2018 Ag_2CrO_4/g-C_3N_4/graphene oxide ternary nanocomposite Z-scheme photocatalyst with enhanced CO_2 reduction activity *Appl. Catalysis* B **231** 368–80

[99] Vacanti J P and Langer R 1999 Tissue engineering: the design and fabrication of living replacement devices for surgical reconstruction and transplantation *Lancet* **354** S32–4

[100] Brown T E, Carberry B J, Worrell B T, Dudaryeva O Y, McBride M K, Bowman C N and Anseth K S 2018 Photopolymerized dynamic hydrogels with tunable viscoelastic properties through thioester exchange *Biomaterials* **178** 496–503

[101] Vedadghavami A, Minooei F, Mohammadi M H, Khetani S, Kolahchi A R, Mashayekhan S and Sanati-Nezhad A 2017 Manufacturing of hydrogel biomaterials with controlled mechanical properties for tissue engineering applications *Acta Biomater.* **62** 42–63

[102] Zhou R and Gao H 2014 Cytotoxicity of graphene: recent advances and future perspective *Wiley Interdiscip. Rev. Nanomed. Nanobiotechnol.* **6** 452–74
[103] Wang K, Ruan J, Song H, Zhang J, Wo Y, Guo S and Cui D 2010 Biocompatibility of graphene oxide *Nanoscale Res. Lett.* **6** 8
[104] Chang Y, Yang S T, Liu J H, Dong E, Wang Y, Cao A, Liu Y and Wang H 2011 In vitro toxicity evaluation of graphene oxide on A549 cells *Toxicol. Lett.* **200** 201–10
[105] Mu Q, Su G, Li L, Gilbertson B O, Yu L H, Zhang Q, Sun Y P and Yan B 2012 Size-dependent cell uptake of protein-coated graphene oxide nanosheets *ACS Appl. Mater. Interfaces* **4** 2259–66
[106] Vecitis C D, Zodrow K R, Kang S and Elimelech M 2011 Electronic-Structure-Dependent Bacterial Cytotoxicity of Single-Walled Carbon Nanotubes *ACS Nano* **4** 5471–9
Liu S, Zeng T H, Hofmann M, Burcombe E, Wei J, Jiang R, Kong J and Chen Y 2011 Antibacterial activity of graphite, graphite oxide, graphene oxide, and reduced graphene oxide: membrane and oxidative stress *ACS Nano* **5** 6971–80
[107] Akhavan O and Ghaderi E 2010 Toxicity of graphene and graphene oxide nanowalls against bacteria *ACS Nano* **4** 5731–6
[108] Wang A, Pu K, Dong B, Liu Y, Zhang L, Zhang Z, Duan W and Zhu Y 2013 Role of surface charge and oxidative stress in cytotoxicity and genotoxicity of graphene oxide towards human lung fibroblast cells *J. Appl. Toxicol.* **33** 1156–64
[109] Seabra A B, Paula A J, de Lima R, Alves O L and Durán N 2014 Nanotoxicity of graphene and graphene oxide *Chem. Res. Toxicol.* **27** 159–68
[110] Zhang W, Yan L, Li M, Zhao R, Yang X, Ji T, Gu Z, Yin J J, Gao X and Nie G 2015 Deciphering the underlying mechanisms of oxidation-state dependent cytotoxicity of graphene oxide on mammalian cells *Toxicol. Lett.* **237** 61–71
[111] Liu Z, Robinson J T, Sun X and Dai H 2008 PEGylated nanographene oxide for delivery of water-insoluble cancer drugs *J. Am. Chem. Soc.* **130** 10876–7
[112] Park Y J, Park S Y and In I 2011 Preparation of water soluble graphene using polyethylene glycol: comparison of covalent approach and noncovalent approach *J. Ind. Eng. Chem.* **17** 298–303
[113] Engler A J, Sen S, Sweeney H L and Discher D E 2006 Matrix elasticity directs stem cell lineage specification *Cell* **126** 677–89
[114] Lee J, Kim J, Kim S and Min D H 2016 Biosensors based on graphene oxide and its biomedical application *Adv. Drug Delivery Rev.* **105** 275–87
[115] Rahmati M, Mills D K, Urbanska A M, Saeb M R, Venugopal J R, Ramakrishna S and Mozafari M 2021 Electrospinning for tissue engineering applications *Prog. Mater Sci.* **117** 100721
[116] Lee-Liu D, Edwards-Faret G, Tapia V S and Larraín J 2013 Spinal cord regeneration: lessons for mammals from non-mammalian vertebrates *Genesis* **51** 529–44
[117] Bedir T, Ulag S, Ustundag C B and Gunduz O 2020 3D bioprinting applications in neural tissue engineering for spinal cord injury repair *Mater. Sci. Eng.* C **110** 110741
[118] Park S Y, Park J, Sim S H, Sung M G, Kim K S, Hong B H and Hong S 2011 Enhanced differentiation of human neural stem cells into neurons on graphene *Adv. Mater.* **23** H263–7
[119] Li N, Zhang Q, Gao S, Song Q, Huang R, Wang L, Liu L, Dai J, Tang M and Cheng G 2013 Three-dimensional graphene foam as a biocompatible and conductive scaffold for neural stem cells *Sci. Rep.* **3** 1–6

[120] Abematsu M, Tsujimura K, Yamano M, Saito M, Kohno K, Kohyama J, Namihira M, Komiya S and Nakashima K 2010 Neurons derived from transplanted neural stem cells restore disrupted neuronal circuitry in a mouse model of spinal cord injury *J. Clin. Invest.* **120** 3255–66

[121] Shah S, Yin P T, Uehara T M, Chueng S T D, Yang L and Lee K B 2014 Guiding stem cell differentiation into oligodendrocytes using graphene-nanofiber hybrid scaffolds *Adv. Mater.* **26** 3673–80

[122] Song J, Gao H, Zhu G, Cao X, Shi X and Wang Y 2015 The preparation and characterization of polycaprolactone/graphene oxide biocomposite nanofiber scaffolds and their application for directing cell behaviors *Carbon* **95** 1039–50

[123] Magaz A, Li X, Gough J E and Blaker J J 2021 Graphene oxide and electroactive reduced graphene oxide-based composite fibrous scaffolds for engineering excitable nerve tissue *Mater. Sci. Eng.* C **119** 111632

[124] Qing H, Jin G, Zhao G, Huang G, Ma Y, Zhang X, Sha B, Luo Z, Lu T J and Xu F 2018 Heterostructured silk-nanofiber-reduced graphene oxide composite scaffold for SH-SY5Y cell alignment and differentiation *ACS Appl. Mater. Interfaces* **10** 39228–37

[125] Girao A F *et al* 2020 3D reduced graphene oxide scaffolds with a combinatorial fibrous-porous architecture for neural tissue engineering *ACS Appl. Mater. Interfaces* **12** 38962–75

[126] Huang C T, Shrestha L K, Ariga K and Hsu S H 2017 A graphene–polyurethane composite hydrogel as a potential bioink for 3D bioprinting and differentiation of neural stem cells *J. Mater. Chem.* B **5** 8854–64

[127] Cha C, Shin S R, Gao X, Annabi N, Dokmeci M R, Tang X and Khademhosseini A 2014 Controlling mechanical properties of cell-laden hydrogels by covalent incorporation of graphene oxide *Small* **10** 514–23

[128] Park J, Choi J H, Kim S, Jang I, Jeong S and Lee J Y 2019 Micropatterned conductive hydrogels as multifunctional muscle-mimicking biomaterials: graphene-incorporated hydrogels directly patterned with femtosecond laser ablation *Acta Biomater.* **97** 141–53

[129] Jing X, Mi H Y, Napiwocki B N, Peng X F and Turng L S 2017 Mussel-inspired electroactive chitosan/graphene oxide composite hydrogel with rapid self-healing and recovery behavior for tissue engineering *Carbon* **125** 557–70

[130] Yi J, Choe G, Park J and Lee J Y 2020 Graphene oxide-incorporated hydrogels for biomedical applications *Polym. J.* **52** 823–37

[131] Liu X, Miller A L, Park S, Waletzki B E, Zhou Z, Terzic A and Lu L 2017 Functionalized carbon nanotube and graphene oxide embedded electrically conductive hydrogel synergistically stimulates nerve cell differentiation *ACS Appl. Mater. Interfaces* **9** 14677–90

[132] Xavier Mendes A, Moraes Silva S, O'Connell C D, Duchi S, Quigley A F, Kapsa R M and Moulton S E 2021 Enhanced electroactivity, mechanical properties, and printability through the addition of graphene oxide to photo-cross-linkable gelatin methacryloyl hydrogel *ACS Biomater. Sci. Eng.* **7** 2279–95

[133] Barrientos S, Stojadinovic O, Golinko M S, Brem H and Tomic-Canic M 2008 Growth factors and cytokines in wound healing *Wound Repair Regen.* **16** 585–601

[134] Kim S H, Turnbull J and Guimond S 2011 Extracellular matrix and cell signalling: the dynamic cooperation of integrin, proteoglycan and growth factor receptor *J. Endocrinol.* **209** 139–51

[135] Aaronson S A 1991 Growth factors and cancer *Science* **254** 1146–53

[136] Zhang Y, Wu C, Guo S and Zhang J 2013 Interactions of graphene and graphene oxide with proteins and peptides *Nanotechnol. Rev.* **2** 27–45
[137] Yoon H H, Bhang S H, Kim T, Yu T, Hyeon T and Kim B S 2014 Dual roles of graphene oxide in chondrogenic differentiation of adult stem cells: cell-adhesion substrate and growth factor-delivery carrier *Adv. Funct. Mater.* **24** 6455–64
[138] Zhou M, Lozano N, Wychowaniec J K, Hodgkinson T, Richardson S M, Kostarelos K and Hoyland J A 2019 Graphene oxide: a growth factor delivery carrier to enhance chondrogenic differentiation of human mesenchymal stem cells in 3D hydrogels *Acta Biomater.* **96** 271–80
[139] Shen H, Lin H, Sun A X, Song S, Wang B, Yang Y, Dai J and Tuan R S 2020 Acceleration of chondrogenic differentiation of human mesenchymal stem cells by sustained growth factor release in 3D graphene oxide incorporated hydrogels *Acta Biomater.* **105** 44–55
[140] La W G, Jung M J, Yoon J K, Bhang S H, Jang H K, Lee T J, Yoon H H, Shin J Y and Kim B S 2014 Bone morphogenetic protein-2 for bone regeneration—dose reduction through graphene oxide-based delivery *Carbon* **78** 428–38
[141] Fu C, Yang X, Tan S and Song L 2017 Enhancing cell proliferation and osteogenic differentiation of MC3T3-E1 pre-osteoblasts by BMP-2 delivery in graphene oxide-incorporated PLGA/HA biodegradable microcarriers *Sci. Rep.* **7** 1–13
[142] Wang Q, Wang M, Wang K, Sun Y, Zhang H, Lu X and Duan K 2020 Molecular mechanisms of interactions between BMP-2 and graphene: effects of functional groups and microscopic morphology *Appl. Surf. Sci.* **525** 146636
[143] Emadi F, Amini A, Gholami A and Ghasemi Y 2017 Functionalized graphene oxide with chitosan for protein nanocarriers to protect against enzymatic cleavage and retain collagenase activity *Sci. Rep.* **7** 1–13
[144] Zhang W, Yang G, Wang X, Jiang L, Jiang F, Li G, Zhang Z and Jiang X 2017 Magnetically controlled growth-factor-immobilized multilayer cell sheets for complex tissue regeneration *Adv. Mater.* **29** 1703795
[145] Magaz A, Ashton M D, Hathout R M, Li X, Hardy J G and Blaker J J 2020 Electroresponsive silk-based biohybrid composites for electrochemically controlled growth factor delivery *Pharmaceutics* **12** 742
[146] Yang K, Wan J, Zhang S, Tian B, Zhang Y and Liu Z 2012 The influence of surface chemistry and size of nanoscale graphene oxide on photothermal therapy of cancer using ultra-low laser power *Biomaterials* **33** 2206–14
[147] Yang K, Zhang S, Zhang G, Sun X, Lee S T and Liu Z 2010 Graphene in mice: ultrahigh *in vivo* tumor uptake and efficient photothermal therapy *Nano Lett.* **10** 3318–23
[148] Zhao X, Liu Y, Shao C, Nie M, Huang Q, Li J, Sun L and Zhao Y 2019 Photoresponsive delivery microcarriers for tissue defects repair *Adv. Sci.* **6** 1901280
[149] Halperin A, Kroger M and Winnik F M 2015 Since LCST defined as the temperature of the minimum of binodal, the cloud point temperature should be used for the transition temperature measured by turbidimetry or DSC: poly(*N*-isopropylacrylamide) phase diagrams: fifty years of research *Angew. Chem. Int. Ed.* **54** 15342–67

IOP Publishing

Anoop Chandran, N V Unnikrishnan, M K Jayaraj, Reenu Elizabeth John and Justin George

Chapter 7

Applications of graphene in electronics: graphene field effect transistors

L Chandrasekar and K P Pradhan

Undoped pristine graphene with zero bandgap and B/N-substitution doped graphene with non-zero bandgap are explored in this chapter. The predominant prerequisites in designing of graphene field effect transistors (GFETs) have been precisely modeled, such as: (1) the electrostatics as energy–momentum dispersion relation, density of states, individual 2D carrier statistics, total carrier sheet charge density, quantum capacitance for undoped pristine graphene, B- and N-substitution doped graphene have been examined extensively; (2) the electronic transport properties like interaction parameter and semi-classical drift-diffusive transport mobility have been modeled and examined for undoped pristine graphene, B- and N-substitution doped graphene. With the aid of the above-mentioned prerequisites, the phenomenological all region I_D for GFETs has been modeled by utilizing the metal insulator–graphene (MIG) equivalent circuit and solving self-consistently through first order Newton–Raphson method. Also, the theoretical mobility model and drift-diffusive I_D model have been validated with experimentally measured data and proven accuracy. In addition to that, the Verilog-A implementation of GFETs has been studied. Utilizing the Verilog-A symbol of GFET in Cadence Spectre circuit simulator, the memoryless linearity and nonlinearity performance of both undoped and B-doped GFET have been analyzed and the improved reliability in B-substitution doped GFET over undoped GFET is reported.

7.1 Introduction

The triumphant isolation of graphene—single carbon atom thick layer—from graphite in 2004, energized the research to explore the influence of other two-dimensional (2D) materials on nanoelectronics. The particular electronic and exceptional transport properties of absolutely 2D single-layer graphene [1] have

doi:10.1088/978-0-7503-3999-5ch7

stimulated research interest over the past decade in exploring GFETs and revealed them to be a novel channel material to replace traditional semiconductors. Recently, numerous research works [2–4] have been extensively explored on GFETs with various architectures, like top-gated, bottom-gated, and double-gated, and the suitability of different substrates has been investigated. These reports are largely focused on both analog/RF and digital applications of GFETs, which emphasize exploring the physical insights in device modeling and linearity performance extraction of GFETs. Hence, the carrier statistics, electronic transport properties, I_D modeling, and memoryless linearity and nonlinearity performance have been presumably manifested in the following sections.

7.2 The carrier statistics and quantum capacitance

When developing the electrostatics for low-dimensional devices, the key quantities of concern are the carrier sheet density and quantum capacitance [5]. The individual 2D carrier sheet concentration, total sheet charge density, and quantum capacitance of single-layer graphene are explored analytically in [5, 6] and observed experimentally in [7, 8]. Recently, quantum capacitance of GFETs has been explored theoretically in gas sensor application [9]. A compact modeling of GFET is established with the aid of a mathematical model of quantum capacitance explored in [6, 10, 11], which emphasizes the significance of quantum capacitance in designing and modeling of nanoscale devices.

7.2.1 The electrostatics: undoped pristine graphene

The two-dimensional nature of graphene and its two inter-leaving triangular sublattice structure make the band structure of pristine graphene distinct from conventional semiconductors. The linear energy–momentum (E–k) relation with zero bandgap (at the point of zero momentum) is one of the exceptional features of pristine graphene [1, 5, 12–14], which is shown in figure 7.1(a). The dispersion

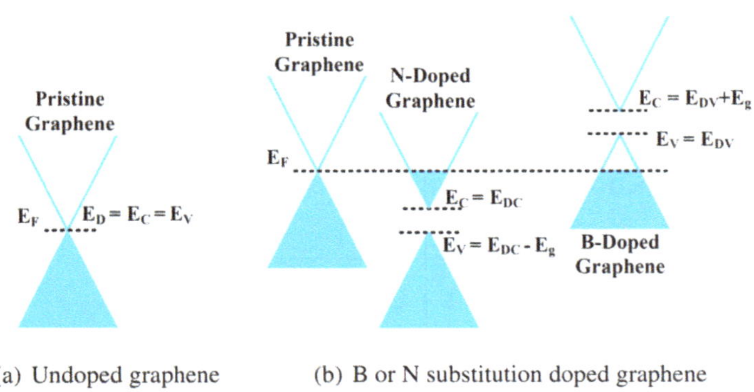

Figure 7.1. E–k dispersion relation of (a) undoped pristine graphene and (b) B- or N-doped graphene—illustrating about Dirac point and E_g. Reproduced with permission [16]. Copyright 2021, *IEEE Trans. Electron Devices.*

relation for single-layer undoped graphene has been abundantly surveyed in the literature [5, 6, 12, 13, 15] and is presented in equation (7.1) as follows:

$$E - E_D = s\hbar v_f \mid k \mid \tag{7.1}$$

where E indicates the energy, E_D represents the Dirac point, $\hbar$ denotes the reduced Planck's constant, for the energy in conduction band $s = 1$ and for the energy in valence band $s = -1$, v_f denotes the Fermi velocity of 10^8 cm s^{-1}, additionally $|k|$ indicates the wave vector computed around the Brillouin zone, which is expressed as [5], $|k| = (k_x^2 + k_y^2)^{1/2}$. For the undoped pristine graphene, the minimum of the conduction band edge (E_C) and the maximum of the valence band edge (E_V) make contact exactly at E_D during the zero momentum state (i.e., $E_D = E_C = E_V$). Also, for undoped graphene E_D has been positioned at 0 eV. Consequently, the single-layer undoped graphene possesses a unique zero E_g and categorized as semi-metal (neither metal nor semiconductor), which is apparent from figure 7.1(a).

The density of states (DOS) per unit area per unit energy for 2D single-layer undoped graphene can be obtained with the aid of E–k dispersion expression from equation (7.1). The DOS in conduction band and in valance band for undoped graphene are presented in equations (7.2) and (7.3), respectively, as follows [15]:

$$D_{CB} \geqslant (E) = \frac{2(E - E_D)}{\pi(\hbar v_f)^2} \quad E \geqslant E_D \tag{7.2}$$

$$D_{VB} \geqslant (E) = \frac{2(E_D - E)}{\pi(\hbar v_f)^2} \quad E \leqslant E_D \tag{7.3}$$

From equations (7.2) and (7.3), it can be observed that DOS varies linearly with respect to energy in both conduction band and valance band due to E_D lying at 0 eV for undoped graphene.

In general, the electron concentration (2D sheet—n) in conduction band can be obtained analytically by integrating the product of DOS in conduction band and Fermi function (probability of occupancy of a state) over the corresponding energy range, which is reppresented as:

$$n = \int_{E_C}^{\infty} D_{CB}(E) f(E)\, dE \tag{7.4}$$

where the Fermi function, $f(E)$ is given by:

$$f(E) = \frac{1}{1 + \exp\left(\dfrac{E - E_F}{k_B T}\right)} \tag{7.5}$$

where E_F represents the Fermi energy, k_B is the Boltzmann constant and T denotes the temperature in kelvin. Since, E_C and E_D are the same for undoped pristine graphene, $D_{CB}(E)$ in equation (7.2) can be exploited in equation (7.4) to compute n, which can be obtained as:

$$n = \int_{E_D}^{\infty} \frac{2(E - E_D)}{\pi(\hbar v_f)^2} \frac{1}{1 + \exp(\frac{E - E_F}{k_B T})} dE \tag{7.6}$$

Using the integration by substitution method, equation (7.6) can be recomputed as:

$$n = \frac{2(k_B T)^2}{\pi(\hbar v_f)^2} \int_0^{\infty} \frac{E}{1 + e^{E - \eta_F}} dE \tag{7.7}$$

where η_F is the dimensionless parameter, which measures the Fermi energy (E_F) relatively with respect to Dirac point energy (E_D) i.e., $\eta_F = (E_F - E_D)/k_B T$. With the aid of complete Fermi–Dirac integral formula, equation (7.7) can be solved analytically and n for single-layer 2D undoped graphene is represented as:

$$n = N_G\, \Im_1(\eta_F) \tag{7.8}$$

where $\Im_1(\eta_F)$ represents the Fermi–Dirac integral over η_F with the order of 1 and N_G denotes the effective single-layer 2D graphene sheet DOS [15], and is represented as: $N_G = 2(k_B T)^2/\pi(\hbar v_f)^2$. Similarly, hole concentration (2D sheet—p) in valance band for single-layer undoped graphene is also analytically derived by using the above-mentioned procedure and is represented as:

$$p = N_G\, \Im_1(-\eta_F) \tag{7.9}$$

The total sheet charge density (Q_{sh}) for single-layer graphene can be quantified by counting electron (n) and hole (p) sheet concentrations with considering the corresponding electronic charge (q) i.e., $Q_{sh} = q(p - n)$. Hence, utilizing equations (7.8) and (7.9), Q_{sh} for undoped graphene is represented as:

$$Q_{sh} = qN_G\left[\Im_1(-\eta_F) - \Im_1(\eta_F)\right] \tag{7.10}$$

The relative measure of E_F in respect of E_D has been modeled as the potential drop inside the graphene and fundamentally archived as local channel potential—V_{ch} in the literature [6, 10, 11, 15] which is represented as $V_{ch} = (E_F - E_D)/q$. The quantum capacitance (C_q) of single-layer graphene is interpreted as variation in total 2D sheet charge density in respect of variation in graphene local channel potential (V_{ch}), which is given by $C_q = -\partial Q_{sh}/\partial V_{ch}$ [5, 6]. In addition to that, the dimensionless parameters η_F and V_{ch} are related as: $V_{ch} = (k_B T/q)\,\eta_F$. Therefore, C_q can be represented as:

$$C_q = -\frac{\partial Q_{sh}}{\partial V_{ch}} = -\frac{q}{k_B T}\frac{\partial Q_{sh}}{\partial \eta_F} \tag{7.11}$$

Utilizing equation (7.10) in equation (7.11), the C_q for single-layer undoped graphene is analytically modeled as:

$$C_q = \frac{q^2 N_G}{k_B T}[\ln(1 + e^{-\eta_F}) + \ln(1 + e^{\eta_F})] \tag{7.12}$$

Further, equation (7.12) is simplified analytically as follows:

$$C_{\mathrm{q}} = \frac{q^2 N_{\mathrm{G}}}{k_{\mathrm{B}}T} \ln \{2[1 + \cosh(\eta_{\mathrm{F}})]\} \tag{7.13}$$

7.2.2 The electrostatics: B-substitution doped graphene

The drastic drawback of single-layer undoped pristine graphene is the lack of E_{g} (i.e., zero bandgap) in it when encountering it with traditional semiconductors. Due to the absence of E_{g} in undoped graphene, the channel becomes conductive even when the device is in the OFF state. Many methods have been archived in the literature to overcome this drawback by the manual introduction of E_{g} in graphene. The first method is cutting the larger area graphene sheet into graphene nanoribbons (GNRs) [17]. However, in this method, the carrier mobility in GNR has been diminished drastically [18]. In the second method, compression and extention of the larger area graphene sheet is done [19]. The third method is the application of electric field (**E**-field) perpendicular to the plane of bilayer graphene (BLG) [20]. This method, however, is not suitable for monolayer graphene. In the fourth method, the graphene is doped with impurities such as boron (B) or nitrogen (N) [21, 22], in which the intrinsic behaviour of graphene would be preserved.

The unusual linear *E–k* relation of single-layer undoped graphene around the Dirac point energy and its hexagonal lattice structure have been preserved by incorporating a dopant such as B or N to the identical sublattice points of graphene. This doping procedure is termed as substitution doping in graphene [21, 22].

The *E–k* relation of B- or N-substitution doped graphene (labelled as sample S_{A} ans S_{B}, respectively) is investigated along with *E–k* relation of undoped pristine graphene in a comparative manner, which is clearly depicted within figure 7.1. Also, it can be noticed that, E_{C} and E_{V} for undoped graphene are exactly positioned at Dirac point energy, which leads to zero E_{g}. On the other hand, B- or N-doped graphene are exhibiting non-zero E_{g}, which is increasing in respect of dopant concentration and is enumerated in table 7.1. In addition to that, B- or N-doping in single-layer graphene shifts the Dirac point energy above or below the Fermi energy level (E_{F}), respectively, which is manifested in figure 7.1(b).

Since the electron-deficient essence of B atoms as compared with carbon (C) atoms, the notable shifting of Dirac point energy above E_{F} has been established in sample S_{A}. Therefore, B-doping in graphene exhibits a behaviour similar to degenerately p-type doped as in traditional semiconductors, and is shown in figure 7.1(b). The *E–k* dispersion relation of sample S_{A} can be modeled analytically in equations (7.14) and (7.15) for conduction band and valence band, respectively, by incorporating doping effects such as: Dirac point energy shifting with non-zero E_{g} [22, 23].

$$E - (E_{\mathrm{DV}} + E_{\mathrm{g}}) = \hbar v_f |k| \qquad E \geqslant E_{\mathrm{DV}} + E_{\mathrm{g}} \tag{7.14}$$

$$E_{\mathrm{DV}} - E = \hbar v_f |k| \qquad E \leqslant E_{\mathrm{DV}} \tag{7.15}$$

Table 7.1. Induced bandgap (E_g) and shifted Dirac points (E_D) in doped graphene with nitrogen (N) and boron (B).

Doping percent		0%	6.25%	12.5%	18.75%	25%
E_g (eV)	B-doped graphene	0	0.35	0.60	0.87	1.18
	N-doped graphene	0	0.36	0.62	0.88	1.14
E_D (eV)	B-doped graphene (E_{DV})	0	0.52	1.12	1.03	1.23
	N-doped graphene (E_{DC})	0	−0.45	−0.93	−0.84	−0.76

where E_{DV} denotes the shift in Dirac point energy due to B-doping. The maximum point of valence band energy (E_V) and E_{DV} are identical for sample S_A i.e., $E_{DV} = E_V$. Therefore, the minimum point of conduction band energy (E_C) for sample S_A can be modeled as a sum of shifted E_{DV} and induced E_g i.e., $E_C = E_{DV} + E_g$, which is deduced in figure 7.1(b).

The DOS per unit area per unit energy in conduction band and valence band for 2D single-layer sample S_A can be obtained with the aid of E–k dispersion expression reported in equations (7.14) and (7.15), respectively, and the corresponding DOS are presented in equations (7.16) and (7.17), respectively, as follows [16, 23]:

$$D_{CB}(E) = \frac{2\left[E - (E_{DV} + E_g)\right]}{\pi(\hbar v_f)^2} \qquad E \geqslant E_{DV} + E_g \tag{7.16}$$

$$D_{VB}(E) = \frac{2(E_{DV} - E)}{\pi(\hbar v_f)^2} \qquad E \leqslant E_{DV} \tag{7.17}$$

Utilizing the DOS and Fermi function, the 2D sheet hole (p) and electron (n) concentration in valence band and conduction band, respectively, for sample S_A has been modeled mathematically as follows:

$$p = N_G \Im_1(-\eta_{FV}) \tag{7.18}$$

$$n = N_G \Im_1(\eta_{FV} - \eta_g) \tag{7.19}$$

where η_{FV} and η_g are dimensionless parameters and η_{FV} relatively measures E_F from the shifted Dirac point energy E_{DV} due to B-dopant, which are denoted as: $\eta_{FV} = (E_F - E_{DV})/k_BT$ and $\eta_g = E_g/k_BT$. $\Im_1(-\eta_{FV})$ denotes the first order complete Fermi–Dirac integral over $-\eta_{FV}$. Subsequently, the total 2D charge sheet density (Q_{sh}) for mono-layered sample S_A is acquired by summing p from equation (7.18) and n from equation (7.19) with counterpart electronic charge (q), and is indicated as:

$$Q_{sh} = qN_G[\Im_1(-\eta_{FV}) - \Im_1(\eta_{FV} - \eta_g)] \tag{7.20}$$

From equation (7.20), it is clear that Q_{sh} is primarily dependent on η_{FV} and η_g, which can be disclosed by the change in E_F, E_{DV} and E_g, respectively. The E_g is established non-zero E_g due to B-dopant and it can be customized by the amount of dopant concentration, and it is constant about any external source of bias. On the other hand, the change in $(E_F - E_{DV})$/q can be modeled as local channel potential (V_{ch}) of B graphene, which is manifested as $V_{ch} = (E_F - E_{DV})/q$. In addition to that, the dimensionless parameter η_{FV} and V_{ch} are related as: $V_{ch} = (k_B T/q)\,\eta_{FV}$. Consequently, the quantum capacitance (C_q) for 2D single-layer sample S_A can be modeled as:

$$C_q = -\frac{q}{k_B T}\frac{\partial Q_{sh}}{\partial \eta} \tag{7.21}$$

Using (7.20) in (7.21), it can be obtained as:

$$C_q = \frac{q^2 N_G}{k_B T} \ln\left[(1 + e^{-\eta_{FV}})(1 + e^{\eta_{FV}-\eta_g})\right] \tag{7.22}$$

Equation (7.22) manifests the all region complete C_q model for sample S_A. When E_g is zero, the E_{DV} and η_{FV} of sample S_A become the same as E_D and η_F in undoped pristine graphene, respectively. Therefore, the C_q model presented in equation (7.22) is compatible for both sample S_A and undoped graphene.

7.2.3 The electrostatics: N-substitution doped graphene

For the electron-rich attribute of N atoms as compared with C atoms, the notable shifting of Dirac point energy below the E_F has been manifested in sample S_B. Therefore, N-doping in graphene is disclosing a behaviour similar to degenerately n-type doped as in traditional semiconductors, and is demonstrated in figure 7.1(b). The E–k dispersion relation of sample S_B can be modeled analytically in equations (7.23) and (7.24) for energy in conduction band and valence band, respectively, by incorporating doping effects such as: Dirac point energy shifting with non-zero E_g [22, 23].

$$E - E_{DC} = \hbar v_f |k| \qquad E \geqslant E_{DC} \tag{7.23}$$

$$(E_{DC} - E_g) - E = \hbar v_f |k| \qquad E \leqslant E_{DC} - E_g \tag{7.24}$$

where E_{DC} denotes the shift in Dirac point energy due to N-doping. The minimum point of conduction band energy (E_C) and E_{DC} are identical to sample S_B, i.e., $E_{DC} = E_C$. Therefore, the maximum point of valence band energy (E_V) for sample S_B can be modeled as a difference of shifted E_{DC} and induced E_g i.e., $E_V = E_{DC} - E_g$, which is clearly depicted in figure 7.1(b).

The DOS per unit area per unit energy in conduction band and valence band for 2D single-layer sample S_B can be obtained with the aid of E–k dispersion expression reported in equations (7.23) and (7.24), respectively, and the corresponding DOS are presented in equations (7.25) and (7.26), respectively, as follows [16, 23]:

$$\mathrm{D_{CB}}(E) = \frac{2(E - E_{\mathrm{DC}})}{\pi(\hbar v_f)^2} \qquad E \geqslant E_{\mathrm{DC}} \tag{7.25}$$

$$\mathrm{D_{VB}}(E) = \frac{2\left[(E_{\mathrm{DC}} - E_{\mathrm{g}}) - E\right]}{\pi(\hbar v_f)^2} \qquad E \leqslant E_{\mathrm{DC}} - E_{\mathrm{g}} \tag{7.26}$$

Utilizing the DOS and Fermi function, the 2D sheet electron (n) and hole (p) concentration in conduction band and valence band, respectively, for sample $\mathrm{S_B}$ has been modeled analytically as follows:

$$n = N_{\mathrm{G}}\mathfrak{I}_1(\eta_{\mathrm{FC}}) \tag{7.27}$$

$$p = N_{\mathrm{G}}\mathfrak{I}_1(-\eta_{\mathrm{FC}} - \eta_{\mathrm{g}}) \tag{7.28}$$

where η_{FC} is dimensionless parameter, which relatively measures E_{F} from the shifted Dirac point energy E_{DC} due to N-dopant, and is termed as: $\eta_{\mathrm{FC}} = (E_{\mathrm{F}} - E_{\mathrm{DC}})/k_{\mathrm{B}}T$. $\mathfrak{I}_1(\eta_{\mathrm{FC}})$ denotes the first order complete Fermi–Dirac integral over η_{FC}. Subsequently, the total 2D charge sheet density (Q_{sh}) for mono-layered sample $\mathrm{S_B}$ is acquired by summing n from equation (7.27) and p from equation (7.28) with counterpart electronic charge (q), and is indicated as:

$$Q_{\mathrm{sh}} = qN_{\mathrm{G}}[\mathfrak{I}_1(-\eta_{\mathrm{FC}} - \eta_{\mathrm{g}}) - \mathfrak{I}_1(\eta_{\mathrm{FC}})] \tag{7.29}$$

From equation (7.29), it is clear that Q_{sh} is primarily dependent on η_{FC} and η_{g}, which can be disclosed by the change in E_{F}, E_{DC} and E_{g}, respectively. The E_{g} is established non-zero E_{g} due to N-substitution dopant and it can be customized by the amount of dopant concentration, and it is constant about any external source of bias. On the other hand, the change in $(E_{\mathrm{F}} - E_{\mathrm{DC}})/q$ can be modeled as local channel potential (V_{ch}) of sample $\mathrm{S_B}$, which is manifested as $V_{\mathrm{ch}} = (E_{\mathrm{F}} - E_{\mathrm{DC}})/q$. In addition to that, the dimensionless parameters η_{FC} and V_{ch} are related as: $V_{\mathrm{ch}} = (k_{\mathrm{B}}T/q)\eta_{\mathrm{FC}}$. Consequently, the quantum capacitance (C_{q}) for 2D single-layer sample $\mathrm{S_B}$ can be modeled as:

$$C_{\mathrm{q}} = -\frac{q}{k_{\mathrm{B}}T}\frac{\partial Q_{\mathrm{sh}}}{\partial \eta} \tag{7.30}$$

Using (7.29) in (7.30), it can be obtained as:

$$C_{\mathrm{q}} = \frac{q^2 N_{\mathrm{G}}}{k_{\mathrm{B}}T}\ln\left[(1 + e^{-\eta_{\mathrm{FC}}-\eta_{\mathrm{g}}})(1 + e^{\eta_{\mathrm{FC}}})\right] \tag{7.31}$$

Equation (7.31) manifests the all region complete C_{q} model for sample $\mathrm{S_B}$. When E_{g} is zero, the E_{DC} and η_{FC} of N-substitution doped graphene become the same as E_{D} and η_{F} in undoped pristine graphene, respectively. Therefore, the C_{q} model presented in equation (7.31) is compatible for both sample $\mathrm{S_B}$ and undoped graphene.

7.3 Electronic transport properties

In the design of nanoscale device models for field effect applications, it is crucial to investigate the fundamental electronic transport features, such as interaction parameter and mobility. The interaction parameter for 2D systems has been studied in [24, 25], where it is investigated for both parabolic and linear dispersion. The literature [26, 27] proposes static and dynamic screening functions for 2D mono-layered graphene with regard to charged impurities. In this section, modeling and comparative analyses of the interaction parameter and mobility for single-layer undoped pristine graphene and B- or N-substitution doped graphene have been conducted.

7.3.1 Interaction parameter

The ratio of average inter-electron Coulomb interaction energy to Fermi energy is expressed as the interaction parameter (r_s) [24] also termed as Wigner–Seitz radius. The interaction parameter for 2D single-layer graphene is analytically represented as follows:

$$r_s = \frac{\dfrac{q^2\sqrt{\pi N_T}}{4\pi\epsilon_0\epsilon_r}}{E_F} \tag{7.32}$$

where N_T denotes the total 2D carrier sheet density (involving both holes and electrons, i.e., $N_T = |p - n|$) in the system, ϵ_0 represents the permittivity of vacuum, ϵ_r denotes the dielectric constant. For SiO_2 substrate-supported single-layer graphene, ϵ_r is defined as $\epsilon_r = (\epsilon_{SiO_2} + 1)/2 \approx 2.5$ [24]. For single-layer undoped pristine graphene, the Fermi energy can be identified with the aid of carrier statistics and is represented as: $E_F = \hbar v_f\sqrt{\pi N_T}$ [24]. Utilizing the above-mentioned E_F in equation (7.32), the r_s of undoped graphene has been constant and varies only with respect to dielectric constant of different supporting substrate material. However, for the case of B- or N-substitution doped graphene, E_F is altered by shifted Dirac point energy and induced E_g due to B- or N-dopant. From the 2D carrier statistics, the position of E_F in B- and N-substitution doped graphene are related as in equations (7.33) and (7.34), respectively as follows:

$$E_F = \begin{cases} E_{DV} + E_g + \hbar v_f\sqrt{\pi N_T} & E_F > E_{DV} + E_g, \\ E_{DV} + \dfrac{E_g}{2} + \dfrac{kT}{2}\ln\left(\dfrac{n}{p}\right) & E_{DV} \leqslant E_F \leqslant E_{DV} + E_g, \\ E_{DV} - \hbar v_f\sqrt{\pi N_T} & E_F < E_{DV}. \end{cases} \tag{7.33}$$

$$E_F = \begin{cases} E_{DC} + \hbar v_f\sqrt{\pi N_T} & E_F > E_{DC}, \\ E_{DC} - \dfrac{E_g}{2} + \dfrac{kT}{2}\ln\left(\dfrac{n}{p}\right) & E_{DC} - E_g \leqslant E_F \leqslant E_{DC}, \\ E_{DC} - E_g - \hbar v_f\sqrt{\pi N_T} & E_F < E_{DC} - E_g. \end{cases} \tag{7.34}$$

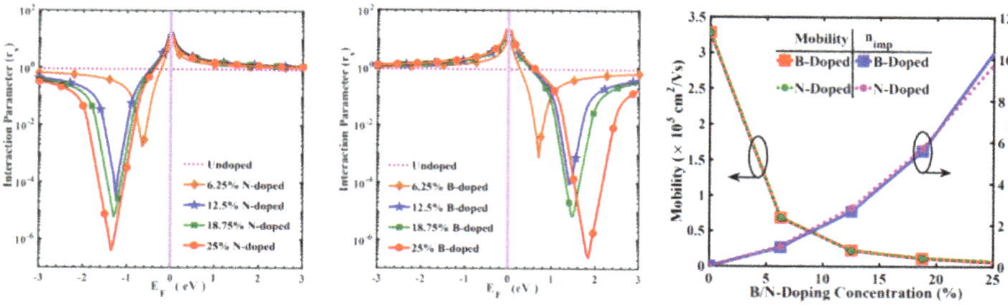

(a) r_s for N-doped graphene (b) r_s for B-doped graphene (c) Mobility and impurity

Figure 7.2. (a) The interaction parameter (r_S) for N and B-substitution doped graphene in respect of different arbitrary values of E_F is plotted in (a) and (b), respectively, and various doping percentages from 0% (undoped) to 25% with 6.25% of step size are considered. Reproduced with permission [28], Copyright 2021, *IEEE Proc.* (c) With regard to various doping concentrations, the mobility and impurity concentration of graphene doped with B- and N-substitutions have been illustrated. Reproduced with permission [16]. Copyright 2021, *IEEE Trans. Electron Devices.*

From equations (7.34) and (7.33), it can be observed that the immense effect of B- or N-substitution doping such as the shift in Dirac point energy (E_{DV} or E_{DC}), induced non-zero E_g, and 2D carrier sheet density (both p and n) are fundamentally paramount in obtaining E_F, and hence, become essential in acquiring the r_S. In relation to various arbitrary values of E_F, the interaction parameter for N and B-substitution doped graphene is shown in figures 7.2(a) and (b), respectively, and compared with r_S of pristine graphene. From figures 7.2(a) and (b), it is evident that, the r_S is constant for pristine graphene, which is equal to 0.875 for graphene on SiO_2 supported substrate. However, the N and B-substitution doped graphene have been disclosing more interactive attributes away from the Dirac points and less interactive near the Dirac points and inside the forbidden gap.

7.3.2 Mobility

The semi-classical drift-diffusive transport mobility (μ) for 2D single-layer graphene has been explored analytically in [24, 29, 30], and it is given as:

$$\mu = \frac{q}{h}\frac{2}{G[2r_S]}\frac{1}{n_i} \tag{7.35}$$

where h represents the Planck's constant, n_i denotes the concentration of impurities in single-layer graphene, and $G[x]$ represents the function of x, which is utilized from the literature [29]. From equation (7.35), it can be noticed that the semi-classical drift-diffusive transport mobility (μ) primarily depends on the interaction parameters r_S and n_i. The interaction parameter is already explored for both pristine and doped graphene in the previous section.

On the other hand, the n_i is composed of two parts: (1) impurities associated while preparing the single-layer graphene samples [29]—n^*; (2) the charged impurities associated due to B- or N-substitution dopant [16]—n_{imp}. Therefore, the effective concentration of impurities (n_i) is represented as in equation (7.36).

$$n_{\mathrm{i}} = \frac{n^* + n_{\mathrm{imp}}}{2} \tag{7.36}$$

where n^* is constant, which ranges from 2.5×10^{12} cm^{-2} to 4.0×10^{12} cm^{-2} for the dirty samples of single-layer graphene [29]. And n_{imp} represents the impurity concentration associated with substitution doping in graphene. The impurity concentration associated with doping in graphene (n_{imp}) can be expressed analytically as:

$$n_{\mathrm{imp}} = \frac{2(kT)^2}{\pi(\hbar v_f)^2}\Im_1(\eta) \tag{7.37}$$

Hence, the elemental parameters of interest to model the semi-classical drift-diffusive transport mobility such as r_{s} and n_{i} are explicitly demonstrated from equation (7.32) to equation (7.37) for both B- or N-substitution doped and undoped pristine graphene. Also, it is evident that the r_{s} and n_{i} are predominantly modulated by shifted Dirac points and induced non-zero E_{g} due to substitution doping in single-layer graphene.

The mobility (μ) and impurity concentration associated with n_{imp} are illustrated in figure 7.2(c) against different B- or N-doping concentration from 0% (pristine undoped graphene) to 25% with increment of 6.25%. The respective shift in Dirac points and induced E_{g} have been exploited from table 7.1 [22] to plot figure 7.2. From figure 7.2(c), it is observed that μ is reducing with increase in B- or N-doping concentration. This is due to the increase in induced E_{g} in respect of B- or N-doping concentration, which in turn expands the n_{imp}. Since both N and B replacement doping consider the same amount of doping concentration, the induced E_{g} in both forms is roughly equivalent. Subsequently, n_{imp} of B- or N-substitution doped graphene are less distinguishable and hence, μ coincides for B- or N-substitution doped graphene.

7.4 Modeling of monolayer GFETs

To examine the large signal features of GFETs, a variety of I_{D} models have been researched in the literature [6, 10, 11]. These models, however, are only applicable to GFETs with zero E_{g}. With respect to diverse substrates, mobility is derived as a constant and imported as an external parameter [6, 10, 11]. However, mobility is solely dependent on the impurity concentration and interaction parameter, which vary depending on the substrate material. In this section, an extensive drift-diffusion I_{D} for GFET has been modeled, which approaches the Newton–Raphson method self-consistent solution by capturing all the physical phenomenon such as total 2D carrier sheet charge density, quantum capacitance, and mobility in respect of bandgap engineering.

7.4.1 The extensive drain current model for GFETs

To incorporate the drift current and diffusion current, the quasi-Fermi potential based all region drain current (I_{D}) model is investigated in this section, which is analytically presented as:

$$I_{\mathrm{D}} = -qW\rho_{\mathrm{sh}}v_{\mathrm{d}} \tag{7.38}$$

where W represents the channel width, ρ_{sh} is the total 2D carrier sheet density and is related as: $\rho_{\mathrm{sh}} = \frac{|Q_{\mathrm{sh}}|}{q}$, and v_{d} denotes the carrier drift velocity [6]. Either 2D single-layer undoped pristine graphene with zero E_{g} or B/N-substitution doped graphene with induced non-zero E_{g} and shifted Dirac point are acting as the channel, which is extensively modeled as a complete bottom-gated GFET device. Also, the schematic picture of bottom-gated GFET is depicted in figure 7.3(a). The carrier drift velocity for the bottom-gated GFET is modeled mathematically in equation (7.39).

$$v_{\mathrm{d}} = \frac{\mu E}{\left[1 + \left(\frac{\mu|E|}{v_{\mathrm{sat}}}\right)^{\gamma}\right]^{\frac{1}{\gamma}}} \tag{7.39}$$

where v_{sat} denotes saturation velocity, γ represents dimensionless parameter for fitting with measured data, and it is ranging from 0.7 to 1.2 [31], and E denotes electric field through the graphene channel, which is expressed as: $E = -\mathrm{d}V/\mathrm{d}x$. And, the external application of drain and source bias is modeled as the quasi-Fermi potential split in the graphene channel, which is denoted as $V(x)$. Utilizing the physical significance of gradual channel approximation technique, the linear variation of $V(x)$ is assumed along the graphene channel, which can be modeled as $V(x = 0) = 0$ V at S terminal and $V(x = L) = V_{\mathrm{DS}}$ at drain terminal. Utilizing v_{d} from equation (7.39), the I_{D} model for both undoped pristine and B- or N-substitution doped GFET is presented as:

$$I_{\mathrm{D}} = \frac{qW\int_0^{V_{\mathrm{DS}}} \mu\rho_{\mathrm{sh}}\, dV}{\left[L^{\gamma} + \left(\int_0^{V_{\mathrm{DS}}} \frac{\mu}{v_{sat}}\, dV\right)^{\gamma}\right]^{\frac{1}{\gamma}}} \tag{7.40}$$

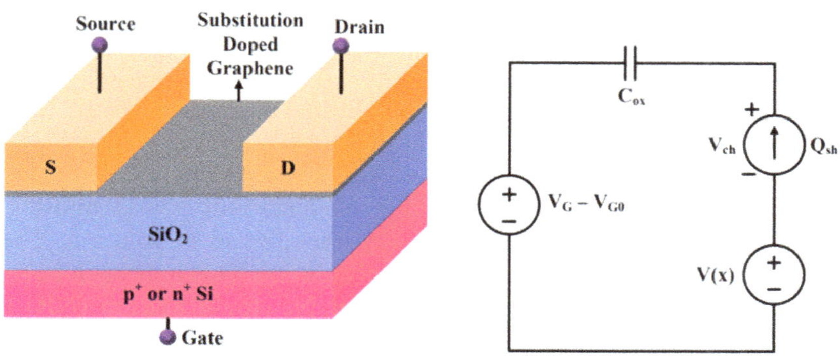

(a) Schematic of GFET (b) MIG equivalent model

Figure 7.3. (a) The schematic structure of bottom-gated graphene FET with SiO_2 as both supporting substrate and bottom gate insulator. (b) MIG electrical equivalent circuit model. Reproduced with permission [16]. Copyright 2021, *IEEE Trans. Electron Devices*.

where L denotes the channel length and v_{sat} has been computed by using the model presented in [31], which is noted as: $v_{sat} = 2\omega/\pi\sqrt{\pi\rho_{sh}}$, and $\hbar\omega$ represents phonon energy of the surface. For single-layer graphene on SiO_2 supported substrate, $\hbar\omega = 99$ meV [10]. Also, the I_D model in equation (7.40) is compatible for both B- or N-substitution doped (both n-type doped and p-type doped) and undoped pristine GFET, when carrier sheet density and mobility are modeled explicitly with respect to substitution doping. The 2D carrier sheet density and semi-classical mobility for both undoped and B/N-substitution doped graphene have been already modeled precisely in the previous sections.

7.4.2 Metal insulator–graphene (MIG) equivalent circuit

The metal insulator–graphene (MIG) electrical equivalent circuit model is illustrated in figure 7.3(b). In this, C_{ox} denotes the gate dielectric capacitance, V_G represents the applied voltage at gate terminal, V_{G0} accounts the shifted Dirac point as a result of the work function potential difference between the metal gate and single-layer graphene channel, which is same as the flat band potential in traditional semiconductor FETs. The external application of drain-to-source potential shifts the position of E_F inside the graphene channel, which can be analytically related as: $E_F = -qV(x)$, where $V(x) = 0$ at $x = 0$ (at S terminal) and $V(x) = V_{DS}$ at $x = L$ (at D terminal) are considered based on the widely accepted gradual channel approximation technique. As discussed in the previous sections, the single-layer graphene local channel potential V_{ch} is defined as: $V_{ch} = (E_F - E_D)/q$ for undoped pristine graphene; $V_{ch} = (E_F - E_{DV})/q$ for B-substitution doped graphene; $V_{ch} = (E_F - E_{DC})/q$ for N-substitution doped graphene. The MIG electrical equivalent circuit model is playing an immense role in obtaining the relation between V_{ch} and Q_{sh}, which is obtained by the application of Kirchoff's laws to the MIG model presented in figure 7.3(b), and is denoted as:

$$V_{ch} = V_G - V_{G0} - V(x) + (Q_{sh}/C_{ox}) \tag{7.41}$$

7.4.3 Self-consistent model

The total 2D carrier sheet charge density (Q_{sh}) and quantum capacitance C_q for both undoped pristine and substitution doped (either B- or N-doped) are poly-logarithmically varying in respect of V_{ch}, which have been explored in the previous sections. Consequently, equations (7.41) and (7.10) or equations (7.41) and (7.20) or equations (7.41) and (7.29) will not have any closed-form of solution. Therefore, these equations demand a numerical method of solution. One approach to achieve a numerical solution is establishing the first order Newton–Raphson (NR) method and solving self-consistently by assuming the initial values of V_{ch} or Q_{sh}. The flow-algorithmic procedure to implement the first order NR method self-consistent solution is briefly demonstrated in figure 7.4. For a given gate and drain bias situation, the root solution such as V_{ch} and Q_{sh} have been manifested from the self-consistent method. Ultimately, the requisite ρ_{sh} in equation (7.40) can be acquired as

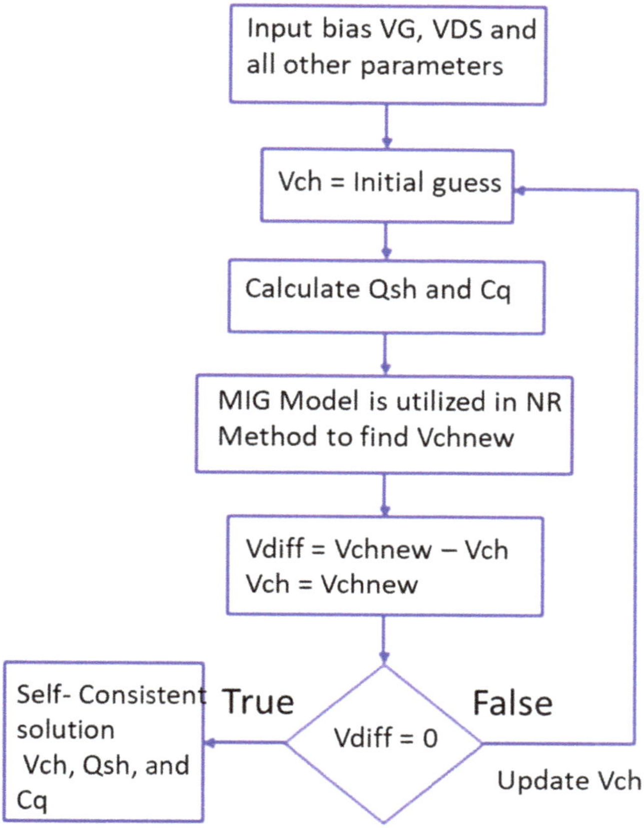

Figure 7.4. The algorithmic flow graph showing the procedure of self-consistent approach by employing first order Newton–Raphson (NR) numerical method. Reproduced with permission [16]. Copyright 2021, *IEEE Trans. Electron Devices*.

$\rho_{sh} = |Q_{sh}|/q$. Although, the variable of integration used in the I_D model of equation (7.40) (dV) can be reordered as: $dV = (dV/dV_{ch})dV_{ch}$. Applying differentiation on equation (7.41) in respect of V and rearranging, which has the outcome:

$$\frac{dV}{dV_{ch}} = -\left(1 + \frac{C_q}{C_{ox}}\right) \tag{7.42}$$

Utilizing equation (7.42) and changing the integration variable into V_{ch} and changing the integration limit in equation (7.40), the most generalized I_D model for single-layer GFET is presented as follows:

$$|I_D| = \frac{qW\int_{V_{chs}}^{V_{chd}} \mu\rho_{sh}\left(1 + \frac{C_q}{C_{ox}}\right)dV_{ch}}{\left[L^{\gamma} + \left(\int_{V_{chs}}^{V_{chd}} \frac{\mu}{v_{sat}}\left(1 + \frac{C_q}{C_{ox}}\right)dV_{ch}\right)^{\gamma}\right]^{\frac{1}{\gamma}}} \tag{7.43}$$

where V_{chs} represents the single-layer graphene local channel potential at S terminal, which can be achieved from equation (7.41) by substituting $V(x) = 0$ and V_{chd} is the single-layer graphene local channel potential at D terminal, which can be acquired from equation (7.41) by incorporating $V(x) = V_{DS}$. Further, for both undoped and substitution doped GFETs, μ, ρ_{sh}, and C_q are directly depend on V_{ch} and hence, the integration is computed in respect of V_{ch} and the appropriate limits are applied.

7.4.4 Model validation

Figure 7.5(a) validates the theoretical model of mobility and experimentally measured data [32] of mobility for both undoped and B-doped GFETs. The five different B-doping concentrations from [32] have been considered: 0 (undoped pristine graphene), 3.63%, 7.52%, 11.31%, and 13.85% to validate the theoretical mobility model. Additionally, the theoretical mobility model has been given the corresponding shift in Dirac points and induced non-zero E_g from [32] as inputs. From figure 7.5(a), it is evident that the analytical mobility model discloses significant matching with experimentally measured mobility for the case of both undoped and B-doped GFETs.

The complete self-consistent I_D model of GFET has been validated with experimentally measured data [32] for the case of bottom-gated 7.5% of boron-doped GFET, which is manifested in figure 7.5(b). From the experiment [32], for 7.5% of boron-doped GFET, the shift of 1.1 eV is observed in Dirac point and induced non-zero E_g of 0.32 eV have been extracted and supplied as a input for the I_D model. Besides, the parameters such as: $L = 1.7$ μm, $W = 1.2$ μm, $V_{G0} = -1.787$ V, $C_{top} = 11.51$ nF cm^{-2}, $\mu = 530$ cm^2 V^{-1} s^{-1} have also been considered for validation. The complete I_D model of B-doped GFET exhibits an excellent agreement with experimentally measured data for $I_D - V_G$ transfer characteristics with various drain–source potential such as: $V_{SD} = 0.1$ V, 0.3 V, and 0.5 V.

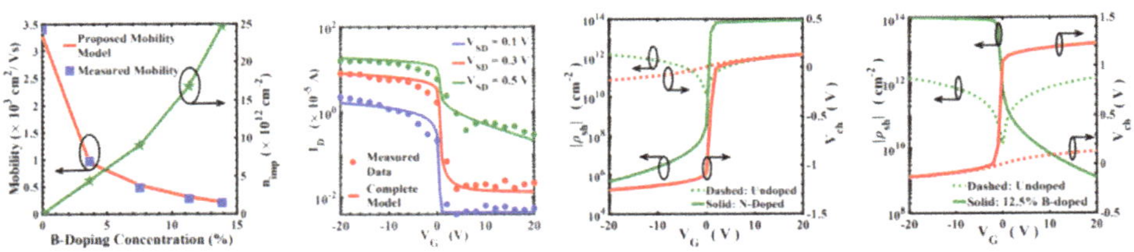

(a) Mobility validation (b) I_D-V_G validation (c) ρ_{sh}:12.5% N-doped (d) ρ_{sh}:12.5% B-doped

Figure 7.5. (a) Justification of the mobility model using data from bottom-gated GFETs that were both undoped and B-doped at various B-doping concentrations [31]. (b) Validation of complete model with measured data from 7.5% B-doped GFET [31]. Total 2D carrier sheet density (ρ_{sh}) and V_{ch} in respect of applied gate bias (V_G) for (c) 12.5% N-doped GFET, and (d) 12.5% B-doped GFET, which are demonstrating the amelioration of ambipolar behaviour in B/N-substitution doped GFETs in comparison with undoped GFET. Reproduced with permission [16]. Copyright 2021, *IEEE Trans. Electron Devices*.

7.4.5 Characteristics of GFETs

7.4.5.1 Carrier statistics of bottom-gated undoped, B- and N-substitution doped GFETs

Figure 7.5 depicts the plot of total 2D carrier sheet density (ρ_{sh}) and V_{ch} for a wide range of bias conditions using the method of self-consistent interpretation. The variations of total 2D carrier sheet density and V_{ch} with respect to gate voltage (V_G) for 12.5% N and 12.5% B-doped GFETs are depicted in figures 7.5(c) and (d), respectively, and investigated against undoped pristine GFET in a comparative manner. From figure 7.5, it is evident that the ambipolar behaviour is apparent in undoped pristine GFETs, which becomes the hurdle to turn off the transistor absolutely. Alternatively, n-type conductivity behaviour in N-substitution doped GFET and p-type conductivity attribute in B-substitution GFET have been manifested instead of ambipolar behaviour, which are transparently illustrated in figures 7.5(c) and (d), respectively. For GFETs with B- or N-doping, the transport carriers are totally drained out in the negative V_G bias region or in the positive V_G bias region, which greatly benefits the OFF current. The displaced Dirac point and generated non-zero E_g from B- or N-substitution doping are the culprits behind this.

7.4.5.2 Transfer characteristics of undoped, B- and N-substitution doped GFETs

Undoped pristine GFET, B- and N-substitution doped GFETs have been simulated with the aid of a complete I_D model by considering various doping percentages from 0% to 25% with 6.25% of step size. The $I_D - V_G$ transfer characteristics for N and B-substitution doped GFETs have been explored and relatively examined with undoped pristine GFET. From the figures 7.6(a) and (c), it can be noticed that, the shifting of Dirac points about negative/positive values of V_G have been observed for N/B-substitution doped GFET with respect to increase in dopant concentration. In addition to that, the OFF current of B- or N-substitution doped GFETs close to Dirac point has been reduced significantly as compared to undoped pristine GFET. Subsequently, B- or N-substitution doping in graphene enhances the OFF current at the cost of reduced ON current due to the diminished mobility by B- or N-dopant. Nevertheless, the central quantity of interest, i.e., ON/OFF ratios of B- or N-substitution doped GFET have been enhanced enormously as compared to undoped pristine GFET. The ON/OFF ratio for undoped GFET is ≈ 1, while for B-substitution doped GFET, it has shown a value $\approx 8 \times 10^4$ for 25% B-doped. Likewise, for N-substitution doped GFET the ON/OFF ratio has been reported $\approx 7.7 \times 10^4$ for 25% N-doped. Also, for B- or N-substitution doped GFET, it is evident that the ambipolar characteristics in undoped pristine graphene have been suppressed notably as compared with undoped GFET, which makes it clear that the former is highly suitable for digital applications.

7.4.5.3 Output characteristics of bottom-gated undoped, B- and N-substitution doped GFETs

Figures 7.6(b) and (d) investigate the output behaviour of N and B-substitution doped GFET, respectively, and relatively compared with undoped pristine GFET.

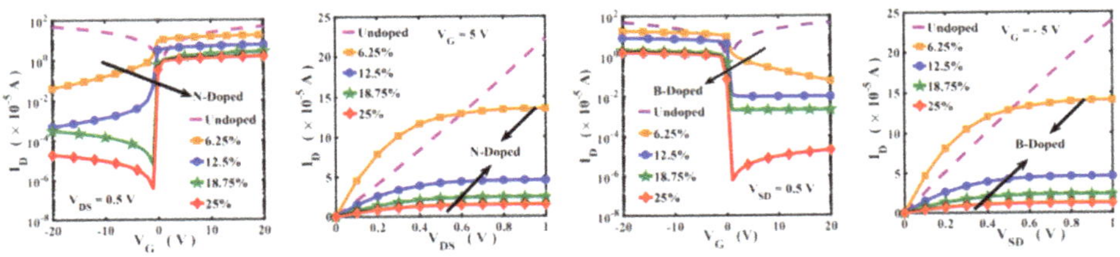

(a) I_D-V_G:N-doped (b) I_D-V_D:N-doped (c) I_D-V_G:B-doped (d) I_D-V_D:B-doped

Figure 7.6. DC characteristics of N- and B-doped GFET with L = 1.7 μm, W = 1.2 μm, V_{G0} = −1.787 V, C_{top} = 11.51 nF cm^{-2}, and γ = 1. The percentage of doping is changed from 0% to 25% with a step size of 6.25%. (a), (c) $I_D - V_G$ at fixed $|V_{DS}|$ = 0.5 V. (b), (d) $I_D - V_{DS}$ at fixed $|V_G|$ = 5 V. Reproduced with permission [16]. Copyright 2021, *IEEE Trans. Electron Devices.*

The saturation behaviours of B- or N-substitution doped GFETs have been enriched, whereas the undoped pristine GFET discloses linear $I_D - |V_{DS}|$ attributes, which is demonstrated in figures 7.6(b) and (d). Since, the carrier transport mobility is decreased from 3282 cm^2 V^{-1} s^{-1} to 60 cm^2 V^{-1} s^{-1} for undoped pristine (0%) GFET to 25% B-substitution doped GFET, the I_D is also reduced. Nevertheless, in B/N-substitution doped GFET, the linear I–V behaviour has been absolutely suppressed when $|V_{DS}| \geqslant$ 0.5 V and saturation has been established when compared to undoped pristine GFET, which is extreamly useful in analog/RF applications.

7.5 Static linearity and nonlinearity analysis of GFETs

The vital RF parameters such as transconductance, output resistance, gain, and f_T of GFETs have been examined, which are important while considering analog/RF applications. At the same time, analyzing the linearity and nonlinearity performance is essential for all RF systems. The device's linearity behaviour reveals that intermodulation distortions and higher order harmonics are minor, indicating improved reliability. It also provides an estimate of the upper limit of the applied input signal that can be handled by an RF system without distortion. However, the existence of nonlinearities in an RF system will have detrimental effects such as gain compression, cross modulation, and harmonic and intermodulation distortion, all of which affect the performance and dependability of the device [33]. Therefore, rather than at the system level, it is essential to study the device's nonlinearity behaviour at the transistor level.

7.5.1 Modeling of V_N for doped GFETs

Figure 7.7(a) depicts the equivalent circuit model B-doped GFETs obtained using a Verilog-A code from the Cadence Spectre circuit simulator [10]. The corresponding symbol is as shown in figure 7.7(b). Source contact resistance R_S and drain contact resistance R_D are both taken to be 172 Ω [10]. Furthermore, C_{GS} and C_{GD} are the gate-source and the gate-drain capacitance, respectively. In the proposed model, the error-free region is considered when $V_{eff} > V_{DSi}/2$. Hence, C_{GS} and C_{GD} can be modeled as $C_{OX} \times WL$ and $C_{OX} \times WL/2$, respectively [11].

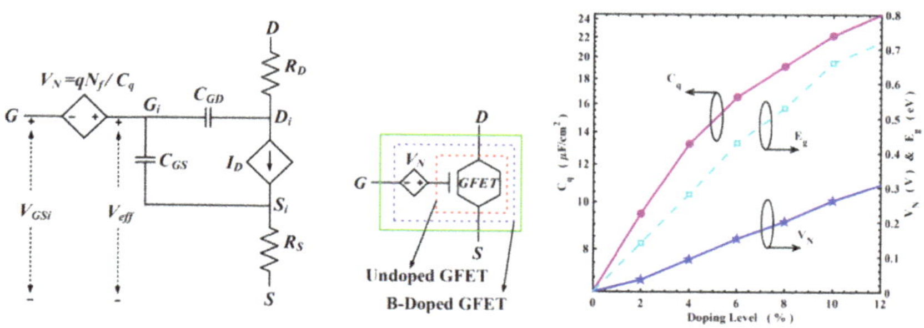

(a) GFET-equivalent model (b) GFET symbol (c) V_N, E_g, and C_q model

Figure 7.7. (a) A compact electrical equivalent circuit for undoped and boron-doped GFET. (b) Representation of equivalent circuit model as obtained from the Cadence Spectre circuit simulator using Verilog-A code. (c) The variations of induced potential (V_{N}), E_{g} and quantum capacitance (C_{q}) with B-doping concentration. Reproduced from [34], copyright 2021, with permission from Elsevier.

The induced potential (V_{N}) due to substitution doping will be investigated with expression and figure. This potential, which is a result of the Dirac point shift and E_{g}triggered by B-doping in single-layer graphene could be modeled based on [23, 35, 36]:

$$V_{\mathrm{N}} = \frac{qN_{\mathrm{f}}}{C_{\mathrm{q}}} \tag{7.44}$$

where N_{f} is concentration of donor or acceptor impurity atoms and C_{q} is quantum capacitance. Boron-doping introduces a E_{g} and considerably moves the Dirac point in single-layer graphene. The different doping levels and their associated induced E_{g} and shifting of the Dirac point are taken from [20]. Both N_{f} and C_{q} can be modeled as in our earlier work based on distinct E_{g} and Dirac points [23]. Figure 7.7(c) shows the change in induced V_{N}, E_{g}, and C_{q} for B-doping with concentration 0, 2, 4, 6, 8,10 and 12 percentages. Boron-doped GFET with doping concentrations 0, 2, 4, 6, 8, 10 and 12% are labelled as sample S1, S2, S3, S4, S5, S6 and S7, respectively, throughout the chapter. As the Dirac point shift and E_{g} for undoped graphene are zero, their N_{f} and V_{N} also remain zero. In contrast, E_{g} for B-doped graphene increases with increase in doping percentage, which in turn increases N_{f} and V_{N} as evident from figure 7.7(c). The values of these parameters for different doping concentrations are listed in table 7.2.

7.5.2 Verilog-A implementation of GFETs

B-doped single-layer graphene is used as the channel, source, and drain with doping concentrations ranging from 2% to 12% with step size of 2%. The doped and undoped GFET can both be studied with the analytical model of the simplified I_{D} from [11], which is expressed as follows:

$$I_{\mathrm{D}} \approx \frac{\mu W C_{\mathrm{OX}}\left(V_{\mathrm{eff}} - \dfrac{V_{\mathrm{DSi}}}{2}\right)}{\dfrac{L}{V_{\mathrm{DSi}}} + \dfrac{\mu}{\omega}\sqrt{\dfrac{\pi C_{\mathrm{OX}}}{q}}\sqrt{V_{\mathrm{eff}} - \dfrac{V_{\mathrm{DSi}}}{2}}} \tag{7.45}$$

Table 7.2. Modelled memoryless linearity and nonlinearity figures of merit (FOMs), E_g, and V_N in B-doped and undoped GFET with different doping concentrations. Reproduced from [34], copyright 2021, with permission from Elsevier.

B-doping Percentage	0 SI	2 S2	4 S3	6 S4	8 SS	10 S6	12 S7
E_g (eV)	0	0.14	0.28	0.43	0.53	0.66	0.72
V_N (V)	0	0.0346	0.0931	0.1523	0.2015	0.2613	0.3082
HD_2 (dB)	−55.66	−55.87	−56.2	−56.53	−56.79	−57.1	−57.33
HD_3 (dB)	−105.03	−105.45	−106.13	−106.79	−107.32	−107.95	−108.42
IM_2 (dB)	−49.64	−49.85	−50.18	−50.51	−50.77	−51.08	−51.31
IM_3 (dB)	−95.49	−95.91	−96.59	−97.25	−97.78	−98.41	−98.88
A_{IIP2} (dBV)	15.66	15.87	16.2	16.53	16.79	17.1	17.33
A_{IIP3} (dBV)	13.77	13.97	14.32	14.65	14.91	15.22	15.46
$A_{in,\ 1\ dB}$ (dBV)	4.13	4.34	4.68	5.01	5.28	5.59	5.82

where L and W are length and width of the channel, respectively. C_{OX} represents the gate-oxide capacitance per unit area and q, the electronic charge (1.6×10^{-19} C). $\hbar\omega$ and V_{DSi} denotes the surface phonon energy and the internal potential between drain (D) and source (S) terminal, respevitvely. V_{eff} is labelled as the effective potential between gate (G) and S terminal. It is given by $V_{eff} = V_{GSi} + V_N$, where V_{GSi} is the sum of internal potential between G and S terminal V_N is the potential induced by doping. Also, V_N for B-doped GFET has been already modeled in the previous section. Doping single-layer graphene with B or N by swapping out carbon atoms with the same sublattice positions is referred to as doping [22, 37]. Furthermore, the effective mass (m^*) of the charge carriers is also changed by B-doping in graphene, and it rises as the dopant concentration rises. The primary factor influencing carriers' movement is their effective mass. It has been observed that the increase in m^* is negligible [22] in relation to the concentration of B-dopant, in contrast to the shift in Dirac point and induced E_g. Therefore, for B-doped GFET, the constant undoped GFET mobility ($\mu = 7000$ cm^2 V^{-1} s^{-1} [11]) is taken into account.

7.5.3 Static nonlinearity transconductance model

From the compact equivalent model (figure 7.7(a)), the I_D, R_S, and R_D are the memoryless components that serve as the primary generator of memoryless (static) nonlinearities. The causes of dynamic nonlinearities are memory elements like C_{GS} and C_{GD}. Because C_{GS} and C_{GD} have relatively low values, their effects on nonlinearity are only readily visible at high frequencies and almost nonexistent at low frequencies. According to the literature [37] and [38], the impacts of R_S and R_D induced nonlinearity are negligible. However, because of its direct reliance on input and memoryless behaviour, I_D should be extensively modeled. For B-doped GFET, the static nonlinear attributes of I_D are modeled analytically using [33] as:

$$I_D \approx g_{m1}x(t) + g_{m2}x^2(t) + g_{m3}x^3(t) + \cdots + g_{mn}x^n(t) \tag{7.46}$$

where $x(t) = A\cos(2\pi ft)$ is the input signal applied to the G–S terminal of the FET with amplitude A and frequency f. It is taken into account in memoryless form so as to extract out the static nonlinearity in B-doped GFET. Hence it could be understood that current I_D depends only on the amplitude of $x(t)$ and is unaffected by the frequency band f of $x(t)$. Therefore, I_D in equation (7.46) can be modified as:

$$I_D \approx g_{m1}A + g_{m2}A^2 + g_{m3}A^3 + \cdots + g_{mn}A^n \tag{7.47}$$

where $g_{m1}, g_{m2},\ldots, g_{mn}$ are the transconductance coefficients. The higher order coefficients, such as g_{m4}, g_{m5}, and others, are essentially nonexistent for small amplitude operation and only become visible at very large amplitudes, where the GFET device's amplifying behaviour is lost. Therefore, only the g_{m1}, g_{m2}, and g_{m3}, which are determined using equation (7.47), are the main quantities of importance for extracting memoryless nonlinearity. These are obtained directly from the relation

$$g_{mn} = \frac{1}{n!}\frac{\mathrm{d}^n I_D}{\mathrm{d}A^n(t)} \quad \text{at} \quad A = 0 \tag{7.48}$$

where $n = 1, 2, 3$. It is clear from equation (7.48) that the transconductance coefficients have a direct dependence on I_D, making them the main cause of memoryless nonlinearity.

The harmonic distortion terms, HD_2 (second) and HD_3 (third order), intermodulation distortion terms, IM_2 (second) and IM_3 (third) and intercept points, A_{IIP2} (second) and A_{IIP3} (third) as well as the gain compression point ($A_{in,\ 1dB}$), are memoryless nonlinearity FOMs that are analytically modeled based on [33]. These terms primarily depend on the transconductance (g_{m1}, g_{m2}, g_{m3}) and amplitude (A) of applied small signal input. The transconductance coefficients were obtained with respect to various B-doping concentrations ranging from 0% to 12% in steps of 2 and are used to investigate the static linearity and nonlinearity created by GFET.

7.5.4 Harmonic and intermodulation distortions

7.5.4.1 Analysis with doping level: HD *and* IM

The effects of memoryless linearity and nonlinearity are explored by studying the various FOMs with regard to various B-doping concentrations. The device specifications are $W = 1$ μm, $t_{OX} = 8.5$ nm, $L = 440$ nm, and h-BN as gate dielectric with 3.5 relative permittivity in order to explore the linearity and nonlinearity FOMs. Figures 7.8(a) and (b), respectively, show the HD_2 and HD_3 (harmonic distortion terms) with regard to different B-doping levels. The biased values for both the undoped and B-doped GFET are observed to be $V_{GSi} = 1.5$ V, $V_{DSi} = 0.4$ V and $A = 20$ mV. In sample S1 can be observed HD_2 and HD_3 values of −55.66 dB and −105.03 dB, respectively, while for sample S7 the values are seen as −57.33 dB and −108.42 dB, respectively. In comparison to S1, samples S2–S7 show considerable reductions in HD_2 which are −0.20, −0.54, −0.87, −1.13, −1.44, and

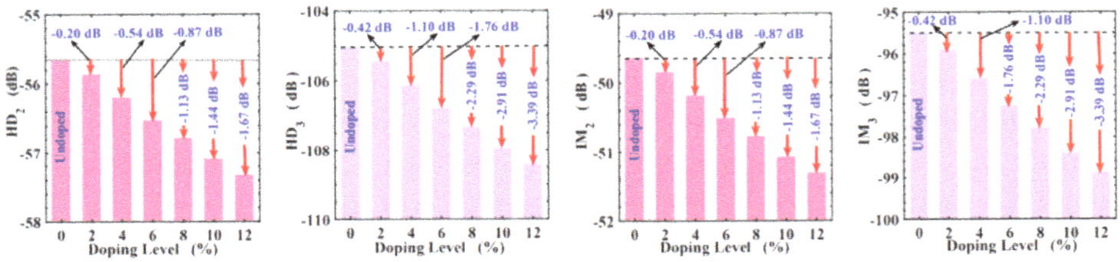

(a) HD_2: B-doping (b) HD_3: B-doping (c) IM_2: B-doping (d) IM_3: B-doping

Figure 7.8. *HD* and *IM* terms for samples S1–S7. (a), (b) H D_2, H D_3. (c), (d) IM_2, IM_3. Reproduced from [34], copyright 2021, with permission from Elsevier.

−1.67 dB, respectively, and also in HD_3 are −0.42, −1.10, −1.76, −2.29, −2.91, and −3.39 dB, respectively. These results are presented in figures 7.8(a) and (b).

The terms IM_2 and IM_3 are depicted with respect to various B-doping levels in figures 7.8(c) and (d), respectively. The observed IM_2 and IM_3 for an undoped GFET are −49.64 and −95.49 dB, respectively. The observed HD_2 and HD_3 for sample S7 are −51.31 dB and −98.88 dB, respectively. The IM_2 values for samples S2–S7 have significantly diminished in comparison to the undoped GFET by 0.20, 0.54, 0.87, 1.13, 1.44, and 1.67 dB, respectively. Similar to this, the lowered IM_3 values for sample S1 are −0.42, −1.10, −1.76, −2.29, −2.91, and −3.39 dB, respectively. These values are clearly displayed in figures 7.8(c) and (d) and are also included in table 7.2 in a comparative manner.

7.5.4.2 Investigation with input amplitude: HD *and* IM

By investigating key static nonlinearity FOMs like HD_2, HD_3, IM_2, and IM_3, figures 7.9(a) and (b) demonstrate the examination of memoryless nonlinearity effects with respect to applied small signal input amplitudes (A). The device parameters are: $W = 1$ μm, $t_{ox} = 8.5$ nm, $L = 440$ nm, and h-BN as gate dielectric with 3.5 relative permittivity. The 12% boron-doped GFET is driven at $V_{GSi} = 1.5$ V, $V_{DSi} = 0.4$ V. Figures 7.9(c) and (d) depict the variations of HD_2, IM_2, HD_3 and IM_3 for B-doping concentration of 12% in regard to various input small signal amplitudes (A). The input signal is adjusted between 5 and 50 mV. From the graphs it can be observerved that HD_2 and IM_2 are curbed by −6 dB and HD_3 and IM_3 are suppressed by −12 dB when A is cut in half from 20 mV to 10 mV. Correspondingly, HD_2 and IM_2 are boosted by 6 dB HD_3 and IM_3 are augmented by 12 dB when A is doubled from 20 mV to 40 mV.

7.5.4.3 Investigation with device parameter: HD *and* IM

For a 12% B-doped GFET, the consequences of memoryless linearity and nonlinearity are investigated by analysing the various FOMs with regard to the device parameter C_{OX}. Due to C_{OX}'s direct dependence on FOMs and intrinsic voltage gain (A_v), linearity and nonlinearity FOMs can be evaluated in relation to C_{OX} regardless of gate-oxide thickness (t_{ox}) and different types of gate insulator. $A = 20$ mV, $W = 1$ μm, $L = 440$ nm, $V_{GSi} = 1.5$ V and $V_{DSi} = 0.4$ V, are the device parameters

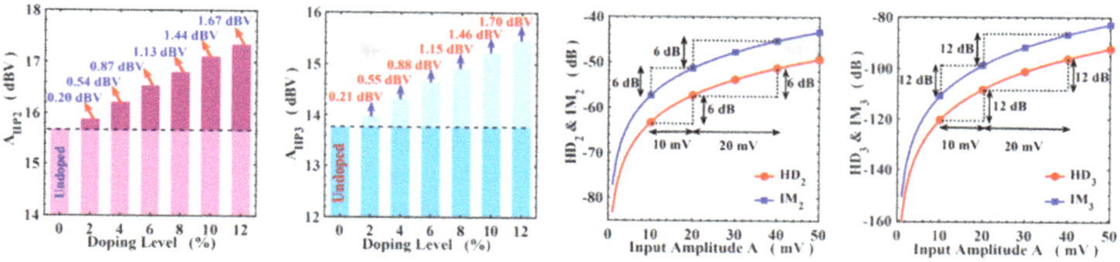

(a) A_{IIP2}: B-doping (b) A_{IIP3}: B-doping (c) HD_2 and IM_2 (d) HD_3 and IM_3

Figure 7.9. (a) and (b) A_{IIP2}, A_{IIP3} for undoped and B-doped GFET (c) HD_2 and IM_2. (d) HD_3 and IM_3. Reproduced from [34], copyright 2021, with permission from Elsevier.

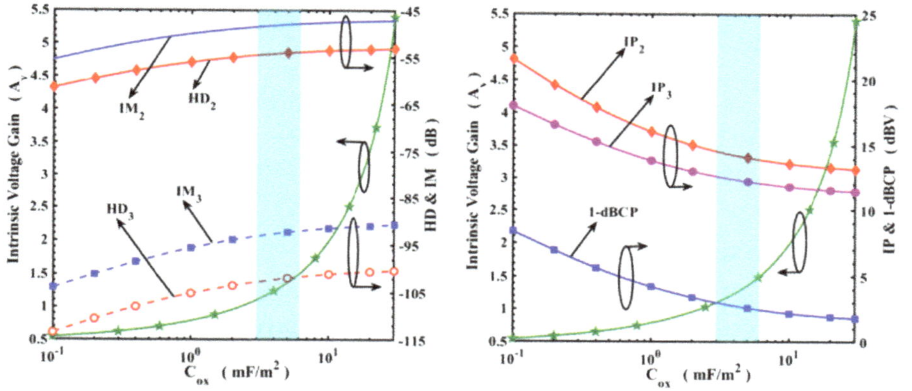

Figure 7.10. The *HD* and *IM* terms, the IP and 1-dBCP points, and A_v for sample S7 with respect to C_{OX} are examined. C_{OX} for sample S7 is traded between A_v and static nonlinearities as a result, the shaded area represents the GFET's reliable zone of operation. (a) *HD* and *IM*. (b) IP and 1-dBCP. Reproduced from [34], copyright 2021, with permission from Elsevier.

that are taken into account. Figure 7.10(a) examines the variations of A_v and distortion terms *HD* and *IM* with regard to varied C_{OX} for sample S7. A_v rises with a rise in C_{OX} because of the field effect principle. In contrast, as g_{m2} and g_{m3} grow in relation to an increase in C_{OX}, due to which the terms HD_2, HD_3, IM_2, and IM_3 also begin to increase, resulting in greater nonlinear behaviour. Because of this, the reliable region of B-doped GFET device operation can be attained by trading C_{OX} between A_v and static nonlinearity FOMs. And in order to achieve more trustworthy operation, the appropriate range of C_{OX} for S7 is 3–6 mF m^{-2}, which is shaded and explicitly explored in figure 7.10(a).

7.5.5 Gain compression and input intercept points

7.5.5.1 Investigation with doping level: IP

Figures 7.9(a) and (b), respectively, show the variations of second and third order input intercept points (A_{IIP2} and A_{IIP3}) with regard to various B-doping levels from 2, 4, 6, 8,10 and 12%. In comparison to those for sample S7, which are 17.33 dBV and

15.46 dBV, respectively, the observed A_{IIP2} and A_{IIP3} for S1 are 15.66 dBV and 13.77 dBV. Figure 7.9 vividly highlights the large improvements in A_{IIP2} and A_{IIP3} samples S2, S3, S4, S6 and S7 are showing in comparison to S1. These enhancements are 0.20, 0.54, 0.87, 1.13, 1.44, and 1.70 dBV, respectively, for A_{IIP3} and 0.20, 0.54, 0.87, 1.13, 1.44, and 1.67 dBV, respectively, for A_{IIP2} as depicted in figures 7.9(a) and (b).

7.5.5.2 Investigation with device parameter: IP and 1-dBCP

The input intercept and gain compression points (IP or A_{IIP} and 1-dBCP or $A_{in,\ 1dB}$) and intrinsic voltage gain (A_v) of sample S7 with respect to various C_{OX} is investigated in figure 7.10(b). Due to the field effect principle, A_v is increasing with increase in C_{OX}. On the other hand, the linearity behaviour weakens as a consequence of the rise in g_{m2} and g_{m3} with regard to the increase in C_{OX}, which causes the input intercept and gain compression points (IP_2, IP_3, and 1-dBCP) to begin lowering. In order to attain the trustworthy zone of B-doped GFET device operation, C_{OX}, is exchanged between A_v and static linearity FOMs. In order to operate in a more dependable zone, the appropriate C_{OX} range S7 is 3–6 mF m^{-2}, which is shaded and plainly demonstrated in figure 7.10(b).

7.5.6 Simulation setup

By creating a Verilog-A code for the compact equivalent model in figure 7.7(a), the symbols for undoped GFET and B-doped GFET have been constructed. The simulation test is implemented on the intended symbol using the Cadence Spectre simulator to study the various linearity and nonlinearity effects. To look at the device's static nonlinearity, numerous analyses have been run in the simulator. $V_{GSi} = 1.5$ V, $V_{DSi} = 0.4$ V, and $A = 20$ mV are the standard biassing parameters. Additional conditions are transparently provided for the specific investigation.

7.6 Conclusions and future prospects

In summary, the electrostatics and electronic transport properties of undoped, B- and N-substitution doped graphene have been explored extensively. Utilizing these effects, the phenomenological all region I_D model for GFETs has been established using the first order NR method of self-consistent solution for MIG equivalent model and corresponding 2D carrier sheet charge density and also its characteristics have been examined. Additionally, the memoryless linearity and nonlinearity performance of GFETs have been extracted using the Verilog-A implementation of GFETs in Cadence Spectre circuit simulator. The paramount linearity and nonlinearity FOMs have been examined with respect to different doping concentrations, different small signal amplitudes, and different gate-oxide capacitances and the improved reliability in B-substitution doped GFET is reported. The fruitful key attributes of B- or N-substitution doping in graphene can be outlined as follows:

1. Induced non-zero bandgap with shifted Dirac point energy and enhancement in quantum capacitance have been achieved.

2. The bipolar gating effect has been suppressed significantly as compared with undoped pristine GFET, which is disclosed from the transfer characteristics.
3. The OFF current of B- or N-substitution doped GFETs has been reduced significantly as compared to undoped pristine GFET.
4. The ON/OFF ratio for undoped GFET is ≈ 1, whereas for B- and N-substitution doped GFET, it has been extremely enhanced $\approx 8 \times 10^4$ and $\approx 7.7 \times 10^4$ for 25% B-doped and N-doped GFET, respectively.
5. The second- and third-order distortion FOMs for B-doped GFET have been suppressed significantly by a factor of 1.67 dB and 3.38 dB, respectively, with respect to undoped GFET, whcih reveals the ameliorated nonlinearity behaviour in B-doped GFET.
6. The second- and third-order linearity FOMs of B-doped GFET have been improved by a factor of 1.69 dBV, when compared to undoped GFET, which manifests enhanced linearity behaviour in B-doped GFET.

Therefore, B- or N-substitution doping in graphene prove the potential of B/N-doped GFETs and make them highly suitable in digital applications, also fulfilling the superior linearity requirement in analog/RF applications.

Further, in future perspective, the exploration of other vital transport properties such as mean free path and energy level broadening at the contacts will play a major role in designing the ballistic transport limit of graphene channel with respect to its length. For the case of ballistic transport, the quantum transport based non-equilibrium Green's function model will be fruitful for GFET with channel length less than the mean free path.

References

[1] Neto A C, Guinea F, Peres N M, Novoselov K S and Geim A K 2009 The electronic properties of graphene *Rev. Mod. Phys.* **81** 109

[2] Asad M, Jeppson K O, Vorobiev A, Bonmann M and Stake J 2021 Enhanced high-frequency performance of top-gated graphene FETs due to substrate-induced improvements in charge carrier saturation velocity *IEEE Trans. Electron Devices* **68** 899–902

[3] Feijoo P C, Pasadas F, Iglesias J M, Rengel R and Jiménez D 2019 Radio frequency performance projection and stability tradeoff of h-BN encapsulated graphene field-effect transistors *IEEE Trans. Electron Devices* **66** 1567–73

[4] Rawat B and Paily R 2015 Analysis of graphene tunnel field-effect transistors for analog/RF applications *IEEE Trans. Electron Devices* **62** 2663–9

[5] Fang T, Konar A, Xing H and Jena D 2007 Carrier statistics and quantum capacitance of graphene sheets and ribbons *Appl. Phys. Lett.* **91** 092109

[6] Thiele S A, Schaefer J A and Schwierz F 2010 Modeling of graphene metal-oxide-semiconductor field-effect transistors with gapless large-area graphene channels *J. Appl. Phys.* **107** 094505

[7] Xia J, Chen F, Li J and Tao N 2009 Measurement of the quantum capacitance of graphene *Nat. Nanotechnol.* **4** 505–9

[8] Xu H, Zhang Z and Peng L M 2011 Measurements and microscopic model of quantum capacitance in graphene *Appl. Phys. Lett.* **98** 133122

[9] Pourasl A H, Ariffin S H S, Ahmadi M T, Gharaei N, Rashid R A and Ismail R 2019 Quantum capacitance model for graphene FET-based gas sensor *IEEE Sens. J.* **19** 3726–32
[10] Fregonese S, Magallo M, Maneux C, Happy H and Zimmer T 2013 Scalable electrical compact modeling for graphene FET transistors *IEEE Trans. Nanotechnol.* **12** 539–46
[11] Rodriguez S, Vaziri S, Smith A, Fregonese S, Ostling M, Lemme M C and Rusu A 2014 A comprehensive graphene FET model for circuit design *IEEE Trans. Electron Devices* **61** 1199–206
[12] Novoselov K S, Geim A K, Morozov S V, Jiang D E, Zhang Y, Dubonos S V, Grigorieva I V and Firsov A A 2004 Electric field effect in atomically thin carbon films *Science* **306** 666–9
[13] Novoselov K S, Geim A K, Morozov S V, Jiang D, Katsnelson M I, Grigorieva I, Dubonos S and Firsov A 2005 Two-dimensional gas of massless Dirac fermions in graphene *Nature* **438** 197–200
[14] Wallace P R 1947 The band theory of graphite *Phys. Rev.* **71** 622
[15] Champlain J G 2011 A first principles theoretical examination of graphene-based field effect transistors *J. Appl. Phys.* **109** 084515
[16] Chandrasekar L and Pradhan K P 2021 Self-consistent modeling of B or N substitution doped bottom gated graphene FET with nonzero bandgap *IEEE Trans. Electron Devices* **68** 3658–64
[17] Kim P, Han M Y, Young A F, Meric I and Shepard K L 2009 Graphene nanoribbon devices and quantum heterojunction devices *2009 IEEE Int. Electron Devices Meeting (IEDM)* (Piscataway, NJ: IEEE) pp 1–4
[18] Schwierz F 2010 Graphene transistors *Nat. Nanotechnol.* **5** 487–96
[19] Ni Z H, Yu T, Lu Y H, Wang Y Y, Feng Y P and Shen Z X 2008 Uniaxial strain on graphene: Raman spectroscopy study and band-gap opening *ACS Nano* **2** 2301–5
[20] Zhang Y, Tang T T, Girit C, Hao Z, Martin M C, Zettl A, Crommie M F, Shen Y R and Wang F 2009 Direct observation of a widely tunable bandgap in bilayer graphene *Nature* **459** 820–3
[21] Rani P and Jindal V K 2013 Designing band gap of graphene by B and N dopant atoms *RSC Adv.* **3** 802–12
[22] Sharma P, Singh S, Gupta S and Kaur I 2017 Enhancing linearity in *I–V* characteristics by B/N doping in graphene for communication devices *J. Mater. Sci., Mater. Electron.* **28** 7668–76
[23] Chandrasekar L and Pradhan K P 2020 Carrier density and quantum capacitance model for doped graphene *2020 4th IEEE Electron Devices Technology and Manufacturing Conf. (EDTM)* (Piscataway, NJ: IEEE) pp 1–4
[24] Sarma S D, Adam S, Hwang E H and Rossi E 2011 Electronic transport in two-dimensional graphene *Rev. Mod. Phys.* **83** 407
[25] Sarma S D, Hwang E H and Li Q 2009 Valley-dependent many-body effects in two-dimensional semiconductors *Phys. Rev.* B **80** 121303
[26] Ando T 2006 Screening effect and impurity scattering in monolayer graphene *J. Phys. Soc. Jpn.* **75** 074716
[27] Hwang E H and Sarma S D 2007 Dielectric function, screening, and plasmons in two-dimensional graphene *Phys. Rev.* B **75** 205418
[28] Chandrasekar L and Pradhan K P 2021 Modeling and investigation of electronic transport properties of boron or nitrogen substitution doped single layer graphene *2021 IEEE 16th Nanotechnology Materials and Devices Conf. (NMDC)* (Piscataway, NJ: IEEE) pp 1–4

[29] Adam S, Hwang E H, Galitski V M and Sarma S D 2007 A self-consistent theory for graphene transport *Proc. Natl Acad. Sci. USA* **104** 18392–7

[30] Chen J H, Jang C, Adam S, Fuhrer M S, Williams E D and Ishigami M 2008 Charged-impurity scattering in graphene *Nat. Phys.* **4** 377–81

[31] Yamoah M A, Yang W, Pop E and Goldhaber-Gordon D 2017 High-velocity saturation in graphene encapsulated by hexagonal boron nitride *ACS Nano* **11** 9914–9

[32] Tang Y B *et al* 2012 Tunable band gaps and p-type transport properties of boron-doped graphenes by controllable ion doping using reactive microwave plasma *ACS Nano* **6** 1970–8

[33] Razavi B 2012 *RF Microelectronics* 2nd edn (Upper Saddle River, NJ: Prentice-Hall)

[34] Chandrasekar L and Pradhan K P 2021 Memoryless linearity in undoped and B-doped graphene FETs: a relative investigation to report improved reliability *Microelectron. Reliab.* **125** 114363

[35] Chandrasekar L, Pradhan K P and Kumar P 2019 Comparative study on nonlinearity of doped and undoped GFET using DC characteristics *2019 IEEE 16th India Council Int. Conf. (INDICON)* (Piscataway, NJ: IEEE) pp 1–4

[36] Lakshumanan C and Pradhan K P 2021 Memoryless non-linearity in B-substitution doped and undoped graphene FETs: a comparative investigation *IET Circuits Devices Syst.* **15** 641–8

[37] Rodriguez S, Smith A, Vaziri S, Ostling M, Lemme M C and Rusu A 2014 Static nonlinearity in graphene field effect transistors *IEEE Trans. Electron Devices* **61** 3001–3

[38] Alam A U, Holland K D, Wong M, Ahmed S, Kienle D and Vaidyanathan M 2015 RF linearity performance potential of short-channel graphene field-effect transistors *IEEE Trans. Microwave Theory Tech.* **63** 3874–87

Chapter 8

Applications of graphene in electronics: graphene for energy storage applications

Dohun Kim, Jihyun Park, Subramani Surendran, Chang-Soo Park, Joon Young Kim, Jung Kyu Kim and Uk Sim

This chapter reviews the use of graphene for electrochemical energy storage devices, including metal-ion batteries, metal–air batteries, and supercapacitors. In recent years, the demand for high-performance electrochemical energy storage applications has significantly increased; hence, numerous studies have been performed to develop advanced materials. In particular, graphene-based materials have drawn substantial interest for energy storage devices due to their good chemical stability, high electrical conductivity, and excellent physical properties. This chapter presents a comprehensive literature review on the development of graphene-based materials for energy storage systems and applications. Research trends and electrochemical performances of graphene-based materials in various electrochemical energy storage devices are also summarized based on recently reported works.

8.1 Introduction

With the rapid improvement of renewable energy technologies, electrochemical energy storage applications, including metal-ion batteries (MIBs), metal–sulfur batteries, metal–air batteries, and supercapacitors, have been extensively investigated in recent years [1–5]. Despite the numerous studies performed, obtaining materials that reveal high energy density, high power density, and long lifespan still remains a challenge [6, 7]. Although there have been many studies for electrochemical energy storage applications, they still hardly meet the demands of energy density, power density, and lifespan [8–10]. Nanomaterials and nanostructure play a key role in electrochemical energy storage applications [11, 12]. Physical and/or

doi:10.1088/978-0-7503-3999-5ch8

chemical interactions at the surface of these materials, such as specific surface area, surface energy, and surface chemistry, are crucial to enhancing the performance of energy storage applications [13–15].

Graphene is a two-dimensional (2D) mono-sheet material composed of carbon atoms in a hexagonal structure. When used as an electrode, the hexagonal structure reduces the volume change between the charge and discharge reactions and increases the electrode's electrical conductivity [16–18]. The unique structure of graphene provides abundant electrochemical active sites due to its large surface area, high electrical conductivity to accelerate electron transport, and excellent chemical stability under harsh conditions [19, 20]. In this chapter, we focus on the latest developments of graphene-based materials for electrochemical energy storage devices.

8.2 Properties of graphene

Graphene, owing to its unique chemical structure and geometry, exhibits excellent properties: high flexibility, high transparency (97.7%), excellent electrical and thermal conductivity, rapid charge mobility (2×10^5 cm^2 V^{-1} s^{-1}), large specific surface area (2630 m^2 g^{-1}), high specific capacity (744 mAh g^{-1}), high surface-to-volume ratio, high Young's modulus (1.1 TPa), and good biocompatibility [21, 22]. As such, graphene is used as the building block for various graphene-based materials such as zero-dimensional (0D) fullerenes (C_{60}), one-dimensional (1D) carbon nanotubes (CNTs), and three-dimensional (3D) graphite, as shown in figure 8.1 [23–28]. Graphene oxide (GO), reduced graphene oxide (rGO), and graphene aerogel have been considered potential candidates for various energy conversion and storage applications due to their impressive properties [29–31].

8.2.1 Physical properties

In the graphene honeycomb structure, a monolayer of carbon atoms is held together by a backbone of overlapping sp^2 hybridization bonds. The sp^2 hybridization bond of carbon consists of one pi (π) bond vertical to the planes and three strong sigma (σ) in-plane bonds. The unique properties of graphene originate from the 2p orbitals, which form delocalized π-bands on the carbon sheet [32]. Also, graphene reveals a high tensile strength (130 GPa), Young's modulus (1.1 TPa), and thermal conductivity (5300 W m^{-1} K^{-1}), it is highly suitable for use in electrochemical energy storage applications, including graphene-based supercapacitors with high theoretical capacity (550 F g^{-1}) and Li-ion batteries with high energy and power output [33, 34].

8.2.2 Electrical properties

Graphene is a semiconductor with high charge mobility and conductivity. Due to the distinct arrangement of the Fermi energy level and the Dirac-like spectrum of graphene, the band overlap of graphene at 1.6 meV makes it a semimetal material containing a single type of exciton [35, 36]. In a normal carbon atom, two and four

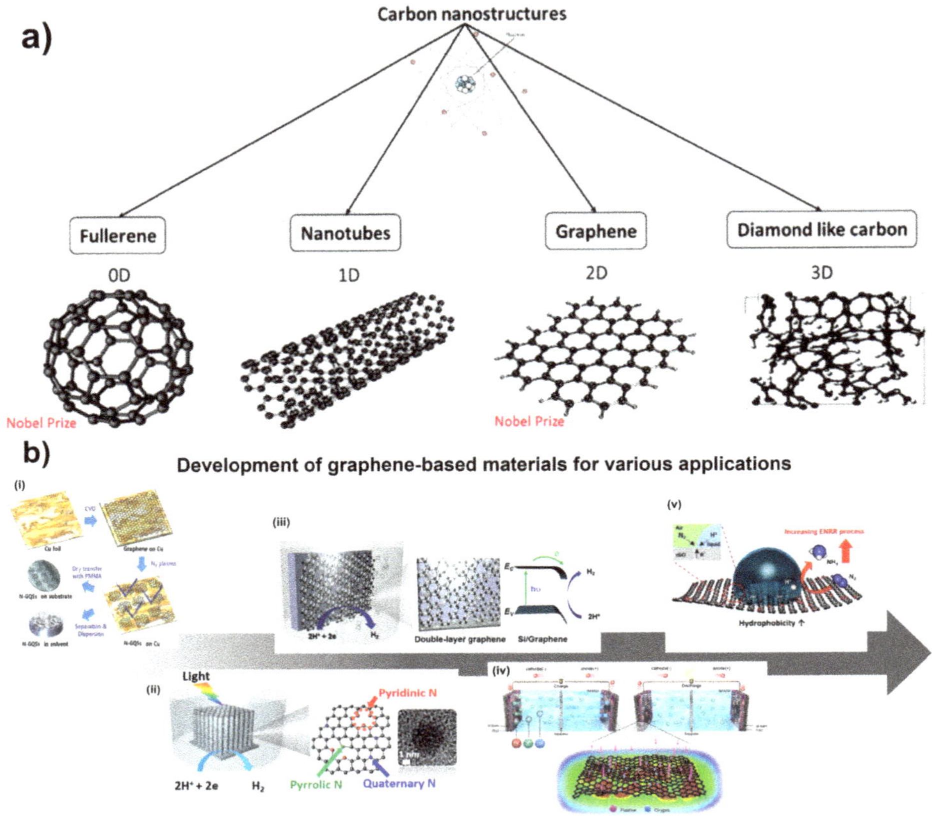

Figure 8.1. (a) Graphene-based materials: 0D fullerene, 1D CNT, 2D graphene, and 3D graphite. Reproduced from [23] under the terms of the Creative Commons Attribution 4.0 International, Copyright 2017, MDPI. (b) Development of graphene-based materials for various applications. (i) Reproduced from [24], John Wiley & Sons, Copyright 2014 Wiley-VCH Verlag GmbH & Co. KGaA, Weinheim; (ii) Reproduced from [25], Copyright 2015, Royal Society of Chemistry; (iii) Reproduced from [26], Copyright 2017, American Chemical Society; (iv) Reproduced from [27], Copyright 2022, with permission from Elsevier; (v) Reproduced from [28], Copyright 2023, with permission from Elsevier.

electrons are located in the inner and outer shell of the atom, respectively, where the four electrons in the outer shell can chemically bond with other carbon atoms. On the other hand, in the graphene carbon atom, only three electrons in the outer shell participate in chemical bonding. The remaining electron does not participate in chemical bonding and is freely available in the 3D area, increasing the electrical conductivity of graphene. Because these highly mobile electrons, known as π electrons, are located above and below the graphene sheet, the electrical properties of graphene can be evaluated through the bonding and antibonding of π orbitals [37–39]. Based on previous reports, the experimental and theoretical values of charge mobility of graphene were recorded as 1.5×10^4 cm^2 V^{-1} s^{-1} and 2×10^5 cm^2 V^{-1} s^{-1}, respectively. Besides, graphene has a low sheet resistance (30 Ω sq^{-1} cm^{-2}) owing to the high charge mobility [40, 41].

8.2.3 Chemical properties

Each graphene carbon atom is connected by three σ in-plane bonds and one π orbital, making possible additional interactions with other molecules. The π–π stacking structure in graphene allows graphene surfaces to react easily with other molecules. Chemical functionalization of the surface further enhances chemical reactions [42–44]. For example, the oxygen functionalization of graphene to form a graphene oxide (GO) enhances its reaction with water, polymer, and organic solvents. Because surface-functionalized graphene allows the control of chemical reaction activities, it is identified as a potential candidate material for various electrochemical energy storage applications [45, 46].

8.3 Graphene in metal-ion batteries

Electrochemical MIBs are the most favorable energy storage applications due to their good flexibility, high energy density, and simple maintenance, allowing efficient electricity storage [47]. Energy and power density, cycle stability, rate capability, safety, and production cost are the main parameters in evaluating the performance of MIBs. Therefore graphene is a promising material for high-performance MIBs due to its excellent carrier mobility, large specific surface area, and flexible structural change. This section focuses on the properties of MIBs using graphene and its derivatives [48, 49].

8.3.1 Lithium-ion batteries

Lithium-ion batteries (LIBs), first marketed by Sony in 1991, are currently one of the most conventional types of batteries in the world. The technologies of rechargeable LIBs are continuously being improved due to the increasing demand for batteries with high energy density and outstanding rate capability [50]. The various cathode materials for LIBs can be categorized into layered cathode materials such as $LiCoO_2$ (LCO), $LiNiO_2$ (LNO), $LiMnO_2$, $LiNi_{1/3}Co_{1/3}Mn_{1/3}O_2$ (NCM); spinel cathode materials such as $LiMn_2O_4$ (LMO); and olivine cathode materials such as $LiFePO_4$ (LFP), have been widely commercialized [51–54]. On the other hand, intercalation materials such as graphite and lithium titanate oxide ($Li_4Ti_5O_{12}$, LTO) are commonly used as anode materials. Graphite is commercially cast-off as the anode material for LIBs due to its high coulombic efficiency and cyclability. However, conventional intercalation-type graphite exhibits a low storage capacity of lithium ions, with a theoretical specific capacity lower than 372 mAh g^{-1}, leading to the formation of intercalation compounds like LiC_6 [55, 56].

To solve this problem, the use of graphene and its derivatives as both the anode and cathode in LIBs has been thoroughly investigated. The theoretical specific capacity of graphene at 744 mAh g^{-1} is approximately twice that of graphite [57–59]. The high theoretical capacity of graphene materials is attributed to the intercalation and adsorption of lithium on each side of graphene. When adsorbed on both edges of the graphene sheet, lithium ions are arranged like a 'house of cards' owing to the formation of the intercalation compound Li_2C_6 [60, 61].

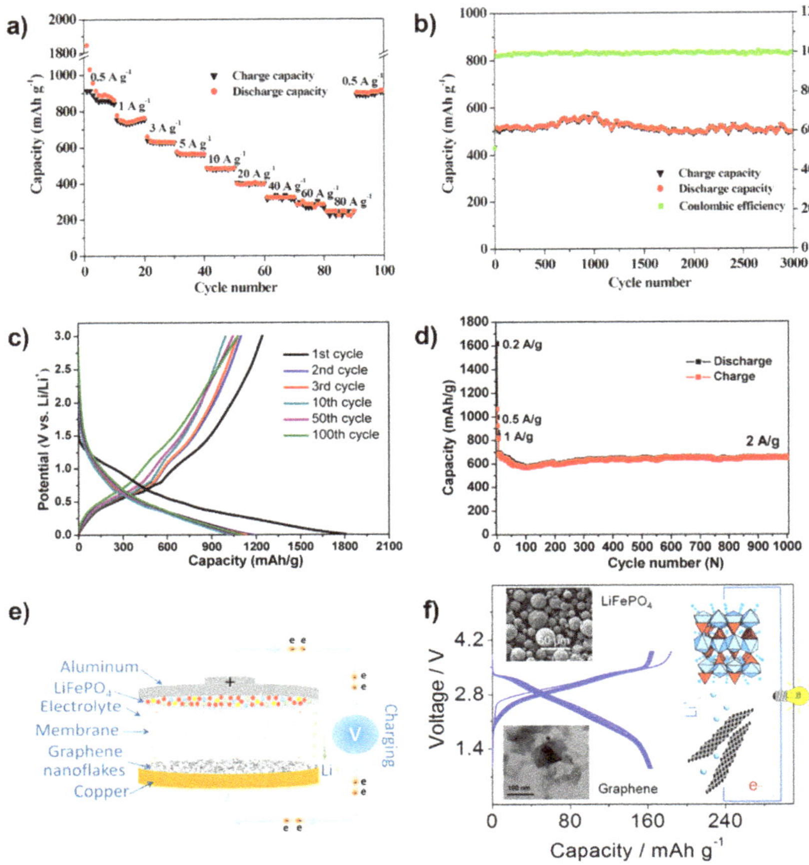

Figure 8.2. (a) Rate capability at current densities of 0.5–80 A g^{-1} and (b) cycling stability at a current density of 5 A g^{-1} over 3000 cycles for porous graphene. Reproduced with permission from [63], Copyright 2013, American Chemical Society; (c) galvanostatic charge discharge (GCD) curve at a current density of 0.2 A g^{-1} and (d) cycling performance at a current density of 2 A g^{-1} over 1000 cycles for Sn@G. Reproduced with permission from [68], Copyright 2013, American Chemical Society; (e) schematic and (f) GCD curve of graphene//LFP full cell. Reproduced with permission from [69], Copyright 2014, American Chemical Society.

Excellent cyclability and rate capability have been demonstrated when using graphene as the anode in LIBs. Hu *et al* [62] investigated the performance of LIBs in two different potential windows of 0.01–3.0 V and 0.01–1.5 V under the same conditions. In the expanded potential window of 0.01–3.0 V, N-doped graphene demonstrated improved cyclability and rate capability. Wang *et al* [63] have demonstrated the excellent storage capacity of porous graphene. Figure 8.2(a) shows reversible capacities of 560 mAh g^{-1} at a current density of 5 A g^{-1} and 220 mAh g^{-1} at an ultrahigh current density of 80 A g^{-1}. The excellent cell cyclability of up to 3000 cycles at a current density of 5 A g^{-1} is shown in figure 8.2(b).

Other anode materials include alloying-type materials, such as Si, Sn, and Ge and conversion-type materials, such as transition metal oxides. Although the alloying-type anode materials are desirable due to their superior specific capacity, they cause large volume expansion and electrode aggregation during cell operation, degrading the cyclability of LIBs [64]. Graphene, which forms a composite with alloying-type materials, can disrupt the aggregation of the electrode and provide high electric conductivity and mechanical strength for LIBs [65, 66]. Liang *et al* [67] fabricated binder-free self-supporting C@Si@graphene (C@Si@G) composite for flexible LIBs. The C@Si@G composite showcased an excellent reversible capacity of 1070 mAh g^{-1} at 0.2 A g^{-1} after 300 cycles. The rate capabilities of C@Si@G were recorded as 1734, 1680, 1611, 1519, 1330, 1055, and 597 mAh g^{-1} at current densities of 0.05, 0.1, 0.2, 0.5, 1, 2 and 5 A g^{-1}, respectively. Qin *et al* [68] synthesized 3D Sn@graphene (Sn@G), which demonstrated long-term cyclability at a high current density of 2 A g^{-1}, as shown in figure 8.2(d). The synthesized composite also possessed a high specific capacity of 682 mAh g^{-1} and a high capacity retention of 96.3% after 1000 cycles.

Hassoun *et al* [69] peeled off graphite through wet-chemical dispersion. Figures 8.2(e) and (f) show a full LIB, in which the graphene anode was coupled with a commercial LFP cathode at a potential window of 0.9–3.9 V. When operated at approximately 2.3 V, the cell showed a reversible capacity of 165 mAh g^{-1}. The electrochemical performances of graphene-based materials for LIBs are summarized in table 8.1.

8.3.2 Sodium-ion batteries

As the price of lithium continues to rise due to the limited lithium sources relative to the growing demand for LIBs, sodium-ion batteries (SIBs) have received substantial interest as an ideal substitute for LIBs [80, 81]. Sodium has chemical properties similar to those of lithium but is available at a much lower price, making it an easy replacement for LIBs [82, 83]. However, sodium ions have a larger ionic radius (0.97 Å) and higher standard electrode potential (−2.71 V versus SHE) compared to lithium ions. Figure 8.3(a) depicts the cell configuration of SIB system with graphene and its derivatives [84]. Several outstanding anode materials for LIBs are unsuitable in SIBs [85, 86]. For instance, commercial graphite cannot fully store sodium ions due to the narrow interlayer distances. Therefore, graphene would be a good substitute for graphite anodes in SIBs, as shown in figure 8.3(b) [87]. Xu *et al* [88] designed an N-doped graphene foam to enhance the performance of SIBs. The N-doped graphene foam anode showed a capacity of 1057.1 mAh g^{-1} at a current density of 0.1 A g^{-1}, which was higher than that of bare N-doped graphene. The N-doped graphene foam also demonstrated good cyclability of 0.5 A g^{-1} with a capacity retention of 69.7%. Yan *et al* [89] proved the enhanced cyclability of porous carbon/graphene composite at a current density of 1 A g^{-1}, where a specific capacity of 250 mAh g^{-1} was retained over 1000 cycles. Yan *et al* [90] developed an rGO/CNT sponge with excellent rate capability for high-performance SIBs. The rGO/CNT sponge exhibited excellent specific capacities of 217 mAh g^{-1} at a high current density of 10 A g^{-1} and 138 mAh g^{-1} after 10 000 cycles, as shown in figure 8.3(c).

Table 8.1. Electrochemical performances of graphene-based materials for LIBs.

Materials	Cyclability (mAh g^{-1})	Rate capability (mAh g^{-1})	Voltage (V)	Electrode	Reference
Graphene nanoflakes	650 at 0.7 A g^{-1} after 150 cycles	1500 at 0.1 A g^{-1}; 1000 at 0.4 A g^{-1}	0.01–2.0	Anode	[69]
Graphene sheets	848 at 0.1 A g^{-1} after 40 cycles	936 at 0.3 A g^{-1}; 445 at 1 A g^{-1}	0.01–3.5	Anode	[70]
N-doped graphene	180 at 10 A g^{-1} after 2000 cycles	798 at 0.5 A g^{-1}; 179 at 30 A g^{-1}	0.01–3.0	Anode	[62]
N-doped graphene	163 at 5 A g^{-1} after 800 cycles	335 at 0.1 A g^{-1}; 40 at 20 A g^{-1}	0.01–1.5	Anode	[62]
Porous graphene	470 at 10 A g^{-1} after 3000 cycles	860 at 0.5 A g^{-1}; 220 at 80 A g^{-1}	0.01–3.0	Anode	[63]
N-doped porous graphene	496 at 0.4 A g^{-1} after 200 cycles	672 at 0.1 A g^{-1}; 317 at 1.5 A g^{-1}	0.005–3.0	Anode	[71]
Vertical graphene sheets/graphite	105.4 at 5 C after 3000 cycles	518.8 at 0.2 C; 146.7 at 5 C	0.0–3.0	Anode	[72]
Cu@N-doped rGO/CNT composite	420 at 1 A g^{-1} after 500 cycles	1250 at 0.1 A g^{-1}; 230 at 10 A g^{-1}	0.002–3.0	Anode	[73]
C@Si@G	1070 at 0.2 A g^{-1} after 300 cycles; 972 at 1 A g^{-1} after 120 cycles	1734 at 0.05 A g^{-1}; 597 at 5 A g^{-1}	0.01–1.5	Anode	[74]
rGO/Si/graphene aerogel	916 at 0.2 C after 50 cycles	1071 at 0.2 C; 819 at 10 C	0.01–2.0	Anode	[67]
Sn@G	1089 at 0.2 A g^{-1} after 100 cycles; 682 at 2 A g^{-1} after 1000 cycles	1022 at 0.2 A g^{-1}; 270 at 10 A g^{-1}	0.005–3.0	Anode	[68]
ZnS/graphene	759 at 0.1 A g^{-1} after 100 cycles	595 at 0.1 A g^{-1}; 376 at 1 A g^{-1}	0.01–3.0	Anode	[75]
Ni_3S_2/GO	1122.3 at 1 A g^{-1} after 500 cycles; 233.2 at 5 A g^{-1} after 1000 cycles	644.4 at 0.1 A g^{-1}; 197.7 at 5 A g^{-1}	0.01–3.0	Anode	[76]
$NiTiO_3$-GO composite	506 at 0.1 A g^{-1} after 100 cycles	505 at 0.1 A g^{-1}; 212 at 3.2 A g^{-1}	0.01–3.0	Anode	[77]
Graphene-wrapped commercial LFP	117 at 2 C after 1000 cycles	160.7 at 0.5 C; 76.6 at 20 C	2.5–4.0	Cathode	[78]
Graphene/V_2O_5-coated NCM	193.3 at 0.1 C after 100 cycles; 165.9 at 1 C after 200 cycles	207.7 at 0.1 C; 108.5 at 10 C	3.0–4.5	Cathode	[79]

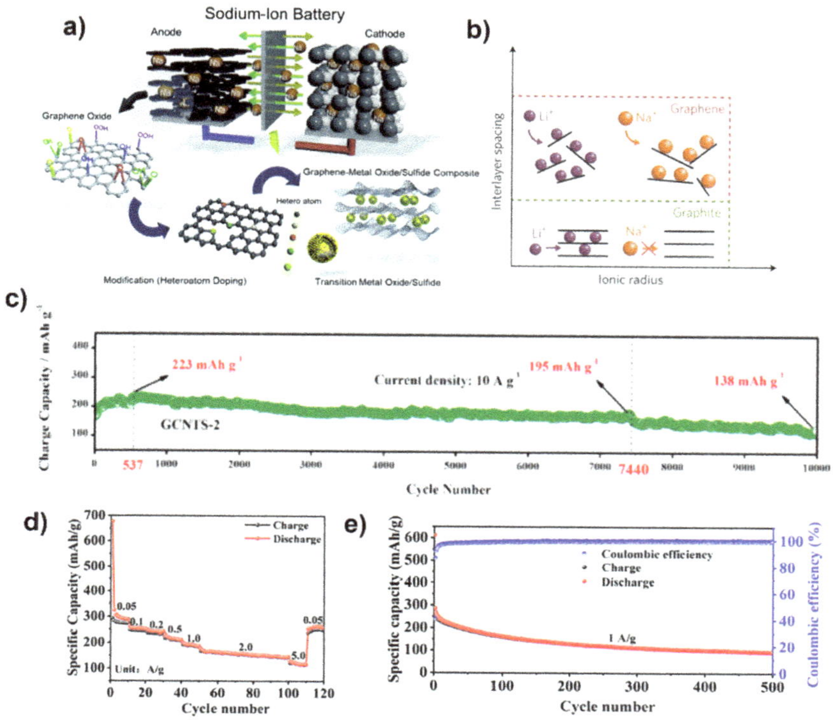

Figure 8.3. (a) Cell configuration of SIB system with graphene and its derivatives. Reproduced with permission from [84], Copyright 2013, American Chemical Society, (b) Li^+ and Na^+ ion intercalation process in graphite and graphene. Reproduced from [87], copyright 2014, with permission of Springer Nature, (c) cycling performance of rGO/CNT sponge at a current density of 10 A g^{-1} over 10 000 cycles. Reproduced with permission from [90], Copyright 2016, American Chemical Society; (d) rate performance at current densities of 0.05–5 A g^{-1} and (e) cycling performance at a current density of 1 A g^{-1} over 500 cycles for Ni_3S_2/GO. Reproduced with permission from [77], Copyright 2020, American Chemical Society.

When using alloying-type materials for SIBs, similar problems to those encountered in LIBs were observed. Therefore, graphene-based materials for SIBs are continuously being researched to solve these problems [91, 92]. For example, the sandwich-like SnS_2/graphene/SnS_2 composites prepared by Jiang *et al* [93] to prevent large volume expansion in SIBs achieved a reversible capacity of 1295 mAh g^{-1} at 0.1 A g^{-1}. At a high current density of 10 A g^{-1}, the specific capacity remained at 765 mAh g^{-1}. Zhang *et al* [76] prepared small-sized Ni_3S_2 nanoparticles deposited on GO using the typical one-pot wet-chemical method. As shown in figure 8.3(d), the Ni_3S_2/GO composite demonstrated capacities of 274, 243, 236, 207.4, 184.2, 144.2, and 115.2 mAh g^{-1} at 0.05, 0.1, 0.2, 0.5, 1.0, 2.0, and 5.0 A g^{-1}, respectively. When the current density returned to 0.05 A g^{-1}, Ni_3S_2/GO demonstrated a high-capacity recovery to 250 mAh g^{-1} with a capacity retention of more than 86%. Ni_3S_2/GO also delivered good cyclability with a capacity of 76.3 mAh g^{-1} over 500 cycles at 5 A g^{-1}, as shown in figure 8.3(e). The electrochemical performances of graphene-based materials for SIBs are summarized in table 8.2.

Table 8.2. Electrochemical performances of graphene-based materials for SIBs.

Materials	Cyclability (mAh g^{-1})	Rate capability (mAh g^{-1})	Voltage (V)	Electrode	Reference
N-doped graphene foam	605.6 at 0.5 A g^{-1} after 150 cycles	1057.1 at 0.1 A g^{-1}; 137.7 at 5 A g^{-1}	0.02–3.0	Anode	[89]
S-doped graphene	150 at 1 A g^{-1} after 200 cycles; 133 at 2 A g^{-1} after 100 cycles	262 at 0.1 A g^{-1}; 83 at 5 A g^{-1}	0.001–3.0	Anode	[95]
P-doped graphene	374 at 0.025 A g^{-1} after 120 cycles	374 at 0.025 A g^{-1}; 210 at 0.5 A g^{-1}	—	Anode	[96]
Porous carbon/graphene	250 at 1 A g^{-1} after 1000 cycles	—	0.01–2.7	Anode	[90]
rGO/CNT	49.2 at 0.2 A g^{-1} after 300 cycles	166.8 at 0.05 A g^{-1}; 9 at 5 A g^{-1}	0.001–3.0	Anode	[97]
rGO/CNT sponge	138 at 10 A g^{-1} after 10 000 cycles	440 at 0.05 A g^{-1}; 217 at 10 A g^{-1}	0.005–3.0	Anode	[91]
$NiTiO_3$–GO composite	290 at 0.1 A g^{-1} after 100 cycles	366 at 0.1 A g^{-1}; 60 at 1.6 A g^{-1}	0.01–3.0	Anode	[77]
SnO_2–CoO–graphene composite	302.8 at 0.1 A g^{-1} after 200 cycles	377.8 at 0.1 A g^{-1}; 228.0 at 2 A g^{-1}	0.01–3.0	Anode	[71]
SnO_2@graphene	650 at 0.2 A g^{-1} after 90 cycles	—	0.005–2.0	Anode	[98]
SnS_2/rGO/SnS_2	1133 at 0.1 A g^{-1} after 100 cycles	1295 at 0.1 A g^{-1}; 765 at 10 A g^{-1}	0.01–3.0	Anode	[94]
Ni_3S_2/GO	93.6 at 1 A g^{-1} after 500 cycles; 76.3 at 5 A g^{-1} after 500 cycles	274 at 0.05 A g^{-1}; 115.2 at 5 A g^{-1}	0.01–3.0	Anode	[76]
ZnS/graphene	491 at 0.1 A g^{-1} after 100 cycles	575 at 0.1 A g^{-1}; 317 at 1 A g^{-1}	0.01–3.0	Anode	[75]

8.4 Metal–air batteries

Recently, metal–air batteries have garnered significant attention due to their high specific energy density, low volume and weight, and safety. Metal–air batteries, including zinc– and lithium–air batteries that have a higher theoretical energy storage density than LIBs, are promising alternatives to metal-ion batteries [98–100]. Electrochemical oxidation and reduction reactions of oxygen are crucial parameters that determine the performance of metal–air batteries. Therefore, the advancement of bifunctional electrocatalysts for the oxygen evolution reaction (OER) and oxygen reduction reaction (ORR) is important for fabricating efficient rechargeable metal–air batteries [101, 102].

8.4.1 Lithium–air batteries

In lithium–air batteries, the overall electrochemical reaction could be expressed as 2 $Li^+ + O_2 + 2e^- \leftrightarrow Li_2O_3$, in which the ORR of discharge occurs in the forward direction, whereas the OER of charge occurs in the reverse direction [103, 104]. However, the low rates of OER and ORR have led to problems of low round-trip efficiency (53%–70%), low charge/discharge rate (0.1–1 mA cm^{-2}), short cycle life (10–100 cycles), and electrolyte instability [105–107]. Because carbon-based materials have low weight, large surface area, high electrical conductivity, and excellent chemical stability, various approaches have been reported of enhancing the performance of Li–air batteries using carbon-based materials, such as nitrogen-doped carbon, bimetal nanoparticles loaded on carbon, and modified graphene catalysts [108–110].

Lin *et al* [111] fabricated an ultrahigh-capacity Li–air battery using dry-pressed holey graphene (hG) as the cathode. The electrode was easily prepared under solvent- and binder-free conditions via the compression-molding method, as shown in figure 8.4(a). The presence of in-plane holes in the prepared binder-free dry-pressed hG allowed facile through-thickness mass transport. The discharge curve of the dry-pressed hG cathode showed high stability during discharge, with an areal mass loading (m_A) of 5 mg cm^{-2} at an areal current density (I_A) of 0.1 mA cm^{-2}, as shown in figure 8.4(b). The exceptional performance of the hG cathode was also demonstrated by the 0.22 V discharge overpotential. Figures 8.4(c) and (d) suggest that the specific gravity depends on the mass loading. The highest specific capacity was exhibited by the hG cathode at I_A of 0.2 mA cm^{-2} and m_A of 5 mg cm^{-2}. The time of full discharge of the 10 mg cm^{-2} sample was 12% higher than that of the 5 mg cm^{-2} sample (figure 8.4(c) inset). Although the two samples with different masses showed similar discharge curves until 150 h, the hG cathode of the 10 mg cm^{-2} sample was stable for 20 h longer than the hG cathode of the 5 mg cm^{-2} sample.

Metal catalysts can be combined with other catalysts via heteroatom-doping. Heteroatom-doped carbon-based materials used on various electrodes have revealed great catalytic performance in OERs and ORRs [112, 113]. Ionescu *et al* [114] reported that utilizing nitrogen-doped CNTs on metallic foams as the cathode improved the performance of Li–air batteries. The scanning electron microscopy (SEM) image shown in figure 8.4(d) demonstrates the overall morphology of carpet-

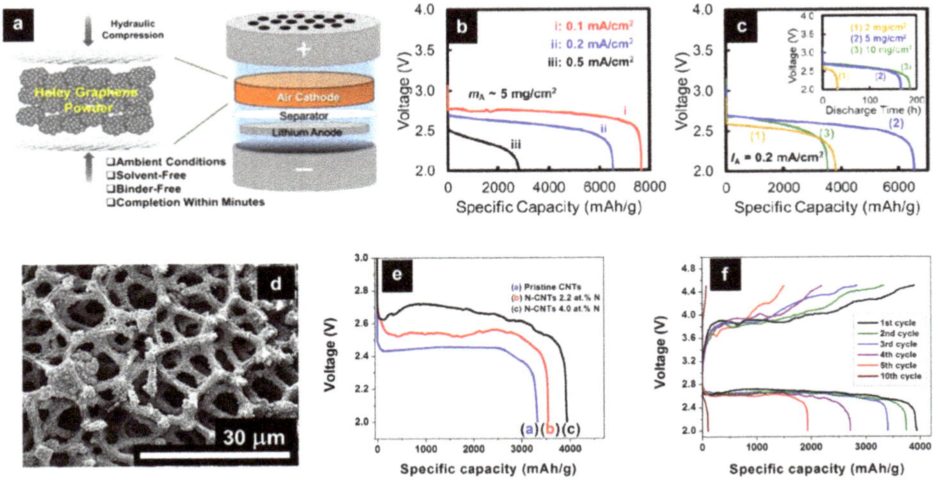

Figure 8.4. (a) Scheme of hG powder to directly deposit hG on the cathode, (b) discharge curves at current densities of 0.1, 0.2, and 0.5 mA cm^{-2} for hG cathode with mass loading of 5 mg cm^{-2}, (c) discharge curves for hG cathode with mass loadings of 2, 5, and 10 mg cm^{-2} at a current density of 0.2 mA cm^{-2}. Reproduced with permission from [111], Copyright 2017, American Chemical Society; (d) SEM image of CNTs deposited on Ni foam, (e) discharge curves for cathodes coated by pristine and N-doped CNTs, (f) full discharge cycle performance of 4.0 at% N-doped CNT electrodes. Reproduced with permission from [114], Copyright 2020, American Chemical Society.

like deposits, including high-density aligned CNTs grown on Ni foam. The cathode fabricated by pristine CNTs revealed a specific capacity of 3320 mAh g^{-1} and a voltage plateau of 2.45 V, as shown in figure 8.4(e). When 2.2 and 4.0 at% nitrogen-doped CNT electrodes were used, the specific capacity was increased to 3511 and 3934 mAh g^{-1}, and the voltage plateau to 2.52 and 2.68 V, respectively, confirming the enhancement of catalytic activity and performance of Li–air batteries with the increase in nitrogen ratio. The cyclability of 4.0 at% N-CNTs was also measured to evaluate the stability of Li–air batteries, as shown in figure 8.4(f). In the first cycle, the specific capacity and voltage plateau were approximately 3542 mAh g^{-1} and 3.8 V, respectively, with a coulombic efficiency of 90%. However, after 10 cycles, the efficiency was significantly reduced to 32% due to the accumulation of solid and non-conductive discharge product (Li_2O_2) in the interface between the electrolyte and electrode, causing the poor overall cycle stability of Li–air batteries. Table 8.3 highlights the advancement of graphene-based cathodes for Li–air batteries.

8.4.2 Zinc–air batteries

Similar to Li–air batteries, zinc metal is the anode, and an air electrode is a cathode in Zn–air batteries. The cathode is a combination of a catalyst layer and a gas diffusion layer. The overall reaction of Zn–air batteries in alkaline conditions is expressed as follows [126]:

Table 8.3. Electrochemical performances of graphene-based cathode for Li–air batteries.

Material	Electrode preparation method	Current density (mA g^{-1})	Specific capacity (mAh g^{-1})	Reference
Graphene with mesoporous carbon	Slurry	100	9000	[115]
Vacuum-promoted, thermal-expanded graphene	Slurry	300	19 800	[116]
rGO	Slurry	100	10 000	[117]
GNP/GO paper	Vacuum filtration	200	6910	[118]
Graphene foam	Hydrogel self-assembly	100	3120	[119]
Porous graphene	Hydrothermal self-assembly	100	13 000	[120]
Porous graphene paper	Vacuum filtration	200	10 176	[121]
Graphene aerogel	Aerogel templated	200	21 507	[122]
3-Dimensional graphene gel	Hydrothermal self-assembly	50	2250	[123]
Vertically aligned carbon nanosheet	Solution growth	75	1248	[124]
CVD graphene on Ni	CVD	200	89	[125]
Holey graphene	Dry compression	20	7667	[111]
Holey graphene	Dry compression	40	6540	[111]
Holey graphene	Dry compression	10	3737	[111]
Holey graphene	Dry compression	20	3529	[111]
N-doped CNTs on Ni foam	Spray pyrolysis CVD	23.8	3934	[114]

$$\begin{array}{lll}
\text{Anode:} & Zn \leftrightarrow Zn^{2+} + 2e^- & \\
 & Zn^{2+} + 4OH^- \leftrightarrow Zn(OH)_4^{2-} & E^o = 1.25\ V \\
 & Zn(OH)_4^{2-} \leftrightarrow ZnO + H_2O + 2OH^- & \\
\text{Cathode:} & 1/2O_2 + H_2O + 2e^- \leftrightarrow 2OH & E^o = 0.4\ V \\
\text{Overall reaction:} & Zn + 1/2O_2 \leftrightarrow ZnO & E^o = 1.65\ V
\end{array}$$

During the discharge reaction, oxygen diffuses in the porous air cathode due to the difference in oxygen pressure and the occurrence of ORR on the catalyst surface, resulting in the production of hydroxyl ions (OH^-) [127, 128]. This three-phase point, which includes catalyst (solid), electrolyte (liquid), and oxygen (gas), becomes an active site for gas-involving electrochemical reactions, such as nitrogen reduction reaction (NRR), carbon dioxide reduction reaction (CO_2RR), and ORR [129–132]. As such, OH^- would react with Zn^{2+} to complete the overall cell reaction, producing a working potential of 1.65 V. However, the practical working potential measured was lower than the theoretical potential due to the combination of activation, ohmic,

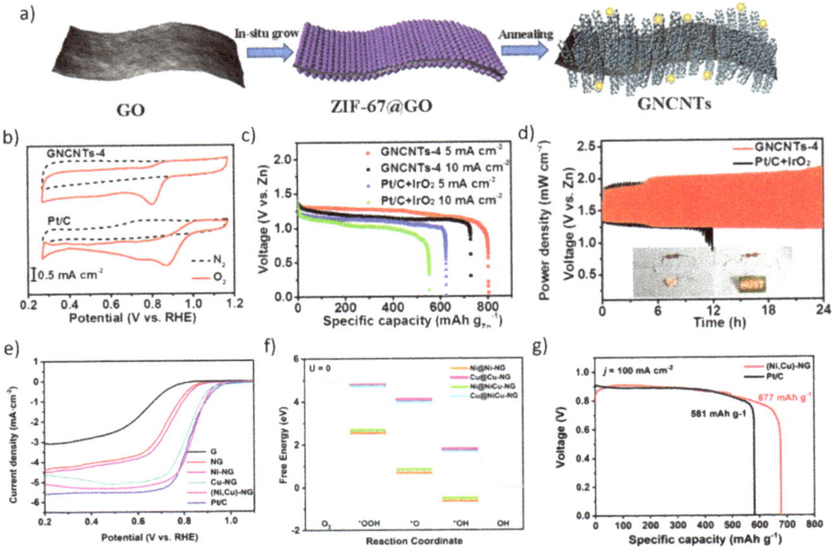

Figure 8.5. (a) Preparation for 2D N-doped CNTs/graphene; (b) CV curves of 2D N-doped CNTs/graphene-4 and Pt/C under N_2- and O_2-saturated conditions in 0.1 M KOH electrolyte; (c) discharge curves of 2D N-doped CNTs/graphene-4, Pt/C, and IrO_2 measured at 5 and 10 mA cm^{-2}; (d) discharge/charge cycling curves at 1 mA cm^{-2}. Reproduced with permission from [138], John Wiley & Sons, copyright 2019 WILEY-VCH Verlag GmbH & Co. KGaA, Weinheim; (e) ORR polarization curves for graphene, NG, Ni–NG, Cu–NG, (Ni, Cu)–NG, and Pt/C in O_2-saturated 0.1 M KOH electrolyte at a rotating speed of 1600 rpm with a scan rate of 100 mV s^{-1}; (f) free energy diagram of ORR; and (g) discharge curves of (Ni, Co)–NG- and Pt/C-based Zn–air batteries at a current density of 100 mA cm^{-2}. Reproduced with permission [141], Copyright 2009, Royal Society of Chemistry.

and concentration losses. Cathode instability during the charging reaction also resulted in low ORR and OER efficiency [133, 134]. Owing to high catalytic activities toward ORR and OER, graphene is considered a suitable material for Zn–air batteries, and defect engineering strategies can be utilized to optimize the catalytic activities of graphene-based materials [135–137].

Xu *et al* [138] fabricated an efficient bifunctional 2D N-doped CNTs/graphene derived from metal–organic framework (MOF) catalysts for air cathodes in rechargeable Zn–air batteries. Figure 8.5(a) shows the synthesis process of the 2D CNTs/graphene, which involves annealing the precursor of zeolitic imidazolate framework (ZIF)-67 on the GO layer. The cyclic voltammogram (CV) in figure 8.5(b) shows the catalytic activity of 2D CNTs/graphene-4 compared to commercial Pt/C, where the number 4 represents the repetition number of depositions of the sample. The peak potential of the 2D CNTs/graphene was 0.8 V versus reversible hydrogen electrode (RHE), corresponding with the ORR peak. The linear sweep voltammetry (LSV) curves show a more positive onset and half-wave potentials of 0.92 and 0.85 versus RHE, respectively, for the 2D CNTs/graphene-4. These results demonstrated that the 2D CNTs/graphene-4 sample exhibited the highest catalytic activity toward ORR among the samples, including the commercial

Pt/C catalyst. The OER performance of the 2D CNTs/graphene-4 was also similar to that of the commercial IrO_2 catalyst. From figure 8.5(c), the Zn–air battery with 2D CNTs/graphene-4 showcased a higher capacity than the Pt/C- and IrO_2-based Zn–air batteries at various current densities. Besides the excellent performance even after a long time, the charge/discharge cycle ability of the 2D CNTs/graphene-4-based Zn–air battery was also more stable than that of the Pt/C- and IrO_2-based batteries, as shown in figure 8.5(d).

Because graphene-based materials show increased catalytic activities toward OER and ORR after heteroatom-doping with metal catalysts, numerous studies have been conducted on various catalysts to enhance OER and ORR activities [5, 139, 140]. Cheng *et al* [141] reported the synthesis of N-doped nanoporous graphene (NG) with atomically dispersed Cu and Ni, (Ni, Cu)–NG, prepared via CVD and high-temperature gas transportation. Figure 8.5(e) shows the ORR electrocatalytic performance of (Ni, Cu)–NG compared to the commercial Pt/C catalyst. Graphene and NG exhibited poor ORR performance, whereas Cu–NG and (Ni, Cu)–NG revealed excellent electrocatalytic activities toward ORR, suggesting that the combination of single-atom transition metals with graphene could boost ORR activities. Density functional theory (DFT) calculations were conducted to evaluate the intrinsic properties of the ORR activity of (Ni, Cu)–NG and to discover the active sites. Synergistic effects were observed in $Cu–N_2–C_2$ and $Ni–N_3$ sites based on the Gibbs free energy diagrams and calculated onset potentials shown in figure 8.5(f). In the $Cu–N_2–C_2$ site, the presence of $Ni–N_3$ reduced the energy required to drive the rate-determining step. The discharge curves of (Ni, Cu)–NG-based Zn–air battery showed a specific capacity of 677 mAh g^{-1} at 100 mA cm^{-2}, as shown in figure 8.5(g). The (Ni, Cu)–NG-based Zn–air battery revealed similar performances to the Pt/C-based battery due to the high electrocatalytic properties of the (Ni, Cu)–NG catalyst. The (Ni, Cu)–NG-based Zn–air battery also demonstrated higher stability and rechargeability than the Pt/C- and IrO_2-based batteries. Table 8.4 summarizes the advancement of graphene-based cathodes for Zn–air batteries.

8.5 Supercapacitors

Supercapacitors (SCs) are emerging as excellent energy storage systems owing to their high power density compared to conventional MIBs, fast charging time (within seconds), low self-discharge rate, and 100 times longer lifespan than conventional batteries. Currently, SCs have several applications, such as electric vehicles, trucks, and portable electronics [149–152]. In an SC, a stable electric double layer (EDL) is formed between the electrolyte and the electrode to store electrical energy through ion adsorption at the electrode interface of an EDL. Electrical energy is also stored through electron transfer between the electrode and the electrolyte as a consequence of a Faradic reaction [153, 154]. Therefore, the functional electrode material used in SCs should possess a large specific surface area with appropriate pore sizes, high density for good stability, and high electrical conductivity. Activated carbon, carbon aerogel, porous carbon and carbon nanotubes are commonly used as electrode

Table 8.4. Electrochemical performances of graphene-based materials for Zn–air batteries.

Material	Electrode preparation method	Current density (mA g^{-1})	Specific capacity (mAh g^{-1})	Reference
2D N-doped CNTs/graphene	Annealing process of ZIF-67	166	801	[138]
(Ni, Cu)-NG	CVD and high temperature gas transportation	250	677	[141]
$NiCo_2S_4$@g-C_3N_4-CNT	Two-step hydrothermal and vacuum filtration	80	493	[142]
3DOM-Co@TiO_xN_y	Thermal treating	95	697	[143]
DN-CP@G	*In situ* exfoliating graphene from carbon paper and high temperature NH_3 treatment	100	591	[144]
$CoIn_2S_4$/S-rGO	*In situ* solvothermal growth process	80	745	[145]
$CuCo_2O_4$/N-CNTs	Electrostatic attraction	48	817	[146]
CoNi/BCF	Hydrothermal	80	711	[147]
CoP@mNSP-C	*In situ* incorporation	80	—	[148]

materials in SCs. Among them, graphene, with 2D nanostructures, is highly suitable due to its large specific surface area and high stability [155, 156].

Figure 8.6(a) shows the main features of a graphene electrode in an SC. The in-plane 2D design provides active sites which enhance the interaction between the ions present in the electrolyte with all carbon layers. The intrinsic electrochemical properties of graphene edges along with the basal plane of graphene in the in-plane 2D design also maximize the usable electrochemical surface area, allowing graphene to be used in extremely small devices [156, 157].

Chen *et al* [158] prepared N-doped interpenetrating porous carbon/graphene networks for SCs (figure 8.6(b)). The prepared carbon-based electrodes showed long cycle stability at an ultrahigh current density of 20 A g^{-1}, with approximately 92.3% high-capacity retention. The N-doped porous carbon/graphene networks also achieved energy densities of 27.6 and 11.7 Wh kg^{-1} at power densities of 600 and 12 000 W kg^{-1}, respectively.

Chen *et al* [159] developed an N, F co-doped holey graphene for high-performance SCs. As shown in the SEM image of the N, F co-doped holey graphene shown in figure 8.6(c), an interconnection of pores ranging from hundreds of nanometers to several micrometers in diameter was observed. This symmetrical SC demonstrated great cyclability with a capacity retention of 87.8% after 10 000 cycles at a current density of 2 A g^{-1}, as shown in figure 8.6(d). Khandelwal *et al* [160] prepared an N, B co-doped graphene, which demonstrated excellent rate capability at current densities ranging from 0.05 to 1 mA cm^{-2}, and high cycle stability after 10 000 cycles with a capacity retention of 87.6%. From their study, multi-heteroatom-

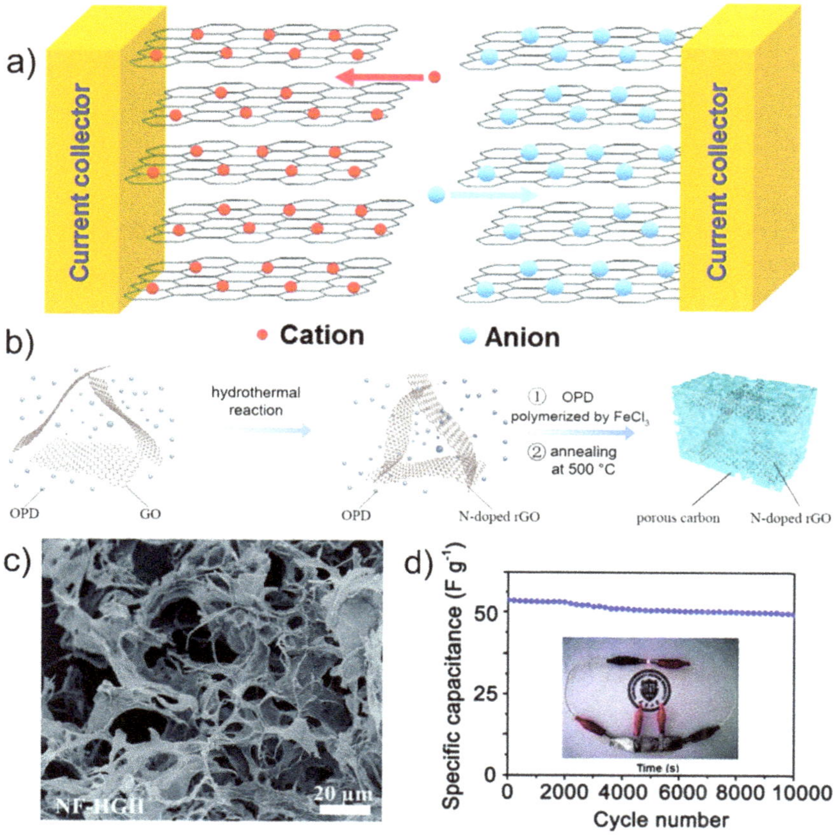

Figure 8.6. (a) Working mechanism of graphene as electrodes in SCs. Reproduced with permission from [156], Copyright 2011, American Chemical Society;, (b) schematic of synthesis process for N-doped interpenetrating porous carbon/graphene networks. Reproduced with permission from [158], Copyright 2021, Elseiver; (c) SEM image and (d) cyclability after 10 000 cycles at a current density of 2 A g^{-1} for N, F co-doped holey graphene. Reproduced with permission from [159], Copyright 2016, Royal Society of Chemistry.

doping into graphene-based materials was proven to be more useful in enhancing SC performance compared to monoatomic doping. Double doping of the graphene lattice creates more defects and favorable electron structures, improving the overall electrode performance. The various graphene-based materials used for SCs are reviewed in table 8.5.

8.6 Conclusions

This chapter introduces graphene and graphene-based materials as potential candidate materials for electrochemical energy storage devices such as metal-ion batteries, metal–air batteries, and supercapacitors. Various studies have suggested using graphene to improve electrochemical energy storage performance. In LIBs, N-doped graphene or GO complexes are used as anodes: in SIBs, N-, S-, or P-doped

Table 8.5. Electrochemical performances of graphene-based materials for SCs.

Materials	Specific capacitance (F g^{-1} or F cm^{-2})	Cycle	Electrolyte	Reference
2D graphene	100–205	1500	$BMIMBF_4$	[161]
Ppy-graphene	165	2000	1.0 M NaCl	[162]
Graphene/PANI	210	1000	1.0 M H_2SO_4	[163]
F-modified graphene	134	5000	$EMIMBF_4$/PVDF-HFP	[164]
N-doped porous carbon/graphene	673	10 000	1 M H_2SO_4	[158]
N, F-doped holey graphene hydrogel	67.7	10 000	PVA and KOH gel	[159]
N, B-doped graphene	104.3	10 000	PVA-H_2SO_4 gel	[160]
rGO aerogel	182.0	10 000	5 M KOH	[165]
S, N-doped three-dimensional graphene aerogel	399	6000	6 M KOH	[166]

graphene; rGO/CNT complexes; and metal oxide/graphene complexes are used as anodes. In metal–air batteries, porous graphene and holey graphene-based materials with a large surface area are used as cathodes for efficient OER and ORR reactions. In SCs, multi-heteroatom-doped graphene, 2D graphene, and rGO aerogel are used as electrodes to enhance specific capacity and lifespan. The unique properties of graphene, such as high accessibility, large surface area, high charge mobility, and chemical stability, graphene make it an attractive material for energy storage applications.

References

[1] Surendran S, Shanmugapriya S, Sivanantham A, Shanmugam S and Kalai Selvan R 2018 Electrospun carbon nanofibers encapsulated with NiCoP: a multifunctional electrode for supercapattery and oxygen reduction, oxygen evolution, and hydrogen evolution reactions *Adv. Energy Mater.* **8** 1800555

[2] Kim H, Surendran S, Chae Y, Lee H Y, An T Y, Han H S, Park W, Kim J K and Sim U 2020 Fabrication of an ingenious metallic asymmetric supercapacitor by the integration of anodic iron oxide and cathodic nickel phosphide *Appl. Surf. Sci.* **511** 145424

[3] Lin J, Zeng C, Lin X, Xu C, Xu X and Luo Y 2021 Metal–organic framework-derived hierarchical MnO/Co with oxygen vacancies toward elevated-temperature Li-Ion battery *ACS Nano* **15** 4594–607

[4] Zhang L, Zhu Y, Nie Z, Li Z, Ye Y, Li L, Hong J, Bi Z, Zhou Y and Hu G 2021 Co/MoC nanoparticles embedded in carbon nanoboxes as robust trifunctional electrocatalysts for a Zn–air battery and water electrocatalysis *ACS Nano* **15** 13399–414

[5] Wang A, Zhao C, Yu M and Wang W 2021 Trifunctional Co nanoparticle confined in defect-rich nitrogen-doped graphene for rechargeable Zn–air battery with a long lifetime *Appl. Catal.* B **281** 119514
[6] Cheng S *et al* 2021 Towards large-scale electrochemical energy storage in the marine environment with a highly-extensible 'paper-like' seawater supercapacitor device *J. Mater. Chem.* A **9** 622–31
[7] Li W, Ma Y, Li P, Jing X, Jiang K and Wang D 2021 Synergistic effect between S and Se enhancing the electrochemical behavior of Se_xS_y in aqueous Zn metal batteries *Adv. Funct. Mater.* **31** 2101237
[8] Tang Z K, Xue Y F, Teobaldi G and Liu L M 2020 The oxygen vacancy in Li-ion battery cathode materials *Nanoscale Horiz.* **5** 1453–66
[9] Liu J, Wang C, Sun H, Wang H, Rong F, He L, Lou Y, Zhang S, Zhang Z and Du M 2020 CoO_x/CoN_y nanoparticles encapsulated carbon-nitride nanosheets as an efficiently trifunctional electrocatalyst for overall water splitting and Zn–air battery *Appl. Catal.* B **279** 119407
[10] Tebyetekerwa M, Marriam I, Xu Z, Yang S, Zhang H, Zabihi F, Jose R, Peng S, Zhu M and Ramakrishna S 2019 Critical insight: challenges and requirements of fibre electrodes for wearable electrochemical energy storage *Energy Environ. Sci.* **12** 2148–60
[11] Meng C, Das P, Shi X, Fu Q, Müllen K and Wu Z S 2021 In situ and operando characterizations of 2D materials in electrochemical energy storage devices *Small Sci.* **1** 2000076
[12] Huang D, Chen Y, Cheng M, Lei L, Chen S, Wang W and Liu X 2021 Carbon dots-decorated carbon-based metal-free catalysts for electrochemical energy storage *Small* **17** 2002998
[13] Li L, Liu W, Dong H, Gui Q, Hu Z, Li Y and Liu J 2021 Surface and interface engineering of nanoarrays toward advanced electrodes and electrochemical energy storage devices *Adv. Mater.* **33** 2004959
[14] Vijayan B L, Misnon I I, Karuppaiah C, Kumar G M A, Yang S, Yang C C and Jose R 2021 Thin metal film on porous carbon as a medium for electrochemical energy storage *J. Power Sources* **489** 229522
[15] Baig M M and Gul I H 2021 Conversion of wheat husk to high surface area activated carbon for energy storage in high-performance supercapacitors *Biomass Bioenergy* **144** 105909
[16] Sun J, Sadd M, Edenborg P, Grönbeck H, Thiesen P H, Xia Z, Quintano V, Qiu R, Matic A and Palermo V 2021 Real-time imaging of Na^+ reversible intercalation in 'Janus' graphene stacks for battery applications *Sci. Adv.* **7** eabf0812
[17] Zhu W, Kierzek K, Wang S, Li S, Holze R and Chen X 2021 Improved performance in lithium ion battery of CNT-Fe_3O_4@ graphene induced by three-dimensional structured construction *Colloids Surf.* A **612** 126014
[18] Mo R *et al* 2020 Tin-graphene tubes as anodes for lithium-ion batteries with high volumetric and gravimetric energy densities *Nat. Commun.* **11** 1–11
[19] Cheng Q, Okamoto Y, Tamura N, Tsuji M, Maruyama S and Matsuo Y 2017 Graphene-like-graphite as fast-chargeable and high-capacity anode materials for lithium ion batteries *Sci. Rep.* **7** 1–14
[20] Shi M, Nie P, Fan Z, Fu R, Fang S, Dou H and Zhang X 2020 Tubular graphene nano-scroll coated silicon for high rate performance lithium-ion battery *Front. Energy Res.* **8** 2

[21] Liu Y, Xie B, Zhang Z, Zheng Q and Xu Z 2012 Mechanical properties of graphene papers *J. Mech. Phys. Solids* **60** 591–605

[22] Fuchs J N and Goerbig M O 2008 11–2 Introduction to the physical properties of graphene *Lecture Notes* University of California, Santa Barbara

[23] Al-Jumaili A, Alancherry S, Bazaka K and Jacob M V 2017 Review on the antimicrobial properties of carbon nanostructures *Materials* **10** 1066

[24] Moon J, An J, Sim U, Cho S-P, Kang J H, Chung C, Seo J-H, Lee J, Nam K T and Hong B H 2014 One-step synthesis of N-doped graphene quantum sheets from monolayer graphene by nitrogen plasma *Adv. Mater.* **26** 3501–5

[25] Sim U, Moon J, An J, Kang J H, Jerng S E, Moon J, Cho S P, Hong B H and Nam K T 2015 N-doped graphene quantum sheets on silicon nanowire photocathodes for hydrogen production *Energy Environ. Sci.* **8** 1329–38

[26] Sim U, Moon J, Lee J, An J, Ahn H Y, Kim D J, Jo I, Jeon C, Han S, Hong B H and Nam K T 2017 Double-layer graphene outperforming monolayer as catalyst on silicon photocathode for hydrogen production *ACS Appl. Mater. Interfaces* **9** 3570–80

[27] Sim Y, Surendran S, Cha H, Choi H, Je M, Yoo S, Seok D C, Jung Y H, Jeon C, Kim D J and Han M K 2022 Fluorine-doped graphene oxide prepared by direct plasma treatment for supercapacitor application *Chem. Eng. J.* **428** 132086

[28] Kim D, Alam K, Han M K, Surendran S, Lim J, Kim J Y, Moon D J, Jeong G, Kim M G, Kwon G and Yang S 2023 Manipulating wettability of catalytic surface for improving ammonia production from electrochemical nitrogen reduction *J. Coll. Interface Sci.* **633** 53–9

[29] Zhu X, Zhu Y, Murali S, Stoller M D and Ruoff R S 2011 Nanostructured reduced graphene oxide/Fe_2O_3 composite as a high-performance anode material for lithium ion batteries *ACS Nano* **5** 3333–8

[30] Chang J, Huang X, Zhou G, Cui S, Hallac P B, Jiang J, Hurley P T and Chen J 2014 Multilayered Si nanoparticle/reduced graphene oxide hybrid as a high-performance lithium-ion battery anode *Adv. Mater.* **26** 758–64

[31] Zhou G, Paek E, Hwang G S and Manthiram A 2016 High-performance lithium-sulfur batteries with a self-supported, 3D Li_2S-doped graphene aerogel cathodes *Adv. Energy Mater.* **6** 1501355

[32] Pierson H O 2012 *Handbook of Carbon, Graphite, Diamonds and Fullerenes: Processing, Properties and Applications* (Norwich, NY: William Andrew)

[33] Şenel M C, Gürbüz M and Koc E 2018 Fabrication and characterization of synergistic Al-SiC-GNPs hybrid composites *Composites* B **154** 1–9

[34] Kumar H P and Xavior M A 2014 Graphene reinforced metal matrix composite (GRMMC): a review *Procedia Eng.* **97** 1033–40

[35] Partoens B and Peeters F M 2006 From graphene to graphite: electronic structure around the K point *Phys. Rev.* B **74** 075404

[36] Avetisyan A A, Partoens B and Peeters F M 2009 Electric-field control of the band gap and Fermi energy in graphene multilayers by top and back gates *Phys. Rev.* B **80** 195401

[37] Gai Y, Wang W, Xiao D, Tan H, Lin M and Zhao Y 2018 Reversible conversion between graphene nanosheets and graphene nanoscrolls at room temperature *RSC Adv.* **8** 9749–53

[38] de Heer W A, Berger C, Conrad E, First P, Murali R and Meindl J 2007 Pionics: the emerging science and technology of graphene-based nanoelectronics *2007 IEEE Int. Electron Devices Meeting* (Piscataway, NJ: IEEE) pp 199–202

[39] Zerbi G, Tommasini M M S, Fustella G, Ferrugiari A and Castiglioni C 2008 pi-Electron delocalization and raman spectra in graphite and graphene molecules: the role of the edges in graphene sheets *The 8th Int. Symp. on Functional Pi-Electron Systems* 1

[40] Sreeprasad T S and Berry V 2013 How do the electrical properties of graphene change with its functionalization? *Small* **9** 341–50

[41] Vicarelli L, Heerema S J, Dekker C and Zandbergen H W 2015 Controlling defects in graphene for optimizing the electrical properties of graphene nanodevices *ACS Nano* **9** 3428–35

[42] Hernández Rosas, J.J., Ramírez Gutiérrez, R.E., Escobedo-Morales, A. and Chigo Anota E 2011 First principles calculations of the electronic and chemical properties of graphene, graphane, and graphene oxide *J. Mol. Model.* **17** 1133–9

[43] Liu L, Qing M, Wang Y and Chen S 2015 Defects in graphene: generation, healing, and their effects on the properties of graphene: a review *J. Mater. Sci. Technol.* **31** 599–606

[44] Wassei J K, Cha K C, Tung V C, Yang Y and Kaner R B 2011 The effects of thionyl chloride on the properties of graphene and graphene–carbon nanotube composites *J. Mater. Chem.* **21** 3391–6

[45] Lei Z, Lu L and Zhao X S 2012 The electrocapacitive properties of graphene oxide reduced by urea *Energy Environ. Sci.* **5** 6391–9

[46] Hu X, Yu Y, Hou W, Zhou J and Song L 2013 Effects of particle size and pH value on the hydrophilicity of graphene oxide *Appl. Surf. Sci.* **273** 118–21

[47] Xu X, Xiong F, Meng J, Wang X, Niu C, An Q and Mai L 2020 Vanadium-based nanomaterials: a promising family for emerging metal-ion batteries *Adv. Funct. Mater.* **30** 1904398

[48] Jing Y, Zhou Z, Cabrera C R and Chen Z 2014 Graphene, inorganic graphene analogs and their composites for lithium ion batteries *J. Mater. Chem.* A **2** 12104–22

[49] Peigney A, Laurent C, Flahaut E, Bacsa R R and Rousset A 2001 Specific surface area of carbon nanotubes and bundles of carbon nanotubes *Carbon* **39** 507–14

[50] Moshtev R and Johnson B 2000 State of the art of commercial Li ion batteries *J. Power Sources* **91** 86–91

[51] Wang B, Bates J B, Hart F X, Sales B C, Zuhr R A and Robertson J D 1996 Characterization of thin-film rechargeable lithium batteries with lithium cobalt oxide cathodes *J. Electrochem. Soc.* **143** 3203

[52] Abraham D P, Twesten R D, Balasubramanian M, Kropf J, Fischer D, McBreen J, Petrov I and Amine K 2003 Microscopy and spectroscopy of lithium nickel oxide-based particles used in high power lithium-ion cells *J. Electrochem. Soc.* **150** A1450

[53] Rossouw M H, Liles D C, Thackeray M M, David W I F and Hull S 1992 Alpha manganese dioxide for lithium batteries: a structural and electrochemical study *Mater. Res. Bull.* **27** 221–30

[54] Liu S, Xiong L and He C 2014 Long cycle life lithium ion battery with lithium nickel cobalt manganese oxide (NCM) cathode *J. Power Sources* **261** 285–91

[55] Striebel K A, Sierra A, Shim J, Wang C W and Sastry A M 2004 The effect of compression on natural graphite anode performance and matrix conductivity *J. Power Sources* **134** 241–51

[56] Xie Z, Li X, Li W, Chen M and Qu M 2015 Graphene oxide/lithium titanate composite with binder-free as high capacity anode material for lithium-ion batteries *J. Power Sources* **273** 754–60

[57] Wang S, Yang B, Chen H and Ruckenstein E 2019 Reconfiguring graphene for high-performance metal-ion battery anodes *Energy Storage Mater.* **16** 619–24

[58] Chang Y, Li J, Wang B, Luo H, He H, Song Q and Zhi L 2013 Synthesis of 3D nitrogen-doped graphene/Fe_3O_4 by a metal ion induced self-assembly process for high-performance Li-ion batteries *J. Mater. Chem.* A **1** 14658–65
[59] Dobrota A S, Pašti I A and Skorodumova N V 2015 Oxidized graphene as an electrode material for rechargeable metal-ion batteries—a DFT point of view *Electrochim. Acta* **176** 1092–9
[60] Ji K *et al* 2019 Lithium intercalation into bilayer graphene *Nat. Commun.* **10** 1–10
[61] Zhou L J, Hou Z F and Wu L M 2012 First-principles study of lithium adsorption and diffusion on graphene with point defects *J. Phys. Chem.* C **116** 21780–7
[62] Hu T, Sun X, Sun H, Xin G, Shao D, Liu C and Lian J 2014 Rapid synthesis of nitrogen-doped graphene for a lithium ion battery anode with excellent rate performance and super-long cyclic stability *Phys. Chem. Chem. Phys.* **16** 1060–6
[63] Wang Z L, Xu D, Wang H G, Wu Z and Zhang X B 2013 In situ fabrication of porous graphene electrodes for high-performance energy storage *ACS Nano* **7** 2422–30
[64] Rönnebro E, Yin J, Kitano A, Wada M and Sakai T 2005 Comparative studies of mechanical and electrochemical lithiation of intermetallic nanocomposite alloys for anode materials in Li-ion batteries *Solid State Ion.* **176** 2749–57
[65] Goli P, Legedza S, Dhar A, Salgado R, Renteria J and Balandin A A 2014 Graphene-enhanced hybrid phase change materials for thermal management of Li-ion batteries *J. Power Sources* **248** 37–43
[66] Evanoff K, Magasinski A, Yang J and Yushin G 2011 Nanosilicon-coated graphene granules as anodes for Li-ion batteries *Adv. Energy Mater.* **1** 495–8
[67] Kim S Y, Kim C H and Yang C M 2021 Binder-free silicon anodes wrapped in multiple graphene shells for high-performance lithium-ion batteries *J. Power Sources* **486** 229350
[68] Qin J, He C, Zhao N, Wang Z, Shi C, Liu E Z and Li J 2014 Graphene networks anchored with Sn@ graphene as lithium ion battery anode *ACS Nano* **8** 1728–38
[69] Hassoun J *et al* 2014 An advanced lithium-ion battery based on a graphene anode and a lithium iron phosphate cathode *Nano Lett.* **14** 4901–6
[70] Lian P, Zhu X, Liang S, Li Z, Yang W and Wang H 2010 Large reversible capacity of high quality graphene sheets as an anode material for lithium-ion batteries *Electrochim. Acta* **55** 3909–14
[71] Qiu H, Zheng H, Jin Y, Yuan Q, Zhang X, Zhao C, Wang H and Jia M 2021 Mesoporous cubic SnO_2–CoO nanoparticles deposited on graphene as anode materials for sodium ion batteries *J. Alloys Compd.* **874** 159967
[72] Mu Y, Han M, Li J, Liang J and Yu J 2021 Growing vertical graphene sheets on natural graphite for fast charging lithium-ion batteries *Carbon* **173** 477–84
[73] Wang H, Que X, Liu Y, Wu X, Yuan Q, Lu J and Gan W 2021 Facile synthesis of yolk–shell structured SiOx/C@ Void@ C nanospheres as anode for lithium-ion batteries *J. Alloys Compd.* **874** 159913
[74] Liang M, Wang W, Jiang Y, Liao C, Long Q, Lai X and Liao L 2021 Fabrication of C@ Si@ G for flexible lithium-ion batteries *J. Alloys Compd.* **878** 160357
[75] Zhang R, Wang Y, Jia M, Xu J and Pan E 2018 One-pot hydrothermal synthesis of ZnS quantum dots/graphene hybrids as a dual anode for sodium ion and lithium ion batteries *Appl. Surf. Sci.* **437** 375–83

[76] Zhang Y H, Liu R H, Xu L J, Zhao L J, Luo S H, Wang Q and Liu X 2020 One-pot synthesis of small-sized Ni_3S_2 nanoparticles deposited on graphene oxide as composite anode materials for high-performance lithium-/sodium-ion batteries *Appl. Surf. Sci.* **531** 147316
[77] Majumder T, Das D and Majumder S B 2021 Nickel titanate–GO composite as negative electrode for lithium and sodium ion batteries *Mater. Lett.* **301** 130293
[78] Tao R *et al* 2021 High-conductivity–dispersibility graphene made by catalytic exfoliation of graphite for lithium-ion battery *Adv. Funct. Mater.* **31** 2007630
[79] Li Y, Xu C, Dang M, Yu C, He Y, Liu W, Jin H, Li W, Zhu M and Zhang J 2021 Improved electrochemical performances of $LiNi_{0.5}Co_{0.2}Mn_{0.3}O_2$ modified by Graphene/ V_2O_5 co-coating *Ceram. Int.* **47** 21759–68
[80] Kim H, Kim J C, Bianchini M, Seo D H, Rodriguez-Garcia J and Ceder G 2018 Recent progress and perspective in electrode materials for K-ion batteries *Adv. Energy Mater.* **8** 1702384
[81] Zhang W, Yan Y, Xie Z, Yang Y, Xiao Y, Zheng M, Hu H, Dong H, Liu Y and Liang Y 2020 Engineering of nanonetwork-structured carbon to enable high-performance potassium-ion storage *J. Colloid Interface Sci.* **561** 195–202
[82] Vaalma C, Buchholz D, Weil M and Passerini S 2018 A cost and resource analysis of sodium-ion batteries *Nat. Rev. Mater.* **3** 1–11
[83] Bauer A, Song J, Vail S, Pan W, Barker J and Lu Y 2018 The scale-up and commercialization of nonaqueous Na-ion battery technologies *Adv. Energy Mater.* **8** 1702869
[84] Thakur A K, Ahmed M S, Oh G, Kang H, Jeong Y, Prabakaran R, Vikram M P, Sharshir S W, Kim J and Hwang J Y 2021 Advancement in graphene-based nanocomposites as high capacity anode materials for sodium-ion batteries *J. Mater. Chem.* A **9** 2628–61
[85] Nayak P K, Yang L, Brehm W and Adelhelm P 2018 From lithium-ion to sodium-ion batteries: advantages, challenges, and surprises *Angew. Chem. Int. Ed.* **57** 102–20
[86] Zheng X, Bommier C, Luo W, Jiang L, Hao Y and Huang Y 2019 Sodium metal anodes for room-temperature sodium-ion batteries: applications, challenges and solutions *Energy Storage Mater.* **16** 6–23
[87] Raccichini R, Varzi A, Passerini S and Scrosati B 2015 The role of graphene for electrochemical energy storage *Nat. Mater.* **14** 271–9
[88] Xu J, Wang M, Wickramaratne N P, Jaroniec M, Dou S and Dai L 2015 High-performance sodium ion batteries based on a 3D anode from nitrogen-doped graphene foams *Adv. Mater.* **27** 2042–8
[89] Yan Y, Yin Y X, Guo Y G and Wan L J 2014 A sandwich-like hierarchically porous carbon/graphene composite as a high-performance anode material for sodium-ion batteries *Adv. Energy Mater.* **4** 1301584
[90] Yan D, Xu X, Lu T, Hu B, Chua D H and Pan L 2016 Reduced graphene oxide/carbon nanotubes sponge: a new high capacity and long life anode material for sodium-ion batteries *J. Power Sources* **316** 132–8
[91] Liang S, Cheng Y J, Zhu J, Xia Y and Müller-Buschbaum P 2020 A chronicle review of nonsilicon (Sn, Sb, Ge)-based lithium/sodium-ion battery alloying anodes *Small Methods* **4** 2000218
[92] Kim D, Zhang K, Cho M and Kang Y M 2019 Critical design factors for kinetically favorable P-based compounds toward alloying with Na ions for high-power sodium-ion batteries *Energy Environ. Sci.* **12** 1326–33

[93] Jiang Y, Song D, Wu J, Wang Z, Huang S, Xu Y, Chen Z, Zhao B and Zhang J 2019 Sandwich-like SnS_2/graphene/SnS_2 with expanded interlayer distance as high-rate lithium/sodium-ion battery anode materials *ACS Nano* **13** 9100–11

[94] Wang X, Li G, Hassan F M, Li J, Fan X, Batmaz R, Xiao X and Chen Z 2015 Sulfur covalently bonded graphene with large capacity and high rate for high-performance sodium-ion batteries anodes *Nano Energy* **15** 746–54

[95] Yang Y *et al* 2017 'Protrusions' or 'holes' in graphene: which is the better choice for sodium ion storage? *Energy Environ. Sci.* **10** 979–86

[96] Hao Y and Wang C 2020 Free-standing reduced graphene oxide/carbon nanotube paper for flexible sodium-ion battery applications *Molecules* **25** 1014

[97] Chen W, Song K, Mi L, Feng X, Zhang J, Cui S and Liu C 2017 Synergistic effect induced ultrafine SnO_2/graphene nanocomposite as an advanced lithium/sodium-ion batteries anode *J. Mater. Chem.* A **5** 10027–38

[98] Girishkumar G, McCloskey B, Luntz A C, Swanson S and Wilcke W 2010 Lithium– air battery: promise and challenges *J. Phys. Chem. Lett.* **1** 2193–203

[99] Jung H G, Hassoun J, Park J B, Sun Y K and Scrosati B 2012 An improved high-performance lithium–air battery *Nat. Chem.* **4** 579–85

[100] Jiang Y, Deng Y P, Liang R, Fu J, Gao R, Luo D, Bai Z, Hu Y, Yu A and Chen Z 2020 d-Orbital steered active sites through ligand editing on heterometal imidazole frameworks for rechargeable zinc–air battery *Nat. Commun.* **11** 1–11

[101] Zhang Y L *et al* 2020 Advanced non-noble materials in bifunctional catalysts for ORR and OER toward aqueous metal–air batteries *Nanoscale* **12** 21534–59

[102] Liu Q, Pan Z, Wang E, An L and Sun G 2020 Aqueous metal–air batteries: fundamentals and applications *Energy Storage Mater.* **27** 478–505

[103] Kondori A *et al* 2020 Kinetically stable oxide overlayers on Mo_3P nanoparticles enabling lithium–air batteries with low overpotentials and long cycle life *Adv. Mater.* **32** 2004028

[104] Wang L, Wang Y, Qiao Y, Wu S, Lu X, Zhu J J, Zhang J R and Zhou H 2020 Superior efficient rechargeable lithium–air batteries using a bifunctional biological enzyme catalyst *Energy Environ. Sci.* **13** 144–51

[105] Kang J H *et al* 2020 Lithium–air batteries: air-breathing challenges and perspective *ACS Nano* **14** 14549–78

[106] Song S, Qin X, Ruan Y, Li W, Xu Y, Zhang D and Thokchom J 2020 Enhanced performance of solid-state lithium–air batteries with continuous 3D garnet network added composite polymer electrolyte *J. Power Sources* **461** 228146

[107] Lai J, Xing Y, Chen N, Li L, Wu F and Chen R 2020 Electrolytes for rechargeable lithium–air batteries *Angew. Chem. Int. Ed.* **59** 2974–97

[108] Yi X, Liu X, Dou R, Wen Z and Zhou W 2021 Unraveling the control mechanism of carbon nanotubes on the oxygen reduction reaction and product growth behavior in lithium–air batteries *ACS Appl. Energy Mater.* **4** 2148–57

[109] Al-Ogaili A W M, Cetinkaya T, Pakseresht S and Akbulut H 2021 Graphene-based nanocomposite cathodes architecture with palladium and α-MnO_2 for high cycle life lithium-oxygen batteries *J. Alloys Compd.* **854** 157293

[110] Novčić K A, Dobrota A S, Petković M, Johansson B, Skorodumova N V, Mentus S V and Pašti I A 2020 Theoretical analysis of doped graphene as cathode catalyst in Li-O_2 and Na-O_2 batteries–the impact of the computational scheme *Electrochim. Acta* **354** 136735

[111] Lin Y, Moitoso B, Martinez-Martinez C, Walsh E D, Lacey S D, Kim J W, Dai L, Hu L and Connell J W 2017 Ultrahigh-capacity lithium–oxygen batteries enabled by dry-pressed holey graphene air cathodes *Nano Lett.* **17** 3252–60

[112] Zahoor A, Faizan R, Elsaid K, Hashmi S, Butt F A and Ghouri Z K 2021 Synthesis and experimental investigation of δ-MnO_2/N-rGO nanocomposite for Li–O_2 batteries applications *Chem. Eng. J. Adv.* **7** 100115

[113] Cui X, Luo Y, Zhou Y, Dong W and Chen W 2021 Application of functionalized graphene in Li–O_2 batteries *Nanotechnology* **32** 132003

[114] Ionescu M I and Laforgue A 2020 Synthesis of nitrogen-doped carbon nanotubes directly on metallic foams as cathode material with high mass load for lithium–air batteries *Thin Solid Films* **709** 138211

[115] Kim J H, Kannan A G, Woo H S, Jin D G, Kim W, Ryu K and Kim D W 2015 A bifunctional metal-free catalyst composed of dual-doped graphene and mesoporous carbon for rechargeable lithium–oxygen batteries *J. Mater. Chem.* A **3** 18456–65

[116] Zhou W, Zhang H, Nie H, Ma Y, Zhang Y and Zhang H 2015 Hierarchical micron-sized mesoporous/macroporous graphene with well-tuned surface oxygen chemistry for high capacity and cycling stability Li–O_2 battery *ACS Appl. Mater. Interfaces* **7** 3389–97

[117] Wu F *et al* 2016 Facile synthesis of boron-doped rGO as cathode material for high energy Li–O_2 batteries *ACS Appl. Mater. Interfaces* **8** 23635–45

[118] Kim M, Kim D W, Suk J, Park O O and Kang Y 2015 Flexible binder-free graphene paper cathodes for high-performance Li-O_2 batteries *Carbon* **93** 625–35

[119] Yang Y, Yin W, Wu S, Yang X, Xia W, Shen Y, Huang Y, Cao A and Yuan Q 2016 Perovskite-type LaSrMnO electrocatalyst with uniform porous structure for an efficient Li–O_2 battery cathode *ACS Nano* **10** 1240–8

[120] Luo W B, Chou S L, Wang J Z, Zhai Y C and Liu H K 2015 A metal-free, free-standing, macroporous graphene@ g-C_3N_4 composite air electrode for high-energy lithium oxygen batteries *Small* **11** 2817–24

[121] Kim M, Kim D W, Suk J, Park J J, Park O O and Kang Y 2016 Graphene paper with controlled pore structure for high-performance cathodes in Li–O_2 batteries *Carbon* **100** 265–72

[122] Yu C, Zhao C, Liu S, Fan X, Yang J, Zhang M and Qiu J 2015 Polystyrene sphere-mediated ultrathin graphene sheet-assembled frameworks for high-power density Li–O_2 batteries *Chem. Commun.* **51** 13233–36

[123] He M, Zhang P, Liu L, Liu B and Xu S 2016 Hierarchical porous nitrogen doped three-dimensional graphene as a free-standing cathode for rechargeable lithium-oxygen batteries *Electrochim. Acta* **191** 90–7

[124] Zhu J, Metzger M, Antonietti M and Fellinger T P 2016 Vertically aligned two-dimensional graphene-metal hydroxide hybrid arrays for Li–O_2 batteries *ACS Appl. Mater. Interfaces* **8** 26041–50

[125] Wang G, Tu F, Xie J, Du G, Zhang S, Cao G and Zhao X 2016 High-performance Li–O_2 batteries with controlled Li_2O_2 growth in graphene/Au-nanoparticles/Au-nanosheets sandwich *Adv. Sci.* **3** 1500339

[126] Ren S, Duan X, Liang S, Zhang M and Zheng H 2020 Bifunctional electrocatalysts for Zn–air batteries: recent developments and future perspectives *J. Mater. Chem.* A **8** 6144–82

[127] Song Z, Ding J, Liu B, Liu X, Han X, Deng Y, Hu W and Zhong C 2020 A rechargeable Zn–air battery with high energy efficiency and long life enabled by a highly water-retentive gel electrolyte with reaction modifier *Adv. Mater.* **32** 1908127

[128] Zhou T, Zhang N, Wu C and Xie Y 2020 Surface/interface nanoengineering for rechargeable Zn–air batteries *Energy Environ. Sci.* **13** 1132–53
[129] Zhang J, Zhao B, Liang W, Zhou G, Liang Z, Wang Y, Qu J, Sun Y and Jiang L 2020 Three-phase electrolysis by gold nanoparticle on hydrophobic interface for enhanced electrochemical nitrogen reduction reaction *Adv. Sci.* **7** 2002630
[130] Matter P H, Wang E, Arias M, Biddinger E J and Ozkan U S 2007 Oxygen reduction reaction activity and surface properties of nanostructured nitrogen-containing carbon *J. Mol. Catal. A: Chem.* **264** 73–81
[131] Li A, Cao Q, Zhou G, Schmidt B V, Zhu W, Yuan X, Huo H, Gong J and Antonietti M 2019 Three-phase photocatalysis for the enhanced selectivity and activity of CO_2 reduction on a hydrophobic surface *Angew. Chem.* **131** 14691–97
[132] Lu Z, Xu W, Ma J, Li Y, Sun X and Jiang L 2016 Superaerophilic carbon-nanotube-array electrode for high-performance oxygen reduction reaction *Adv. Mater.* **28** 7155–61
[133] Pan J, Xu Y Y, Yang H, Dong Z, Liu H and Xia B Y 2018 Advanced architectures and relatives of air electrodes in Zn–air batteries *Adv. Sci.* **5** 1700691
[134] Zhang J, Zhou Q, Tang Y, Zhang L and Li Y 2019 Zinc–air batteries: are they ready for prime time? *Chem. Sci.* **10** 8924–9
[135] Qiu H J, Du P, Hu K, Gao J, Li H, Liu P, Ina T, Ohara K, Ito Y and Chen M 2019 Metal and nonmetal codoped 3D nanoporous graphene for efficient bifunctional electrocatalysis and rechargeable Zn–air batteries *Adv. Mater.* **31** 1900843
[136] Tang C, Wang B, Wang H F and Zhang Q 2017 Defect engineering toward atomic Co–N_x–C in hierarchical graphene for rechargeable flexible solid Zn–air batteries *Adv. Mater.* **29** 1703185
[137] Wang Q, Ji Y, Lei Y, Wang Y, Wang Y, Li Y and Wang S 2018 Pyridinic-N-dominated doped defective graphene as a superior oxygen electrocatalyst for ultrahigh-energy-density Zn–air batteries *ACS Energy Lett.* **3** 1183–91
[138] Xu Y, Deng P, Chen G, Chen J, Yan Y, Qi K, Liu H and Xia B Y 2020 2D nitrogen-doped carbon nanotubes/graphene hybrid as bifunctional oxygen electrocatalyst for long-life rechargeable Zn–air batteries *Adv. Funct. Mater.* **30** 1906081
[139] Tam T V, Kang S G, Kim M H, Lee S G, Hur S H, Chung J S and Choi W M 2019 Novel graphene hydrogel/B-doped graphene quantum dots composites as trifunctional electrocatalysts for Zn–air batteries and overall water splitting *Adv. Energy Mater.* **9** 1900945
[140] Xu L, Tian Y, Deng D, Li H, Zhang D, Qian J, Wang S, Zhang J, Li H and Sun S 2020 Cu nanoclusters/FeN_4 amorphous composites with dual active sites in N-doped graphene for high-performance Zn–air batteries *ACS Appl. Mater. Interfaces* **12** 31340–50
[141] Cheng Y *et al* 2021 Atomic Ni and Cu co-anchored 3D nanoporous graphene as an efficient oxygen reduction electrocatalyst for zinc–air batteries *Nanoscale* **13** 10862–70
[142] Han X, Zhang W, Ma X, Zhong C, Zhao N, Hu W and Deng Y 2019 Identifying the activation of bimetallic sites in $NiCo_2S_4$@ g-C_3N_4-CNT hybrid electrocatalysts for synergistic oxygen reduction and evolution *Adv. Mater.* **31** 1808281
[143] Liu G *et al* 2019 An oxygen-vacancy-rich semiconductor-supported bifunctional catalyst for efficient and stable zinc–air batteries *Adv. Mater.* **31** 1806761
[144] Hang C, Zhang J, Zhu J, Li W, Kou Z and Huang Y 2018 In situ exfoliating and generating active sites on graphene nanosheets strongly coupled with carbon fiber toward self-standing bifunctional cathode for rechargeable Zn–air batteries *Adv. Energy Mater.* **8** 1703539

[145] Fu G, Wang J, Chen Y, Liu Y, Tang Y, Goodenough J B and Lee J M 2018 Exploring indium-based ternary thiospinel as conceivable high-potential air-cathode for rechargeable Zn–Air batteries *Adv. Energy Mater.* **8** 1802263

[146] Cheng H, Li M L, Su C Y, Li N and Liu Z Q 2017 Cu–Co bimetallic oxide quantum dot decorated nitrogen-doped carbon nanotubes: a high-efficiency bifunctional oxygen electrode for Zn–air batteries *Adv. Funct. Mater.* **27** 1701833

[147] Wan W, Liu X, Li H, Peng X, Xi D and Luo J 2019 3D carbon framework-supported CoNi nanoparticles as bifunctional oxygen electrocatalyst for rechargeable Zn–air batteries *Appl. Catal.* B **240** 193–200

[148] Ahn S H and Manthiram A 2017 Cobalt phosphide coupled with heteroatom-doped nanocarbon hybrid electroctalysts for efficient, long-life rechargeable zinc–air batteries *Small* **13** 1702068

[149] Kötz R and Carlen M J E A 2000 Principles and applications of electrochemical capacitors *Electrochim. Acta* **45** 2483–98

[150] Deka B K, Hazarika A, Kim J, Park Y B and Park H W 2017 Recent development and challenges of multifunctional structural supercapacitors for automotive industries *Int. J. Energy Res.* **41** 1397–411

[151] Raza W, Ali F, Raza N, Luo Y, Kim K H, Yang J, Kumar S, Mehmood A and Kwon E E 2018 Recent advancements in supercapacitor technology *Nano Energy* **52** 441–73

[152] Ray A, Roy A, Ghosh M, Ramos-Ramón J A, Saha S, Pal U, Bhattacharya S K and Das S 2019 Study on charge storage mechanism in working electrodes fabricated by sol-gel derived spinel $NiMn_2O_4$ nanoparticles for supercapacitor application *Appl. Surf. Sci.* **463** 513–25

[153] Najib S and Erdem E 2019 Current progress achieved in novel materials for supercapacitor electrodes: mini review *Nanoscale Adv.* **1** 2817–27

[154] Zhang L L and Zhao X S 2009 Carbon-based materials as supercapacitor electrodes *Chem. Soc. Rev.* **38** 2520–31

[155] Pan H, Li J and Feng Y 2010 Carbon nanotubes for supercapacitor *Nanoscale Res. Lett.* **5** 654–68

[156] Yoo J J *et al* 2011 Ultrathin planar graphene supercapacitors *Nano Lett.* **11** 1423–7

[157] Tan Y B and Lee J M 2013 Graphene for supercapacitor applications *J. Mater. Chem.* A **1** 14814–43

[158] Chen Z, Zhao S, Zhao H, Zou Y, Yu C and Zhong W 2021 Nitrogen-doped interpenetrating porous carbon/graphene networks for supercapacitor applications *Chem. Eng. J.* **409** 127891

[159] Chen Y, Li Y, Yao F, Peng C, Cao C, Feng Y and Feng W 2019 Nitrogen and fluorine co-doped holey graphene hydrogel as a binder-free electrode material for flexible solid-state supercapacitors *Sustain. Energy Fuels* **3** 2237–45

[160] Khandelwal M, Van Tran C, Lee J and In J B 2022 Nitrogen and boron co-doped densified laser-induced graphene for supercapacitor applications. *Chem. Eng. J.* **428** 131119

[161] Lehtimaki S, Suominen M, Damlin P, Tuukkanen S, Kvarnstrom C and Lupo D 2015 Preparation of supercapacitors on flexible substrates with electrodeposited PEDOT/graphene composites *ACS Appl. Mater. Interfaces* **7** 22137–47

[162] Biswas S and Drzal L T 2010 Multilayered nanoarchitecture of graphene nanosheets and polypyrrole nanowires for high performance supercapacitor electrodes *Chem. Mater.* **22** 5667–71

[163] Wu Q, Xu Y, Yao Z, Liu A and Shi G 2010 Supercapacitors based on flexible graphene/polyaniline nanofiber composite films *ACS Nano* **4** 1963–70
[164] Zhou F *et al* 2018 Electrochemically scalable production of fluorine-modified graphene for flexible and high-energy ionogel-based microsupercapacitors *J. Am. Chem. Soc.* **140** 8198–205
[165] Lee S P, Ali G A, Hegazy H H, Lim H N and Chong K F 2021 Optimizing reduced graphene oxide aerogel for a supercapacitor *Energy Fuels* **35** 4559–69
[166] Liu Q, Zhang L, Chen H, Jin J, Wang N, Wang Y and Sui D 2021 Sulfur and nitrogen co-doped three-dimensional graphene aerogels for high-performance supercapacitors: a head to head vertical bicyclic molecule both as pillaring agent and dopant *Appl. Surf. Sci.* **565** 150453

IOP Publishing

Recent Advances in Graphene and Graphene-Based Technologies

Anoop Chandran, N V Unnikrishnan, M K Jayaraj, Reenu Elizabeth John and Justin George

Chapter 9

Photonic and optoelectronic applications of graphene: nonlinear optical properties of graphene and its applications

Arpana Agrawal and Gyu-Chul Yi

Implementation of any material for the successful fabrication of optoelectronic devices warrants the precise knowledge of their nonlinear optical properties including the nonlinear optical absorption/refraction coefficients, carrier dynamics, etc. The magnitude and sign of such parameters are the decisive factor for specific device applications such as optical limiters, saturable absorbers, mode lockers, etc. For such applications, graphene-based nanomaterials have engendered immense scientific research interest, owing to their various fascinating and unique nonlinear optical properties which can be further tailored by integrating them with several other materials, and hence add another dimension to their utility. Accordingly, the present chapter provides a comprehensive overview of the nonlinear optical properties of various graphene-based nanomaterials including metals decorated graphene-based nanomaterials, dispersion of graphene-based nanomaterials in various solvents, thin films of graphene-based nanomaterials, 2D materials/graphene-based nanomaterials and semiconductor/graphene-based nanomaterials. Outcomes based on the nonlinear optical properties from recent studies were compiled and are discussed in detail.

9.1 Introduction

Invention of intense laser light sources and the possibility of strong light–matter interactions give birth to a new branch of optics termed as nonlinear optics. Nonlinear optics is the optics of intense light which examines the behavior of light in the nonlinear media that enables the realization of new phenomena in all aspects of photonics including absorption, transmission, photon generation and manipulation, imaging and detection. Material scientists are in search of excellent materials that can exhibit excellent optical

doi:10.1088/978-0-7503-3999-5ch9

nonlinearity so that they can be utilized in making various optoelectronic devices including optical limiters, optical switches, saturable absorbers, mode lockers, etc. Under such class of materials, novel two-dimensional (2D) graphene (G)-based nanomaterials have proven themselves as a potential candidate material and since their discovery in 2004, this material has been immensely explored for several of its fascinating properties including the nonlinear optical (NLO) properties [1]. Its inimitable electronic band structure with the presence of delocalized π-electrons and ultrafast carrier dynamics have led to stupendous linear as well as NLO properties. Also, the linear dispersion of massless Dirac fermions entails nearly unvarying absorption over the entire optical spectrum signifying its resonant temperament for any optical excitation wavelength.

Utilization of graphene-based nanomaterial for the fabrication of optoelectronic devices (optical limiters, saturable absorbers or optical switches etc) warrants the precise knowledge of their NLO parameters including nonlinear absorption (NLA) coefficient and nonlinear refractive (NLR) index. So far, several experimental techniques have been employed for the determination of NLO properties including self-phase modulation [2], second/third harmonic generation [3], four-wave mixing [4], z-scan technique [5, 6] etc. Among all these, z-scan technique, based on the principles of spatial beam distortion is the simplest and most sensitive technique to determine the magnitude and sign of the NLO parameters [7–10]. Herein, the sample is translated in the positive and negative *z*-direction with respect to the focal point of the tight focusing lens along the axis of the focused Gaussian laser beam. While the input energy is kept constant, the sample experiences different incident field at different z-positions. This experiment can be performed in two geometries: open aperture (OA) z-scan geometry utilized to determine the intensity dependent NLA coefficient and closed aperture (CA) z-scan geometry to determine the NLR index.

Figure 9.1 schematically illustrates the z-scan experimental setup in both the geometries where a collimated laser beam from the source L is tightly focused on the sample

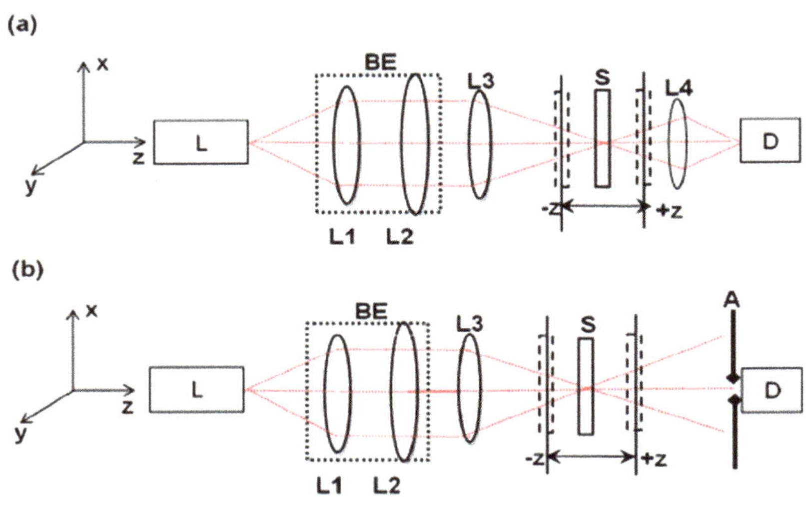

Figure 9.1. Schematic illustration of z-scan techniques in OA (a) and CA (b) geometries.

using a tight focusing convex lens L3 after passing through a beam expander BE (combination of lens L1 and L2) and the sample S is mounted on a translation stage translating along the z-axis with respect to the focal point of lens L3. Transmitted beam is then collected by a detector D. For OA experiment (figure 9.1(a)), another large aperture convex lens (L4) is used to collect whole of the transmitted light to the detector D which is placed at the focal plane of lens L4 while for CA experiment (figure 9.1(b), an aperture A is placed in front of the detector to allow only the on-axis beam to enter the detector.

It is worth mentioning here either individual saturation absorption (SA), reverse saturation absorption (RSA) or a combination of these two effects is responsible for the absorptive optical nonlinearity within the material. In the first case where the nonlinear transmission was described by purely considering individual SA or RSA processes, the sign of the NLA coefficient (β) can be numerically obtained by solving the wave propagation equation of the form [9];

$$\frac{dI}{dz'} = \alpha_o I + \beta I^2 \tag{9.1}$$

where α_o, z' and I are the unsaturated linear absorption coefficient, propagation distance within the material and laser intensity, respectively. In such situation, the total absorption coefficient ($\alpha(I)$) and transmittance as a function of transverse distance ($T(z)$) can be expressed as [9];

$$\alpha(I) = \alpha_o + \beta I \tag{9.2}$$

$$T(z) \approx 1 - \frac{q_o}{2\sqrt{2}(1 + z^2/z_R^2)} \tag{9.3}$$

Here, $q_o = \beta I_o L_{eff}$, with I_o and L_{eff} being the maximum intensity at the focal point ($z = 0$) and linear absorption compensated length, respectively, and z_R is the Rayleigh range. From equation (9.3), it can be clearly observed that the transmittance from the material can either increase or decrease with increasing light intensity depending upon the sign of β and, consequently, the practical application of the material. Positive (negative) absorptive optical nonlinearity implies decreases (increases) in transmittance with increasing laser light intensity and this phenomenon is known as RSA (SA). Both positive and negative optical absorptive nature has practical applications in optical limiters to provide safety to detectors, sensors, human eye etc and amplifiers, saturable absorbers used in Q-switching and mode locking, respectively. Optical limiters (saturable absorbers) are the smart materials that are transparent (opaque) for low intensities and become opaque (transparent) for high intensities and thus provide safety to detectors, sensors, human eye, etc and are based on positive (negative) absorptive optical nonlinearity. On the other hand, optical switches require the knowledge of NLO refraction.

Contrary to individual SA or RSA effect, it is generally observed that the transmittance signal obtained from graphene-like materials exhibits both SA and RSA behavior and under such a case, $T(z)$ can be numerically obtained by solving the following wave propagation equation [10] given by;

$$\frac{dI}{dz'} = \frac{\alpha_o I}{1 + I/I_s} + \beta I^2 \tag{9.4}$$

here I_s is the saturation intensity which characterizes the saturation behavior and α (I) will be given by,

$$\alpha(I) = \frac{\alpha_o}{1 + I/I_s} + \beta I \tag{9.5}$$

Consequently, $T(z)$ after considering both RSA and SA processes can be approximated as [10];

$$T(z) \approx 1 - \frac{\alpha_o L_{eff}}{1 + I/I_s} + \frac{q_o}{2\sqrt{2}\left(1 + z^2/z_R^2\right)^2} \tag{9.6}$$

The above equation takes care of the combined effect of SA and RSA.

Refractive optical nonlinearity arises as a result of modification in the electron distribution upon irradiating the sample with an intense laser light beam and consequently, the total refractive index is a function of intensity and can be expressed as [9];

$$n(I) = n_o + n_2 I \tag{9.7}$$

where n_o and n_2 represents linear and NLR index, respectively. It should be noted here that because of the Gaussian beam profile of laser beam, there will be higher intensity at the center of the beam spot and hence the refractive index of the material can never be considered as constant over the beam cross-section. The central part of the beam can either propagate slower or faster depending upon the sign of the n_2, which is the intrinsic property of the material. Consequently, the sample acts as a lens possessing 'variable focal length' giving rise self-focusing and self-defocusing phenomenon and just by visualization of $T(z)$ signals, one can predict the sign of the NLR index. The appearance of pre-focal peak (valley) followed by a post-focal valley (peak) feature in $T(z)$ curve obtained in CA assembly signalizes the negative (positive) sign of NLR index. In case of positive NLR index, the sample behaves as a convex lens (converging lens) exhibiting positive focal length that converges the beam leading to self-focusing effect. Thus there is larger refractive index at the center of beam than its periphery. Contrary to this, in the case of negative refraction, self-defocusing effect, the sample behaves as a concave lens having negative focal length (diverging lens) and the beam passing through the sample tends to spread out. Thus there is larger refractive index at the periphery than its center and the refractive index is negative.

NLO response of graphene-based nanomaterials is one of the decisive factors for its utilization for specific optoelectronic device. Additionally, their optical nonlinearity can be further enhanced by coupling them with several other materials including several metals decorated graphene-based nanomaterials, graphene-based nanomaterials dispersed in various solvents, thin films of graphene-based nanomaterials, organic glasses made up of other 2D materials (MoS_2, $MoSe_2$, etc) and graphene-based nanomaterials, or semiconductor/graphene-based nanomaterials heterostructures. In view of this, the present chapter provides a comprehensive overview of the NLO responses and possible applications of several graphene-based nanomaterials. However, it is unfeasible to cover all the aspects of optical nonlinearity of graphene-based nanomaterial in a single chapter; efforts are being made to discuss few of them along with the outcomes of the recent studies.

9.2 Nonlinear optical properties of graphene-based materials

Graphene (G)-based nanomaterials exhibit excellent optical nonlinearity owing to the linear dispersion of massless Dirac fermions and have been extensively employed for the fabrication of outstanding optoelectronic devices. Such nanomaterials include: (i) graphene-based nanomaterials decorated with metals; (ii) graphene-based nanomaterials dispersed in various solvents; (iii) thin films of graphene-based nanomaterials grown on various substrates; (iv) 2D materials/graphene-based nanomaterials; and (v) semiconductor/graphene-based nanomaterials. Detailed discussion of all these graphene-based nanomaterials is provided below.

9.2.1 Metals/graphene-based nanomaterials

Several metals such as gold (Au), silver (Ag), platinum (Pt), Pd, Ni, etc are being employed so far for decorating graphene-based nanomaterials and can enhance the NLO responses of these materials and hence their utility. Yue *et al* [11] have reported an encouraging optical nonlinearity in Ag decorated graphene oxide (GO) nanoparticles (NPs) (AgNPs/GO) which were prepared by femtosecond (fs) laser ablation process. Optical nonlinearity has been investigated using z-scan technique employing fs laser source. Saturation intensity of 18.5 MW cm^{-2} and enhanced nonlinear refractivity (1.1×10^{-12} m^2 W^{-1}) was obtained for AgNPs/GO. This enhancement as compared to reduced graphene oxide (rGO) was attributed to the interaction of energy states of AgNPs. Enhancement in thermal nonlinearity of functionalized graphene by Au nanorods (NRs) has also been examined by Abdulaal *et al* [12] using continuous wave (CW) argon laser (λ = 514 nm). A negative NLO refractivity was found, suggesting the existence of self-defocusing effect.

Sadrolhesseini *et al* [13] have examined the NLO properties, particularly nonlinear refraction of GO, with different concentration and AuNPs decorated GO (AuNPs/GO) nanocomposites. AuNPs/GO nanocomposites were prepared using laser ablation technique with varying ablation time. The complete setup of laser ablation process is depicted in figure 9.2(a) where second harmonic of Q-switched Nd:YAG laser (λ: 532 nm; energy: 1200 mJ; pulse duration: 10 ns; repetition rate: 40 Hz) is employed to ablate the Au plate dipped in 10 ml GO solution for different ablation times varying from 10 min to 40 min, resulting in the formation of AuNPs/GO nanocomposites. Figure 9.2 shows the GO sheets before (figure 9.2(b)) and after (figure 9.2(c)) the laser ablation of Au plate. The particle shapes of the AuNPs ablated at 10, 20, 30 and 40 min have also been analyzed and the corresponding transmission electron microscope (TEM) micrographs are shown in figures 9.2(d)–(g), respectively.

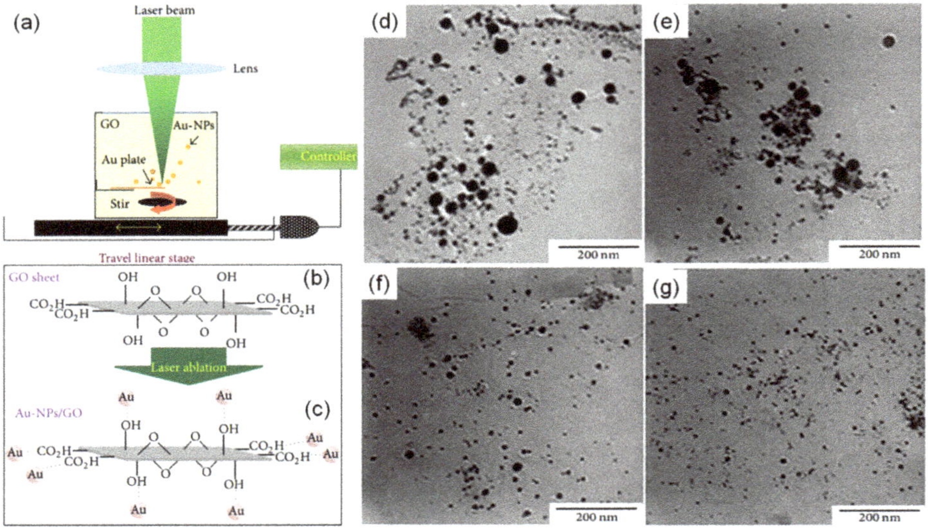

Figure 9.2. (a) Fabrication of AuNPs in GO using laser ablation method. GO sheet before (b) and after (c) the ablation of gold plate, TEM image of AuNPs in GO for 10, 20, 30 and 40 min ablation times showing the particle size of 16.55 nm (d), 13.4 nm (e), 9.51 nm (f), and 5.18 nm (g), respectively. Reproduced from [13] under a Creative Commons Attribution License, Copyright 2014, Hindawi.

A decrease in the particle size has been observed from 16.55 nm to 5.18 nm with increase in ablation times.

NLO properties of GO and AuNPs/GO nanocomposites have been investigated using a diode laser (180 mW) having a wavelength of 532 nm. Figures 9.3(a)–(d) show the z-scan experimental results for different concentrations of GO (0.1–0.8 mg ml^{-1}), where a peak-valley has been found, suggesting the existence of self-defocusing effect and hence negative NLR index. Furthermore, magnitude of nonlinear refraction is also found to increase with increase in concentration. The z-scan transmittance signal for AuNPs/GO nanocomposites prepared at different ablation times has also been shown in figures 9.3(e)–(h), which also illustrates negative optical refractivity with increased magnitude upon increasing the ablation time. Kalanoor *et al* [14] have also investigated the NLO properties of AgNPs decorated functionalized multilayer graphene (AgNPS/fG) at 1064 nm. A change in the sign of NLA coefficient from negative to positive and hence from SA behavior to RSA behavior has been observed upon increasing the input intensity. Table 9.1 summarizes few experimentally obtained NLO parameters (nonlinear refraction (NLR), NLA, limiting threshold F_{th}) for metal-decorated graphene-based nanomaterials.

9.2.2 Graphene-based nanomaterials dispersed in various solvents

Graphene-based nanomaterials dispersed in various solvents including *N*,*N*-dimethylformamide (DMF), *N*-methyl-2-pyrrolidone (NMP), γ-butyrolactone (GBL),

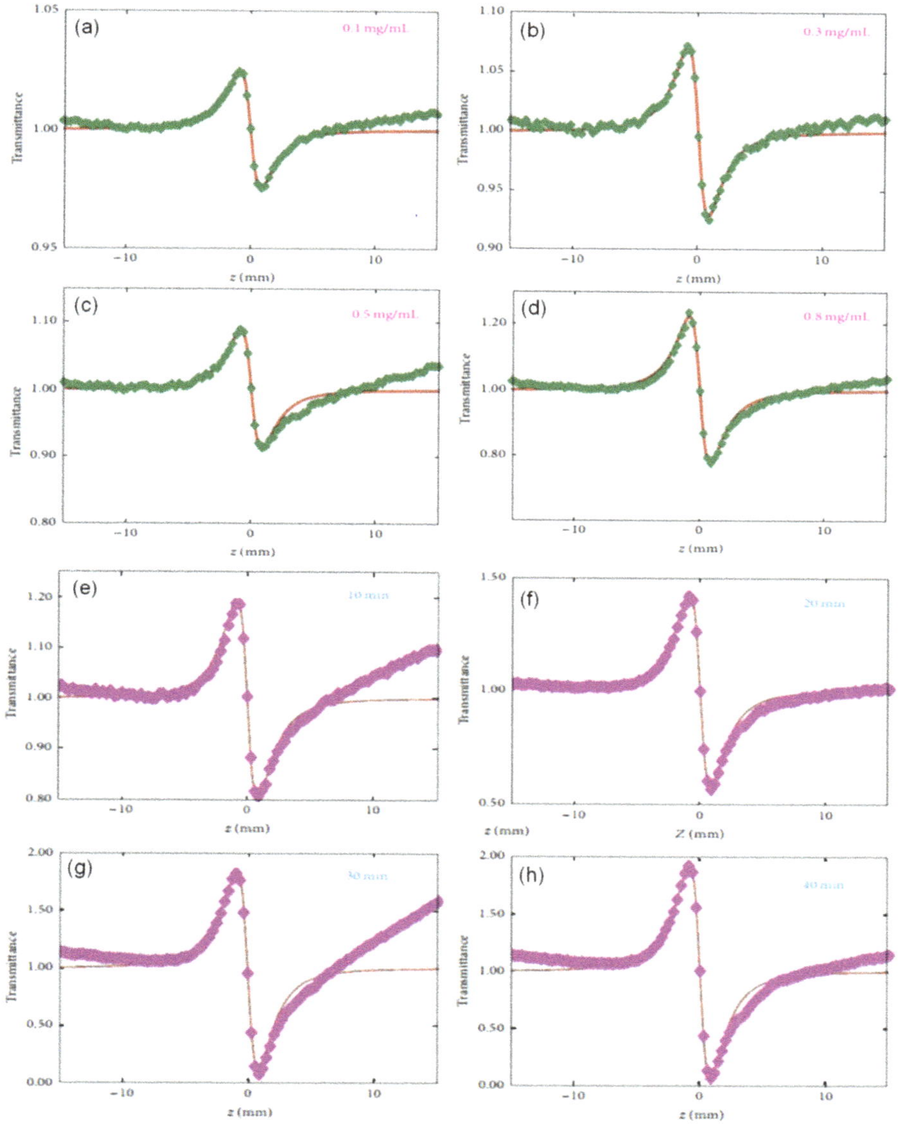

Figure 9.3. (a)–(d) CA z-scan signal for the GO solution for various concentrations of GO in water. (e)–(h) CA z-scan signal for AuNPs/GO nanocomposite for various ablation times. Reproduced from [13], under a Creative Commons Attribution License, Copyright 2014, Hindawi.

dimethyl sulfoxide (DMSO), *N,N*-dimethylacetamide (DMA), tetrahydrofuran (THF), acetonitrile (CAN), ethanol, water, etc, are also reported to exhibit excellent NLO properties. Liu *et al* [19] have investigated the NLO properties of GO dispersed in DMF by performing z-scan measurement using second harmonic of Q-switched Nd:YAG laser generating 5 ns pulses and mode locked Nd:YAG laser under picoseconds regime (35 ps pulses). Under ps regime, a symmetrical peak

Table 9.1. Experimentally obtained NLO parameters for metal-decorated graphene-based nanomaterials.

Metal/graphene-based nanomaterials	Laser parameters	NLO parameters	Reference
AuNPs (16.55 nm)/GO	532 nm; 180 mW	NLR = -1.85 cm^2 GW^{-1}	[13]
AuNPs (13.41 nm)/GO	532 nm; 180 mW	NLR = -2.7 cm^2/ GW^{-1}	[13]
AuNPs (9.52 nm)/GO	532 nm; 180 mW	NLR = -4.1 cm^2 GW^{-1}	[13]
AuNPs (5.18 nm)/GO	532 nm; 180 mW	NLR = -5.8 cm^2 GW^{-1}	[13]
AgNPs/rGO	—	I_s = 18.5 MW cm^{-2} NLR = -1.1×10^{-12} m^2 W^{-1}	[11]
AgNPs/fG composite	532 nm; 40 ps; 10 Hz	NLA = 812 cm GW^{-1} I_s = 3.7 GW cm^{-2}	[14]
AgNPs/fG composite	1064 nm; 40 ps; 10 Hz	NLA = 600 cm GW^{-1}	[14]
Ag-GO	532 nm; 10 ns	NLA = 45.4 cm GW^{-1} at 0.20 GW cm^{-2} NLA = 39.7 cm GW^{-1} at 0.16 GW cm^{-2} NLA = 32.4 cm GW^{-1} at 0.10 GW cm^{-2} NLA = 30.0 cm GW^{-1} at 0.08 GW cm^{-2}	[15]
Non-covalent f-rGO/AgNPs (1 M)	532 nm; 5 ns	NLA = 13.9 m GW^{-1}	[16]
Non-covalent f-rGO/AgNPs (1 M)	800 nm; 100 fs	NLA = 5.8×10^{-15} m W^{-1}	[16]
NiNPs/rGO	532 nm; 1 Hz; 4 ns; 50 μJ	NLA = 1.29 cm GW^{-1}	[17]
PtNPs/rGO	532 nm; 1 Hz; 4 ns; 50 μJ	NLA = 1.38 cm GW^{-1}	[17]
Pt/functionalized hydrogen exfoliated-G (f-HEG)	532 nm; 5 ns; 0.5 Hz	F_{th} = 13.7 J cm^{-2}	[18]
Pt/f-HEG	800 nm; 100 fs; 0.5 Hz	F_{th} = 1.8 J cm^{-2}	[18]
Pt–Ni NP/rGO	532 nm; 1 Hz; 4 ns; 50 μJ	NLA = 1.64 cm GW^{-1}	[17]
Pd/f-HEG	532 nm; 5 ns; 0.5 Hz	F_{th} = 8.8 J cm^{-2}	[18]
Pd/f-HEG	800 nm; 100 fs; 0.5 Hz	F_{th} = 1.5 J cm^{-2}	[18]

indicating SA behavior was observed at an input intensity of 2.1×10^9 W cm^{-2} which turns to valley-feature at 1×10^{10} W cm^{-2} and 3.4×10^{10} W cm^{-2}, suggesting RSA behavior. Furthermore, there is a strong dependence of NLA coefficient upon input intensity under ns regime but a weaker dependence under ps regime. Wang *et al* [20] have reported the NLO properties of graphene dispersed in various dispersion solvents including DMA, NMP, GBL under an excitation of 532 nm and 1064 nm, with repetition rate of 10 Hz and pulse duration of 6 ns.

Fang *et al* [21] have demonstrated the fabrication of tunable optical limiting optofluidic device filled with GO dispersion in ethanol. The optofluidic device was fabricated using photolithographic technique where the optical nonlinearity is manipulated by adjusting the concentration of GO and flow rate ratio of injection. Figure 9.4(a) schematically illustrates the design of the optofluidic device which has two inlets—one for GO/ethanol solution and the other for ethanol solvent; one outlet for effluent; a mixture area with zigzag geometry and an optical cavity (radius = 10 nm) to store the fully mixed liquid. Figure 9.4(b) shows the optofluidic fabricated on polydimethylsiloxane (PDMS) which is transparent within the 1550 nm band. The width, height and total length of channel in the mixture area are 200 μm, 100 μm and 180 nm, respectively. Z-scan technique and pump probe measurements in fs regime within 1500 nm wavelength band were employed to examine the optical nonlinearity and dynamic excitation relaxation of GO/ethanol solution, respectively. OA z-scan experimental results for 100 μg mL^{-1} and 200 μg mL^{-1} GO/ethanol

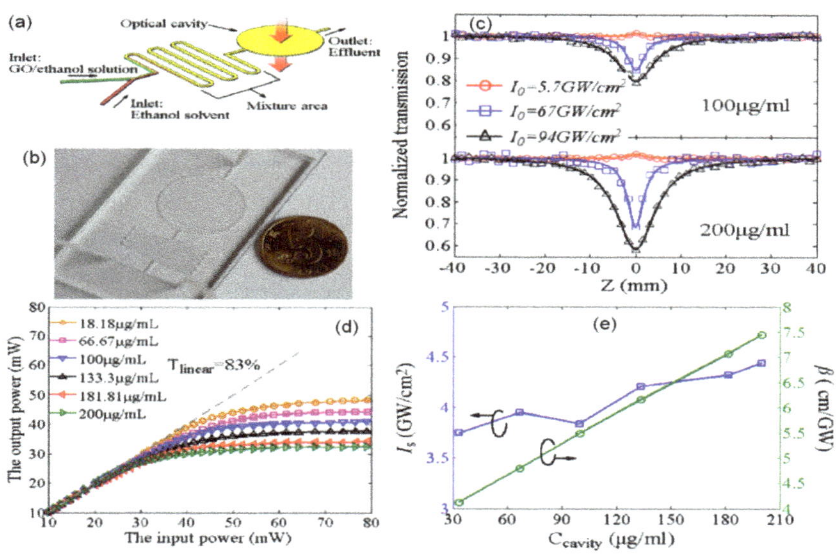

Figure 9.4. (a) Schematically illustration of the design of GO/ethanol based optofluidic device. (b) Optofluidic fabricated on PDMS. (c) OA z-scan experimental results for 100 μg mL^{-1} and 200 μg mL^{-1} GO/ethanol solutions obtained at different input intensities. (d) Optical limiting behavior of the fabricated device at various concentrations of GO/ethanol solution. (e) Dependence of saturation intensity and NLA coefficient on the concentration of GO/ethanol solution. Reproduced from [21], under the terms of the Creative Commons Attribution 4.0 International License, Copyright 2015, Springer Nature.

solutions are depicted in figure 9.4(c) which clearly shows a dip at the focal point of the transmittance versus distance signal, representing the occurrence of RSA behavior which increases with increasing the input fluence.

Optical limiting behavior at various concentrations of GO/ethanol solution is shown in figure 9.4(d) where the linear transmittance is 83% and upon increasing the input power, output power increases linearly and then becomes constant, suggesting the optical limiting behavior. Figure 9.4(e) demonstrates the dependence of saturation intensity and NLA coefficient on the concentration of GO/ethanol solution filled within the cavity. The NLO properties and hence the optical limiting behavior of GO/water, fluorinated GO/water have been reported by Chantharasupawong *et al* [22]. Few experimental results of NLO parameters obtained for several graphene-based nanomaterials dispersed in various solvents are listed in table 9.2.

9.2.3 Thin films of graphene-based nanomaterials

It should be noted that graphene-based nanomaterial in the form of thin films (single-, bi-, few-, multilayer graphene) grown on transparent quartz substrate can be easily employed for optoelectronic device applications. Hence, it is crucial to study the NLO responses of thin films of graphene-based nanomaterial. Thakur *et al* [35] have demonstrated the tunable optical Kerr nonlinearity in graphene films grown on quartz substrate using pump probe integrated z-scan measurement, employing a Ti:Sapphire laser exhibiting a tunable wavelength (690–1050 nm), repetition rate of 8 MHz at 100 fs. Optical limiting properties of rGO/polymethylmethacrylate (PMMA) films, poly[(9,9-dihexyl-9*H*-fluorene)-alt-(1,1,2,2-tetraphenylethane)] (PFTP)-/rGO/PMMA and annealed PFTP-/rGO/PMMA films has also been investigated by Liu *et al* [36]. Demetrious *et al* [37] have reported the NLO properties of multilayer graphene using z-scan technique employing 100 fs laser source at various operating wavelengths (1150, 1550, 1900 and 2400 nm).

Our group has also investigated the NLO absorptivity and refractivity in as-grown and annealed graphene multilayers grown on quartz substrate following the chemical vapor deposition (CVD) method where graphene layers were initially grown on Cu foil employing methane and hydrogen gases as graphene precursor and ambient gas [10]. Figure 9.5(a) depicts the complete growth process with four stages, showing the proper gas flows (in sccm) along with the time of flow. After the growth of graphene layers on Cu foil, they are then transferred onto quartz substrate following the wet-transfer technology [10]. The as-grown graphene layers was also annealed at 250 °C in a vacuum chamber with base pressure and working pressure of 1.8×10^{-6} Torr and 3.1×10^{-6} Torr, respectively. Figures 9.5(b) and (c) compare the Raman spectra for as-grown and annealed graphene layers at the excitation wavelengths of 633 nm and 785 nm, respectively. The figure clearly shows the predominance of Raman-G band ($\approx$1578 cm^{-1}) at both the excitation wavelengths, which are associated with the bond stretching of sp^2 atoms. Second strongest Raman 2D band (~2660 cm^{-1}) is related with the double resonance enhanced two-phonon processes. Apart from this, Raman D-band ($\approx$1330 cm^{-1}) and weak D′ band also

Table 9.2. Experimentally obtained NLO parameters for graphene-based nanomaterials dispersed in various solvents.

Sample	Solvent	Laser parameters	NLO properties	Reference
GO	DMF	532 nm; 5 ns	NLA = 5.6×10^{-8} cm W^{-1}; Is = 0.12 GW cm^{-2}	[19]
GO	DMF	532 nm; 6 ns; 10 Hz	NLA = 30.22 cm GW^{-1}	[23]
GO	DMF	1064 nm; 6 ns; 10 Hz	NLA = 6.19 cm GW^{-1}	[23]
G sheets	DMF	532 nm; 8 ns	F_{th} = 0.15/1.9 (mJ/J cm^{-2})	[24]
GF	DMF	532 nm; 4 ns	NLA = 20.71 cm GW^{-1}	[25]
GF	DMF	1064 nm; 4 ns	NLA = 12.53 cm GW^{-1}	[25]
GO	DMF	800 nm; 120 fs; 1 kHz; 82.1 GW cm^{-2}	NLA = 2.5×10^{-11} cm W^{-1} NLR = -5.3×10^{-16} cm^2 W^{-1}	[26]
G	DMF	532 nm; 6 ns; 2 Hz; 57~166 nJ	NLA = 2.89 cm GW^{-1}	[27]
FLGO	DMF	532 nm; 4 ns	NLA = 220 cm GW^{-1}	[28]
GF	DMF	532 nm	NLR = 0.16×10^{-18} m^2 W^{-1} at 35 ps NLR = 14.7×10^{-18} m^2 W^{-1} at 4 ns	[28]
G-Cu porphyrin	DMF	532 nm; 6 ns; 10 Hz	NLA = 3570 cm GW^{-1}	[29]
GO-PcZn	DMF	532 nm; 6 ns; 10 Hz	NLA = 51.16 cm GW^{-1}	[23]
GO-PcZn	DMF	1064 nm; 6 ns; 10 Hz	NLA = 31.04 cm GW^{-1}	[23]
G	DMA	532 nm; 6 ns; 10 Hz	F_{th} = 2 J cm^{-2}	[20]
G	DMA	1064 nm; 6 ns; 10 Hz	F_{th} = 4.2 J cm^{-2}	[20]
GO	DMSO	532 nm; 4 ns; 10 Hz; 0.43 J cm^{-2}	NLA = 50 cm GW^{-1}	[30]
GO-ZnPc	DMSO	532 nm; 4 ns; 10 Hz; 0.43 J cm^{-2}	NLA = 300 cm GW^{-1}	[30]
R-GO-ZnPc	DMSO	532 nm; 4 ns; 0.43 J cm^{-2}	NLA = 1500 cm GW^{-1}	[30]
G	NMP	532 nm; 6 ns; 10 Hz	F_{th} ~ 2.5 J cm^{-2}	[20]
G	NMP	1064 nm; 6 ns; 10 Hz	F_{th} ~ 7.8 J cm^{-2}	[20]
G	NMP	532 nm; 6 ns; 2 Hz; 57~166 nJ	NLA = 2.6 cm GW^{-1}	[27]
HF-GO	NMP	532 nm; 5 ns	NLA = 0.7 nm W^{-1}	[22]
G	GBL	532 nm; 6 ns; 10 Hz	F_{th} ~ 3 J cm^{-2}	[20]
G	GBL	1064 nm; 6 ns; 10 Hz	F_{th} ~ 10 J cm^{-2}	[20]
G sheet	THF	532 nm; 8 ns	F_{th} = 0.097/1.2 (mJ/J cm^{-2})	[24]
G sheet	ACN	532 nm; 8 ns	F_{th} = 0.11/1.4 (mJ/J cm^{-2})	[24]
G	Alcohol	800 nm; 50 fs; 1 kHz; 46 GW cm^{-2}	NLA = 1.96 × 10−2 cm GW^{-1}	[31]
G sheet	Water	532 nm; 8 ns	F_{th} = 0.25/3.2 (mJ/J cm^{-2})	[24]

(Continued)

Table 9.2. *(Continued)*

Sample	Solvent	Laser parameters	NLO properties	Reference
GO	Water	532 nm; 5 ns	NLA = 0.35 nm W^{-1}	[22]
GO	Water	800 nm; 50 fs; 1 kHz; 46 GW cm^{-2}	NLA = −6.84 × 10−3 cm GW^{-1}	[31]
GO NPs	Water	810 nm; 150 fs; 80 MHz	NLA (2PA) ~0.045 cm GW^{-1} (740 nm) to ~0.023 cm GW^{-1} (820–850 nm)	[32]
GO sheets	Water	532 nm; 4 ns	NLA = 130.10 cm GW^{-1}	[33]
GO sheets	Water	1064 nm; 4 ns	NLA = 20.01 cm GW^{-1}	[33]
1–3 layers GO	Water	532 nm; different pulse duration	NLA = 108.4 cm GW^{-1} at 4 ns NLA = 2.08 cm GW^{-1} at 35 ps	[28]
GO	Water	534 nm	NLR = −9.971 cm^2 GW^{-1}	[34]
fG fluoride	Water	532 nm	NLR = −34.2 × 10−18 m^2 W^{-1} at 4 ns NLR = −1.2 × 10−18 m^2 W^{-1} at 35 ps	[28]
Fluorinated GO	Water	532 nm; 5 ns	NLA = 1.40 nm W^{-1}	[22]

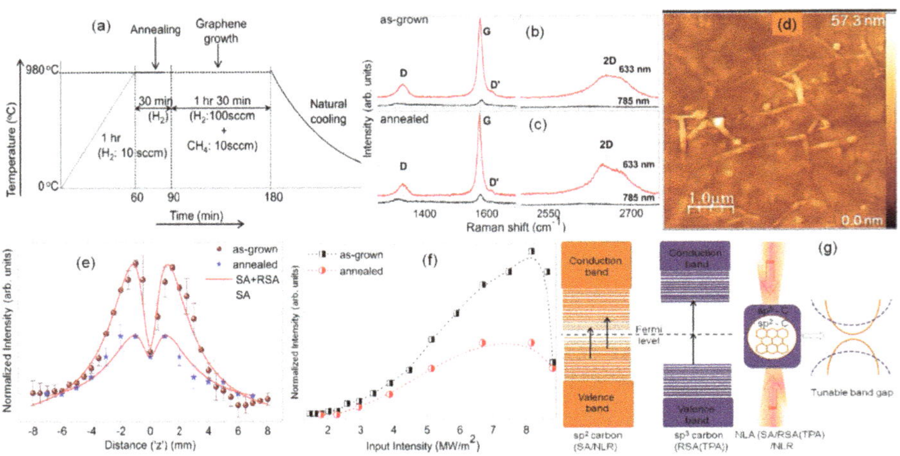

Figure 9.5. (a) Temperature–time profile indicating four stage heating process for CVD growth of graphene. Raman spectra of as-grown (b) and annealed graphene multilayers (c) obtained at 633 nm and 785 nm. (d) AFM image for as-grown graphene multilayers. (e) OA z-scan results for as-grown and annealed graphene multilayers. (f) Variation of normalized transmittance as a function of input intensity for as-grown and annealed graphene multilayers. (g) Schematic illustration for the tuning of energy bandgap by embedding sp^2 hybridized carbon atoms into sp^3 matrix. Reproduced from [10], copyright 2020, with permission from Elsevier.

appear and are defect/disorder induced bands, suggesting the contributions of sp^3 hybridized carbon atoms in graphene.

An atomic force microscope (AFM) image of the as-grown sample is shown in figure 9.5(d) which indicates a piled-up growth of graphene layers. The NLA response of the grown samples was investigated using OA z-scan experiment employing CW He–Ne laser source (633 nm), as shown in figure 9.5(e), where combined SA and RSA behavior was observed. Here, symbols as experimental data points and red thick solid lines are theoretical curves fitted using equation (9.6). NLA coefficients and the saturation intensity for as-grown (annealed) graphene layers was found to be $\beta = 15.4 \times 10^{-3}$ m W^{-1}; $I_s = 1.64$ MW m^{-2} and $\beta = 4.2 \times 10^{-3}$ m W^{-1} and $I_s = 1.27$ MW m^{-2}, respectively, which clearly indicates a reduction in optical nonlinearity upon annealing and is attributed to the presence of sp^2 hybridized carbon atoms into sp^3 matrix. Figure 9.5(f) shows the transmittance signal as a function of input fluence where initially the transmittance increases with increasing input fluence and then drops down after a limiting threshold and hence represents the saturable absorber as well as optical limiting behavior within the same material. The concept of embedded sp^2 hybridized carbon atoms into sp^3 matrix along with their energy band profile is schematic illustrated in figure 9.5(g). It is worth mentioning here that the energy bandgap of graphene layers can be modified via manipulating the sp^2/sp^3 ratio which can result in the occurrence of combined SA and RSA behavior within the same material.

Yang *et al* [38] have also reported the NLR in graphene layers grown onto silicon/silica substrate via a facile bottom-up approach. A valley followed by a peak was observed in the CA z-scan transmittance signal and hence positive NLO refractivity. Ren *et al* [39] have also explored giant third-order optical nonlinearity in electrochemical GO under fs regime using the same z-scan technique. A valley-kind feature is observed in the OA geometry which increases with increasing the input laser fluences, suggesting RSA behavior and hence the utility in making optical limiters. At low input fluence (80 mJ cm^{-2}), SA behavior was initially observed which switches to RSA behavior. Under CA geometry, pre-focal peak followed by a post-focal valley was observed, signifying the negative sign of NLO refraction. A few experimentally obtained NLO parameters for thin films of graphene-based nanomaterials are listed in table 9.3.

9.2.4 2D nanomaterials/graphene-based nanomaterials

Optical nonlinearity of graphene-based nanomaterials can also be modified by integrating them with other 2D materials such as MoS_2, h-BN, $MoSe_2$ etc. Ouyang *et al* [46] have demonstrated an enhanced RSA behavior in G/MoS_2 organic glasses. G/MoS_2 composite was first prepared by hydrothermal method followed by dispersion in methyl-methacrylate (MMA) and then polymerized to synthesize G/MoS_2 organic glass. OA z-scan technique was employed to determine the NLA coefficient and a valley-feature was observed in the transmittance signal suggesting RSA behavior with NLA coefficient ~2110 cm GW^{-1}, which was significantly larger than G/PMMA (~242 cm GW^{-1}) and MoS_2/PMMA (~365 cm GW^{-1}). Such an

Table 9.3. Experimentally obtained NLO parameters for thin films of graphene-based nanomaterials.

Sample	Laser parameters	NLO properties	Reference
G/quartz	690–1050 nm; 80 MHz; 100 fs	NLR = 9.07×10^{-9} to 1.76×10^{-8} cm^2 W^{-1}	[35]
MLG/quartz	633 nm; 20 mW	NLR = −14.5 m^2 GW^{-1}	[10]
Annealed MLG/ quartz (250 °C)	633 nm; 20 mW	NLR = −1.4 m^2 GW^{-1}	[10]
G layers/quartz	1550 nm; 3.8 ps; 10 MHz	NLR ~ 10^{-7} cm^2 W^{-1}	[40]
G/quartz	733 nm;100 fs; 80 MHz; 94 GW cm^{-2}	NLA = 6 cm MW^{-1} NLR = 1.4 cm^2 GW^{-1}	[41]
rGO/PMMA	532 nm; 2 Hz; 300 μJ	NLA = 129.01 cm GW^{-1}	[36]
PFTP-rGO/PMMA	532 nm; 2 Hz; 300 μJ	NLA = 215.77 cm GW^{-1}	[36]
rGO/PMMA	1064 nm; 2 Hz; 300 μJ	NLA = 148.42 cm GW^{-1}	[36]
PFTP-rGO/PMMA	1064 nm; 2 Hz; 300 μJ	NLA = 300.13 cm GW^{-1}	[36]
G/PMMA [50]	532 nm; 6 ns; 1 Hz	NLA = 242 cm GW^{-1}	[42]
Multilayer G (5–7 layers)	2400 nm; 100 fs; 1 kHz; 22 GW cm^{-2}	NLA = 19 000 cm W^{-1} NLR= −2.5 cm^2 GW^{-1}	[37]
Electrochemically derived GO film	800 nm; 10 kHz; 85 fs; 80 mJ cm^{-2}	NLA = −2 cm GW^{-1} NLR = 3.63 cm^2 GW^{-1}	[39]
Electrochemically derived GO film	800 nm; 10 kHz; 85 fs; 400 mJ cm^{-2}	NLA = +7 cm GW^{-1} NLR = 0.57 cm^2 GW^{-1}	[39]
GO films/glass	400 nm;100 fs; 1 kHz	NLA ~ 41 000 cm GW^{-1}	[43]
GO films/glass	800 nm;100 fs; 1 kHz	NLA(2PA) ~31 cm GW^{-1} NLA(3PA) ~0.47 cm^3 GW^{-2}	[43]
GO thin film	800 nm ; 100 fs; 70 GW cm^{-2}	NLA = 0.47×10^{-18} cm^3 GW^{-2}	[44]
GO films	1560 nm; 67 fs; 20 MHz	NLA ~ 103 cm GW^{-1} NLR = 0.45 cm^2 GW^{-1}	[45]

enhanced optical nonlinearity suggests its utility in optical limiter and optical shutters. Improved NLO response of MoS_2/G nanocomposite for ultrafast photonic applications has also been observed by Jiang *et al* [47]. The nanocomposite was fabricated using hydrothermal method. TEM image of the prepared nanocomposite is shown in figure 9.6(a), which reveals a homogeneous spread of MoS_2 sheet on graphene. The inset of figure 9.6(a) shows selected area electron diffraction (SAED) pattern of MoS_2/G nanocomposite. Figure 9.6(b) depicts the ultraviolet–visible (UVVis) spectra of the grown nanocomposite and clearly signatures the spectral peaks of MoS_2/G at 610 nm and 670 nm and two absorption peaks of MoS_2. Raman spectra has also been shown in figure 9.6(c), showing Raman D-band and Raman G band at 1342 cm^{-1} and 1593 cm^{-1}, respectively, along with in-plane $E_{2g}{}^1$, $A^1{}_g$ and 2LA bands which are associated with two sulfur atoms vibrating in opposite direction with respect to Mo atom, out-of-plane vibration of sulfur atoms in opposite direction and second-order longitudinal acoustic mode at M point, respectively.

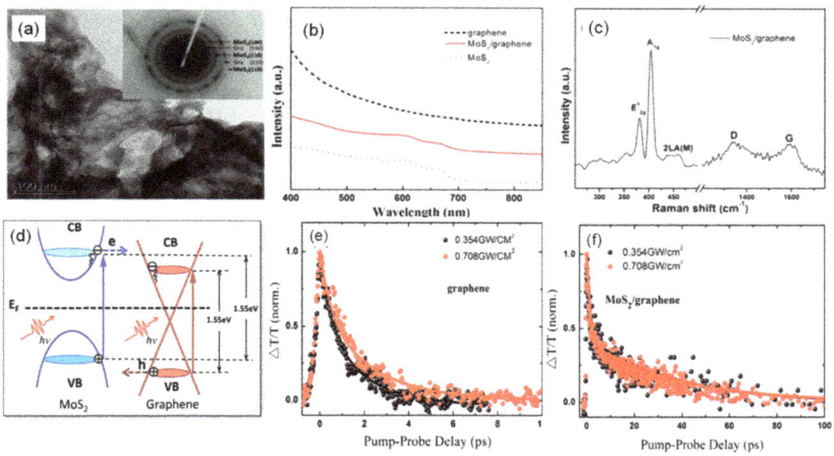

Figure 9.6. (a) TEM image of MoS_2/graphene nanocomposites; Inset: SAED pattern of MoS_2/graphene nanocomposites. (b) UV–Vis spectra of graphene, MoS_2/graphene nanocomposites and MoS_2. (c) Raman spectrum of the MoS_2/graphene nanocomposites. (d) Schematically illustration of interaction between MoS_2 and graphene. Carrier dynamics of graphene (e) and MoS_2/graphene nanocomposite (f) under different pump-power intensities. Solid thick lines are the theoretical fit to the experimental data using equation (9.8). Reproduced from [47], under the terms of Creative Commons Attribution 4.0 International License, Copyright 2015, Springer.

Carrier dynamics and NLO response of MoS_2/G nanocomposite was investigated using pump-probe and OA z-scan measurements, respectively, employing a Ti: Sapphire amplified laser system (pulse repetition rate ~1 kHz; centre wavelength = 800 nm; pulse width ~100 fs). Figure 9.6(d) schematically illustrates the interaction between MoS_2 and graphene which plays a vital role in the relaxation dynamic mechanism where photon excitation causes an electron transfer from MoS_2 to graphene. Pump-probe ultrafast carrier dynamics for graphene and MoS_2/G nanocomposites for different pump intensities were shown in figures 9.6(e) and (f), respectively, where symbols are the experimental data points and solid thick lines are the theoretical fits using the equation [47];

$$\frac{\Delta T}{T} = A_1 \exp(-t/\tau_1) + A_2 \exp(-t/\tau_2) \tag{9.8}$$

where τ_1 and τ_2 are the decay time with amplitude weights A_i and slower constant related to cooling time of thermalized carrier–phonon coupling system and electron–hole recombination. Figures 9.6(e)–(f) clearly suggest a dependence of carrier dynamics on the pump-power intensity in case of graphene but no relaxation-time dependence on pump-power intensity in the case of the nanocomposite. The fast relaxation-time constants τ_1 for graphene and MoS_2/G are found to be 0.47 ± 0.03 ps and 1.3 ± 0.2 ps at a power of 0.354 GW cm^{-2}, whereas slower relaxation-time constant τ_2 at the same power was found to be 1.7 ± 0.1 ps and 36 ± 2 ps for graphene and MoS_2, respectively.

Figures 9.7(a)–(c) illustrates the OA z-scan results for MoS_2/G nanocomposite at the wavelengths of 400 nm, 800 nm and 1562 nm and different input fluencies, depicting a peak-feature, suggesting SA behavior and an increase in optical nonlinearity has also been observed with increase in input fluence. Figures 9.7(d) and (e) compare the NLO response of MoS_2/G and graphene (wavelength = 800 nm; input fluence = 20.4 GW cm^{-2}) and MoS_2/G and MoS_2 (wavelength = 1562 nm; input fluence = 104.5 MW cm^{-2}), respectively, where better optical nonlinearity has been observed in MoS_2/G nanocomposite as compared to bare graphene and MoS_2, suggesting its utility for fabricating optical limiters.

He *et al* [48] have also reported an improved nonlinear SA in MoS_2/G nanocomposites films which were grown by vacuum filtering and liquid phase exfoliated MoS_2/G suspension. SA behavior was examined using OA z-scan experiment at 800 nm. An enhancement in nonlinear SA has been observed in MoS_2/G with NLA coefficient of −1217.8 cm GW^{-1}, which is larger as compared to MoS_2 (−136.1 cm GW^{-1}) and graphene (−961.6 cm GW^{-1}). Such an encouraging SA response of the nanocomposite suggests the applicability of this nanocomposite for optical switching and mode locking applications. Apart from MoS_2/G, NLO response in h-BN-GO heterostructure has also been explored by Biswas *et al* [49], by performing the same z-scan experiment in the nanosecond regime. An increased two-photon absorption coefficient has been observed in BN nanosheet/GO as compared to bare BN nanosheets.

It is worth stressing here that the combination of semiconductors (ZnO, TiO_2, MgO etc) with graphene-based nanomaterials can improve the NLO properties of

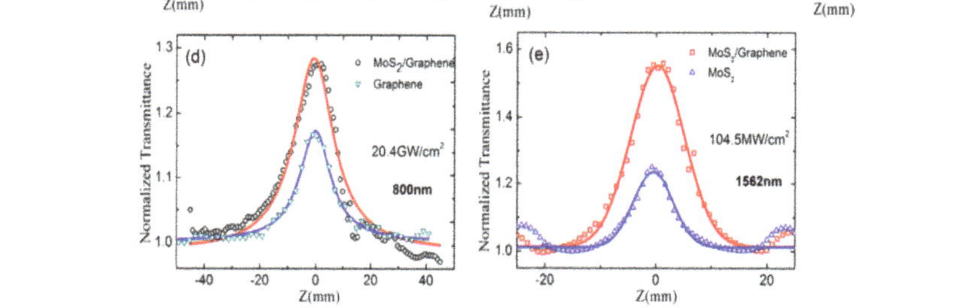

Figure 9.7. OA z-scan curves for MoS_2/graphene nanocomposites at different input fluences at 400 nm (a), 800 nm (b) and 1562 nm (c). (d) OA z-scan curves for MoS_2/graphene nanocomposites and graphene at the same input fluences of 20.4 GW/cm^2 at 800 nm. (e) MoS_2/graphene nanocomposites and MoS_2 at input fluences of 104.5 mW cm^{-2} at 1562 nm. Reproduced from [47], under the terms of Creative Commons Attribution 4.0 International License, Copyright 2015, Springer.

semiconductors owing to the superior optical nonlinearity of graphene. So far, several semiconductor/graphene-based nanomaterials have been investigated for their NLO properties. Tong *et al* [50] have explored the NLO properties and multi-photon absorption (PA) in G-ZnO nanocomposites by performing OA z-scan experiment employing fs laser with operating wavelength of 1030 nm. The G-ZnO nanocomposites were prepared using modified solvothermal method with GO and zinc nitrate as starting materials. They have reported the occurrence of 5PA, 3PA and 2PA at 46.8 GW cm^{-2}, 28.1 GW cm^{-2} and 18.7 GW m^{-2}, respectively, where a switching from SA feature to RSA has been observed upon increasing the incident intensity. Superior NLO properties along with encouraging optical limiting behavior in G/ZnO hybrid organic glasses have also been investigated by Ouyang *et al* [51]. The G/ZnO composites were prepared using wet chemical method and then dispersed in MMA polymerized at 75 °C to prepare G/ZnO/PMMA organic glass. Modified z-scan technique was employed to examine the optical nonlinearity and optical limiting behavior of the organic glass where the NLA coefficient was found to be 1530 cm GW^{-1} which was significantly larger than that of G/PMMA and ZnO/PMMA organic glasses. Kavitha *et al* [52] have also reported enhanced optical limiting properties in GO–ZnO hybrid where facile solution methods (solution precipitation method and hydrothermal method) were employed for the *in situ* growth of GO. OA z-scan experiment was performed to examine the NLA coefficient and hence the optical limiting properties of the grown hybrid where a 15 times enhancement has been observed in the NLA coefficient as compared to bare ZnO. Kimiagar and Abrinaei [53] have investigated the effect of temperature on the NLO properties of MgO doped GO nanocomposites, which were synthesized via hydrothermal method at different temperatures. The NLO response of the grown MgO–GO nanocomposite was investigated by performing z-scan experiment using second harmonic of nanosecond Nd:YAG laser. The absorptive and refractive optical nonlinearity was found to be of the order of 10^{-7} cm W^{-1} and -10^{-12} cm^2 W^{-1}, respectively, at a laser intensity of 1.1×10^8 W cm^{-2}. Absorptive optical nonlinearity and the optical limiting behavior in TiO_2-rGO nanocomposites have also been considered by Wang *et al* [54]. The nanocomposite was again synthesized using hydrothermal method where GO was first reduced to rGO in water:ethanol (2:1) mixture with varying amount of tetrabutyl-titanate which results in the deposition of TiO_2 on rGO sheets. The absorptive optical nonlinearity of TiO_2 doped rGO nanocomposite as solution in DMF was examined at 532 nm and a dip at the focal point of the transmittance versus distance signal has been observed indicating RSA behavior. Table 9.4 provides a compilation of few experimentally obtained NLO parameters for 2D material/graphene-based nanomaterials and semiconductor/graphene-based nanomaterials.

9.3 Conclusions

In conclusion, this chapter details the NLO properties of various graphene-based nanomaterials including metals decorated graphene-based nanomaterials, 2D materials/graphene-based nanomaterials, dispersion of graphene-based nanomaterials in

Table 9.4. Experimentally obtained NLO parameters for 2D material/graphene-based nanomaterials and semiconductor/graphene-based nanomaterials.

Sample	Laser parameters	NLO properties	Ref.
$MoSe_2$/G (Rt = 6 h)	532 nm; 30 ps; 10 Hz; 6.6 GW/cm^2	NLA = -6.46×10^{-12} m W^{-1} NLR = 1.54×10^{-11} esu	[55]
$MoSe_2$/G (Rt = 12 h)	532 nm; 30 ps; 10 Hz; 6.6 GW/cm^2	NLA = -6.12×10^{-12} m W^{-1} NLR = 1.35×10^{-11} esu	[55]
$MoSe_2$/G (Rt=18 h)	532 nm; 30 ps; 10 Hz; 6.6 GW/cm^2	NLA = -3.90×10^{-12} m W^{-1} NLR = 1.18×10^{-11} esu	[55]
MoS_2/G	800 nm; 1 Hz	NLA ~ −1217.8 cm GW^{-1}	[48]
MoS_2/G	400 nm; 100 fs	I_s = 1.427 GW cm^{-2}	[47]
MoS_2/G	800 nm; 100 fs	I_s = 2.02 GW cm^{-2}	[47]
MoS_2/G	1562.6 nm; 565 fs	I_s = 2.44 mW cm^{-2}	[47]
G/MoS_2/PMMA	532 nm; 6 ns; 1 Hz; 66 μJ	NLA = 2110 cm GW^{-1}	[46]
Bi_2S_3/rGO	532 nm; 30 ps; 10 Hz; 4.5 GW cm^{-2}	NLA = 2.48×10^{-11} m W^{-1} NLR = 4.03×10^{-11} esu	[56]
Bi_2S_3/rGO	532 nm; 30 ps; 10 Hz; 4.5 GW cm^{-2}	NLA = 2.28×10^{-11} m W^{-1} NLR = 3.48×10^{-11} esu	[56]
Bi_2S_3/rGO	532 nm; 30 ps; 10 Hz; 4.5 GW cm^{-2}	NLA = 1.77×10^{-11} m W^{-1} NLR = 1.81×10^{-11} esu	[56]
CdSe-rGO	532 nm; 30 ps; 10 Hz	NLA = 224.42 cm GW^{-1} NLR = 48.90×10^{-11} esu	[57]
TiO_2/rGO	532 nm; 4 ns	NLA = 6.0×10^{-10} m W^{-1}	[54]
GO-Fe_3O_4	532 nm; 5 ns; 10 Hz	NLA = 26 cm GW^{-1} NLR = 2.83×10^{-13} cm^2 W^{-1}	[58]
S-rGO(10 mg)-ZnO	532 nm; 5 ns; 10 Hz	NLA = 5.8 cm GW^{-1} I_s = 1.5 GW cm^{-2}	[52]
GO-Fe_3O_4	532 nm; 5 ns	RSA: enhanced by Fe_3O_4	[59]
$ZnFe_2O_4$–rGO (40 wt%)	532 nm; 5 ns; 10 Hz; 100 μJ	NLA = 1.26 × 10–10 m W^{-1} NLR = 1.29×10^{-38} esu	[60]
MgO–GO (T = 60 °C)	532 nm; 10 ns; 200 Hz	NLA = 97 cm GW^{-1} NLR = -1.95×10^{-12} cm^2 W^{-1}	[53]

various solvents, thin films of graphene-based nanomaterials, and semiconductor/graphene-based nanomaterials. Results based on the NLO properties from recent studies were compiled and discussed in detail. Overall, it is manifested from the discussions that graphene-based nanomaterials offer prodigious potential as an optoelectronic material, owing to their various fascinating and unique NLO properties which can be further improved by their integration with several other materials, and hence add another dimension to their utility. However, there's still inevitability for more and more novel graphene-based materials so that their commercialization can be realized for today's technological optoelectronic applications. Consequently, combined efforts are strongly recommended to meet the speedily growing demand for graphene-based devices.

References

[1] Agrawal A and Yi G-C 2020 Sample pretreatment with graphene materials *Analytical Applications of Graphene for Comprehensive Analytical Chemistry* vol 91 (New York: Elsevier) p 21

[2] Hendry E, Hale P J, Moger J, Savchenko A K and Mikhailov S A 2010 Coherent nonlinear optical response of graphene *Phys. Rev. Lett.* **105** 097401

[3] Wu R, Zhang Y, Yan S, Bian F, Wang W, Bai X, Lu X, Zhao J and Wang E 2011 Purely coherent nonlinear optical response in solution dispersions of graphene sheets *Nano Lett.* **11** 5159–64

[4] Kumar N, Kumar J, Gerstenkorn C, Wang R, Chiu H Y, Smirl A L and Zhao H 2013 Third harmonic generation in graphene and few-layer graphite films *Phys. Rev.* B **87** 121406

[5] Agrawal A, A Dar T, Solanki R, Phase D M and Sen P 2015 Study of nonlinear optical properties of pure and Mg-doped ZnO films *Phys. Status Solidi* B **252** 1848–53

[6] Dar T A, Agrawal A, Sen P K, Choudhary R J and Sen P 2016 Thermo-optic coefficients of pure and Ni doped ZnO thin films *Thin Solid Films* **603** 115–8

[7] Agrawal A, Dar T A, Andrews J T, Sen P K and Sen P 2016 Negative thermo-optic coefficients and optical limiting response in pulsed laser deposited Mg-doped ZnO thin films *J. Opt. Soc. Am.* B **33** 2015–9

[8] Agrawal A, Saroj R K, Dar T A, Baraskar P, Sen P and Dhar S 2017 Insight into the effect of screw dislocations and oxygen vacancy defects on the optical nonlinear refraction response in chemically grown ZnO/Al_2O_3 films *J. Appl. Phys.* **122** 195303

[9] Sheik-Bahae M, Said A A and Van Stryland E W 1989 High-sensitivity, single-beam $n(2)$ measurements *Opt. Lett.* **14** 955–7

[10] Agrawal A, Park J Y, Sen P and Yi G C 2020 Unraveling absorptive and refractive optical nonlinearities in CVD grown graphene layers transferred onto a foreign quartz substrate *Appl. Surf. Sci.* **505** 144392

[11] Yue M, Si J, Yan L, Yu Y and Hou X 2018 Enhanced nonlinear optical properties of reduced graphene oxide decorated with silver nanoparticles *Opt. Mater. Express* **8** 698–703

[12] Al Abdulaal T and Salamo G 2017 Thermal nonlinearity of functionalized graphene enhanced by gold nanorods *J. Electr. Eng.* **5** 171–80

[13] Sadrolhosseini A R, Noor A S M, Faraji N, Kharazmi A and Mahdi M 2014 Optical nonlinear refractive index of laser-ablated gold nanoparticles graphene oxide composite *J. Nanomater.* **2014** 962917

[14] Kalanoor B S, Bisht P B, Ali S A, Baby T T and Ramaprabhu S 2012 Optical nonlinearity of silver-decorated graphene *J. Opt. Soc. Am.* B **29** 669–75
[15] Biswas S, Kole A K, Tiwary C S and Kumbhakar P 2016 Enhanced nonlinear optical properties of graphene oxide–silver nanocomposites measured by Z-scan technique *RSC Adv.* **6** 10319–25
[16] Oluwafemi O S, Sreekanth P, Philip R, Thomas S and Kalarikkal N 2016 Improved nonlinear optical and optical limiting properties in non-covalent functionalized reduced graphene oxide/silver nanoparticle (NF-RGO/Ag-NPs) hybrid *Opt. Mater.* **58** 476–83
[17] Zheng C, Lei L, Huang J, Chen W, Li W, Wang H, Huang L and Huang D 2017 Facile control of metal nanoparticles from isolated nanoparticles to aggregated clusters on two-dimensional graphene to form optical limiters *J. Mater. Chem.* C **5** 11579–89
[18] Anand B, Kaniyoor A, Sai S S S, Philip R and Ramaprabhu S 2013 Enhanced optical limiting in functionalized hydrogen exfoliated graphene and its metal hybrids *J. Mater. Chem.* C **1** 2773–80
[19] Liu Z, Wang Y, Zhang X, Xu Y, Chen Y and Tian J 2009 Nonlinear optical properties of graphene oxide in nanosecond and picosecond regimes *Appl. Phys. Lett.* **94** 021902
[20] Wang J, Hernandez Y, Lotya M, Coleman J N and Blau W J 2009 Broadband nonlinear optical response of graphene dispersions *Adv. Mater.* **21** 2430–5
[21] Fang C, Dai B, Hong R, Tao C, Wang Q, Wang X, Zhang D and Zhuang S 2015 Tunable optical limiting optofluidic device filled with graphene oxide dispersion in ethanol *Sci. Rep.* **5** 1–10
[22] Chantharasupawong P, Philip R, Narayanan N T, Sudeep P M, Mathkar A, Ajayan P M and Thomas J 2012 Optical power limiting in fluorinated graphene oxide: an insight into the nonlinear optical properties *J. Phys. Chem.* C **116** 25955–61
[23] Zhu J, Li Y, Chen Y, Wang J, Zhang B, Zhang J and Blau W J 2011 Graphene oxide covalently functionalized with zinc phthalocyanine for broadband optical limiting *Carbon* **49** 1900–5
[24] Zhao B, Cao B, Zhou W, Li D and Zhao W 2010 Nonlinear optical transmission of nanographene and its composites *J. Phys. Chem.* C **114** 12517–23
[25] Liaros N, Bourlinos A B, Zboril R and Couris S 2013 Fluoro-graphene: nonlinear optical properties *Opt. Express* **21** 21027–38
[26] Zhang X L, Liu Z B, Li X C, Ma Q, Chen X D, Tian J G, Xu Y F and Chen Y S 2013 Transient thermal effect, nonlinear refraction and nonlinear absorption properties of graphene oxide sheets in dispersion *Opt. Express* **21** 7511–20
[27] Cheng X, Dong N, Li B, Zhang X, Zhang S, Jiao J, Blau W J, Zhang L and Wang J 2013 Controllable broadband nonlinear optical response of graphene dispersions by tuning vacuum pressure *Opt. Express* **21** 16486–93
[28] Liaros N, Orfanos I, Papadakis I and Couris S 2016 Nonlinear optical response of some Graphene oxide and Graphene fluoride derivatives *Optofluid. Microfluid. Nanofluid.* **3** 53–8
[29] Krishna M B M, Kumar V P, Venkatramaiah N, Venkatesan R and Rao D N 2011 Nonlinear optical properties of covalently linked graphene–metal porphyrin composite materials *Appl. Phys. Lett.* **98** 081106
[30] Song W, He C, Zhang W, Gao Y, Yang Y, Wu Y, Chen Z, Li X and Dong Y 2014 Synthesis and nonlinear optical properties of reduced graphene oxide hybrid material covalently functionalized with zinc phthalocyanine *Carbon* **77** 1020–30

[31] Fan Y, Jiang Z G and Yao L F 2012 Large nonlinear optical absorption response in monolayer graphene and graphene oxide *Adv. Mater. Res.* **510** 768–71

[32] Qian J, Wang D, Cai F H, Xi W, Peng L, Zhu Z F, He H, Hu M L and He S 2012 Observation of multiphoton-induced fluorescence from graphene oxide nanoparticles and applications in *in vivo* functional bioimaging *Angew. Chem.* **124** 10722–7

[33] Liaros N, Aloukos P, Kolokithas-Ntoukas A, Bakandritsos A, Szabo T, Zboril R and Couris S 2013 Nonlinear optical properties and broadband optical power limiting action of graphene oxide colloids *J. Phys. Chem.* C **117** 6842–50

[34] Jiménez-Pérez J L, Gutiérrez-Fuentes R, López-Gamboa G, Sánchez-Ramírez J F, Correa-Pacheco Z N and Carbajal-Valdéz R 2018 Measurement of optical nonlinear refractive index response of graphene nanoparticles dispersed in an aqueous solution by Z scan technique *Opt. Mater.* **84** 236–41

[35] Thakur S, Semnani B, Safavi-Naeini S and Majedi A H 2019 Experimental characterization of the ultrafast, tunable and broadband optical kerr nonlinearity in graphene *Sci. Rep.* **9** 1–10

[36] Liu Z, Dong N, Jiang P, Wang K, Wang J and Chen Y 2018 Reduced graphene oxide chemically modified with aggregation-induced emission polymer for solid-state optical limiter *Chem. Eur. J.* **24** 19317–22

[37] Demetriou G, Bookey H T, Biancalana F, Abraham E, Wang Y, Ji W and Kar A K 2016 Nonlinear optical properties of multilayer graphene in the infrared *Opt. Express* **24** 13033–43

[38] Yang Y *et al* 2015 Bottom-up fabrication of graphene on silicon/silica substrate via a facile soft-hard template approach *Sci. Rep.* **5** 1–7

[39] Ren J, Zheng X, Tian Z, Li D, Wang P and Jia B 2016 Giant third-order nonlinearity from low-loss electrochemical graphene oxide film with a high power stability *Appl. Phys. Lett.* **109** 221105

[40] Zhang H, Virally S, Bao Q, Ping L K, Massar S, Godbout N and Kockaert P 2012 Z-scan measurement of the nonlinear refractive index of graphene *Opt. Lett.* **37** 1856–8

[41] Chen W, Wang G, Qin S, Wang C, Fang J, Qi J, Zhang X, Wang L, Jia H and Chang S 2013 The nonlinear optical properties of coupling and decoupling graphene layers *AIP Adv.* **3** 042123

[42] Ouyang Q, Yu H, Xu Z, Zhang Y, Li C, Qi L and Chen Y 2013 Synthesis and enhanced nonlinear optical properties of graphene/CdS organic glass *Appl. Phys. Lett.* **102** 031912

[43] Jiang X F, Polavarapu L, Neo S T, Venkatesan T and Xu Q H 2012 Graphene oxides as tunable broadband nonlinear optical materials for femtosecond laser pulses *J. Phys. Chem. Lett.* **3** 785–90

[44] Roy S and Yadav C 2013 Femtosecond all-optical parallel logic gates based on tunable saturable to reverse saturable absorption in graphene-oxide thin films *Appl. Phys. Lett.* **103** 241113

[45] Xu X, Zheng X, He F, Wang Z, Subbaraman H, Wang Y, Jia B and Chen R T 2017 Observation of third-order nonlinearities in graphene oxide film at telecommunication wavelengths *Sci. Rep.* **7** 1–7

[46] Ouyang Q, Yu H, Wu H, Lei Z, Qi L and Chen Y 2013 Graphene/MoS_2 organic glasses: fabrication and enhanced reverse saturable absorption properties *Opt. Mater.* **35** 2352–6

[47] Jiang Y, Miao L, Jiang G, Chen Y, Qi X, Jiang X F, Zhang H and Wen S 2015 Broadband and enhanced nonlinear optical response of MoS_2/graphene nanocomposites for ultrafast photonics applications *Sci. Rep.* **5** 1–12

[48] He M, Quan C, He C, Huang Y, Zhu L, Yao Z, Zhang S, Bai J and Xu X 2017 Enhanced nonlinear saturable absorption of MoS_2/graphene nanocomposite films *J. Phys. Chem.* C **121** 27147–53

[49] Biswas S, Tiwary C S, Vinod S, Kole A K, Chatterjee U, Kumbhakar P and Ajayan P M 2017 Nonlinear optical properties and temperature dependent photoluminescence in hBN-GO heterostructure 2D material *J. Phys. Chem.* C **121** 8060–9

[50] Tong Q *et al* 2018 Nonlinear optical and multi-photon absorption properties in graphene–ZnO nanocomposites *Nanotechnology* **29** 165706

[51] Ouyang Q, Xu Z, Lei Z, Dong H, Yu H, Qi L, Li C and Chen Y 2014 Enhanced nonlinear optical and optical limiting properties of graphene/ZnO hybrid organic glasses *Carbon* **67** 214–20

[52] Kavitha M K, John H, Gopinath P and Philip R 2013 Synthesis of reduced graphene oxide–ZnO hybrid with enhanced optical limiting properties *J. Mater. Chem.* C **1** 3669–76

[53] Kimiagar S and Abrinaei F 2018 Effect of temperature on the structural, linear, and nonlinear optical properties of MgO-doped graphene oxide nanocomposites *Nanophotonics* **7** 243–51

[54] Wang A, Yu W, Fang Y, Song Y, Jia D, Long L, Cifuentes M P, Humphrey M G and Zhang C 2015 Facile hydrothermal synthesis and optical limiting properties of TiO_2-reduced graphene oxide nanocomposites *Carbon* **89** 130–41

[55] Liu G, Dai S, Zhu B, Li P, Wu Z and Gu Y 2019 Third-order nonlinear optical properties of $MoSe_2$/graphene composite materials *Opt. Laser Technol.* **120** 105746

[56] Liu G, Dai S, Cao F, Zhu B, Li P and Gu Y 2019 Preparation and enhanced nonlinear optical properties of Bi_2S_3/RGO composite materials *Opt. Mater.* **89** 112–7

[57] Li P, Zhu B, Li P, Zhang Z, Li L and Gu Y 2019 A facile method to synthesize CdSe-reduced graphene oxide composite with good dispersion and high nonlinear optical properties *Nanomaterials* **9** 957

[58] Zhang X L, Zhao X, Liu Z B, Shi S, Zhou W Y, Tian J G, Xu Y F and Chen Y S 2011 Nonlinear optical and optical limiting properties of graphene oxide–Fe_3O_4 hybrid material *J. Opt.* **13** 075202

[59] Zhang X, Yang X, Ma Y, Huang Y and Chen Y 2010 Coordination of graphene oxide with Fe_3O_4 nanoparticles and its enhanced optical limiting property *J. Nanosci. Nanotechnol.* **10** 2984–7

[60] Saravanan M and TC S G 2017 Enhanced nonlinear optical absorption and optical limiting properties of superparamagnetic spinel zinc ferrite decorated reduced graphene oxide nanostructures *Appl. Surf. Sci.* **392** 904–11

IOP Publishing

Recent Advances in Graphene and Graphene-Based Technologies

Anoop Chandran, N V Unnikrishnan, M K Jayaraj, Reenu Elizabeth John and Justin George

Chapter 10

Photonic and optoelectronic applications of graphene: Applications of graphene in surface-enhanced Raman scattering

K Hasna, K S Anju and M K Jayaraj

Surface-enhanced Raman scattering (SERS) has become a dominant tool in material science and life science, owing to its core features such as high sensitivity and unique detection abilities. It is a surface sensitive technique which arises when the molecules are adsorbed on metal nanoparticles and an enhancement of the order of 10^{10} is observed in the Raman signals of adsorbed species. The enhancement effect of SERS is ascribed to the electromagnetic mechanism (EM) and the chemical enhancement mechanism (CM). The EM originates due to the presence of localized surface plasmons and surface roughness present in metallic nanostructures, while CM arises mainly due to the charge transfer (CT) between the substrate and molecule. EM plays a superior role in metallic substrates and has been well studied over the past few decades, but the mechanism of CM is still not understood entirely. Major drawbacks of metallic SERS substrates are poor stability and reproducibility, making its application strenuous in practical scenarios. Therefore, development of a non-metallic SERS substrate not only helps to overcome the drawbacks of traditional metallic SERS substrates, but also helps to understand the CM. Graphene, with its unique properties is an excellent choice to overcome the drawbacks of traditional SERS substrates and to elevate enhancement of Raman signal through chemical enhancement. Exploration of graphene and graphene-based SERS substrates has intensified to such an extent that it finds applications in many scientific branches.

This chapter gives an overview of progress in graphene enhanced Raman spectroscopy over the past decade. The chapter begins with an insight into the general aspects of SERS following the role of graphene in SERS. General understanding of Raman spectrum of graphene and how it changes when coupled with metal nanoparticles is also given. A subsequent section gives general perception of enhancement mechanism when graphene itself is used as a SERS substrate. Succeeding sections elaborately describe how research has progressed in making use of the unique properties of graphene in developing graphene–metal hybrid structures as SERS substrates following diverse strategies.

10.1 Introduction to surface-enhanced Raman spectroscopy

Raman spectroscopy is a unique spectroscopic tool to identify the molecular structure and characterize the chemical compounds. Each band in the Raman spectrum is associated with a specific vibrational mode of the molecule. Each Raman active molecule has unique Raman spectrum making this technique highly specific [1]. Yet, the sensitivity of this technique is insufficient at lower concentrations due to the small Raman scattering cross-section of the molecules. This problem may be tackled with surface-enhanced Raman spectroscopy (SERS)—a technique where the Raman signal from molecules adsorbed on the metal surface gets enhanced allowing the detection of organic molecules at sub-femtomolar concentration and even single molecular detection [2].

Experimental observation of SERS was first done by Fleischmann *et al* in 1974 during their attempts to develop a spectroscopic probe to study electrochemical processes. They observed an exceptional increase in the intensity of the Raman signal of pyridine molecule when adsorbed on rough silver electrode [3]. Later on, this phenomenon was much explored and is confirmed using other nanoscale featured metallic systems such as colloidal metal nanoparticles, metallic nanoparticle assemblies, metallic electrodes, etc. In general, the metallic systems which provide Raman enhancement are known as SERS substrate, and the adsorbed molecule whose Raman spectra get enhanced is known as probe molecule. The SERS phenomenon is much explored on the hybrid system consisting of more than one entity, such as semiconductor–metallic systems, polymer–metallic systems, metal–metal systems [4]. Also, functionalisation of metal nanoparticles and formation of metal nanocomposites are offering improved SERS performance [5]. SERS substrates of various morphologies are also being investigated. A variety of synthesis methods is being developed to fabricate SERS substrates in order to tackle specific problems [6].

10.2 Enhancement mechanism

SERS mainly arises from the interaction between the adsorbed samples and the SERS substrate. In the early stages of development of SERS, the enhancement in the Raman signal of the adsorbed species was attributed to the increased surface area resulting from the increase in the surface roughness. Presently, two mechanisms are widely accepted for the enhancement:

(a) electromagnetic enhancement mechanism;
(b) chemical enhancement mechanism.

The governing equation that determines Raman scattering is, polarization $\mu = \alpha E$, E (amplitude of the electric field), and α (molecular polarizability).

EM theory is formulated from the 'E' factor and the CM theory focuses on 'α'. Both mechanisms are explained in detail in the following sections.

10.2.1 The electromagnetic enhancement mechanisms

Jeanmaire and Van Duyne proposed the EM enhancement mechanism in 1977 based on the properties of metallic nanostructures [7]. The collective oscillation of free electrons on the metal surface is known as plasmons and the surface plasmons are confined near the surface of metal nanoparticles. The propagating plasmon on a flat surface is known as surface plasmon polaritons. On the other hand, the plasmons gets localized on a rough surface with nanoscale roughness. When the frequency of incident field coincides with the plasmon frequency, the metal becomes extremely polarizable and large local fields are produced on the metal surface because of localized surface plasmon resonance (LSPR), as shown in figure 10.1. The Raman intensity greatly depends on the square of applied field at the molecule, and these local fields cause an increase in the Raman intensity when the molecule is placed in its vicinity. The governing equation for the electric field enhancement during LSPR is given in the following equation (10.1)

$$E_{\text{out}} = E_0\,\hat{z} - \alpha E_0\left[\frac{\hat{z}}{r^3} - \frac{3z}{r^5}(x\hat{x} + y\hat{y} + z\hat{z})\right] \tag{10.1}$$

where α is metallic polarizability and expressed as $\alpha = ga^3$, where 'a' is radius of sphere and 'g' is defined as in equation (10.2)

$$g = \frac{\epsilon_{\text{in}} - \epsilon_{\text{out}}}{\epsilon_{\text{in}} + 2\epsilon_{\text{out}}} \tag{10.2}$$

where ϵ_{in} and ϵ_{out} represent dielectric function of metal and external environment, respectively. Maximum electric field enhancement occurs when $\epsilon_{\text{in}} = -2\epsilon_{\text{out}}$. Mie theory suggests that, the condition $\epsilon_{\text{in}} = -2\epsilon_{\text{out}}$ satisfies during LSPR and which occurs in visible region for metals such as silver (Ag), gold (Au), copper (Cu) etc. EM is a long-range effect which rapidly falls off with distance.

Nie *et al* and Kneipp *et al* studied single molecular SERS and came to the conclusion that isolated single particles could not achieve high enhancement factors [8, 9]. Experimentally observed field enhancements were very high in nanoparticle clusters compared to that on single nanoparticles. This is due to the formation of

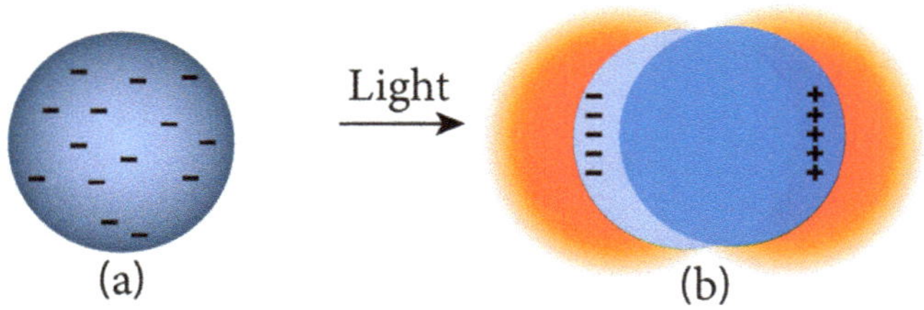

Figure 10.1. Schematic of (a) metal nanoparticle charge distribution. (b) Charge separation and formation of enhanced electric field when the light falls on the metal nanoparticle.

regions with intense fields known as 'hotspots' in the junction between two metallic particles [10]. Shape of the nanoparticle also contributes to the enhancement, as electric field is enhanced profoundly on the tip or sharp edges of certain nanostructures. This effect is known as 'lightning rod effect'.

10.2.2 The chemical enhancement mechanism

The chemical enhancement arises when the molecule is directly adsorbed onto the SERS substrate. This is a short-range process, and the enhancement is up to 10^2 which is much weaker than the EM enhancement. Although many theories are being proposed to explain chemical enchantment mechanism, CT mechanism is the most accepted explanation [11, 12]. When a molecule is adsorbed on to a metal surface a metal–adsorbate complex is formed. This induce CT resonance (μ_{CT}) either from the Fermi level of metal cluster to lowest unoccupied molecular orbital (LUMO) of the molecules, or from the highest occupied molecular orbital (HOMO) of the molecules to the Fermi level of metal cluster is shown in figure 10.2(a) [13]. This leads to an increase in molecular polarizability and hence the enhancement in Raman signals.

In semiconductor-based SERS substrate, the CT interaction between semiconductor and probe molecule is like that of metals, apart from the existence of bandgap between conduction band and valence band. The CT between semiconductors and molecules depends on the conduction band and the valence band energies, E_c and E_v of semiconductor with respect to the molecular energies. The CT scheme is as shown in figure 10.2(b) [14, 15]. In semiconductor–metal heterostructure-based SERS substrates, CT direction and the type depend on the probe molecules, nature of metals and the arrangement of energy levels. In most of the cases the following pathways are found: semiconductor-to-molecule-to-metal, metal-to-semiconductor-

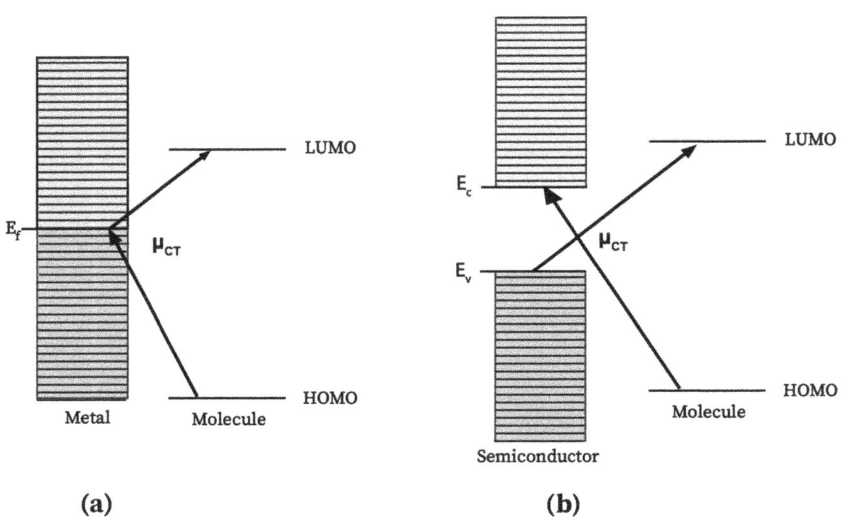

Figure 10.2. CT interaction in (a) metal–molecule system and (b) semiconductor–metal system.

to-molecule, metal-to-molecule-to-semiconductor and all depending on the relative energy positions and excitation wavelength [16].

10.3 Qualitative analysis of SERS substrate

The enhancement factor (EF) is the significant aspect to measure the performance of SERS substrate. It is the measure of enhancement in the Raman intensity in the presence of SERS substrate compared to normal Raman intensity of the molecule under the same environment. Simply, analytical enhancement factor is calculated using equation 10.3

$$EF = \frac{I_{SERS}}{I_{RS}} \times \frac{C_{RS}}{C_{SERS}} \tag{10.3}$$

where I_{SERS} and I_{RS} are SERS and Raman signal intensity of analyte molecule, respectively. C_{SERS} and C_{RS} are concentrations of the same analyte used to measure SERS and Raman signal intensity. However, I_{SERS} is the Raman signal from the analyte molecules adsorbed on the surface. So a more accurate equation (equation (10.4)) used to calculate the enhancement factor is [17, 18]

$$EF = \frac{I_{SERS}}{I_{RS}} \times \frac{N_{RS}}{N_{SERS}} \tag{10.4}$$

where N_{SERS} and N_{RS} are the number of molecules within the excitation volume on the sensing region of the SERS substrate and on the reference substrate, respectively.

Another factor which determines the quality of substrate is the lowest concentration of the molecule which can be detected using the SERS substrate and is known as the sensitivity of a SERS substrate. In general, the lower the detection limit, the better the efficiency of the substrate. Apart from that, multi-analyte detection capability, substrate-to-substrate reproducibility and uniformity defines the quality of SERS substrate.

10.4 Applications of SERS

The trace level detection capability of SERS led to the tremendous applications in detection of chemicals, biology and medicine, monitoring chemical reactions etc [19]. SERS has been used for the detection of presence of nutrients, anti-oxidants and flavonoids in food samples. Also, SERS has been employed to study water contamination and pesticide detection [20]. Brackx *et al* employed SERS method for detecting the presence of Zn^{2+} in water quantitatively, which indicates the contamination of water [21]. Chen *et al* decorated Ag nanoparticles (AgNPs) on nanocellulose and fabricated jelly-like flexible SERS substrate to detect pesticides thiram and thiabendazole on apple peels and cabbages [22]. Food adulterant detection is another important area where SERS technique can contribute extensively. For example, realisation of rapid and sensitive detection of melamine in milk was done by Li and Chin down to 0.5 ppm using Ag nanocube arrays [23]. Another advancement is the detection of gaseous substances by designing SERS sensors in appropriate configuration. Park *et al* studied volatile organic compounds emitted from plant materials using Tenax-TA polymer film decorated with Ag nanospheres [24].

SERS finds application in spectro-electrochemistry, catalysis and pH sensing [6]. Also, SERS is widely used in archaeology, forensics, chemical warfare agents, and in the detection of pharmaceuticals [25]. Another challenging area is the identification of colorants in traditional art works with minimum invasion. Choffel *et al* followed a SERS-based method to identify the presence of organic pigments like madder lake, carmine lake, indigo, gambode etc in 18th century portraits [26]. SERS sensing of common forensic substances like cyclosarin, RDX, amphetamine and picric acid was done by Hakonen *et al* using a hand held Raman spectrometer and commercially available SERS substrate down to femtograms [27]. SERS is also used for the label-free detection of biomolecules, pathogens, living cells, tissues, body fluids, nucleic acids and proteins, making major advancements in the medical field [28]. Another remarkable application of SERS-based sensors are real-time monitoring of alteration in a cellular environment such as pH, temperature and ion concentration, which affects the cell functioning and life processes. Indirect SERS sensors are also used for this purpose which monitors the alterations in SERS tags [29]. For example, in the work of Wang *et al* an optical fibre tip decorated with AgNPs and Raman reporter 4-mercptopyridine are employed to monitor the pH changes in between normal cells and cancer cells [30]. These advances open up a minimally invasive pathway for the diagnosis and treatment of cancer, which is a major breakthrough in health care field. Recently, Mao *et al* developed SERS-based sensor for the successful detection of multiple biomarkers miR-21 and miR-196a-5p in human urine with high sensitivity as low as 3.31pM and 2.18pM [31]. Also He *et al* fabricated octahedral Ag oxide nanoparticles for the detection of circulating tumour cells [32].

In more advanced scenarios, SERS is used for the continuous monitoring and treatment of cancer cells. Sujai *et al* employed SERS substrate based on a Au–MnO_2 core–shell nano-envelope for monitoring the photothermal therapy of pancreatic cancer cells [33]. A study of estrogen receptor alpha (ERα) expression in breast cancer cells was done by Kapara *et al* using SERS. The effect of drug treatment was monitored using functionalised Au nanoparticles (AuNPs) [34]. Among the plethora of applications of SERS, only a few are mentioned here. All these above cited applications of SERS point towards the significance of SERS and need for developing a variety of SERS substrates that are appropriate for specific applications.

10.5 Graphene-based surface-enhanced Raman spectroscopy

Graphene with sp^2-bonded carbon atoms can be used to enhance Raman signal of the adsorbed molecule, by virtue of chemical enhancement. The chemical enhancement in graphene arises mainly from the charge transfer interaction between graphene and the adsorbed analyte molecule. The enhancement is also a function of number of graphene layers which invokes the function of graphene-based SERS substrate [35]. Since π electrons are abundant in graphene the CT can easily happen, as shown in figure 10.3. The continuous energy band of graphene also contributes to the CT in the graphene–molecule system.

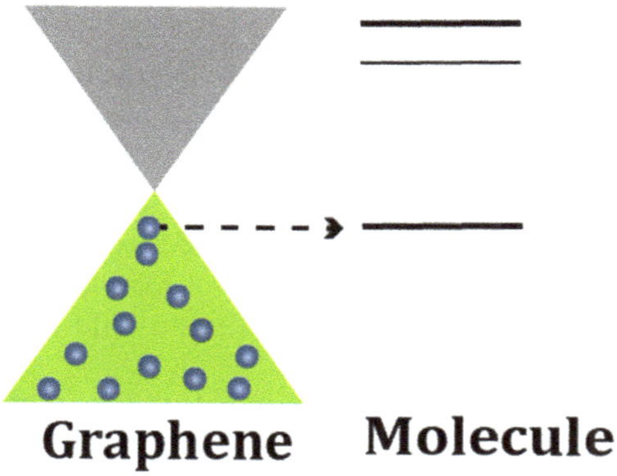

Figure 10.3. Schematic analysis of CT in graphene and probe molecule.

Although SERS has many advantages over other spectroscopic techniques, the high local fields produced during the measurement lead to photo-induced damage to the analyte molecule. Also, the florescent background produced by the analyte molecules obstructs the Raman signal enhancement. Furthermore, metal nanoparticles experience agglomerations and instability due to oxide formation which compounds the limitations of conventional metallic SERS substrates.

These limitations may be circumvented by using graphene-mediated SERS also known as graphene-mediated surface-enhanced Raman scattering (G-SERS). First of all, background fluorescence is avoided in G-SERS. Fluorescence quenching can be achieved by the non-radiative transition of electron in excited state to the ground state through energy transfer or electron transfer. In graphene, the π–π structure is favourable for the stacking and functionalization of most of the organic dyes due to their structural similarities. This leads to the movement of excited electrons from the molecule to graphene and thereby reduction of fluorescence [36].

To overcome the problem of photo-induced damage to the analyte molecule, a spacer could be introduced between analyte and the substrate. This may lead to a reduction in Raman signal enhancement as the plasmonic enhancement rapidly decreases with distance. Graphene may be used as a spacer between the metal nanostructure and molecule to tackle this problem, owing to the atomic thickness and chemical inertness of graphene. Also, graphene can be used as a sub-nanospacer for metal film–metal nanoparticle system for multiple plasmonic coupling and sensitive detection of molecules [37].

Graphene can be used as a protective layer for preventing the agglomeration and oxidation of metal nanoparticles. The optical response of nanostructures like graphene may be probed by analyzing the SERS of graphene since it can be easily integrated with plasmonic nanostructures. Moreover, the atomic uniformity, biocompatibility and atomic thickness make it suitable for promising SERS applications. Likewise, graphene itself is widely used as SERS substrate to detect other probe molecules. Graphene plasmonic systems are also extensively used as SERS substrates since they overcome the problems associated with conventional SERS substrates and get higher enhancement factors due to chemical enhancement, which will be discussed in the following sections.

10.5.1 Raman spectroscopy of graphene

Raman spectroscopy is used predominantly to determine the number of layers of graphene. It also used to identify the structural alterations, addition of functional groups, formation of by-products, and chemical changes brought during the development of graphene [38]. Even though the spectra of all carbon based materials are similar, considerable information can be derived from the intensity, position and shape of the peaks [39]. The prominent features of Raman spectra of an amorphous carbon, carbon nanotube, and doped graphene are different, making them distinguishable from one another [40].

Graphene has a single atomic layer with sp^2-bonded carbon atoms with a two-dimensional structure. Raman spectrum of graphene with defect is as shown in figure 10.4(a). The peaks at ~1585 cm^{-1} and ~2685 cm^{-1} namely G band and G′ band, respectively, which are the most intense Raman peaks, show the vibrations of pristine graphene [41]. D band (~1345 cm^{-1}) and D′ band ~1625 cm^{-1} appear when the graphene contains defects or when measured at the edge of graphene [42, 43]. Oher bands, namely D_2 and D_3, are also reported at 1450 cm^{-1} and 1500 cm^{-1}, respectively [44, 45]. The G band arises due to the in-plane vibrations of sp^2 hybridized C atoms, which is a first-order process corresponding to the degenerate iTO and iLO phonons at the Γ point and is independent of the excitation energy [36]. D band, which corresponds to the breathing modes of six-atom rings, requires the presence of a defect to activate it. A double resonance intervalley Raman scattering process with two iTO points at K point gives rise to the G′ band and is referred to as 2D band in common literature [46], as shown in figure 10.4(b). Two more low intensity peaks are observed at ~42 cm^{-1} arises due to weak interlayer interactions and at ~3250 cm^{-1} which is the second-order mode of D′ band [40, 47].

Many graphene properties can be studied by analysing the Raman spectrum of graphene. The changes in Raman spectrum with the number of layers are well studied throughout the literature. It is commonly found that as the number of layers in graphene increases, due to the newly aroused interactions between layers, the 2D peak becomes wider and shorter and shifts to higher frequency due to the splitting produced from increased number of modes. A redshift is also observed for G peak as the number of layer increases. The ratio of I_{2D}/I_G gives the number of layers for

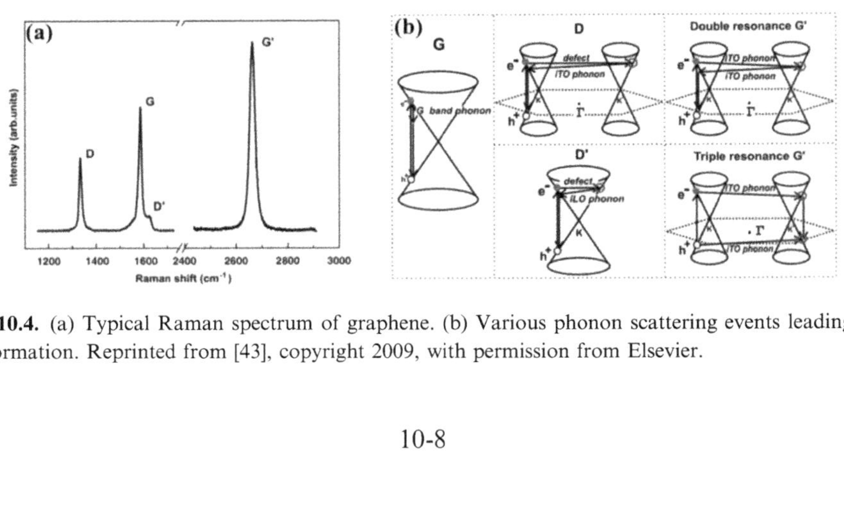

Figure 10.4. (a) Typical Raman spectrum of graphene. (b) Various phonon scattering events leading to the band formation. Reprinted from [43], copyright 2009, with permission from Elsevier.

AB-stacked graphene. For more than five layers, the Raman spectrum becomes hardly distinguishable from that of bulk graphite [45, 48]. The level of disorder in graphene can be determined by ratio of peak intensities of D and G peak (I_D/I_G ratio).

In low defect density regime, I_D/I_G ratio increases with increase in defects due to more elastic scattering. In the high defect density regime, the I_D/I_G ratio decreases with increase in defects due to the presence of amorphous carbon structure produced. Also, increased electron concentration leads to a decreased 2D peak position and decrease in intensity of 2D peak [49, 50]. It is also important to note that D and 2D are dispersive, i.e., the peak positions depend on the excitation energy [40]. The effect of doping in graphene is also reflected in the Raman spectrum of graphene. In general, n-type doping results in the blue shift of D peak and redshift of 2D band. Both G and 2D bands get blue shifted as a result of p-type doping of graphene. The intensity ratio of peaks G and 2D peaks also gets affected by the doping [50].

10.5.2 Graphene as a probe

Graphene is used as a probe for SERS studies either to determine the substrate properties or to study the interactions of graphene with metal nanoparticles. In a typical experimental setup, plasmonic–graphene systems are fabricated either by transferring metal nanoparticles on the graphene layer or by growing metal nanoparticles of various morphologies on graphene layer and recording the Raman spectra. Metal nanoparticles of Au or Ag are used widely to study the doping effects since coupling with AgNPs invokes n-doping and AuNPs result in p-doing.

In general, maximum enhancement is obtained when the SPR frequency of metal nanoparticles matches with the frequency of excitation source. This effect is examined by Lee *et al* by depositing AuNPs on single- and few-layer graphene by thermal evaporation. The SERS measurement was done with three different excitation sources namely, 532 nm, 633 nm and 785 nm laser and maximum enhancement was obtained when 633 nm excitation source was used, which is close to the surface plasmon resonance of AuNPs [51]. The wavelength dependency on the enhancement factor was also studied by Zhou *et al.* It is reported that Ag deposited on n-layer graphene shows more Raman signal enhancement when excited with laser wavelength of 514 nm, which matches with the SPR of AgNPs [52]. So in most of the cases the excitation wavelength is chosen so as to match the SPR of the metal nanoparticles to provide maximum plasmonic enhancement of Raman signals of graphene.

Recently, Almokhtar *et al* investigated the dependence of Raman spectra on the incident laser power on a graphene/AgNP system. A redshift is observed for G and 2D peak with increase in laser power. I_D/I_G ratio increased from 1.27 to 1.34 with an increase in laser power from 1 mW to 7 mW. The higher value of I_D/I_G ratio in graphene/AgNP system compared to graphene on SiO_2/Si indicates the photo-induced CT from Ag. Decrease in G peak half width with increase in laser power is also indicative of the CT in graphene/AgNP system [53].

When a graphene–metal interface is formed, it affects the properties of both metal surface and graphene. Osvath *et al* studied the properties of graphene on AuNPs. LSPR of AuNPs at 597 nm is broadened and red shifted by 9 nm when a graphene layer is transferred onto AuNP, which shows the coupling between localized surface plasmons and graphene. With a 633 nm excitation on these graphene/Au systems, an enhancement in Raman signal of graphene is observed. When this sample is annealed, the separation between graphene layer and AuNPs decreases and better enhancement is observed [54]. Another observed change is with the doping of graphene. A decrease I_{2D}/I_G was observed by Iyer *et al* as a result of the p-doping of graphene. Also, the blue shift in G and 2D peak confirmed this effect [55]. Carles *et al* studied the interaction of graphene with buried 3D printed AgNPs. The distance between graphene and AgNP layer is estimated to be 4 nm and they observed a tunnelling of electrons from AgNPs to graphene when excited with laser [56].

From the studies of Zhou *et al*, the number of graphene layers affects the AgNP morphologies. More dense AgNPs are formed when AgNPs are deposited on graphene with lower thickness. Also, the size of the AgNPs increased with the increase in layer number. More enhancement in Raman signal of graphene was observed for less number of layers. Also, it was found that the enhancement factor for G band is higher than that of 2D band [52].

Zhou *et al* studied the effect of surface roughness of plasmonic film on the SERS of graphene. When graphene is placed on corrugated Ag film, a tensile strain is induced in graphene resulting from the bending of graphene. This results in the loss of symmetry of graphene peaks due to the loss of symmetry of graphene hexagonal structure [57]. Quenching of 2D peak was observed by Ranjan *et al* when graphene was covered with AuNPs. Due to the uniaxial strain developed by the presence of AuNPs, a peak arose at ~1546 cm^{-1}. Another peak arising at ~1600 cm^{-1} is due to the hybridization of graphene p orbitals with the 5d orbitals of Au and this is the reason for quenching 2D peak. A split in D band also is observed due to defects present in the sample [58].

10.5.3 Graphene as SERS substrate

Graphene is widely used as SERS substrate, to study the interaction between substrate and probe molecules and thereby to get more understanding of chemical enhancement mechanism, investigation of which is difficult using usual metallic nanostructures. Exploration of the SERS properties of graphene increased after the observation of fluorescence quenching of dye molecules adsorbed on graphene by Xie *et al* during the measurement of Raman signals from these molecules [59]. The theoretical observation by Swathi *et al* proved the existence of a resonant energy transfer which paves a way to the clean measurement of Raman spectra of these molecules without hindrance of fluorescence [60]. Later on, Ling *et al* discovered the emergence of many bands on the exfoliated graphene treated with organic dyes [35]. The presence of smooth surface, high visible range optical transmission and absence of SPR in visible range rules out the possibility of EM. Many works suggest that the π–π interaction between graphene and molecules, leads to the CT interaction between adsorbed molecules and graphene and is responsible for the Raman signal enhancement [36].

More insight into the CT interaction was obtained by Xu *et al* by modulation of CT interaction in graphene–cobalt phthalocyanine (CoPc) system by applying external electric field and various doping conditions. Fermi level of graphene is at −4.6 eV and HOMO–LUMO energies of CoPc lie at −5.02 and −3.41 eV, respectively. The Fermi level changes from −4.98 to −4.22 eV when gate voltage sweeps from −150 V to +150 V and consequently the energy gap between LUMO of molecule and graphene Fermi level changes from 1.57 to 0.81 eV. When the energy gap is closer to excitation energy of laser (1.96 eV) the Raman intensity increases. Under n-doping, the Fermi level is upshifted above Dirac point and variation range of Fermi level with applied electric field is closer to the LUMO of molecule. Compared to the excitation energy the energy gap is small, which makes CT process sensitive to Fermi level variation. The opposite effect is observed on p-doping, where the Fermi level variation is downshifted, making the energy gap between LUMO and Fermi level closer to the excitation energy, which makes the CT process insensitive to the energy gap variation [61].

The CT interaction in graphene is highly sensitive to molecular orientation. Ling *et al* studied this effect on Cu phthalocyanine (CuPc)–graphene system. The CuPc molecules where adsorbed on the top of graphene with an upstanding orientation. On annealing the sample the CuPc molecules take a lying down configuration. On comparing the enhancement of Raman signal of CuPc molecule, more enhancement was observed for the lying down configuration of CuPc molecule. In lying down configuration, the π-orbital cloud of CuPc molecule overlaps with graphene, and electron transition takes place readily. Also, it induces larger interfacial dipole, which magnifies the molecular polarizability. This shows that different molecular interactions are highly dependent on molecular orientation and stronger interaction gives rise to stronger enhancement [62].

10.5.4 Graphene–metal hybrid SERS substrate

As graphene does not exhibit SPR in visible wavelength region, it does not show EM effect and thus graphene itself cannot form an efficient SERS-active substrate with single molecule detection capability. Therefore, the effort to develop a highly sensitive SERS-active substrate with stable, reproducible and clean fingerprint Raman peaks is now progressing towards the development of graphene–metal hybrid SERS substrate. Graphene–metal SERS substrate can take advantage of both metal substrates (EM) and graphene (CM). As graphene provides large surface area and can prevent metal nanostructures from oxidation, these hybrid SERS possess additional qualities such as high chemical stability, superior molecule adsorption and suppressed fluorescence background. The sensitivity of the substrate mainly depends on distribution density and inter-particle distance of metal nanostructures. Depending on the geometry of the hybrid substrate, they are mainly divided into graphene-supported, graphene-encapsulated and graphene sandwiched hybrid substrates [63]. Various forms of graphene such as graphene oxide (GO), graphene sheets, single-layer graphene, etc have been utilised for the fabrication of G-SERS substrate.

10.5.4.1 Graphene sheet–metal nanocomposites SERS substrates

Graphene oxide/graphene sheet–metal nanocomposites can be developed through simple chemical reactions. AgNPs/graphene sheets based SERS substrate is developed by a galvanic displacement process in which Cu nanoparticles (CuNPs) loaded onto graphene sheets are replaced by AgNPs [64]. AgNPs of small size (5 nm) enable the formation of highly sensitive SERS substrate. These graphene sheet/Ag based SERS substrates are employed for detection of prohibited colorants. A schematic illustration of synthesis of Ag/graphene sheet hybrid SERS substrate is shown in figure 10.5 and quantitative determination of 2-thiouracil is demonstrated using as-prepared hybrid SERS substrates [65]. SERS substrate based on Ag nanoplates decorated on graphene sheets has been employed for ultrasensitive SERS detection of organic pesticides [66]. Jiang *et al* dispersed and anchored AgNPs onto GO surface and proved that SERS activity for rhodamine 6G is increased by 7.8-fold in comparison to bare AgNPs-based SERS substrate [67]. Zhang *et al* reported Ag–graphene composite nanosheets with adjustable size and well-controlled densities of AgNPs by one-pot method using poly(*N*-vinyl-2-pyrrolidone) as a reductant and

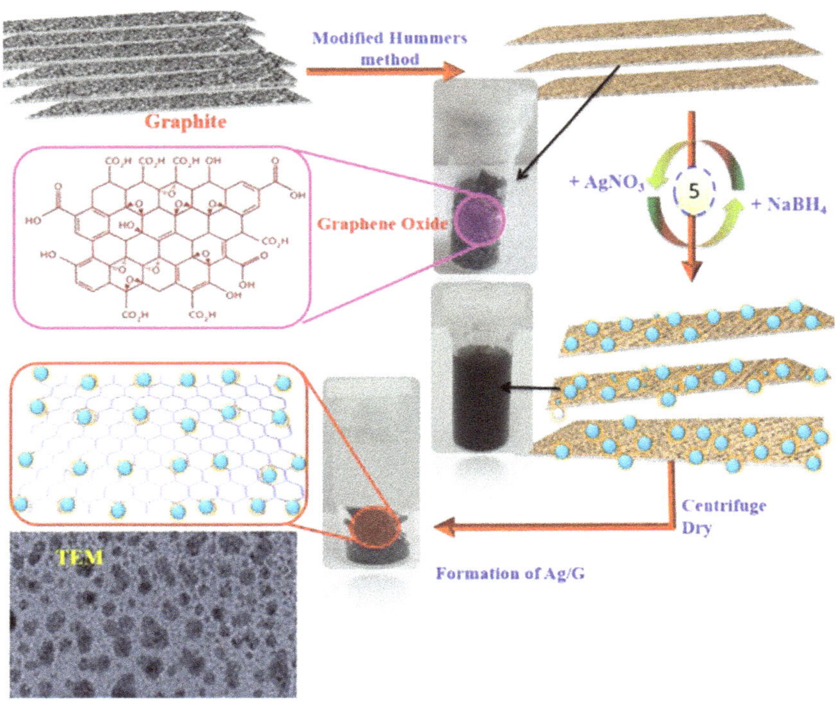

Figure 10.5. Schematic illustration of synthesis of Ag/graphene sheets. Reproduced from [65], copyright 2017 with permission from Elsevier.

stabilizer [68]. Work by Huang and co-workers shows that modifying GO with Au, improved SERS activity and stability of the substrate. At the same time, it exhibited enhanced catalytic activity too [69]. Cu/GO hybrid substrate is developed by facile chemical reduction method. A list of different GO/Ag and GO/Au hybrids substrates are tabulated elsewhere in a review article [70]. SERS enhancement mechanism in graphene–Ag hybrid substrate is investigated using rhodamine 6G (R6G) adsorbed onto the substrate with different sizes of AgNPs. It is observed that larger metal nanoparticles play major roles in SERS performance of the substrate [71]. A free-standing hybrid substrate is then developed by vacuum filtration followed by annealing which enables decreasing the inter-particle distance and hence increased density of 'hotspots'.

But these hybrid substrates do not hold the capability to perform as an efficient SERS substrate with reproducible signal over a larger area and mono/few-layer graphene can offer many advantages over graphene sheets.

10.5.4.2 Chemical vapour deposition graphene–metal hybrids

10.5.4.2.1 Graphene-supported SERS substrates

Graphene can be easily employed as a supporter and as a molecule enricher by simply depositing with a metal thin film or decorating graphene with metal nanostructures. Graphene–metal hybrids using Au, Ag and Cu nanostructures have been widely developed as SERS substrates [72]. Morphology, arrangement and location of metal nanostructure in the composites and the connection type of metal and graphene play a significant role in performance of SERS substrate. Metal nanostructures can be deposited easily via *in situ* method, self-assembly, thermal evaporation etc. Wang *et al* reported dependence of Au film thickness on SERS activity of G-SERS [73]. Chemical vapour deposition (CVD) grown single-layer graphene on Si/SiO_2 substrate is deposited with Au film with different thickness. It is optimized that 7 nm thick Au film over graphene produces maximum SERS activity and weakest photoluminescence background, which proves the effectiveness of graphene for PL quenching and for boosting SERS activity. Wang *et al* observed single molecule detection capability of Au nanopyramids–graphene hybrids [74]. Au nanopyramids with high density of 'hotspots' are fabricated via wafer-scale templating technology and single-layer graphene is then transferred onto sharp Au nanotips.

AuNPs are decorated onto a graphene layer by dipping the graphene layer in Au chloride solution, and these graphene–Au SERS substrates act as a high-sensitivity substrate for sensitive and multiplex DNA detection [75]. DNA can be easily immobilised onto a graphene surface with the help of AuNPs, and DNA with a concentration of 10 nM is detected. The large planar surface of CVD-grown graphene film opens up the possibility of simultaneous detection of many more types of DNA on the same SERS substrate with single-laser excitation. Figures 10.6(a)–(c) illustrate the preparation of Au/graphene hybrid SERS substrate and its use for multiplex DNA detection.

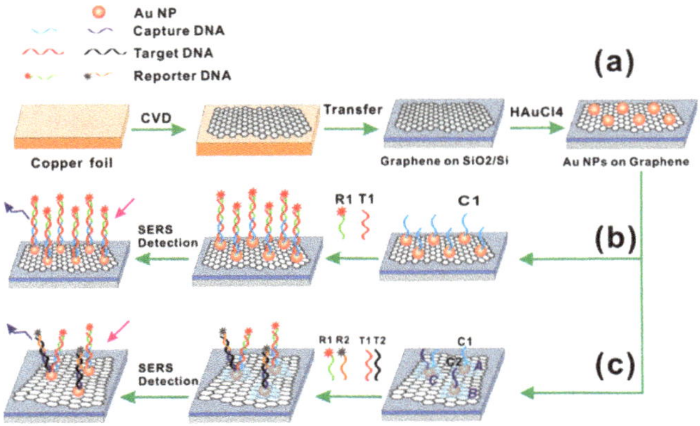

Figure 10.6. Schematic illustration of (a) decoration of AuNPs onto CVD-grown graphene, (b) detection of target DNA by graphene–Au hybrid SERS substrate, and (c) multiplexing detection of two different target types of DNA by graphene–Au hybrid SERS substrate. Reprinted with permission from [75], copyright 2012, American Chemical Society.

SPR of Au nanorods can be easily tuned along the visible spectrum and can be matched with laser excitation wavelength for Raman spectroscopy. Therefore, Au nanorods have great potential to be used as SERS substrate, in graphene–metal hybrids.

Three types of pesticides are detected using an Au nanorod–graphene-based SERS substrates [76]. In this work by Nguyen *et al*, SERS activity of graphene–Au film–Au nanorod (G–Au–AuNR) substrate, Au film–Au nanorod (Au–AuNR) substrate, and graphene coupled with Au nanorods (G–AuNR) are compared and it is verified that SERS activity is highest for G–Au–AuNR system. These G–Au–AuNR substrate offers a lightweight and flexible platform for SERS application in food safety. CVD-grown graphene monolayer is transferred onto AuNP arrays synthesized by electron beam evaporation through anodic aluminium oxide. It was identified that graphene boosted the density of hotspots, enabled SERS analysis with well-defined molecular interaction and improved the reproducibility of SERS signal [77]. Development of a large area quasi-continuous SERS substrate is challenging using traditional top-down approach and that limits its practical applications. This is overcome by employing cold-etching method and Au–graphene hybrids are developed through a controllable break-up of a thin Au film and then transfer of graphene on to the periodically distributed nanoscale Au tips formed by cold-etching method [78].

Srichan *et al* proposed two distinct mechanisms, namely multiple cascaded amplification and collective excitation, to explain SERS enhancement on AgNPs over graphene. In their study AgNPs are sputtered onto graphene surface for different sputtering time to control the density of AgNPs [79]. The enhancement factor estimated using multiple cascaded amplification model is given by equation (10.5):

$$\mathrm{EF} = A_{\mathrm{m}}(\nu_1)^4 A_{\mathrm{g}}(\nu_1)^2 \tag{10.5}$$

where ν_l is the incident light frequency and A_m and A_g are, respectively, SERS enhancement factor for metal and graphene. The electric field enhancement of metal nanoparticle on metallic surface (E_{SP}) and electric field enhancement of metal nanoparticles on graphene (E_G) are related as follows:

$$\left| \frac{E_G}{E_{SP}} \right|_{max} = 3\epsilon_G \frac{(2\pi)^2}{(\lambda_{SP}^G)^2} rd \tag{10.6}$$

where ϵ_G the dielectric constant of graphene, r the radius of the metallic nanoparticle and d is the distance from analyte to silver nanoparticle surface, (λ_{SP}^G) is given by,

$$\frac{(\lambda_{SP}^G)}{\lambda_0} = \frac{8\pi\alpha E_F}{h\omega\ (\varepsilon_r + 1)} \tag{10.7}$$

where λ_0 is the wavelength of incident light, α fine constant, E_F is the Fermi energy of graphene and ε_r is the dielectric constant of medium. Putting all the constant values, it is estimated that $A_g \sim 100$. Hence the contribution of graphene to total enhancement in metal hybrid substrate is estimated to be 10^4.

A SERS-based chemical sensor is developed using three-dimensional microporous graphene foam (GF) decorated with AgNPs. It is demonstrated that EF of AgNPs/GF sensor is four orders of magnitude larger than that of AgNPs/Si substrate. The SERS sensitivity is also tuned by controlling AgNPs size [79]. In the work by Sidorov *et al*, graphene layers are deposited from highly oriented pyrolytic graphite (HOPG) onto Si/SiO_2, and the Ag and Au nanostructures with different morphologies are grown by seed-mediated method [80]. Hexanethiolate-coated Au monolayer acts as seed for the growth of Ag nanostructures.

A more sensitive graphene–metal hybrid substrate can be developed using metal alloys as it can utilize the advantage of both metal nanostructures. For example, Ag nanostructure has higher scattering cross-section as compared to Au and Cu and uniform sized/shaped AuNPs can be more easily synthesized. Therefore, a metal alloy structure can be a potential candidate for hybrid SERS substrate with ideal performance. Ag–CuNPs are wrapped by a monolayer graphene layer grown on SiO_2 and highly reproducible SERS signal are observed, and detection of nucleosides extracted from human urine also demonstrated using these hybrid substrates [81]. Even though electron beam lithography offers fabrication of SERS substrates using uniform shaped and sized metal nanostructures, controlling the inter-particle space to nano or sub-nanometer dimension to maximise EF is still challenging. Also, the density of 'hotspots' that can effectively contribute to total EM enhancement is much less in these 2D patterned substrates. Placing atomic flat graphene layer between two metal layers offers flexible integration of graphene and metal. Li *et al* proposed a novel nanoparticle–film gap system by adding graphene as a sub-nanospacer between AgNP and Ag film [37]. Zhao *et al* introduced a monolayer graphene between AuNPs and Ag nanohole arrays to develop an efficient 3D SERS substrate, and these substrates utilize the large scattering cross-section of Ag, high chemical stability of Au and atomic flatness, biocompatibility and flexibility of graphene to promote SERS for practical applications [82]. A detection limit of 10^{-13} M R6G molecules is achieved using these substrates and good reproducibility and stability is achieved. A flexible G-SERS substrate based on AuNP/graphene/AuNP (AuNP/G/AuNP) sandwich geometry is fabricated on polyethylene (PE) film through self-assembly. Detection of thiram at very low concentration on the surface of orange peels is also carried out using these flexible substrates that proves the effectiveness of G-SERS tape substrate for *in situ* detection in food safety and environmental monitoring (figures 10.7(a)–(c)) [83].

Graphene hybrid SERS substrates are not only limited to metal nanostructures but can be developed using semiconductors. Graphene–semiconductor hybrid composites, exhibit superior optical and electrical properties that lead to enhanced performance of batteries, solar cells and sensors [70, 84, 85]. Kuo and Chen developed a graphene–semiconductor hybrid SERS substrate using TiO_2 and Au and demonstrated the graphene thickness controls the CT site and hence SERS enhancement [86]. The CVD-grown graphene is transferred to TiO_2/ZnO panel and

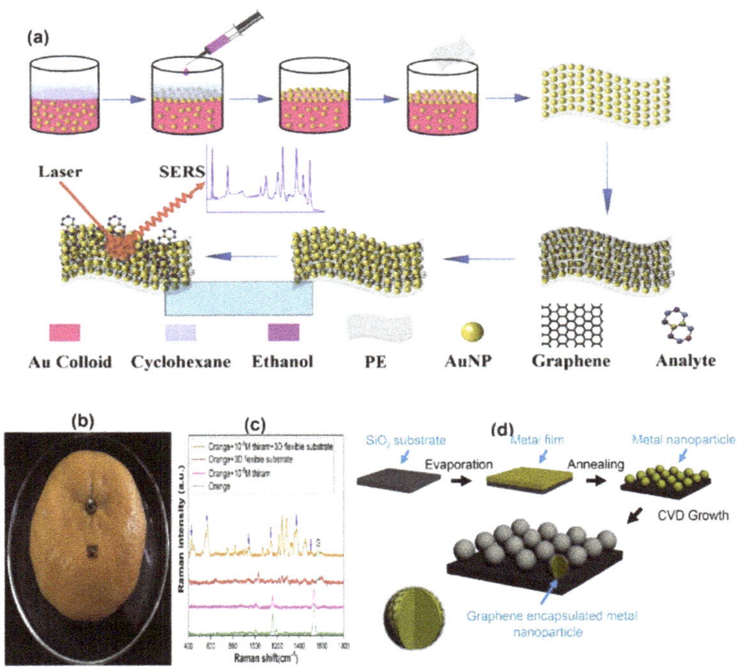

Figure 10.7. (a) Schematic of preparation of AuNP/graphene/AuNP 'sandwich' SERS substrate. (b) Demonstration of thiram detection on orange surface. (c) *In situ* detection of Raman spectrum of thiram under different conditions using as-prepared substrate. Reprinted from [83], copyright 2019, with permission from Elsevier. (d) Schematic diagram fabrication of graphene-encapsulated SERS substrate. Reprinted with permission from [90], copyright 2014 American Chemical Society.

then it is deposited with AuNPs to form an efficient SERS substrate. A hybrid SERS structure using three innovative materials, graphene, AuNP dimer and DNA is developed by Ilko Bald and his group [87]. DNA enables precise arrangement of metal nanoparticles and analyte molecule, which is an essential criterion for an efficient SERS substrate. As-prepared substrate is very efficient to suppress fluorescence and to reduce fluorescent background. The photo-bleaching rate is also decreased considerably. A substrate that integrates two plasmonic nanostructures of AuNPs and WS_2–nanodomes (WS_2–ND) is demonstrated in AuNP/WS_2–ND/graphene heterostructures [88]. The enhanced sensitivity of these heterostructures is due to the dipole–dipole interaction at the interface. The same group explored a high sensitive non-metallic SERS substrate based entirely on 2D atomic materials using a rationale design of intermixed WS_2 nanodiscs (WS_2NDs) and MoS_2 nanodiscs (MoS_2NDs) on graphene [89].

10.5.4.2.2 Graphene-encapsulated SERS substrate

Stability of SERS substrate can be improved by coating the metal nanostructures with a ultra-thin shell, termed as shell-isolated surface-enhanced Raman scattering (shell-isolated SERS), which prevents metal–molecule interactions. The inert shell should be very thin so that there won't be any loss of EM enhancement from metal and pinhole-free for complete isolation of metal and molecule to avoid unwanted interactions between them. Due to its unique properties, graphene is identified as a good candidate for shell-isolated SERS substrate and can offer a Raman spectrum with a clear baseline due to the reduced photo-carbonization and photo-bleaching.

A typical preparation technique for graphene-encapsulated SERS substrate is given in figure 10.7(d). Graphene with a controlled number of layers can be grown on the surface of metal nanoparticles via CVD [90]. AgNPs are encapsulated by hydrogen-terminated chemical vapour deposited single-layer graphene, and the high resistivity of hydrogenated graphene increases EM enhancement in the AgNPs due to smaller screening effect of the localized electric field. Graphene also protected AgNPs from reactive and harsh environments and that offers a chemically inert and biocompatible surface for SERS applications. Graphene-veiled Au substrate is fabricated using thermal evaporation of Au on SiO_2/Si and graphene prepared from mechanically exfoliated Kish graphite [91]. Cu based hybrid substrate can be developed using graphene as it can prevent Cu from oxidation which inhibits the cost-effective Cu nanostructures from employment as SERS substrate. Core–shell graphene@Cu is deposited onto graphene@Cu via CVD, which possess enhanced stability and higher sensitivity as CM from graphene also contributing to the total enhancement. An excellent SERS substrate is fabricated by transferring monolayer graphene onto CuNPs synthesised on SiO_2 by liquor-phase reduction synthesis (LPRS) method [92]. A SERS-active substrate that can detect 10^{-16} M human lactoferrin molecules is developed using silvered porous silicon in which the negative surface potential of AgNPs helps to attract positively charged lactoferrin molecules. Graphene is used as a protective layer to prevent the overheating of analyte molecules and also it improves the SERS sensitivity so that lactoferrin molecules with a concentration as low as 10^{-18} M are carried out [93]. A reproducible ultrasensitive SERS substrate is fabricated using graphene/CuNP hybrids (G/CuNPs) system in which CuNPs are wrapped by graphene and adenosine in serum was detected at a lower concentration of 5 nM [94]. The linear response curve with the concentration of adenosine is achieved and adenosine detection in a real human urine sample also carried out. In this system, graphene shell enriches and fixes the adenosine molecule to the surface and improves its chemical and optical stability. It is identified that direct-growth method is an efficient way to fabricate ultrasensitive reproducible graphene-based SERS substrates. Because the graphene layers can tightly and seamlessly wrap the nanoparticles, that can reduce the loss of EM enhancement from the metal structures and magnify the CM from graphene layer. But the direct-growth method is usually carried out at high temperature (1000 °C) and that can cause the heterogenous size and distribution of metal nanoparticles which will ultimately affect the reproducibility of SERS signal. Liu *et al* introduced a low-temperature growth method to fabricate Cu/graphene double nanocaps array-based SERS substrate. First, Cu is evaporated onto the surface of a close-packed array of polystyrene spheres, and then graphene is grown by CVD in the absence of CH_4 using polystyrene as a template and also a source of carbon [95].

10.5.4.2.3 Free-standing flat graphene SERS substrates

Observation of homogeneous SERS signal and SERS quantification on flat thin polymer layer over metal nanostructures leads to the development of G-SERS substrate with atomic flatness. As the ‘hotspot’ sites are pushed to the flat graphene surface, its SERS activity is high as compared to graphene-supported SERS substrates. The surface adsorption on G-SERS substrate is possible through the π–π interactions of graphene–molecule and which serve as a new driving force for surface adsorption. As compared with traditional roughened surface, these flat surfaces are preferred for many analytical situations. The behaviour of molecules on a traditional SERS substrate and flat graphene substrate are illustrated in figure 10.8(a) and (b). Graphene acts as supporter for more controllable molecular arrangement and helps to reduce the possible disturbances due to metal–molecular interactions. It is also observed that arrayed hotspot sites should be located close to flat graphene surface and hence hemisphere structures are more apt for flat graphene substrate than spheres. This effect is proven by 3D-FDTD simulations on Au hemispheres and results show that presence of graphene layer causes a one-fold increase in electric field enhancement and electromagnetic field will not be decayed due to the presence of graphene layer. Effect of atomic flat G-SERS is also experimentally verified using Au and Ag based G-SERS substrate [96]. A new form of flat G-SERS substrate that is flexible, transparent and free-standing (G-SERS tape) is developed for extended applications. Analytical applications using G-SERS tape is demonstrated by detection of CuPc molecule adsorbed onto cauliflower surface (figure 10.8(c) and (d)) [96]. This offers almost freedom from sample pre-treatments, quicker data acquisition and analysis, less contamination and is more economical. It is also found that they are reusable and with all these advantages, G-SERS substrates offer a ‘greener’ way to do SERS.

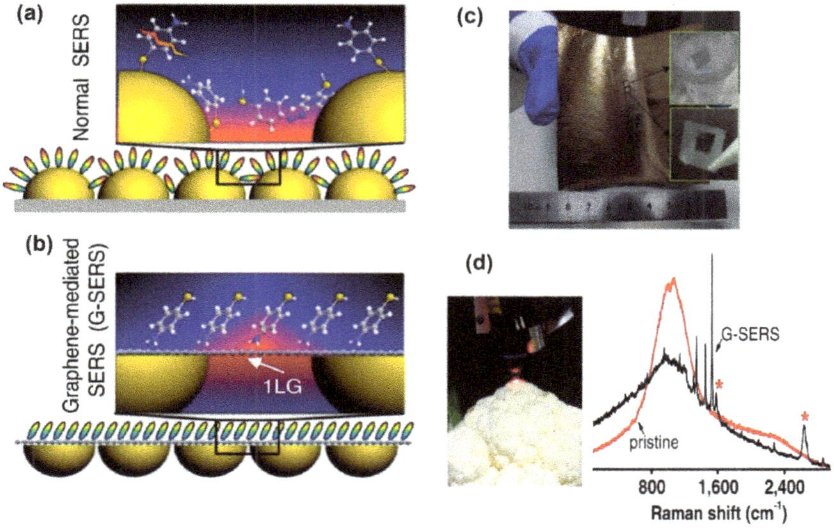

Figure 10.8. Schematic illustration of molecular arrangement (a) on a normal SERS substrate and (b) on a G-SERS flat substrate. (c) Photograph of a G-SERS substrate before the removal of Cu. Insets on the right are two free-standing G-SERS tapes. (d) Pristine and G-SERS measurements of a cauliflower surface with adsorbed CuPc molecule. Reproduced with permission [96]. Copyright 2012, National Academy of Sciences.

10.6 Conclusions

This chapter discusses how graphene works as multi-functional agent for the better performance of SERS. The chapter introduces SERS technology, its enhancement mechanism and some SERS prominent applications are given. Then, the role of graphene in SERS and recent progress in development of G-SERS substrates are discussed. G-SERS substrate using graphene only, graphene–metal hybrids and their advantage over traditional metal-based SERS substrates are discussed in detail. A review on G-SERS substrates that have been used as supporter, nanospacer and as a protecting layer to enhance the sensitivity, stability and suppress fluorescent background has been included. It is identified that G-SERS has become an important material in SERS technology and is progressing towards applications in both fundamental science and in real-life applications. Hence, the study and development of G-SERS substrates would advance SERS technology as a facile and vital tool in future.

References

[1] Jones R R, Hooper D C, Zhang L, Wolverson D and Valev V K 2019 Raman techniques: fundamentals and frontiers *Nanoscale Res. Lett.* **14** 1–34

[2] Le R E C and Etchegoin P G 2012 Single-molecule surface-enhanced Raman spectroscopy *Annu. Rev. Phys. Chem.* **63** 65–87

[3] Fleischmann M, Hendra P J and McQuillan A J 1974 Raman spectra of pyridine adsorbed at a silver electrode *Chem. Phys. Lett.* **26** 163–6

[4] Ouyang L, Ren W, Zhu L and Irudayaraj J 2017 Prosperity to challenges: recent approaches in SERS substrate fabrication *Rev. Anal. Chem.* **36** 20160027
[5] Kubackova J, Fabriciova G, Miskovsky P, Jancura D and Sanchez-Cortes S 2015 Sensitive surface-enhanced Raman spectroscopy (SERS) detection of organochlorine pesticides by alkyl dithiol-functionalized metal nanoparticles-induced plasmonic hot spots *Anal. Chem.* **87** 663–9
[6] Schlücker S 2014 Surface-enhanced Raman spectroscopy: concepts and chemical applications *Angew. Chem. Int. Ed.* **53** 4756–95
[7] Jeanmaire D L and Duyne, R.P. V 1977 Surface Raman spectroelectrochemistry: part I. Heterocyclic, aromatic, and aliphatic amines adsorbed on the anodized silver electrode *J. Electroanal. Chem. Interfacial Electrochem.* **84** 1–20
[8] Kneipp K, Wang Y, Kneipp H, Perelman L T, Itzkan I, Dasari R R and Feld M S 1997 Single molecule detection using surface-enhanced Raman scattering (SERS) *Phys. Rev. Lett.* **78** 1667
[9] Nie S and Emory S R 1997 Probing single molecules and single nanoparticles by surface-enhanced Raman scattering *Science* **275** 1102–6
[10] Le R E C, Meyer M, Blackie E and Etchegoin P G 2008 Advanced aspects of electromagnetic SERS enhancement factors at a hot spot *J. Raman Spectrosc.* **39** 1127–34
[11] Lombardi J R, Birke R L, Lu T and Xu J 1986 Charge-transfer theory of surface enhanced Raman spectroscopy: Herzberg–Teller contributions *J. Chem. Phys.* **84** 4174–80
[12] Le R E and Etchegoin P 2008 *Principles of Surface-Enhanced Raman Spectroscopy: And Related Plasmonic Effects* (Amsterdam: Elsevier)
[13] Lombardi J R and Birke R L 2009 A unified view of surface-enhanced Raman scattering *Acc. Chem. Res.* **42** 734–42
[14] Lombardi J R 2017 The theory of surface-enhanced Raman scattering on semiconductor nanoparticles; toward the optimization of SERS sensors *Faraday Discuss.* **205** 105–20
[15] Lombardi J R and Birke R L 2014 Theory of surface-enhanced Raman scattering in semiconductors *J. Phys. Chem.* C **118** 11120–30
[16] Han X, Ji W, Zhao B and Ozaki Y 2017 Semiconductor-enhanced Raman scattering: active nanomaterials and applications *Nanoscale* **9** 4847–61
[17] Le R E C, Blackie E, Meyer M and Etchegoin P G 2007 Surface enhanced Raman scattering enhancement factors: a comprehensive study *J. Phys. Chem.* C **111** 13794–803
[18] Le R E C and Etchegoin P G 2013 Quantifying SERS enhancements *MRS Bull.* **38** 631–40
[19] Sun J, Gong L, Wang W, Gong Z, Wang D and Fan M 2020 Surface-enhanced Raman spectroscopy for on-site analysis: a review of recent developments *Luminescence* **35** 808–20
[20] Perumal J, Wang Y, Attia A B E, Dinish U S and Olivo M 2021 Towards a point-of-care SERS sensor for biomedical and agri-food analysis applications: a review of recent advancements *Nanoscale* **13** 553–80
[21] Brackx G, Guinoiseau D, Duponchel L, Gélabert A, Reichel V, Zrig S, Di Meglio J M, Benedetti M F, Gaillardet J and Charron G 2020 A frugal implementation of surface enhanced Raman scattering for sensing Zn^{2+} in freshwaters—in depth investigation of the analytical performances *Sci. Rep.* **10** 1–11
[22] Chen J, Huang M, Kong L and Lin M 2019 Jellylike flexible nanocellulose SERS substrate for rapid *in situ* non-invasive pesticide detection in fruits/vegetables *Carbohydrate Polym.* **205** 596–600
[23] Li L and Chin W S 2021 Rapid and sensitive SERS detection of melamine in milk using Ag nanocube array substrate coupled with multivariate analysis *Food Chem.* **357** 129717

[24] Park J, Thomasson J A, Gale C C, Sword G A, Lee K M, Herrman T J and Suh C P C 2020 Adsorbent-SERS technique for determination of plant VOCs from live cotton plants and dried teas *ACS Omega* **5** 2779–90

[25] Sharma B, Frontiera R R, Henry A I, Ringe E and Van Duyne R P 2012 SERS: materials, applications, and the future *Mater. Today* **15** 16–25

[26] Choffel M A, Farling C G, Frano K A, Matecki M K, Zheng Z, Svoboda S A and Wustholz K L 2021 Revealing the fugitive palette of the early American South: a SERS study of eighteenth-century oil paintings *Stud. Conserv.* **67** 556–68

[27] Hakonen A, Wu K, Schmidt M S, Andersson P O, Boisen A and Rindzevicius T 2018 Detecting forensic substances using commercially available SERS substrates and handheld Raman spectrometers *Talanta* **189** 649–52

[28] Zheng X S, Jahn I J, Weber K, Cialla-May D and Popp J 2018 Label-free SERS in biological and biomedical applications: recent progress, current challenges and opportunities *Spectrochim. Acta* A **197** 56–77

[29] Li P, Long F, Chen W, Chen J, Chu P K and Wang H 2020 Fundamentals and applications of surface-enhanced Raman spectroscopy–based biosensors *Curr. Opin. Biomed. Eng.* **13** 51–9

[30] Wang J, Geng Y, Shen Y, Shi W, Xu W and Xu S 2019 SERS-active fiber tip for intracellular and extracellular pH sensing in living single cells *Sens. Actuators* B **290** 527–34

[31] Mao Y, Sun Y, Xue J, Lu W and Cao X 2021 Ultra-sensitive and high efficiency detection of multiple non-small cell lung cancer-related miRNAs on a single test line in catalytic hairpin assembly-based SERS-LFA strip *Anal. Chim. Acta* **1178** 338800

[32] He M, Lin J, Akakuru O U, Xu X, Li Y, Cao Y, Xu Y and Wu A 2021 Octahedral silver oxide nanoparticles enabling remarkable SERS activity for detecting circulating tumor cells *Sci. China Life Sci.* **65** 561–71

[33] Sujai P T, Shamjith S, Joseph M M and Maiti K K 2021 Elucidating gold–MnO_2 core–shell nanoenvelope for real time SERS-guided photothermal therapy on pancreatic cancer cells *ACS Appl. Bio Mater.* **4** 4962–72

[34] Kapara A, Brunton V G, Graham D and Faulds K 2020 Characterisation of estrogen receptor alpha (ERα) expression in breast cancer cells and effect of drug treatment using targeted nanoparticles and SERS *Analyst* **145** 7225–33

[35] Ling X, Xie L, Fang Y, Xu H, Zhang H, Kong J, Dresselhaus M S, Zhang J and Liu Z 2010 Can graphene be used as a substrate for Raman enhancement? *Nano Lett.* **10** 553–61

[36] Xu W, Mao N and Zhang J 2013 Graphene: a platform for surface-enhanced Raman spectroscopy *Small* **9** 1206–24

[37] Li X, Choy W C, Ren X, Zhang D and Lu H 2014 Highly intensified surface enhanced Raman scattering by using monolayer graphene as the nanospacer of metal film–metal nanoparticle coupling system *Adv. Funct. Mater.* **24** 3114–22

[38] Bonaccorso F, Lombardo A, Hasan T, Sun Z, Colombo L and Ferrari A C 2012 Production and processing of graphene and 2D crystals *Mater. Today* **15** 564–89

[39] Ferrari A C and Robertson J 2004 Raman spectroscopy of amorphous, nanostructured, diamond-like carbon, and nanodiamond *Philos. Trans. R. Soc. London, Ser.* A **362** 2477–512

[40] Ferrari A C 2007 Raman spectroscopy of graphene and graphite: disorder, electron–phonon coupling, doping and nonadiabatic effects *Solid State Commun.* **143** 47–57

[41] Vidano R P, Fischbach D B, Willis L J and Loehr T M 1981 Observation of Raman band shifting with excitation wavelength for carbons and graphites *Solid State Commun.* **39** 341–4

[42] Cancado L G, Pimenta M A, Neves B R A, Dantas M S S and Jorio A 2004 Influence of the atomic structure on the Raman spectra of graphite edges *Phys. Rev. Lett.* **93** 247401
[43] Malard L M, Pimenta M A, Dresselhaus G and Dresselhaus M S 2009 Raman spectroscopy in graphene *Phys. Rep.* **473** 51–87
[44] Ferrari A C and Basko D M 2013 Raman spectroscopy as a versatile tool for studying the properties of graphene *Nat. Nanotechnol.* **8** 235–46
[45] Gupta A, Chen G, Joshi P, Tadigadapa S and Eklund P C 2006 Raman scattering from high-frequency phonons in supported n-graphene layer films *Nano Lett.* **6** 2667–73
[46] Jorio A, Dresselhaus M S, Saito R and Dresselhaus G 2011 *Raman Spectroscopy in Graphene Related Systems* (New York: Wiley)
[47] Gupta A K, Russin T J, Gutiérrez H R and Eklund P C 2009 Probing graphene edges via Raman scattering *ACS Nano* **3** 45–52
[48] Ferrari A C *et al* 2006 Raman spectrum of graphene and graphene layers *Phys. Rev. Lett.* **97** 187401
[49] Lucchese M M, Stavale F, Ferreira E M, Vilani C, Moutinho M V D O, Capaz R B, Achete C A and Jorio A 2010 Quantifying ion-induced defects and Raman relaxation length in graphene *Carbon* **48** 1592–7
[50] Das A *et al* 2008 Monitoring dopants by Raman scattering in an electrochemically top-gated graphene transistor *Nat. Nanotechnol.* **3** 210–5
[51] Lee J, Shim S, Kim B and Shin H S 2011 Surface-enhanced Raman scattering of single-and few-layer graphene by the deposition of gold nanoparticles *Chem. Eur. J.* **17** 2381–7
[52] Zhou H, Qiu C, Yu F, Yang H, Chen M, Hu L and Sun L 2011 Thickness-dependent morphologies and surface-enhanced Raman scattering of Ag deposited on n-layer graphenes *J. Phys. Chem.* C **115** 11348–54
[53] Almokhtar M, Fares H, Inoue K and Matsumoto K 2021 Graphene on single Ag nanoparticles for nanoscale and quantum applications *Appl. Surf. Sci.* **541** 148390
[54] Osváth Z, Deák A, Kertész K, Molnár G, Vértesy G, Zámbó D, Hwang C and Biró L P 2015 The structure and properties of graphene on gold nanoparticles *Nanoscale* **7** 5503–9
[55] Iyer G R S, Wang J, Wells G, Guruvenket S, Payne S, Bradley M and Borondics F 2014 Large-area, freestanding, single-layer graphene–gold: a hybrid plasmonic nanostructure *ACS Nano* **8** 6353–62
[56] Carles R, Bayle M and Bonafos C 2018 Plasmon-enhanced scattering and charge transfer in few-layer graphene interacting with buried printed 2D-pattern of silver nanoparticles *Nanotechnology* **29** 175301
[57] Zhao Y, Liu X, Lei D Y and Chai Y 2014 Effects of surface roughness of Ag thin films on surface-enhanced Raman spectroscopy of graphene: spatial nonlocality and physisorption strain *Nanoscale* **6** 1311–7
[58] Ranjan P, Laha R and Balakrishnan J 2017 Au concentration-dependent quenching of Raman 2D peak in graphene *J. Raman Spectrosc.* **48** 586–91
[59] Xie L, Ling X, Fang Y, Zhang J and Liu Z 2009 Graphene as a substrate to suppress fluorescence in resonance Raman spectroscopy *J. Am. Chem. Soc.* **131** 9890–1
[60] Swathi R S and Sebastian K L 2008 Resonance energy transfer from a dye molecule to graphene *J. Chem. Phys.* **129** 054703
[61] Xu H, Chen Y, Xu W, Zhang H, Kong J, Dresselhaus M S and Zhang J 2011 Modulating the charge-transfer enhancement in GERS using an electrical field under vacuum and an n/p-doping atmosphere *Small* **7** 2945–52

[62] Ling X, Wu J, Xu W and Zhang J 2012 Probing the effect of molecular orientation on the intensity of chemical enhancement using graphene-enhanced Raman spectroscopy *Small* **8** 1365–72

[63] Lai H, Xu F, Zhang Y and Wang L 2018 Recent progress on graphene-based substrates for surface-enhanced Raman scattering applications *J. Mater. Chem.* B **6** 4008–28 May

[64] Zhao H, Fu H, Zhao T, Wang L and Tan T 2012 Fabrication of small-sized silver NPs/graphene sheets for high-quality surface-enhanced Raman scattering *J. Colloid Interface Sci.* **375** 30–4

[65] Saleh T A, Al-Shalalfeh M M and Al-Saadi A A 2018 Silver loaded graphene as a substrate for sensing 2-thiouracil using surface-enhanced Raman scattering *Sens. Actuators* B **254** 1110–7

[66] Wang X, Zhu C, Hu X, Xu Q, Zhao H, Meng G and Lei Y 2019 Highly sensitive surface-enhanced Raman scattering detection of organic pesticides based on Ag-nanoplate decorated graphene-sheets *Appl. Surf. Sci.* **486** 405–10

[67] Jiang Y, Wang J, Malfatti L, Carboni D, Senes N and Innocenzi P 2018 Highly durable graphene-mediated surface enhanced Raman scattering (G-SERS) nanocomposites for molecular detection *Appl. Surf. Sci.* **450** 451–60

[68] Zhang Z, Xu F, Yang W, Guo M, Wang X, Zhang B and Tang J 2011 A facile one-pot method to high-quality Ag–graphene composite nanosheets for efficient surface-enhanced Raman scattering *Chem. Commun.* **47** 6440–2

[69] Huang J, Zhang L, Chen B, Ji N, Chen F, Zhang Y and Zhang Z 2010 Nanocomposites of size-controlled gold nanoparticles and graphene oxide: formation and applications in SERS and catalysis *Nanoscale* **2** 2733–8

[70] Wang Z, Wu S, Ciacchi L C and Wei G 2018 Graphene-based nanoplatforms for surface-enhanced Raman scattering sensing *Analyst* **143** 5074–89

[71] Zhou Y, Cheng X, Du D, Yang J, Zhao N, Ma S, Zhong T and Lin Y 2014 Graphene–silver nanohybrids for ultrasensitive surface enhanced Raman spectroscopy: size dependence of silver nanoparticles *J. Mater. Chem.* C **2** 6850–8

[72] Prasad A, Chaichi A, Mahigir A, Sahu S P, Ganta D, Veronis G and Gartia M R 2020 Ripple mediated surface enhanced Raman spectroscopy on graphene *Carbon* **157** 525–36

[73] Wang Y, Ni Z, Hu H, Hao Y, Pei Wong C and Yu Tothers 2010 Gold on graphene as a substrate for surface enhanced Raman scattering study *Appl. Phys. Lett.* **97** 163111 October

[74] Wang P, Liang O, Zhang W, Schroeder T and Xie Y-H 2013 Ultra-sensitive graphene-plasmonic hybrid platform for label-free detection *Adv. Mater.* **25** 4918–424

[75] He S, Liu K K, Su S, Yan J, Mao X, Wang D, He Y, Li L J, Song S and Fan C 2012 Graphene-based high-efficiency surface-enhanced Raman scattering-active platform for sensitive and multiplex DNA detection *Anal. Chem.* **84** 4622–7

[76] Nguyen T H, Zhang Z, Mustapha A, Li H and Lin M 2014 Use of graphene and gold nanorods as substrates for the detection of pesticides by surface enhanced Raman spectroscopy *J. Agric. Food Chem.* **62** 10445–51

[77] Han Y, Wang H, Qiang L, Gao Y, Li Q, Pang J, Liu H, Han L, Wu Y and Zhang Y 2020 Fabrication of a uniform Au nanodot array/monolayer graphene hybrid structure for high-performance surface-enhanced Raman spectroscopy *J. Mater. Sci.* **55** 591–602

[78] Wu T *et al* 2019 Ultrawideband surface enhanced Raman scattering in hybrid graphene fragmented-gold substrates via cold-etching *Adv. Opt. Mater.* **7** 1900905

[79] Srichan C, Ekpanyapong M, Horprathum M, Eiamchai P, Nuntawong N, Phokharatkul D, Danvirutai P, Bohez E, Wisitsoraat A and Tuantranont A 2016 Highly-sensitive

surface-enhanced Raman spectroscopy (SERS)-based chemical sensor using 3D graphene foam decorated with silver nanoparticles as SERS substrate *Sci. Rep.* **6** 23733
[80] Sidorov A N, Sławiński G W, Jayatissa A H, Zamborini F P and Sumanasekera G U 2012 A surface-enhanced Raman spectroscopy study of thin graphene sheets functionalized with gold and silver nanostructures by seed-mediated growth *Carbon* **50** 699–705
[81] Zhang X *et al* 2015 Growth graphene on silver–copper nanoparticles by chemical vapor deposition for high-performance surface-enhanced Raman scattering *Appl. Surf. Sci.* **353** 63–70
[82] Zhao Y, Yang D, Li X, Liu Y, Hu X, Zhou D and Lu Y 2017 Toward highly sensitive surface-enhanced Raman scattering: the design of a 3D hybrid system with monolayer graphene sandwiched between silver nanohole arrays and gold nanoparticles *Nanoscale* **9** 1087–96
[83] Dong J *et al* 2020 Flexible and transparent Au nanoparticle/graphene/Au nanoparticle 'sandwich' substrate for surface-enhanced Raman scattering *Mater. Today Nano* **9** 100067
[84] Liang Y, Li Y, Wang H, Zhou J, Wang J, Regier T and Dai H 2011 Co_3O_4 nanocrystals on graphene as a synergistic catalyst for oxygen reduction reaction *Nat. Mater.* **10** 780–6
[85] Qiu B, Xing M and Zhang J 2014 Mesoporous TiO_2 nanocrystals grown *in situ* on graphene aerogels for high photocatalysis and lithium-ion batteries *J. Am. Chem. Soc.* **136** 5852–5
[86] Kuo C and Chen C 2014 Graphene thickness-controlled photocatalysis and surface enhanced Raman scattering *Nanoscale* **6** 12805–13
[87] Prinz J, Matković A, Pešić J, Gajić R and Bald I 2016 Hybrid structures for surface-enhanced Raman scattering: DNA origami/gold nanoparticle dimer/graphene *Small* **12** 5458–67
[88] Ghopry S A, Alamri M, Goul R, Cook B, Sadeghi S M, Gutha R R, Sakidja R and Wu J Z 2020 Au nanoparticle/WS_2 nanodome/graphene van der Waals heterostructure substrates for surface-enhanced Raman spectroscopy *ACS Appl. Nano Mater.* **3** 2354–63
[89] Ghopry S A, Sadeghi S M, Farhat Y, Berrie C L, Alamri M and Wu J Z 2021 Intermixed WS_2+MoS_2 nanodisks/graphene van der Waals heterostructures for surface-enhanced Raman spectroscopy sensing *ACS Appl. Nano Mater.* **4** 2941–51
[90] Liu Y, Hu Y and Zhang J 2014 Few-layer graphene-encapsulated metal nanoparticles for surface-enhanced Raman spectroscopy *J. Phys. Chem.* C **118** 8993–8
[91] Xu W, Xiao J, Chen, Yanfeng, Chen, Yabin, Ling X and Zhang J 2013b Graphene-veiled gold substrate for surface-enhanced Raman spectroscopy *Adv. Mater.* **25** 928–33
[92] Qiu H W, Xu S C, Chen P X, Gao S S, Li Z, Zhang C, Jiang S Z, Liu M, Li H S and Feng D J 2015 A novel surface-enhanced Raman spectroscopy substrate based on hybrid structure of monolayer graphene and Cu nanoparticles for adenosine detection *Appl. Surf. Sci.* **332** 614–9
[93] Zavatski S *et al* 2019 Surface enhanced Raman spectroscopy of lactoferrin adsorbed on silvered porous silicon covered with graphene *Biosensors* **9** 34
[94] Gao S, Liu H, Li Z, Li H and Qiu H 2015 Graphene/Cu nanoparticle hybrids fabricated by chemical vapor deposition as surface-enhanced Raman scattering substrate for label-free detection of adenosine *ACS Appl. Mater. Interfaces* **7** 10977–87
[95] Zhu H, Liu A, Li D, Zhang Y, Wang X, Yang W, Gooding J J and Liu J 2017 Wafer-scale fabrication of a Cu/graphene double-nanocap array for surface-enhanced Raman scattering substrates *Chem. Commun.* **53** 3273–6
[96] Xu W, Ling X, Xiao J, Dresselhaus M S, Kong J, Xu H, Liu Z and Zhang J 2012 Surface enhanced Raman spectroscopy on a flat graphene surface *Proc. Natl Acad. Sci. USA* **109** 9281–6

IOP Publishing

Recent Advances in Graphene and Graphene-Based Technologies

Anoop Chandran, N V Unnikrishnan, M K Jayaraj, Reenu Elizabeth John and Justin George

Chapter 11

Photonic and optoelectronic applications of graphene: applications of graphene in solar cells

Susana Sério, Paulo A Ribeiro and Maria Raposo

Graphene's unique characteristics, together with the simplicity associated with its synthesis and functionalization, demonstrate immense potential in energy conversion systems and therefore can tackle the increasing energy demands of global society. This chapter presents an updated revision of the relevant achievements in the performance of solar cell devices with organic, inorganic or hybrid architectures, as result of the integration of graphene or its derivatives in their design. It further highlights the difficulties in practical applications of graphene-based materials as anode or cathode, hole or electron transport layer, or active layer in a solar cell device. The studies carried out so far demonstrate a high potential for photovoltaic applications, showing that graphene-based energy conversion devices are more efficient compared to those using conventional materials, and point out other approaches for new applications in this area.

11.1 Introduction

Currently, the global population is almost completely reliant on the use of fossil fuels such as coal, oil, and natural gas for energy. However, theses type of fuels substantially enhance the emission of greenhouse gases thus increasing global warming and therefore are considered to be a threat to global society. Furthermore, the fossil reserves are drying up fast due to which these types of fuels are not capable of meeting the future energy demands of mankind. This is why the scientific community is looking for alternate strategies to effectively generate and store energy with less or almost no negative impact on the environment, which is imperative for sustainable development. A review of the literature shows considerable efforts to achieve this task from renewable sources like the Sun, water, wind etc [1]. Of them, electricity generation and storage using solar energy has attracted much attention recently. Solar cells and photo-electrochemical cells have the potential to be replacements for fossil fuels in many sectors. Considering the economic aspects and better efficiency compared to other renewable energy resources, photovoltaic technology will continue to thrive in the coming era as well.

Solar cells were first developed in 1839, when Alexandre-Edmond Becquerel witnessed the photovoltaic effect by coating platinum electrodes with silver chloride. When these electrodes were illuminated, it was observed that a potential difference and an electric current were generated between them [2]. In 1873, Willoughby Smith discovered that when an electric current was passed through a selenium bar, its resistance decreased when the bar was exposed to light. This work motivated W G Adams and R E Day to further analyze this phenomenon, leading to the observation of photovoltaic effect on solid selenium in 1877. The aforementioned projects culminated in the production of the first functional photovoltaic solar cell—consisting of gold-coated selenium—with an efficiency less than 1%, by Charles Fritts, in 1883 [3]. In 1904, Wilhelm Hallwachs also presented a solar cell, with semiconductor behavior, formed by copper and copper oxide. The most important milestone in the development of solar cells occurred in 1954, with the announcement of the production of the first silicon-based solar cell carried out by a group of researchers from Bell Laboratories. D M Chapin, S Fuller and G L Pearson managed to achieve an efficiency of approximately 6%, which exceeded by a large scale all projects developed so far, with the first generation of solar cells being implemented. This drastic improvement in cell efficiency was only achieved due to an intensive study of the properties of these devices, particularly of the p–n junctions.

11.2 Working principle of a solar cell

Solar cells correspond to devices that convert energy from solar radiation into electrical energy through the photovoltaic effect, and in most cases semiconductor materials are used in their production. In general terms, this energy conversion is based on the absorption of the energy of a photon, which leads to the formation of electron–hole pairs in the semiconductor, and subsequently, the separation of charge carriers, based on p–n junctions [4]. Creation of p–n junctions is pivotal for the functioning of solar cells, allowing the photovoltaic effect to occur. This type of junction is achieved by doping one part of a semiconductor material (such as silicon) with atoms of a p-type dopant and another part of the same material with atoms of n-type dopant creating, therefore, two distinct regions.

The n-type region is characterized by having a considerable number of free electrons, this number (and charge) being equalled by positive ions from the dopant. Similarly, the p-type region has a certain number of holes, and an equal number of negative dopant ions. When these two types of materials are linked, electrons migrate from the n-region to the p-region, and holes migrate from the p-region to the n-region. This diffusion process will cause an accumulation of positive and negative ions at the interface of the two regions and, consequently, the creation of an intrinsic electric field. In turn, the generated electric field triggers the electrons to be attracted to the positive ions in the n-region and the holes to the negative ions in the p-region, creating a flow with opposite direction to that created by diffusion. Figure 11.1 shows a representative scheme of the electrons and holes movement due to diffusion and intrinsic electric field, which occur between the n-region and p-region in a solar cell device. These opposite movements of electrons and holes give rise to the so-called depletion region at the interface of the two materials, where the number of moving electrons and holes is reduced. Although an electric field is present, the depletion region is characterized by being highly resistive. When a silicon cell is exposed to light, each incident photon on the depletion region gives rise to an electron–hole pair which, due to the cell's intrinsic electric field, is separated. The separation of these charge carriers causes a potential difference in the solar cell itself, in a process called the photovoltaic effect.

A solar cell device is made up of numerous parts, including electrodes, a photoactive layer, and carrier transport layers. Figure 11.2 depicts a representative structure of a solar cell device. Each component should embrace particular characteristics to be a potential candidate for application in a photovoltaic device. In this regard, materials presenting characteristics such as low sheet resistance, high

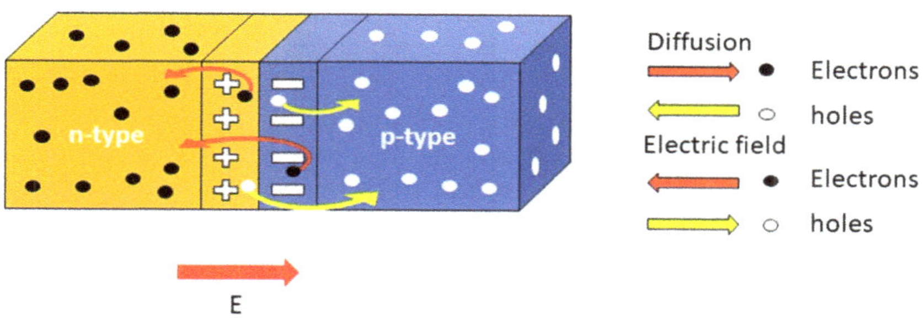

Figure 11.1. Electrons and holes movement due to diffusion and intrinsic electric field, which occur between the n-region and p-region in a solar cell device.

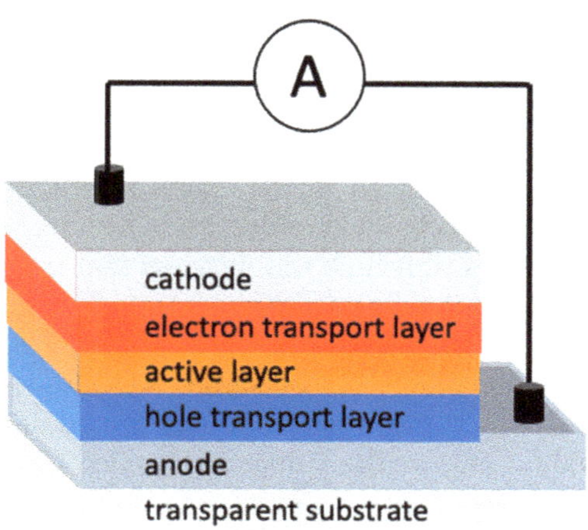

Figure 11.2. Schematic representation of a solar cell device showing the main components.

transparency and mechanical stability are generally employed as electrodes in photovoltaic (PV) devices. For the photoactive layers, the materials should present the capability to capture a wider light spectrum, have superior charge carrier mobility and be able to improve the thermal and photochemical stability. The carrier transport layers, namely hole and electron transport layers, are used to boost the carrier transport, which is accomplished by reduced hole–electron recombination rate.

In short, when solar light shines on a photovoltaic device several steps occur in the regions shown in figure 11.2, which are briefly described below.

(i) **Exciton formation**

Inorganic semiconductors yield free carriers immediately after light absorption, whereas organic semiconductors require an additional process called exciton formation. In the case of this latter group of semiconductors, the absorption of a photon from solar radiation should present energy greater than the difference between the two energy levels HOMO (highest occupied molecular orbital) and LUMO (lowest unoccupied molecular orbital), to excite an electron from HOMO to LUMO. This process is analogous to the excitation of an electron from the valence band to the conduction band in inorganic semiconductors. This photoexcitation leads to the generation of electron–hole pairs called excitons.

(ii) **Exciton transport**

After their formation, the excitons will be diffused to the donor–acceptor interface where they can dissociate and produce separated negative and positive charges, with the exciton diffusion length being defined by the maximum distance that the exciton can travel before decaying via radiative or non-radiative emission, losing its energy.

(iii) **Charge separation**

To separate the excitons, a donor–acceptor interface must be defined. This allows the creation of local electric fields, which are formed due to a potential energy difference between the donor and the acceptor. Knowing that this excitonic dissociation occurs at the donor–acceptor interface, it is necessary that the organization of these materials is well defined, in order to obtain a device with a good performance.

(iv) **Charge transport**

After separating the charges, it is necessary to transport them to the electrodes before the charge carriers start to recombine and thus obtain high-efficiency solar cells. Therefore, it is necessary that the donor and acceptor materials exhibit high mobility, so that the charge carriers can be transported towards the electrodes.

(v) **Charge collection**

The last step includes the collection of charge carriers at electrodes, which is done by a transparent conductive oxide (TCO) such as indium doped tin oxide (ITO) or fluor doped tin oxide (FTO), at one side and a metal contact on the other side. Thus, a built-in-potential barrier in the cell activates the electrons to generate a voltage, to drive a current through a circuit.

Another important issue in the photovoltaic technology is the evaluation of the solar cell performance, denominated power conversion efficiency (PCE), which is assessed by the ratio between absorbed photons and amount of generated photocurrent, and can be obtained using the following expression:

$$\mathrm{PCE} = \frac{\mathrm{FF} \times V_{\mathrm{OC}} \times J_{\mathrm{CC}}}{P_{\mathrm{inc}}} \tag{11.1}$$

where P_{inc} corresponds to the incident photon power, V_{OC} denotes maximum open circuit voltage, J_{CC} maximum close circuit current and FF is the filling factor which can be calculated by the expression:

$$\mathrm{FF} = \frac{V_{\max} \times J_{\max}}{V_{\mathrm{OC}} \times J_{\mathrm{SC}}} \tag{11.2}$$

where $V_{\max}$, $J_{\max}$, V_{OC} and J_{SC} are maximum voltage, maximum current, open circuit voltage and short circuit current, respectively [5].

11.3 Types of solar cells

Following the discovery of the photovoltaic effect, various types of solar cells have been investigated and developed in order to achieve high-energy conversion performance systems. The conversion efficiency of the photovoltaic device depends on two key factors: solar cell category and materials used in its manufacture. In the literature, it has been claimed that single-junction crystalline materials like Si and GaAs, which constitute the first generation of photovoltaic (PV) cells, can accomplish PCEs around 30% [6]. Despite the fact that there has been substantial progress in the technology and large-scale production of the traditional silicon-based solar modules which allowed significant cost savings, this technology has certain drawbacks, such as materials consumption, low production speed and limited flexibility. Therefore, the development of novel solar cell technologies ironing out the above-mentioned drawbacks is of huge interest to the industry.

Two decades ago, dye sensitized solar cells (DSSCs) were projected as a low-priced option to the traditional Si solar cells, especially due to the simplicity in the fabrication processes. These types of cells correspond to the third generation PV cells, where innovative approaches in the photovoltaic technology are employed [7, 8]. In the first generation PV cells, which presently exhibit the highest conversion efficiency, the electric interface is generated between doped n-type and p-type bulk silicon [9]. In the case, of second generation PV cells, the devices are based on thin film technology and consequently the fabrication cost decreases, because these cells use less material. Nonetheless, they offer lower power conversion efficiency in comparison with the bulk cells. Moreover, these two types of PV cells are composed of opaque materials, which require front-face illumination and moving supports to track the Sun position. The third generation solar cells are implemented by using nanostructured materials, of pure organic or a mixture of organic and inorganic components, which constitute a more environmentally friendly and cheaper generation of solar cells. The DSSCs, quantum dot solar cells (QDSCs) and organic solar cells (OSCs) fall under this category [10–12]. This PV cells generation exhibits lower PCE and stability in comparison with the other generations [13]. Hence, more investigation is being performed to improve the efficiency and stability of these PV devices.

The performance of a solar cell is mostly determined by the materials employed in its production. As a result, numerous materials have been investigated for use in the production of solar cells. Graphene (Gr), a carbon sheet with a hexagonally packed honeycomb structure, stands out as having promising features for solar cell applications. Wrapped graphene is a zero-dimensional (0D) structure called fullerene, rolled graphene is a two-dimensional (2D) namely carbon nanotube (CNT), and ultimately stacked graphene leads to three-dimensional (3D) graphite. Figure 11.3 depicts the evolution of graphene to fullerene, CNT, and graphite structures [14].

Several synthesis techniques, like mechanical exfoliation, chemical vapor deposition (CVD), solution-processing, among others, have been used to produce graphene for diverse applications. These methods are extremely important because they can affect significantly the yield and structure of graphene and consequently may play a key role on defining its electrical performance. The principal techniques to produce graphene are represented in figure 11.4. The graphene end products and reduced graphene oxide (rGO) are also utilized in a variety of applications due to their outstanding properties.

Graphene and its various composites exhibit unique characteristics that make them the best choice to address today's need for more effective materials for photovoltaic technology, namely:

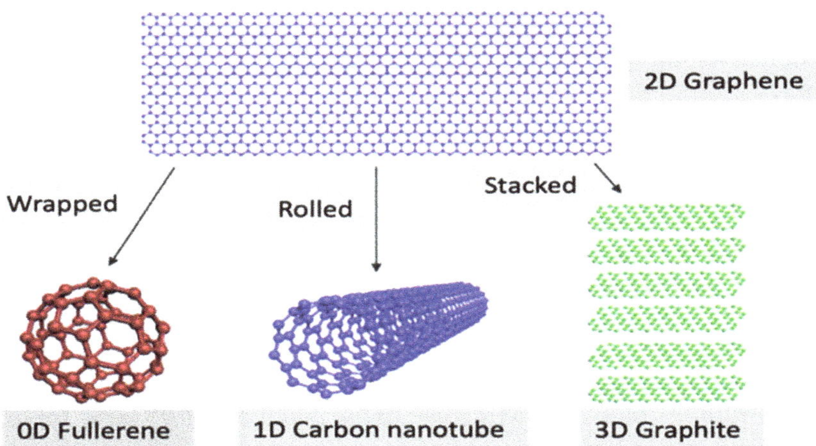

Figure 11.3. From 2D graphene to 0D fullerene, 1D carbon nanotube and 3D graphite. Reproduced from [14], copyright 2020, with permission from Elsevier.

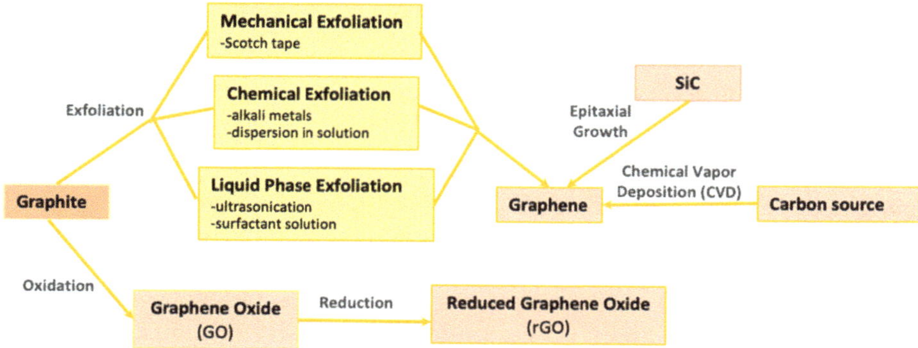

Figure 11.4. Summary of the main methods used for synthesis of graphene.

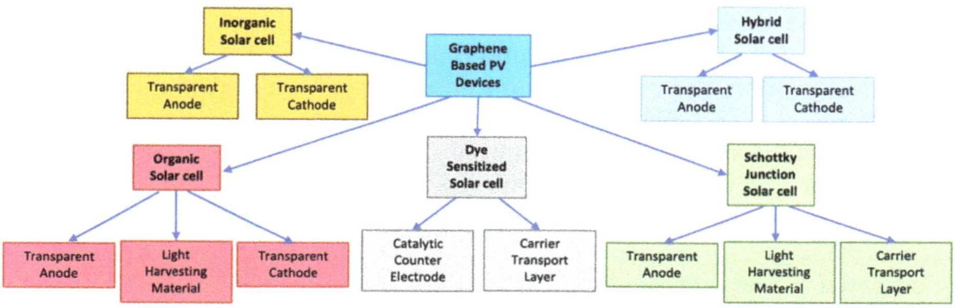

Figure 11.5. Graphene inclusion schematic for numerous photovoltaic device applications.

- capture larger spectrum—2.3% of irradiated light is absorbed [12];
- low sheet resistance [15];
- intrinsic electron mobility as high as 25 000 cm^2 V^{-1} s^{-1} [16];
- high specific surface area (2630 m^2 g^{-1})—facilitates charge collection and release process and therefore enhances the photovoltaic performance [17];
- easy chemical activity;
- easy functionalization—suitable for photoactive layer in solar cells [18, 19]; and
- high transparency and superior flexibility and stretchability—graphene is the thinnest material that can transmit up to 98% of light; it can be stretched up to 120% of its length and return its original shape.

Moreover, the carrier separation and diffusion process is sped up by graphene/semiconductor Schottky junctions. Consequently, it can be concluded that these chemical properties of graphene and its various composites can potentiate the photovoltaic technology performance [20] and therefore are consider as promising candidates for application in PV devices such as electrodes (anode, cathode, counter electrode), photoactive layer, Schottky junction and carrier transporter. Figure 11.5 presents the roles that could be played by graphene in several solar cell technologies.

11.4 Graphene applications in photovoltaic devices

The main achievements found in the literature regarding the incorporation of graphene and its derivatives in several photovoltaic devices are reviewed in this section.

11.4.1 Transparent conducting anode

When graphene is incorporated into a solar cell device as transparent conducting anode, it can present two functions: window for the photoactive layer and ohmic contact [21]. The requisites for this PV device component, include high transmittance (T_r) (over 90%) and low sheet resistance (R_s) preferably less than 10 Ω/□ [22]. Typically used ITO and FTO electrodes are expensive in comparison with the other components used in dye sensitized solar cells [23]. Other disadvantages of ITO and FTO are the reduced flexibility and stability, which constrains the utilization of those materials as electrodes [15]. Hence, other materials were investigated to overcome these drawbacks, such as metallic nanowires and carbon nanotubes [24–26]. Nevertheless, their surface properties hinder large-scale use. Graphene, on the other hand, is one of the primary possibilities for applications as electrodes that might improve the efficiency of a PV system due to its unique properties which perfectly meets the aforementioned requisites. Although the most perfect combination of R_s and T_r has not yet been achieved, graphene-based electrodes have been successfully incorporated into organic, inorganic, dye sensitized, and hybrid solar cells. [25, 27, 28].

Compared to inorganic solar cells (ISCs), organic solar cells (OSCs) have definite advantage in terms of flexibility and price. There are also well-established industrial methods for their large-scale production. However, an OSC's performance is limited by the fact that, the ITO which is used as electrode permits the diffusion of metal ions into the active layer.

Arco *et al* reported a transparent anode formed of few-layer graphene (1–5 layers) grown by CVD (CVD-Gr) with Gr/PEDOT:PSS/CuPc/C60/BCP/Al architecture [28]. This work revealed that the CVD-Gr exhibiting lower R_s (~230 Ω/□) and T_r (~72%) have potentiated an efficiency increase of 1.8% as compared with ITO anode (PCE 1.27%) [28]. Furthermore, the flexibility of graphene aided PCE because it can work at a bending angle of 138°, as opposed to ITO, which performed poorly at bending angles around 60°. The explanation, of the reduced ITO-based device performance, is due to the fast decline of FF to zero under mechanical stress, because of the appearance of micro cracks in ITO layer as a result of the bending.

The optical transparency of electrodes has a significant effect on the performance of flexible OSCs [29]. In that work rGO was used as anode in the rGO/PEDOT:PSS/P3HT:PCBM/TiO_2/Al device. The rGO-based OSCs showed a performance increase with 1000 bending cycles under 2.9% tensile strain. In another work it is demonstrated that a hybrid single-layer graphene/ITO (SLG/ITO) electrode displays greater electromechanical properties in comparison to normal ITO [30]. When tensile strain surpasses 10%, ITO becomes unstable, whereas the graphene electrode preserves its excellent features even at 20% tensile strain. The relative resistance of graphene ($\Delta R/R_0$) is approximately 3.28, whereas for ITO it is around 125.91 for similar strain value. R_0 corresponds to the original lateral resistance for a 1×1 cm^2 sample and ΔR denotes the resistance variation. The R_s, mobility (μ), and carrier density (ρ) for various electrode materials are given in table 11.1.

The work of Bi *et al* shows the use of Gr electrode on CdTe-based photovoltaic devices, which was produced by the economic CVD method and transferred on a glass support [31]. The number of CVD-Gr layer(s) was monitored by inlet H_2 flow rate through the process. The CVD-Gr exhibited $R_s \sim 220$ Ω/□ and $T_r \sim 84\%$ and glass/Gr/ZnO/CdS/CdTe/(graphite paste) device accomplished a PCE of 4.17%. In the work of Bae *et al*, 30 CVD-Gr monolayers were transferred from a copper foil to a flexible polyethylene terephthalate (PET) substrate employing the roll-to-roll technique. In this study the obtained CVD-Gr showed high transparency ($T_r \sim 97.4\%$) and low sheet resistance ($R_s \sim 125$Ω/□). Subsequently, four Gr layers were attached by layer-by-layer technique, followed by a doping with HNO_3, which allowed $R_s \sim 30$ Ω/□ and $T_r \sim 90\%$ and thus permitted a CVD-Gr with superior properties in comparison to ITO for PV technology purposes [32]. The PCE values of solar cell devices with graphene as electrode are summarized in table 11.2. The reported studies reveal that graphene, and its derivatives offer high transparency and low sheet resistance together with high mechanic stability, which constitute important properties for this type of application. It further, shows that the photovoltaic devices' performance is constantly enhancing with the introduction of graphene and its derivatives in the devices. Hence, graphene and related materials are advantageous in comparison with the conventional materials normally used as anodes in photovoltaic technology.

11.4.2 Transparent conducting cathode (TCC)

The TCCs play an important role in defining the performance of PV devices. Graphene and graphene-based materials are potential candidates as cathodes on solar cells due to their distinctive characteristics already mentioned, as high conductivity, ultrahigh mobility and transparency. Recently, reduced graphene oxide (rGO) was used as cathode in a hybrid photovoltaic device [38]. In this work, rGO was produced by hydrazine vapor followed by annealing and exhibited a R_s roughly equal to 420 Ω/□ and T_r close to 61%. After this, the n-type ZnO nanorod-coating was given to rGO electrode electrochemically. This was followed by the spin coating of p-type poly 3-hexylthiophene (P3HT) resulting in the

Table 11.1. Sheet resistance, carrier density and mobility for several electrodes. Adapted from [30].

Electrode type	μ (cm^2 V^{-1} s^{-1})	R_s (Ω/□)	ρ (cm^{-2})
Gr[a]/ITO[b]	56.35	76.46	2.551×10^{15}
ITO	32.01	78.34	2.295×10^{15}
ITO/ITO	32.71	78.25	2.366×10^{15}
Gr	1.403×10^4	440.34	5.37×10^{13}

[a]Gr—graphene.

[b]ITO—indium doped tin oxide.

Table 11.2. An overview of graphene-based solar cells' PCE.

Materials	Device architecture	PCE (%)	Reference
rGO	Glass/rGO/TiO_2/dye/spiro-OMeTAD/Au	0.26	[27]
rGO	PET/rGO/PEDOT:PSS/P3HT:PCBM/TiO_2/Al	0.78	[33]
rGO-CNT	Glass/rGO-CNT/PEDOT:PSS/P3HT:PCBM/Ca:Al	0.85	[34]
CVD-Gr	Quartz/Gr/PEDOT:PSS/CuPc:C_{60}/BCP/Ag	0.85	[35]
CVD-Gr	PET/Gr/PEDOT:PSS/CuPc:C_{60}/BCP/Al	1.18	[28]
CVD-Gr	Quartz/Gr/MoO_3+PEDOT:PSS/P3HT:PCBM/LiF/Al	2.5	[36]
TCNQ-Gr	Glass/TCNQ-Gr/PEDOT:PSS/P3HT:PCBM/Ca/Al	2.58	[37]
Au-doped Gr	Au-Gr/PEDOT:PSS/P3HT:PCBM/ZnO/ITO	3.04	[37]
rGO	Quartz/rGO/ZnO/P3HT/PEDOT:PSS/Au	0.31	[38]
Al–TiO_2–Gr	Au/Gr/Al–TiO_2/P3HT:PCBM/MoO_3/Ag	2.58	[39]
CVD-Gr	Glass/Gr/ZnO/CdS/CdTe/graphite paste	4.17	[31]
CVD-Gr	Glass/Gr/PEDOT:PEG/ZnO/PbS/P3HT/MoO_3/Au	4.2	[40]
rGO	FTO/TiO_2/dye/I_3-/I_1-mediated electrolyte/rGO	4.99	[41]
CVD-Gr	FTO/TiO_2/dye/I_3-/I_1-mediated electrolyte/Gr	5.73	[42]
TiN-rGO	FTO/TiO_2/dye/I_3-/I_1-mediated electrolyte/TiN-rGO	5.78	[43]
CNT-rGO	FTO/TiO_2/dye/I_3-/I_1-mediated electrolyte/CNT-rGO	6.05	[37]
Gr	FTO/TiO_2/dye/Co(III)/(II) mediated electrolyte/Gr	9.3	[44]
CVD-Gr	P(VDF-TrFE)/Gr/PEDOT/PSS/P3HT/PCBM/Ca/Al	2.07	[45]
CVD-Gr	Glass/Au-doped SLGNRs/PEDOT:PSS/WO_3/SMPV1: $PC_{71}BM$/ZNO/PEDOT:PSS/PTTBDT-FTT: $PC_{71}BM$/CA/Al	8.48	[46]
CVD-Gr	Glass/Pt grids/Gr/TiO_2/dye/electrolyte/counter electrode	0.4	[47]
rGO	Glass/ITO/HTL(rGO)/active layer/LiF/Al	3.71	[48]
c-Si/GQDs	In/Ga/c-Si/GQDs/Au	6.63	[49]

[a]PCBM—phenyl-C_{61}-butyric acid methyl ester.
[b]PCDTBT—poly[*N*-9″-heptadecanyl-2,7-carbazole-alt-5,5-(4′,7′-di-2-thienyl-2′,1′,3′-benzothiadiazole)].
[c]$PC_{71}BM$—[6,6]-phenyl C 71 butyric acid methyl ester.
[d]PEDOT:PSS—poly(3,4-ethylenedioxythiophene) polystyrene sulfonate.

formation of a hybrid photovoltaic device. The developed device with quartz/rGO/ZnONRs/P3HT/PEDOT:PSS/Au architecture accomplished an efficiency of 0.31%. Both, R_s and T_r, are fundamental factors of this type of electrodes and can determine the solar cell performance. In fact, it was demonstrated that the efficiency of quartz/rGO/ZnONRs/P3HT/PEDOT:PSS/Au device improved from 0.31% to 0.5% for quartz/Gr/ZnONRs/P3HT/MoO_3/Au device [38, 40]. Thus, it is clear that the inclusion of CVD-Gr has improved the device properties [50, 51]. Therefore, graphene has a high potential to substitute the conventional transparent materials used for photovoltaic applications.

11.4.3 Catalytic counter electrodes (CCEs)

The DSSCs, has a disadvantage in that their catalytic counter electrode is often Pt, which is highly chemically active and expensive. Graphene is a good substitute to Pt, as it can overcome the above-mentioned issues in addition to absorbing a wider spectrum of wavelengths [27, 52–54]. In table 11.2 are summarized some of the reported results of graphene application as CCEs. The studies of Mayhew *et al*, substituted Pt with functionalized graphene sheet (FGrS) in DSSCs [41]. The efficiency of FGrS-based solar cell is around 5% which is almost 10% lower in comparison with Pt-based cell. By applying a bias, the development of the catalytic effect of the functionalized graphene sheet occurs due to the resistance to charge transfer and the functional group containing oxygen. Low charge transfer resistance is achieved by functionalization-induced surface-modification of graphene sheets, leading to highly conductive flexible CCEs It was also discovered that FGrS ink casted plastic substrate can act as an effective counter electrode and it could be of use even in the absence FTO substrate.

Hong *et al* employed a counter electrode made of thin (60 nm) Gr/PEDOT:PSS film in DSSC, which was obtained by depositing Gr/PEDOT:PSS aqueous mixture onto ITO using spin coating technique [55]. The formed Gr/PEDOT: PSS counter electrode displayed greater graphene electrolytic activity along with higher PEDOT:PSS conductivity. The developed DSSC with this counter electrode, when exposed to a white light of 100 mW power, showed a PCE around 4.5%, which close to the PCE (~6.3%) obtained when a Pt counter electrode was used. One decade ago, Zhang *et al* investigated the application of graphene nanosheets' (GrNs) as CCE in DSSCs, to allow the reduction of tri-iodide and further compared to a Pt-based device [56]. To prepare GrNs counter electrode, graphite oxide was exposed to hydrazine reduction followed by an annealing treatment. Subsequently, electrochemical impedance spectroscopy was performed to explore the electrochemical properties of GrNs and Pt electrodes, annealed at distinct temperatures. This study further revealed that FF and PCE, were negatively affected by the charge transfer resistance, Nernst diffusion impedance and FTO sheet resistance. The charge transfer resistance decreased as a result of the reduced organic binding initiated by annealing. Moreover, the annealed GrNs showed a new 3D structure with augmented charge transfer reaction in the electrolyte by increasing the reaction area and enhancing the adherence among GrNs and FTO substrate. An inferior PV device performance by the sample annealed at 450 °C in comparison with the one annealed at 400 °C is due to lower charge transfer resistance of the former This is triggered by the thermally induced separation of GrNs and the reduced GrNs–FTO contact. Table 11.3 presents solar cells performances with GrNs as counter electrode obtained at different annealing temperatures. Despite, GrNs based devices exhibited higher open circuit voltage, low PCE are observed for lower close circuit current density and FF values. It can be concluded that the observed high performance of photovoltaic devices is mainly related to the simplicity of synthesis procedure, reduction of charge transfer resistance, and the improvement in both catalytic activity and flexibility of graphene and its related materials [57].

Recently, in 2019, Wahyuono *et al* fabricated Gr and MWCNTs bilayers by the double self-assembly method to be used as counter electrode for a DSSC [58]. This work revealed that the MWCNTs created random porous networks onto the Gr

Table 11.3. Comparison of GrNs and Pt-based devices at various annealing temperatures. Adapted from [56].

Counter electrode	Annealing temperature (°C)	R_{ct} (Ω cm^2)	J_{sc} (mA cm^{-2})	V_{oc} (V)	FF (%)	PCE (%)
Pt	N/A	0.76	18.507	0.714	57.51	7.59
GrNs	250	337.18	14.478	0.697	32.56	3.29
GrNs	300	274.19	16.147	0.680	32.56	4.44
GrNs	350	3.01	16.358	0.742	44.34	5.38
GrNs	400	1.20	16.988	0.747	53.62	6.81
GrNs	450	16.40	14.525	0.672	45.74	4.46

[a] R_{ct}—redox reaction resistance, R_{ct}, at the counter electrode.
[b] J_{sc}—short-circuit current density.
[c] V_{oc}—open circuit voltage.

flakes surface and the MWCNT/Gr showed a network of CNTs which are randomly oriented (84% of the area covered by a graphene monolayer). This study further showed that the DSSCs based on the Gr/MWCNT counter electrode displayed superior performance parameters ($V_{oc} = 0.62$ V, $J_{sc} = 13.3$ mA cm^{-2}, FF = 0.49, and PCE = 4.1%), in comparison to those of the Pt-based solar cells [58]. Moreover, it was claimed that double self-assembled counter electrodes are mechanically stable and this allows their recycling for DSSC fabrication without losing the solar cell performance. In 2020, Wang *et al* used rGO together with nickel phosphide (Ni_xP_y) as counter electrode in DSSCs to enhance solar cell performance and obtained higher PCE for Ni_xP_y/rGO composites compared to those of the commercial Pt. This result can be due to the high density of Ni_xP_y catalytic sites and superior electrical conductivity of rGO [59]. In the same year, Oh *et al* proposed a nanohybrid counter electrode of graphene-based Cu_2Zn–$NiSe_4$ (CZNS) with tungsten trioxide (WO_3) nanorods, which was fabricated by the hydrothermal method and exhibited an exceptional PV performance of 12.16%, which was superior to the one obtained with a Pt electrode under the same environmental conditions [60].

As evident from the above discussion there are several reports on substantial and incremental improvements on the performance of graphene-based counter electrodes in DSSCs suggesting that new ideas in this field would continue to emerge, which will eventually provide the required properties for the large-area fabrication of DSSCs.

11.4.4 Active layer

Graphene can be easily functionalized and therefore is a promising candidate as active layer in solar cells devices [18, 19]. Moreover, the work function of Gr can be tuned by the addition of a semiconducting material with similar band structure to create a Schottky junction [25, 61–64]. Both Gr and rGO can be used as charge carrier transporting layers due to high carrier mobility [20, 65–69] and interfacial layers for tandem solar cells [70–72]. Next, are briefly discussed the graphene applications within active layer components.

11.4.4.1 Light-harvesting materials

The functionalization of graphene oxide (GO) can be realized using several functional groups in such a way that it enhances the light-harvesting applications. In the work of Liu *et al*, the GO functionalization was carried out with phenyl iso-cyanate and the surface was modified from hydrophilic to hydrophobic [18]. The P3OT/Gr composite was employed as active layer in a bulk heterojunction (BHJ) solar cell. The crystallinity of P3OT can be varied through elimination of functional groups from graphene increasing the annealing temperature. Consequently, was used $T = 160$ °C for 20 min to develop the organic solar cell device, which attained a PCE of 1.4%. The work further demonstrated that functionalized graphene is a perfect to substitute [6,6]-phenyl C_{61}-PCBM as electron acceptor, allowing the development of highly efficient OSCs. Fan *et al* explored the use of graphene and silicon nanowires (SiNWs) as a light-harvesting material in a Schottky junction solar cell type [73]. The Schottky junction was formed by coating N-type SiNWs with graphene layers. In contrast to the Gr/planner silicon device, the Gr/SiNWs Schottky junction solar cell showed exceptional optical absorption and rapid charge transfer. The Gr/SiNWs Schottky junction solar cell reached a PCE of 2.86%. The reported studies found in literature, allowed concluding that Gr based heterostructures can be a potential candidate for light-harvesting materials to obtain highly efficient PV devices.

11.4.4.2 Schottky junction (SJ)

Schottky junctions can be formed through the combination of metallic Gr with semiconductors. Graphene-based Schottky junctions are chosen over traditional ITO-based Schottky junctions as they allow better flexibility and improved performance. In this context, electron beam lithography followed by a lift-off process was employed to developed a CdSe nanobelt SJ utilizing patterned Gr [74]. The obtained photovoltaic device with open circuit voltage of 0.51 V and a close circuit current of 5.75 mA cm^{-2} reached a PCE of 1.25%. Miao *et al* reported Gr/Si Schottky junction solar cells with high PCE (8.6%) [62], which were obtained by adding a p-type impurity-bis (trifluoro methane sulfonyl) amide [$(CF_3SO_2)_2NH$] (TFSA) to graphene via spin coating. The doping led to 70% Gr R_s diminution and a work function rise, without affecting the transparency. The results showed an increased Gr/Si efficiency from 1.9% to 8.6%.

Zhang *et al* employed SiNWs and silicon nano-holes (SiNHs) to engineer Gr/Si Schottky junction solar cells [75]. SiNHs-based devices outperformed those made of SiNWs in terms of effective area at similar light absorption rates and giving 8.71% and 10.30% of PCE, respectively. Recently, Gnisci *et al* developed n-type crystalline Si with graphene and graphene-based derivative layer stacks by CVD to obtain a high-efficiency Schottky junction solar cell [76].

In short, graphene/semiconductor heterojunction for Schottky junctions solar cells enables charge separation and diffusion, resulting in performance that is comparable to that of materials that are typically used.

11.4.4.3 Electron transport layer (ETL)

Enhanced mobility as well as conductivity are characteristics that ETLs require, and graphene and its composites exhibit these qualities. For rGO–TiO_2 composite-based DSSCs, Yang *et al* used rGO as the electron transport layer, which demonstrated improved performance in comparison to CNT–TiO_2 composite-based DSSCs [69]. Despite the photogenerated electrons in graphene, composite-based electrodes can be easily seized and transported, one-dimensional (1D) CNT–TiO_2 composite shows reduced adherence between both materials, leading to a large barrier and therefore high electron–hole recombination rate. The graphene in the device acts as electron transport layer, accepting electrons from TiO_2 and transferring them to FTO and thus a low electron–hole recombination rate is accomplished. The best result was attained with 0.4% rGO coating over TiO_2 [69]. Wang *et al* produced a bulk heterojunction (BHJ) solar cell with PCDTBT:PC_{71}BM architecture through the use of flexible GO as ETL [77]. The CVD-Gr was fixed to thermal release tape and an etching procedure was performed to detach the copper foil. After, it was reacted with HNO_3 to produce GO and subsequently applied as ETL in BHJ solar cell. The GO layer was stamped with a pressure of 0.2 MPa at a temperature around 80 °C–90 °C to accomplish good adherence among GO layer and BHJ, leading to an increase in PCE from 5.9% to 7.5% for PCDTBT/PC_{71}BM device.

Graphene-based flexible ETL oxides produced by atomic layer deposition (ALD) have huge impact in high-performance perovskite solar cells. Nevertheless, ALD oxide growth on graphene is challenging. According to Xu *et al*, flexible perovskite solar cells made with ALD oxides are extremely effective. They also claimed that molecular layer deposition with ethylene glycol (EG) could contribute active sites to the graphene surface through van der Waals forces [78]. The ALD–ZnO film was produced onto EG-functionalized graphene surface, which presented high transmittance, low energy levels, and a superior electron injection capability. The EG–graphene/ZnO-based perovskite solar cells displayed a PCE of 133.9%, which was much higher than that of the graphene/ZnO-based perovskite solar cells.

In this way, graphene-based materials enhance electron transport mechanism, leading to an increase of mobility and conductivity and therefore allowing a better performance of PV devices.

11.4.4.4 Hole transport layer (HTL)

As graphene displays outstanding electron transport layer properties, GO and its composites also constitute promising candidates as HTLs in PV devices. Kim *et al*, for P3HT:PCBM-based solar cells explored the combination of a few micron of GO with single-walled carbon nanotubes (SWCNTs), as HTL [79]. The hole transport conductivity exhibited remarkable increase because of the incorporation of SWCNTs to GO. The hole transport resistance was reduced as a result of the GO:SWCNTs arrangement, improving the transfer rate from the photoactive layer to the anode. Another team employed a spray process to grow rGO to be used as HTL in a polymer solar cell and reached a PCE of 3.71% at 400 °C [48], revealing a potential alternative for application as transport layers in polymer solar cells. Wu *et al* studied the ambipolar transport properties of graphene as HTL in an inverted planar heterojunction perovskite solar cell [80]. The coating of the perovskite film onto Gr enabled the extraction of holes, enhancing crystallinity, surface coverage area, crystal orientation and a PCE of 12.40% was found.

The work of Li *et al* showed a decrease in the recombination rate of electrons and holes as well as the current leakage, which was accomplished with a GO film with 2 nm of thickness inserted among ITO and P3HT:PCBM as HTL, instead of PEDOT:PSS [24]. The GO was produced with a neutral aqueous solution, which eliminated the likelihood of substrate corrosion at elevated temperatures along with water contamination in the active layer, leading to a better performance in comparison to a PEDOT:PSS-based device. It was found that GO thickness is a key factor in the performance device, being higher for lower thickness. This research further revealed, that the recombination rate evidences a decreasing tendency in comparison to ITO under different lighting powers and therefore the device life is prolonged. More recently, in 2016, Lee *et al* explored the possibility of GO/PEDOT:PSS as hole transport layer in a planar heterojunction perovskite photovoltaic device with ITO/HTL(GO/PEDOT:PSS)/$CH_3NH_3PbI_3$/PCBM architecture attaining a PCE of 9.7%, which is greater than the one exhibited by the conventional PEDOT:PSS (PCE = 8.2%) [81]. The GO large bandgap (~3.6 eV) leads to the blocking of the electrons, increasing the shunt resistance and consequently the PCE.

All the results published up to now have revealed that graphene and its derivatives are promising candidates for photovoltaic generation, because of the increase of hole extraction and electron blocking, reduced electron–hole recombination rate, and improved PCE of photovoltaic devices.

11.5 Conclusions

The main advances made in photovoltaic technology by incorporating graphene and its derivatives into the photovoltaic devices in a bid to make them suitable for future applications are highlighted in this chapter. Several studies have been published regarding the role of graphene as transparent electrodes, CCEs, SJs and material for active layer in different solar cells architectures. It is clear from the review that graphene is an ideal material for transparent electrodes with low R_s and high T_r in organic, inorganic and hybrid solar cells devices. For DSSC devices, the functions of graphene as a CCE and a carrier transporting layer have been investigated. Furthermore, graphene has been seen to be used as HTL in perovskite solar cells to enhance the hole transfer rate. Therefore, it is clear from the literature that adding graphene to photovoltaic devices improves their performance and increases their ability to convert sunlight into power. It has also been investigated whether graphene could be used to create flexible and robust photovoltaic devices.

A thorough examination of graphene-based materials and their properties, as was covered in this chapter, presents a tremendous opportunity to address the difficulties associated with energy conversion. There have already been many discoveries regarding the fundamental characteristics of graphene, but there are still many new discoveries to be made regarding the properties of materials based on graphene and their use in photovoltaic technology.

Acknowledgments

This work was partially supported by project through FEDER and FCT (Portugal) programs, and by the research centre Grant UID/FIS/04559/2020 to Laboratory for Instrumentation, Biomedical Engineering and Radiation Physics (LIBPhys-UNL) from the FCT/MCTES/PIDDAC (Portugal).

References

[1] Jurasz J, Canales F A, Kies A, Guezgouz M and Beluco A 2020 A review on the complementarity of renewable energy sources: concept, metrics, application and future research directions *Sol. Energy* **195** 703–24

[2] Becquerel M E 1839 Mémoire sur les effets électriques produits sous l'influence des rayons solaires *C. R. Hebd. Séances Acad. Sci.* **9** 561–7

[3] Fraas L M 2014 *Low-Cost Solar Electric Power* (New York: Springer)

[4] Soga T (ed) 2006 *Nanostructured Materials for Solar Energy Conversion* (Amsterdam: Elsevier)

[5] Green M A, Emery K, Hishikawa Y, Warta W and Dunlop E D 2015 Solar cell efficiency tables (2015) *Prog. Photovolt., Res. Appl* **23** 1–9

[6] Green M A, Hishikawa Y, Dunlop E D, Levi D H, Hohl-Ebinger J, Yoshita M, Ho-Baillie and Anita W Y 2018 Solar cell efficiency tables (Version 53) *Prog. Photovolt., Res. Appl.* **26** 3–12

[7] Alaaeddin M H, Sapuan S M, Zuhri M Y M, Zainudin E S and Al-Oqla F M 2019 Photovoltaic applications: status and manufacturing prospects *Renew. Sustain. Energy Rev.* **102** 318–32

[8] Brown G F and Wu J 2009 Third generation photovoltaics *Laser Photonics Rev.* **3** 394–405

[9] Shockley W and Queisser H J 1961 Detailed balance limit of efficiency of p–n junction solar cells *J. Appl. Phys.* **32** 510–9

[10] Hoppe H and Sariciftci N S 2004 Organic solar cells: an overview *J. Mater. Res.* **19** 1924–45

[11] O'regan B and Grätzel M 1991 A low-cost, high-efficiency solar cell based on dye-sensitized colloidal TiO_2 films *Nature* **353** 737–40

[12] Radich E J, Dwyer R and Kamat P V 2011 Cu_2S reduced graphene oxide composite for high-efficiency quantum dot solar cells. Overcoming the redox limitations of S_2^-/S_n^{2-} at the counter electrode *J. Phys. Chem. Lett.* **2** 2453–60

[13] Mathew S, Yella A, Gao P, Humphry-Baker R, Curchod B F, Ashari-Astani N, Tavernelli I, Rothlisberger U, Nazeeruddin M and Grätzel M 2014 Dye-sensitized solar cells with 13% efficiency achieved through the molecular engineering of porphyrin sensitizers *Nat. Chem.* **6** 242–7

[14] Zhao S, Zhao Z, Yang Z, Ke L, Kitipornchai S and Yang J 2020 Functionally graded graphene reinforced composite structures: a review *Eng. Struct.* **210** 110339

[15] Bonaccorso F, Sun Z, Hasan T and Ferrari A C 2010 Graphene photonics and optoelectronics *Nat. Photon.* **4** 611–22

[16] Zhu D, Wang Z and Zhu D 2020 Highly conductive graphene electronics by inkjet printing *J. Electron. Mater.* **49** 1765–76
[17] Gupta S 2020 Recent development in Graphene technology for multidisciplinary properties and its applications: A Review *Int. J. Sci. Res. Publ.* **10** 199–211
[18] Liu Z, Liu Q, Huang Y, Ma Y, Yin S, Zhang X, Sun W and Chen Y 2008 Organic photovoltaic devices based on a novel acceptor material: graphene *Adv. Mater.* **20** 3924–30
[19] Yan X, Cui X, Li B and Li L S 2010 Large, solution-processable graphene quantum dots as light absorbers for photovoltaics *Nano Lett.* **10** 1869–73
[20] Zhang X, Xie C, Jie J, Zhang X, Wu Y and Zhang W 2013 High-efficiency graphene/Si nanoarray Schottky junction solar cells via surface modification and graphene doping *J. Mater. Chem.* A **1** 6593–601
[21] Liu J, Xue Y, Gao Y, Yu D, Durstock M and Dai L 2012 Hole and electron extraction layers based on graphene oxide derivatives for high-performance bulk heterojunction solar cells *Adv. Mater.* **24** 2228–33
[22] Rowell M W and McGehee M D 2011 Transparent electrode requirements for thin film solar cell modules *Energy Environ. Sci.* **4** 131–4
[23] Kalowekamo J and Baker E 2009 Estimating the manufacturing cost of purely organic solar cells *Sol. Energy* **83** 1224–31
[24] Li S S, Tu K H, Lin C C, Chen C W and Chhowalla M 2010 Solution-processable graphene oxide as an efficient hole transport layer in polymer solar cells *ACS Nano* **4** 3169–74
[25] Li X, Zhu H, Wang K, Cao A, Wei J, Li C, Jia Y, Li Z, Li X and Wu D 2010 Graphene-on-silicon Schottky junction solar cells *Adv. Mater.* **22** 2743–8
[26] Lin T, Huang F, Liang J and Wang Y 2011 A facile preparation route for boron-doped graphene, and its CdTe solar cell application *Energy Environ. Sci.* **4** 862–5
[27] Wang X, Zhi L and Müllen K 2008 Transparent, conductive graphene electrodes for dye-sensitized solar cells *Nano Lett.* **8** 323–7
[28] Gomez De Arco L, Zhang Y, Schlenker C W, Ryu K, Thompson M E and Zhou C 2010 Continuous, highly flexible, and transparent graphene films by chemical vapor deposition for organic photovoltaics *ACS Nano* **4** 2865–73
[29] Yin Z, Sun S, Salim T, Wu S, Huang X, He Q, Lam Y M and Zhang H 2010 Organic photovoltaic devices using highly flexible reduced graphene oxide films as transparent electrodes *ACS Nano* **4** 5263–8
[30] Liu J, Yi Y, Zhou Y and Cai H 2016 Highly stretchable and flexible graphene/ITO hybrid transparent electrode *Nanoscale Res. Lett.* **11** 1–7
[31] Bi H, Huang F, Liang J, Xie X and Jiang M 2011 Transparent conductive graphene films synthesized by ambient pressure chemical vapor deposition used as the front electrode of CdTe solar cells *Adv. Mater.* **23** 3202–6
[32] Bae S *et al* 2010 Roll-to-roll production of 30-inch graphene films for transparent electrodes *Nat. Nanotechnol.* **5** 574–8
[33] Huang X, Yin Z, Wu S, Qi X, He Q, Zhang Q, Yan Q, Boey F and Zhang H 2011 Graphene-based materials: synthesis, characterization, properties, and applications *small* **7** 1876–902
[34] Tung V C, Chen L M, Allen M J, Wassei J K, Nelson K, Kaner R B and Yang Y 2009 Low-temperature solution processing of graphene–carbon nanotube hybrid materials for high-performance transparent conductors *Nano Lett.* **9** 1949–55

[35] Eletskii A V, Zitserman V Y and Kobzev G A 2015 Graphene for solar energy *Nanotechnol. Russ.* **10** 181–91
[36] Wang H, Yang Y, Liang Y, Cui L F, Sanchez Casalongue H, Li Y, Hong G, Cui Y and Dai H 2011 $LiMn_{1-x}Fe_xPO_4$ nanorods grown on graphene sheets for ultrahigh-rate-performance lithium ion batteries *Angew. Chem.* **123** 7502–6
[37] Li S, Luo Y, Lv W, Yu W, Wu S, Hou P, Yang Q, Meng Q, Liu C and Cheng H M 2011 Vertically aligned carbon nanotubes grown on graphene paper as electrodes in lithium-ion batteries and dye-sensitized solar cells *Adv. Energy Mater.* **1** 486–90
[38] Yin Z, Wu S, Zhou X, Huang X, Zhang Q, Boey F and Zhang H 2010 Electrochemical deposition of ZnO nanorods on transparent reduced graphene oxide electrodes for hybrid solar cells *Small* **6** 307–12
[39] Yin Z, Zhu J, He Q, Cao X, Tan C, Chen H, Yan Q and Zhang H 2014 Graphene-based materials for solar cell applications *Adv. Energy Mater.* **4** 1300574
[40] Park H *et al* 2013 Graphene cathode-based ZnO nanowire hybrid solar cells *Nano Lett.* **13** 233–9
[41] Roy-Mayhew J D, Bozym D J, Punckt C and Aksay I A 2010 Functionalized graphene as a catalytic counter electrode in dye-sensitized solar cells *ACS Nano* **4** 6203–11
[42] Kavan L, Yum J H and Grätzel M 2011 Optically transparent cathode for dye-sensitized solar cells based on graphene nanoplatelets *ACS Nano* **5** 165–72
[43] Wen Z, Cui S, Pu H, Mao S, Yu K, Feng X and Chen J 2011 Metal nitride/graphene nanohybrids: general synthesis and multifunctional titanium nitride/graphene electrocatalyst *Adv. Mater.* **23** 5445–50
[44] Kavan L, Yum J H, Nazeeruddin M K and Grätzel M 2011 Graphene nanoplatelet cathode for Co (III)/(II) mediated dye-sensitized solar cells *ACS Nano* **5** 9171–8
[45] Kim K, Bae S H, Toh C T, Kim H, Cho J H, Whang D, Lee T W, Özyilmaz B and Ahn J H 2014 Ultrathin organic solar cells with graphene doped by ferroelectric polarization *ACS Appl. Mater. Interfaces* **6** 3299–304
[46] bin Mohd Yusoff A R, Kim D, Schneider F K, da Silva W J and Jang J 2015 Au-doped single layer graphene nanoribbons for a record-high efficiency ITO-free tandem polymer solar cell *Energy Environ. Sci.* **8** 1523–37
[47] Dong P *et al* 2014 Graphene on metal grids as the transparent conductive material for dye sensitized solar cell *J. Phys. Chem.* C **118** 25863–8
[48] Jeon Y J, Yun J M, Kim D Y, Na S I and Kim S S 2014 Moderately reduced graphene oxide as hole transport layer in polymer solar cells via thermal assisted spray process *Appl. Surf. Sci.* **296** 140–6
[49] Gao P, Ding K, Wang Y, Ruan K, Diao S, Zhang Q, Sun B and Jie J 2014 Crystalline Si/graphene quantum dots heterojunction solar cells *J. Phys. Chem.* C **118** 5164–71
[50] Olson D C, Lee Y J, White M S, Kopidakis N, Shaheen S E, Ginley D S, Voigt J A and Hsu J W 2007 Effect of polymer processing on the performance of poly (3-hexylthiophene)/ZnO nanorod photovoltaic devices *J. Phys. Chem.* C **111** 16640–5
[51] Cheng M, Yang R, Zhang L, Shi Z, Yang W, Wang D, Xie G, Shi D and Zhang G 2012 Restoration of graphene from graphene oxide by defect repair *Carbon* **50** 2581–7
[52] Murakami T N *et al* 2006 Highly efficient dye-sensitized solar cells based on carbon black counter electrodes *J. Electrochem. Soc.* **153** A2255

[53] Reina A, Jia X, Ho J, Nezich D, Son H, Bulovic V, Dresselhaus M S and Kong J 2009 Large area, few-layer graphene films on arbitrary substrates by chemical vapor deposition *Nano Lett.* **9** 30–5
[54] Verma V P, Das S, Lahiri I and Choi W 2010 Large-area graphene on polymer film for flexible and transparent anode in field emission device *Appl. Phys. Lett.* **96** 203108
[55] Hong W, Xu Y, Lu G, Li C and Shi G 2008 Transparent graphene/PEDOT–PSS composite films as counter electrodes of dye-sensitized solar cells *Electrochem. Commun.* **10** 1555–8
[56] Zhang D W, Li X D, Li H B, Chen S, Sun Z, Yin X J and Huang S M 2011 Graphene-based counter electrode for dye-sensitized solar cells *Carbon* **49** 5382–8
[57] Liu Z, Lau S P and Yan F 2015 Functionalized graphene and other two-dimensional materials for photovoltaic devices: device design and processing *Chem. Soc. Rev.* **44** 5638–79
[58] Wahyuono R A *et al* 2019 Self-assembled graphene/MWCNT bilayers as platinum-free counter electrode in dye-sensitized solar cells *Chem. Phys. Chem.* **20** 3336–45
[59] Wang T, Jing L C, Zhu Q, Ethiraj A S, Tian Y, Zhao H, Yuan X T, Wen J G, Li L K and Geng H Z 2020 Fabrication of architectural structured polydopamine-functionalized reduced graphene oxide/carbon nanotube/PEDOT: PSS nanocomposites as flexible transparent electrodes for OLEDs *Appl. Surf. Sci.* **500** 143997
[60] Oh W C, Cho K Y, Jung C H and Areerob Y 2020 Hybrid of graphene based on quaternary $Cu_2ZnNiSe_4$–WO_3 nanorods for counter electrode in dye-sensitized solar cell application *Sci. Rep.* **10** 1–10
[61] Lin Y *et al* 2013 Graphene/semiconductor heterojunction solar cells with modulated antireflection and graphene work function *Energy Environ. Sci.* **6** 108–15
[62] Miao X, Tongay S, Petterson M K, Berke K, Rinzler A G, Appleton B R and Hebard A F 2012 High efficiency graphene solar cells by chemical doping *Nano Lett.* **12** 2745–50
[63] Song J, Yin Z, Yang Z, Amaladass P, Wu S, Ye J, Zhao Y, Deng W Q, Zhang H and Liu X W 2011 Enhancement of photogenerated electron transport in dye-sensitized solar cells with introduction of a reduced graphene oxide–TiO_2 junction *Chem. Eur. J.* **17** 10832–7
[64] Wang Y, Tong S W, Xu X F, Özyilmaz B and Loh K P 2011 Interface engineering of layer-by-layer stacked graphene anodes for high-performance organic solar cells *Adv. Mater.* **23** 1514–8
[65] Guo C X, Yang H B, Sheng Z M, Lu Z S, Song Q L and Li C M 2010 Layered graphene/quantum dots for photovoltaic devices *Angew. Chem. Int. Ed.* **49** 3014–7
[66] Tang Y B *et al* 2010 Incorporation of graphenes in nanostructured TiO_2 films via molecular grafting for dye-sensitized solar cell application *ACS Nano* **4** 3482–8
[67] Wang S, Goh B M, Manga K K, Bao Q, Yang P and Loh K P 2010 Graphene as atomic template and structural scaffold in the synthesis of graphene–organic hybrid wire with photovoltaic properties *ACS Nano* **4** 6180–6
[68] Zhou W *et al* 2011 A general strategy toward graphene@ metal oxide core–shell nanostructures for high-performance lithium storage *Energy Environ. Sci.* **4** 4954–61
[69] Yang N, Zhai J, Wang D, Chen Y and Jiang L 2010 Two-dimensional graphene bridges enhanced photoinduced charge transport in dye-sensitized solar cells *ACS Nano* **4** 887–94
[70] Liu J, Durstock M and Dai L 2014 Graphene oxide derivatives as hole-and electron-extraction layers for high-performance polymer solar cells *Energy Environ. Sci.* **7** 1297–306
[71] Tong S W, Wang Y, Zheng Y, Ng M F and Loh K P 2011 Graphene intermediate layer in tandem organic photovoltaic cells *Adv. Funct. Mater.* **21** 4430–5

[72] Jeong G, Koo D, Seo J, Jung S, Choi Y, Lee J and Park H 2020 Suppressed interdiffusion and degradation in flexible and transparent metal electrode-based perovskite solar cells with a graphene interlayer *Nano Lett.* **20** 3718–27

[73] Fan G, Zhu H, Wang K, Wei J, Li X, Shu Q, Guo N and Wu D 2011 Graphene/silicon nanowire Schottky junction for enhanced light harvesting *ACS Appl. Mater. Interfaces* **3** 721–5

[74] Ye Y, Gan L, Dai L, Dai Y, Guo X, Meng H, Yu B, Shi Z, Shang K and Qin G 2011 A simple and scalable graphene patterning method and its application in CdSe nanobelt/graphene Schottky junction solar cells *Nanoscale* **3** 1477–81

[75] Zhang D, Xie F, Lin P and Choy W C 2013 Al–TiO_2 composite-modified single-layer graphene as an efficient transparent cathode for organic solar cells *ACS Nano* **7** 1740–7

[76] Gnisci A, Faggio G, Messina G, Lancellotti L, Bobeico E, Veneri P D, Capasso A, Dikonimos T and Lisi N 2018 Graphene-based derivative as interfacial layer in graphene/n-Si Schottky barrier solar cells *Adv. Model. Anal.* A **55** 144–50

[77] Wang D H, Kim J K, Seo J H, Park I, Hong B H, Park J H and Heeger A J 2013 Transferable graphene oxide by stamping nanotechnology: electron-transport layer for efficient bulk-heterojunction solar cells *Angew. Chem. Int. Ed.* **52** 2874–80

[78] Xu X *et al* 2020 Surface functionalization of a graphene cathode to facilitate ALD growth of an electron transport layer and realize high-performance flexible perovskite solar cells *ACS Appl. Energy Mater.* **3** 4208–16

[79] Kim J, Tung V C and Huang J 2011 Water processable graphene oxide: single walled carbon nanotube composite as anode modifier for polymer solar cells *Adv. Energy Mater.* **1** 1052–7

[80] Wu Z, Bai S, Xiang J, Yuan Z, Yang Y, Cui W, Gao X, Liu Z, Jin Y and Sun B 2014 Efficient planar heterojunction perovskite solar cells employing graphene oxide as hole conductor *Nanoscale* **6** 10505–10

[81] Lee D Y, Na S I and Kim S S 2016 Graphene oxide/PEDOT: PSS composite hole transport layer for efficient and stable planar heterojunction perovskite solar cells *Nanoscale* **8** 1513–22

IOP Publishing

Recent Advances in Graphene and Graphene-Based Technologies

Anoop Chandran, N V Unnikrishnan, M K Jayaraj, Reenu Elizabeth John and Justin George

Chapter 12

Photonic and optoelectronic applications of graphene: graphene-based transparent conducting electrodes for LED/OLED

Pradeep Kumar, Monika Gupta, Nurul Nadia Norhakim, Mohamed Shuaib Mohamed Saheed, Huzein Fahmi Hawari and Zainal Arif Burhanudin

Graphene is a two-dimensional (2D) carbon material that has outstanding electronic, optical, thermal, chemical, and mechanical properties. Its application such as in transparent conducting electrodes (TCEs) has been extensively investigated for the realization of a new class of inorganic and organic light-emitting diodes (LEDs/OLEDs). In this chapter, a brief overview of commercially available and recently demonstrated TCEs will be given. Then, a discussion on the metrics of the TCEs will be presented. Next, several methods of TCE fabrication will be reviewed. This is followed by a detailed review of hybrid TCEs made of graphene and other materials such as carbon nanotubes, conducting polymers, and metal nanostructures. The effects of different graphene-based electrodes on the performance of LEDs/OLEDs will be discussed in detail. Finally, a summary with a brief discussion on the prospects of graphene-based TCEs will be presented.

12.1 Introduction

Inorganic and organic light-emitting diodes (LEDs/OLEDs) are attracting huge attention as the source of light for next-generation lighting and display applications. As the display technology, the LEDs and OLEDs offer numerous advantages such as brilliant colors, lower power consumption, faster response, light weight, and wide viewing angle over the existing liquid crystal display technology [1, 2]. In the case of lighting applications, the LEDs/OLEDs are thin, glare-free, and emit light in a wide spectrum covering from ultraviolet to near-infrared regions [3–7]. In comparison to the performance of the general-purpose lighting (efficiency~4% from incandescent

doi:10.1088/978-0-7503-3999-5ch12

and~17% from fluorescent lamps), the reported efficiency of ~20% from LEDs/OLEDs has revolutionized the lighting industry and their market is aimed to reach around 95 936 million USD in 2025 [8]. The cost and the performance of the LEDs and OLEDs are influenced by their device structures. In the device structure, the electrodes are a crucial component. In LED/OLEDs, the electrodes are placed in the form of transparent conducting films made of nanomaterials. The transparency and electrical conductivity of these TCEs affect the device performance [9].

Conventionally, ITO (indium tin oxide; In_2O_3-SnO_2; 74% In, 18% O_2, 8% Sn) thin films (~140 nm) are being used as transparent conducting electrodes (TCEs) in LED/OLEDs due to their high electrical conductivity (sheet resistance~15–20 Ω/□) and high transparency (~90%). However, they have some limitations, for example: (i) the diffusion of indium into the active layers of the devices, which degrades the device performance over time; (ii) high refractive index that causes high trapping of light in the waveguide mode; (iii) high processing cost; (iv) chemical instability in acids and alkalis; (v) fragile nature; and (vi) poor transparency (20%) in the UV region [10–14]. Due to these issues with conventional ITO TCEs, alternative TCE materials having similar optoelectrical properties are demanded.

As the ITO-substitutes, various materials like conducting polymers, metallic nanowires (NWs), carbon nanotubes (CNTs), and graphene have been widely investigated [9, 15–20]. Among these materials, graphene has attracted attention as the viable TCE for existing as well as next-generation LEDs/OLEDs, because of its outstanding properties such as very high transmittance (~98%), high stability, high flexibility, high quantum electronic transports, high thermal conductivity (~5000 W $(m\cdot K)^{-1}$), high mobility (~200 000 $cm^2\ V^{-1}\ s^{-1}$) at room temperature, and electrical resistivity in the range of 1–6 kΩ/□ [12, 21–25].

In this chapter, we cover the graphene-based TCEs for LEDs and OLEDs applications. We highlight the TCE metrics at the beginning. Afterwards, we discuss the fabrication methods such as chemical vapor deposition (CVD), solution-processing, layer-by-layer (LBL) assembly to produce graphene TCEs with different structural and optoelectrical properties. We also present some strategies to produce highly conductive graphene-based hybrid TCEs for achieving high device performance. Later, we review the device performance of inorganic and organic LEDs with different graphene-based TCEs. A summary is presented in the end with the prospects to be considered for future opportunities.

12.2 Metrics of transparent conducting electrodes

12.2.1 Optoelectronic property

High transmittance and high electrical conductivity are the crucial requirements for an ideal TCE for optoelectronic applications. In addition, the TCE materials must be practically colorless, low-priced, and non-toxic. The formation of tailored interfaces between TCE and active electronic material is also an essential requirement. From the physics point of view, a material with high carrier concentration (n) and mobility ($\mu_{n,p}$) must exhibit high conductivity as given by equation (12.1),

$$\sigma = qn\mu_{n,\,p} \tag{12.1}$$

where q is the elementary charge. The electrons are more mobile than the ions due to their small effective mass. This indicates that the good conducting materials have electrons as the charge carriers. The equation (12.1) empirically suggests that the mobility and carrier concentration can be increased to achieve enhanced conductivity however, n is limited by the absorption of light by free carriers (electrons in good conductors), which can be understood by classical Drude theory as equations (12.2) and (12.3) [26],

$$\lambda = \frac{2\pi\hbar}{\omega} \tag{12.2}$$

$$\omega = \left(\frac{ne^2}{m^*\varepsilon}\right)^{1/2} \tag{12.3}$$

where λ, ω, ε, $\hbar$, and m^* are the wavelength, plasma frequency, material permittivity, modified Plank's constant, and effective mass of the carriers. So, the strategy of limiting the carrier concentration and increasing the carrier mobility can be very effective to develop high-performance electrodes. Low surface roughness and high adhesion should also be considered for the development of high-performance graphene-based TCEs [14, 27, 28].

12.2.2 Figure of merit (FoM)

FoM is a quantified trade-off relationship between the optical property (transmittance, T) and electrical property (sheet resistance, R_S) of a TCE. In other words, the FoM sets a standard that could quantify and compare the optoelectrical properties of different TCEs [29]. FoM relates the DC conductivity (σ_{DC}) to optical conductivity (σ_{op}) and the transmittance in the visible wavelength range is displayed as a function of sheet resistance as defined in equation (12.4) [30, 31] and shown in figure 12.1.

$$\text{FoM} = \frac{\sigma_{DC}}{\sigma_{op}} = \frac{Z_0}{2R_S}\frac{1}{(T^{-0.5} - 1)} \tag{12.4}$$

where Z_0 is vacuum impedance (377 Ω) [32]. Transmittance is typically measured in the visible range at a wavelength of 550 nm. The σ_{DC}/σ_{op} ratio can be maximized for high transmittance at a particular sheet resistance and high values of σ_{DC}/σ_{op} results in the desired properties (high T, low R_S). The minimum optoelectronic industry standard for a substitute material to replace conventional ITO-based electrode is a sheet resistance lower than 100 Ω/□ coupled with a transmittance higher than 90% in the visible spectrum. This indicates that material thin films showing FoM higher than 35 can be viable alternatives of ITO TCEs for optoelectronic applications. However, for most current-driven applications, a more stringent condition of R_S that should be lower than 10 Ω/□ (at $T = 85\%$) is necessary. This substantiates the requirement that FoM should be greater than 220 [30].

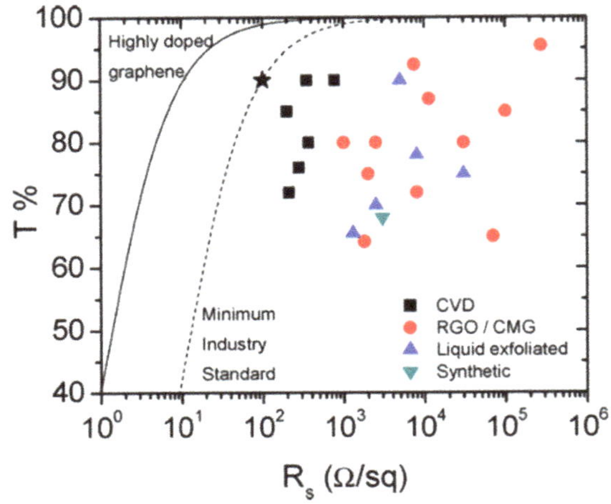

Figure 12.1. Transmittance versus sheet resistance of different graphene-based TCEs. Reproduced with permission from [30], Copyright 2010, American Chemical Society.

12.3 Fabrication of graphene-based transparent conducting electrodes

12.3.1 CVD-grown graphene TCEs

CVD is a technique used to deposit the gaseous reactant onto the transition metal substrates like nickel (Ni), palladium (Pd), and copper (Cu) [33]. Using CVD, impervious and very high-quality graphene thin films can be achieved [34]. Thin foil sheets of copper or nickel are used as the catalytic substrate to grow the graphene nanostructure. Methane (CH_4) is used as the precursor gas. In a typical synthesis process, the catalytic substrate is annealed up to 1000 °C in the presence of H_2 gas, as shown in figure 12.2(a) [35]. The precursor gas is then flowed into the CVD quartz tube chamber to begin the graphene growth on the catalytic surface. During cooling, the carbon precipitates on catalytic substrate to form graphene nanostructure (monolayer or multilayer). The synthesized graphene films are later transferred to transparent substrates using the polymer-assisted wet-transfer method to prepare TCEs, as shown in figure 12.2(b). For single-layer graphene (SLG), copper is a highly preferred substrate because carbon has very low solubility in the copper substrate, whereas carbon solubility in nickel or palladium catalytic sheet is very high and causes the formation of a combination of monolayer, bilayer, and multilayer nanostructure in the grown graphene film.

High-quality SLG was first demonstrated on the polycrystalline copper substrate by Ruoff *et al* [36]. This attracted the attention of the research community because of the advantages of good control of graphene layers, ability to transfer and low cost. Today, a variety of different CVD methods are available that can be employed to synthesize graphene-based materials. The optoelectrical properties of CVD-grown graphene TCEs vary according to the processing parameters like temperature, precursor nature, pressure, growth duration, gas flow state, and activation

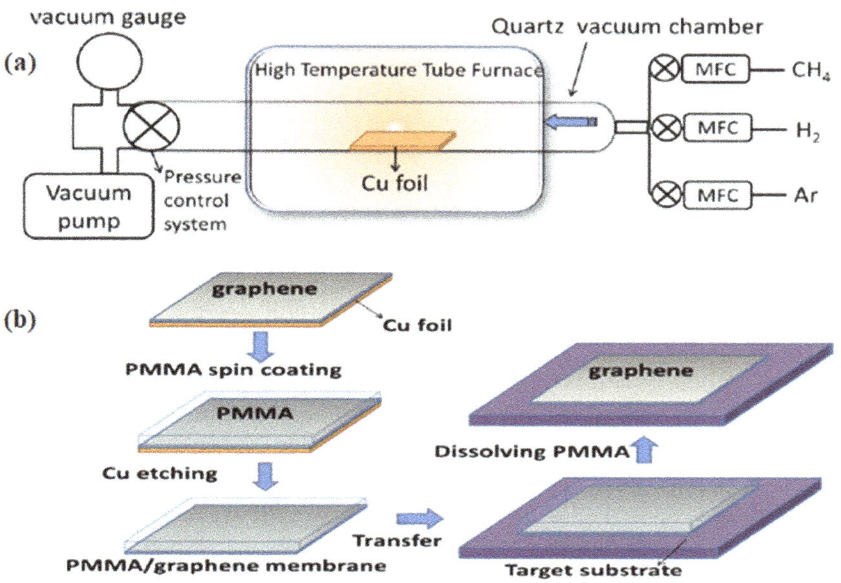

Figure 12.2. (a) CVD growth and (b) transfer processes for fabricating graphene TCE [35]. Reproduced [35] with Creative Commons Attribution 3.0 Unported, Copyright 2013, InTech.

manner [37]. The SLG TCEs have shown high optical transmittance (~98%) and high intrinsic carrier mobility (~2×10^5 cm^2 V^{-1} s^{-1}) that make them excellent TCE choices for LED/OLED applications [38]. However, atomic thin SLG has a high R_S (~1000 Ω/□) [39].

Ease of setup in research laboratories, successful long-term use in industrial settings, and the potential to scale up fabrication make CVD a popular approach for producing high-quality graphene [40, 41]. Moreover, the CVD technique is highly compatible with the existing complementary metal-oxide-semiconductor (CMOS) technology but it has some limitations. For example, controlling the thickness of graphene film is challenging and secondary crystals are formed during the growth process. In the CVD method, expensive substrate materials are required for growing graphene. This limits its applicability for large-scale production. Nevertheless, the CVD technique has emerged as a significant method for the mass production of graphene with fewer electronic and structural defects or disorders. In consideration of both ecological and cost factors, the CVD approach is among the best available routes for the synthesis of graphene-based materials [42].

12.4 Solution-processed graphene derivative-based TCEs

A graphene derivative like graphene oxide (GO) is commonly synthesized by modified Hummer's method [43–45]. Various other graphene derivatives such as reduced graphene oxide (rGO), functionalized graphene oxide, can be produced by the chemical reduction of GO. In this chemical reduction process, different organic and inorganic reducing agents, for example hydrazine hydrate, ascorbic acid, sodium borohydride, glucose, hydroxylamine, hydroquinone, amino acids, pyrrole, urea, and alkaline solution have been explored [46–51]. Chemical reduction of GO, offers easy surface modification and surface functionalization of thin films and their coatings. Easy dispersion and deposition on different target substrates make the solution-processed graphene TCEs suitable for light-emitting device applications.

The large-scale production of electrodes made of graphene derivatives can be done by depositing the GO sheets on target substrates followed by its reduction by thermal and chemical methods [48, 52, 53]. The easy dispersion of GO facilitates a large-area film. Various deposition methods such as spin coating, spray coating, dip coating, electrophoretic deposition, and Langmuir–Blodgett (LB) assembly have been used for developing TCEs. Among them, spin coating, and spray coating are the most convenient methods. However, the partial agglomerations in solution-processed GO suspension and wrinkles in coated thin films are unavoidable. The TCEs-based on solution-processed graphene are easy to process, low in cost, and have excellent stability, but they have higher R_S (0.8–43 kΩ/□) than that of metal-based thin TCEs that exhibit the same level of transparency (71%–95%) [43, 50, 54, 55].

12.5 Doped and layered graphene TCEs

In doped graphene TCEs, the graphene thin films are doped or decorated with chemicals, for example, gold chloride ($AuCl_3$), nitric acid (HNO_3), or triethyloxonium hexachloroatimonate (OA). In layered graphene TCEs, the LBL assembly or stacking of undoped or doped single-layer or multilayer graphene TCEs are prepared to develop the TCEs. These electrodes have shown high optoelectronic properties. For example, Li *et al* prepared doped graphene electrodes by soaking CVD-grown and wet transferred graphene thin film into triethyloxonium hexachloroantimonate (OA)/dichloroethene solution [12]. This p-type chemical doping of graphene significantly reduced the sheet resistance of CVD-SLG (80% reduction, ~200 Ω/□) with negligible deformation in its structure, as illustrated in figure 12.3(a). The transmittance of used doped TCE was found to be ~95% at 550 nm wavelength as depicted in figure 12.3(b). The work function of graphene

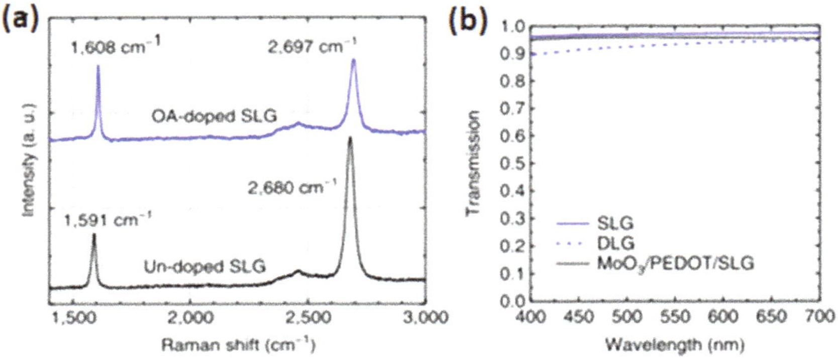

Figure 12.3. (a) Raman spectra and (b) transmittance of graphene TCE doped with OA chemical. Reproduced from [12], copyright 2013, with permission from Springer Nature.

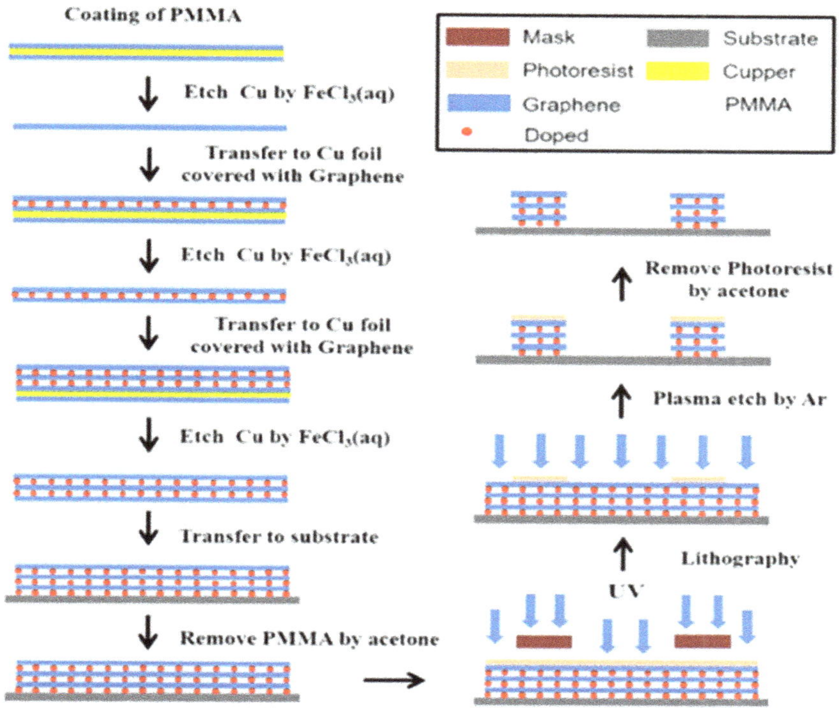

Figure 12.4. The procedure of preparing the doped multilayer graphene TCE. Reproduced [57] under Creative Commons Attribution License 4.0, Copyright 2017, Springer.

TCEs can also be modified by such chemical doping. The enhancement of ~0.4 eV in the work function of SLG was observed after OA doping. In another study, Han *et al* developed the layered TCEs by doping the multilayer graphene (four layers) using nitric acid HNO_3 [56]. Such chemical doping of layered graphene is found to be very effective in achieving high optoelectrical (R_S~54 Ω/□ and T~88%) and electronic properties (work function~4.6 eV).

Recently, Xu *et al* developed a chemically modified multilayer graphene (MLG) TCE [57]. They developed the graphene-based TCE by assembling three layers of $AuCl_3$-doped SLG. In the development process, the CVD-grown graphene film is repetitively transpired to newly grown graphene after etching the catalytic Cu substrate using ferric chloride and doping the film using gold chloride solution, as shown in figure 12.4. The $AuCl_3$ doped MLG TCE was found to be very effective in achieving high sheet resistance (150 Ω/□) and optical transparency (~91%). 82% enhancement in the electrical conductivity was observed from such a TCE arrangement.

12.6 Graphene-based hybrid transparent conducting electrodes

In order to improve the optoelectrical and electronic properties of CVD-grown SLG, its various hybrid structures with inorganic and organic materials have been widely explored. The graphene-based hybrid TCEs are categorized as follows.

12.6.1 Hybrid TCEs of graphene with metal oxides and ultrathin metals

The hybrid structures of CVD-SLG with oxide semiconductors have been found efficient in achieving high optoelectrical properties. The introduction of a thin layer of metal or metal oxides improves the electrical conductivity and work function of SLG with a trade-off of optical properties. Ag (silver), MoO_x (molybdenum oxide), WO_3 (tungsten trioxide) have been studied to improve the optoelectronic properties of SLG TCEs. Li *et al* developed the hybrid TCE of graphene with Ag, and aluminum-doped zinc oxide (AZO) by depositing thin films of AZO and Ag onto CVD-grown graphene [58]. For preparing the hybrid structure, thin layers of Ag (~10 nm) and AZO (~50 nm) were deposited on the graphene using magnetron sputtering, as shown in figure 12.5(a). The hybrid TCE of graphene/Ag/AZO exhibited optoelectrical properties (T~80%, R_S~10 Ω/□, FoM~160) as presented in figure 12.5(b). Another advantage of the presence of thin metallic film and oxide layer is that they smoothen the surface of hybrid TCE (see figures 12.5(c) and (d)). By such hybridization, around a 20% reduction in the surface roughness was noticed.

In another case, graphene and metal oxide-based hybrid TCE was developed by depositing the tungsten trioxide (WO_3) thin film of graphene thin film surface. Kuruvila *et al* introduced a WO_3 thin layer (3–5 nm) of using thermal evaporation on graphene TCE to improve the electrical conductivity of graphene TCE [59]. By this hybrid scheme, ~60% reduction in the sheet resistance (300 Ω/□ at 90% transparency) of graphene TCE was noticed (see figures 12.6(a) and (b)) and showed its TCE FoM~11.62. A strong p-doping was also evident when comparing the work

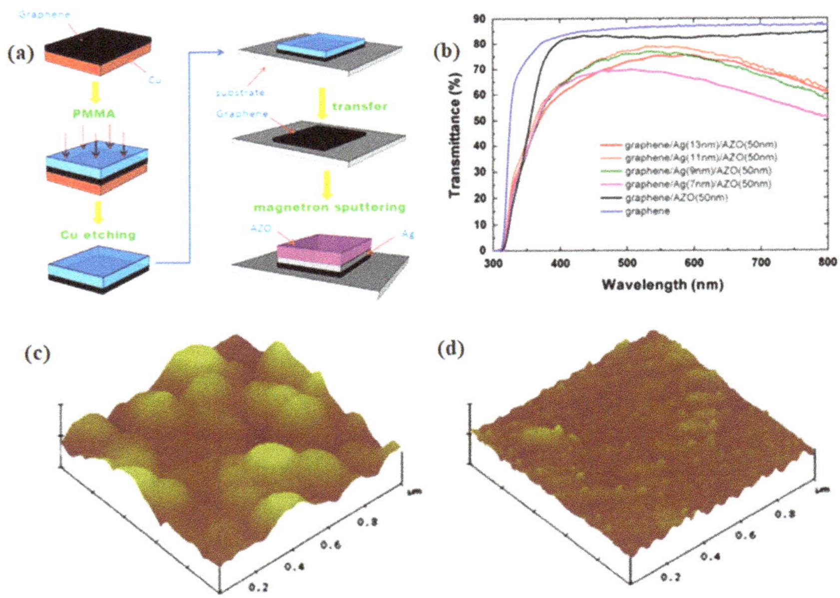

Figure 12.5. (a) Development process, (b) transmittance, and (c) surface profile of graphene/Ag/AZO hybrid TCE. Reproduced from [58], copyright 2013 with permission from Elsevier.

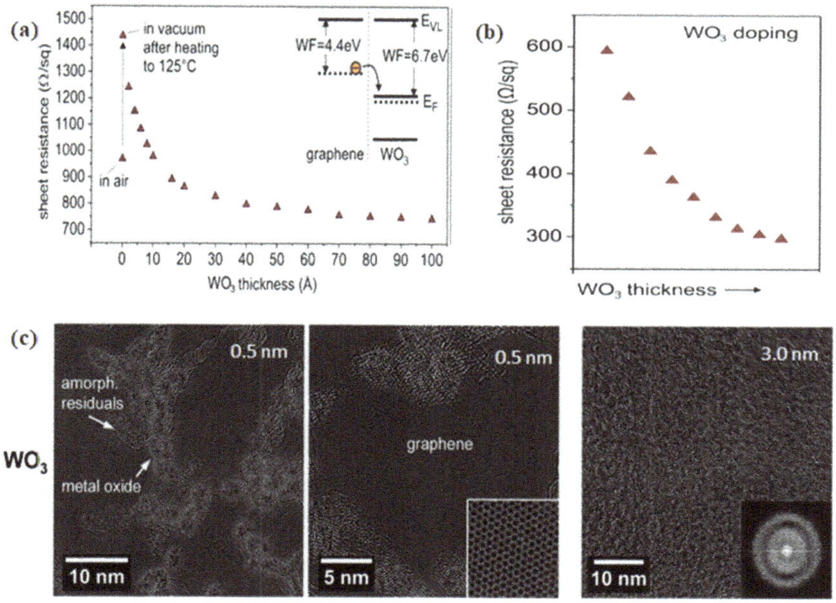

Figure 12.6. (a) and (b) Sheet resistance variation of SLG on various tungsten trioxide thin film thicknesses. (c) SEM image of graphene/WO_3 hybrid TCE. Reproduced with permission from [59] copyright 2014, Royal Society of Chemistry.

function difference. After employing a thin layer of transition metal oxide on a graphene film, the work function of hybrid graphene/WO_3 TCE was enhanced by ~1 eV. However, the amorphous character of the metal oxide layer at different thicknesses (0.5 nm, 3 nm) was shown during the microscopic study, as displayed in figure 12.6(c).

Recently, Lee *et al* used a different technique to develop the hybrid TCE of graphene and metal oxide. They spin-coated molybdenum oxide (MoO_x) nanoparticle suspension onto CVD-grown graphene thin films with a post-annealing treatment (100 °C, 30 min) for preparing the hybrid TCE, as shown in figure 12.7 [60]. The optoelectrical properties of the hybrid TCE in the LBL assembly of graphene (two layers)/MoO_x (12 nm)/graphene (two layers)/MoO_x (12 nm) were found to be very promising (R_S~96 Ω/□, T~85%, FoM~23.2) for electrode applications. The work function of hybrid TCE was also increased by 0.6 eV, indicating the reduction of in the energy barrier at the interface of graphene/MoO_x and the upper polymeric layer. Such reduction in the energy barrier can achieve efficient hole injection into the active layer of OLED.

12.6.2 Hybrid TCEs of graphene with conducting polymers and metal nanostructures

Combinations of graphene with some nanomaterials such as conducting polymers and metal nanowires have been investigated to prepare hybrid TCEs for LED/OLED applications. The synergistic effect of a unified hybrid structure plays an

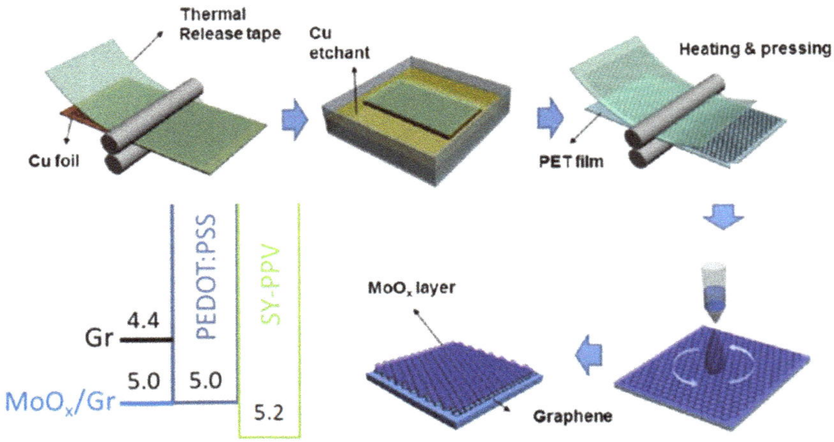

Figure 12.7. Development process and energy levels of MoO_x/graphene hybrid TCE. Reproduced with permission [60], Copyright 2017, IOP Publishing Ltd, all rights reserved.

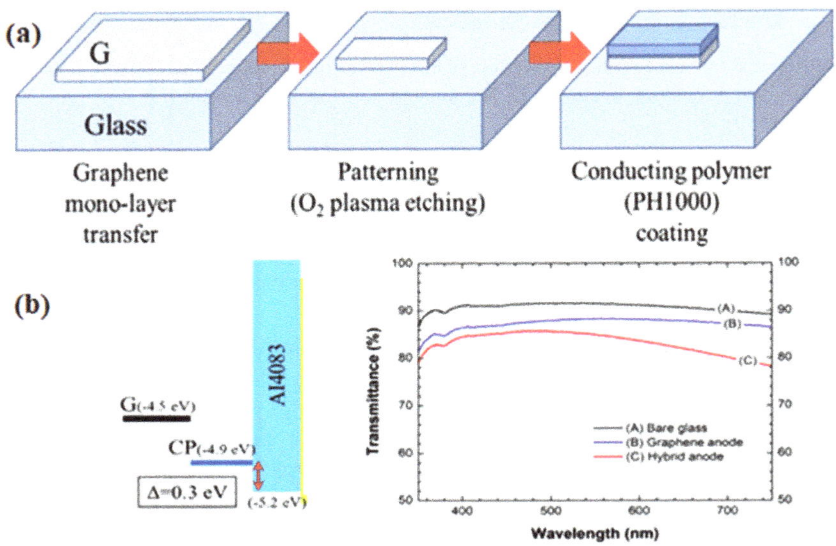

Figure 12.8. (a) Development process, (b) transmittance and energy levels of graphene/conducting polymer hybrid TCE. Reproduced from [61], copyright 2013, with permission from Elsevier.

efficient function in realizing the high electrode properties of graphene-based TCEs. For instance, Shin *et al* developed the hybrid TCE structure of graphene and conducting polymer by depositing PEDOT:PSS (poly(3,4-ethylenedioxythiophene): polystyrene sulfonate) on the patterned graphene grown by CVD [61], as shown in figure 12.8(a). After applying the conducting polymer, the uniform surface morphology of the hybrid TCE was observed with a lower sheet resistance (~90 Ω/□ at a T~92.8%) and FoM~37.56. This suggests a 15-fold increase in the electrical

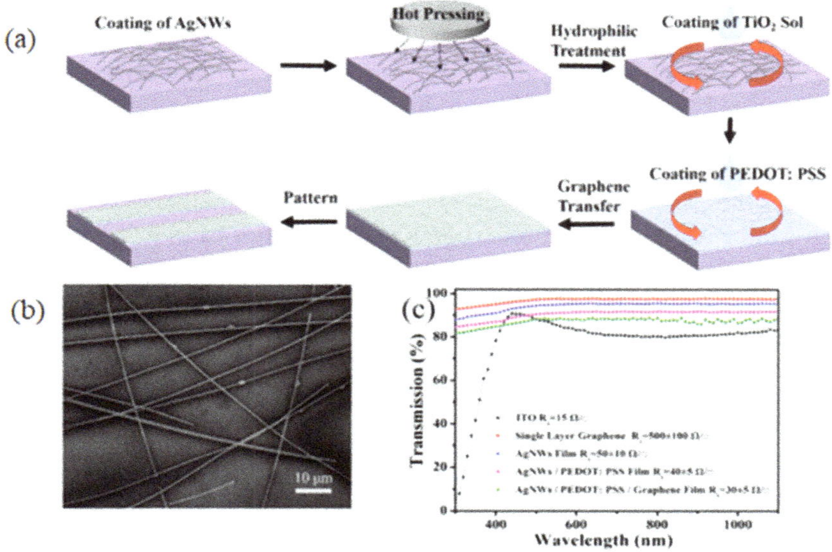

Figure 12.9. (a) Preparation process, (b) SEM image, and (c) transmittance of graphene/conducting polymer/silver nanowires hybrid TCE. Reproduced [62] with Creative Commons Attribution 4.0 International license, Copyright 2017, Nature Publishing Group.

conductivity of graphene-based TCE after introducing the conducting polymer in their LBL structure. The work function of graphene-based TCE was also increased to ~5.0 eV after hybridization of it with the PEDOT:PSS as depicted in figure 12.8(b).

Xu *et al* employed a different strategy to improve the electrode properties of graphene-based TCEs [62]. They prepared a hybrid of graphene, PEDOT:PSS, and Ag nanowires (NWs) by transferring the CVD-grown SLG onto the conducting polymer deposited, hot-pressed silver nanowires-based thin film. A schematic of this hybridization scheme is illustrated in figure 12.9(a). The surface properties of TCE were improved by heat treatment and metal oxide coating. A smooth surface (~9 nm) with higher wettability was achieved when the TCE was passing through hot-press followed by TiO_2 coating, as shown in figure 12.9(b). The graphene/PEDOT:PSS/AgNWs TCE (~100 nm) demonstrated low R_S~30 Ω/□ at T~88%, indicating TCE FoM~95.2 as shown in figure 12.9(c). In comparison to SLG TCE, ~94% improvement in electrical conductivity can be achieved by this hybridization scheme.

12.6.3 Hybrid TCEs of graphene with carbon materials

Both graphene and CNT have good thermal stability, remarkable electronic and mechanical properties. The hybridization of graphene with CNTs can prevent the aggregation of carbon materials and realize the synergistic effect between graphene and CNTs [63]. In hybrid nanostructure, graphene fuses with CNTs. By such fusion (chemically bonded sp^2 carbons), an interconnected conducting network of totally carbon nanostructures is formed, wherein, graphene provides a two-dimensional conducting channel for charge carrier movement [27, 63, 64]. This improves the electrical conductivity of TCE and inhibits the possible oxidation of CNTs [65, 66].

As illustrated by Kumar *et al* the graphene/CNT hybrid transparent thin film can be prepared by a one-step method using thermal CVD (T-CVD) [27]. In the process, the catalytic copper template coated with CNTs is placed in a T-CVD chamber for graphene synthesis. At high temperatures (~1080 °C), the methane gas which acts as a carbon source decomposes into carbon ions which later form a 2D surface across the dispersed CNT network on the catalytic copper template. After the wet-transfer process, the TCE of graphene/CNT is prepared on the transparent substrate, as shown in figure 12.10. Such CVD-based hybrid TCE preparation offers the high optical property of the electrode. The graphene/CNT hybrid TCE showed high transparency (96.6%) with good electrical conductivity (R_S~300 Ω/□) and TCE FoM~36. Additional surface modification like partial decoration of hybrid TCE can enhance the electrical conductivity with a little compromise with the electrode's transparency (R_S~100 Ω/□, T~96%, FoM~88). In another case, Kang *et al* also obtained good optoelectrical properties of CNT/graphene hybrid TCEs (T~88%, R_S~533 Ω/□) and FoM~6 [64]. The CNT/graphene junction contacts are improved by directly synthesizing the graphene nanostructure across the CNT network using CVD and this expedites the charge transport that improves the electrical conductivity of hybrid TCEs [67, 68].

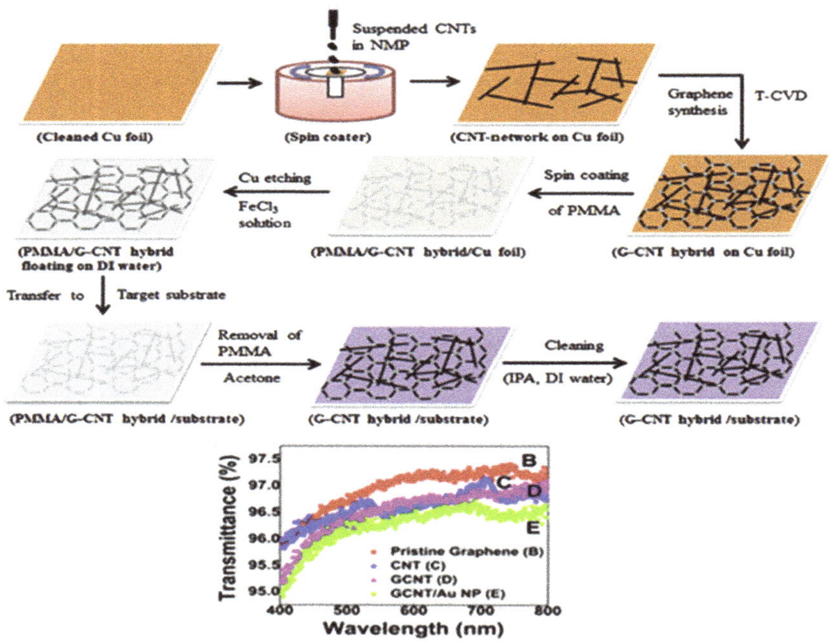

Figure 12.10. Development process and transmittance of graphene/CNT hybrid TCE. Reproduced from [27], copyright 2019, with permission from Elsevier.

12.7 LEDs and OLEDs with graphene-based TCEs

A substantial understanding of the fundamental properties of graphene-related materials (GRMs) and TCEs has led to the fabrication and integration of numerous state-of-the-art device configurations of LEDs and OLEDs. The present research trend is headed for developing several sustainable lighting and display technologies that aim for higher emission efficiency and lower power consumption. The focal point has been on the integration of graphene-based TCEs with LED structures, or bottom-up/top-down fabrication of OLED structures on these TCEs in order to boost the device performances, owing to their unique properties of high transparency, conductivity, high carrier mobilities, high current transport and high stability.

In the development of inorganic LEDs (table 12.1), Tae Hoon *et al* [69] were among the early groups to demonstrate graphene TCEs for GaN-based LEDs. The structure of the LED is as illustrated in figure 12.11(a). At an injection current of 20 mA, the fabricated LED with graphene TCE showed ~6 V forward voltage (V_F) much higher than the 3.5 V forward voltage of LED with conventional ITO TCE (see figure 12.11(b)). The higher forward voltage of graphene-LED is attributed to the higher sheet resistance of graphene-based TCE. However, graphene-LED showed light output power (LOP) that was ~25% higher than the ITO-LED (see figure 12.11(c)). The higher LOP of graphene-LED than that of ITO-LED is attributed to higher transparency and more uniform current spreading of graphene-based TCE. Relatively higher and more uniform electroluminance (EL) was observed in graphene-LED than ITO-LED (see figure 12.11(d)).

Subsequently, MLG films were investigated as TCEs for GaN-based LED [70]. The MLG film (R_S~108 Ω/□) was utilized as the effective current spreading layer in LED. However, the LOP of MLG-LED was affected due to its low transparency. Among SLG, MLG, and Ni/Au-based TCEs, MLG schemes showed the lowest V_F of 3.1 V at 20 mA. In order to improve the device performance of GaN-based LEDs, doped graphene TCEs have also been explored. Choe *et al* [71] demonstrated MLG doped with gold nanoparticles (AuNPs) as a TCE for GaN-based LED. AuNPs doping of graphene-based TCE was found to be essential in terms of achieving lower sheet resistance (20 mM AuNP/MLG~105 Ω/□, MLG~1 kΩ/□) and high work function (AuNP/MLG~5.12 eV, MLG~4.42 eV). These improvements in graphene-based TCE resulted in the increased hole injection and higher efficiency of graphene-LED compared to the LED with undoped MLG TCE. The doped MLG TCE-based LED showed a V_F of 3.86 V that was found to be lower than the undoped MLG TCE-based LED (V_F~5 V). In another case, HNO_3-doped MLG TCE was explored in realizing high optical power for GaN-based LED [72]. The work function of graphene-based TCE was also increased to ~5 eV after doping with HNO_3 solution. Reduction in the work function difference between graphene TCE and p-GaN layer further reduced the forward voltage of the LED, resulting in the attainment of lower V_F~5.4 V (at 80 mA).

Apart from doped graphene TCEs, graphene-based hybrids TCEs have also been explored for GaN LEDs. For example, GaN LED with hybrid TCE of graphene/

Table 12.1. Graphene-based TCEs for inorganic LEDs.

Graphene-based TCE					Inorganic LED with graphene-based TCE		
TCE type	TCE nanostructure	Sheet resistance (Ω/□)	Transmittance (%)	FoM	Forward voltage (V) at 20 mA	Light output power (mW)	Reference
CVD graphene	SLG	—	~97	—	5.87	65 at 100 mA	[69]
	MLG (3 layers)	300	87	8.71	3.1	—	[70]
Doped graphene	AuNPs doped MLG (~9 layer graphene)	105	78	13.6	3.86	—	[71]
	HNO_3 doped CVD-MLG	—	80	—	~4	~20 at 80 mA	[72]
Hybrid graphene	MLG (~3 layers)/AgNWs	300	92.8	16.51	6.6	~3 at 20 mA	[73]
	CVD-grown CNT-graphene hybrid	500	90	6.97	~5	~2 at 4 mA	[64]
	CVD-grown MLG (~6 nm thick)/Cu NWs TCE	37	89	84.92	4.6	—	[74]

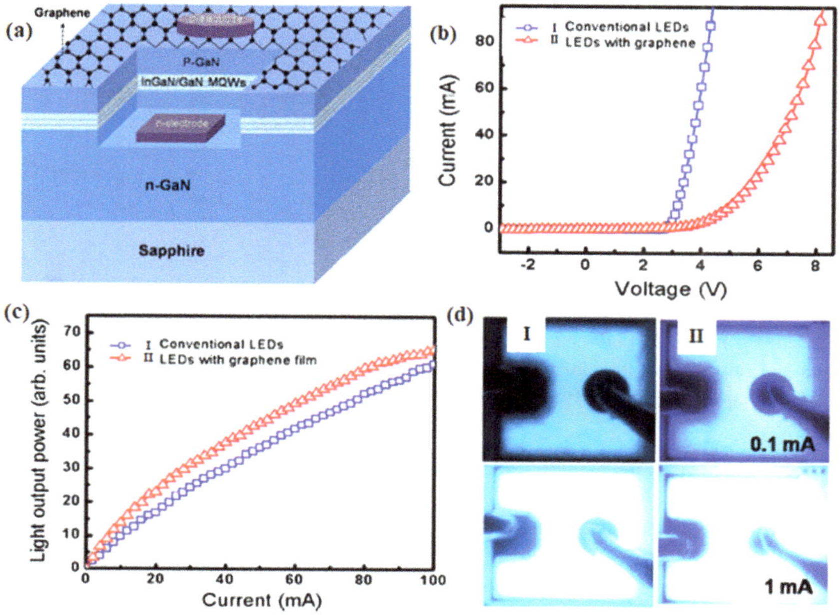

Figure 12.11. (a) Illustration of GaN-based LEDs with graphene TCE. (b) Current–voltage (*I–V*) and (c) LOP—injection current characteristics of LEDs with conventional ITO and graphene TCEs. (d) Photographs of EL from LEDs with ITO (I) and graphene TCEs (II) at different injection currents (0.1, 1 mA). Reproduced with permission from [69] The Japan Society of Applied Physics. Copyright IOP Publishing Ltd. All rights reserved.

silver nanowires (AgNWs) showed two-fold higher LOP and lower V_F (~6.5 V at 20 mA) than the LED with pristine graphene TCE (~12V) [73]. The improved device performance is attributed to the barrier height reduction between hybrid TCE and p-GaN as well as ~50% reduction in sheet resistance of hybrid TCE (~365 Ω/□) in comparison to pristine graphene TCE (~800 Ω/□). In graphene/AgNWs hybrid nanostructures, AgNWs bridge the gaps between the graphene defects and this improves the electrical conductivity. In another case, Kang *et al* investigated the performance of hybrid TCE of graphene and CNT in GaN LED by fabricating the hybrid TCE using CVD, wherein the grown graphene was fused across CNT networks [64]. A four-fold enhancement in LOP and lower V_F (~5 V) was obtained for LED with graphene/CNT hybrid TCE in comparison to the LED with pristine SLG (see figure 12.12). The improved device performance is attributed to the enhanced current spreading (lateral) across the graphene/CNT hybrid TCE (R_S~500 Ω/□, T~90%). Recently, graphene/copper (Cu) NWs TCE was investigated for achieving high device performance of GaN-based LEDs [74]. The hybrid TCE fabricated using low-temperature CVD showed high optoelectrical properties (R_S~37 Ω/□, T~89%). In addition, a remarkable oxidation barrier was observed in hybrid TCE wherein, the multilayer graphene sheets work as the passivation layers to prevent Cu NWs from the possible oxidation. 2.6 times higher EL intensity

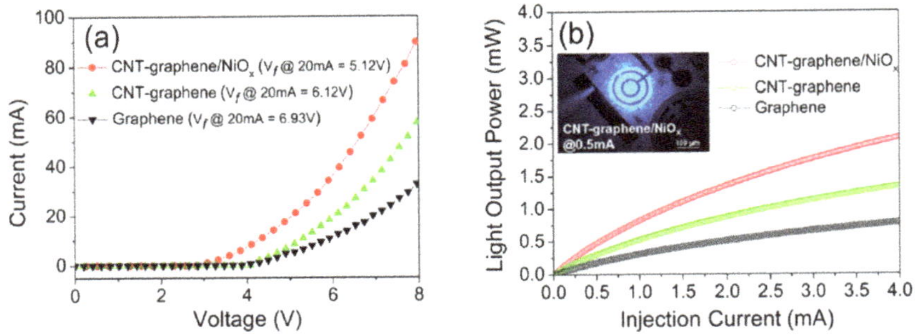

Figure 12.12. (a) *I–V* and (b) LOP characteristics of LEDs with CNT/graphene hybrid and graphene TCEs. Inset shows the optical microscopic image of LED with CNT-graphene hybrid TCE. Reproduced with permission [64], Copyright 2016, AIP Publishing.

was achieved by the blue LED (V_F~4.6 V) with graphene/Cu NWs hybrid TCE compared to the LED (V_F~4.5) with conventional ITO TCE.

In OLEDs developments (table 12.2), Wu *et al* were one of the early groups demonstrating that graphene-based TCEs can be effectively used for OLEDs [54]. The graphene-based TCE was developed by the reduction of GO using the solution-process technique (Hummers method) followed by spin coating the material on the transparent substrate. The basic OLED structure with ~7 nm thick solution-processed graphene TCE (R_S~800 Ω/□, transparency ~82%) realized ~300 cd m^{-2} luminance (at 11.7 V) and turn-on voltage (V_{on}) ~4.5 V, power efficiency ~0.5 lm W^{-1}, and external quantum efficiency (EQE)~0.2. The performance of the OLED with graphene TCE (solution-processed) was observed to be comparable to that of commercial ITO-OLED (luminance ~300 cd m^{-2} at 9.9 V, V_{on}~3.8 V, power efficiency ~0.3 lm W^{-1}, EQE~0.3), possibly due to the optoelectrical properties differences in both TCEs as illustrated in figures 12.13(a) and (b).

After the successful demonstration of solution-processed graphene TCEs for OLED, different graphene-based TCEs were also explored [12, 56, 75]. Sun *et al* fabricated OLEDs with CVD-grown MLG (R_S~310 Ω/□, T~85%) as anode TCE without any polymer interface layer [75]. MLG TCE-based OLED device showed the V_{on}~4 V, maximum luminous current efficiency ~0.75 cd A^{-1}, and maximum power efficiency ~0.38 lm W^{-1}, which were found to be comparable with conventional ITO-OLEDs. High leakage current was noticed from this OLED without polymer interface due to (i) the surface roughness, and (ii) mismatching in the work functions of anode TCE and active polymeric layer. Further, some other groups also worked on the performance-improvement of OLED devices with graphene-based TCEs. For example, Han *et al* used doped graphene TCE as anode and devised the gradient hole injection (GraHIL) layer in OLED for reducing the work functions mismatch between graphene-based anode (work function ~4.4 eV) and hole transport layer (NPB, work function ~5.4 eV). The GraHIL was prepared by mixing the PEDOT:PSS and perfluorinated ionomer (PFI) such as tetrafluoroethylene-perfluoro-3,6-dioxa-4-methyl-7-octenesulphonic acid copolymers [56]. The advantage of using the GraHIL in OLED is that it provides a work function gradient

Table 12.2. Graphene-based transparent conducting electrodes for OLEDs.

Graphene-based TCE					OLED with graphene-based TCE				Reference
TCE type	TCE nanostructure	Sheet resistance (Ω/□)	Transmittance (%)	FoM	OLED structure (TCE structure/thickness, device layers/cathode)	Maximum luminance (cd m^{-2})	Turn-on voltage (V)	Maximum luminous efficiency	
Solution-processes graphene	7 nm thick reduced graphene oxide	800	82	2.26	Graphene TCE/40 nm, PEDOT:PSS/50 nm, NPD/50 nm Alq_3/0.5 nm LiF/100 nm Al	300	4.5	0.5 lm W^{-1}	[54]
	rGO (chemically, thermally reduced)	3200	80	0.5	20 nm rGO/20 nm CuPc/40 nm β-NPB/ 50 nm Alq_3/120 nm Al	—	3.5.0V	—	[55]
CVD graphene	MLG (~20 layers, total thickness 6–7 nm)	310	85	7.2	60 nm Al/glass/MLG/5 nm V_2O_5/40 nm NPB/40 nm CBP:Ir (ppy)$_2$(acac) (10 wt %)/25 nm Bphen/15 nm Bphen: Cs_2CO_3 (mass rate 1:1)/15 nm Sm/15 nm Au	1000	4	0.75 cd A^{-1} 0.38 lm W^{-1}	[75]
	MLG (5–8 layers, total thickness~2–3 nm)	289	~90	12.1	2–3 nm graphene/30 nm TAPC/10 nm HAT-CN/30 nm TAPC/10 nm HAT-CN/30 nm TAPC/5 nm TCTA: FIrpic/5 nm DCzPPy: FIrpic/40 nm BmPyPB/1 nm LiF/ 100 nm Al	1000	3.9	24.1 lm W^{-1}	[77]

(*Continued*)

Table 12.2. (*Continued*)

Graphene-based TCE					OLED with graphene-based TCE				Reference
TCE type	TCE nanostructure	Sheet resistance (Ω/□)	Transmittance (%)	FoM	OLED structure (TCE structure/thickness, device layers/cathode)	Maximum luminance (cd m^{-2})	Turn-on voltage (V)	Maximum luminous efficiency	
Doped and layered graphene	$AuCl_3$ and HNO_3-doped MLG (4 layers)	30 ($AuCl_3$-doped 4layer graphene) 54 (HNO_3-doped 4layer graphene)	~87	87.1 48.4	Doped graphene/50 nm GraHIL/20 nm NPB/30 nm $Bebq_2$:C545T/20 nm Beq_2/1 nm LiF/130 nm Al	~8000	~2.5	30.2 cd A^{-1}, 37.2 lm W^{-1} (HNO_3) 27.4 cd A^{-1}, 28.1 lm W^{-1} ($AuCl_3$)	[56]
	Triethyloxonium hexachloro antimonate (OA)-doped SLG	200	95	36.3	OA-doped SLG/MoO_3: PEDOT/CBP:MoO_3/CBP/CBP:Ir$(ppy)_2$(acac)/TPBi/LiF/Al for green OLEDOA-doped SLG/MoO_3: PEDOT/CBP:MoO_3/CBP/CBP:Ir$(ppy)_2$(acac): Ir$(MDQ)_2$(acac)/CBP: Fripic/TPBi/LiF/Al for white OLED	20 000 (green OLED) 1000 (white OLED)	2.6 V	240 cd A^{-1} (green OLED) 80 cd/A (white OLED)	[12]

	$AuCl_3$-doped trilayer graphene (layer-by-layer assembly of $AuCl_3$-doped CVD-grown SLG)	150	91	26.1	$AuCl_3$-doped trilayer graphene/60 nm NPB/60 nm Alq_3/ 1 nm LiF/80 nm Al	150	4.8	0.263	[57]
Graphene-based hybrid	Graphene/metal thin layer (Ag~11 nm)/ doped semiconductor oxide (AZO~50 nm)	~10	~80	159.7	Hybrid TCE/30 nm α-NPB/105 nm Alq_3/0.5 nm LiF/ 200 nm Al	1150	6	1.43 cd A^{-1}	[58]
	CVD-grown SLG/ PEDOT:PSS/ AgNWs	30	88	95.2	hybrid TCE/60 nm NPB/60 nm Alq_3/ 1 nm LiF/100 nm Al	~5000	4.4	2 cd A^{-1}	[62]
	CVD-grown SLG-CNT hybrid TCE partially decorated with gold nanoparticles	100	96.5	104.9	Graphene-CNT-AZO-AuNPs/80 nm, PEDOT:PSS/80 nm SYPPV/0.8 nm, LiF/ 100 nm, Al	650	~5	2.1 cd/A	[27]
	CVD-grown MLG (five layers)/DMSO-doped highly conducting polymer PEDOT:PSS (35 nm)	80	83	24.1	Hybrid TCE/10 nm, HAT-CN/65 nm, TAPC/15 nm, Ir $(ppy)_2$(acac):$Bepp_2$/50 nm, $Bepp_2$/1 nm, LiF/ 120 nm, Al	19 000	~3.5	76 cd A^{-1}61 lm W^{-1}	[76]
	CVD-grown SLG/WO_3 (5 nm)	300	90	11.6	Hybrid TCE/135 nm, CBP:MoO_3/5 nm, CBP/15 nm, CBP:Ir (ppy) (acac)/60 nm, TPBi/1 nm, LiF/100 nm, Al	2000	2.8	60 lm W^{-1}	[59]

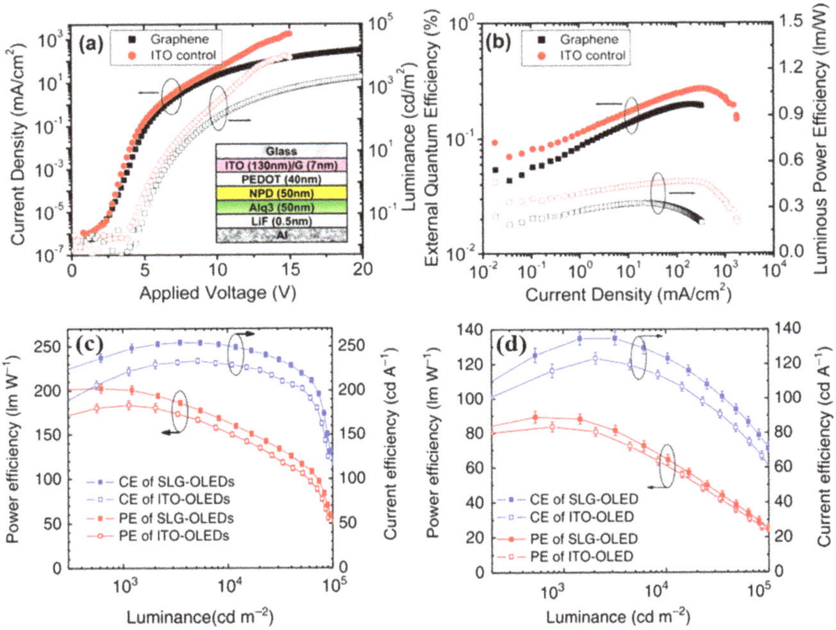

Figure 12.13. (a) Current density–applied bias–luminance (*J*–*V*–*L*) characteristics for an OLED with graphene and ITO TCEs. Inset shows the fabricated OLED structure. (b) External quantum and luminous power efficiencies. Reproduced with permission [54], Copyright 2010, American Chemical Society. (c) and (d) Power efficiency and current efficiency of green and white OLEDs with OA-doped SLG and ITO, respectively. Reproduced from [12], Copyright 2013, with permission from Springer Nature.

throughout the hole injection layer, that increases the efficient hole injections. The performance of the fabricated OLED (advanced structure) with MLG (four layers, T~87%) doped with $AuCl_3$ (R_S~30 Ω/□) or HNO_3 (R_S~54 Ω/□) and GraHIL was observed to be superior (fluorescent OLED—37.2 lm W^{-1}, phosphorescent OLED —102.7 lm W^{-1}) to that of ITO-OLEDs (fluorescent OLED—24.1 lm W^{-1}, phosphorescent OLED—85.6 lm W^{-1}). Similarly, Li *et al* demonstrated the white and green OLEDs (advanced structures) with triethyloxonium hexachloroantimonate (OA)-doped SLG TCEs (R_S~200 Ω/□, T~95%) [12]. The green (current efficiency >240 cd A^{-1}) and white (current efficiency >80 cd/A) OLEDs with doped SLG TCEs outperformed the OLEDs with conventional ITO TCEs as shown in figures 12.13(c) and (d). The improved performance of these OLEDs is credited to high hole injection from doped graphene-based TCE to the emission layer and reduction in the efficiency roll-off.

Moreover, several graphene-based hybrids as TCEs have been discovered to achieve high OLED performance. The OLED with hybrid TCE of graphene/metal thin layer/doped semiconductor oxide was demonstrated to realize a stable light emission with maximum current efficiency of ~1.5 cd A^{-1}, maximum luminance (L_{max}) of ~1150 cd m^{-2}, and V_{on}~6 V [58]. In the graphene/Ag/AZO hybrid TCE, the 11 nm thin layer of Ag enhances the electrical conductivity of graphene, and the

50 nm thick AZO layer provides the smoothening of the hybrid TCE surface. In another example, Xu *et al* developed the graphene/PEDOT:PSS/AgNWs hybrid TCE for OLED (basic structure) that achieved maximum luminous efficiency of ~2 cd A^{-1} with V_{on}~4.5 V and L_{max}~5000 cd m^{-2} [62]. The graphene/PEDOT:PSS/AgNWs hybrid TCE-based OLED demonstrated higher device performance than pristine graphene-based OLED. The high device performance of hybrid TCE-based OLED is attributed to the reduced sheet resistance of the hybrid TCE and the additional reduction in surface roughness offered by the PEDOT:PSS layer.

Later, Kumar *et al* devised the hybrid TCE of SLG and CNTs to achieve high device performance of graphene-OLEDs [27]. The graphene/CNT hybrid TCE partially decorated with AuNPs showed high electrical conductivity (~100 Ω/□) while maintaining its high transparency (96.5%). After introducing the AZO buffer layer with graphene/CNT hybrid TCE, the fabricated OLED (basic structure) demonstrated current efficiency ~2.1 cd A^{-1}, V_{on}~5 V, L_{max}~650 cd m^{-2}, as presented in figures 12.14(a) and (b). The combination of fusion of graphene with CNTs and AuNP decoration accumulatively reduces the sheet resistance and improves the work function of graphene-based TCEs. This leads to attainment of higher device performance of hybrid TCE-OLED compared to pristine graphene-OLED. The thin AZO layer in hybrid TCE provides high surface smoothening and

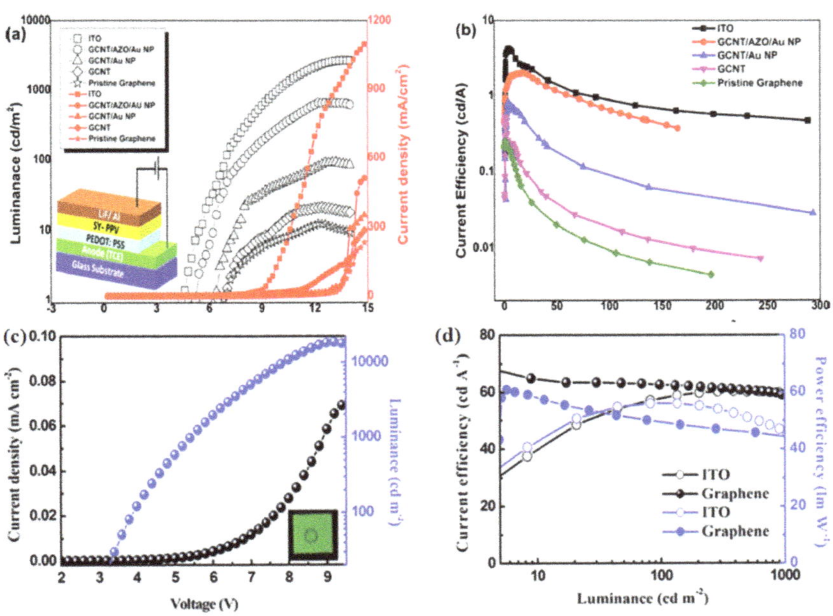

Figure 12.14. (a) *J–V–L* and (b) current efficiency–current density characteristics of graphene-CNT hybrid TCEs-based OLEDs (inset: device structure). Reproduced from [27], Copyright 2019, with permission from Elsevier. (c) *J–V–L* characteristics and (d) efficiencies as a function of luminance for OLEDs with hybrid TCEs of graphene (five layers)/DMSO-doped highly conductive polymers (inset: illuminated OLED device with hybrid TCE at low brightness). Reproduced with permission [76], Copyright 2021, The Royal Society of Chemistry.

high wettability to CVD-grown graphene hybrid TCE. Recently, the hybrid structure of multilayer graphene (five layers) and doped conducting polymer (DMSO-doped PEDOT:PSS) has been investigated as TCE for OLED [76]. The OLED with CVD-grown graphene with highly conductive polymer (R_S~80 Ω/□, T~83%) demonstrated high device performance (L_{max}~19 000 cd m^{-2}, V_{on}~3.5 V, current efficiency ~76 cd A^{-1}, power efficiency ~61 lm W^{-1}) as illustrated in figures 12.14(c) and (d). The device performance of OLED with graphene (five layers)/35 nm DMSO-doped PEDOT:PSS hybrid TCE outperformed the ITO-based OLED. The higher device performance of hybrid TCE-based OLED is possibly due to the high hole injection that is caused by the high work function and low sheet resistance of hybrid TCE.

1,1-bis (di-4-tolylamino)phenyl (cyclohexane) (TAPC); 1,4,5,8,9,11-hexaazatriphenylene hexacarbonitrile (HAT-CN); 4′-tri(N-carbazolyl) triphenylamine:iridium (III) bis(4,6-difluorophenyl)-pyridinato-(N,C2)(picolinate) (TCTA:FIrpic); 2,6-bis [3′-(*N*-carbazole)phenyl]pyridine:iridium(III) bis(4,6-difluorophenyl)-pyridinato-(N,C2) (picolinate) (DCzPPy: FIrpic); 1,3-bis(3,5-dipyrid-3-yl-phenyl)benzene (BmPyPB); lithium fluoride (LiF); aluminum (Al); Copper(II) phthalocyanine (CuPc), *N,N′*-bis(naph- thalen-2-yl)-*N,N*0-bis(phenyl)-benzidine (b-NPB); tris-(8-hydroxyquinoline) aluminum (Alq_3).

12.8 Summary and prospects

In this chapter, several graphene-based TCEs for LED/OLED devices are discussed. For a viable TCE, among the important design parameters that need to be optimized are surface conductivity and transparency. These parameters are captured inside a FoM to ease the evaluation of high performing TCEs. Based on the FoM that we have extracted, high quality graphene-based TCEs can be produced using CVD technique. Furthermore, the FOM can be improved by hybridizing the CVD graphene with nanomaterials such as metal thin films, oxide semiconductors, conducting polymers, and other carbon nanomaterials. By interchanging the nanolayer structure, the FOM of graphene-based TCE can reach nearly the FoM of a commercially available ITO-based TCE. It is clear that the performance of LEDs/OLEDs with graphene-based TCEs has approached that of conventional ITO-based TCE. However, there are some practical points that need to be considered before graphene-based TCEs are accepted for commercial use. Most important of these are the techniques to consistently mass produce high-quality large-area graphene-based TCEs at low cost. These need to be addressed in order for the use of graphene-based TCE to take off in the near future.

References

[1] Chen Y, Zhang N, Li Y F, Bi Y G, Yue Y Y, Feng J and Sun H B 2018 Microscale-patterned graphene electrodes for organic light-emitting devices by a simple patterning strategy *Adv. Opt. Mater.* **6** 1701348

[2] Park I J, Kim T I, Yoon T, Kang S, Cho H, Cho N S, Lee J I, Kim T S and Choi S Y 2018 Flexible and transparent graphene electrode architecture with selective defect decoration for organic light-emitting diodes *Adv. Funct. Mater.* **28** 1704435

[3] Song N, Jia J, Wang W, Gao Y, Zhao Y and Chen Y 2016 Green production of pristine graphene using fluid dynamic force in supercritical CO_2 *Chem. Eng. J.* **298** 198–205

[4] Wu T L, Yeh C H, Hsiao W T, Huang P Y, Huang M J, Chiang Y H, Cheng C H, Liu R S and Chiu P W 2017 High-performance organic light-emitting diode with substitutionally boron-doped graphene anode *ACS Appl. Mater. Interfaces* **9** 14998–5004

[5] Lee H, Cho H, Byun C W, Han J H, Kwon B H, Choi S, Lee J and Cho N S 2018 Color-tunable organic light-emitting diodes with vertically stacked blue, green, and red colors for lighting and display applications *Opt. Express* **26** 18351–61

[6] Ràfols-Ribé J, Will P A, Hänisch C, Gonzalez-Silveira M, Lenk S, Rodríguez-Viejo J and Reineke S 2018 High-performance organic light-emitting diodes comprising ultrastable glass layers *Sci. Adv.* **4** eaar8332

[7] Cui Y, Jiang Y, Zhou L, Liu W, Zhu Q and Zhang H 2018 High performance organic light-emitting diodes with tunable color range from orange to warm white based on single thermally activated delayed fluorescence emitter *Opt. Mater.* **85** 349–55

[8] QY R 2019 Global LED & OLED Displays and Lighting Products Market Insights, Forecast to 2025. USA *Market Research* Available at: https://marketresearch.com/QYResearch-Group-v3531/Global-LED-OLED-Displays-Lighting-12393802/

[9] Guo H *et al* 2013 Copper nanowires as fully transparent conductive electrodes *Sci. Rep.* **3** 1–8

[10] Lee J Y, Connor S T, Cui Y and Peumans P 2008 Solution-processed metal nanowire mesh transparent electrodes *Nano Lett.* **8** 689–92

[11] Saxena K, Jain V K and Mehta D S 2009 A review on the light extraction techniques in organic electroluminescent devices *Opt. Mater.* **32** 221–33

[12] Li N, Oida S, Tulevski G S, Han S J, Hannon J B, Sadana D K and Chen T C 2013 Efficient and bright organic light-emitting diodes on single-layer graphene electrodes *Nat. Commun.* **4** 1–7

[13] Liu Y F, Feng J, Zhang Y F, Cui H F, Yin D, Bi Y G, Song J F, Chen Q D and Sun H B 2015 Improved efficiency of indium-tin-oxide-free organic light-emitting devices using PEDOT: PSS/graphene oxide composite anode *Org. Electron.* **26** 81–5

[14] Wang L, Liu W, Zhang Y, Zhang Z H, Tan S T, Yi X, Wang G, Sun X, Zhu H and Demir H V 2015 Graphene-based transparent conductive electrodes for GaN-based light emitting diodes: challenges and countermeasures *Nano Energy* **12** 419–36

[15] Bansal M, Srivastava R, Lal C, Kamalasanan M N and Tanwar L S 2009 Carbon nanotube-based organic light emitting diodes *Nanoscale* **1** 317–30

[16] Song T B and Li N 2014 Emerging transparent conducting electrodes for organic light emitting diodes *Electronics* **3** 190–204

[17] Ahn Y, Jeong Y, Lee D and Lee Y 2015 Copper nanowire–graphene core–shell nanostructure for highly stable transparent conducting electrodes *ACS Nano* **9** 3125–33

[18] Jeong G J, Lee J H, Han S H, Jin W Y, Kang J W and Lee S N 2015 Silver nanowires for transparent conductive electrode to GaN-based light-emitting diodes *Appl. Phys. Lett.* **106** 031118

[19] Lee J *et al* 2016 Synergetic electrode architecture for efficient graphene-based flexible organic light-emitting diodes *Nat. Commun.* **7** 1–9

[20] Sim H, Kim C, Bok S, Kim M K, Oh H, Lim G H, Cho S M and Lim B 2018 Five-minute synthesis of silver nanowires and their roll-to-roll processing for large-area organic light emitting diodes *Nanoscale* **10** 12087–92

[21] Balandin A A, Ghosh S, Bao W, Calizo I, Teweldebrhan D, Miao F and Lau C N 2008 Superior thermal conductivity of single-layer graphene *Nano Lett.* **8** 902–7

[22] Chen J H, Jang C, Xiao S, Ishigami M and Fuhrer M S 2008 Intrinsic and extrinsic performance limits of graphene devices on SiO_2 *Nat. Nanotechnol.* **3** 206–9

[23] Ellmer K 2012 Past achievements and future challenges in the development of optically transparent electrodes *Nat. Photon.* **6** 809–17

[24] Jin Y, Hu B, Wei Z, Luo Z, Wei D, Xi Y, Zhang Y and Liu Y 2014 Roles of H_2 in annealing and growth times of graphene CVD synthesis over copper foil *J. Mater. Chem.* A **2** 16208–16

[25] Cho H, Shin J W, Cho N S, Moon J, Han J H, Kwon Y D, Cho S and Lee J I 2015 Optical effects of graphene electrodes on organic light-emitting diodes *IEEE J. Sel. Top. Quantum Electron.* **22** 48–53

[26] Pankove J I 1971 *Optical Processes in Semiconductors* (New York: Dover) p 320

[27] Kumar P, Woon K L, Wong W S, Saheed M S M and Burhanudin Z A 2019 Hybrid film of single-layer graphene and carbon nanotube as transparent conductive electrode for organic light emitting diode *Synth. Met.* **257** 116186

[28] Hwang Y, Choi J, Kim J W, Lee J W, Kim J G, Hwang H, Choi K W, Lee W and Ju B K 2020 Ag-fiber/graphene hybrid electrodes for highly flexible and transparent optoelectronic devices *Sci. Rep.* **10** 1–8

[29] Haacke G 1976 New figure of merit for transparent conductors *J. Appl. Phys.* **47** 4086–9

[30] De S and Coleman J N 2010 Are there fundamental limitations on the sheet resistance and transmittance of thin graphene films? *ACS Nano* **4** 2713–20

[31] Hojati-Talemi P, Bächler C, Fabretto M, Murphy P and Evans D 2013 Ultrathin polymer films for transparent electrode applications prepared by controlled nucleation *ACS Appl. Mater. Interfaces* **5** 11654–60

[32] Cao Y, Parker I D, Yu G, Zhang C and Heeger A J 1999 Improved quantum efficiency for electroluminescence in semiconducting polymers *Nature* **397** 414–7

[33] Shi Q *et al* 2020 Substrate developments for the chemical vapor deposition synthesis of graphene *Adv. Mater. Interfaces* **7** 1902024

[34] Lisi N, Dikonimos T, Buonocore F, Pittori M, Mazzaro R, Rizzoli R, Marras S and Capasso A 2017 Contamination-free graphene by chemical vapor deposition in quartz furnaces *Sci. Rep.* **7** 1–11

[35] Kumar A, Kumar A and Lee C H 2013 Synthesis and biomedical applications of graphene: present and future trends *Advances in Graphene Science* (Rijeka: In Tech Open), pp 5772–8

[36] Li X *et al* 2009 Large-area synthesis of high-quality and uniform graphene films on copper foils *Science* **324** 1312–4

[37] Ferrari A C *et al* 2015 Science and technology roadmap for graphene, related two-dimensional crystals, and hybrid systems *Nanoscale* **7** 4598–810

[38] Bonaccorso F, Sun Z, Hasan T and Ferrari A C 2010 Graphene Photonics and Optoelectronics *Nat. Photon.* **4** 611–22

[39] Kim H, Song I, Park C, Son M, Hong M, Kim Y, Kim J S, Shin H J, Baik J and Choi H C 2013 Copper-vapor-assisted chemical vapor deposition for high-quality and metal-free single-layer graphene on amorphous SiO_2 substrate *ACS Nano* **7** 6575–82

[40] Warner J H *et al* 2013 Methods for obtaining graphene *Graphene: Fundamentals and Emergent Applications* (New York: Elsevier) pp 129–228

[41] Chen Y, Gong X L and Gai J G 2016 Progress and challenges in transfer of large-area graphene films *Adv. Sci.* **3** 1500343

[42] Yang X, Zhang G, Prakash J, Chen Z, Gauthier M and Sun S 2019 Chemical vapour deposition of graphene: layer control, the transfer process, characterisation, and related applications *Int. Rev. Phys. Chem.* **38** 149–99

[43] Kholmanov I N, Domingues S H, Chou H, Wang X, Tan C, Kim J Y, Li H, Piner R, Zarbin A J and Ruoff R S 2013 Reduced graphene oxide/copper nanowire hybrid films as high-performance transparent electrodes *ACS Nano* **7** 1811–6

[44] Yang J, Ma M, Li L, Zhang Y, Huang W and Dong X 2014 Graphene nanomesh: new versatile materials *Nanoscale* **6** 13301–13

[45] Wu X, Lian L, Yang S and He G 2016 Highly conductive PEDOT: PSS and graphene oxide hybrid film from a dipping treatment with hydroiodic acid for organic light emitting diodes *J. Mater. Chem.* C **4** 8528–34

[46] Muszynski R, Seger B and Kamat P V 2008 Decorating graphene sheets with gold nanoparticles *J. Phys. Chem.* C **112** 5263–6

[47] Moon I K, Lee J, Ruoff R S and Lee H 2010 Reduced graphene oxide by chemical graphitization *Nat. Commun.* **1** 1–6

[48] Zhou X, Zhang J, Wu H, Yang H, Zhang J and Guo S 2011 Reducing graphene oxide via hydroxylamine: a simple and efficient route to graphene *J. Phys. Chem.* C **115** 11957–61

[49] Sharma K, Maiti K, Kim N H, Hui D and Lee J H 2018 Green synthesis of glucose-reduced graphene oxide supported Ag–Cu_2O nanocomposites for the enhanced visible-light photocatalytic activity *Composites* B **138** 35–44

[50] Zhi M, Liu Q, Chen H, Feng S and He Y 2019 Study on efficient and green reduction of graphene oxide by a one-step hydrothermal method *J. Phys.: Conf. Ser.* **1213** 052047

[51] Gupta M, Hawari H F, Kumar P, Burhanudin Z A and Tansu N 2021 Functionalized reduced graphene oxide thin films for ultrahigh CO_2 gas sensing performance at room temperature *Nanomaterials* **11** 623

[52] Wang Y, Chen Y, Lacey S D, Xu L, Xie H, Li T, Danner V A and Hu L 2018 Reduced graphene oxide film with record-high conductivity and mobility *Mater. Today* **21** 186–92

[53] Junaid M, Md Khir M H, Witjaksono G, Ullah Z, Tansu N, Saheed M S M, Kumar P, Hing Wah L, Magsi S A and Siddiqui M A 2020 A review on graphene-based light emitting functional devices *Molecules* **25** 4217

[54] Wu J, Agrawal M, Becerril H A, Bao Z, Liu Z, Chen Y and Peumans P 2010 Organic light-emitting diodes on solution-processed graphene transparent electrodes *ACS Nano* **4** 43–8

[55] Lima A H, Mendonça J P, Duarte M, Stavale F, Legnani C, De Carvalho G S G, Maciel I O, Sato F, Fragneaud B and Quirino W G 2017 Reduced graphene oxide prepared at low temperature thermal treatment as transparent conductors for organic electronic applications *Org. Electron.* **49** 165–73

[56] Han T H, Lee Y, Choi M R, Woo S H, Bae S H, Hong B H, Ahn J H and Lee T W 2012 Extremely efficient flexible organic light-emitting diodes with modified graphene anode *Nat. Photon.* **6** 105–10

[57] Xu Y, Yu H, Wang C, Cao J, Chen Y, Ma Z, You Y, Wan J, Fang X and Chen X 2017 Multilayer graphene with chemical modification as transparent conducting electrodes in organic light-emitting diode *Nanoscale Res. Lett.* **12** 1–7

[58] Li F, Lin Z, Zhang B, Zhang Y, Wu C and Guo T 2013 Fabrication of flexible conductive graphene/Ag/Al-doped zinc oxide multilayer films for application in flexible organic light-emitting diodes *Org. Electron.* **14** 2139–43
[59] Kuruvila A, Kidambi P R, Kling J, Wagner J B, Robertson J, Hofmann S and Meyer J 2014 Organic light emitting diodes with environmentally and thermally stable doped graphene electrodes *J. Mater. Chem.* C **2** 6940–5
[60] Lee D, Kim D, Lee Y and Jeon D Y 2017 Solution processible MoO_x-incorporated graphene anode for efficient polymer light-emitting diodes *Nanotechnology* **28** 235201
[61] Shin S, Kim J, Kim Y H and Kim S I 2013 Enhanced performance of organic light-emitting diodes by using hybrid anodes composed of graphene and conducting polymer *Curr. Appl. Phys.* **13** S144–7
[62] Xu Y, Wei X, Wang C, Cao J, Chen Y, Ma Z, You Y, Wan J, Fang X and Chen X 2017 Silver nanowires modified with PEDOT: PSS and graphene for organic light-emitting diodes anode *Sci. Rep.* **7** 1–7
[63] Fan W, Zhang L and Liu T 2017 Strategies for the Hybridization of CNTs with graphene *Graphene-Carbon Nanotube Hybrids for Energy and Environmental Applications* (Singapore: Springer), pp 21–51
[64] Kang C H, Shen C, Saheed M S, Mohamed N M, Ng T K, Ooi B S and Burhanudin Z A 2016 Carbon nanotube-graphene composite film as transparent conductive electrode for GaN-based light-emitting diodes *Appl. Phys. Lett.* **109** 081902
[65] Kong H X 2013 Hybrids of carbon nanotubes and graphene/graphene oxide *Curr. Opin. Solid State Mater. Sci.* **17** 31–7
[66] Kumar P, Kang C H, Burhanudin Z A, Saheed M S M, Irshad M I and Mohamed N M 2016 Graphene-based hybrid thin films as transparent conductive electrode for optoelectronic devices *2016 IEEE Int. Conf. on Semiconductor Electronics (ICSE)* (Piscataway, NJ: IEEE) pp 216–9
[67] Zhao M Q, Liu X F, Zhang Q, Tian G L, Huang J Q, Zhu W and Wei F 2012 Graphene/single-walled carbon nanotube hybrids: one-step catalytic growth and applications for high-rate Li–S batteries *ACS Nano* **6** 10759–69
[68] Shim W, Kwon Y, Jeon S Y and Yu W R 2015 Optimally conductive networks in randomly dispersed CNT: graphene hybrids *Sci. Rep.* **5** 1–10
[69] Seo T H, Oh T S, Chae S J, Park A H, Lee K J, Lee Y H and Suh E K 2011 Enhanced light output power of GaN light-emitting diodes with graphene film as a transparent conducting electrode *Jpn. J. Appl. Phys.* **50** 125103
[70] Chandramohan S, Bok Ko K, Han Yang J, Deul Ryu B, Katharria Y S, Yong Kim T, Jin Cho B and Hong C H 2014 Performance evaluation of GaN light-emitting diodes using transferred graphene as current spreading layer *J. Appl. Phys.* **115** 054503
[71] Choe M, Cho C Y, Shim J P, Park W, Lim S K, Hong W K, Hun Lee B, Lee D S, Park S J and Lee T 2012 Au nanoparticle-decorated graphene electrodes for GaN-based optoelectronic devices *Appl. Phys. Lett.* **101** 031115
[72] Youn D H, Yu Y J, Choi H, Kim S H, Choi S Y and Choi C G 2013 Graphene transparent electrode for enhanced optical power and thermal stability in GaN light-emitting diodes *Nanotechnology* **24** 075202
[73] Li Z, Kang J, Liu Z, Du C, Lee X, Li X, Wang L, Yi X, Zhu H and Wang G 2013 Enhanced performance of GaN-based light-emitting diodes with graphene/Ag nanowires hybrid films *AIP Adv.* **3** 042134

[74] Yoon H, Shin D S, Kim T G, Kim D and Park J 2018 Facile synthesis of graphene on Cu nanowires via low-temperature thermal CVD for the transparent conductive electrode *ACS Sustain. Chem. Eng.* **6** 13888–96

[75] Sun T, Wang Z L, Shi Z J, Ran G Z, Xu W J, Wang Z Y, Li Y Z, Dai L and Qin G G 2010 Multilayered graphene used as anode of organic light emitting devices *Appl. Phys. Lett.* **96** 55

[76] Zhang Z, Xia L, Liu L, Chen Y, Wang Z, Wang W, Ma D and Liu Z 2021 Ultra-smooth and robust graphene-based hybrid anode for high-performance flexible organic light-emitting diodes *J. Mater. Chem.* C **9** 2106–14

[77] Hwang J *et al* 2012 Multilayered graphene anode for blue phosphorescent organic light emitting diodes *Appl. Phys. Lett.* **100** 82

Chapter 13

Graphene-based sensors

Anshul Kumar Sharma, Sagar Sardana and Aman Mahajan

The environment we are living in contains various kinds of potentially dangerous pollutants, volatile organic compounds (VOCs) and toxic gases such as ammonia (NH_3), nitrogen dioxide (NO_2), carbon monoxide (CO), nitic oxide (NO) and chlorine (Cl_2) etc, and there have been increasing demands for real-time monitoring of these toxic gases, which can be acquired by designing sensors that are flexible, cost-effective, with ease of use and integration. The fabrication of effective and efficient devices for monitoring harmful and toxic gases present in the environment is a challenge in the sensor industry. As a member of the carbon family, graphene having hexagonal arrangement of carbon atoms making a one-atom-thick planar sheet, along with its outstanding electronic and mechanical properties, has captured significant research interest in the development of different types of graphene sensors for environmental and health monitoring. However, from the viewpoint of practical applications of graphene, economical process is required in the synthesis of high-quality graphene. It is worth noting that the expensive cost of large-scale graphene synthesis has led to the development of reduced graphene oxide (rGO), a two-dimensional carbon-based material with adjustable optical and electrical properties. Furthermore, combining graphene with metal oxides, noble metal nanoparticles (NPs), and organic semiconductors enhances gas sensing properties while also increasing selectivity for a certain type of target analyte. This book chapter focuses on the advancements done in the development and fabrication of different kinds of graphene sensors and their application in detection of different hazardous gases.

13.1 Introduction

Concerns about air pollution, medical diagnostics and national defence have prompted the creation of sensors that can detect chemical species in the atmosphere [1]. When it comes to identification, the human nasal sensory system possesses a

doi:10.1088/978-0-7503-3999-5ch13

remarkable range of flexibility and sensitivity. Humans are capable of recognising and distinguishing hundreds of various odours fast. However, smell detection can be difficult at times due to the lack of uniqueness in the chemical basis of most odours and the toxicity of chemicals. To protect people's safety and health, gas sensors are required to manage and monitor these dangerous gases. Without sensors, significant advancements in industrial and environmental monitoring and control will be unattainable. The ultimate goal is to create sensors that have enough selectivity and sensitivity to detect a wide range of compounds at appropriate concentration levels. A sensor is a device that comprises an active sensing material and a signal transducer that detects events in the physical environment and replies with an electrical, mechanical, or optical signal as an output. The most basic example is a domestic mercury thermometer, which detects temperature and responds with measured liquid expansion. Sensors are utilised in a variety of applications, including touch-sensitive elevator buttons and touch-sensitive lamp dimmer surfaces, as well as countless unseen sensors used in medicine, robotics, aircraft, and other fields. Gas sensors are among the various types of sensors that are used to record the air quality in the environment and to spot harmful pollutants [2]. It's worth noting that a gas sensor is a device that converts gas concentration data into electrical impulses. Sensitivity, selectivity, detection limit, response and recovery are different types of parameters that can be used to define the performance of a gas sensor. Sensor sensitivity is measured in Hz ppm^{-1} or Hz/vol percent, and represents the degree of change in response to a specific analyte concentration [3]. Furthermore, sensor selectivity refers to a sensor's capacity to identify one species from another in the presence of several analytes. The detection limit, also known as the limit of detection (LOD), of gas sensors is the smallest amount of gas that the sensor can detect. The adsorption and desorption speeds of a sensor with regard to the detected analyte are referred to as response and recovery time, respectively. These characteristics differ depending on the type of sensing material employed in the gas sensor and the detecting technique [3]. Further, in sensor technology, sensing materials play an important role in detecting and distinguishing different types of pollutants in environmental processes. A variety of materials have been used in the creation of gas sensors. As sensing materials, a wide range of materials such as semiconductor metal oxide [2], carbon nanotubes (CNTs) [4] and graphene [5] have recently been investigated. Due to their high sensitivity and quick reaction time, metal oxide semiconductor (MOS) gas sensors are the most widely used gas sensors in the world [2]. The MOS gas sensors' detecting process is ascribed to changes in electric charge carriers produced by oxidation or reduction reactions at the metal oxide's surface. They do, however, have the drawbacks of a short life, low selectivity, and a high working temperature. Sensitivity isn't enough when it comes to precise measurement. The specific surface area, which is the total surface area of a material per unit of mass, is a crucial indication of material gas sensitivity. Nanomaterials' high particular surface area promotes gas molecule adsorption, increasing the sensitivity of gas detection. Among carbon-based materials, graphene due to its two-dimensional (2D) structure and extraordinarily high surface area has shown excellent detecting capability in gas sensors. The adsorption and desorption of gaseous molecules on the graphene surface, results in changes in graphene's electrical conductivity and has a significant impact on its electrical properties (depending upon the oxidizing and reducing nature of gas molecule) [6]. However, due to its high sensitivity to many gaseous molecules, graphene has been used as a gas sensor to detect hydrogen, carbon monoxide, ammonia, ethanol and oxygen. However, as discussed earlier, the electrical properties of graphene are highly influenced by the molecules that are adsorbed on its surface, and it has very low signal disturbance compared to typical gas sensors due to its unique structure. Graphene has a high specific surface area and strong electron mobility, making it a suitable gas sensing material. As a p-type semiconductor, graphene has numerous holes and exhibits the pull electron effect in a gaseous environment. Gas molecules will undergo mild hybridization and coupling with the electron on the surface after being adsorbed by graphene, causing the fermi level to go up and down in tiny increments. The condition of electron or hole doping alters the fermi level, resulting in graphene conductivity variations. As a result, graphene is especially good at detecting adsorbed small molecule gases. The relative location of the electron energy level orbit of the system determines the donor and acceptor. The gas molecules function as a donor for electrons if the valence band of the adsorbed gas is higher than the fermi surface of graphene; on the other hand, if the valence band is lower than the fermi surface of graphene, the gas molecules work as a sink for electrons. As a result, graphene is an excellent choice for gas sensing, particularly for identifying single molecules. However, because pristine graphene has a small number of active groups on its surface, chemisorption of gas molecules onto the surface are limited, and recovery is sluggish. Another strategy for improving recovery speed and selective sensing toward a target gas is graphene surface modification [5, 7]. Graphene, as an active sensing material or as the channel material in a field-effect transistor, is a potential material for the development of high-sensitivity sensing devices. In this chapter, a comprehensive review on the most essential aspects of graphene like sensor fabrication process, sensing mechanism and its performance for various types of gas sensors, such as strain sensors, gas sensors, electrochemical sensors, and optical sensors, are highlighted (figure 13.1).

13.2 Graphene-based chemiresistive sensor

Chemical sensors are devices that detect the presence of analytes, which are specific chemical compounds. The chemical sensor's recognition element detects the presence of certain chemical compounds, which are turned into a quantifiable signal by a transducer, with the resulting signal's value directly proportional to the analyte's concentration. Chemical sensors are divided into chemiresistive and electrochemical sensors, which are developed to promote widespread use of chemicals in a variety of industries, including food, biomedical, pharmaceuticals, and industrial safety, as well as environmental protection and security. For practical applications, an ideal gas sensor must have high responsiveness, excellent selectivity, quick response/recovery, high stability/repeatability, room-working temperature, cheap cost, and ease of fabrication. Chemiresistors have the advantage of being able

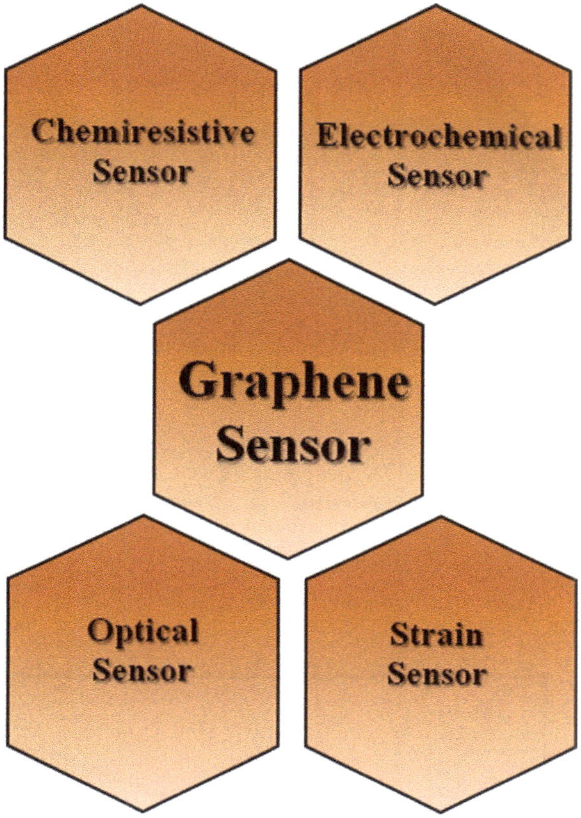

Figure 13.1. Presentation of different applications of graphene sensors.

to detect a wide range of analytes due to the sensor's adjustable layer characteristics. A schematic depiction of the chemiresistive gas sensor and typical response curve showing the variation of resistance of sensor as a function of time are shown in figures 13.2 and 13.3, respectively

Furthermore, chemiresistive sensors are easier to manufacture, use less power, and have a longer operational life than electrochemical detectors [8, 9]. They show promise as candidates for ultra-low molecular concentration detection in a non-invasive, continuous healthcare monitoring strategy, with the potential to detect and indicate significant physiological states by recognising target gaseous analytes [10]. Nevertheless, a wide range of materials such as carbon-based materials like CNTs and graphene, conducting polymers, inorganic and organic semiconductors have extensively been utilized for gas sensor fabrication. Metal oxide (MOx) sensitive films, such as tin or titanium oxide, are often used in chemiresistors, and they require a high temperature for gas detection [11]. Furthermore, operating at high temperatures might cause structural reorganization, affecting the repeatability of such sensors. In addition to the high power consumption, adding a heater and

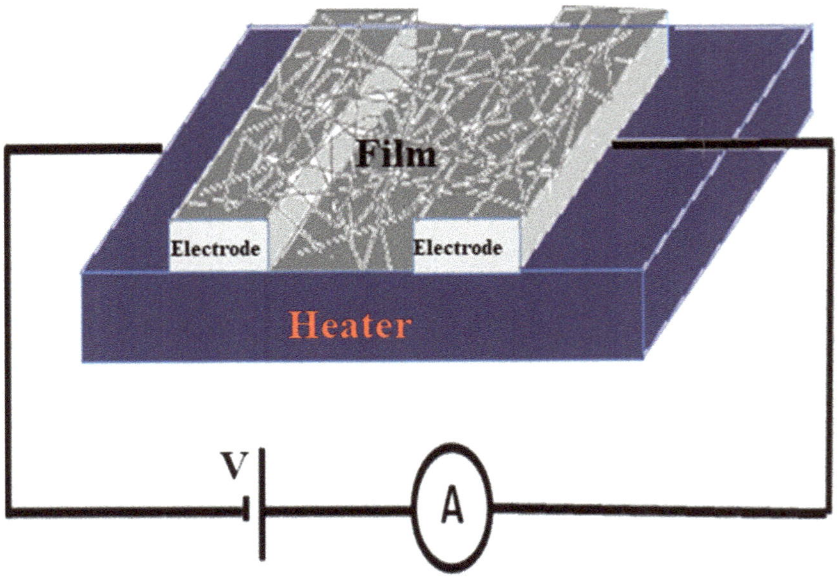

Figure 13.2. Schematic of a chemiresistive gas sensor.

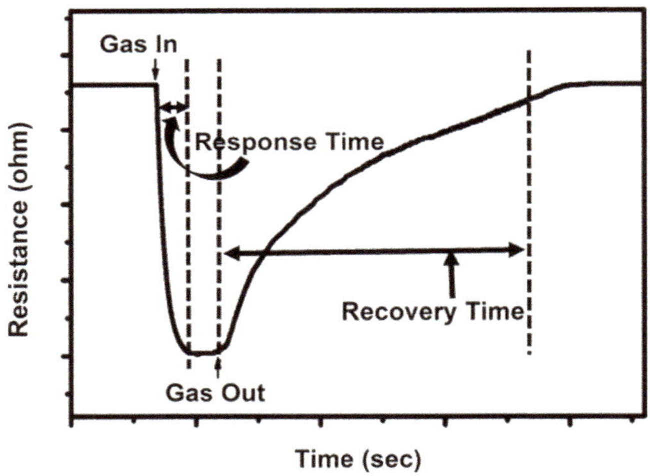

Figure 13.3. Schematic response curve of a chemiresistive gas sensor.

temperature sensor to the device raises the net cost of the gadget. In addition, when subjected to high temperatures, some test gases can burst into flames, posing a fire hazard. These high temperature gas sensors are not recommended for use in the home or office due to the considerable risk connected with them. As a result, operating temperature has a significant impact on sensing qualities and market demand. To counter these drawbacks, room temperature (RT) sensors have been created, which are significantly safer, more cost-effective, use less power, and can reproduce for longer periods of time. Other carbonaceous materials, such as CNTs, fullerenes, and graphene, have recently sparked interest in gas sensing technology due to their superior structural, optical, and electrical capabilities, as well as their low operating temperatures. Graphene for gas sensors has recently attracted a lot of attention, owing to its exceptional electrical conductivity, great thermal and chemical durability, and high mechanical strength. It's worth noting that the graphene gas sensor's mechanism is based on a change in conductance induced by gas molecules adsorption on the graphene surface, which alters the carrier concentration/density of the graphene surface. Pointedly, oxidizing gas molecules remove electrons from the graphene surface and hence increase the majority carrier concentration on its surface that results in drop in resistance due to lower recombination probability, whereas electron–hole recombination increases in the presence of electron-donating gas/vapor molecules, lowering effective hole concentration on the graphene surface and increasing resistance value. A lot of work has recently resulted in a significant advancement in the development of graphene gas sensors for ammonia detection. Gautam and his colleagues performed ammonia gas sensing characteristics of graphene produced by chemical vapor deposition (CVD), and have found that sensitivity and recovery were enhanced by deposition of gold (Au) NPs on the surface [12]. Furthermore, hydrogen (H_2) has sparked a lot of interest as one of the most significant industrial chemicals and a possible source of clean energy in the future. Johnson and colleagues have reported a new Pd-functionalized multilayered graphene nanoribbon network having exceptional sensitivity to H_2 at ppm levels. However, a rapid response and recovery time at room temperature were attributed to noble metal alteration and porous material structure. Khurshid *et al* [14] under different humidity conditions have studied NH_3 gas adsorption properties on graphene oxide-based sensor with sensitivity device about 45%–58% as the NH_3 concentration is changed from 100 to 400 ppm. Furthermore, noble metal nanostructures hybridized with graphene exhibit high catalytic activity due to their size effect and graphene accelerates the catalytic process by transferring of the electrons. Phan *et al* [15] have synthesized Pd nanocube–graphene hybrid via a simple chemical method which exhibits improved sensitivity due to the controlled morphology and the size of noble metals. The as-developed resistive-type sensor can detect hydrogen in the range from 6 to 1000 ppm at room temperature. The response values at room temperature, 50 °C and 100 °C for 1000 ppm of H_2 are 13%, 10.4%, and 9.2%, respectively. Also, metal-oxide/graphene hybrids are found to be a promising material because graphene can control the size and the morphology of metal oxides during synthesis [16]. Meng *et al* [17] have demonstrated SnO_2/graphene hybrids using chemical method which exhibit a very attractive improved response to benzene and ethanol compared to a traditional SnO_2. Shi *et al* [18] have developed a chemically modified graphene-based chemiresistor sensor for NO_2 detection. The fabricated sensor exhibits 4–16 times enhanced response with high sensitivity, selectivity and reversibility toward NO_2 in comparison to that of rGO, indicating that the presence of functional groups plays a significant role in the sensing process. Mackin *et al* [19] have demonstrated a compact sensor system competent in monitoring hundreds of graphene sensors in a fast and easy fashion. The functionalization of graphene is observed to increase the sensitivity towards ammonia four-fold over pristine graphene sensors. Smith *et al* [6] have demonstrated graphene-based carbon dioxide (CO_2) sensing and its cross-sensitivity with humidity (figure 13.4). They have shown that cross-sensitivity of CO_2 is negligible at typical CO_2 concentration present in air. Moreover, high selectivity to water (H_2O) in the presence of other gases is attractive for execution of graphene-based humidity sensing in ambient air environment. A schematic of binding of CO_2 molecules with the graphene surface upon diffusion into the chamber is shown in figure 13.4(a). The graphene resistance response is found to be varied linearly with R^2 value of 0.9661 with the variation in CO_2 concentration (figure 13.4(b)). Upon

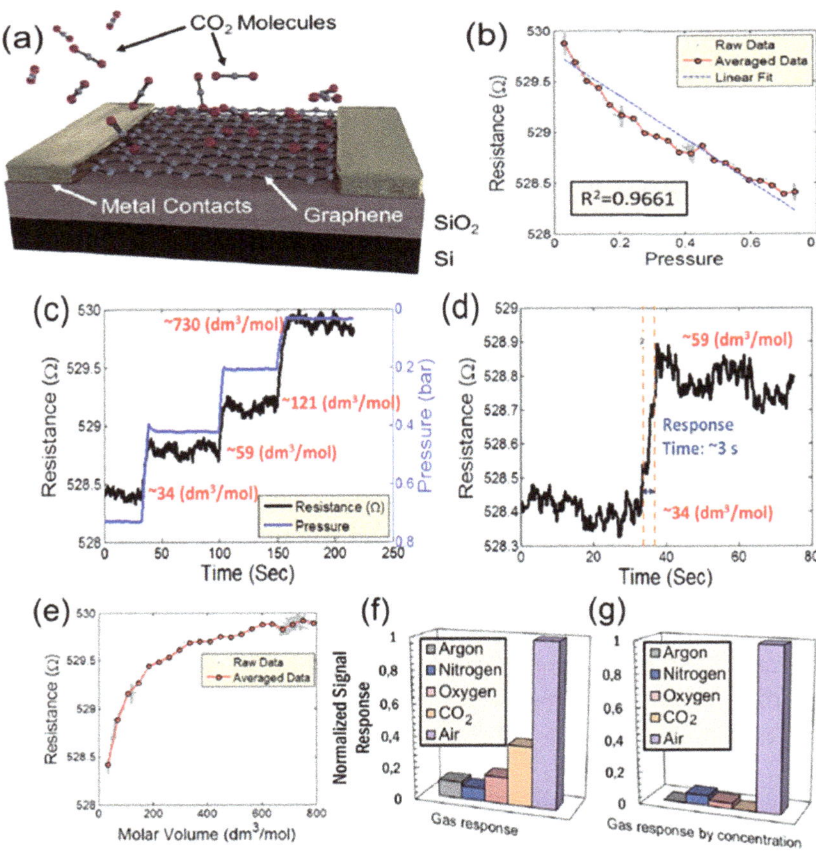

Figure 13.4. (a) Schematic of the active graphene sensing area of the device. (b) Resistance response of the device to changes in pressure. (c) Resistance response to CO_2. (d) Response time of the CO_2 sensor. (e) Relationship of the resistance versus the molar volume of CO_2 in the chamber. (f) Device sensitivity of various constituent gases of air. (g) Normalized sensitivity of each gas with respect to its relative concentration in air. Reproduced [6] under Creative Common Attribution 3.0 Unported License, Copyright 2017, Royal Society of Chemistry.

Table 13.1. rGO/graphene-based chemiresistive gas sensors.

Sl. No.	Sensing material	Analyte	Concentration	Working temperature (°C)	Response (%)	Reference
1.	Graphene	NH_3	75 ppm	150	3.8	[20]
2.	AgNPs/rGO	NO_2	15 ppm	RT	40	[21]
3.	Pd/Graphene	NH_3	100 ppm	150	2.04	[22]
	Al/Graphene	NO_2	1.2 ppm		2.89	
4.	Ag/rGO	NH_3	1 ppm	RT	6.52	[23]
5.	SnO_2/rGO	NH_3	50 ppm	RT	0.92	[24]
6.	SnO_2 NRs/rGO	NH_3	200 ppm	RT	1.3	[25]
7.	TiO_2/rGO	NH_3	10 ppm	RT	0.62	[26]
8.	TiO_2/rGO	CO	100 ppm	RT	1.12	[27]
9.	NiO/rGO	CH_4	1000 ppm	260	15.2	[28]
10.	ZnO/rGO	CH_4	1000 ppm	190	12.1	[29]
11.	ZnO NWs/rGO	NO_2	5 ppm	RT	3.5	[30]
12.	ZnO/rGO	NO_2	1.5 ppm	RT	1.1	[30]
		H_2S	8.1 ppm		0.87	
13.	WO_3/rGO	NO_2	0.5 ppm	RT	1.19	[31]
14.	Cu_2O NRs/rGO	NH_3	200 ppm	RT	2.04	[32]
15.	CuO/rGO	NO_2	1 ppm	RT	14	[33]
16.	PTh/rGO	NO_2	10 ppm	RT	26.36	[34]
17.	P3HT/rGO	NH_3	50 ppm	RT	12.63	[35]
18.	PANI/rGO	NH_3	100 ppm	RT	344.2	[36]

constant chamber pressure, CO_2 response of the graphene is found to be very stable (figure 13.4(c)). Moreover, the device exhibits quick response time as demonstrated in figure 13.4(d). Figure 13.4(e) shows the variation of the resistance with the molar volume of CO_2 in the chamber. Figures 13.4(f) and (g) describe the sensitivity for different gases and normalized sensitivity. Table 13.1 list different rGO based chemiresistive gas sensors.

13.3 Graphene-based strain sensor

Strain sensors have recently received a lot of attention in a variety of applications, including human motion detection [37], healthcare [38], damage detection [39] and structural characterization [40] etc. Nevertheless, well established strain sensors are required for these applications. A strain gauge sensor is a form of sensor in which the resistance changes as a result of the application of force, and this change results in a different electrical output. This approach was used to measure pressure, force, weight, and tension with strain gauges. The strain sensors can simply deduce the effective force by monitoring the surface strain. Indirect force measurement with strain sensors may determine huge forces with a small strain sensor at a low cost. Figure 13.5 shows the schematic of a graphene strain sensor.

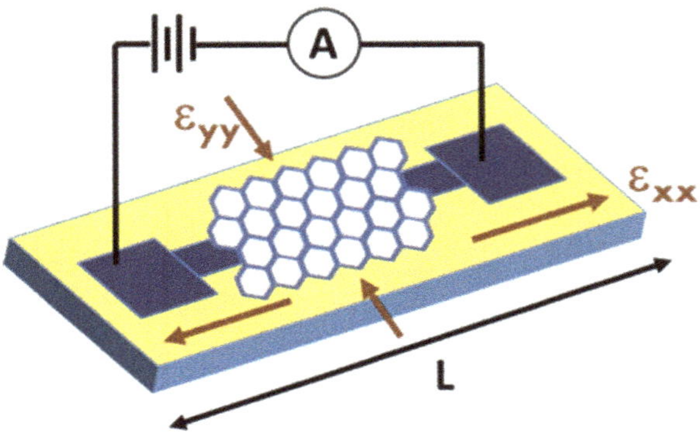

Figure 13.5. Schematic diagram of a strain sensor based on graphene.

Strain sensors are commonly constructed using a variety of metals and semiconductors. The graphene-based strain sensor works on the idea that when a mechanical strain is applied, the electrical structure of graphene is altered, resulting in a change in resistance. Graphene could be a potential option for a highly sensitive strain sensing material based on this idea. Graphene-based strain sensors also have unique possibilities for new types of smart wearable applications. The most significant attribute of a strain sensor is sensitivity, which is commonly given as gauge factor (GF). The GF is a ratio of the relative resistance change as a function of applied strain that is used to evaluate the sensitivity of a strain sensor, given by equation (13.1).

$$\mathrm{GF} = \frac{\Delta R/R_{\mathrm{o}}}{\varepsilon} \tag{13.1}$$

Here, R_{o} represents the sensor's resistance in its original state, ΔR represents the difference between the stretched state's resistance (R) and the original state's resistance (R_{o}), and ε represents the mechanical strain. The strain sensor's sensitivity increases as the GF grows larger. Diverse types of flexible and very sensitive strain sensors based on various phenomena such as piezo-resistive, piezoelectric, capacitive, and optical have been reported in recent decades. Among such strain sensors, piezo-resistive sensors transduce applied mechanical strain into electrical resistance are widely used for monitoring minute structural deformations due to their feasible preparation, flexibility, mechanical robustness and the ease of signal processing. Graphene, on the other hand, has been discovered to be a good material for piezo-resistive strain sensor because its resistance changes dramatically with applied pressure, and it has superior elasticity and stiffness with adjustable transparency for use in flexible wearable devices. Tang *et al* [41] have demonstrated a low-cost piezo-resistive sensor fabricated using graphene-nanowalls wrinkle and polydimethylsiloxane (PDMS) elastomer, which exhibits high sensitivity and a low limit of detection. Hassan *et al* [42] have demonstrated self-healable strain sensor developed using graphene and iron oxide composite that exhibits excellent performance towards applied strain along with outstanding self-healing characteristics. The operating sensitivity of about 94% has been recovered even after cutting the sensor with stretchability up to 54.5%. They found that with the increase in amount of the iron oxide, the performance of the sensor increases and the resistance error decreases. Figure 13.6 shows the performance of developed sensor with graphene as the active layer and behaving as a stain sensor along with *I*–*V* curves at different bending diameters.

Fu *et al* [43] have studied strain dependent resistance in graphene developed using CVD and observed that resistance first decreases with increase in strain and is then found to be increased as strain increases with a sensitivity value of 151, which exhibits the potential towards promising applications as sensor, switches, gauges, etc. Table 13.2 lists different graphene-based strain sensors.

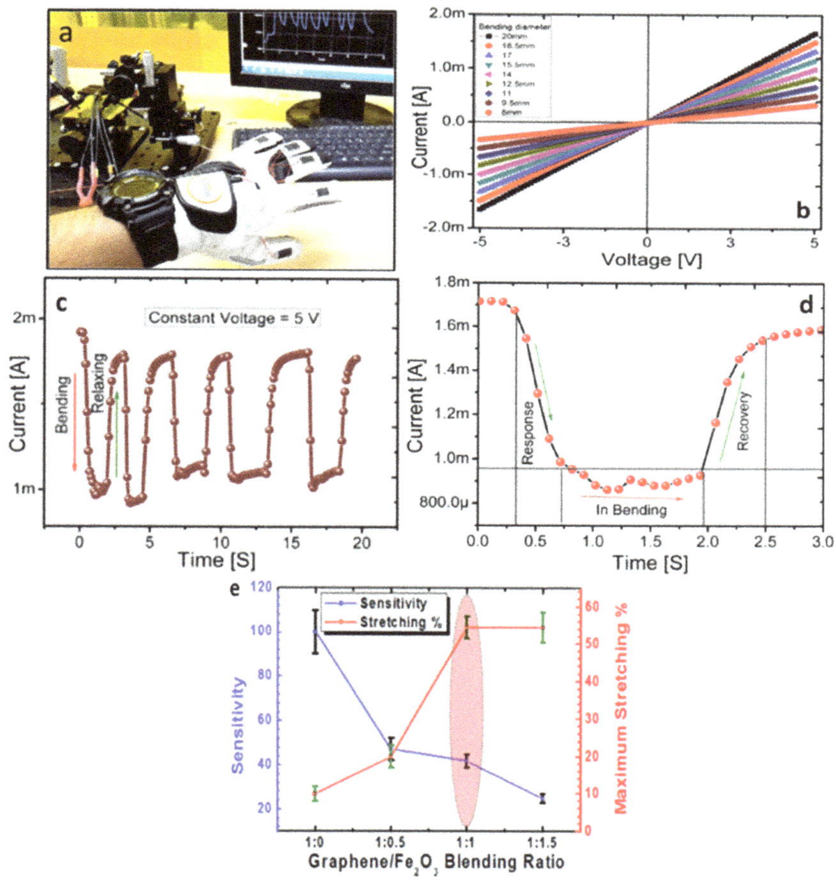

Figure 13.6. (a) Strain sensor with graphene as the active layer. (b) Current–voltage characteristics. (c) Bending and relaxing cycles. (d) Response and recovery time. (e) Sensitivity and stretching percentage of composite. Reproduced [42] under Creative Common Attribution 4.0 International License, Copyrights 2020, Springer Nature.

Table 13.2. Different graphene-based strain sensors.

Sl. No.	Materials	Technique	Gauge factor	Maximum strain/ stretchability (%)	Reference
1.	Graphene/PDMS	CVD	151	5	[43]
2.	Graphene/paper	Inkjet printing	125	1.25	[44]
3.	PDMS/CNT mesh	Adhesion	4	30	[45]
4.	PDMS/Carbon black	Spin coating	12	30	[46]
5.	GWF	CVD	35	0.2	[47]
6.	Graphene/PDMS	Spraying method	1054	26	[48]
7.	Graphene woven fabric/PDMS	—	223	3	[49]

13.4 Graphene-based electrochemical sensor

Electrochemical sensor is a type of sensor in which current is produced by an electrochemical reaction which occurs due to reaction of an electrolyte with target gas. An electrochemical sensor works on the idea of detecting the electron that is transported during an electrochemical reaction. Electrochemical sensors for analytes in solution are made up of electrochemical cells that are sandwiched between an active material of the working electrode (sensory unit), the analyte, the reaction medium, and the electrodes that are employed. Interface techniques are electrolyte processes in which the electrodes' polarisation is controlled (cathode and anode). Two or three electrodes can be used in the electrochemical gas sensor. Figure 13.7 exhibits the principle of a simple two-electrode. A three-electrode electrochemical cell consists of a working electrode, on which the species involved undergo oxidation and reduction reactions, a counter electrode (or auxiliary electrode), which completes the electrochemical system's electric circuit, and a reference electrode, which serves as a reference for the evaluation of other measured parameters. Due to their wide range tunability, strong sensitivity, low cost, and ease of application, electrochemical sensors have proven to be superior to other types of sensors. Nevertheless, potentiometric (change in potential), conductometric (change in conductance), impedimetric (change in impedance), and voltammetric or amperometric electrochemical sensors may all be distinguished based on the electrical magnitude used for transduction of the recognition event (change of current for an electrochemical reaction with the applied voltage in the first case, or with time at a fixed applied potential in the latter). Further, voltammetric methods have one thing in common: they all entail applying a voltage (E) to an electrode and measuring the current (I) that flows through the electrochemical cell as a result. The applied potential is frequently altered or the current is controlled for a set amount of time (t). As a result, all voltammetric procedures are functions of potential, current, and time (E, I, and t).

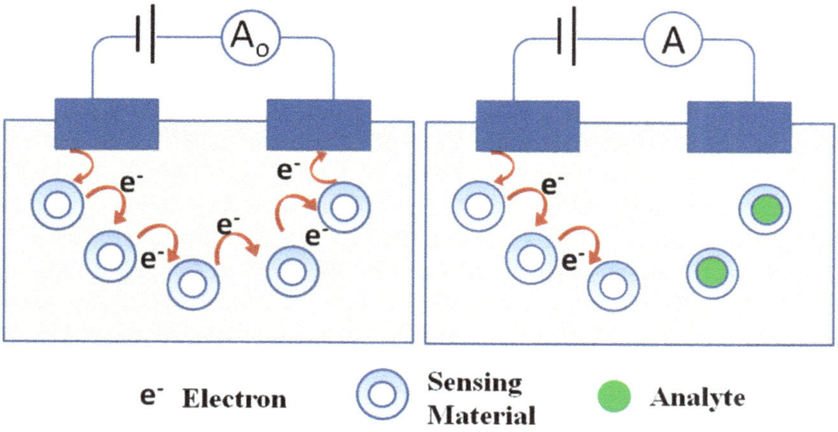

Figure 13.7. Schematic of electrochemical sensor without gas analytes (left figure) and with gas analytes (right figure).

The graphene on interacting with oxidizing analytes results in the depletion of charge carriers, whereas the opposite effect occurs as it interacts with reducing analytes [50]. There has been a lot of interest in graphene as an electrochemical gas sensor in recent years. By putting graphene's unique features into the sensor, it improves the properties of traditional sensing materials. Weak van der Waals interaction, hydrogen bonding and charge transfer are all ways through which graphene atoms might interact with adsorbed gas molecules [51, 52]. The Dirac point shift and electron concentration variation occurs from these interactions, which can be detected using an electronic system [52]. It is worth noting that surface imperfections in modified graphene boost the gas molecule absorption efficiency by allowing anchoring capacity for a variety of chemical functions [53]. The incorporation of AuNPs and rGO sheets in molybdenum disulfide (MoS_2) to form AuNPs@rGO–MoS_2 hybrid has been reported to enhance the voltammetric detection of hydroquinone (HQ), catechol (CC) or resorcinol (RC) [54]. They utilized the cyclic voltammetry (CV) and differential pulse voltammetry (DPV) on the glassy carbon electrode (GCE) to investigate the oxidation of hybrid towards HQ, CC and RC. The electrochemical impedance spectroscopy (EIS) study showed the decreasing semicircle diameter pattern in order of MoS_2/GCE > MoS_2–rGO/GCE > GCE > Au@MoS_2–rGO/GCE. This implies that integration of AuNPs and rGO reduced the electron transfer resistance of MoS_2 due to their synergistic interaction. The sensors showed the linear ranges of 0.1–950 μM, 3–560 μM, and 40–960 μM with LOD 0.04 μM, 0.95 μM, and 14.6 μM for HQ, CC and RC, respectively. Hasani *et al* [55] have demonstrated cadmium sulfide (CdS)/graphene oxide (GO) (1%, 3% and 5%)-based CO_2 gas sensor fabricated through photo reduction of GO using highly efficient photocatalytic CdS NPs under visible irradiation for 2 h. The prepared CdS/GO (5%) achieved excellent sensitivity at a level of 200–1000 ppm concentration due to its higher conductivity. Moreover, Bai *et al* [56] have demonstrated the effect of copper sulfide (CuS) NPs in CuS/rGO nanocomposite. They have prepared the nanocomposite by heating the mixture of $CuCl_2$ and Na_2S aqueous solutions. The independence of cathodic peak potential (E_{pc}) and the anodic peak potential (E_{pa}) with the scan rate (−0.8 to +0.2 V) have revealed the electrochemical reaction ability of CuS at the CuS/rGO/GC electrode. The as-synthesized CuS/rGO composite has shown excellent sensitivity and stability for determination of H_2O_2. It is worth mentioning that graphene-based electrochemical sensors are mostly fabricated using rGO instead of GO due to the presence of abundant chemical moieties on the rGO surface that make an easy charge transfer process which is responsible for the enhance electrochemical activity. Ibrahim *et al* [57] have fabricated a photo-electrochemical (PEC) sensor based on the CdS–rGO nanocomposite using aerosol-assisted CVD (AACVD) and dip-coating strategies. The CdS–rGO nanocomposite showed the highest PEC activity compared to that of bare GO, rGO, CdS, and CdS–GO electrodes under visible-light illumination. This is because the presence of rGO facilitated the transfer of electrons from the conduction band of the CdS, thus manifesting the photocurrent performance of the nanocomposite. The sensor exhibits excellent sensitivity and detection towards copper (II) ions (Cu^{2+}) with photoelectrode response for Cu^{2+} ion detection in a linear range of 0.5–120 μM, with an LOD of 16 nM [57]. Furthermore, the influence of cationic composition in $Cd_xZn_{1-x}S$–rGO nanocomposite has been investigated [58]. Their research results showed that the photoelectrode response increased with x value, achieved optimal composition $Cd_{0.5}Zn_{0.5}S$–rGO with linear range of 0.02–20 mM and LOD of 6.7 nM. Pandhi *et al* [59] have studied electrochemical performance of multilayered graphene (MLG) electrodes on Kapton substrates. CV studies have shown electrochemical reversibility of an MLG electrode and inkjet-printed (IJP) three-electrode device using the ferric/ferrocyanide redox couple as the analyte. They have observed that IJP three-electrode sensors exhibit superior electrochemical response. Figure 13.8 exhibits the observed electrochemical performance using CV curves having distinct redox peaks.

Furthermore, 30 printed passes of MLG are found to exhibit a higher current and lower sheet resistance in comparison to the 25 printed passes along with a higher peak-to-peak separation (figure 13.8(b)). They have suggested that additional printed passes provide uniformity and decrease resistance with denser packing of the graphene flakes which slows the redox reaction. Yang *et al* [60] have studied sensing properties of CuS/rGO nanocomposite for the development of hydrogen peroxide (H_2O_2) and hydrazine sensor and the resulting sensor exhibits promising results in the range from 1 μM to 10 mM at −0.2 V and LOD of 100 nM towards H_2O_2 and sensor also exhibits good performance in the range of 1 μM–10 mM at +0.4 V and LOD of 300 nM for hydrazine. Jin *et al* [61] have performed *in situ* growth of CuS decorated graphene oxide-multiwalled carbon nanotubes for their application as H_2O_2 amperometric sensor which shows response in the linear range of 0.45–60 mM with LOD of 600 nM. It has been found that the synergetic interaction of CNTs and graphene-based materials accelerate the electron transfer

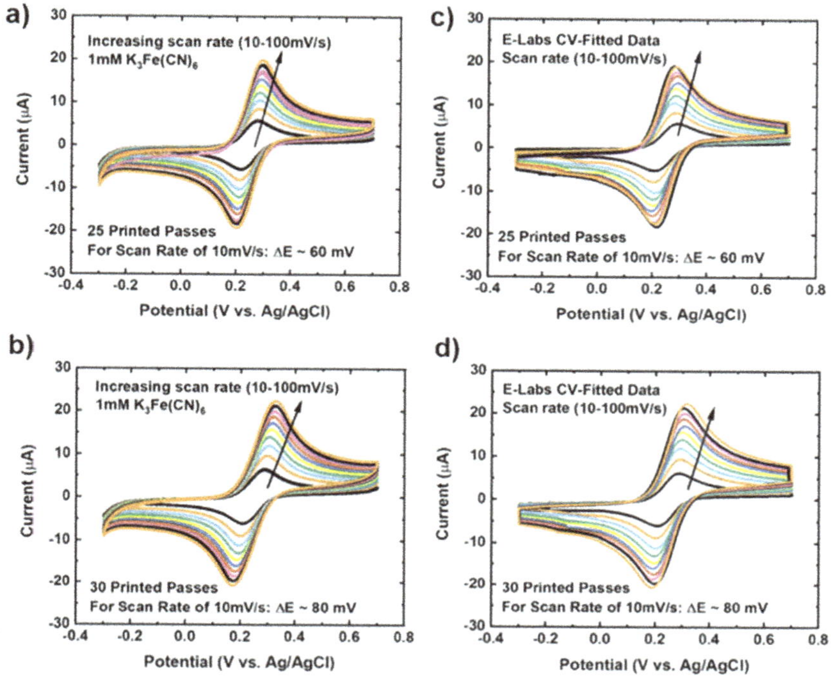

Figure 13.8. CV scan rate data of graphene for (a) 25 printed passes and (b) 30 printed passes of graphene. E-labs CV data for (c) 25 printed passes of graphene and (d) 30 printed passes of graphene. Reproduced [59] under Creative Common Attribution 3.0 Unported Licence, Copyrights 2020, Royal Society of Chemistry.

mechanism between the solution and electrodes and thus achieved a fast response time. Hu *et al* [62] have investigated rGO–MoS_2 heterostructures for electrochemical performance towards nitrite which exhibits high sensitivity in the linear range of 0.2–4800 μM and LOD of 170 nM. Table 13.3 list different graphene-based electrochemical gas sensors.

13.5 Graphene-based optical sensors

The aforesaid sensor methods, have a distinct benefit, such as simple and inexpensive manufacturing using common microelectronic or micromechanical processes, small size, or high operating temperature, however, they suffer from a lack of broad range selectivity due to a lack of sensitive materials for each gas and/or crosstalk. In the following part, outstanding properties of optical gas sensing schemes are discussed. When compared to the concepts discussed above, optical gas sensors have a number of advantages. One key reason is that the species used for detection, photons, are fundamentally different from the species to be detected, gas atoms/molecules. In contrast to the other notions discussed thus far, the species employed for detection have no effect on the detection method. Because photons have no residual mass and no charge, they have no effect on charge- or mass-based detection systems. In addition, because IR spectroscopy-based sensors do not rely on catalytic or

Table 13.3. Graphene-based electrochemical gas sensors.

Sl. No.	Electrode material	Analyte	Linear range	LOD	Reference
1.	AuNPs/graphene	H_2O_2	5–8600 μM	2000 nM	[63]
2.	PtNPs/graphene	H_2O_2	0–2500 μM	200 nM	[64]
3.	Ag/NCs/graphene	H_2O_2	20–10 000 μM	3000 nM	[64]
4.	AuNPs–porous graphene film	H_2O_2	0.5–4900 μM	100 nM	[65]
5.	Cube like Pt/GO	NO_2^-	0.5–227 780 μM	200 nM	[66]
6.	Ag/Fe_3O_4/GO	NO_2^-	0.5–720 720–8150 μM	170 nM	[67]
7.	AuNP/rGO/MWCNT/GCE	NO_2^-	0.05–2200 μM	14 nM	[68]
8.	Fe_2O_3–rGO/GCE	NO_2^-	0.05–780 μM	15 nM	[69]
9.	Graphene–copper NP	Glucose	400–500 μM	500 nM	[70]
10.	Palladium NP–graphene nanosheet film	H_2O_2	0.1–1000 μM	500 nM	[71]
11.	AuNP–graphene nanosheet film	H_2O_2	0.3–1800 μM	1100 nM	[72]
12.	Graphene–palladium NP	Hydrogen	20–1000 ppm	20 ppm	[73]

electrochemical activities, they are not susceptible to surface depletion or contamination. However, an optical gas sensor is a device which monitors the change of optical properties (absorbance, fluorescence, reflectance etc) of a sensing material on interacting with gas molecules. The recognition of gas species can be done from the wavelength dependent absorption/reflectance spectra of sensing material. The absorbance of sensing layer may increase or decrease depending on the oxidizing (SO_2, NO_2, O_3, etc) or reducing (CO, H_2, etc) nature of target gas. The graphene covering is reported to boost sensing performance by offering a high surface-to-volume ratio, greater analyte absorption, and superior plasmonic characteristics. Graphene-based devices, on the other hand, benefit from the extraordinary features of graphene, which offer high photocurrent production and a huge active area. This makes it a good choice for infrared (IR) detection in the IR regimes of near-IR and middle-wave infrared regime [74]. Despite the fact that numerous research groups have revealed several novel features of graphene, there is still much work to be done to improve device fabrication and performance. Lai *et al* [74] have studied the optical properties of graphene-based optical detectors and found that at room temperature in the presence of IR, devices can create photocurrent signals and this increases as devices are composed of multilayer graphene. Zhang *et al* [75] have studied reflectance properties of graphene film-based optical sensor synthesized using CVD method for acetone vapor detection. The prepared sensor exhibited a significant decrease in reflectance under the exposure of 44–352 ppm acetone vapor. Yao *et al* [76] have described ammonia sensor using optical interferometric sensor based on graphene/microfiber hybrid waveguide (GMHW). The resolution of as fabricated sensor is ~0.3 ppm. The adsorption of NH_3 changes the conductivity of graphene and thus the effective refractive index of the GMHW, and the transmitting light along the GMHW is particularly sensitive to NH_3 gas concentration. Further, synergistic interplay by combining the semiconducting and photocatalytic activity behavior of the partially reduced GO with the localized surface plasmon resonance (LSPR) of AuNPs result in an enhancement of the photocatalytic properties of GO has been studied by Cittadini *et al* [77]. The charge transfer mechanism between the AuNPs and GO flakes induced SPR effect, which leads to an increase in absorbance for the case of H_2 (reducing gas) and decrease in absorbance for NO_2 (oxidizing gas). Further, Some *et al* [78] have studied the sensor array fabricated using hydrophilic GO and hydrophobic rGO flakes coated on polymer optical fiber tip that exhibits the selective sensing for VOCs gases. They reported the selective ability of hydrophilic GO and hydrophobic rGO flakes towards tetrahydrofuran (THF) and dichloromethane (MC), respectively, and high stability of as-developed sensor in humid conditions. High response towards methane with good selectivity has been observed in graphene-carbon nanotubes–poly methyl methacrylate (GCNT/PMMA) hybrid based SPR fiber optic sensor [79]. The authors carried out the testing of a sensor based on rGO, CNT, GCNT, and GCNT/PMMA hybrid nanocomposite towards different gases methane, ammonia, hydrogen, nitrogen, carbon dioxide, hydrogen sulfide, and chlorine. Out of all the four sensors, GCNT/PMMA nanocomposite showed a maximal red shift (0.33 nm ppm^{-1}) in resonance wavelength for methane. The doping concentration of GCNT has been optimized as 5 wt% to increase the defect levels efficiently in nanocomposite structure, which further increase the active adsorption sites of GCNT/PMMA nanocomposite towards methane. The same group has also studied the sensing properties of PMMA, rGO and PMMA/rGO nanocomposites based SPR based fiber optic gas sensor towards ammonia, nitrogen, hydrogen, chlorine and hydrogen sulfide [80]. The PMMA/rGO nanocomposite-based sensor showed higher response towards 10–20 ppm ammonia than pristine PMMA and rGO based sensor. The polar nature of ammonia is favourable for the formation of hydogen bond with the oxygen functionalized rGO sheets resulting in decreasing the dielectric constant of nanocomposite. The increase in resistance has been observed due to the adsorption of ammonia molecules on the PMMA matrix. These observed variations were responsible for the high sensitivity (1 nm ppm^{-1}) and selectivity of PMMA/rGO nanocomposite in favour of ammonia. To demonstrate the effect of inorganic coatings of MFI zeolite crystals in fiber Bragg grating (FBG) chemical sensors, Snelder *et al* [81] have demonstrated fiber sensors coated with silicalite of different thicknesses. It has been observed that ZSM-5 coated optical sensor exhibited higher responses towards alkanes C_2H_6, C_3H_8, C_4H_{10}, octane and isopropanol under ambient conditions. Moreover, the prepared sensors showed stable responses towards C_2H_6 and C_3H_8 at extreme conditions. They attributed excellent sensing performance to the high adsorption of analyte at zeolite framework controlled by gaseous molecular affinity. Lai *et al* [74] have demonstrated the development of optical detectors using multilayer graphene for IR sensing. They have shown that at ambient temperature their devices are capable of producing photocurrent when illuminated in the presences of IR signals and have observed that photocurrent is increased as fabrication of devices is done using multilayer graphene. Abid *et al* [82]

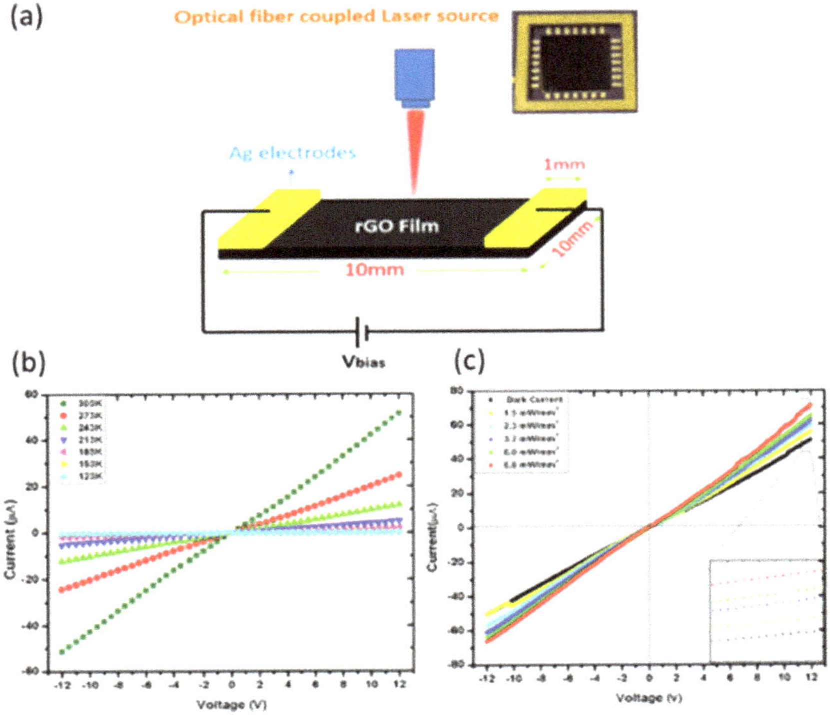

Figure 13.9. (a) Schematics of the developed optical sensor along with current–voltage characteristics measured at various (b) temperature (c) laser power density. Reproduced [82] under Creative Common Attribution 4.0 International License, Copyrights 2018, Springer Nature.

have demonstrated optical sensor based on rGO and observed that defect states play an important role in tuning the sensor parameters.

Figure 13.9(a) exhibits the schematic diagram of the developed sensor and figures 13.9(b) and (c) show *I–V* characteristics recorded at different temperature and laser power density. Sridevi *et al* [83] have shown improved sensitivity for strain and temperature using rGO coated with an etched fiber Bragg grating (eFBG) sensor which exhibits strain sensitivity of 5.5 pm and temperature sensitivity of 33 pm. Further, Kavinkumar *et al* [84] have studied the role of hydroxyl, carboxyl functional groups on dielectric and optical gas sensing properties of GO, heat-treated GO for ammonia, ethanol and methanol vapors is reported. High surface area, exfoliation of GO provides better vapor sensing ability than the heat-treated GO. The sensitivities of GO sensor are found to be 0.32, 0.26 and 0.20 counts per ppm for ammonia, ethanol and methanol vapors, respectively. Table 13.4 lists different configurations of graphene-based gas sensors.

13.6 Conclusions

In this book chapter, we have discussed the advancements made in the area of graphene sensors along with their different types. The sensing properties of sensors

Table 13.4. Various configurations of graphene-based gas sensors.

Sl. No.	Materials	Analyte	Detection tool	SR	References
1.	Graphene	(NH_3)	MBG	6 pm ppm^{-1}	[85]
2.	Graphene	—	GBMR	200+ 0.2 kHz ppm^{-1}	[86]
3.	Graphene	Ag	SPR	340%	[87]
4.	rGO/Polyimide	(H_2S)	Resistance monitoring	0.238% (20 ppm)	[88]

have also been well described for various graphene-based hybrid materials. The challenge for researchers, however, is to develop materials that can be used to fabricate stable, high selective, rapid, and high response gas sensors, and there has been significant progress in gas sensors based on carbonaceous materials, such as graphene, as evidenced by the extensive literature survey discussed in this chapter. Graphene due to its applicability to perform at comparatively lower operating temperature along with lower detection limit has resulted in its application as advanced gas sensors. Furthermore, we have outlined the interesting approaches to solve sensitivity and selectivity difficulties by developing composites or hybrids of graphene with semiconductor metal oxides, NPs, polymer, etc or by heteroatom doping. Summing up, as there are no global standards and procedures for determining graphene quality, the quality of graphene generated varies greatly between countries and even from batch to batch within the same company. Using graphene of unknown quality in a sensor could be a major problem, especially since the sensors must be extremely accurate and exact. Finally, understanding the correct graphene material and how to optimize its use will lead to graphene being commercialized as next-generation gas sensors. We expect that by combining existing knowledge with new research, researchers will be able to develop new ways of making gas sensors.

References

[1] Zhang T, Mubeen S, Myung N V and Deshusses M A 2008 Recent progress in carbon nanotube-based gas sensors *Nanotechnology* **19** 332001

[2] Tian W, Liu X and Yu W 2018 Research progress of gas sensor based on graphene and its derivatives: a review *Appl. Sci.* **8** 1118

[3] Wang T, Huang D, Yang Z, Xu S, He G, Li X, Hu N, Yin G, He D and Zhang L 2016 A review on graphene-based gas/vapor sensors with unique properties and potential applications *Nano-Micro Lett.* **8** 95–119

[4] Liang X, Chen Z, Wu H, Guo L, He C, Wang B and Wu Y 2014 Enhanced NH_3-sensing behavior of 2,9,16,23-tetrakis (2,2,3,3-tetrafluoropropoxy) metal (II) phthalocyanine/multi-walled carbon nanotube hybrids: An investigation of the effects of central metals *Carbon* **80** 268–78

[5] Zhou X, Wang X, Wang B, Chen Z, He C and Wu Y 2014 Preparation, characterization and NH_3-sensing properties of reduced graphene oxide/copper phthalocyanine hybrid material *Sens. Actuators* B **193** 340–8

[6] Smith A D *et al* 2017 Graphene-based CO_2 sensing and its cross-sensitivity with humidity *RSC Adv.* **7** 22329–39

[7] Shafiei M, Spizzirri P G, Arsat R, Yu J, du Plessis J, Dubin S, Kaner R B, Kalantar-Zadeh K and Wlodarski W 2010 Platinum/graphene nanosheet/SiC contacts and their application for hydrogen gas sensing *J. Phys. Chem.* C **114** 13796–801

[8] Sanchez J B, Berger F, Fromm M and Nadal M H 2006 A selective gas detection micro-device for monitoring the volatile organic compounds pollution *Sens. Actuators* B **119** 227–33

[9] Lin C, Gillespie J, Schuder M D, Duberstein W, Beverland I J and Heal M R 2015 Evaluation and calibration of Aeroqual series 500 portable gas sensors for accurate measurement of ambient ozone and nitrogen dioxide *Atmos. Environ.* **100** 111–6

[10] Majumder S, Mondal T and Deen M J 2017 Wearable sensors for remote health monitoring *Sensors* **17** 130

[11] Wang C, Yin L, Zhang L, Xiang D and Gao R 2010 Metal oxide gas sensors: sensitivity and influencing factors *Sensors (Basel)* **10** 2088–106

[12] Gautam M and Jayatissa A H 2012 Ammonia gas sensing behavior of graphene surface decorated with gold nanoparticles *Solid-State Electron.* **78** 159–65

[13] Johnson J L, Behnam A, Pearton S J and Ural A 2010 Hydrogen sensing using Pd-functionalized multi-layer graphene nanoribbon networks *Adv. Mater.* **22** 4877–80

[14] Khurshid F, Jeyavelan M, Hussain T, Hudson M S L and Nagarajan S 2020 Ammonia gas adsorption study on graphene oxide based sensing device under different humidity conditions *Mater. Chem. Phys.* **242** 122485

[15] Phan D T and Chung G S 2014 A novel Pd nanocube–graphene hybrid for hydrogen detection *Sens. Actuators* B **199** 354–60

[16] Meng F L, Guo Z and Huang X J 2015 Graphene-based hybrids for chemiresistive gas sensors *TrAC Trends Anal. Chem.* **68** 37–47

[17] Meng F L, Li H H, Kong L T, Liu J Y, Jin Z, Li W, Jia Y, Liu J H and Huang X J 2012 Parts per billion-level detection of benzene using SnO_2/graphene nanocomposite composed of sub-6 nm SnO_2 nanoparticles *Anal. Chim. Acta.* **736** 100–7

[18] Yuan W, Liu A, Huang L, Li C and Shi G 2013 High-performance NO_2 sensors based on chemically modified graphene *Adv. Mater.* **25** 766–71

[19] Mackin C, Schroeder V, Zurutuza A, Su C, Kong J, Swager T M and Palacios T 2018 Chemiresistive graphene sensors for ammonia detection *ACS Appl. Mater. Interfaces* **10** 16169–76

[20] Gautam M and Jayatissa A H 2011 Gas sensing properties of graphene synthesized by chemical vapor deposition *Mater. Sci. Eng.* C **31** 1405–11

[21] Li W, Qi W, Cai L, Li C, Sun Y, Sun M, Yang X, Xiang L, Xie D and Ren T 2020 Enhanced room-temperature NO_2-sensing performance of AgNPs/rGO nanocomposites *Chem. Phys. Lett.* **738** 136873

[22] Cho B *et al* 2014 Graphene-based gas sensor: metal decoration effect and application to a flexible device *J. Mater. Chem.* C **2** 5280–5

[23] Karaduman I, Er E, Çelikkan H, Erk N and Acar S 2017 Room-temperature ammonia gas sensor based on reduced graphene oxide nanocomposites decorated by Ag, Au and Pt nanoparticles *J. Alloys Compd.* **722** 569–78

[24] Feng Q, Li X and Wang J 2017 Percolation effect of reduced graphene oxide (rGO) on ammonia sensing of rGO–SnO_2 composite based sensor *Sens. Actuators* B **243** 1115–26

[25] Chen Y, Zhang W and Wu Q 2017 A highly sensitive room-temperature sensing material for NH_3: SnO_2-nanorods coupled by rGO *Sens. Actuators* B **242** 1216–26
[26] Ye Z, Tai H, Xie T, Su Y, Yuan Z, Liu C and Jiang Y 2016 A facile method to develop novel TiO_2/rGO layered film sensor for detecting ammonia at room temperature *Mater. Lett.* **165** 127–30
[27] Bandi S, Hastak V, Peshwe D and Srivastav A 2018 In situ TiO_2–rGO nanocomposites for CO gas sensing *Bull. Mater. Sci.* **41** 115
[28] Zhang D, Chang H, Li P and Liu R 2016 Characterization of nickel oxide decorated-reduced graphene oxide nanocomposite and its sensing properties toward methane gas detection *J. Mater. Sci., Mater. Electron.* **27** 3723–30
[29] Zhang D, Yin N and Xia B 2015 Facile fabrication of ZnO nanocrystalline-modified graphene hybrid nanocomposite toward methane gas sensing application *J. Mater. Sci., Mater. Electron.* **26** 5937–45
[30] Ugale A D, Umarji G G, Jung S H, Deshpande N G, Lee W, Cho H K and Yoo J B 2020 ZnO decorated flexible and strong graphene fibers for sensing NO_2 and H_2S at room temperature *Sens. Actuators* B **308** 127690
[31] Su P G and Peng S L 2015 Fabrication and NO_2 gas-sensing properties of reduced graphene oxide/WO_3 nanocomposite films *Talanta* **132** 398–405
[32] Meng H, Yang W, Ding K, Feng L and Guan Y 2015 Cu_2O nanorods modified by reduced graphene oxide for NH_3 sensing at room temperature *J. Mater. Chem.* A **3** 1174–81
[33] Li Z, Liu Y, Guo D, Guo J and Su Y 2018 Room-temperature synthesis of CuO/reduced graphene oxide nanohybrids for high-performance NO_2 gas sensor *Sens. Actuators* B **271** 306–10
[34] Bai S, Guo J, Sun J, Tang P, Chen A, Luo R and Li D 2016 Enhancement of NO_2 sensing performance at room temperature by graphene-modified polythiophene *Ind. Eng. Chem. Res.* **55** 5788–94
[35] Ye Z, Jiang Y, Tai H and Yuan Z 2014 The investigation of reduced graphene oxide/P3HT composite films for ammonia detection *Integr. Ferroelectr.* **154** 73–81
[36] Bai S, Zhao Y, Sun J, Tian Y, Luo R, Li D and Chen A 2015 Ultrasensitive room temperature NH_3 sensor based on a graphene–polyaniline hybrid loaded on PET thin film *Chem. Commun.* **51** 7524–7
[37] Yamada T, Hayamizu Y, Yamamoto Y, Yomogida Y, Izadi-Najafabadi A, Futaba D N and Hata K 2011 A stretchable carbon nanotube strain sensor for human-motion detection *Nat. Nanotechnol.* **6** 296–301
[38] Kang I, Schulz M J, Kim J H, Shanov V and Shi D 2006 A carbon nanotube strain sensor for structural health monitoring *Smart Mater. Struct.* **15** 737
[39] Kim J *et al* 2014 Stretchable silicon nanoribbon electronics for skin prosthesis *Nat. Commun.* **5** 1–11
[40] Zhao D, Liu T, Zhang M, Liang R and Wang B 2012 Fabrication and characterization of aerosol-jet printed strain sensors for multifunctional composite structures *Smart Mater. Struct.* **21** 115008
[41] Tang X, Yang W, Yin S, Tai G, Su M, Yang J, Shi H, Wei D and Yang J 2021 Controllable graphene wrinkle for a high-performance flexible pressure sensor *ACS Appl. Mater. Interfaces* **13** 20448–58

[42] Hassan G, Khan M U, Bae J and Shuja A 2020 Inkjet printed self-healable strain sensor based on graphene and magnetic iron oxide nano-composite on engineered polyurethane substrate *Sci. Rep.* **10** 1–12

[43] Fu X W *et al* 2011 Strain dependent resistance in chemical vapor deposition grown graphene *Appl. Phys. Lett.* **99** 213107

[44] Casiraghi C, Macucci M, Parvez K, Worsley R, Shin Y, Bronte F, Borri C, Paggi M and Fiori G 2018 Inkjet printed 2D-crystal based strain gauges on paper *Carbon* **129** 462–7

[45] Guo F M, Cui X, Wang K L and Wei J Q 2016 Stretchable and compressible strain sensors based on carbon nanotube meshes *Nanoscale* **8** 19352–8

[46] Chang X, Sun S, Sun S, Liu T, Xiong X, Lei Y, Dong L and Yin Y 2018 ZnO nanorods/carbon black-based flexible strain sensor for detecting human motions *J. Alloys Compd.* **738** 111–7

[47] Wang Y, Wang L, Yang T, Li X, Zang X, Zhu M, Wang K, Wu D and Zhu H 2014 Wearable and highly sensitive graphene strain sensors for human motion monitoring *Adv. Funct. Mater.* **24** 4666–70

[48] Yang Y F, Tao L Q, Pang Y, Tian H, Ju Z Y, Wu X M, Yang Y and Ren T L 2018 An ultrasensitive strain sensor with a wide strain range based on graphene armour scales *Nanoscale* **10** 11524–30

[49] Liu X *et al* 2017 A highly sensitive graphene woven fabric strain sensor for wearable wireless musical instruments *Mater. Horiz.* **4** 477–86

[50] Dan Y, Lu Y, Kybert N J, Luo Z and Johnson A C 2009 Intrinsic response of graphene vapor sensors *Nano Lett.* **9** 1472–5

[51] Latif U and Dickert F L 2015 Graphene hybrid materials in gas sensing applications *Sensors* **15** 30504–24

[52] Zhu Y, Murali S, Cai W, Li X, Suk J W, Potts J R and Ruoff R S 2010 Graphene and graphene oxide: synthesis, properties, and applications *Adv. Mater.* **22** 3906–24

[53] Hu N, Wang Y, Chai J, Gao R, Yang Z, Kong E S W and Zhang Y 2012 Gas sensor based on p-phenylenediamine reduced graphene oxide *Sens. Actuators* B **163** 107–14

[54] Ma G, Xu H, Wu M, Wang L, Wu J and Xu F 2019 A hybrid composed of MoS_2, reduced graphene oxide and gold nanoparticles for voltammetric determination of hydroquinone, catechol, and resorcinol *Microchim. Acta.* **186** 1–9

[55] Hasani A, Sharifi Dehsari H, Amiri Zarandi A, Salehi A, Taromi F A and Kazeroni H 2015 Visible light-assisted photoreduction of graphene oxide using CdS nanoparticles and gas sensing properties *J. Nanomater.* **2015** 930306

[56] Bai J and Jiang X 2013 A facile one-pot synthesis of copper sulfide-decorated reduced graphene oxide composites for enhanced detecting of H_2O_2 in biological environments *Anal. Chem.* **85** 8095–101

[57] Ibrahim I, Lim H N, Huang N M and Pandikumar A 2016 Cadmium sulphide-reduced graphene oxide-modified photoelectrode-based photoelectrochemical sensing platform for copper (II) ions *PLoS One* **11** e0154557

[58] Yan J, Wang K, Liu Q, Qian J, Dong X, Liu W and Qiu B 2013 One-pot synthesis of Cd_xZn_{1-x} S-reduced graphene oxide nanocomposites with improved photoelectrochemical performance for selective determination of Cu^{2+} *RSC Adv.* **3** 14451–7

[59] Pandhi T, Cornwell C, Fujimoto K, Barnes P, Cox J, Xiong H, Davis P H, Subbaraman H, Koehne J E and Estrada D 2020 Fully inkjet-printed multilayered graphene-based flexible electrodes for repeatable electrochemical response *RSC Adv.* **10** 38205–19

[60] Yang Y J, Li W and Wu X 2014 Copper sulfide|reduced graphene oxide nanocomposite for detection of hydrazine and hydrogen peroxide at low potential in neutral medium *Electrochim. Acta.* **123** 260–7
[61] Jin W, Fu Y and Cai W 2019 In situ growth of CuS decorated graphene oxide-multiwalled carbon nanotubes for ultrasensitive H_2O_2 detection in alkaline solution *New J. Chem.* **43** 3309–16
[62] Hu J, Zhang J, Zhao Z, Liu J, Shi J, Li G, Li P, Zhang W, Lian K and Zhuiykov S 2018 Synthesis and electrochemical properties of rGO–MoS_2 heterostructures for highly sensitive nitrite detection *Ionics* **24** 577–87
[63] Xiao F, Song J, Gao H, Zan X, Xu R and Duan H 2012 Coating graphene paper with 2D-assembly of electrocatalytic nanoparticles: a modular approach toward high-performance flexible electrodes *ACS Nano* **6** 100–10
[64] Liang B, Fang L, Hu Y, Yang G, Zhu Q and Ye X 2014 Fabrication and application of flexible graphene silk composite film electrodes decorated with spiky Pt nanospheres *Nanoscale* **6** 4264–74
[65] Xi Q, Chen X, Evans D G and Yang W 2012 Gold nanoparticle-embedded porous graphene thin films fabricated via layer-by-layer self-assembly and subsequent thermal annealing for electrochemical sensing *Langmuir* **28** 9885–92
[66] Bai W, Sheng Q and Zheng J 2016 Morphology controlled synthesis of platinum nanoparticles performed on the surface of graphene oxide using a gas–liquid interfacial reaction and its application for high-performance electrochemical sensing *Analyst* **141** 4349–58
[67] Li B Q, Nie F, Sheng Q L and Zheng J B 2015 An electrochemical sensor for sensitive determination of nitrites based on Ag–Fe_3O_4–graphene oxide magnetic nanocomposites *Chem. Pap.* **69** 911–20
[68] Yu H, Li R and Song K L 2019 Amperometric determination of nitrite by using a nanocomposite prepared from gold nanoparticles, reduced graphene oxide and multi-walled carbon nanotubes *Microchim. Acta.* **186** 1–9
[69] Radhakrishnan S, Krishnamoorthy K, Sekar C, Wilson J and Kim S J 2014 A highly sensitive electrochemical sensor for nitrite detection based on Fe_2O_3 nanoparticles decorated reduced graphene oxide nanosheets *Appl. Catal.* B **148** 22–8
[70] Justino C I, Gomes A R, Freitas A C, Duarte A C and Rocha-Santos T A 2017 Graphene based sensors and biosensors *TrAC Trends Anal. Chem.* **91** 53–66
[71] Chen X M, Cai Z X, Huang Z Y, Oyama M, Jiang Y Q and Chen X 2013 Ultrafine palladium nanoparticles grown on graphene nanosheets for enhanced electrochemical sensing of hydrogen peroxide *Electrochim. Acta.* **97** 398–403
[72] Song H, Ni Y and Kokot S 2013 A novel electrochemical biosensor based on the hemin-graphene nano-sheets and gold nano-particles hybrid film for the analysis of hydrogen peroxide *Anal. Chim. Acta.* **788** 24–31
[73] Chung M G, Kim D H, Seo D K, Kim T, Im H U, Lee H M, Yoo J B, Hong S H, Kang T J and Kim Y H 2012 Flexible hydrogen sensors using graphene with palladium nanoparticle decoration *Sensors Actuators* B **169** 387–92
[74] Kar Man Fung C, Xi N, Lai K, Lou J and Chen H 2011 Noncryogenic, cooled IR sensors based on carbon nanotubes *SPIE Newsroom*
[75] Zhang H, Kulkarni A, Kim H, Woo D, Kim Y J, Hong B H, Choi J B and Kim T 2011 Detection of acetone vapor using graphene on polymer optical fiber *J. Nanosci. Nanotechnol.* **11** 5939–43

[76] Yao B, Wu Y, Cheng Y, Zhang A, Gong Y, Rao Y J, Wang Z and Chen Y 2014 All-optical Mach–Zehnder interferometric NH_3 gas sensor based on graphene/microfiber hybrid waveguide *Sens. Actuators* B **194** 142–8

[77] Cittadini M, Bersani M, Perrozzi F, Ottaviano L, Wlodarski W and Martucci A 2014 Graphene oxide coupled with gold nanoparticles for localized surface plasmon resonance based gas sensor *Carbon* **69** 452–9

[78] Some S, Xu Y, Kim Y, Yoon Y, Qin H, Kulkarni A, Kim T and Lee H 2013 Highly sensitive and selective gas sensor using hydrophilic and hydrophobic graphenes *Sci. Rep.* **3** 1–8

[79] Gupta B D and Verma R K 2009 Surface plasmon resonance-based fiber optic sensors: principle, probe designs, and some applications *J. Sens.* **2009** 979761

[80] Mishra S K, Tripathi S N, Choudhary V and Gupta B D 2014 SPR based fibre optic ammonia gas sensor utilizing nanocomposite film of PMMA/reduced graphene oxide prepared by *in situ* polymerization *Sens. Actuators* B **199** 190–200

[81] Snelders D J, Mackenzie F O V, Boersma A and Peeters R H 2016 Zeolites as coating materials for Fiber Bragg Grating chemical sensors for extreme conditions *Sens. Actuators* B **235** 698–706

[82] Sehrawat P, Islam S S, Mishra P and Ahmad S 2018 Reduced graphene oxide (rGO) based wideband optical sensor and the role of temperature, defect states and quantum efficiency *Sci. Rep.* **8** 1–13

[83] Vasu K S, Asokan S and Sood A K 2016 Enhanced strain and temperature sensing by reduced graphene oxide coated etched fiber Bragg gratings *Opt. Lett.* **41** 2604–7

[84] Kavinkumar T, Sastikumar D and Manivannan S 2015 Effect of functional groups on dielectric, optical gas sensing properties of graphene oxide and reduced graphene oxide at room temperature *RSC Adv.* **5** 10816–25

[85] Darafsheh A 2021 Photonic nanojets and their applications *J. Phys. Photon.* **3** 022001

[86] Yao B *et al* 2017 Graphene-enhanced Brillouin optomechanical microresonator for ultrasensitive gas detection *Nano Lett.* **17** 4996–5002

[87] Maharana P K, Jha R and Padhy P 2015 On the electric field enhancement and performance of SPR gas sensor based on graphene for visible and near infrared *Sens. Actuators* B **207** 117–22

[88] Choi S J, Kim S J and Kim I D 2016 Ultrafast optical reduction of graphene oxide sheets on colorless polyimide film for wearable chemical sensors *NPG Asia Mater.* **8** e315

IOP Publishing

Recent Advances in Graphene and Graphene-Based Technologies

Anoop Chandran, N V Unnikrishnan, M K Jayaraj, Reenu Elizabeth John and Justin George

Chapter 14

Graphene-based biosensors

Paulo M Zagalo, Susana Sério, Paulo A Ribeiro and Maria Raposo

Graphene and its derivatives present unique electrical and optical properties that have led to a rising interest in these materials over the past years. Currently, graphene and its derivatives are claimed for the fabrication of molecular electronics devices, namely sensor devices. This chapter presents an updated overview of relevant achievements on biosensors devices based on graphene molecules. Both electrical and optical measurements performed with these devices demonstrate that the developed graphene-based biosensors can detect a panoply of molecules, macromolecules, and viruses, thus making them an invaluable tool in many fields (e.g. medicine, industry, genetics, criminology). However, despite having been revealed to be adequate for bio-measurements, even at the edge of the detection limit, the overall device handling is not straightforward, which might hinder potential applications.

14.1 Introduction

The development of sensor devices based on electrical transducing has at its core the need to find and develop efficient ways through which the process of applying an external bias to a material becomes more optimized for relevant features such as sensitivity, linearity, resolution, hysteresis behavior, drift features, and durability, all contributing to device reliability. This process involves materials processing, electrical interfacing and encapsulation. These needs have been reinforced by the relatively recent growth of technologies based on nanotubes, hybrid conductors, and organic conductors [1, 2], particularly considering the rapid approach to the non-passable edge set by the limitations of performance and optimization of the widespread traditional silicon-based devices, which holds a quasi-monopoly of both the market and this type of technologies.

doi:10.1088/978-0-7503-3999-5ch14

These novel molecular organic structures present remarkable features that can be used for the development of novel sensor devices. An example has been reported by K S Novoselov *et al* with the occurrence of an electric field effect in few-layer graphene (FLG), a material with an innate two-dimensional (2D) nature [3]. The electric field effect in most semiconductors is responsible for the change of the carrier concentration, thus allowing a change in the electric current. It is possible to attain 2D graphene by isolating a single graphene layer, which holds in its composition a planar, sheet-like formation of carbon atoms that are tightly arranged into a benzene ring structure, displaying the thickness of a single atom (approx. 0.34 nm), a hexagonal repeating arrangement and is a common denominator among many carbon-based materials in nature, such as graphite (3D), nanotubes (1D), and fullerenes (0D) [4].

Aside from being approximately two hundred times stronger than steel and having zero effective mass, graphene also presents a myriad of properties that have shaped it into a highly sought after material, often being called a 'wonder material', such as large surface area (2630 m^2 g^{-1}); high electrical conductivity (approx. 202 S cm^{-1}); outstanding mechanical properties; high room-temperature electron mobility (250 000 cm^2 V^{-1} s^{-1}); good thermal conductivity (5000 W mK^{-1}); a modulus of elasticity of around 1 TPa; impermeability to gases and good optical transmittance (97.7%) [5–12]. All these featured properties have allowed it to be explored and put into practice through a multitude of applications including: supercapacitors; semiconductors; field-effect transistors (FETs); electronics; lithium-ion batteries; low-cost, highly efficient, transparent, and flexible solar cells; top of the line composite materials; strain sensors; and biosensors [13–27].

In the processes that brought about the inception, adaptation, and implementation of graphene devices, a number of graphene-derived subproducts have also been envisioned and developed, as is the case of graphene oxide (GO), graphene nanoribbons, graphene nanotubes (single- or multi-walled), and reduced graphene (be it through chemical, UV radiation, thermal, electrochemical procedures), polymers, graphene nano-sheets, quantum dots, amongst [28–34].

As mentioned, one of the applications of graphene-derived materials is strongly applied to the development of biosensors, in which different kinds of biosensors are welcome, namely for the detection of biological molecules or even virus associated with pathologies or pernicious to life, detection of specific DNA sequences, which is currently an area of tremendous interest since more and more research has shown that the mutations of genes are responsible for numerous inherited human disorders [35, 36]. The different applications found in the literature of graphene-based biosensors are summarized in figure 14.1. Interestingly, both electrical and optical techniques have been used to transduce the biosensors' properties. This chapter presents an overview of the use of these graphene-based biosensors.

14.2 Electrical biosensors

Electrochemical type biosensors are the most common ones as they are easy-to-implement, accurate, quick, and can be even sensitive biomarkers [37–39].

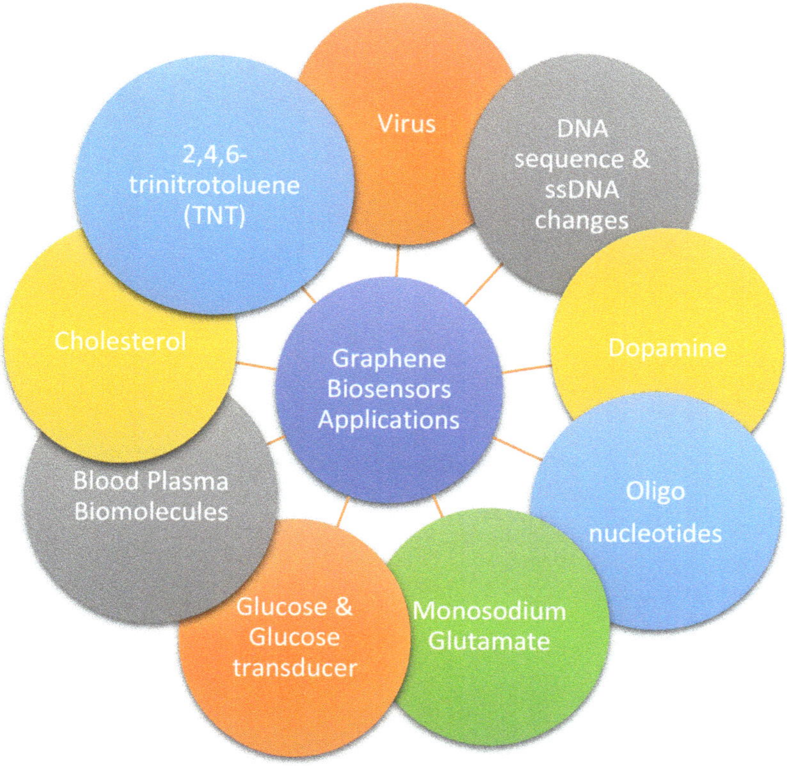

Figure 14.1. Different applications of graphene-based biosensors.

In addition, the sensitivity and selectivity of the electrochemical sensors can be significantly enhanced simply by the development of nanostructured modified electrodes. This section will demonstrate that the graphene nanostructures are interesting for the development of biosensors as evidenced by electrochemical methods.

14.2.1 Electrical biosensors: types and characteristics

Despite their promising applications in the biomedical sciences, the development of electrochemical biosensors with improved sensitivity and specificity together with a low detection limit is still a great challenge. As a remark, in what concerns biomolecular detection, the limits of detection can be extremely low as traces, say ppm or ppt, of a given molecular species often must be detected. Furthermore, the detection is to take place in complex biological media such as blood, urine, or tears.

In particular, because of the shown usefulness in diagnostic analysis of different molecules or entities associated with diseases found in biological complex media, electrochemical biosensors based on nanomaterials, including graphene-based films have been extensively investigated. An electrochemical sensor is based on the measurement of the current–voltage characteristic of an electrolyte media because

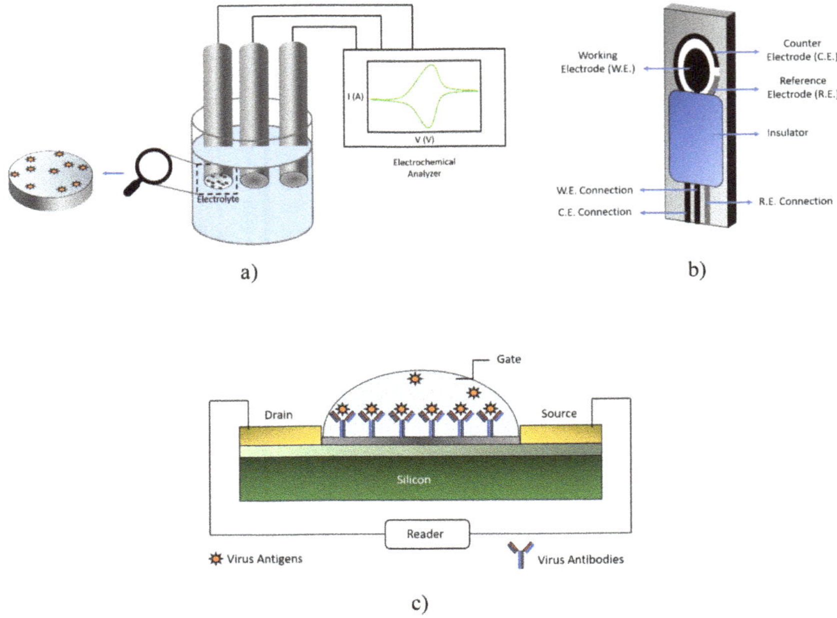

Figure 14.2. (a) Scheme of an electrochemical sensor. (b) A typical design of a screen-printed electrode. (c) FET biosensor.

of an applied voltage. The analyte to be detected is normally introduced in the electrolyte and the applied voltage induces an electrochemical reaction at the electrodes. Generally, the electrochemical sensor's response is the overall I–V characteristic curve with some features taken from it, which can be translated, via adequate calibration, into an applicable qualitative or quantitative output.

Thus, the overall process involves transducing and a chemical recognition system. Normally, the physical systems consist of a sensing electrode, a counter electrode, and often an extra reference electrode, as shown in figure 14.2(a). Currently, the electrodes are condensed in strips, as depicted in figure 14.2(b), which are low-cost and disposable devices specially designed to work with samples volumes of the order of microliters. These electrodes are screen-printed and are based on carbon, gold, platinum, silver, or carbon nanotubes inks. From the point of view of electrical measurements, the electrochemical sensors are divided into potentiometric, voltammetric, amperometry, and impedance spectroscopy types, depending on the physical variable being measured. In potentiometry, evaluation is normally done under flow or batch conditions, the change in potential is used to evaluate or quantify the amount of an ion in solution through the electrical current. A potentiometric sensor type, measures the difference in potential between the working electrode and a reference electrode, operating under conditions of near-zero current flow. Several electroanalytical voltammetry techniques variants have been implemented, namely, cyclic voltammetry (CV); differential pulse voltammetry (DPV); square wave voltammetry (SWV); linear sweep voltammetry (LSV), working at a controlled variable potential, the current being measured as a function of applied potential and electrochemical impedance spectroscopy (EIS).

Selectivity towards several analytes, sensitivity within a broad concentration range, fast analysis times (few seconds), and the ability to determine kinetic parameters and estimate the mechanisms of chemical and/or electrochemical reactions are the main advantages of the voltammetry [40]. Amperometry is based on the detection of an analyte by measuring the electrical current because of its reduction or oxidation. In the case of amperometry, at a constant applied potential, the measured current is related to the analyte concentration. As the mass transport of electroactive species should be governed only by convection, the amperometric techniques are generally run under a stirred solution or in a flow system and, therefore, can be integrated into microfluidic systems for continuous and real-time analysis. EIS technique is simple, inexpensive, fast, and has proved to be a useful tool for the characterization and analysis of materials and work as a transducer. In this technique, a sinusoidal potential over a wide frequency range is applied and both current response and phase between the electrical potential, V, and current, I, are measured over the frequency range [41]. This allows one to obtain the electrical impedance which is the ratio between voltage and current:

$$Z = \frac{v}{i} \tag{14.1}$$

Z is frequency dependent, and the result is an impedance spectrum $Z(w)$, which can be seen in several forms. The impedance in a complex form can be written as:

$$Z(w) = Z'(w) + i\,Z''(w) \tag{14.2}$$

Several graphical spectrum representations can come through either representing the module of impedance ($|Z(w)|$), real part of impedance ($Z'(w)$) and imaginary part of impedance ($Z''(w)$). In addition, other impedance features can be used to characterize the sensor impedance, namely the phase difference and the loss tangent. The shape of the attained spectrum can be modeled through a resistance capacitor (RC) equivalent circuit representing the electrolyte. This will consist of a series. The impedance spectrum features allow us to assess parallel RC circuits to both intrinsic material properties and the processes involved in the conductivity/resistivity (in-phase) or the capacity (not in phase) of the electrochemical system.

FETs have also been used as biosensors. A schematic of these devices is shown in figure 14.2(c).

14.2.2 Graphene-based electrical biosensors: applications

14.2.2.1 Glucose

The development of glucose sensors is of great relevance due to their potential for clinical utility for health care monitoring. Different kinds of sensors have been proposed with graphene and graphene derivatives. For example, a glucose biosensor was developed by depositing a solution (5:1) of GO and polyaniline (PANI) onto a 3 mm diameter polished glassy carbon electrode (GCE) [42]. This approach, which consists of a PANI/GO composite membrane, was demonstrated to improve the sensitivity and limit of detection of this glucose biosensor. Cyclic voltammetry was used to analyze the electrochemical responses of the biosensor when exposed to varying glucose phosphate-buffered saline (PBS) concentrations of solutions saturated with O_2. The PANI/GO/GCE biosensor displayed an overall sensitivity in a range between 0.1 and 10 mM concerning glucose. It was able to reach a limit of detection (LOD) of 0.1 mM, while exhibiting both a rather satisfactory long-term stability and good reproducibility. Other approaches for glucose biosensors were achieved by combining graphene sheets with metallic nanocomposites namely plain graphene (Gr); graphene–nickel (Gr–Ni); graphene–platinum (Gr–Pt); graphene–copper (Gr–Cu). And posteriorly attaching these to glucose oxidase enzymes and glassy carbon paste electrodes [43]. The syntheses of these nanocomposites all share similar processes, differing just in the starting material. These Gr sensors used graphite powder, Gr–Ni and Gr–Cu used graphene powder while Gr–Pt sensors used GO powder. Each of these biosensors when used as electrochemical sensors were revealed to have wide linear ranges and low limits of detection (magnitude of μM), with Gr–Cu reaching the lowest LOD at 2.87 μM. Gr–Ni, Gr, and Gr–Pt reached LOD values of 24.71, 7.2, and 3.06 μM, respectively.

14.2.2.2 Glutamate

Given glutamate's widespread use in the food industry, mainly in the form of monosodium glutamate (MSG), the need to detect and monitor its presence and concentration due to its potential pernicious consequences in human health [44, 45] becomes imperative. To detect this molecule and quantify its concentration, a biosensor has been developed by immobilizing glutamate oxidase (GLOx) on a tricobalt (Co_3O_4), chitosan (CS), and Gr nanocomposite film layer (Co_3O_4/CS/Gr), which was, in turn, deposited onto the surface of a glassy carbon electrode (GCE) [46]. An amperometric study was then conducted to ascertain the biosensor's sensitivity and LOD whilst detecting glutamate. The results showed that the Co_3O_4/CS/Gr/GCE sensor displayed a good reproducibility (and stability), a rather satisfactory response time (25 s), had a lower cost when compared to other existing biosensors and its development process was simple. The reported values of sensitivity and LOD were 0.73 μA mM^{-1} and 2.0 μM, respectively.

14.2.2.3 Dopamine

There has been an increase in interest and demand for wearable (ideally flexible), biodegradable and eco-friendly biosensors to be used in several applications, such as the detection of dopamine. A flexible biosensor based on a film layer of GO and CS (GO/CS) was developed to detect dopamine (DA) [47]. So, to achieve this objective, chitosan was functionalized with GO which caused it to change from a non-conductive material into a film layer that exhibits conductivity, a much desired and essential property for electrochemical biosensors. The developed biosensor showed it could reach a LOD of $\leqslant$20 μM while displaying the properties it set out to achieve by being flexible, eco-friendly, and biodegradable, which allows for its usage and subsequent disposal to be easier and seamless.

14.2.2.4 Cholesterol

Detecting cholesterol in human blood samples is paramount to diagnosing and identifying numerous food products. To this end, nanoplatelets of Gr were functionalized, and through chemical reduction gold nanoparticles (Au NPs) were decorated onto the surface of the Gr nanoplatelets [48]. This layer was deposited over a glassy carbon support electrode (GCE) and then cholesterol oxidase (Cox) was immobilized on its surface to work as a detection mechanism for cholesterol. The biosensor showed outstanding stability over time as well as an enhanced anti-interference ability against other biological species, which proved to be an advantage when detecting cholesterol. The Au/Gr biosensor displayed a sensitivity of 31.4 nA μM^{-1}.

14.2.2.5 Riboflavin

Riboflavin or vitamin B2 is a water-soluble vitamin and an essential constituent of flavoenzymes andh plays a vital role in the biochemical reactions in the human body and its deficiency is associated with eye lesions and skin disorders. Recently, Cioatesz [49] published a review about electrochemical sensors of riboflavin from different samples such as pharmaceutical products, cow and powder milk, non-alcoholic beer, or human plasma, using different kinds of electrodes including graphene-based films. Also, Sriramprabha *et al* [50] developed an electrochemical selective sensor of riboflavin by preparing a tin oxide and reduced graphene oxide (SnO_2)/rGO nanocomposite by a simple hydrothermal process. The reason for using SnO_2 is that it presents suitable oxidation states that allow a higher catalytic activity in sensor applications. Figure 14.3 demonstrates the interfacial electron transfer reaction facilitated by the active redox sites on the rGO matrix. This sensor based on SnO_2/rGO/GCE nanocomposite exhibited significant linearity in the range of 0.1–150 μmol l^{-1} from the SWV responses. This sensor demonstrated a 30 days

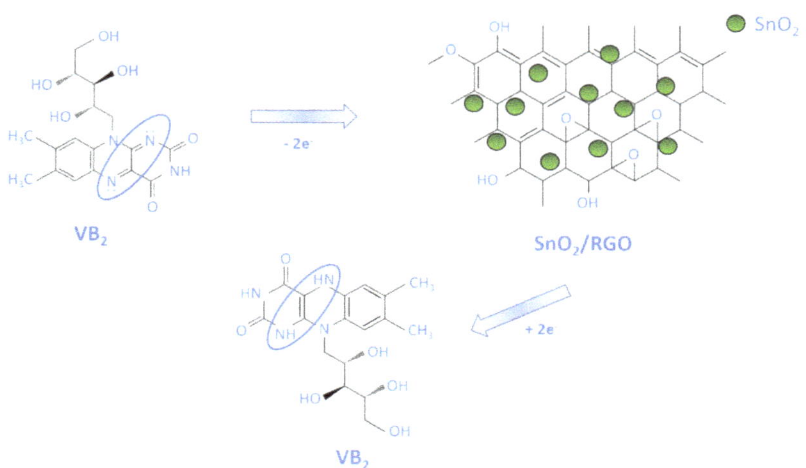

Figure 14.3. Redox mechanism of riboflavin on SnO_2/rGO modified electrode suggested by Sriramprabha *et al* [50].

electrode's stability and high sensitivity and selectivity towards riboflavin sensing and quantification in the presence of various common interferences like L-tryptophan, L-proline, L-cysteine, lysine, L-glutamic acid, L-ascorbic acid, glucose, cholesterol, dopamine hydrochloride, uric acid, thiamine, sodium sulfate, and ammonium chloride.

14.2.2.6 DNA biosensors

A myriad of areas, as is the case of the food industry, genetics, forensic science, criminology, and pharmacology has been relying evermore on diagnostic techniques based on the detection of DNA and by extension on the development and use of DNA-focused biosensors. A few examples of this development and growth are presented below.

A thionine–graphene (Th–Gr) based DNA electrochemical biosensor was created to detect complementary oligonucleotide, whilst having its fabrication procedure presented and its detection capabilities analyzed [51]. The Th–Gr component (layer) was achieved by adding an aqueous solution of thionine to a previously prepared graphene solution, the resulting solution was then stirred, heated, centrifuged, washed, and finally dissolved in distilled water. A gold electrode (AuE) was prepared and treated, to have deposited onto its surface a layer of Th–Gr (among other compounds) and a DNA probe (figure 14.4). The attained biosensor, device response was characterized through differential pulse voltammetry as to its ability to detect oligonucleotides. It was then demonstrated that this DNA biosensor displays good stability, a high degree of reproducibility, and good detection capabilities, having reached an LOD of 12.6 pM.

Similarly, to the already mentioned above biosensor, a Gr and polyaniline (PANI)-based electrochemical biosensor was developed [52]. However, the major difference lies in the sensor structures that have been implemented: PANI nanowires (PANIws) were used instead of PANI in their regular form. To this end, a GCE had its surface coated with graphene and later a layer of PANIw was deposited on the

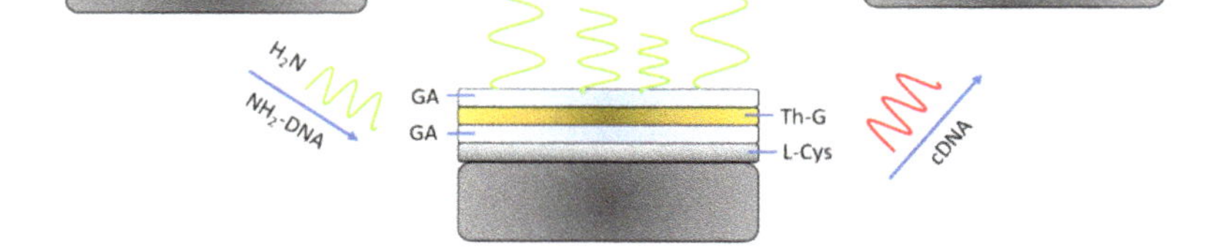

Figure 14.4. Schematic diagram of covalent immobilization of NH_2-substituted probe ssDNA on Th–G nanocomposite using glutaraldehyde as an arm linker.

graphene modified GCE. The final step of the preparation of this biosensor consisted of immobilizing DNA probes on its surface, which had the function of detecting DNA in the analyzed samples. The resulting sensor (ssDNA/PANIw/Gr/GCE) was subjected to several electrochemical studies and displayed high selectivity, good stability and reproducibility, high sensitivity, swift amperometric response, and an LOD of 32.5 pM.

14.2.2.7 Virus

A graphene-based biosensor was created to study its response regarding selectivity and sensitivity while detecting a pathogenic virus (rotavirus) [53]. The graphene film was prepared using a speed vacuum concentrator and a subsequent process of thermal annealing, having as a source material GO colloidal suspension (3 mg ml^{-1}). The result consisted of a free-standing graphene film (thickness of 400 nm) displaying a high conductivity (56 Ω/sq), as well as a rapid electron transfer property, which was posteriorly deposited on a glass substrate with an Au/Cr patterned configuration, thus obtaining a functioning graphene-based electrode. This electrode's surface was then functionalized and had rotavirus-specific antibodies immobilized on it, hence making this new biosensor capable of capturing rotavirus. The LOD was investigated and for several input cells of the virus corresponding to 10^3 pfu/mL, a sensitivity of about 1.3% was achieved.

After the most recent Zika virus outbreak in South and North America in 2015, the necessity to detect this virus became increasingly urgent, particularly with improved devices capable of achieving lower LODs and sensitivity. A graphene-based field-effect biosensor (FEB) was used as a means to detect in real time the presence of Zika antigens [54]. This biosensor consisted of a FET in which the gate dielectric was composed of an underlying layer of graphene on top where anti-Zika NS1 antibodies were immobilized. This biosensor allowed the detection of the virus through the variation of current flow and charge distribution in the FET. In this analysis, several virus contents were tested and from the signal information it was possible to observe that the LOD attained was 0.45 nM. It has also been demonstrated that the FEB displayed an outstanding selectivity to the target molecules, which in this case were the anti-Zika NS1 antigens. The present device not only has a low cost but is also portable, making it an invaluable tool when detecting this virus' presence in its early stages and allowing for an enhanced diagnosis.

Recently, El-Said *et al* [55] reported the synthesis of reduced porous graphene oxide (rPGO) ornated with Au NPs to modify the ITO electrode. This highly uniform ITO/Au NPs@rPGO structure was used as a surface-enhanced Raman spectroscopy-active surface and as a working electrode, since the use of the Au NPs and porous graphene enhance the Raman signals and the electrochemical conductivity, respectively. Anti-COVID-19 antibodies were immobilized onto the ITO modified electrode to capture the COVID-19 protein. The developed biosensor showed the capability of monitoring the COVID-19 protein within a concentration range from 100 nmol l^{-1} to 1 pmol l^{-1} with an LOD of 75 fmol l^{-1}. Furthermore, COVID-19 protein was detected based on electrochemical techniques within a concentration range from 100 nmol l^{-1} to

500 fmol l^{-1} that showed an LOD of 39.5 fmol l^{-1}, investigating three concentrations of COVID-19 protein spiked in human serum.

14.2.2.8 Other

Through the processes of self-assembly and photopolymerization, a biosensor composed of rGO layer with a thickness around 20 nm and hybridized with polydiacetylene (PDA) layer, was developed [56]. PDAs have been commonly used in some areas, namely in sensor development science and organic nano-electronics owing to some of their advantageous properties both of electrical and optical nature. Graphene's outstanding biocompatibility is also one of the reasons around which electrochemical biosensors and varied different related studies have emerged over the past years. The PDA/rGO biosensor's detection ability was analyzed through cyclic voltammetry and the results revealed that it not only can succeed in this task, but it also allows the detection of low-level analytes concentration, thus rendering it a strong candidate for future applications in the biomedical fields. Some of the detection features of such a type of device are summarized in table 14.1.

Table 14.1. Responses of graphene biosensors in the literature using electrical detection.

Detection	Measurement	Sensor	Range	Detection limit	Long-term	Sensitivity
Glucose [41]	Cyclic voltammetry	PANI/GO/GCE	0.1–10 mM	0.1 mM		
Glucose [42]	Amperometric	Gr	50–800 μM	7.2 μM		
		Gr–Ni	25–600 μM	24.71 μM		
		Gr–Cu		2.87 μM		
		Gr–Pt	10–600 μM	3.06 μM		
Glutamate	amperometric	Co_3O_4/CS/Gr		0.73 μA mM^{-1} and 2.0 μM		
Dopamine		GO/CS		⩽20 μM		
Cholesterol						31.4 nA/μM
Riboflavin	CV, ASV, and SWV	SnO_2/rGO/GCE	0.1–150 μmol l^{-1}	34 nmol l^{-1}	Stability for 30 days	
DNA	Cyclic voltammetry	Th–Gr		12.6 pM		
DNA	Cyclic voltammetry	ssDNA/PANIw/Gr/GCE		32.5 pM		
Rotavirus [52]	Cyclic voltammetry	Graphene film+Au/Cr		10^3 pfu ml^{-1}		1.3%
Zika virus [53]	FET	ITO/Au NPs@rPGO		0.45 nM		
COVID-19 [54]	Electrochemical techniques	ITO/Au NPs@rPGO/Anti-COVID-19 antibodies	100 nmol l^{-1} to 1 pmol l^{-1}	75 fmol l^{-1}		

14.3 Optical biosensors

Despite the interesting properties of electrochemical biosensors, they are not without limitations or weaknesses, which only spurs the need to design and develop graphene-based sensing devices whose core principle differs from already existing ones. A fine example of this multifaceted need for innovation is the biosensors that rely on the use of optical transducing principles. When compared to electrochemical devices, optical biosensors allow their users to conduct studies on living organisms without having to worry about sample damage (using electric current), while simultaneously collecting data with a high spatial and time resolution [57]. Graphene-based optical biosensors can make use of total internal reflection features at a graphene interface. In fact, under adequate conditions reflection at a graphene interface can exhibit both heightened broadband and polarization absorption, in such a way that the detection of the refractive index variation close to its surface can be performed. This type of sensor also allows for a deep and rather wide detection range, as well as high precision and sensitivity [58]. Several cases studies conducted and published pertinent to this subject will now be analyzed as fine examples of this technology.

14.3.1 Optical biosensors: types and characteristics

Surface plasmon resonance (SPR) relies on the detection of shifts in the refractive indexes brought about by molecular interactions that occur at the surface of metals, through surface plasmon waves. The phenomenon described as surface plasmon is manifested via the interaction between the incident photons and the electrons present in the metal's surface, causing the plasmon to travel parallelly to the surface, thus forming a surface plasmon wave which in turn propagates through the interface between the metal and dielectric parts and ends up being greatly attenuated (exponential behavior) in the dielectric [59] (figure 14.5). By having an uninterrupted incident beam, both the reflection angle and SPR angle are strictly dependent on the refractive index of the sensing medium, which is in tandem with the metal and thus means that variations on the surface's refractive index will trigger fluctuations on the

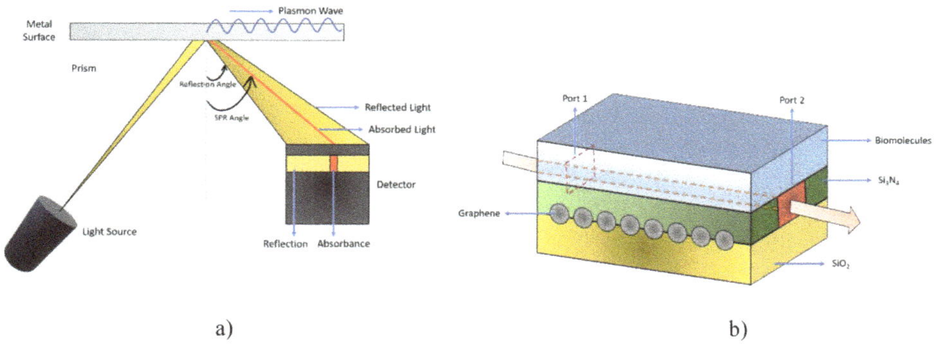

Figure 14.5. Schematic representation of the experimental setups for optical detection: (a) surface plasmon resonance apparatus; and (b) waveguide apparatus.

SPR measurements making it feasible to detect analytes (figure 14.5(a)). Therefore, analyte detection is made possible via the signal that reaches the detector containing information such as the reflected light intensity, shifts in the resonance angle, and absorbed light [60].

Waveguide configuration can also be used to develop a biosensor (figure 14.5(b)). The transduction principle can also rely on total internal reflection (TIR), which can be achieved by assuring that the incident light hits an optical waveguide at a specific angle (critical angle), thus resulting in a near-perfect reflection of the light inside the material. Usually, waveguides such as these are composed of the sample and substrate films, as well as an optically transparent layer through which light can travel, typically possessing a higher refractive index compared to the other layers of films present [61]. By taking advantage of TIR, the incident light can travel long stretches of distance but, as stated before, this process is not perfect, which means that a minor percentage of that light, also called an evanescent field, leaks through into the substrate and sample layers. The intensity of the evanescent field reaches its maximum peak at the sensing layer/surface and consequently, the overall sensitivity of the biosensor is greatly enhanced, which translates into a sensing device capable of high precision and sensitivity, making this kind of biosensor very desirable and coveted in myriads of applications [62].

14.3.2 Graphene-based optical biosensors: applications

14.3.2.1 Optical biosensors as an electronic nose

Biosensors that are based around the concept of an electronic nose [63] frequently have in their composition materials that seldom possess long periods of stability, despite being able to maintain their sensitivity, or in the case that it is possible to prolong their stability, the means to do so can be rather expensive and not easily implemented. Hence, the need arose to develop and test new biosensors that possess high sensitivity, good and long stability, and are cost-effective. The detection of explosives for example is of great relevance for applications ranging from environmental conservation to public safety and has received increasing attention throughout the scientific community. Many approaches, especially various sensors based on different technologies, have been proposed for the detection of explosives. Among them, biosensors emulating the olfactory systems have attracted tremendous attention for their portability and high sensitivity. This necessity birthed the idea of designing an optical biosensor based on GO with functionalized peptides aiming at the detection of explosives such as 2,4,6-trinitrotoluene (commonly referred to as TNT). As seen GO is a rather versatile material that not only exhibits outstanding optical and electrical properties but also can be simply altered by NPs, proteins, and other species making it suitable to be exploited to incorporate an electronic nose sensor device system. This is particularly relevant to confer specificity to the sensor device. These desirable properties allow for the functionalization of GO with remarkably useful compounds such as peptides, which possess different biological sensitivities and have an inherent biorecognition nature. Besides being possible to chemically engineer these molecules to bind only to very specific targets, they can also be tailored to improve stability over time and sensitivity which prolongs their usage lifetime. As an example, the optical detection of TNT has been performed through ultraviolet spectroscopy (200–400 nm) and by plotting absorbance spectra it was possible to demonstrate that the biosensor presented a great selectivity towards TNT when compared with 2,6-dinitrotoluene (DNT), which was used as a means of comparison. An LOD of 4.4 pM was achieved [64].

14.3.2.2 Waveguide biosensors

By studying optical responses such as fluorescence, transmittance, absorption, reflectance (among others) one can demonstrate the effectiveness, sensitivity, and detection power of an optical biosensor. These sensors can then be applied as detection devices in several areas, such as the food industry or medicine. It is in the latter that a graphene metasurface based sensitive infrared biosensor was designed to detect biomolecules in blood plasma form. The biosensor was created by depositing a waveguide layer of Si_3N_4 onto the top of a silica substrate and then by adding graphene perturbations in the waveguide, it became a leaky wave radiating structure. Several measurements were then performed by placing either the target blood plasma biomolecules or air molecules onto the biosensor's surface, to have a means of comparison when analyzing the results. A sensitivity of 431.42 nm/RIU was obtained in an infrared range of 250–320 THz [65].

To improve SPR sensor-based devices a simple base guided-wave SPR biosensor (GWSPR) and afterward a step-by-step procedure is normally used. For example, starting with a conventional SPR biosensor based on Kretschmann geometry [66] with a thin-film layer of silver (Ag), used to excite the surface plasmons of the sensing medium, thus achieving a sensitivity of 35.6°/RIU. A thin-film layer of silicon (Si) was then deposited onto the Ag layer to function as a waveguide, which in turn improved sensitivity to 41.8°/RIU. Then, by finally adding a hybrid layer of molybdenum disulfide (MoS_2) and Gr, the proposed GWSPR biosensor was complete and sensitivity of 45.2°/RIU was achieved. However, by adding more layers (where N denotes the number of layers) of MoS_2 the refraction index of the sensor suffers variations that alter the resonance angle, which in turn changes the overall sensitivity's value. It was then concluded that an improvement of 98.2% of the sensitivity was possible with $N = 6$, thus increasing the GWSPR biosensor's detection abilities [66].

14.3.2.3 Surface plasmon resonance biosensors

SPR biosensors based on Gr and GO with multiple layers have also been proposed and investigated as a means of detection of single-stranded (ss) DNA changes during the processes of formation of double-stranded (ds) DNA. To achieve this end, a conventional SPR biosensor configuration was adapted from a gold (Au) thin-film layer on top of a prism. Then two different types of sensors were prepared: one in which a layer of Gr coated the Au surface, increasing its electrical properties while simultaneously protecting processes of oxidation and contamination; and another type consisting of a deposition of a GO layer onto the Au layer, working as a biomolecular recognition element (BRE) for DNA molecules. To study the influence of multiple layers of Gr and GO in the sensitivities of the abovementioned sensors, experiments were performed in which the number of these layers varied from one up to ten and then the resulting sensitivity values were compared with that of a conventional Au/prism biosensor (91.11°/RIU). The Gr biosensor displayed an increase in sensitivity with the addition of layers, reaching its maximum at 10 Gr layers with a value of 120°/RIU. The GO sensor reached a maximum sensitivity value of 112°/RIU also for 10 GO layers. Given that the Gr and GO biosensors attained an improvement of 32% and 23% (respectively) regarding the conventional SPR biosensor, which revealed that using these materials in the development of devices of this nature is an important and invaluable step [67].

A graphene plasmonic-based infrared biosensor has also been developed and presented. Its configuration consisted of a nanofluidic biosensor chip, which was achieved by etching a nanochannel on a silicon substrate and having a layer of graphene placed on top of the substrate's surface, in such a way that the nanochannel remained free to let the nanofluid pass through it without any obstacles. The detection of the nanofluid passing through the nanochannel was made possible by using SPR. Through this method, a light source is aimed at the biosensor while the nanofluid is flowing beneath the graphene layer and changing the refractive index as a direct consequence of the absorption of its molecules in the channel. The polarized light emitted from the light source interacts with the surface plasmons causing them to undergo an excitation process and by having a detector be irradiated by the reflected light, it becomes possible to detect collective oscillations of the plasmons and to analyze the bonding processes that are occurring. Further simulations demonstrate that by optimizing this system sensitivity values up to 1920 nm/RIU can be attained [68]. Table 14.2 summarizes some responses of graphene-based biosensors reported in the literature.

Table 14.2. Responses of graphene-based biosensors described in the literature using optical detection.

Detection	Measurement	Sensor	Range	Detection Limit	Sensitivity
2,4,6-Trinitrotoluene (TNT)	Absorbance spectra	GO functionalized with peptides	0.4 μM–4.4 pM	4.4 pM	N/A
Biomolecules in blood plasma	Infrared	$SiO_2/Gr/Si_3N_4$	250 THz–320 THz	N/A	431.42 nm/RIU
Refractive index (RI) changes	Guided-wave SPR	MoS_2/Gr hybrid	RI between 1.33 and 1.8	N/A	89.6°/RIU
Single-stranded DNA	SPR	Au/Gr Au/GO	N/A	N/A	120°/RIU 112°/RIU
Nanofluid	SPR	Gr plasmonic	N/A	N/A	1920 nm/RIU
FOM	SPR	Au/Gr	N/A	N/A	33.05 RIU^{-1}
Biomolecules	SPR	Au/Si/MoS_2/Au/Gr	N/A	N/A	210°/RIU

N/A means not available.

To evaluate and categorize the quality and performance of an SPR biosensor, a set of properties and parameters need to be analyzed and studied. These comprise the biosensor's sensitivity, its sensibility, the angle through which the minimum value of reflectivity (resonance angle) is obtained, the full width at half maximum (FWHM) of the SPR dip, observed in the plots of reflectance versus incident angle and that represents the direct correspondence between the propagation constants of the incident beam of light and the surface plasmon wave. There is, however, another important tool commonly used to assess the sensor's performance which consists of the ratio between the sensitivity and the FWHM, which is dubbed figure of merit (FOM). These parameters have been theoretically investigated for a gold/graphene (Au/Gr) SPR biosensor [69]. The first step involved optimizing the minimum reflectivity attained at resonance angle by tuning the thickness of the Au layer, which was done by characterizing the response of different Au thicknesses between a set interval of wavelengths (580–1000 nm). Results revealed that the optimal Au layer thickness was 52.6 nm (at 710 nm). The performance of the biosensor when the number of graphene layers was changed was then investigated and it was observed that to achieve the optimal FOM value of 33.05 RIU^{-1}, there should only be present a single layer of Gr [69].

Conventional prism-metal based SPR biosensors are commonly used in a variety of fields when it comes to optical sensing. And although this configuration displays good results, it is always important to strive for higher standards. This has been investigated in a multilayer biosensor of silicon, molybdenum disulfide (MoS_2), and graphene, aside from the conventional layer/s of Au that is used in this type of device. Each one of these materials by itself exhibits an increased absorption rate, which then leads to an enhanced excitation of the surface plasmons and which in turn results in an increased value of sensitivity. Therefore, by building a biosensor with several layers of these materials (and optimizing it) it can be possible to attain an enhanced and improved SPR biosensor capable of achieving a sensitivity value of 210°/RIU [70].

Although it has been demonstrated that there is a linear dependency between the increase of Gr layers and the overall sensitivity in SPR biosensors, Bari *et al* [71] have managed to show that the increase of graphene layers on an SPR biosensor exhibits, in fact, a non-linear behavior. There is, however a linear behavior, associated with the gradual increase of Gr layers, between the concentration of targeted molecules and the SPR frequency (SPRF), and this linear dependency remains unchanged up to six layers of Gr. It was then shown that the sensitivity curves between the first and four Gr layers display a third-degree equation behavior, while between five and eight Gr layers the curves present a fourth-degree equation tendency. These results demonstrate that by increasing the number of Gr layers in an SPR biosensor, its sensitivity increases even further than previously shown and expected.

14.4 Conclusions

Graphene-based electrochemical biosensors present a panoply of advantages and useful properties such as swift response time, ability to remain stable for long periods, versatility and depending on the technique used, low production cost, to name just a few. Despite their promising applications in biomedical research, the development of electrochemical biosensors with improved sensitivity and low detection limits has remained a great challenge. Especially, because of the usefulness in diagnostic analysis of different molecules or entities associated with diseases and found in complex media (e.g. blood, urine, etc), electrochemical biosensors based on nanomaterials including graphene-based films have been studied.

The search for novel and more efficient sensor devices is an evermoving train braving new trails into the unknown, regardless of the base principle from which these biosensors branch out. When it comes to graphene-based optical biosensors, not only are new studies being conducted and published, but also research teams around the world seek to improve and upgrade already existing concepts. Due to their multiple advantages regarding signal-to-noise ratio, long stability, high sensitivity, selectivity, insusceptibility to external influences, optical biosensors have conquered a solid foothold on the vanguard of sensor science. And as the study and development of such devices continue to grow, their potential applications also keep growing and evolving, such as in the fields of environmental monitoring, food quality assessment, biomedical advances, medical diagnostics, or drug control.

Acknowledgments

This work was partially supported by the project through FEDER and FCT (Portugal) programs and by the research center Grant UID/FIS/04559/2020 to Laboratory for Instrumentation, Biomedical Engineering and Radiation Physics (LIBPhys-UNL) from the FCT/MCTES/PIDDAC (Portugal). The authors acknowledge also FCT, through research project grants PTDC/FIS-NAN/0909/2014 and the Bilateral Project entitled 'Deteção de Estrogénio-um Contaminante Emergente-em Corpos Hídricos' within the scope of 'Cooperação Transnacional_FCT (Portugal)-CAPES (Brazil) 2018'. PM Zagalo acknowledges his fellowship PD/BD/142767/2018 from RABBIT Doctoral Programme.

References

[1] Baughman R H, Zakhidov A A and De Heer W A 2002 Carbon nanotubes—the route toward applications *Science* **297** 787–92

[2] Dimitrakopoulos C D and Mascaro D J 2001 Organic thin-film transistors: a review of recent advances *IBM J. Res. Dev.* **45** 11–27

[3] Novoselov K S, Geim A K, Morozov S V, Jiang D E, Zhang Y, Dubonos S V, Grigorieva I V and Firsov A A 2004 Electric field effect in atomically thin carbon films *Science* **306** 666–9

[4] Spain. I 1981 Chemistry and physics of carbon *Carbon* **19** 119–304

[5] Balandin A A, Ghosh S, Bao W, Calizo I, Teweldebrhan D, Miao F and Lau C N 2008 Superior thermal conductivity of single-layer graphene *Nano Lett.* **8** 902–7

[6] Novoselov K S, Geim A K, Morozov S V, Jiang D, Katsnelson M I, Grigorieva I, Dubonos S and Firsov A 2005 Two-dimensional gas of massless Dirac fermions in graphene *Nature* **438** 197–200

[7] Zhu Y, Murali S, Cai W, Li X, Suk J W, Potts J R and Ruoff R S 2010 Graphene and graphene oxide: synthesis, properties, and applications *Adv. Mater.* **22** 3906–24

[8] Lee C, Wei X, Kysar J W and Hone J 2008 Measurement of the elastic properties and intrinsic strength of monolayer graphene *Science* **321** 385–8
[9] Sheehy D E and Schmalian J 2009 Optical transparency of graphene as determined by the fine-structure constant *Phys. Rev.* B **80** 193411
[10] Khan M E, Khan M M and Cho M H 2018 Recent progress of metal–graphene nanostructures in photocatalysis *Nanoscale* **10** 9427–40
[11] Xiang Q and Yu J 2013 Graphene-based photocatalysts for hydrogen generation *J. Phys. Chem. Lett.* **4** 753–9
[12] Chowdhury S and Balasubramanian R 2014 Graphene/semiconductor nanocomposites (GSNs) for heterogeneous photocatalytic decolorization of wastewaters contaminated with synthetic dyes: a review *Appl. Catal.* B **160** 307–24
[13] Yoo J J *et al* 2011 Ultrathin planar graphene supercapacitors *Nano Lett.* **11** 1423–7
[14] Brownson D A and Banks C E 2012 Fabricating graphene supercapacitors: highlighting the impact of surfactants and moieties *Chem. Commun.* **48** 1425–7
[15] Hu Y *et al* 2021 Recent advance of graphene/semiconductor composite nanocatalysts: synthesis, mechanism, applications and perspectives *Chem. Eng. J.* **414** 128795
[16] Huang J, Hines D R, Jung B J, Bronsgeest M S, Tunnell A, Ballarotto V, Katz H E, Fuhrer M S, Williams E D and Cumings J 2011 Polymeric semiconductor/graphene hybrid field-effect transistors *Org. Electron.* **12** 1471–6
[17] Schwierz F 2010 Graphene transistors *Nat. Nanotechnol.* **5** 487–96
[18] Eda G and Chhowalla M 2010 Chemically derived graphene oxide: towards large-area thin-film electronics and optoelectronics *Adv. Mater.* **22** 2392–415
[19] Kumar P, Singh A K, Hussain S, Hui K N, San Hui K, Eom J, Jung J and Singh J 2013 Graphene: synthesis, properties and application in transparent electronic devices *Rev. Adv. Sci. Eng.* **2** 238–58
[20] Le Xie J, Guo C X and Li C M 2014 Construction of one-dimensional nanostructures on graphene for efficient energy conversion and storage *Energy Environ. Sci.* **7** 2559–79
[21] Patil K, Rashidi S, Wang H and Wei W 2019 Recent progress of graphene-based photoelectrode materials for dye-sensitized solar cells *Int. J. Photoenergy* **2019** 1812879
[22] Kong X, Zhang L, Liu B, Gao H, Zhang Y, Yan H and Song X 2019 Graphene/Si Schottky solar cells: a review of recent advances and prospects *RSC Adv.* **9** 863–77
[23] Das S, Pandey D, Thomas J and Roy T 2019 The role of graphene and other 2D materials in solar photovoltaics *Adv. Mater.* **31** 1802722
[24] Potts J R, Dreyer D R, Bielawski C W and Ruoff R S 2011 Graphene-based polymer nanocomposites *Polymer* **52** 5–25
[25] Papageorgiou D G, Kinloch I A and Young R J 2015 Graphene/elastomer nanocomposites *Carbon* **95** 460–84
[26] Raju A P A, Lewis A, Derby B, Young R J, Kinloch I A, Zan R and Novoselov K S 2014 Wide-area strain sensors based upon graphene-polymer composite coatings probed by Raman spectroscopy *Adv. Funct. Mater.* **24** 2865–74
[27] Szunerits S and Boukherroub R 2018 Graphene-based biosensors *Interface Focus* **8** 20160132
[28] Wu T, Liu S, Li H, Wang L and Sun X 2011 Production of reduced graphene oxide by UV irradiation *J. Nanosci. Nanotechnol.* **11** 10078–81
[29] Yang X, Dou X, Rouhanipour A, Zhi L, Räder H J and Müllen K 2008 Two-dimensional graphene nanoribbons *J. Am. Chem. Soc.* **130** 4216–7

[30] Cai J *et al* 2010 Atomically precise bottom-up fabrication of graphene nanoribbons *Nature* **466** 470–3
[31] Mohan V B, Lau K T, Hui D and Bhattacharyya D 2018 Graphene-based materials and their composites: a review on production, applications and product limitations *Composites* B **142** 200–20
[32] Jia X, Campos-Delgado J, Terrones M, Meunier V and Dresselhaus M S 2011 Graphene edges: a review of their fabrication and characterization *Nanoscale* **3** 86–95
[33] Singh V, Joung D, Zhai L, Das S, Khondaker S I and Seal S 2011 Graphene based materials: past, present and future *Prog. Mater Sci.* **56** 1178–271
[34] Shen J, Zhu Y, Yang X and Li C 2012 Graphene quantum dots: emergent nanolights for bioimaging, sensors, catalysis and photovoltaic devices *Chem. Commun.* **48** 3686–99
[35] O'Malley M and Hutcheon R G 2007 Genetic disorders and congenital malformations in pediatric long-term care *J. Am. Med. Dir. Assoc.* **8** 332–4
[36] Van Borsel J and Tetnowski J A 2007 Fluency disorders in genetic syndromes *J. Fluen. Disord.* **32** 279–96
[37] El-Said W A, Kim T H, Chung Y H and Choi J W 2015 Fabrication of new single cell chip to monitor intracellular and extracellular redox state based on spectroelectrochemical method *Biomaterials* **40** 80–7
[38] Asad M, Zulfiqar A, Raza R, Yang M, Hayat A and Akhtar N 2020 Orange peel derived C-dots decorated CuO nanorods for the selective monitoring of dopamine from deboned chicken *Electroanalysis* **32** 11–8
[39] Khalifa M M, Elkhawaga A A, Hassan M A, Zahran A M, Fathalla A M, El-Said W A and El-Badawy O 2019 Highly specific electrochemical sensing of *Pseudomonas aeruginosa* in patients suffering from corneal ulcers: a comparative study *Sci. Rep.* **9** 1–12
[40] Ghanam A, Mohammadi H, Amine A, Haddour N and Buret F 2021 Chemical sensors: Electrochemical sensors; voltammetry/amperometry *Reference Module in Biomedical Sciences* (Amsterdam: Elsevier)
[41] Ribeiro A B, Mateus E P and Couto N 2016 Electrokinetics across disciplines and continents *New Strategies for Sustainable Development* (Heidelberg: Springer International Publishing)
[42] Ge Y, Chen Z, Jin Q, Mao H, Liu J and Zhao J 2013 Polyaniline/graphenebased electrochemical biosensor for sensitive detection of glucose *Asian J. Chem.* **25** 5036–8
[43] Altuntas D B, Tepeli Y and Anik U 2016 Graphene-metallic nanocomposites as modifiers in electrochemical glucose biosensor transducers *2D Mater.* **3** 034001
[44] Batra B and Pundir C S 2013 An amperometric glutamate biosensor based on immobilization of glutamate oxidase onto carboxylated multiwalled carbon nanotubes/gold nanoparticles/chitosan composite film modified Au electrode *Biosens. Bioelectron.* **47** 496–501
[45] Basu A K, Chattopadhyay P, Roychudhuri U and Chakraborty R 2006 A biosensor based on co-immobilized L-glutamate oxidase and L-glutamate dehydrogenase for analysis of monosodium glutamate in food *Biosens. Bioelectron.* **21** 1968–72
[46] Dalkıran B, Erden P E and Kılıç E 2017 Graphene and tricobalt tetraoxide nanoparticles based biosensor for electrochemical glutamate sensing *Artif. Cells Nanomed. Biotechnol.* **45** 340–8
[47] Kafi M A, Paul A and Dahiya R 2017 Graphene oxide-chitosan based flexible biosensor *2017 IEEE SENSORS* 1–3
[48] Aravind S S J, Baby T T, Arockiadoss T, Rakhi R B and Ramaprabhu S 2011 A cholesterol biosensor based on gold nanoparticles decorated functionalized graphene nanoplatelets *Thin Solid Films* **519** 5667–72

[49] Cioates C N 2020 Electrochemical sensors used in the determination of riboflavin *J. Electrochem. Soc.* **167** 037558
[50] Sriramprabha R, Divagar M, Ponpandian N and Viswanathan C 2018 Tin Oxide/reduced graphene oxide nanocomposite-modified electrode for selective and sensitive detection of riboflavin *J. Electrochem. Soc.* **165** B498
[51] Zhu L, Luo L and Wang Z 2012 DNA electrochemical biosensor based on thionine-graphene nanocomposite *Biosens. Bioelectron.* **35** 507–11
[52] Bo Y, Yang H, Hu Y, Yao T and Huang S 2011 A novel electrochemical DNA biosensor based on graphene and polyaniline nanowires *Electrochim. Acta* **56** 2676–81
[53] Liu F, Choi K S, Park T J, Lee S Y and Seo T S 2011 Graphene-based electrochemical biosensor for pathogenic virus detection *Biochip J.* **5** 123–8
[54] Afsahi S *et al* 2018 Novel graphene-based biosensor for early detection of Zika virus infection *Biosens. Bioelectron.* **100** 85–8
[55] El-Said W A, Al-Bogami A S and Alshitari W 2022 Synthesis of gold nanoparticles@ reduced porous graphene-modified ITO electrode for spectroelectrochemical detection of SARS-CoV-2 spike protein *Spectrochim. Acta* A **264** 120237
[56] Yun J S, Yang K S and Kim D H 2011 Multifunctional polydiacetylene-graphene nano-hybrids for biosensor application *J. Nanosci. Nanotechnol.* **11** 5663–9
[57] Xing F, Liu Z B, Deng Z C, Kong X T, Yan X Q, Chen X D, Ye Q, Zhang C P, Chen Y S and Tian J G 2012 Sensitive real-time monitoring of refractive indexes using a novel graphene-based optical sensor *Sci. Rep.* **2** 1–7
[58] Wu L, Chu H S, Koh W S and Li E P 2010 Highly sensitive graphene biosensors based on surface plasmon resonance *Opt. Express* **18** 14395–4400
[59] Malmqvist M 1993 Biospecific interaction analysis using biosensor technology *Nature* **361** 186–7
[60] Li G 2018 *Nano-Inspired Biosensors for Protein Assay with Clinical Applications* (Amsterdam: Elsevier)
[61] Taitt C R, Anderson G P, Lingerfelt B M, Feldstein M J and Ligler F S 2002 Nine-analyte detection using an array-based biosensor *Anal. Chem.* **74** 6114–20
[62] Kovacs B and Horvath R 2019 Modeling of label-free optical waveguide biosensors with surfaces covered partially by vertically homogeneous and inhomogeneous films *J. Sens.* **2019** 1762450
[63] Capelli L, Sironi S and Del Rosso R 2014 Electronic noses for environmental monitoring applications *Sensors* **14** 19979–20007
[64] Zhang Q, Zhang D, Lu Y, Yao Y, Li S and Liu Q 2015 Graphene oxide-based optical biosensor functionalized with peptides for explosive detection *Biosens. Bioelectron.* **68** 494–9
[65] Patel S K, Parmar J, Kosta Y P, Ladumor M, Zakaria R, Nguyen T K and Dhasarathan V 2020 Design of graphene metasurface based sensitive infrared biosensor *Sens. Actuators* A **301** 111767
[66] Wu L, Guo J, Dai X, Xiang Y and Fan D 2018 Sensitivity enhanced by MoS_2–graphene hybrid structure in guided-wave surface plasmon resonance biosensor *Plasmonics* **13** 281–5
[67] Meshginqalam B, Ahmadi M T, Ismail R and Sabatyan A 2017 Graphene/graphene oxide-based ultrasensitive surface plasmon resonance biosensor *Plasmonics* **12** 1991–7
[68] Wei W, Nong J, Zhang G and Zhu Y 2014 An infrared biosensor based on graphene plasmonic for integrated nanofluidic analysis *Proc. SPIE – Int. Soc. Opt. Eng.* 9278 33–41

[69] Chen S and Lin C 2019 Figure of merit analysis of graphene based surface plasmon resonance biosensor for visible and near infrared *Opt. Commun.* **435** 102–7

[70] Shushama K N, Rana M, Inum R and Hossain M 2017 Sensitivity enhancement of graphene coated surface plasmon resonance biosensor *Opt. Quantum Electron.* **49** 1–13

[71] Bari M M, Sarkar A K and Hossain S 2016 Sensitivity analysis of a graphene based surface plasmon resonance biosensor in terms of number of graphene layers *2016 2nd Int. Conf. on Electrical, Computer & Telecommunication Engineering (ICECTE)* (Piscataway, NJ: IEEE) pp 1–4

Chapter 15

Graphene membranes and coatings

Avishek Chanda and Avishek Mishra

Development of modern membranes and coatings and the quest for advanced techniques and material systems have led to the use of graphene (Gr) and its oxide. Gr-based structures are known to have excellent mechanical strengths, chemical inertness, impermeability to gases, good adsorption, lubricity, antibacterial properties and thermal stability resulting in the huge boost in research for different applications. This chapter outlines the fabrication techniques of Gr-based membranes and coatings, while particularly focusing on the applications and the critical studies on the same. The recent advances in the use of Gr membranes and coatings and their future potentials are also highlighted.

15.1 Introduction

Graphene (Gr) is the newest allotrope of carbon with a 2D hexagonal lattice structure, which results in excellent and unique properties [1]. Gr is an atomically thin material, the thinnest material in existence, with significantly high thermal, $\geqslant$3000 W m^{-1} K^{-1} [2], and electrical, $\geqslant 2.5 \times 10^5$ cm^2 V^{-1} s^{-1} [3], conductivities. It also has a Young's modulus $\geqslant$1TPa and an intrinsic strength of ~130 GPa [4], making it exceptionally strong mechanically due to the C–C bonds which act as beams [5]. The excellent conductivities and mechanical strengths make Gr a versatile material suitable for a wide range of applications [6]. A range of applications, including electrodes, flexible electronics, sensors and energy storage devices has experienced the involvement of Gr and its derivatives [7]. Other unique and interesting characteristics of Gr include resistance to any type of chemical including acids, bases and salts [8, 9], and having antibacterial properties [10], impermeability to gases [11], various other membrane applications, thermal stability [12, 13], high specific surface areas [14] and along with all these, being environmentally friendly [15].

doi:10.1088/978-0-7503-3999-5ch15

These unique aspects significantly help in providing a strong ground to achieve high performance membranes and surface coatings, having various functions.

Nature uses membrane technology to permeate ingredients through its connected networks and to mimic that ability; the inception of synthetic membranes by humankind has led to a new technological advancement. Membranes can be defined as a barrier which selectively permeates material through its pores or channels [16]. Membranes have been improved over the past two decades to provide viable solutions to many challenges faced by mankind with respect to global pollution, large-scale energy efficient separation and they are also key components in advanced equipment systems such as battery packs, solar cells, purification processes etc. Membranes ideally function by providing a higher membrane flux, which is defined as the amount of permeate produced per unit area of membrane, where improved selectivity and their pore dimensions and shapes control the membrane stability [17]. Polymeric and inorganic membranes have shown great potential for industrial applications, but they fail in terms of permeability and selectivity. Membrane technologies are addressed to various filtration types such as microfiltration, ultra-filtration, nanofiltration, reverse osmosis, forward osmosis, dialysis, electrodialysis, gas separation and pervaporation. Metal–organic framework compounds (MOFs), molecular sieves and carbon nanotubes (CNTs) have been extremely successful in showing excellent reverse osmosis membrane permeation flux [18].

Existing membrane technologies have high cost and manufacturing difficulties, but since 2010 the synthesis of 2D Gr helped manufacture atomic thickness and nearly frictionless surfaced membranes with maximum permeate flux and minimum transport resistance. Based on the membrane microstructures, Gr membranes can be classified as porous Gr layer, assembled Gr laminates and Gr-based composites [17]. Manufacturing of Gr membranes can be done by mechanical exfoliation, chemical exfoliation, chemical synthesis, pyrolysis, epitaxial growth and chemical vapor deposition (CVD) [19]. The membranes produced from these manufacturing techniques can fit into various roles depending on the requirements which can be further divided as pressure-driven membranes, concentration-driven membranes and electric-driven membranes [18].

Another important and growing field for Gr application is coating, which can be defined as that material that can cover the surface of a particular bulk material system, giving it unique and specific properties. Engineered coatings, both synthetic and natural, are usually applied on surfaces for functional or decorative purposes and even for both. Major goals can include but are not be limited to resistance from corrosion, mechanical wear, fouling, environmental damage, microorganism growth, bacterial and septic growths, mechanical damage protection and many others [20]. Thus, the use of coatings in everyday life is immense, with considerable usage in military, shipping, aircraft, automobile, medical, electronic and household utility industries. Gr coatings interestingly have another functional applicability where they can be designed to absorb chemical contaminants and hazardous gases, such as NO_x, CO, SO_x, NH_3, H_2S, sulfur containing compounds like organothiols and others [21], from various environments. This extends applications to absorbing

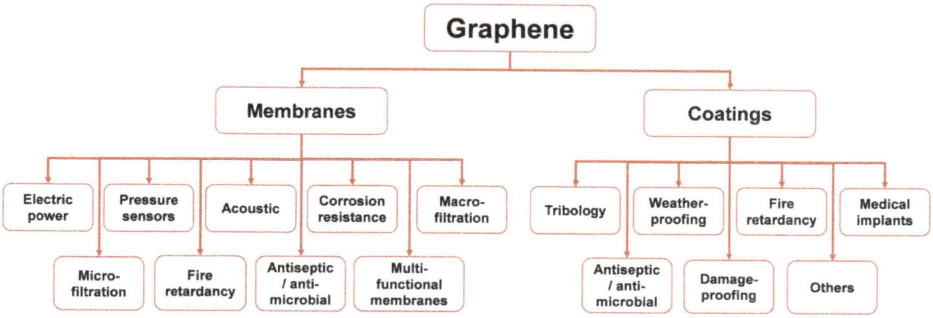

Figure 15.1. Application-oriented of Gr-based membranes and coatings.

toxins, resisting fire/frost/irradiation and killing bacteria, with the possible applications in protection against chemical and biological weaponry [19].

Conventional coatings, which have been used in various applications, often result in weak outcomes or hazardous bi-products or are associated with negative impacts on the environment, leading to many being banned or restricted from being used. Tributylin (TBT) and hexavalent chromium (Cr(VI)), commonly used as metal coatings for many decades, have been restricted due to the carcinogenic risks and biocidal properties of the elements. The imposed limitation includes Cr(VI) percentage to be reduced to only 0.1% in corrosion preventive coatings from 1 July 2007 [20], and TBT being completely prohibited as anti-fouling coatings to protect marine life from 1 January 2008 [22]. Other components, substantially used as protective coatings, namely, cobalt, copper and cadmium, have also been labelled as carcinogens and toxic to the ecosystem. Therefore, it is extremely important for the coating industry to find alternatives that are eco-friendly and useful. This work illustrates the usefulness, applications and recent advancements in Gr membranes and coatings that can be a viable alternative to the conventional materials used, giving the industries a significant environment friendly and strong option. An application-oriented classification of the various Gr-based membranes and coatings is illustrated in figure 15.1.

15.2 Graphene membranes

Membranous structure developed by stacking layers of Gr or Gr oxide on top of each other are known as Gr membranes or Gr oxide membranes. These resemble the structure of nacre (mother of pearl). These membranes have the capability of forming a perfect barrier against liquids and gasses.

Based on their microstructures they can be classified into three main types:

(1) a porous Gr layer,
(2) assembled Gr laminates, and
(3) a Gr-based composite.

Nanopores exist in Gr membranes, and they can be functionalized to facilitate membrane separation which has recently emerged as a critical method for addressing environmental issues. Due to its regulated pore size and shape, it is perfect for increased permeate flux, higher selectivity, and improved stability. To increase permeability and obtain better throughput, the membrane thickness must be reduced, which will improve membrane performance [17].

15.2.1 Graphene membranes and graphene-based membranes

The Gr sheet itself acts as an excellent membrane for separation. The single carbon atom thick Gr shows maximum membrane permeance that is inversely related to the membrane thickness. The dense delocalized π-orbital cloud makes the single-layered Gr impermeable to molecules as it blocks the gap within its aromatic rings. Bunch and colleagues experimentally showed the impermeability of Gr sheets [23]. In addition to its high mechanical strength and inert chemical properties, being one atom thick with a two-dimensional structure has opened up several material applications along with its usefulness in creating highly effective separation membranes. Despite the excellent separation performance, fabrication of large-area single/few-layered Gr membranes with uniformity in structure remains a challenge [24].

On the other hand, the synthesis of Gr-based materials has a major influence on the membrane synthesis, microstructure and permeability of the membranes as it provides a more practical approach for application of Gr-based membranes. Out of all Gr derivatives graphene oxide (GO) is considered to be one of the most important one that exhibits similar properties to Gr but has a significantly large number of oxygen atoms containing functional groups attached to the edges and basal planes of the films, which can be any of the epoxy, hydroxyl, carbonyl and carboxyl groups. In this context, the processes of phase inversion, solvent evaporation, *in situ* polymerization, and melt mixing have produced the most successful high-performance membranes [25].

15.2.2 Synthesis of membranes

Despite the excellent separation performance fabrication of large-area single/few-layered Gr membranes with uniformity in structure remains a challenge, and the physicochemical features of the GO sheet, as well as those of its derivative membranes, may be easily tuned using these oxygen functional groups. On the one hand, the hydrophilic groups provide for stable GO sheet dispersions in aqueous conditions, allowing for easy processing and stacking [26]. Furthermore, the functional groups give reactive handles for diverse GO alterations, allowing for easy control of GO membrane microstructures and chemical characteristics, as well as the development of GO-based hybrid membranes.

Methods for generating Gr can be divided as either top-down or bottom-up methods, as shown in figure 15.2, which are broadly classified into five major types: (1) mechanical exfoliation, solution based exfoliation, electrochemical exfoliation processes of a single Gr sheet from bulk graphite [28]; (2) crystal growth of Gr films; (3) CVD of Gr monolayers; (4) unzipping of CNTs; and (5) reduction of Gr derivatives such as GO high-quality Gr monolayers; however, ultrasonic cleavage of

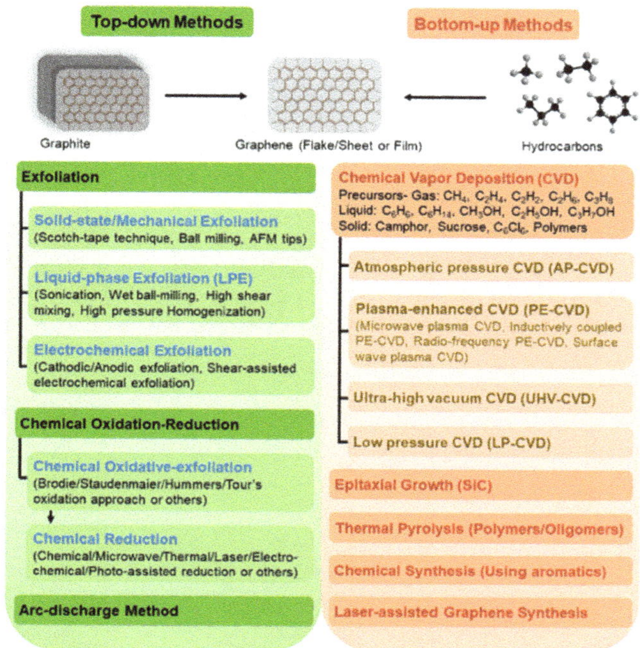

Figure 15.2. Illustration of the current method of preparation of Gr sheets. Reproduced from [27], copyright 2021, with permission from Elsevier.

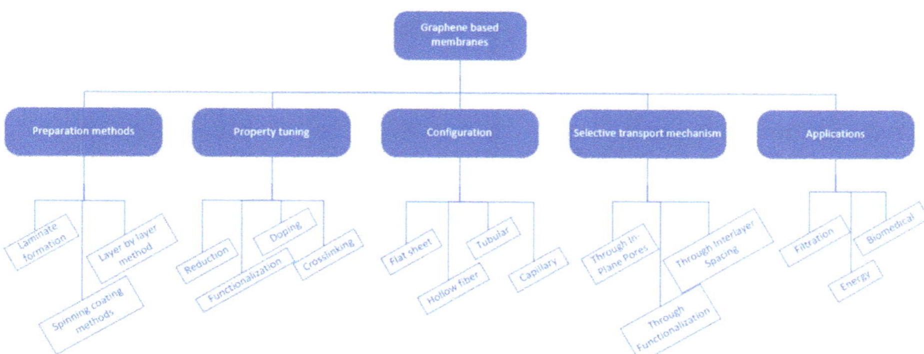

Figure 15.3. Classification of Gr-based membranes.

graphite and reduction of Gr derivatives enable for scale Gr layer manufacturing. In general, GO is made using Brodie [29], Staudenmaier [30], and Hummers *et al* [31] procedures, which involve oxidizing graphite to various degrees. The extent of graphite oxidation, as measured by the carbon/oxygen ratio, is determined by both the synthesis technique and the reaction time. Figure 15.3 shows the classification of Gr-based membranes.

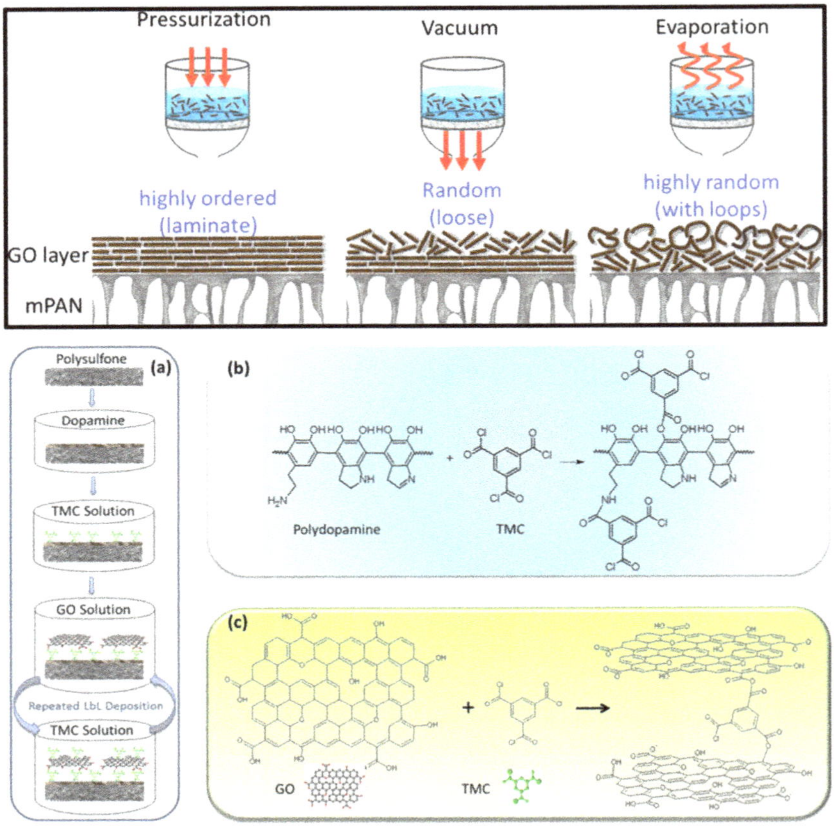

Figure 15.4. Schematic of assembled GO laminates prepared by different methods [24, 32]. Top, reproduced from [24], copyright 2015, with permission from Elsevier. Bottom, reproduced with permission [32], Copyright 2013, American Chemical Society.

Figure 15.4 illustrates assembled GO laminates prepared by different methods [24, 32]. The most common technique for making membranes is layer-by-layer construction, melt technique, solvent/non-solvent coagulation, *in situ* polymerization, solution technique, and spin coating. Hybrid composite and nanocomposite membranes are given particular attention in membrane technology. The type of additive, additive concentration, pore size, pore distribution, polymer concentration, and fabrication method have all been proven to have an impact on the properties of the membrane. Nanofiller content, random stacking of Gr nanosheets, and Gr functionalization all can influence the performance and characteristics of membranes.

15.2.3 Use of graphene as a membrane

With high selective permeability of Gr membrane, the potential use of it in food and pharmaceuticals packaging can stop the transfer of oxygen and water, resulting in improved shelf life. As per the prediction of UN by 2025, 14% of the world's population is likely to experience water scarcity which brings out the question of how can clean water be produced from recycling wastewater. Gr membranes have been considered for this purpose as their nanochannels formed in their porous structure helps eliminate impurities to a great extent. Highly selective permeability of gasses is also possible with Gr membranes, which can help removal of carbon dioxide from power station exhaust gases and Gr has the potential to carry out this economically as well.

They can create membranes with little transport barrier and maximum permeate flux since their surface is only one atom thick and almost frictionless. Because of the significant electron density of its aromatic rings, the ideal single-layer Gr sheet is impenetrable to gases as tiny as helium. Atoms and molecules attempting to pass through these rings are repelled by them [23]. Therefore, early theoretical investigations concentrated on making porous Gr membranes by making holes in Gr sheets to allow for the selective passage of water, ions, and gases [33–35]. Despite a few recent attempts, the enormous issue of perforating selective pores in large area continues despite outstanding separation performance being promised. As an alternative, Gr, particularly GO nanosheets, can be put together into laminar structures via filtration or coating techniques, creating quick and precise 2D nanochannels for the transportation of tiny molecules. The ability to apply Gr-based membranes for a variety of membrane processes, including pressure filtration (such as ultrafiltration, nanofiltration, reverse osmosis, and forward osmosis), pervaporation, and gas separation, are made possible by the membranes' distinctive microstructures and transport pathways [19].

15.2.4 Applications

Gr-based separation membranes or GBSMs make a major contribution in catalyzing selective regulatory transmittance of various molecules through nanopores in single-layer Gr and Gr derivative membranes for applications like water desalination, chemical mixture separation and gas control.

15.2.4.1 Gas transport

Porous Gr membranes have emerged as excellent alternatives for gas-separation applications because of their atomic thickness that enables ultrahigh permeance, but they have low gas selectivity. Ashirov *et al* [36] states that decreasing the pore size below 3 nm is expected to improve gas selectivity due to molecular sieving and the group studied a pore narrowing approach using gold deposition onto porous graphene surface to tune the pore size and thickness of the membrane to increase the number of small pores. In contrast, Yuan *et al* [37] stated the limitation of etching-based pore generation methods that cannot decouple pore nucleation and pore growth which leads to a trade-off between high pore density and high selectivity. In their work, they have highlighted the advantages of CVD that exhibits high pore density while maintaining its gas selectivity. One distinctive feature of stacked Gr or GO nanosheets is the capacity for adaptive and dynamic layer

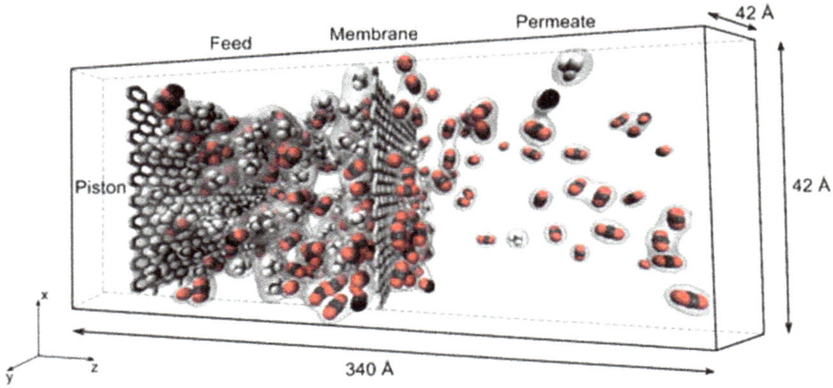

Figure 15.5. Modelling of gas transport behavior through ultrathin GO membranes. The Gr membrane is placed in the middle of the cell while the Gr piston is on the left. Carbon atoms are represented in gray/black, oxygens in red and hydrogens in white. The shaded areas represent the van der Waals surfaces of gas molecules. Reproduced [38] under the terms of the Creative Commons Attribution 4.0 unported license. Copyright 2018, MDPI.

distances. The stacking structures can be carefully controlled to further enhance separation factors and flow rates. In fact, the stacking strategy results in many-layered GO membranes that behave as either molecular-sieving membranes (method two) or nanoporous membranes (method one). Both membranes exhibit CO_2-philic permeability activity, and the addition of water further increases this tendency. As such, these membranes are promising materials for industrial CO_2 separation processes related to petrochemical engineering (CO_2 removal from natural gas), the environment (CO_2 capture from fuel gas), and biomass energy (CO_2 recovery from landfill gas). Figure 15.5 illustrates the gas transport behaviour through GO membranes.

15.2.4.2 Liquid filtration

Unimpeded water permeation through GO laminates encouraged researchers to explore Gr-based membranes for water purification. Though porous Gr monolayer with 2 nm hydrogenated of hydroxylated pores are predicted to efficiently filter NaCl salt from water with ultrahigh flux as stated by Liu *et al*, the obtainability of such a perfect membrane remains a challenge. The ease of preparation for GO membranes through vacuum filtration or layer-by- layer method have been studied experimentally for nanofiltration (NF), ultrafiltration (UF) and forward osmosis (FO) (figure 15.6) [33].

Nanochannels formed within the interlayer spacing of GO membranes were thought to be the primary transport routes, which appeared to be reasonable given the presence of only a very small areal ratio of intrinsic defects within GO flakes that could be simply blocked by the adjacent sp^3 oxidized regions upon mutual stacking. This means that Gr desalting membrane preparation method can be divided into two types (figure 15.7). In the first process, the preparation of single-layer nano-porous Gr (porous Gr membrane, PGM) uses nanopores to achieve desalination.

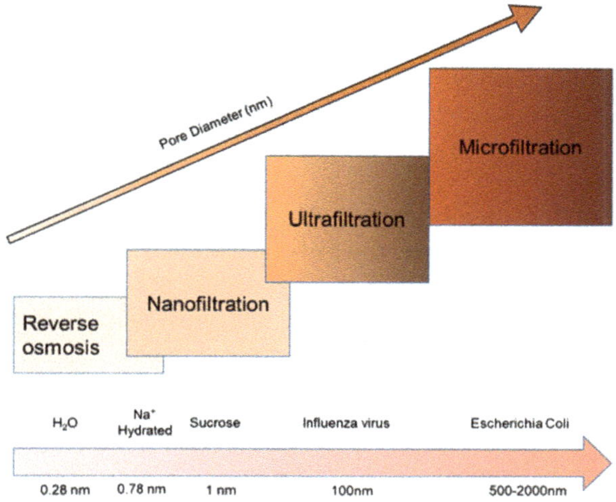

Figure 15.6. Membrane based liquid phase filtrations.

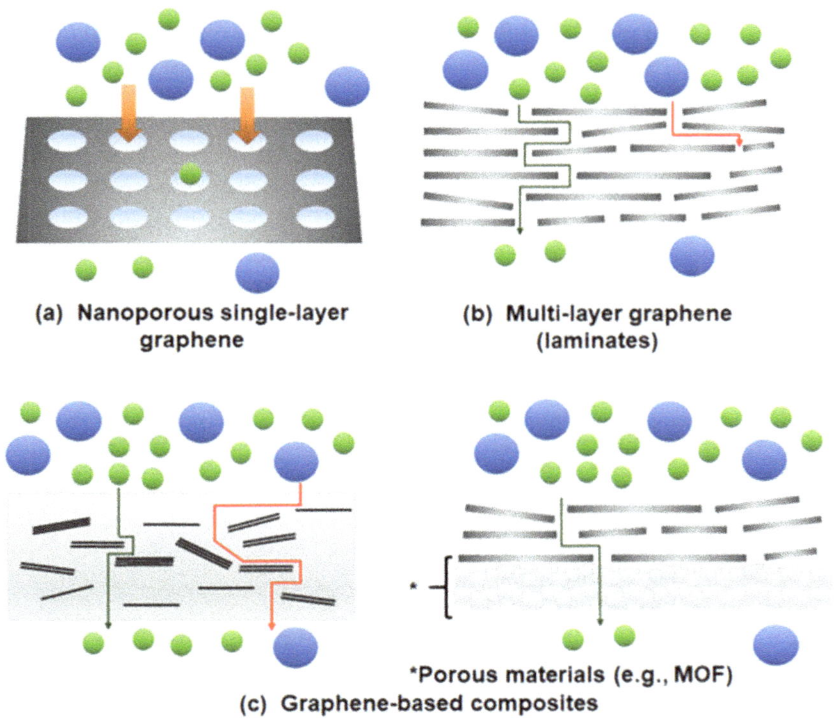

Figure 15.7. Schematic representation of the types of Gr-based membranes. Possible membrane configurations that can be developed with the use of Gr: (a) nanoporous single-layer Gr, (b) multilayer Gr (laminates) and (c) Gr-based composites. MOF: metal–organic frameworks. Reproduced [39] under the terms of the Creative Commons Attribution 4.0 unported license. Copyright 2020 MDPI.

In the second, the assembly of Gr layer by layer (layered GO membrane, LGOM), using a two-dimensional nano-channel between adjacent GO layers achieves desalination. Both can control the separation performance of the Gr separation membrane by modification of the size of nanopores or the interlayer spacing of GO.

15.2.4.3 Biomedical

The applications of Gr membranes in biomedicine include DNA translocations through Gr membrane nanopores towards the fabrication of devices for genomic screening, in particular DNA sequencing (figure 15.8); sub nanometer trans-electrode membranes with high throughput nanopore-based single-molecule detectors; antibacterial activity of Gr, graphite oxide, Gr oxide, and reduced Gr oxide; nanopore sensors for nucleic acid analysis; and use of Gr multilayers as the gates for sequential release. These set the trend for using GMs and GBSMs for biomedical applications.

15.2.4.4 Fuel cell

Fuel cells have been found to be more energy efficient than conventional engines that transform chemical energy of fuel into electrical energy by utilizing some mechanical energy. So, more research is underway to find environment friendly, sustainable and highly efficient energy conversion and storage systems which are economical at the same time. Electrochemical fuel cell technology is important in providing clean and sustainable energy conversion systems for continuous energy supply. Gr dispersion over conductive electrodes provides electrochemically active sites, which improves electrocatalytic activity in oxidation reduction reactions and fuel oxidation. The main issue with fuel cell degradation is corrosion of bipolar plates, which can be reduced by using Gr-based metal composites [41]. Gr and related materials have been used successfully in polymeric fuel cell membranes. An increase in Gr loading to an optimum level has been found to improve proton conductivity, ion transport, and current density at low operating temperatures [42]. Figure 15.9 depicts a typical lithium ion battery (LIB) with graphite as the anode.

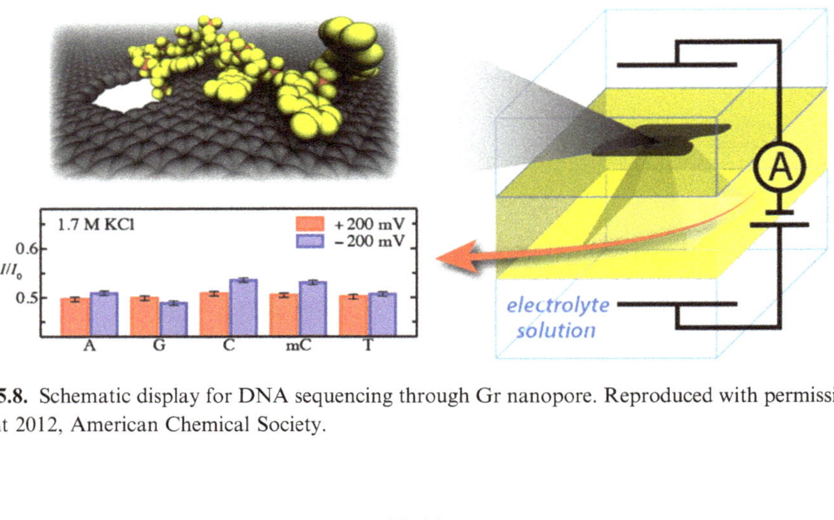

Figure 15.8. Schematic display for DNA sequencing through Gr nanopore. Reproduced with permission [40], copyright 2012, American Chemical Society.

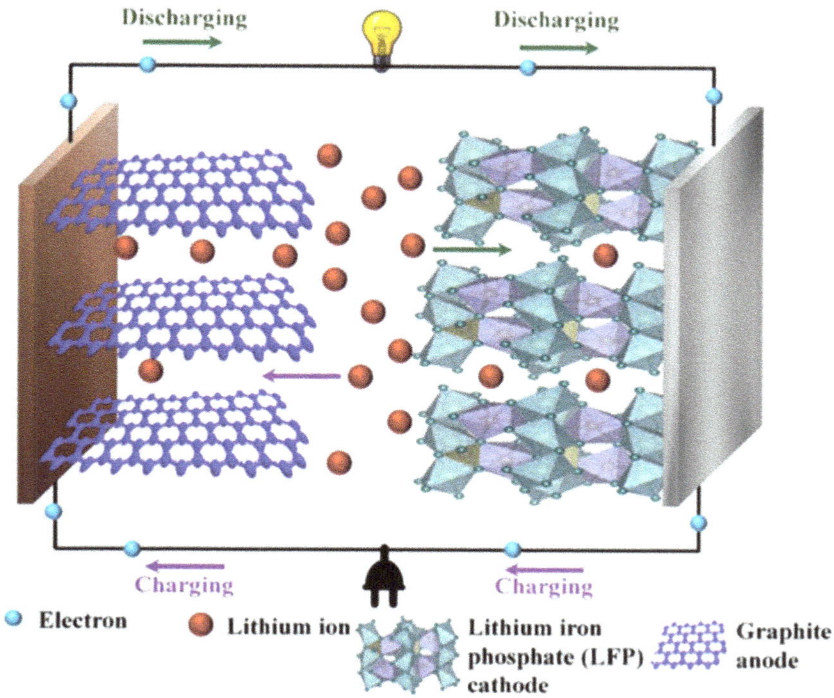

Figure 15.9. Schematic demonstration of typical LIB comprising graphite as anode, lithium iron phosphate as cathode, and lithium salt-based electrolyte. Reproduced [43] under the terms of the Creative Commons Attribution 4.0 unported license. Copyright 2020 MDPI.

15.3 Graphene coatings

15.3.1 Coating techniques

Gr-based functional coatings can be fabricating by tailoring many aspects, out of which the two key parameters are the Gr structure and the surface properties of the corresponding Gr [44, 45]. The modifications of the Gr's surface and structure form the important basis to not only achieve tunable surface properties, but also to ensure that the desired contact disposition and film formation is achieved on the substrate. Inherited primarily from the studies of CNTs, there are numerous studies showing different possible ways of achieving the desired modifications of the aforementioned two aspects [46–48]. These methods have been systematically categorized based on their functionalization (covalent and non-covalent), immobilization (nanoparticle) and substitutional doping [49–52], with the achieved coatings that have several applications in different industries, figure 15.10.

The easiest method for functionalizing the Gr surface is through controlling the oxidation of the graphite flakes, which generates GO, whose basal plane and edges contain the functional groups of epoxy, carbonyl, hydroxyl and carboxyl [25, 45, 53, 54]. Additionally, the GO's stepwise and selective reduction via thermal and chemical

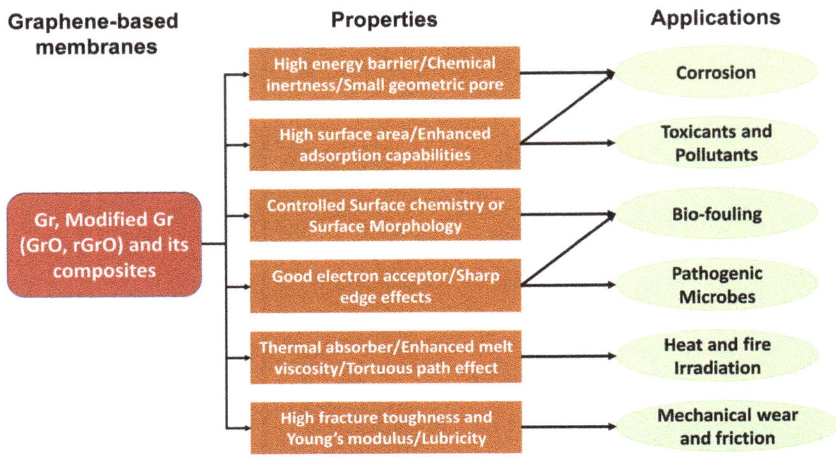

Figure 15.10. Some overarching protecting applications of Gr-based coatings. Adapted from [20] with permission from the Royal Society of Chemistry.

modification methods can result in the removal and reduction of oxygen, which will in turn result in transforming the hydrophilic GO to reduced GO (rGO), which is hydrophobic [55, 56]. The oxygen-containing functional groups, present in the basal planes and edges of the GO structure, play a useful and important role to form covalent bonds with polymers, diazonium compounds, chromophores and other organic molecules [49, 57]. Other significant covalent bonding includes those with halogen groups that are highly reactive, which can result in the enhancement of useful properties, including energy storage, electrochemical sensing and heat spreading [58]. Gr and GO structures, with improved functionalization, have also been observed to substantially help in nano-architectonics, whereas nitrogen (N), oxygen (O), phosphorus (P) and boron (B) doping of the Gr structures has been proven to improve their catalyzing performance. Moreover, significant effects on the Gr surface's properties and structure have been observed due to the non-covalent interactions with cationic and anionic molecules and hydrogen bonds, resulting in Gr π-systems [49, 52]. The various diverse methods of functionalizing Gr thus provide ample opportunities for developing improved coatings and protective layers for different applications, that can help in corrosion, fire retardancy and protecting from the spread of bacterial infection, toxic compounds and fouling. These diverse possible applications of Gr coatings, figure 15.10, will be discussed elaborately in the remaining part of the chapter.

Numerable methods have been implemented for preparing protective coatings from Gr, which include CVD [59–62], rapid thermal annealing (RTA) [43], high temperature pyrolysis of the organic molecules [63, 64], electrostatic powder and plasma spray coatings (powder spray) [65], electrophoretic deposition (EPD) [66–68], dip coating [69, 70], drop casting [71], solution spray [72, 73], spin coating [66, 74], brushing [75] and vacuum filtration [76]. These methods can be broadly classified into two types of coating techniques, dry and wet techniques, some of the important and widely used ones being detailed in table 15.1. The various methods, listed under

Table 15.1. Commonly used dry and wet coating techniques for fabricating Gr-based protective films.

General coating process	Types under the methods	Concepts
Dry	CVD	• Methane and acetylene gases form the precursors. • Argon and hydrogen gases are injected inside a CVD reactor • The reactor is at very high temperatures, around 1000 °C, and synthesis a corrosion-resistant Gr coating.
	Rapid thermal processing	• Organic materials, namely, polyacrylonitrile, coronene, naphthalene and anthracene are applied on the metal substrate • The deposited substrate then is converted into a multilayer Gr coating by pyrolysis at around 1000 °C.
	Powder spray	• Gr-based polymer composites are mixed with a binder and with pigments, granulated in powder form. • The powder mixture is usually sprayed electromagnetically, after which it is cured under heat. • A compact film remains as residue after the drying process.
Wet	Electrophoretic deposition	• Negatively charged GrO is attracted to and deposited at an electrode of opposite charge due to an electric field. • The process once dried results in a compact film.
	Solution spray	• GO or GO mixed with other materials is dispersed in a suitable solvent. • The solvent is sprayed over the substrate. • Finally, the substrate is cured under heat to achieve the coating.
	Dip coating	• The substrate is initially dipped in solution of GO. • It is then dried to achieve the film.
	Spin coating	• The substrate is rotated at high speeds on a disc. • Dispersed GO solution is applied on it while rotating. • Centrifugal force helps in spreading the coating material to form a film.
	Drop casting	• The surface of the substrate is treated with cationic surfactant • Droplets of GO solution is added on it. • The substrate is then dried in air or in an oven to achieve the uniform film.
	Vacuum filtration	• Gr or GO coating dispersions are prepared, and the vacuum filtered using a membrane support. • The coating helps in depositing the Gr/GO sheets fabricating GO-based antibacterial paper and anti-fouling coatings.
	Brushing	• Gr ink and GO-based paints are often used for coatings. • They are prepared using a brush to form corrosion-resistant thin films for metals and alloys.

the two broad processing techniques, have their own degree of usefulness that depends on the sensitivity of the application and also that of the environment. Further details on the techniques can be found in the study carried out by Nine *et al* [20].

15.3.2 Applications and advancements

The various fabrication techniques, as detailed in the previous section, have resulted in numerus applications of Gr coatings, and the field is regularly growing. The applications have been segregated into broader fields, figure 15.1, and the critical ones will be discussed in detail in this section with the primary focus on the application and their utility, advantages and efficiency for the particular field.

15.3.2.1 Tribological aspects

Gr has been successfully used both as micro- and macro-tribological coatings, resulting in positive tribological outcomes in different dynamic loading scenarios, proving that Gr-based coatings have the ability to reduce wear or mechanical failure of sliding surfaces [20]. The application of Gr coatings helps in two ways, firstly by acting as a nanofiller and developing bridging networks and secondly as a solid lubricant that helps in reducing abrasive wear. Often, application of Gr coatings on generally weaker structures results in enhanced mechanical and structural performance along with better tribological properties. One such example is aerogel, commonly used in insulation and lubrication. Aerogel is in general lightweight, fragile and inelastic, however, the application of Gr coating helps them to become super-elastic and stronger, with the ability to resist mechanical fatigue [63]. Additionally, the coating also increases the aerogel's loss modulus by a factor of about 3 and the Young's modulus and energy storage modulus by a factor of about 6.

Steady-state operation of nano- and macro-electro-mechanical systems, which are based on titanium (Ti), can be ensured by applying rGO coatings that act as anti-wear and anti-frictional coatings on the surface [70]. Significant reduction in the frictional and adhesion forces, providing less affinity towards humid environments, can be achieved through self-assembled GO deposited on Ti that results in its hydrothermal reduction and improved anti-wear properties, having wear depths of below 500 nm. Wang *et al* [77] proved that in macroscale tribology, polyurethane nanocomposite coating with only 2 wt.% of functionalized-GO, when applied on polytetrafluoroethylene (PTFE) increases the tensile strength and Young's modulus by 71% and 86%, respectively. The wear-life of polymer-based Gr composite coatings could also be increased by adding small quantities of Gr in polyamide 11, which is about 0.4 wt.%, that when added to even steel (ASTM steel 1045) results in 462%–880% increment in wear resistance under a force of 150 N [20]. Additionally, Gr has been proven to have 10–30 times higher ability of suppressing wear when compared to polymer composites containing graphite [72]. Furthermore, inclusion of a greater amount of Gr (10%) in PTFE results in a wear rate of $\sim 10^{-7}$ $mm^3\ N\ m^{-1}$ from $0.4 \times 10^3\ mm^3\ N\ m^{-1}$ [78], which can be estimated to be about

10–30 times higher than that achieved from adding a similar amount of micrographite in polymer composites.

Substantial success of GO in tribological applications, has resulted in its extension to the bio-tribological aspect with quite a few studies showing positive potential of its application. One such example is the implant bearings for joint implants, where the use of GO-incorporated ultrahigh molecular weight polyethylene coating results in the transfer of fatigue wear to abrasive wear, efficiently reducing the wear rate [79]. Wear and friction in other commonly used biomaterials, namely, calcium silicate and alumina, have also been reported to be significantly reduced by the application of Gr-based coatings [65, 80]. Moreover, the number of adherent surface particles increases with the increase in the Gr content in calcium silicate-based Gr coatings, due to the increased wetting behavior or cell adhesion experienced in the ceramic matrix [65]. Additionally, the inclusion of Gr platelets in silicon nitride also has the potential to have enhanced tribological properties because the nanofillers create an adherent protective tribo-film through exfoliation resulting from the high pressure generated due to the sliding surfaces [81]. Gr reinforced ceramic coatings also form a similar film through exfoliation, playing a significant role in stopping delamination under bearing loads.

Wang *et al* [82] proposed, more than a decade ago, that the possible use of thin films, made from solid lubricants, can potentially increase the lubricating performance in comparison to the commonly used bulk or powder forms. Gr-based coatings helped to prove the hypothesis correct, because they work as a protective layer from solid lubricants and reduce surface wear. Gr-nanoflake coatings are further proven to have even more efficient tribological properties compared to graphite particles, making Gr-based coatings new, emerging and a sort-after lubricant [83]. GO and Gr have also been reported to be efficient as a solid lubricant [84] and as an additive for the base lubricants [85–87]. Multilayered Gr coatings even have the ability to reduce the puckering effect, an aspect usually experienced due to shear between layers in general multilayered systems or laminates on a dynamic surface [20]. The presence of Gr results in a passivation film that helps reduce frictional forces and adhesion between the layers [88]. Thus, a thin tribo-film is formed in the layered Gr structures, under dynamic loading conditions, resisting tribo-corrosion [88, 89].

The temperature sensitive rotational and transitional motions that can be observed between the initial and final states of loading conditions enables the Gr flakes to slide over each other, resulting in achieving a superior lubricity [90]. Substantially less number of Gr layers also has the ability to reduce the frictional coefficients by factor of 6 between steels in both oxygen [91] and dry nitrogen [92]. Moreover, Gr-based materials work as better lubricants in reducing wear rate in ethanol [92], oils [86] and even water [85]. The performance of Gr coatings in water was studied elaborately with a tungsten carbide ball on a stainless-steel plate by Kinoshita *et al* [93], who reported that in water Gr materials form a thin layer of protective coating that has no significantly visible surface wear even after 60 000 cycles of frictional testing under a force of 1.88 N. Song and Li [85] even deduced that multilayer GO coatings in water have superior tribological properties compared to multi-walled CNTs (MWCNTs). Oleic acid modified Gr also showed the capability to reduce scar diameters by about 14% [87]. Overall, it can be concluded that Gr and Gr-based oxides and other modifications as fillers in polymers and ceramic-based protective coatings have superior performance compared to conventional lubricants with many promising applications.

15.3.2.2 Corrosion and weather proofing

Studies on Gr have proved that the impermeable 2D structure exhibits exceptional barrier properties, resulting in excellent corrosion resistance and weather proofing [8]. The physical structure of Gr is principally responsible for providing the excellent resistance or barrier properties. The Gr lattice is comprised of hexagonal structures, which has a pore diameter of 0.246 nm [73, 94]. While considering the van der Waal's radii of carbon atoms, these pore diameters further reduce to 0064 nm. Therefore, it becomes very difficult for even smaller atoms, like helium, to cross through the pores [11]. Additionally, the π-conjugated network of carbon atoms, which is dense and delocalized in Gr molecules, helps blocking the gaps within the closed packed atomic rings and establishing a repulsive field to reactive atoms and molecules. This helps in creating an actual physical separation between the surface that is being protected and the reactants in the environment [20]. Due to these excellent properties, even a single layer of Gr coating has the ability to block oxygen permeation, with studies suggesting that Gr-based coatings are the thinnest known corrosion-resistant coatings [9].

CVD has been the most used technique of fabricating barrier systems for corrosion resistance from Gr. Chen *et al* [59] gave one of the pioneering reports of the concept, where a single-layer Gr-based coating was proved to be sufficient for protecting copper and copper/nickel surfaces from liquid etchants and air-oxidation. A later repetition of the work further proved that Gr-based coatings fabricated through annealing at 500 °C also had improved barrier protection, compared to general coatings [95]. Although effective as a short-term corrosion-resistant barrier, the single layers often resulted in eventual oxidation due to the wrinkles and possible point defects caused due to synthesis and transportation. Therefore, multilayered Gr-based coatings were fabricated, which proved to solve the issues with a single layer, making the coatings long-lasting and even more efficient [60, 61, 96–98]. It is interesting to note that, CVD-based coatings can both be grown on the material where protection is needed or grown externally and then transported mechanically and laid on the surface for protection. Studies have shown that the mechanically transferred layers had five times greater corrosion rate, compared to that of the *in situ* grown one [60]. Additionally, it is important to have defect-free single-layered or multilayered systems with Gr-based coatings because oxygen and water permeance will oxidize the layer to be protected resulting in wet corrosion and crack propagation along the defects. This oxidation is not very pleasing, as the corrosion has been reported to be more severe when compared to normal corrosion without any coating due to the electrochemical reactions occurring all across the surface [99]. Studies have also been conducted to compare the pioneering CVD method with other known methods, as far as corrosion resistance is concerned, such as electrophoretic deposition of GrO [66], layer-by-layer deposition of GrO and polyethyleneimine [100], nano-paint fabricated from GO [75] and templated epoxy-based hydrophobic coating [101]. Detailed comparison can be observed from the review work illustrated by Nine *et al* [20].

Another interesting and useful application of Gr-based coatings is anti-fouling coating. One of the most widespread issues, when it comes to fouling, is the deposition of inorganic and organic materials in various marine vessels and structures. This issue is huge, and the estimated cost for repairing fouling damage in the hulls of the US navy alone is in the range of $180–$260 million per year [102]. Moreover, with the ban on TBT, the fabrication of new anti-fouling coatings has become extremely important [20]. Strategies generally used to fight surface fouling include self-cleaning, biocidal, biomimetic nanostructure and amphiphilic surfaces [47, 102]. Gr can be successfully used as an anti-sticking coating system, where the surface tailoring can help in generating excellent anti-fouling coating with desired wettability and polarity [55, 56]. Two types of Gr-based coatings, superhydrophobic and superhydrophilic, can be used as anti-fouling coatings [74, 103]. Additionally, the impermeable structure further helps in rejecting salts when Gr-based coatings are done on membranes [104]. Carbon-based coatings are commonly used as anti-fouling agents, due to their known anti-settlement, antibacterial and salt rejection abilities [105–110]. Thus, the success with other allotropes of carbon has inspired work with Gr, which also has been proven to have superb antibacterial properties [111], controlled protein adhesion [112], flexible surface chemistry [50] and the ability to form covalent bonds easily with polymers on functionalization [113]. Gr-based coatings have been able to demonstrate excellent anti-fouling capabilities, due to their resistance to dust, marine salt, protein, cell and microorganism depositions on the surfaces.

Titanium dioxide mixed with Gr has been proved to have improved photo-induced self-cleaning ability, which showed increased hydrophilicity and performance even under low UV intensity [74]. Thus, it can also be used for light indoor applications. Water contact angle measurements on oligomeric silsesquioxane (POSS) and CNTs mixed with Gr were observed to be 155°, proving to be a superhydrophobic, bi-functional and conductive coating [103]. The study concluded that Gr-based nanostructured coatings have the potential to resist water and snow deposition on windows and other devices, along with being self-cleaning. Superoleophobic anti-fouling coating can also be possibly made on Gr coated cotton fabrics, through controlled flourosilanization treatment [114]. Another simple way of creating Gr-based foulant resistant coating is to use the Gr molecules as pigments or fillers in common paints. Studies have shown that using GrO pigments with alkyd resin can resist microorganism colonization and biofilm formation [75]. Cell adhesion was also reported to be reduced in Gr-infused gallium nitride nanowire film due to the nanostructure and surface wettability influenced by GO [115].

Wastewater treatment studies have further shown that only 1 wt.% inclusion of GO in polysulfone can result in huge increase in anti-fouling properties, where the formation of biofilm on membranes can be largely reduced [116]. The reason being the increase in negative zeta potential value by GO due to the negatively charged functional groups containing oxygen. Many other studies have shown enhanced performance of Gr-based composite coatings as anti-fouling agents in bioreactors [117], which can even resist follow-up fouling due to the ability of surpassing the deposition of particles that avoids internal clogging of the pores. Other applications include introducing GO into polysulfone to reduce salt deposition [104], GO/polyaniline-modified coating that results in 82% NaCl rejection and many others. In summary, again the chemical structure and the impermeable lettuce structure helps in fabricating various types of composite coatings that have immense potential as corrosion and anti-fouling agents.

15.3.2.3 Fire-retardancy

The general and industrially accepted flame retardants usually contain halogens, phosphorus and melamine compounds, which are generally toxic to humans and also to the environment. Moreover, it has been observed that a high amount of fire-retardant (FR) fillers tends to reduce the strength of the composites [118, 119]. With the introduction of Gr, it has been proved that the loading of FR fillers can be reduced substantially while improving or maintaining the mechanical integrity of the composites [20]. It has been proven that Gr fillers have the potential to increase the composite's operating temperature, reduce the overall moisture content, protect from lightning strikes by improving the compressive strength and can even induce antistatic behavior [1]. The incredible thermal stability, even at elevated temperatures of around 2126 °C, of Gr has helped in exploiting their use as FR fillers, although there are possibilities of experiencing high vibrations at higher temperatures [120]. However, mono-layered structures usually show higher vibrations, whereas multiple layers have been proved to show more stability [20]. Additionally, multilayered Gr coatings prepared via mechanical cleavage method have been proven to be more stable compared to the same multilayered structure prepared through CVD, due the stronger interlayer π–π interaction [121].

Gr and GrO based materials can further be considered ideal for applications concerning high temperatures because of their significantly high thermal conductivities ranging between 2000 W m^{-1} K^{-1} and 5000 W m^{-1} K^{-1}, almost complete impermeability to gases, excellent mechanical properties and proven capability to dissipate heat [2, 4, 11]. Studies have shown the potential of Gr-based materials to reduce peak heat release rate (PHRR) [65], rate and amount of smoke production [122], increase thermal stability [123], reduce limiting oxygen index (LOI) values [124], increase char yield [125] and reduce melt viscosity index (MFI) [126] and dripping properties [126]. All these positive findings have resulted in the growth of Gr-based FR fillers in many studies of recent times. In should also be noted that, even though most of the Gr-based materials have the potential to enhance both fire and mechanical performances of the composite, a high amount of rGO has been found to be mechanically detrimental even though the thermal properties do get enhanced substantially.

Wang *et al* [126] provided one of the first ever studies on Gr-based FR fillers, where the polymer matrix induced with Gr developed a barrier system that was able to delay the time to oxidation. 5% loading of graphite, its oxide and functionalized graphite oxide into epoxy-based coatings have also been proven to reduce the PHRR of the composites [127]. Glass transition temperature, and thus thermal stability, was also reported to increase with the infusion of GO fillers in polyurethane acrylate [123]. Other studies on nanofillers made from expanded graphite have also shown that the thermal absorbing ability and time to ignition increase, whereas the PHRR decreases for the polymer matrix composite. The synergistic effects of Gr with other FR additives, such as cuprous oxide, intumescent FRs, 9,10-dihydro-9-oxa-10-phosphaphenanthrene-10-oxide, multi-walled CNTs zinc sulfide and molybdenum di-sulfide, in polymers have illustrated second degree improvement in fire performance [122, 124–126, 128–131].

The well-established ability of Gr-based nanocomposite filler to take the tortuous path for creating impermeable layers also helps in substantial reduction in smoke production by hindering the escape of degraded materials in the form of volatiles, mainly carbon monoxide that ensures eliminating on the major hazards during fire [122]. Apart from reducing smoke production, as can be imagined, the barrier also helps in reducing the LOI and PHRR values, proving to be an effective FR filler. Gr and multi-walled CNTs have also been reported to work in synergy and reduce the smoke production by suppressing volatile escape [124]. Another way in which Gr-based fillers have been shown to be able to enhance fire performance is by increasing the amount and rate of char production, where Gr flakes infused with zinc sulfide resulted in densifying the char produced in an epoxy-based polymer composite [129]. The studies performed on Gr-based fillers FR materials show great potential in many applications where it is imperative to protect the structures from fire and other forms of thermal damage, including laser irradiation [132].

15.3.2.4 Medical implants

Relatively new alloys, namely, nitinol (NiTi) and good old stainless steel are the two most commonly used metals as medical implants because of their superior mechanical properties [133–135]. However, there are substantial challenges, of which the most critical ones are bio- and hemo-compatibility and cytotoxicity of the exogenous materials. The poor hemo- and biocompatibility can be attributed to the metallic nature of the alloys, which results in leaching, proliferation, thrombosis and reduced cell adhesion once flowing blood comes in contact with the surface [133, 136, 137]. Additionally, strong immunological responses can be caused when the living cells and/or proteins come in direct contact with the metallic alloys, which might also result in a cascade of biochemical reactions affecting the functionality of the device [138]. Thus, it becomes very important to ensure the interactions of the implants with the surrounding environment are under control and within expectations. One of the common ways to reduce and/or prevent such issues in biomedical implants is surface modification or coating. A surface coating that is ideal will have chemical inertness, excellent adhesion strengths, good bio- and hemo-compatibility and high smoothness. Before the use of Gr, many other materials, including a few allotropes of carbon, such as TiO_2, TiN, SiC, diamond-like carbon and various polymeric materials have been employed as possible coatings for the implants to reduce the interaction [139–148]. Unfortunately, none of them have been able to meet the stringent criteria that must be adhered to for a commercial and suitable coating of biomedical implants.

This increasing and everlasting issue saw the light of possible solution with the invention of thick sp^2 carbon allotrope, in the form of Gr, which opened new doors for fabricating novel multifunctional coatings for various applications, including biomedical implants. Gr-based coatings are expected to be the ideal candidates for surface coating of implants due to their chemical inertness, high durability and smooth atomic structure. Podila *et al* [138] proved that Gr-NiTi, NiTi coated with Gr, can actually meet all the required functional needs, while having some more advantages. The additional advantages include supporting endothelial cell growth and smooth muscle growth that leads to excellent cell proliferation. Microscopic analysis revealed no charge transfer between fibrinogen and Gr, which can be possibly caused because of the ability of the Gr coatings to reduce or inhibit platelet activation by implants. Moreover, their biocompatibility was also confirmed by proving that Gr-based coatings do not show any type of substantial *in vitro* toxicity. These observations along with the known aspect of Gr being chemically inert and durable prove that Gr-based coatings are ideal for surface coating biomedical implants, and there is huge potential in this aspect.

15.3.2.5 Antiseptic and antimicrobial

One of the comparatively newer and developing area of Gr-based coatings is antiseptic and antibacterial applications, where it is critical to protect structures and systems from harmful pathogens and bacteria. Proper coatings on surgical instruments can substantially help in decreasing the post-surgery complexities and thus, accelerating the healing process. Other than medical devices, other industries that also demand antiseptic coatings are food and beverage industries required for storing and marketing [149, 150]. Gr-based coatings have still proven to be the most recent antibacterial and antifungal coating that has no substantial or known toxic effects on the human cells or on the ecosystem [111, 151–156].

Studies showing potential in using Gr-based antiseptic coatings have primarily fabricated bandages made from biocompatible materials, namely silver [157] and chitosan–PVA nanofibers [158], coated with Gr and tested successfully on healing wounds of rabbits and mice. GO flakes have also been used as fillers in polyelectrolyte coating, having porous structures, in blood and vascular applications with the coatings having antithrombogenic and antibacterial functions [159]. Bacteriological analyses of the coatings showed that the GO flakes were successful in reducing bacterial presence on the surfaces of the coated structures. Furthermore, the authors also performed dynamic blood analysis, which proved that the composite coating helped in stronger platelet consumption, higher coagulation and stronger immune responses. Santos *et al* [160] were the pioneers in introducing Gr–polymer composite coatings with proven advanced antibacterial effects and were applied in industrial and biomedical fields. It was proven by the authors that the GO and poly-*N*-vinyl carbazole coating had the ability to reduce bacterial colonization by about 90%, when compared to non-modified surfaces. Another study has also pointed to the possibility of using GO-based antibacterial paper inhibiting the growth of bacteria (*Escherichia coli*), which can be further modified and used in coatings for applications in medical devices and food packaging applications.

The antibacterial mechanisms of Gr-based coatings and its oxide can be primarily associated with disruption of the bacteria or bacterial membrane [10] and oxidative stress [151, 161]. These phenomena take place through few unique mechanisms that include insertion of the Gr-based membranes sharp edges into the bacterial membrane [10], charge transfer [158, 162] and through intertwining resulting in the bacterial membrane's perturbation [111]. Interestingly, most of the studies have reported observing that the first unique mechanism of insertion by the Gr-based membrane is the most active mechanism [10, 111, 152, 153]. The mechanism of intertwining has also been reported to be observed in some studies where a wide range of GO sheets intertwine with the pores of pathogens, resulting in electrolyte leakage from the fungal spores, reduction in the potential of the bacterial membrane and eventual lysis and death of the pathogens [111]. The third mechanism has also been reported to be observed between a bacterium and a GO coating [153]. GO is made up of negative charges and can work as a good electron acceptor that in turn damages the bacterial membrane. Another example being the relatively easy damage to the prokaryotic cell's DNA structure by electron from the Gr coating due to the lack of nuclear membrane. Krishnamoorthy *et al* [152] also reported that the antibacterial properties of Gr-based coatings are potentially higher than the commonly used drugs. Therefore, although it is a less ventured field, the potential is immense and proper studies are needed to ensure a smooth and fruitful transition of Gr and GO-based coatings for antiseptic and antimicrobial applications.

15.3.2.6 Mechanical damage-proofing

Gr coatings have been proved to have excellent mechanical damage-proofing ability, an application which has often demonstrated to be even more efficient than fire and corrosion resistance. The reason behind its efficiency in wear and scratch resistance can be attributed to it being the lightest [163], thinnest [163, 164] and strongest known nanomaterial [4]. The monolayer coatings have been reported to be able to withstand extremely high-pressure differences of 6 atm, which indicates their mechanical robustness [11]. Gr is known to have incredible mechanical strength due to the strong C–C covalent bonds, resulting from the sp^2 hybridization observed in the closed pack hexagonal crystal structure that has a bond length of 0.142 nm [20]. The strength can be quantified by the reported elastic modulus of 0.25 TPa in chemically modified GO, through exfoliation [165]. Furthermore, a monolayer of defect-free Gr has been reported to have an elastic modulus of 1 TPa, which gives an intrinsic breaking strength of 42 N m^{-1} [4].

Previous studies [165, 166] have further shown that Gr coatings have the potential to eliminate or reduce different mechanical failures as they can toughen and strengthen the loaded surface and help in stress distribution along the entire structure. They also have proven ability to bridge grain boundaries, especially useful in brittle and fragile materials, that help in adding an extra reinforcement in materials, such as ceramics [1, 3]. Reinforcing with Gr nanofillers helps during crack formation by crack branching, resisting crack-tip propagation, bridging cracks and deflecting cracks [167]. One example is ceramic materials, where Gr coating can help hindering the grain growth phenomenon, by assisting in deflecting, branching and bridging the cracks that usually propagate under severe loading conditions. Furthermore, 2D Gr coatings and fillers also have the potential to enhance the flexural modulus, cold modulus of rapture and force displacements, with proven performance in ceramic matrices [168]. In general, the overall mechanical performance of the host material is substantially enhanced by applying Gr coatings, where the materials can withstand fatigue, structural collapse and permanent deformation due to the operating stress distribution and increased elasticity. Additionally, the improvement in the materials' structural integrity assists in improving the wear resistance of the sliding surfaces [20].

15.4 Concluding remarks

The details outlining the scope and quality of Gr, GO and functionalized GOs prove that they have immense potential in membranes and coatings applications. There has been unprecedented progress in the field of Gr research due to their excellent and superior chemical and physical properties, which outclass most of the commonly and traditionally used materials in the industries. There are broad application areas that can actually benefit from the use of pure Gr, Gr-based or Gr-enhanced composites and GOs that can provide superior protective properties with minimal-to-negligible disadvantages. The literature and the current chapter also outline the possibility of further enhancing the protective and barrier properties by modifying the structure of Gr and the microstructure of the nanomaterials. It is important to note that in many research fronts the use of Gr and Gr-based coatings have reached a technical readiness level of 8 and 9, with many industries contemplating and planning the transformation. One of the most promising among these techniques is polymer matrix composites with Gr nanoparticles, which has immense potential in automobile and aerospace industries, and has already seen a few applications. The authors believe that the current chapter will help in understanding the various possible application fields where Gr coatings and membranes can be applied with potential to outperform the currently used products.

Acknowledgments

The authors would like to acknowledge the contribution of the staff and management of CACM, University of Auckland, and CMEC, Washington State University, Pullman.

References

[1] Novoselov K S, Colombo L, Gellert P R, Schwab M G and Kim K 2012 A roadmap for graphene *Nature* **490** 192–200

[2] Balandin A A 2011 Thermal properties of graphene and nanostructured carbon materials *Nat. Mater.* **10** 569–81

[3] Mayorov A S *et al* 2011 Micrometer-scale ballistic transport in encapsulated graphene at room temperature *Nano Lett.* **11** 2396–9

[4] Lee C, Wei X, Kysar J W and Hone J 2008 Measurement of the elastic properties and intrinsic strength of monolayer graphene *Science* **321** 385–8
[5] Mohan V B, Souri H, Jayaraman K and Bhattacharyya D 2018 Mechanical properties of thin films of graphene materials: a study on their structural quality and functionalities *Curr. Appl Phys.* **18** 879–85
[6] Ferrari A C *et al* 2015 Science and technology roadmap for graphene, related two-dimensional crystals, and hybrid systems *Nanoscale* **7** 4598–810
[7] Compton O C and Nguyen S T 2010 Graphene oxide, highly reduced graphene oxide, and graphene: versatile building blocks for carbon-based materials *Small* **6** 711–23
[8] Su Y, Kravets V G, Wong S L, Waters J, Geim A K and Nair R R 2014 Impermeable barrier films and protective coatings based on reduced graphene oxide *Nat. Commun.* **5** 1–5
[9] Topsakal M, Şahin H and Ciraci S 2012 Graphene coatings: an efficient protection from oxidation *Phys. Rev.* B **85** 155445
[10] Tu Y *et al* 2013 Destructive extraction of phospholipids from *Escherichia coli* membranes by graphene nanosheets *Nat. Nanotechnol.* **8** 594–601
[11] Bunch J S, Verbridge S S, Alden J S, Van Der Zande A M, Parpia J M, Craighead H G and McEuen P L 2008 Impermeable atomic membranes from graphene sheets *Nano Lett.* **8** 2458–62
[12] Campos-Delgado J *et al* 2009 Thermal stability studies of CVD-grown graphene nano-ribbons: defect annealing and loop formation *Chem. Phys. Lett.* **469** 177–82
[13] Kahng Y H, Lee S, Park W, Jo G, Choe M, Lee J H, Yu H, Lee T and Lee K 2012 Thermal stability of multilayer graphene films synthesized by chemical vapor deposition and stained by metallic impurities *Nanotechnology* **23** 075702
[14] Chen G, Weng W, Wu D, Wu C, Lu J, Wang P and Chen X 2004 Preparation and characterization of graphite nanosheets from ultrasonic powdering technique *Carbon* **42** 753–9
[15] Salas E C, Sun Z, Lüttge A and Tour J M 2010 Reduction of graphene oxide via bacterial respiration *ACS Nano* **4** 4852–6
[16] Yoon H W, Cho Y H and Park H B 2016 Graphene-based membranes: status and prospects *Phil. Trans. R. Soc.* **374** 20150024
[17] Liu G, Jin W and Xu N 2015 Graphene-based membranes *Chem. Soc. Rev.* **44** 5016–30
[18] Song N, Gao X, Ma Z, Wang X, Wei Y and Gao C 2018 A review of graphene-based separation membrane: materials, characteristics, preparation and applications *Desalination* **437** 59–72
[19] Mohan V B, Lau K T, Hui D and Bhattacharyya D 2018 Graphene-based materials and their composites: a review on production, applications and product limitations *Composites* B **142** 200–20
[20] Nine M J, Cole M A, Tran D N and Losic D 2015 Graphene: a multipurpose material for protective coatings *J. Mater. Chem.* A **3** 12580–2602
[21] Barea E, Montoro C and Navarro J A 2014 Toxic gas removal – metal–organic frameworks for the capture and degradation of toxic gases and vapours *Chem. Soc. Rev.* **43** 5419–30
[22] Templeton D M and Liu Y 2010 Multiple roles of cadmium in cell death and survival *Chem. Biol. Interact.* **188** 267–75
[23] Berry V 2013 Impermeability of graphene and its applications *Carbon* **62** 1–10
[24] Tsou C H, An Q F, Lo S C, De Guzman M, Hung W S, Hu C C, Lee K R and Lai J Y 2015 Effect of microstructure of graphene oxide fabricated through different self-assembly techniques on 1-butanol dehydration *J. Membr. Sci.* **477** 93–100

[25] Dreyer D R, Park S, Bielawski C W and Ruoff R S 2010 The chemistry of graphene oxide *Chem. Soc. Rev.* **39** 228–40

[26] Dikin D A, Stankovich S, Zimney E J, Piner R D, Dommett G H, Evmenenko G, Nguyen S T and Ruoff R S 2007 Preparation and characterization of graphene oxide paper *Nature* **448** 457–60

[27] Kumar N, Salehiyan R, Chauke V, Botlhoko O J, Setshedi K, Scriba M, Masukume M and Ray S S 2021 Top-down synthesis of graphene: a comprehensive review *Flat Chem.* **27** 100224

[28] Kim H W *et al* 2013 Selective gas transport through few-layered graphene and graphene oxide membranes *Science* **342** 91–5

[29] Brodie B C 1859 XIII. On the atomic weight of graphite *Phil. Trans. R. Soc.* **149** 249–59

[30] Staudenmaier L 1898 Verfahren zur darstellung der graphitsäure *Ber. Dtsch. Chem. Ges.* **31** 1481–7

[31] William S, Hummers J R and Offeman R E 1958 Preparation of graphitic oxide *J. Am. Chem. Soc.* **80** 1339–9

[32] Hu M and Mi B 2013 Enabling graphene oxide nanosheets as water separation membranes *Environ. Sci. Technol.* **47** 3715–23

[33] Cohen-Tanugi D and Grossman J C 2012 Water desalination across nanoporous graphene *Nano Lett.* **12** 3602–8

[34] Sint K, Wang B and Král P 2008 Selective ion passage through functionalized graphene nanopores *J. Am. Chem. Soc.* **130** 16448–49

[35] Jiang D E, Cooper V R and Dai S 2009 Porous graphene as the ultimate membrane for gas separation *Nano Lett.* **9** 4019–24

[36] Ashirov T, Yazaydin A O and Coskun A 2022 Tuning the transport properties of gases in porous graphene membranes with controlled pore size and thickness *Adv. Mater.* **34** 2106785

[37] Yuan Z, He G, Faucher S, Kuehne M, Li S X, Blankschtein D and Strano M S 2021 Direct chemical vapor deposition synthesis of porous single-layer graphene membranes with high gas permeances and selectivities *Adv. Mater.* **33** 2104308

[38] Tronci G, Raffone F and Cicero G 2018 Theoretical study of nanoporous graphene membranes for natural gas purification *Appl. Sci.* **8** 1547

[39] Chuah C Y, Lee J and Bae T H 2020 Graphene-based membranes for H2 separation: recent progress and future perspective *Membranes* **10** 336

[40] Wells D B, Belkin M, Comer J and Aksimentiev A 2012 Assessing graphene nanopores for sequencing DNA *Nano Lett.* **12** 4117–23

[41] Iqbal M Z, Siddique S, Khan A, Haider S S and Khalid M 2020 Recent developments in graphene based novel structures for efficient and durable fuel cells *Mater. Res. Bull.* **122** 110674

[42] Kausar A 2019 Applications of polymer/graphene nanocomposite membranes: a review *Mater. Res. Innovations* **23** 276–87

[43] Joy R, Balakrishnan N T M, Das A, Shafeek S, Thakur V K, Zaghib K, Jaffarali J F M, Reddy M V V and Raghavan P 2022 Graphene: chemistry and applications for lithium-ion batteries *Electro. chem.* **3** 143–83

[44] Pu J, Wan S, Lu Z, Zhang G A, Wang L, Zhang X and Xue Q 2013 Controlled water adhesion and electrowetting of conducting hydrophobic graphene/carbon nanotubes composite films on engineering materials *J. Mater. Chem.* A **1** 1254–60

[45] Shin Y J, Wang Y, Huang H, Kalon G, Wee A T S, Shen Z, Bhatia C S and Yang H 2010 Surface-energy engineering of graphene *Langmuir* **26** 3798–802
[46] Hirsch A 2002 Functionalization of single-walled carbon nanotubes *Angew. Chem. Int. Ed.* **41** 1853–9
[47] Banerjee S, Hemraj-Benny T and Wong S S 2005 Covalent surface chemistry of single-walled carbon nanotubes *Adv. Mater.* **17** 17–29
[48] Balasubramanian K and Burghard M 2005 Chemically functionalized carbon nanotubes *Small* **1** 180–92
[49] Georgakilas V, Otyepka M, Bourlinos A B, Chandra V, Kim N, Kemp K C, Hobza P, Zboril R and Kim K S 2012 Functionalization of graphene: covalent and non-covalent approaches, derivatives and applications *Chem. Rev.* **112** 6156–214
[50] Gogotsi Y 2011 Controlling graphene properties through chemistry *J. Phys. Chem. Lett.* **2** 2509–10
[51] Jiao Y, Zheng Y, Jaroniec M and Qiao S Z 2014 Origin of the electrocatalytic oxygen reduction activity of graphene-based catalysts: a roadmap to achieve the best performance *J. Am. Chem. Soc.* **136** 4394–403
[52] Kuila T, Bose S, Mishra A K, Khanra P, Kim N H and Lee J H 2012 Chemical functionalization of graphene and its applications *Prog. Mater Sci.* **57** 1061–105
[53] Wang S, Zhang Y, Abidi N and Cabrales L 2009 Wettability and surface free energy of graphene films *Langmuir* **25** 11078–81
[54] Wang H and Hu Y H 2011 Effect of oxygen content on structures of graphite oxides *Ind. Eng. Chem. Res.* **50** 6132–7
[55] Xu C, Yuan R S and Wang X 2014 Selective reduction of graphene oxide *New Carbon Mater.* **29** 61–6
[56] Mathkar A, Tozier D, Cox P, Ong P, Galande C, Balakrishnan K, Leela Mohana Reddy A and Ajayan P M 2012 Controlled, stepwise reduction and band gap manipulation of graphene oxide *J. Phys. Chem. Lett.* **3** 986–91
[57] Peng C, Xiong Y, Liu Z, Zhang F, Ou E, Qian J, Xiong Y and Xu W 2013 Bulk functionalization of graphene using diazonium compounds and amide reaction *Appl. Surf. Sci.* **280** 914–9
[58] Poh H L, Šimek P, Sofer Z and Pumera M 2013 Halogenation of graphene with chlorine, bromine, or iodine by exfoliation in a halogen atmosphere *Chem. Eur. J.* **19** 2655–62
[59] Chen S *et al* 2011 Oxidation resistance of graphene-coated Cu and Cu/Ni alloy *ACS Nano* **5** 1321–7
[60] Prasai D, Tuberquia J C, Harl R R, Jennings G K and Bolotin K I 2012 Graphene: corrosion-inhibiting coating *ACS Nano* **6** 1102–8
[61] Krishnamurthy A, Gadhamshetty V, Mukherjee R, Chen Z, Ren W, Cheng H M and Koratkar N 2013 Passivation of microbial corrosion using a graphene coating *Carbon* **56** 45–9
[62] Zhang W *et al* 2014 Use of graphene as protection film in biological environments *Sci. Rep.* **4** 1–8
[63] Kim K H, Oh Y and Islam M F 2012 Graphene coating makes carbon nanotube aerogels superelastic and resistant to fatigue *Nat. Nanotechnol.* **7** 562–6
[64] Mogera U, Kurra N, Radhakrishnan D, Narayana C and Kulkarni G U 2014 Low cost, rapid synthesis of graphene on Ni: an efficient barrier for corrosion and thermal oxidation *Carbon* **78** 384–91

[65] Xie Y, Li H, Zhang C, Gu X, Zheng X and Huang L 2014 Graphene-reinforced calcium silicate coatings for load-bearing implants *Biomed. Mater.* **9** 025009

[66] He W, Zhu L, Chen H, Nan H, Li W, Liu H and Wang Y 2013 Electrophoretic deposition of graphene oxide as a corrosion inhibitor for sintered NdFeB *Appl. Surf. Sci.* **279** 416–23

[67] Singh B P, Nayak S, Nanda K K, Jena B K, Bhattacharjee S and Besra L 2013 The production of a corrosion resistant graphene reinforced composite coating on copper by electrophoretic deposition *Carbon* **61** 47–56

[68] Singh B P, Jena B K, Bhattacharjee S and Besra L 2013 Development of oxidation and corrosion resistance hydrophobic graphene oxide-polymer composite coating on copper *Surf. Coat. Technol.* **232** 475–81

[69] Shateri-Khalilabad M and Yazdanshenas M E 2013 Preparation of superhydrophobic electroconductive graphene-coated cotton cellulose *Cellulose* **20** 963–72

[70] Li P F, Zhou H and Cheng X 2014 Investigation of a hydrothermal reduced graphene oxide nano coating on Ti substrate and its nano-tribological behavior *Surf. Coat. Technol.* **254** 298–304

[71] Guo F, Silverberg G, Bowers S, Kim S P, Datta D, Shenoy V and Hurt R H 2012 Graphene-based environmental barriers *Environ. Sci. Technol.* **46** 7717–24

[72] Pan B, Xu G, Zhang B, Ma X, Li H and Zhang Y 2012 Preparation and tribological properties of polyamide 11/graphene coatings *Polym.-Plast. Technol. Eng.* **51** 1163–6

[73] David L, Feldman A, Mansfield E, Lehman J and Singh G 2014 Evaluating the thermal damage resistance of graphene/carbon nanotube hybrid composite coatings *Sci. Rep.* **4** 1–6

[74] Anandan S, Narasinga Rao T, Sathish M, Rangappa D, Honma I and Miyauchi M 2013 Superhydrophilic graphene-loaded TiO_2 thin film for self-cleaning applications *ACS Appl. Mater. Interfaces* **5** 207–12

[75] Krishnamoorthy K, Jeyasubramanian K, Premanathan M, Subbiah G, Shin H S and Kim S J 2014 Graphene oxide nanopaint *Carbon* **72** 328–37

[76] Hu W, Peng C, Luo W, Lv M, Li X, Li D, Huang Q and Fan C 2010 Graphene-based antibacterial paper *ACS Nano* **4** 4317–23

[77] Wang X, Xing W, Song L, Yang H, Hu Y and Yeoh G H 2012 Fabrication and characterization of graphene-reinforced waterborne polyurethane nanocomposite coatings by the sol–gel method *Surf. Coat. Technol.* **206** 4778–84

[78] Kandanur S S, Rafiee M A, Yavari F, Schrameyer M, Yu Z Z, Blanchet T A and Koratkar N 2012 Suppression of wear in graphene polymer composites *Carbon* **50** 3178–83

[79] Tai Z, Chen Y, An Y, Yan X and Xue Q 2012 Tribological behavior of UHMWPE reinforced with graphene oxide nanosheets *Tribol. Lett.* **46** 55–63

[80] Kim H J, Lee S M, Oh Y S, Yang Y H, Lim Y S, Yoon D H, Lee C, Kim J Y and Ruoff R S 2014 Unoxidized graphene/alumina nanocomposite: fracture-and wear-resistance effects of graphene on alumina matrix *Sci. Rep.* **4** 1–10

[81] Belmonte M, Ramírez C, González-Julián J, Schneider J, Miranzo P and Osendi M I 2013 The beneficial effect of graphene nanofillers on the tribological performance of ceramics *Carbon* **61** 431–5

[82] Wang D Y, Chang C L, Chen Z Y and Ho W Y 1999 Microstructural and tribological characterization of MoS_2–Ti composite solid lubricating films *Surf. Coat. Technol.* **120** 629–35

[83] Kim K S, Lee H J, Lee C, Lee S K, Jang H, Ahn J H, Kim J H and Lee H J 2011 Chemical vapor deposition-grown graphene: the thinnest solid lubricant *ACS Nano* **5** 5107–14

[84] Lee H, Lee N, Seo Y, Eom J and Lee S 2009 Comparison of frictional forces on graphene and graphite *Nanotechnology* **20** 325701
[85] Song H J and Li N 2011 Frictional behavior of oxide graphene nanosheets as water-base lubricant additive *Appl. Phys.* A **105** 827–32
[86] Lin J, Wang L and Chen G 2011 Modification of graphene platelets and their tribological properties as a lubricant additive *Tribol. Lett.* **41** 209–15
[87] Zhang W *et al* 2011 Tribological properties of oleic acid-modified graphene as lubricant oil additives *J. Phys. D: Appl. Phys.* **44** 205303
[88] Lee C, Li Q, Kalb W, Liu X Z, Berger H, Carpick R W and Hone J 2010 Frictional characteristics of atomically thin sheets *Science* **328** 76–80
[89] Berman D, Erdemir A and Sumant A V 2014 Graphene: a new emerging lubricant *Mater. Today* **17** 31–42
[90] Feng X, Kwon S, Park J Y and Salmeron M 2013 Superlubric sliding of graphene nanoflakes on graphene *ACS Nano* **7** 1718–24
[91] Berman D, Erdemir A and Sumant A V 2013 Few layer graphene to reduce wear and friction on sliding steel surfaces *Carbon* **54** 454–9
[92] Berman D, Erdemir A and Sumant A V 2013 Reduced wear and friction enabled by graphene layers on sliding steel surfaces in dry nitrogen *Carbon* **59** 167–75
[93] Kinoshita H, Nishina Y, Alias A A and Fujii M 2014 Tribological properties of monolayer graphene oxide sheets as water-based lubricant additives *Carbon* **66** 720–3
[94] Gass M H, Bangert U, Bleloch A L, Wang P, Nair R R and Geim A K 2008 Free-standing graphene at atomic resolution *Nat. Nanotechnol.* **3** 676–81
[95] Nayak P K, Hsu C J, Wang S C, Sung J C and Huang J L 2013 Graphene coated Ni films: a protective coating *Thin Solid Films* **529** 312–6
[96] Kirkland N T, Schiller T, Medhekar N and Birbilis N 2012 Exploring graphene as a corrosion protection barrier *Corros. Sci.* **56** 1–4
[97] Kousalya A S, Kumar A, Paul R, Zemlyanov D and Fisher T S 2013 Graphene: an effective oxidation barrier coating for liquid and two-phase cooling systems *Corros. Sci.* **69** 5–10
[98] Dlubak B *et al* 2012 Graphene-passivated nickel as an oxidation-resistant electrode for spintronics *ACS Nano* **6** 10930–4
[99] Schriver M, Regan W, Gannett W J, Zaniewski A M, Crommie M F and Zettl A 2013 Graphene as a long-term metal oxidation barrier: worse than nothing *ACS Nano* **7** 5763–8
[100] Yu L, Lim Y S, Han J H, Kim K, Kim J Y, Choi S Y and Shin K 2012 A graphene oxide oxygen barrier film deposited via a self-assembly coating method *Synth. Met.* **162** 710–4
[101] Chang K C *et al* 2014 Room-temperature cured hydrophobic epoxy/graphene composites as corrosion inhibitor for cold-rolled steel *Carbon* **66** 144–53
[102] Callow J A and Callow M E 2011 Trends in the development of environmentally friendly fouling-resistant marine coatings *Nat. Commun.* **2** 1–10
[103] Jin J, Wang X and Song M 2011 Graphene-based nanostructured hybrid materials for conductive and superhydrophobic functional coatings *J. Nanosci. Nanotechnol.* **11** 7715–22
[104] Ganesh B M, Isloor A M and Ismail A F 2013 Enhanced hydrophilicity and salt rejection study of graphene oxide-polysulfone mixed matrix membrane *Desalination* **313** 199–207
[105] Wang G, Zhu L, Liu H and Li W 2011 Zinc-graphite composite coating for anti-fouling application *Mater. Lett.* **65** 3095–7

[106] Camp D R and Zawacky S R 2006 Anti-fouling coating containing copper and graphite *U. S. Patent* 7,022,750PPG Industries Ohio Inc.

[107] Beigbeder A, Degee P, Conlan S L, Mutton R J, Clare A S, Pettitt M E, Callow M E, Callow J A and Dubois P 2008 Preparation and characterisation of silicone-based coatings filled with carbon nanotubes and natural sepiolite and their application as marine fouling-release coatings *Biofouling* **24** 291–302

[108] Lyon D Y, Brown D, Sundstrom E R and Alvarez P J 2008 Assessing the antibiofouling potential of a fullerene-coated surface *Int. Biodeterior. Biodegrad.* **62** 475–8

[109] Su X J, Zhao Q, Wang S and Bendavid A 2010 Modification of diamond-like carbon coatings with fluorine to reduce biofouling adhesion *Surf. Coat. Technol.* **204** 2454–8

[110] Mesarič T, Sepčič K, Piazza V, Gambardella C, Garaventa F, Drobne D and Faimali M 2013 Effects of nano carbon black and single-layer graphene oxide on settlement, survival and swimming behaviour of *Amphibalanus amphitrite* larvae *Chem. Ecol.* **29** 643–52

[111] Chen J, Peng H, Wang X, Shao F, Yuan Z and Han H 2014 Graphene oxide exhibits broad-spectrum antimicrobial activity against bacterial phytopathogens and fungal conidia by intertwining and membrane perturbation *Nanoscale* **6** 1879–89

[112] Russo C J and Passmore L A 2014 Controlling protein adsorption on graphene for cryo-EM using low-energy hydrogen plasmas *Nat. Methods* **11** 649–52

[113] Kuilla T, Bhadra S, Yao D, Kim N H, Bose S and Lee J H 2010 Recent advances in graphene based polymer composites *Prog. Polym. Sci.* **35** 1350–75

[114] Pan S, Guo R and Xu W 2014 Durable superoleophobic fabric surfaces with counter-intuitive superwettability for polar solvents *AlChE J.* **60** 2752–6

[115] Yang R, Zhang Y, Li J, Han Q, Zhang W, Lu C, Yang Y, Dong H and Wang C 2013 Graphene oxide assisted synthesis of GaN nanostructures for reducing cell adhesion *Nanoscale* **5** 11019–27

[116] Lee J, Chae H R, Won Y J, Lee K, Lee C H, Lee H H, Kim I C and Lee J M 2013 Graphene oxide nanoplatelets composite membrane with hydrophilic and antifouling properties for wastewater treatment *J. Membr. Sci.* **448** 223–30

[117] Li N, Liu L and Yang F 2014 Highly conductive graphene/PANi-phytic acid modified cathodic filter membrane and its antifouling property in EMBR in neutral conditions *Desalination* **338** 10–6

[118] Chanda A, Kim N K, Wijaya W and Bhattacharyya D 2021 Fire reaction of sandwich panels with corrugated and honeycomb cores made from natural materials *J. Sandw. Struct. Mater.* **23** 4196–217

[119] Chanda A, Kim N K and Bhattacharyya D 2021 Manufacturing and characterisation of wood-veneer sandwich panels with flame-retardant composite cores *Compos. Commun.* **27** 100870

[120] Barnard A S and Snook I K 2008 Thermal stability of graphene edge structure and graphene nanoflakes *J. Chem. Phys.* **128** 094707

[121] Nan H Y, Ni Z H, Wang J, Zafar Z, Shi Z X and Wang Y Y 2013 The thermal stability of graphene in air investigated by Raman spectroscopy *J. Raman Spectrosc.* **44** 1018–21

[122] Shi Y, Qian X, Zhou K, Tang Q, Jiang S, Wang B, Wang B, Yu B, Hu Y and Yuen R K 2013 CuO/graphene nanohybrids: preparation and enhancement on thermal stability and smoke suppression of polypropylene *Ind. Eng. Chem. Res.* **52** 13654–60

[123] Yu B, Wang X, Xing W, Yang H, Song L and Hu Y 2012 UV-curable functionalized graphene oxide/polyurethane acrylate nanocomposite coatings with enhanced thermal stability and mechanical properties *Ind. Eng. Chem. Res.* **51** 14629–36

[124] Zhuo D, Wang R, Wu L, Guo Y, Ma L, Weng Z and Qi J 2013 Flame retardancy effects of graphene nanoplatelet/carbon nanotube hybrid membranes on carbon fiber reinforced epoxy composites *J. Nanomater.* **2013** 820901

[125] Liao S H, Liu P L, Hsiao M C, Teng C C, Wang C A, Ger M D and Chiang C L 2012 One-step reduction and functionalization of graphene oxide with phosphorus-based compound to produce flame-retardant epoxy nanocomposite *Ind. Eng. Chem. Res.* **51** 4573–81

[126] Wang X, Song L, Yang H, Lu H and Hu Y 2011 Synergistic effect of graphene on antidripping and fire resistance of intumescent flame retardant poly (butylene succinate) composites *Ind. Eng. Chem. Res.* **50** 5376–83

[127] Guo Y, Bao C, Song L, Yuan B and Hu Y 2011 In situ polymerization of graphene, graphite oxide, and functionalized graphite oxide into epoxy resin and comparison study of on-the-flame behavior *Ind. Eng. Chem. Res.* **50** 7772–83

[128] Huang G, Yang J, Gao J and Wang X 2012 Thin films of intumescent flame retardant-polyacrylamide and exfoliated graphene oxide fabricated via layer-by-layer assembly for improving flame retardant properties of cotton fabric *Ind. Eng. Chem. Res.* **51** 12355–66

[129] Jiang S D, Bai Z M, Tang G, Hu Y and Song L 2014 Synthesis of ZnS decorated graphene sheets for reducing fire hazards of epoxy composites *Ind. Eng. Chem. Res.* **53** 6708–17

[130] Wang D, Zhou K, Yang W, Xing W, Hu Y and Gong X 2013 Surface modification of graphene with layered molybdenum disulfide and their synergistic reinforcement on reducing fire hazards of epoxy resins *Ind. Eng. Chem. Res.* **52** 17882–90

[131] Wang X, Xing W, Feng X, Yu B, Lu H, Song L and Hu Y 2014 The effect of metal oxide decorated graphene hybrids on the improved thermal stability and the reduced smoke toxicity in epoxy resins *Chem. Eng. J.* **250** 214–21

[132] Li J, Li J, Meng H, Xie S, Zhang B, Li L, Ma H, Zhang J and Yu M 2014 Ultra-light, compressible and fire-resistant graphene aerogel as a highly efficient and recyclable absorbent for organic liquids *J. Mater. Chem.* A **2** 2934–41

[133] Roy R K and Lee K R 2007 Biomedical applications of diamond-like carbon coatings: a review *J. Biomed. Mater. Res.* B **83** 72–84

[134] Shah A K, Sinha R K, Hickok N J and Tuan R S 1999 High-resolution morphometric analysis of human osteoblastic cell adhesion on clinically relevant orthopedic alloys *Bone* **24** 499–506

[135] Huang N, Yang P, Leng Y X, Chen J Y, Sun H, Wang J, Wang G J, Ding P D, Xi T F and Leng Y 2003 Hemocompatibility of titanium oxide films *Biomaterials* **24** 2177–87

[136] Gutensohn K, Beythien C, Bau J, Fenner T, Grewe P, Koester R, Padmanaban K and Kuehnl P 2000 In vitro analyses of diamond-like carbon coated stents: reduction of metal ion release, platelet activation, and thrombogenicity *Thromb. Res.* **99** 577–85

[137] Gillespie W J, Frampton C M, Henderson R J and Ryan P M 1988 The incidence of cancer following total hip replacement *J. Bone Joint Surg. Br.* **70** 539–42

[138] Podila R, Moore T, Alexis F and Rao A M 2013 Graphene coatings for enhanced hemo-compatibility of nitinol stents *RSC Adv.* **3** 1660–5

[139] Sperling C, Schweiss R B, Streller U and Werner C 2005 In vitro hemocompatibility of self-assembled monolayers displaying various functional groups *Biomaterials* **26** 6547–57

[140] Mikhalovska L I, Santin M, Denyer S P, Lloyd A W, Teer D G, Field S and Mikhalovsky S V 2004 Fibrinogen adsorption and platelet adhesion to metal and carbon coatings *Thromb. Haemost.* **92** 1032–9

[141] Airoldi F *et al* 2004 Comparison of diamond-like carbon-coated stents versus uncoated stainless steel stents in coronary artery disease *Am. J. Cardiol.* **93** 474–7

[142] Allen M, Myer B and Rushton N 2001 In vitro and *in vivo* investigations into the biocompatibility of diamond-like carbon (DLC) coatings for orthopedic applications *J. Biomed. Mater. Res.* **58** 319–28

[143] Butter R, Allen M, Chandra L, Lettington A H and Rushton N 1995 In vitro studies of DLC coatings with silicon intermediate layer *Diam. Relat. Mater.* **4** 857–61

[144] Dearnaley G and Arps J H 2005 Biomedical applications of diamond-like carbon (DLC) coatings: a review *Surf. Coat. Technol.* **200** 2518–24

[145] Kalbacova M, Broz A, Kong J and Kalbac M 2010 Graphene substrates promote adherence of human osteoblasts and mesenchymal stromal cells *Carbon* **48** 4323–9

[146] Li N, Zhang X, Song Q, Su R, Zhang Q, Kong T, Liu L, Jin G, Tang M and Cheng G 2011 The promotion of neurite sprouting and outgrowth of mouse hippocampal cells in culture by graphene substrates *Biomaterials* **32** 9374–82

[147] Grill A 2003 Diamond-like carbon coatings as biocompatible materials—an overview *Diam. Relat. Mater.* **12** 166–70

[148] Hauert R 2003 A review of modified DLC coatings for biological applications *Diam. Relat. Mater.* **12** 583–9

[149] Roozbahani F, Sultana N, Fauzi Ismail A and Nouparvar H 2013 Effects of chitosan alkali pretreatment on the preparation of electrospun PCL/chitosan blend nanofibrous scaffolds for tissue engineering application *J. Nanomater.* **2013** 641502

[150] Guo L, Yuan W, Lu Z and Li C M 2013 Polymer/nanosilver composite coatings for antibacterial applications *Colloids Surf.*, A **439** 69–83

[151] Liu S, Zeng T H, Hofmann M, Burcombe E, Wei J, Jiang R, Kong J and Chen Y 2011 Antibacterial activity of graphite, graphite oxide, graphene oxide, and reduced graphene oxide: membrane and oxidative stress *ACS Nano* **5** 6971–80

[152] Krishnamoorthy K, Veerapandian M, Zhang L H, Yun K and Kim S J 2012 Antibacterial efficiency of graphene nanosheets against pathogenic bacteria via lipid peroxidation *J. Phys. Chem.* C **116** 17280–7

[153] Akhavan O and Ghaderi E 2010 Toxicity of graphene and graphene oxide nanowalls against bacteria *ACS Nano* **4** 5731–6

[154] Sun X, Liu Z, Welsher K, Robinson J T, Goodwin A, Zaric S and Dai H 2008 Nano-graphene oxide for cellular imaging and drug delivery *Nano Res.* **1** 203–12

[155] Varghese N, Mogera U, Govindaraj A, Das A, Maiti P K, Sood A K and Rao C N R 2009 Binding of DNA nucleobases and nucleosides with graphene *Chem. Phys. Chem.* **10** 206–10

[156] Feng L and Liu Z 2011 Graphene in biomedicine: opportunities and challenges *Nanomedicine* **6** 317–24

[157] Fan Z, Liu B, Wang J, Zhang S, Lin Q, Gong P, Ma L and Yang S 2014 A novel wound dressing based on Ag/graphene polymer hydrogel: effectively kill bacteria and accelerate wound healing *Adv. Funct. Mater.* **24** 3933–43

[158] Lu B, Li T, Zhao H, Li X, Gao C, Zhang S and Xie E 2012 Graphene-based composite materials beneficial to wound healing *Nanoscale* **4** 2978–82

[159] Major R, Sanak M, Mzyk A, Lipinska L, Kot M, Lacki P, Bruckert F and Major B 2014 Graphene based porous coatings with antibacterial and antithrombogenous function—materials and design *Arch. Civil Mech. Eng.* **14** 540–9

[160] Santos C M, Tria M C R, Vergara R A M V, Ahmed F, Advincula R C and Rodrigues D F 2011 Antimicrobial graphene polymer (PVK-GO) nanocomposite films *Chem. Commun.* **47** 8892–4

[161] Gurunathan S, Han J W, Dayem A A, Eppakayala V and Kim J H 2012 Oxidative stress-mediated antibacterial activity of graphene oxide and reduced graphene oxide in *Pseudomonas aeruginosa Int. J. Nanomed.* **7** 5901

[162] Li J, Wang G, Zhu H, Zhang M, Zheng X, Di Z, Liu X and Wang X 2014 Antibacterial activity of large-area monolayer graphene film manipulated by charge transfer *Sci. Rep.* **4** 1–8

[163] Wen B, Wang X X, Cao W Q, Shi H L, Lu M M, Wang G, Jin H B, Wang W Z, Yuan J and Cao M S 2014 Reduced graphene oxides: the thinnest and most lightweight materials with highly efficient microwave attenuation performances of the carbon world *Nanoscale* **6** 5754–61

[164] Geim A K and Novoselov K S 2007 The rise of graphene *Nat. Mater.* **6** 183–91

[165] Gómez-Navarro C, Burghard M and Kern K 2008 Elastic properties of chemically derived single graphene sheets *Nano Lett.* **8** 2045–9

[166] Young R J, Kinloch I A, Gong L and Novoselov K S 2012 The mechanics of graphene nanocomposites: a review *Compos. Sci. Technol.* **72** 1459–76

[167] Gao C, Liu T, Shuai C and Peng S 2014 Enhancement mechanisms of graphene in nano-58S bioactive glass scaffold: mechanical and biological performance *Sci. Rep.* **4** 1–10

[168] Wang Q, Li Y, Luo M, Sang S, Zhu T and Zhao L 2014 Strengthening mechanism of graphene oxide nanosheets for Al_2O_3–C refractories *Ceram. Int.* **40** 163–72

Chapter 16

Magnetism in graphene

Sudipta Dutta, Soumya Ranjan Das and Souren Adhikary

Low-dimensional systems offer a plethora of interesting electronic properties owing to the confined electronic degrees of freedom and associated quantum fluctuations. One of the major breakthroughs in the quest for low-dimensional materials has been the discovery of graphene. It is a strictly two-dimensional array of sp^2-hybridized carbon atoms packed into a monolayer honeycomb lattice. Graphene has emerged as a wonder material, demonstrating unprecedented electronic, mechanical, transport and optical properties arising from the confinement of charge carriers in truly two dimensions. The possibility of achieving magnetism in graphene has especially garnered immense attention. It is due to the itinerant behavior of its electrons that is known to prohibit any spin localization and due to the electronic confinement in two dimensions that suppresses any magnetic ordering at finite temperature as described within Mermin–Wagner theory. In spite of these fundamental obstructions, graphene provides numerous ways of inducing magnetism as evident from recent studies. In this chapter, we present a detailed theoretical perspective of magnetic properties of graphene along with recent experimental validations. We extend our discussion towards the general two-dimensional materials beyond graphene in the light of their unprecedented magnetic properties.

16.1 Introduction

Magnetism is a unique property of materials that has its origin from purely quantum mechanical description of matter and can be experimentally realized in bulk materials [1, 2]. It arises from the interacting electron spins and consequent spin ordering that can translate till the macroscopic length scale. This phenomenon has attracted sustained interest over the decades due to its complexity, diversity and its huge application possibilities in electrical and electronic devices for data storage,

doi:10.1088/978-0-7503-3999-5ch16

energy harvesting and quantum information [3]. The advancement of a new era of quantum technologies heavily depends on novel magnetic materials.

Enhanced quantum confinement and fluctuations in low-dimensional systems modify the interactions and dynamics of the spins, leading to unprecedented and novel magnetic phases [4, 5]. This has been tempting extensive fundamental research along with their exploitations in achieving advanced spin and magnetic devices [6–9]. Recent experimental sophistications enable us to fabricate such low-dimensional magnetic materials that can validate a range of theoretical predictions based on quantum condensed matter theory [10, 11]. This field was mainly confined within the transition metal systems, since the localization of the d-electrons has been known to result in magnetic behavior. The successful isolation of graphene by mechanical exfoliation has changed that perspective [12–15]. Graphene has been a major breakthrough in the quest of low-dimensional materials that has allowed researchers to explore the electronic properties in strictly two dimensions [16–19]. It is also one of the foremost candidates for future applications in carbon-based nanoelectronic devices [20–22], owing to its high charge carrier mobility [23] and thermal conductivity [24]. However, such a material with only itinerant p-electrons is not expected to show conventional magnetic behavior.

The planar honeycomb lattice of graphene consists of two nonequivalent sublattices, *A* and *B*, as depicted by dark and light colored lattice points in the shaded unit cell (see figure 16.1(a)). Because of that, graphene is called a bipartite lattice. The electronic states of graphene near the Fermi energy are governed by itinerant π-electrons originating from the p_z orbital of carbon atoms. The motion of these electrons in graphene is mainly described by the massless Dirac equation or Weyl equation. The hexagonal Brillouin zone consists of two nonequivalent *k* points, namely *K* and *K*′, also known as Dirac points (see inset of figure 16.1(b)). The bands from two different sublattices of the 2D bipartite graphene lattice intersect at the Fermi energy to give rise to an unusual conical energy dispersion, known as Dirac cone and consequent zero density of states (DOS) at the Fermi energy

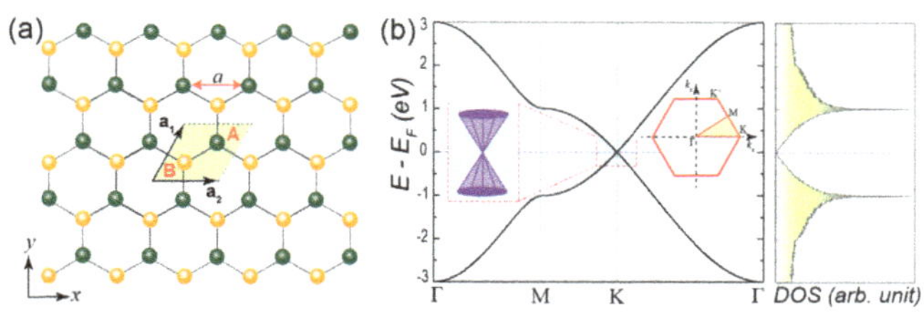

Figure 16.1. (a) Two-dimensional graphene honeycomb lattice (periodic in the *x*–*y*-plane) with dark (green) and light (yellow) circles denoting the nonequivalent *A* and *B* sublattice points, respectively. The unit cell is shown by the shaded rhombus along with the lattice vectors. (b) The scaled energy spectrum of graphene in the irreducible Brillouin zone (shown in the inset along with the high-symmetric points) with the corresponding DOS, calculated using the nearest-neighbor tight-binding model. The Fermi energy is set to zero (horizontal dashed line) and the high-symmetric points are shown by the vertical dashed lines. The linear dispersion near the Fermi energy at high-symmetric point *K*, i.e., at the Dirac point is zoomed in the inset to depict the Dirac conical dispersion.

(see figure 16.1(b)) [16, 25–27]. This unique energy spectrum of semi-metallic graphene leads to highly unconventional electronic properties, such as the half-integer quantum Hall effect [14, 20], the absence of backward scattering [18, 28, 29], Klein tunneling [30] and the π-phase shift of Shubnikov–de Haas oscillations [31]. Furthermore, the magnetic ground states in analogous bipartite lattices and large spin relaxation length in graphene have generated much hope of achieving magnetism in graphene, in spite of its itinerant π-electrons [17, 25, 26, 32–35]. There have been few previous studies on unpaired p-electron induced magnetic ground states in organic materials [36–39]. However, the complete delocalization of π-electrons on pristine graphene monolayer makes it extremely challenging to stabilize any kind of magnetic ordering. Many successful attempts have been made towards this direction in recent times [40, 41].

In the subsequent sections, we systematically develop this intriguing phenomenon. We first present the theoretical background, highlighting the scope of various theoretical frameworks. Then varying defect-induced magnetism in graphene is discussed in the light of non-bipartite lattice and edge states. This is followed by the recent observations of proximity-induced magnetism in graphene. Finally, we present very recent advancements of magnetism in two-dimensional systems beyond graphene, especially in van der Waal's magnetic systems from the perspective of Mermin–Wagner theory followed by summary and outlook.

16.2 Theoretical background

The basic mechanism of magnetism in graphene lies in the bipartite nature of its lattice that satisfies Lieb's theorem [42]. The A and B sublattice points in pristine graphene (see figure 16.1(a)) prefer to localize opposite spins that lead to the ground state with zero magnetization. According to Lieb's theorem, any inequality between the number of sublattice points of A and B, i.e., N_A and N_B, respectively, induces a net magnetization of $(N_A - N_B)/2$. This shows a potential way of making graphene magnetic by creating sublattice imbalance. This will be further discussed in the context of defect and grain boundary induced magnetism in graphene in subsequent sections.

Non-interacting tight-binding Hamiltonian cannot capture such spin localization in lattice systems. This makes the consideration of electron–electron correlations indispensable. We consider the following Hubbard Hamiltonian that considers the on-site Coulomb correlations [43–46].

$$\boldsymbol{H}_{\boldsymbol{Hubbard}} = -\boldsymbol{t}\sum\nolimits_{\langle i,j\rangle,\sigma}\left(\boldsymbol{c}_{i,\sigma}^{\dagger}\boldsymbol{c}_{j,\sigma} + h.\,c.\right) + \boldsymbol{U}\sum\nolimits_{i}\boldsymbol{n}_{i,\uparrow}\boldsymbol{n}_{i,\downarrow} \tag{16.1}$$

The first term in the above equation is the kinetic energy approximated within nearest-neighbor tight-binding model. The operator $c_{i,\sigma}^{\dagger}$ ($c_{i,\sigma}$) creates (annihilates) an electron with spin σ at atomic site i, and 'h.c.' stands for the Hermitian conjugate counterparts. The nearest-neighbor atoms are only considered, as denoted by the angular brackets ($\langle .,. \rangle$). This ensures electron–hole symmetry in the energy dispersion. The hopping integral, t is considered to be the energy scale for the numerical results. The second term includes the on-site electron–electron interactions through U. Here, the number operator, $n_{i,\sigma} = c_{i,\sigma}^{\dagger} c_{i,\sigma}$, denotes the spin-resolved electron occupancy at atomic site i.

The two-electron operator in the second term of the above Hamiltonian makes the numerical calculation extremely challenging, because of exponentially growing many-body basis size. Many approximations have been proposed to tackle such challenges [47]. Mean-field theory has been quite successful to overcome this with close agreements with the experimental observations for weakly interacting systems with negligible quantum fluctuations. Here, the unrestricted Hartree–Fock approximation is considered to map the two-electron operator form into effective one-electron operators by separating the up- and down-spin sectors, as follows [48].

$$H_{\mathrm{MF}} = -t \sum_{\langle i,j\rangle,\sigma} \left(c_{i,\sigma}^{\dagger} c_{j,\sigma} + h.c. \right) + U \sum_{i} \left[n_{i,\uparrow} < n_{i,\downarrow} > + \, n_{i,\downarrow} < n_{i,\uparrow} > - < n_{i,\uparrow} > < n_{i,\downarrow} > \right] \tag{16.2}$$

In the above mean-field (MF) Hamiltonian, an electron of a specific spin belonging to the atomic site i, interacts with the average-field or the mean-field (denoted by the angular brackets) produced by the electron density with an opposite spin orientation at the same site and vice versa. The above Hamiltonian has to be solved self-consistently by minimizing the free energies with respect to the average spin densities. After attaining the self-consistent convergence, the net magnetization M_i (in units of Bohr magneton μ_{B}) at each atomic site i can be calculated via the following expression.

$$M_i = \frac{n_{i,\uparrow} - n_{i,\downarrow}}{2} \tag{16.3}$$

The total spin of the system S is then calculated by adding the contributions of spin density from every site ($S = \sum_i M_i$). The spin moments and consequent net magnetization depends on the strength of the scaled on-site Coulomb correlation U/t.

The mean-field results in graphene and its nanostructures have been validated by comparing them with more accurate theoretical methods, like exact diagonalization and quantum Monte Carlo simulations [49]. Moreover, with a proper choice of the parameter U/t ($\approx$1) in the mean-field Hubbard model, a close agreement with the first-principles methods has been achieved [50–52]. Nevertheless, the empirical parameter U/t has to be derived through comparison with experimental observations. Magnetic resonance experiments of soliton states in quasi-one-dimensional *trans*-polyacetylene gives a close match of U/t ($\approx$1.2–1.3) with that obtained in theory for graphene nanostructures [53, 54]. In particular, for values of U/t above 2.23, non-magnetic graphene undergoes a Mott–Hubbard quantum phase transition into an antiferromagnetic state with bulk spin polarization [55]. Therefore, the magnetic graphene systems discussed in the following sections have been studied theoretically by considering $U/t \approx 1$ in the Hubbard Hamiltonian (equation 16.1).

Note that, the Stoner criterion determines the onset of magnetism in these systems and depends on the competition between the exchange energy gain and the kinetic energy penalty due to spin polarization [56]. The energy splitting (Δ_S) that causes a gain in the magnetic exchange energy [57] is given by,

$$\Delta_S = \frac{U}{2} \sum_i n_i^2 \tag{16.4}$$

where $\sum_i n_i^2$ quantifies the extent of localization of the associated electronic state. Equation (16.4) shows that irrespective of the extent of localization, the system undergoes spin polarization for any non-zero value of the on-site Coulomb potential ($U > 0$) to prevent the instability of low-energy midgap states. This effect becomes significant in reduced dimensions. The sp^2-hybridized carbon systems usually tend to be non-magnetic (NM) as the kinetic energy dominates the exchange energy. However, this can be reversed due to quantum confinement and defects, thereby endowing magnetism in intrinsically non-magnetic graphene nanostructures.

The *ab initio* electronic structure methods based on density functional theory (DFT) have been extremely successful in probing the magnetic properties of graphene and related systems. This is actually an effective mean-field theory that maps the many-electron description of a system onto a single-particle picture, thereby making the computation much easier [58–60]. Within this theoretical framework, the systems are assumed to be weakly interacting with negligible quantum fluctuations. This method shows good agreement with the mean-field Hubbard results for graphene-like systems. We present a series of exciting results and their agreement with experimental observations in the subsequent sections.

16.3 Magnetism due to sublattice inequality

As discussed before, the spin alignment in bipartite graphene lattice is described by Lieb's theorem [42]. Within repulsive Hubbard Hamiltonian, the total spin S of the ground state of a half-filled bipartite system is given by $(N_A - N_B)/2$. Therefore, any inequality of the number of A and B sublattice points leads to net magnetization in graphene lattice. Such inequality can arise due to the vacancy defect or chemical substitution. This can also happen due to the ill-defined sublattice symmetry and associated spin frustrations. Lieb's theorem holds for all dimensions, irrespective of whether the system is periodic or not. In this section, we demonstrate the magnetic properties of graphene systems in the presence of such defects.

The presence of different kinds of defects modifies the chemical, mechanical, and electronic properties of any pristine system. However, such defects or disorders cannot be avoided during the material synthesis or preparation. In the case of low-dimensional systems like graphene, such defects or disorders are more common. These can severely affect the carrier mobility and intrinsic electronic properties of graphene due to unwanted scattering processes and hence has been considered as a major deterrent. However, a growing body of theoretical and experimental investigations indicates that such defects can help in making graphene magnetic by setting a long-range magnetic order.

16.3.1 Vacancy defects

Removal of one or more carbon atoms from a graphene sheet creates a vacancy in the lattice. Such localized and randomly distributed vacancies are very common in any graphene sample. In the case of an isolated single-atom vacancy that is termed as point defect, there is formation of three dangling σ bonds on three neighboring carbon atoms. This leads to lattice reconstruction or formation of non-hexagonal carbon rings around the vacancy center [61]. Moreover, it creates an inequality in the number of A and B sublattice points, thereby inducing net magnetization according to Lieb's theorem [42].

The magnetic ordering due to vacancy defects was first observed in high-energy proton (~2.25 MeV) irradiated graphite samples [62–64]. The knock-on collisions force the carbon atoms to leave their equilibrium position [62]. Proton irradiation and hydrogen plasma treatment also can lead to the chemisorption of hydrogen adatoms on the graphene surface [65]. In such cases, the carbon atom beneath the hydrogen undergoes a change in hybridization from sp^2 to sp^3. Such sp^3 defects break the planarity and the π-electron network of graphene and act as scattering centers. This is also akin to a vacancy defect, albeit without any dangling σ bonds and consequent preservation of the three-fold symmetry of the graphene lattice.

The point defects due to vacancies and hydrogen chemisorption profoundly affect the electronic and magnetic properties of graphene. In particular, they create an imbalance in the π-electrons between the two nonequivalent sublattices of graphene, giving rise to a quasi-localized zero-energy state in the energy dispersion and consequent sharp peak in the DOS [66]. The dangling σ bonds around such defects can also produce sharp DOS peaks arising from the non-bonding electrons [66]. For hydrogen chemisorption, the peak in the DOS fully splits by magnetic exchange and leads to a net magnetic moment of ~1 μ_B at any defect concentration per supercell [66–69]. Moreover, the direct exchange coupling due to Stoner ferromagnetism between the induced itinerant moments results in an unusual long-range spin texture, extending over several nanometers and has been observed in STM experiments [65, 66].

The lattice reconstruction around the vacancy can lead to formation of non-hexagonal odd-membered carbon rings [61, 66]. Consequently, the bipartite nature of graphene collapses, resulting in suppressing the magnetic moments due to spin frustration. However, it is still estimated to vary from 1.12 μ_B to 1.53 μ_B for defect concentration ranging from 20% to 0.5% per supercell [66, 68]. The defect-induced magnetization can be tuned further by a gate voltage or an external electric field [70–73], opening up new avenues for their use in spintronics. In spite of these observations, there is a lack of consensus over the long-range room temperature ferromagnetism predicted in these systems. Experimental studies have claimed that graphene with vacancies or sp^3 defects with fluorine adatoms [74, 75] show paramagnetism and are devoid of any magnetic order at low temperatures [76–78].

Nevertheless, the unique catenation property and consequent range of varying hybridizations of carbon allows the formation of various unique defects in graphene. In particular, the defects can originate from various local dislocations and grain boundary formations in polycrystalline graphene, that can be characterized by unique topological invariants [79]. All these defects bring new functionalities that are not necessarily magnetic always. For example, the Stone–Wales defect in graphene [80] is generated by the 90° rotation of a single C–C bond, resulting in the formation of fused pairs of two five and two seven-membered rings in the hexagonal lattice. The bipartite character of the honeycomb lattice collapses in such defects due to the lack of sublattice clarity. However, due to conserved total number of atoms and the three-fold symmetry of the honeycomb lattice, no magnetic ordering arises from it. In contrast, the formation of a fused single pair of five and seven-membered rings at the finite graphene boundaries has been predicted to show magnetic behavior [81].

16.3.2 Grain boundary

As we discussed before, the defects usually form during the experimental preparation of graphene. The most effective, low-cost, and scalable method to prepare graphene has been the chemical vapor deposition (CVD) technique, which can produce a large-area sample in a short time [82]. In this experimental technique, a carbon-containing precursor molecule, mainly methane gas flows over a metal substrate at very high temperature inside the CVD chamber. At such elevated temperature methane dissociates and the resulting carbon atoms get deposited on the metal substrate in random positions, forming several nucleation centers. With time, those nucleation centers grow with gradual deposition of carbon atoms in honeycomb fashion and form several graphene domains, as shown schematically in figure 16.2(a). Such isolated graphene domains grow randomly in any crystallographic direction [83–85]. When two such gradually grown graphene domains meet each other, the mismatch of crystallographic orientations along their in-plane interface results in grain boundaries that do not necessarily conform with the honeycomb arrangements of the lattice points in individual domains. Such grain boundaries are generally made of non-hexagonal rings spanning over a relatively longer length as compared to the lattice constant of graphene. The bipartite symmetry of the graphene lattice gets broken in such non-hexagonal rings, as shown schematically in figure 16.2(a). Merging of several such non-uniform domains leads to a polycrystalline graphene sample with different grain boundaries that vary in shape and size [83–85].

The network of point defects and grain boundaries in polycrystalline graphene leads to considerable deviation from the characteristic behavior of pristine graphene and can be a severe detriment to electronic transport [86]. However, such defects can also result in localized magnetic states in the proximity of Fermi energy and can significantly affect the transport properties [61, 87–90]. We explore such grain boundaries from the perspective of broken bipartite symmetry and consequent magnetism within DFT framework [61]. We consider large rectangular graphene supercell and the corresponding rectangular Brillouin zone (dashed lines), as shown

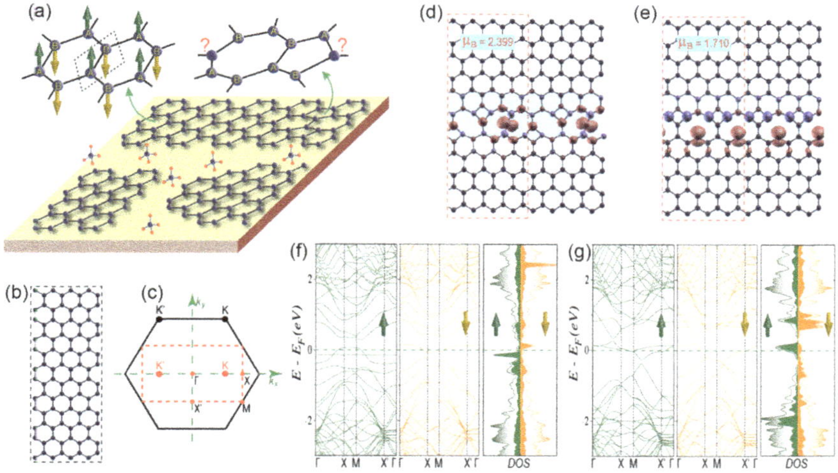

Figure 16.2. (a) Schematic representation of chemical vapor deposition (CVD) technique of preparing graphene on a metal substrate from methane precursor. It shows several graphene domains grown from different nucleation centers. The bipartite nature of the graphene honeycomb lattice with two nonequivalent sublattice points *A* and *B* in the rhombus unit cell with opposite spin localization is shown schematically in the top left side. The fused five-seven-membered ring with lack of sublattice clarity (depicted by the question marks) is shown schematically in the top right side. (b) The rectangular graphene supercell. (c) The two-dimensional Brillouin zone of rhombus graphene unit cell is shown by hexagon with solid line. The rectangular Brillouin zone of rectangular graphene supercell is shown by the dashed lines. The circular dots represent the high-symmetric points. Panels (d) and (e) show the graphene system with two different grain boundary geometries (shaded region) with spin density localizations. The supercell is shown by the dashed rectangles along with the net magnetization values per supercell. Panels (f) and (g) show the spin polarized band structures of graphene supercells in (d) and (e), respectively, along with corresponding total DOS (faded shade) and projected DOS (darker shade) arising from the grain boundary regions. The vertical solid lines (horizontal dashed lines) represent the high-symmetric points (Fermi energy). Reproduced [61] under Creative Common Attribution 4.0 International License.

in figures 16.2(b) and (c), respectively. Note that, the change of supercell shape as compared to the rhombus unit cell of graphene, makes the original hexagonal Brillouin zone folded into a rectangular one with a new set of high-symmetric points and irreducible Brillouin zone Γ–X–M–X′–Γ. Within such a supercell, we force two graphene domains to meet with irregular crystallographic orientations that lead to the formation of different grain boundaries. Here, we present two such periodic grain boundaries containing fused four-eight and four-ten membered rings in figures 16.2(d) and (e), respectively. Due to localized states arising from the dangling bonds, such grain boundaries exhibit localization of spins, as depicted by spin density plots in figures 16.2(d) and (e). Such asymmetric spin localizations give rise to large net magnetization in the graphene supercells that breaks the time-reversal symmetry. Here the grain boundaries in figures 16.2(d) and (e) show net magnetizations of 2.399 μ_B and 1.710 μ_B per supercell, respectively. The linear energy dispersion of pristine graphene near the Fermi energy at high-symmetric point K completely disappears due to these magnetic states, leading to an itinerant magnetic system as shown by the asymmetric up and down-spin band structures and metallic DOS at Fermi energy for both the defect structures (see figures 16.2(f) and (g)). For the system with fused four-eight membered rings (figure 16.2(d)), a completely (partially) dispersion-less flat band appears for the up (down) spin just below (above) the Fermi energy (see figure 16.2(f)). In contrast, both the spins conduct for the four-ten membered fused ring grain boundary (figure 16.2(e)), as shown by the dispersive bands near Fermi energy in figure 16.2(g). The projected DOS clearly shows that the bands near Fermi energy predominantly arise from the grain boundaries, indicating their role in regulating the properties of CVD grown graphene systems. The magnetic ordering along the grain boundaries can further be tuned by doping. These observations agree quite well with the recent experiments on spontaneous time-reversal symmetry breaking across graphene grain boundaries [91]. Therefore, designing novel magnetic grain boundaries in graphene nanostructures will enable the unique properties of these defect systems to be harnessed in various novel spintronic applications, like asymmetric spin transport and spin-filtering.

16.3.3 Elemental substitution

We can achieve similar magnetism by elemental substitution or adatoms that change the number of electrons and the nature of chemical bonding [92–99]. These in turn shift the Fermi energy of the pristine graphene and can result in narrow impurity bands near Fermi energy. It has been observed that in the presence of elemental substituents like nitrogen (N) and sulfur (S), graphene behaves predominantly as a diamagnet, with paramagnetic impurity centers imprinted onto it [92–95]. Upon increase in the concentration of substituents, these paramagnetic centers couple through the itinerant π-electrons, giving rise to Stoner ferromagnetism in graphene. Theoretical studies have also shown that codoping boron enhances ferromagnetism for N-substituted graphene [92]. The magnetization arising from such direct exchange coupling are demonstrated by the Curie temperatures of 62 K and 69 K, respectively, for S- and N-substituted graphene in experiments [93, 94]. Besides non-magnetic atoms, doping transition metal atoms in graphene also showed interesting magnetic behavior [100]. Furthermore, the surface chemical functionalizations of graphane [101, 102], graphone [103], and reduced graphene oxide [104, 105] systems with hydrogen and oxygen atoms and other functional groups have been explored to imprint magnetism. However, their magnetic signatures were demonstrated to be relatively weak at room temperature and strongly dependent on the stoichiometry, internal defects, transition metal impurities, and the level of oxidation. A judicious choice of defect engineering of graphene systems using an amalgamation of vacancies, defects, and doping can thus open new prospects in nanoscale magnetism.

16.4 Magnetism from nanographene edge

The most fascinating way of inducing magnetism in graphene has been through the formation of edge, more specifically zigzag edge. Extensive theoretical studies have been carried out to explore the unprecedented functionalities of graphene through the edge manipulations, modifications and functionalizations. A broad spectrum of electronic, magnetic, optical and transport properties in graphene and related systems originates from the unique boundary conditions and associated quantum confinements in effective quasi-one-dimension [16, 25–27, 106–110]. Many novel theoretical predictions have been motivating enormous experimental explorations in recent times. In this section, we present a detailed discussion on edge-induced magnetism in quasi-one-dimensional and zero-dimensional graphene and related systems.

Phenomenal sophistications in experimental techniques that allow the preparations and characterizations with atomic scale precision have made such reduced length scale accessible to us. They include top-down approaches like lithography [111–113], the unzipping of carbon nanotubes (CNTs) [114, 115], solution dispersion and sonication of chemically exfoliated graphite [116], or the epitaxial growth on silicon carbide sidewalls [117]. In addition, recent experiments using bottom-up synthesis provide graphene nanostructures with smooth edges with minimal disorder and precise control of their edge terminations [118–120]. Depending on the chosen precursor molecule in the bottom-up approach, different unconventional edge geometries, like cove, chevron, and chiral topologies have been prepared, significantly expanding their versatility in device applications [121, 122]. These have huge bearing on the recent upsurge of cutting-edge quantum technologies.

16.4.1 Graphene nanoribbons

Termination of the pristine two-dimensional graphene sheet along the high symmetry crystallographic directions results in two unique edge geometries, e.g., armchair and zigzag, which differ by an angle of 30° in their crystallographic orientations, as shown in figure 16.3(a). The quasi-one-dimensional periodic ribbon geometries with parallel armchair and zigzag edges are famously known as armchair graphene nanoribbons (AGNRs) and zigzag graphene nanoribbons (ZGNRs). Their energy spectrums and consequent electronic properties are remarkably different, owing to their distinct boundary conditions. The AGNRs exhibit metallic or semiconducting behavior depending on their width and subsequent boundary conditions in the folded Brillouin zone [26, 107, 108, 123, 124]. The semiconducting gap reduces with increase in width of the AGNRs. Nevertheless, the absence of localized states in pristine AGNRs makes them intrinsically non-magnetic [25, 107, 108, 124].

In that perspective, the ZGNRs are quite unique in localizing the zero-energy states along the edges. Figure 16.3(b) shows the geometry of the ZGNRs with the width expressed in terms of the number of zigzag lines, N_z along the cross-ribbon direction. The unit cell of periodic ribbon structure is shown by the shaded rectangle. The dangling bonds on the edge atoms are assumed to be passivated by hydrogen atoms that do not show any contribution near the Fermi energy, thereby not altering the electronic properties of the system. Such a unit cell maps the original hexagonal Brillouin zone of pristine graphene along the k_x momentum axis (see figure 16.3(c)). Due to the Brillouin zone folding, the Dirac points (K and K′) now lie on the line

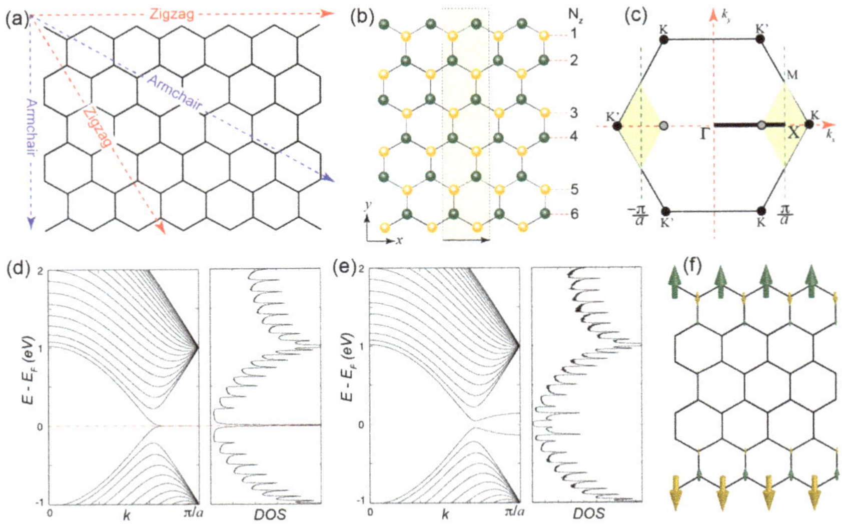

Figure 16.3. (a) Two-dimensional graphene lattice with zigzag and armchair crystallographic directions (shown by dashed arrows), that differ by an angle of 30°. (b) The quasi-one-dimensional zigzag graphene nanoribbon structure that is periodic along the x-axis. The width is defined by the number of zigzag chains, N_z along the y-axis. The unit cell is depicted by the shaded rectangle with the lattice vector along the x-axis shown by the arrow. (c) The hexagonal Brillouin zone of graphene rhombus unit cell. Due to finite termination this gets mapped to the thick solid line spanning from Γ to X. The grey dots along the k_x momentum axis show the location of high-symmetric K and K′ points in the modified one-dimensional Brillouin zone. (d) and (e) present the band structures and corresponding DOS of zigzag graphene nanoribbon with $N_z = 20$, as obtained from tight-binding and mean-field level of calculations, respectively. Note that, the mean-field calculation corresponds to the system, modeled within Hubbard Hamiltonian with $U/t = 1$. (f) The schematic representation of spin ordering in the ribbon structure with opposite spins localizing on either edge with an exponential decay of site-spin magnitudes towards the center of the ribbon.

connecting the two high-symmetric points Γ and X (see figure 16.3(c)). Numerical calculations within non-interacting DFT and tight-binding levels show that all ZGNRs, irrespective of their widths, are gapless and feature an unusual pair of partial flat bands from the Dirac point ($ka = 2\pi/3$, where a is the lattice constant of graphene) to the X point [124–128]. In figure 16.3(d), we present the tight-binding band dispersion of ZGNR with $N_z = 20$. The partially flat top of the valence band and bottom of the conduction band result in diverging DOS at the Fermi energy (see figure 16.3(d)). These zero-energy flat bands arise from the localized edge states that decay exponentially in the bulk [125–131].

Such a sharp peak in the DOS at the Fermi energy in the non-interacting energy spectrum results in Fermi instability that is expected to either induce lattice distortion via the electron–phonon coupling or magnetic polarization due to the electron–electron interaction. As the localized edge states are non-bonding, it is unlikely to cause any lattice distortions near the zigzag edge [124, 132, 133]. Therefore, the consideration of electron–electron interaction becomes inevitable to understand the electronic properties of ZGNRs. These systems are generally modeled within Hubbard Hamiltonian and explored through mean-field approximations [125, 128, 134–137] or more sophisticated theoretical methods like density matrix renormalization group (DMRG) technique [138] or configuration interactions [139–142]. The mean-field energy dispersion of ZGNR with $N_z = 20$ shows the opening of a bandgap at the Fermi energy due to electron–electron interaction with a splitting of DOS at Fermi energy and consequent disappearance of Fermi instability (see figure 16.3(e)). The splitting of DOS originates from the spin-polarized edge states. Although the spatial spin symmetry gets broken upon introduction of electronic correlation, the spin channels remain degenerate. The spin polarized edge states lead to a peculiar spin ordering along the zigzag edge, even in the presence of infinitesimally small electronic correlations (see figure 16.3(f)). The ground state of ZGNRs exhibits parallel (antiparallel) spin alignment along the same (opposite) zigzag edges, with an exponential decay of spin density towards the ribbon center. The overall ground state is antiferromagnetic due to the vanishing net magnetization, consistent with Lieb's theorem of the half-filled Hubbard model for the bipartite lattice [42]. Although the ZGNRs ensure the equality of two nonequivalent sublattice points A and B, the magnetic order solely originates from the edge geometry and consequent zero-energy modes. Similar results can be obtained from the DFT based first-principles calculations. The second nearest-neighbor hopping term breaks the particle–hole symmetry of the energy spectrum but preserves the mirror symmetry of magnetization [143]. The spin degeneracy of the ZGNRs can be lifted by applying an external transverse electric field along the cross-ribbon direction [143–145]. In particular, one of the spin gaps gradually increases along with the decrease of other spin gap with increase in field strength. This finally makes only a single spin channel conducting and consequent half-metallicity beyond a certain critical field strength [25, 26, 143–145]. The ZGNRs can be made magnetic by blocking one of the edges by formation of lateral heterostructures with non-magnetic materials, having minimal lattice mismatch as well [146]. This makes the ZGNRs a promising candidate for obtaining spin-polarized current that is essential for spintronic and magnetic data storage applications. In particular, the critical electric field strength beyond which ZGNRs behave as half-metals decreases with the increase in nanoribbon width [143–149].

The partial flat-band mediated edge magnetism in ZGNRs is also robust towards any changes in the size and geometry of the systems [150, 151]. Furthermore, the edge states in zigzag nanoribbons host an unconventional topological Zak phase [152, 153] due to the non-zero Berry connection despite of the absence of Berry curvature [154, 155]. Such topological edge states are robust towards the chemical substitutions as well [156]. Hence, these intriguing electronic and magnetic properties of ZGNRs can be further tailored to design novel spintronic functionality in nanoscale systems.

The theoretical observation of the zigzag edge-state induced magnetism has been confirmed from scanning tunneling spectroscopy experiments on chiral graphene nanoribbons at low temperatures [157, 158]. In figures 16.4(a) and (b), we present the atomically resolved topography and the corresponding atomic structure of the system under investigation, respectively [158]. The differential conductance

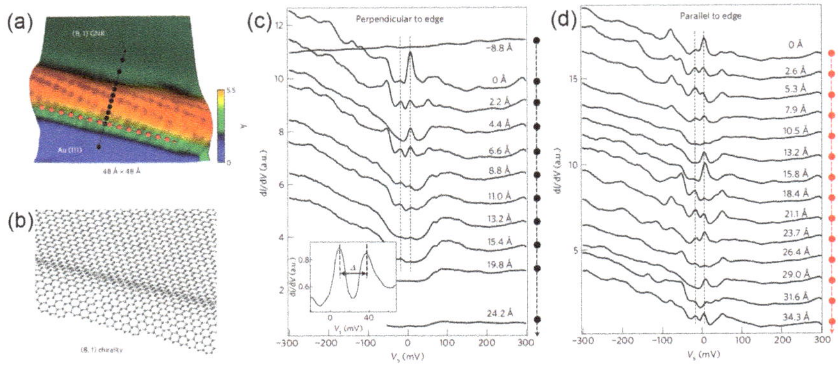

Figure 16.4. (a) Atomically resolved topography and (b) the corresponding structural depiction of a graphene nanoribbon with almost smooth zigzag edge. Panels (c) and (d) show the differential conductance (d*I*/d*V*), measured at different points perpendicular and parallel to the nanoribbon edge as shown in (a), respectively. The vertical lines are guides to eye. The inset in (c) shows the enlarged image of differential conductance for a nanoribbon with shorter width. Reproduced from [158], copyright 2011 with permission from Springer Nature.

measurements have been performed at various places along the perpendicular and parallel direction of the zigzag edge, as shown by the dots in figure 16.4(a). As can be seen from figure 16.4(c), the differential conductance at and near the zigzag edge show a two-peak structure in either side of the charge neutrality point that is corresponding to the Fermi energy. This resembles the DOS peak splitting in the mean-field results in figure 16.3(e). The differential conductance measurements along the parallel direction to the zigzag edge always show such two-peak structure (see figure 16.4(d)), indicating the close agreement of mean-field results with the experiments. This establishes the importance of electronic correlations in graphene nanoribbon systems and the consequent spin structure.

16.4.2 Graphene nanoflakes

Graphene nanoflakes are finite-sized zero-dimensional sp^2-hybridized carbon nanostructures of varying shapes and sizes. Their size may vary from ~1 nm to ~50 nm. These polycyclic aromatic hydrocarbons can be synthesized using on-surface bottom-up approaches by stitching six-membered hexagonal benzene rings in different shapes [159–161]. Due to large edge to bulk ratio in such nanostructures or supermolecules, the edge effects become dominant, giving rise to interesting electronic properties. In general, the unsatisfied carbon valency and resulting dangling bonds make the edges highly reactive towards functionalizations and reconstructions. To avoid such a scenario, in theoretical calculations the edge atoms are assumed to be passivated by hydrogen atoms, just like the nanoribbon edges. A wide range of graphene nanoflakes have been proposed theoretically with experimental realization of quite a few. These graphene nanofragments can appear in varying shapes, e.g., coronene [162], triangulene [163], the '*Star of David*' [163, 164] etc. Very recently, there have been experimental syntheses of *Sierpinski* type quantum fractals as well [165–169]. These nanoflakes can host large spins emerging from their topological edge states and can give rise to unprecedented magnetic ordering in zero-dimensional cluster geometries.

Here we discuss about two distinct classes of such graphene nanoflakes with and without broken *A* and *B* sublattice symmetry. All these nanoflake structures show localization of zigzag edge states and consequent localized spin moments. In the case of sublattice inequality, the net magnetization is anyway expected to arise, as described by Lieb's theorem [42]. We first start with an example of triangular graphene nanoflakes, known as triangulenes that can appear in different sizes. In figure 16.5(a), we present one such system. It has an excess of atoms belonging to one sublattice (N_A = 12 and) in its structure. As can be seen from figure 16.5(a), three edges of a triangulene system belong to the same sublattice, thereby localizing the same type of spins along all three edges. Theoretically it is predicted to exhibit ferromagnetic behavior with a spin-triplet ($S = 1$) ground state along with a pair of zero-energy states where the spins are oriented parallel to each other, according to Hund's rule [25, 41, 68, 163]. Based on the size of triangulene nanoflakes, the number of edge atoms varies, thereby changing the net magnetization in such supermolecules.

For the systems with $N_A = N_B$, magnetic ordering can happen between different domains, however, keeping the net magnetization zero in the antiferromagnetic ground state. To construct such a system, one can fuse two triangulene units in vertically inverted fashion, as shown in figure 16.5(b). This kind of structure is also known as Clar's goblet [170, 171]. Notice that the edges of two such triangulene

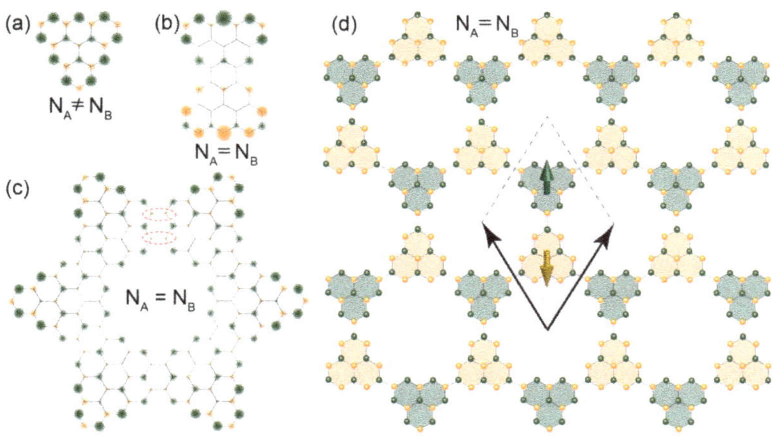

Figure 16.5. Graphene nanoflake systems in (a) triangular, (b) bowtie and (c) star shapes. The number of *A* and *B* sublattice points are the same in the case of bowtie and star structures. The circles with two different shades of colors depict two opposite spins and the relative sizes of those circles are proportional to the relative values of spin densities. The dashed ellipses in (c) are used to emphasize the topological frustration. (d) The two-dimensional periodic structure made of oppositely aligned small triangulene units, connected through their vortices. The unit cell is shown by the shaded rhombus with arrows indicating the lattice vectors. Two triangulene units in the unit cell localize opposite spins, arising from sublattice imbalance in each of them, giving rise to a net zero magnetization. Note that the total number of *A* and *B* sublattice points in each unit cell remain the same.

units are made of two different sublattice points *A* and *B*. As a result, they show non-bonding localized zero-energy edge states of opposite spin polarizations. Therefore, although each triangulene unit shows ferromagnetic spin polarization with a net magnetic moment, the antiferromagnetic ordering between two such units makes the overall ground state antiferromagnetic [172, 173], in accordance with Lieb's theorem of bipartite lattice [42]. A magnetic phase transition between antiferromagnetic and ferromagnetic states can also be induced by carrier doping and by applying an external electric or magnetic field, making them highly desirable for fundamental logic gate applications in novel high-density ultrafast spintronic devices [172, 173].

Furthermore, the fusion of six triangulene units with a 60° relative angle between the adjacent units gives rise to a star-shaped graphene nanoflake, famously known as '*Star of David*', as shown in figure 16.5(c). Here the numbers of *A* and *B* sublattice points remain equal. However, the topological frustration (depicted by the dashed ellipses in figure 16.5(c)) at the fusion points between two adjacent ferromagnetic triangulene units leads to an overall ferromagnetic ground state with large net magnetization [163]. Nevertheless, the consideration of broken spin symmetry eliminates such spin frustrations and induces antiferromagnetic correlation between adjacent intrinsically ferromagnetic triangulene domains, thereby stabilizing the nanoflake system in an antiferromagnetic ground state [164]. There is another widely studied graphene nanoflake system called coronene. This has a hexagonal geometry without any sublattice imbalance and can exist with either zigzag or armchair edges. However, there exist localized edges states induced spin moments along the zigzag edges, the opposite sublattice symmetry in adjacent edges results in overall antiferromagnetic ground state with net zero magnetization. The armchair edge coronene, in contrast, stabilizes in non-magnetic ground state.

Tailoring magnetism in the graphene nanoflakes through experiments can be pretty challenging. The sublattice imbalance in some structures makes them highly reactive due to the presence of unpaired electrons. Moreover, the magnetic stability is limited by weak exchange coupling between p-electrons, which remains below room temperature thermal energy. Nevertheless, contemporary sophistication in the on-surface bottom-up synthesis of graphene open-shell nanostructures has paved new inroads for the successful synthesis of such structures [174–179]. In particular, this has resulted in the experimental realization of the highly reactive and consequently elusive Clar's goblet structure [180, 181].

Edge-state induced magnetism and other novel attributes of these zero-dimensional graphene nanoflake structures motivate the exploration of lattice systems that can be obtained by connecting or fusing such supermolecules in a periodic array [182, 183]. There have been plenty of studies on such covalent organic frameworks that exhibit many interesting functionalities. In figure 16.5(d), we present a two-dimensional array of small triangulene units attached in a honeycomb fashion through their vortices. This can be viewed as a super-graphene system with a rhombus unit cell containing two oppositely aligned triangulene units. Although, the numbers of *A* and *B* sublattice points are unequal in each triangulene unit, their overall counts remain the same in a unit cell. Each of these units exhibits non-zero net magnetization arising from the localized edge states on its more abundant sublattice points. However, localization of opposite spin moments in either of these triangulene units makes the ground state antiferromagnetic. This opens up a direct bandgap at high-symmetric point K, making the system a Mott insulator [184]. This has been predicted to exhibit circular dichroism Hall effect [184]. Further studies are being carried out in such magnetic covalent organic frameworks for their potential applications in qubits.

16.5 Doping induced magnetism in graphene

The magnetic properties of finite graphene systems like graphene nanoribbons or graphene nanoflakes can be further tuned in terms of charge carrier doping by applying a gate bias or by elemental substitution. It can change the location of the Fermi energy with respect to the overall energy dispersions. In the case of elemental substitutions, it can modify the hybridization, thereby changing the energy dispersion profiles altogether along with the shift in Fermi energy. These have huge implications in tuning the electronic, transport, optical and magnetic properties of nanoscale graphene systems.

Replacement of edge carbon atoms in ZGNRs by boron atoms induces spin polarized metallic bands at Fermi energy with non-zero net magnetization, as can be seen in figure 16.6(a). Consequently, the ground state becomes ferromagnetic with parallel alignment of spins between opposite edges, as shown in figure 16.6(b) [145]. The replacements of edge carbon atoms by nitrogen atoms, in contrast, results in

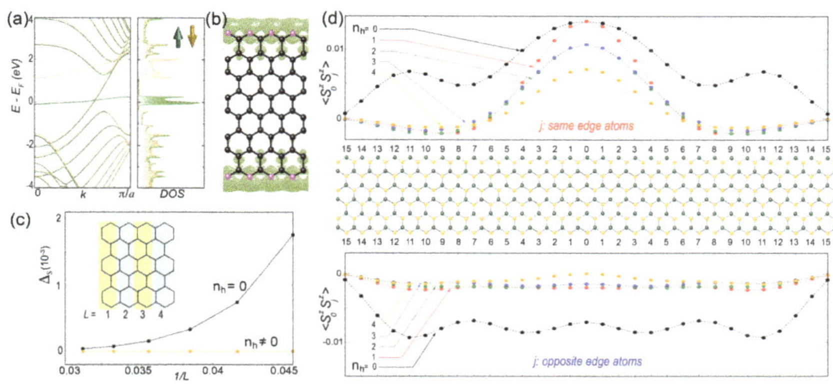

Figure 16.6. (a) Spin polarized band structure and DOS of zigzag edge graphene nanoribbons with edge carbon atoms substituted by boron atoms. The energy is scaled with respect to the Fermi energy (horizontal dashed line). (b) The spin density profile of the same system, showing parallel spin alignments along the boron substituted edges in both sides. Carbon and edge boron atoms are depicted by different color balls. (c) The many-body spin gap of graphene nanoconstriction with long zigzag edge as a function of inverse of length in absence and presence of holes (n_h). The length, L is considered to be the number of unit cells, as shown in the inset. The spin gap value is obtained in the units of hopping integral, t within Hubbard Hamiltonian, using configuration interaction level of calculations. (d) The spin–spin correlation function for a graphene nanoconstriction with $L = 31$ (middle panel), as calculated between the '0'-th atom in one edge with all the other atoms in the same edge (upper panel) and opposite edge (bottom panel) for various hole doping. Reproduced [140] under Creative Common Attribution 4.0 International License.

non-magnetic ground state. Because of the differences of valence electrons, boron replacements make the ZGNRs hole-doped, whereas the nitrogen atom induces one extra electron. The observation of ferromagnetism in boron-edge ZGNRs shows close resemblance with that of ferromagnetism in strongly correlated systems in the presence of holes [185–190]. Quantum many-body treatment of hole-doped graphene nanoconstrictions agrees well with such ferromagnetism as well. For that, we consider graphene quantum dots with long zigzag edges and plot their spin gap variation as a function of the quantum dot length in figure 16.6(c). Spin gap is a measure of energy required to flip a spin in a strongly correlated system. A vanishing spin gap indicates the stabilization of ferromagnetic ground state. As can be seen in figure 16.6(c), in the absence of holes, the graphene quantum dots show non-zero spin gap, indicating an antiferromagnetic ground state. With introduction of holes (n_h), the spin gap drops to zero, showing the stabilization of ferromagnetic ground state [140–142].

To investigate the spin ordering in such ferromagnetic ground state, we further present the spin–spin correlation function along the same edge and between the opposite edges of a graphene quantum dot with long zigzag edges in figure 16.6(d). The correlation function is calculated between the central '0'-th edge atom (middle panel) with all other atoms (denoted by *j*) belonging to the same (top panel) and opposite (bottom panel) edges. A positive (negative) spin–spin correlation indicates a ferrimagnetic (antiferromagnetic) alignment of spins along the same (opposite) edge. As can be seen, the positive correlations decay slowly as compared to the antiferromagnetic correlations upon increase in hole concentration. This indicates a ferromagnetic ground state with parallel spin polarization between opposite edges. Such unique observations make the carrier doping a viable strategy to achieve spin-dependent magneto-transport [191].

The chemical substitution of a bonded pair of carbon atoms by a bonded boron–nitrogen pair also significantly modifies the electronic properties of ZGNRs. Our further studies on lateral heterostructures of ZGNRs with embedded hexagonal boron–nitride nanoribbons with zigzag edges show antiferromagnetic ground state with half-metallic behavior [192]. Such interesting behavior originates from the internal potential gradient arising from the Lewis acid and base characters of boron and nitrogen atoms, respectively and subsequent charge transfers from/to the adjacent carbon atoms [192]. Furthermore, the cyclodehydrogenation step during the on-surface synthesis of ZGNRs by bottom-up methods can result in missing edge m-xylene units, resulting in 'bite' defects in the system. The edge disorder arising from the 'bite' defects was theoretically predicted to give rise to spin-polarized quantum transport [193]. These provide unique ways of achieving spin polarized currents that can find a huge number of applications in spintronics.

16.6 Proximity-induced magnetism in graphene

The magnetic proximity effect imparts magnetic ordering in an intrinsic nonmagnet by coupling it with an adjacent ferromagnet due to the exchange interaction between the spins in the magnetic and non-magnetic materials. As graphene is the prototypical atomically thin material, it presents an ideal platform to study such interactions [194]. Bringing it near a ferromagnetic insulator can magnetize a single sheet of graphene while preserving its unique electronic properties. First-principles theoretical investigations predicted an exchange splitting of the order of tens of meV for ideal lattice matching [195, 196]. Many graphene-on-magnetic insulator systems have been synthesized experimentally by considering several ferromagnetic insulators, e.g., EuO ($T_C = 69$ K) [197], EuS ($T_C = 16.5$ K) [198] and yttrium iron garnet (YIG) thin films ($T_C = 550$ K) [199, 200]. Here, T_C is the Curie temperature of the ferromagnetic insulators. Unlike metals, they do not shunt any current away from graphene and preserve its exceptional charge transport properties. The high T_C, relative chemical stability, and ease in transferring graphene onto its surface make YIG an ideal material for fabricating magnetic graphene-based spintronic devices.

In figure 16.7(a), we show the schematic diagram of the graphene-YIG device. Smooth YIG thin films were grown epitaxially on gadolinium gallium garnet (GGG) substrate by pulsed laser deposition, and monolayer graphene was transferred onto it from SiO_2 substrate [199]. In order to tune the charge density of the system, a thin poly-(methyl methacrylate) (PMMA) top gate was used. The anomalous Hall effect (AHE) measurements were performed to probe the proximity-induced magnetic states in graphene. For ordinary ferromagnets, the Hall resistivity can have contributions from the Lorentz force on the charge carriers due to the external magnetic field (B) and also from the magnetization of the system due to its intrinsic spin–orbit coupling. The AHE resistance at different temperatures for this device arrangement saturates at $B \approx 2300$ G, that is, the saturation magnetization of YIG. This observation indicates that the induced ferromagnetism in graphene arises from the spin polarization due to the proximity coupling of the p-electrons of graphene with the d-electrons of YIG. The interfacial magnetism enhances the typically low spin–orbit coupling of planar graphene, manifested in magnetoresistance measurements and ferromagnetic spin pumping [200]. The AHE resistance decreases with an increase in temperature and stays relatively constant before reaching ~300 K, which defines the Curie temperature of the proximity-induced magnetic graphene systems [199]. Furthermore, the sign of the AHE signal

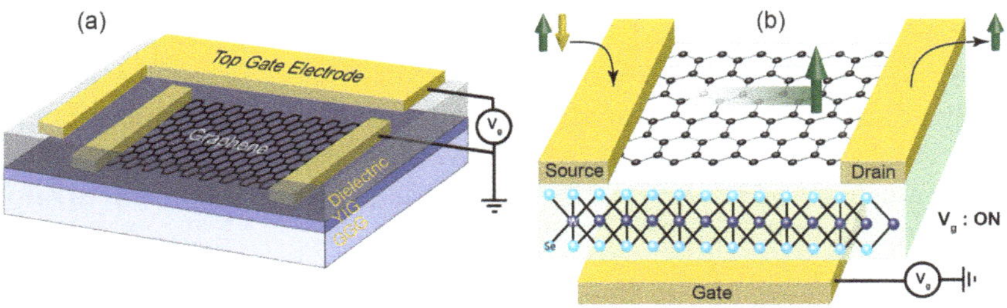

Figure 16.7. (a) Schematic device of single-layer graphene transferred on ferromagnetic and insulating YIG thin film on a GGG substrate with a top gate. (b) Schematic device of bilayer graphene on transition metal dichalcogenide (TMD) system, WSe_2. A back gate is attached to control the carrier doping. This device produces spin polarized current, as depicted schematically.

does not change by changing the carrier type on the application of the top-gate voltage [199].

However, just the observation of AHE signals cannot conclusively establish interfacial ferromagnetism in graphene on YIG thin films. Similar results can be obtained by other effects, such as spin-dependent interfacial scattering or due to magnetic impurities [201]. The observation of a sizeable interfacial exchange field (~14 T) in EuS [198] and the quantum Hall effect at much lower external magnetic fields provides essential information about the interfacial magnetic ordering. Nevertheless, the exchange interaction and the spin–orbit coupling are needed for charge transport. Hence, direct evidence of the exchange interaction comes from the spin–current transport in exfoliated lateral graphene spin valves on YIG substrates [202, 203], where it is probed using only the spin degrees of freedom.

The valley degrees of freedom and intrinsic spin–orbit coupling with anisotropic spin polarizations in two-dimensional materials can also be exploited for inducing spin current in graphene. The two-dimensional TMDs provide a unique platform for this purpose [204–210]. In figure 16.7(b), we present the schematic of a bilayer graphene device on top of TMD system, WSe_2. The underlying bonded WSe_2 layer can tune the anisotropic spin–orbit coupling of bilayer graphene to induce coherent spin precession. The polarity of resulting spin current can be controlled by gate bias in the absence of external magnetic field. This recent study has been a breakthrough in achieving a room temperature spin transport device [211].

The magnetism in graphene is highly sought after for future memory and quantum information processing. So far we have shown various ways of inducing spin-ordering and consequent magnetism in graphene. Because of magnetism due to itinerant p-electrons, the exchange coupling in graphene systems has been very low, which limits its applications in room temperature magnetic and spin transport devices. Although the proximity effect has been predicted to be a viable way of overcoming such limitations, it brings additional challenges of finding an optimal ferromagnetic insulator or TMD and subsequent device fabrication. Therefore, considerable efforts are being invested to explore new two-dimensional materials with unprecedented magnetic properties. We shall discuss further such emergent functional two-dimensional magnetic systems in the next section.

16.7 Magnetism in other two-dimensional materials

Graphene is just the tip of the iceberg in the vast spectrum of two-dimensional materials with novel functionalities, e.g., enhanced spin–orbit coupling, strong electronic correlation, direct semiconducting gap, all of which are capable of inducing magnetism [212–219]. Here we discuss such magnetism arising from electrons from p and d orbitals.

16.7.1 Magnetism in two-dimensional p-electron systems

We have already discussed the edge-state mediated flat-band magnetism in ZGNR systems, arising from the p-electrons. Here, we further discuss our recent findings on magnetism in two-dimensional defect-free SiP_3 system, obtained by silicon (Si) atom replacement in blue phosphorene [220]. The pristine blue phosphorene is a two-dimensional honeycomb lattice made of phosphorus (P) atoms with a rhombus unit cell similar to graphene but having a buckled out-of-plane distortion, as can be seen in figure 16.8(a) [221–223]. The pristine blue phosphorene shows a diverging DOS below the Fermi energy, arising from the flat top of the valence band in its electronic dispersion (see figure 16.8(b)) and a non-magnetic ground state [224].

We have already discussed before how the diverging DOS at the Fermi energy creates instability in ZGNRs, thereby inducing magnetism through electron–electron interactions [225, 226]. Therefore, a band-engineering strategy to shift the Fermi energy downwards to align with such diverging DOS has been proposed to endow magnetic ordering in an otherwise non-magnetic blue phosphorene. A large Fermi energy shift (~2 eV) would be challenging to achieve by conventional electrostatic doping. Therefore, the elemental substitution of P with Si atoms has been proposed [220]. This results in the downshift of the Fermi energy level upon hole doping, as the Si atom has one less electron than the P atom. Similar doping strategies were investigated in other two-dimensional materials, e.g., BC_3 [109, 110, 156], BC_6N [227] and black phosphorene structures [228] that lead to exotic phenomena. The structure of SiP_3 is shown in figure 16.8(c), where two Si dopants are in the 2×2 supercell of blue phosphorene, resulting in a P hexagon surrounded by a Si hexagon. The non-interacting band structure of SiP_3 shows two almost

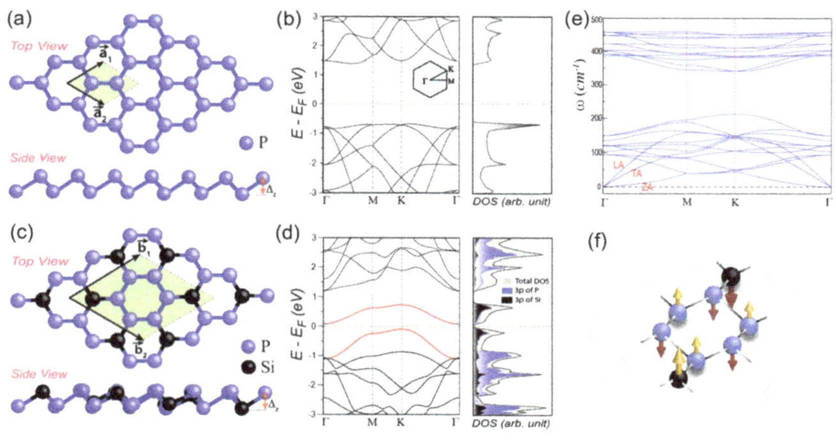

Figure 16.8. (a) The top and side views of pristine blue phosphorene lattice with the rhombus unit cell (shaded region) and the lattice vectors. (b) The band structure and density of states (DOS) of the same in 2×2 unitcell. (c) The top and side views of SiP_3 lattice with eight-atom rhombus unit cell (shaded region). (d) The band structure and DOS of SiP_3 system. The bands near the Fermi energy (horizontal dashed lines) show splitting and consequent opening up of a small gap upon inclusion of electronic correlations. The projected DOS arising from the 3p orbitals of P and Si are highlighted with different color shades. The Δ_z in the side views of the structures is the buckling parameter. (e) The phonon dispersion of SiP_3 system. The vertical dashed lines in electronic and phonon dispersions indicate the location of the high-symmetric points. (f) The spin structure of SiP_3 system with up and down spins shown by arrows of different color shades. The size of the arrows is proportional to the spin moment values. Reproduced [220] under creative common attribution-noncomercial 3.0 unported, Copyright 2021, Royal Society of Chemistry.

dispersion-less bands near the Fermi energy, giving rise to a sharp peak in the DOS. Inclusion of electronic correlation via spin-polarization results in a spin splitting of the P–Si hybrid bands, as can be seen from the band structure and DOS plots in figure 16.8(d).

The dynamical stability of the Si-doped blue phosphorene was confirmed by phonon analysis that shows no imaginary mode and well-separated acoustic and optical modes in the phonon band structure (see figure 16.8(e)) [220].

The resulting spin texture of SiP_3 is shown in figure 16.8(f). The spin moments on each Si atom and its three neighboring P atoms are ferromagnetically coupled. However, the P hexagons show a perfect bipartite nature of opposite spin alignments between nearest neighbors. This creates two triangular tiny ferromagnetic domains with opposite spin orientations and a unique antiferromagnetic ground state [220]. The observation of magnetic ordering in such defect-free layered materials made up entirely of p-block elements is unprecedented, paving new ways for their use in memory storage and qubit applications.

16.7.2 Magnetism in two-dimensional d-electron systems

Among many d-electron systems, the transition metal mono, di and trichalcogenide systems have been widely studied because of their layered structure and novel properties that can be tuned by controlling number of layers, carrier concentration and polarity, valley polarizations and spin–orbit couplings [205, 229–233]. We discuss here mainly the TMD systems.

TMDs consist of transition metal (Mo, W, Nb, V) atoms sandwiched between two layers of chalcogen atoms (S, Se, Te), arranged in a trigonal prismatic geometry [230]. Two-dimensional TMDs, like MoS_2 and WSe_2 are non-magnetic semiconductors with a direct energy bandgap and high carrier mobility at room temperature, possessing interesting electronic, optoelectronic, and valleytronic properties [205, 230–233]. Various other single and few-layered TMDs have been synthesized so far, using the mechanical exfoliation, CVD, or molecular beam epitaxy methods [234]. Except for the chemically active edge or defect sites, pristine layered TMDs are usually passive due to the absence of any dangling bonds [235], making them diamagnetic like graphene. However, previous theoretical and experimental results have shown how layer-selective oxidation of the chalcogen edge atoms gives rise to novel tunable electronic properties in two-dimensional WSe_2 [207–210, 236]. Therefore, room temperature ferromagnetism can be imparted in MoS_2 by creating defects in its honeycomb lattice structure or through its edge states. First-principles investigations have shown the appearance of magnetic states due to a combination of different non-magnetic and transition metal adatoms with different vacancy states, depending upon the position of Mo and S in the system [237]. MoS_2 nanosheets exhibit room temperature ferromagnetism due to unsaturated Mo and S atoms in their edges [238]. Therefore, controlling the edge structure of MoS_2 by creating a ribbon-like geometry can induce magnetism, like zigzag-edge graphene nanoribbons. Depending upon the terminating edge atoms, two zigzag ribbons could be possible. The S-terminated zigzag ribbons have been reported to be more stable than their Mo-terminated counterparts, irrespective of hydrogen passivation [239, 240]. Theoretical calculations show that both edges of MoS_2 nanoribbons are ferromagnetically coupled [239], making them viable for spintronic device applications. However, it is difficult to synthesize the one-dimensional spin-polarized edge states in experiments due to enhanced thermal fluctuations. Nevertheless, edge-reconstructed patterns of zigzag-edge MoS_2 have been reported, where the edge Mo atoms are self-passivated by the S atoms [241], leading to a more stable ferromagnetic state.

16.7.3 Two-dimensional van der Waals magnets

The low-dimensional transition metal systems, especially layered van der Waals materials, e.g., transition metal trihalides and trichalcogenides have been in the forefront of research in recent times due to their unprecedented magnetic behavior [242–245]. The enhanced spin–orbit coupling and consequent uniaxial magnetic anisotropy in these systems have resulted in the recent discovery of long-range magnetic ordering in pristine two-dimensional materials. Therefore, in this last section, we present the recent developments in this emergent and exciting research field.

The fundamental question of the existence of long-range magnetic order at finite temperature in low dimensions has been tempting intense research in condensed matter physics and materials science over last few decades. The seminal theories on long-range order in superfluids, superconductors, and various magnetic model systems were first worked out by Onsager [246], Hohenberg [247], and Mermin–Wagner [248]. Experimentally, quasi-two-dimensional in-plane magnetism enabled by anisotropy was first observed in certain bulk systems [249] and monolayer thick magnetic films [250]. However, the major impediment in discovering low-dimensional magnetic materials has been the Mermin–Wagner theorem [248] that clearly states the suppression of magnetic ordering in a one- or two-dimensional Heisenberg model at finite temperature. The isotropic nature implies the rotational invariance of exchange coupling, which preserves the continuous $SU(2)$ symmetry. Any long-range magnetic ordering breaks the continuous rotational symmetry and results in infinitely degenerate ground states with gapless spin-wave excitations, i.e., Goldstone modes. At finite temperature, the thermally activated magnon excitations therefore destroy any long-range magnetic ordering. However, an arbitrarily small anisotropy breaks the continuous rotational symmetry and consequently opens up a spin-wave excitation gap. This eliminates the possibility of logarithmic divergence of the thermal fluctuations, thereby stabilizing the long-range magnetic order in two-dimension [251, 252]. The anisotropy can arise due to magnetic dipole–dipole interaction or by the shape anisotropy created by demagnetizing fields at the surface. For such anisotropies in the system, the magnetic scale is tiny as compared to the exchange interaction, resulting in a very small Curie temperature (in mK). However, the newly discovered van der Waals magnetic materials possess uniaxial magnetic anisotropy due to their enhanced intrinsic spin–orbit coupling, thus allowing preferential orientation of spins in a particular direction. Therefore, this opens up a spin-wave excitation gap that resists the thermal agitation, resulting in a considerably high Curie temperature (T_C). In particular, we discuss here the properties of two-dimensional van der Waals ferromagnetic semiconductors CrI_3 and $CrGeTe_3$ and itinerant ferromagnet Fe_3GeTe_2, whose magnetic ordering has been investigated experimentally using magneto-optic Kerr microscopy [253–255].

In CrI_3 structure, each chromium (Cr) atom is surrounded by six iodine (I) ligands arranged in an octahedral fashion. Such edge-sharing octahedra create the extended two-dimensional lattice where the Cr atoms form honeycomb pattern, as can be seen in figure 16.9(a). A strong uniaxial magnetic anisotropy in CrI_3 arises from the enhanced spin–orbit interactions of the iodine ligands that mediate the superexchange between neighboring Cr ions [256]. Therefore, CrI_3 is considered as an Ising ferromagnet with out-of-plane spin orientation and a spin moment of 3/2 μ_B per Cr atom [253]. It exhibits spin-polarized indirect bandgap semiconducting property (see figure 16.9(d)) [257]. The Curie temperatures of monolayer and bulk CrI_3 have been reported to be 45 K and 61 K, respectively. Such low difference in their Curie temperatures indicates a weak interlayer coupling [253]. In the case of bulk CrI_3, the interlayer ferromagnetic coupling in the bulk changes to an antiferromagnetic one for surface layers [258]. Such interlayer antiferromagnetic coupling is predominantly visible in the case of bilayer CrI_3, where a metamagnetic phase transition can be induced on application of external magnetic field of ~0.6 T, with the ferromagnetic saturation happening within a short span of 100 mT [253, 259]. The layer-dependent magnetism in CrI_3 can also be tuned by changing the orbital occupancy through an applied electric field or electrostatic gating that modifies the exchange interaction parameters and magneto-anisotropy of the system [259–261].

Intrinsic long-range ferromagnetism has also been reported in the transition metal trichalcogenide, $CrGeTe_3$ (CGT) [254]. The two-dimensional periodic CGT structure is formed by edge-sharing trigonal bipyramidal cages of telurium (Te) atoms, which host the Cr atoms in their centers, thereby forming a honeycomb Cr

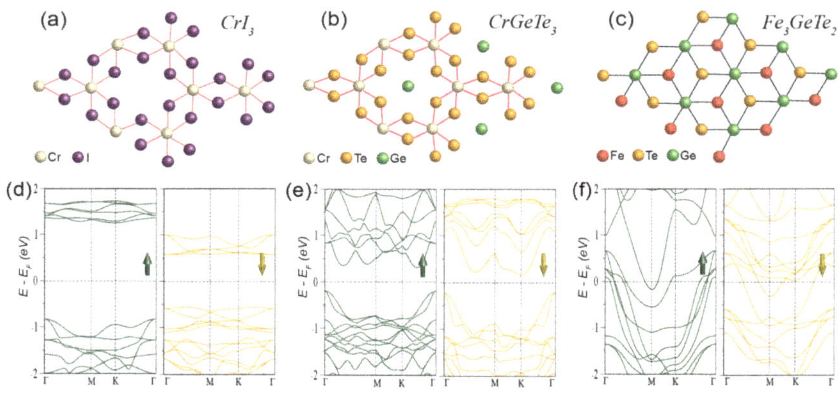

Figure 16.9. The top view of (a) CrI_3, (b) $CrGeTe_3$ and (c) Fe_3GeTe_2 lattices. The corresponding spin-polarized band structures are shown in (d), (e) and (f), respectively. The horizontal (vertical) dashed lines depict the location of the Fermi energy (high-symmetric points).

arrangement much like the CrI_3 system, as can be seen in figure 16.9(b). However, unlike CrI_3, The Te atoms are attached to a group-IV element germanium (Ge) that modifies the superexchange path significantly. So far, the CGT has been experimentally synthesized up to bilayer. The spin–orbit coupling of Cr ions, together with a slight distortion of Te octahedra into trigonal bipyramidal geometry, lead to a small out-of-plane magnetocrystalline anisotropy in this system. The ferromagnetic saturation has been observed at ~0.6 T, with a non-vanishing remanence near zero magnetic field. Because of that, CGT has been classified as a soft Heisenberg ferromagnet [254]. This also shows spin-polarized indirect bandgap semiconducting behavior (see figure 16.9(e)) [262]. Interestingly, a strong interlayer coupling in these materials leads to a significant layer dependence of the Curie temperature. The Curie temperature for bulk and bilayer CGT has been reported to be 68 K and 30 K, respectively. Therefore, the application of an external magnetic field can tune the ferromagnetic order by enhancing the uniaxial magnetic anisotropy and consequently leading to an increase in the Curie temperature of the system [254].

Furthermore, the application of biaxial compressive strain has also been proposed as an essential parameter to tune the magnetic properties of this class of TMDs [263]. The superexchange between two adjacent Cr atoms mediated through the bridging Te atoms gets modified due to presence of side group elements from group-IV and therefore the superexchange phenomenon obeys modified Goodenough–Kanamori–Anderson rule [264–268]. The $CrCTe_3$ (CCT) and $CrSiTe_3$ (CST) magnetic layers are the outset crystals of CGTs where the superexchange between the Cr atoms depends significantly on the atomic radius of the group-IV element and consequent changes in bond-angles. The presence of carbon (C) atoms in CCT makes the ground state antiferromagnetic that can sustain up to ±16% biaxial strain [269]. In the case of CST, the presence of Si atoms results in a ferromagnetic ground state and an antiferromagnetic phase transition can be induced by application of only 0.5% biaxial compressive strain [270, 271]. The lattice mismatch in CCT–CST vertically stacked hetero-bilayers and additional biaxial compressive strain can lead to various intra and interlayer magnetic ordering [263]. These soft two-dimensional van der Waals magnets have great potential in electronic and spintronic device applications where magnetic anisotropic exchange can be tuned with external perturbations.

The discovery of itinerant ferromagnet Fe_3GeTe_2 (FGT) has been a major breakthrough [265]. Its structure is shown in figure 16.9(c). In the monolayer FGT, three of the quintuple sublayers are occupied by iron (Fe) atoms in varied crystallographic environments. Such crystallographic anisotropy along with high spin–orbit coupling of Fe atoms result in a sizeable uniaxial anisotropy that stabilizes the long-range magnetic ordering in two dimensions [272]. The Curie temperature of FGT has been found to be higher than the ferromagnetic semiconductors CrI_3 and CGT and can be easily tuned by changing the Fe concentration [265]. This system shows metallic behavior with spin-polarized band structures, as can be seen from figure 16.9(f) [272]. This class of metallic ferromagnets constitutes a promising platform of two-dimensional materials, where the interplay of the itinerant electrons and magnetic ordering can lead to various emergent phenomena. Moreover, the room temperature ferromagnetism observed in bilayer FGTs can be tailored by electrostatic gating, thereby opening new avenues for voltage-controlled magnetoelectronics and applications in magnetic tunnel junction heterostructures while combined with other layered materials [273].

A plethora of two-dimensional magnetic van der Waals materials with novel properties have been realized in recent times [274–281]. However, industrial scale device applications demand stability of magnetic states at higher temperature. This is currently being undertaken by strengthening the exchange interaction and uniaxial magnetic anisotropy of these systems employing external fields, strain, and interface engineering by coupling other layered materials to create novel heterostructures, opening up huge possibilities in atomically precise spintronic, magnonic, valleytronic and spin-orbitronics device applications [255, 282].

16.8 Summary and outlook

To summarize, we have systematically classified and illustrated the various ways to imprint magnetism in graphene and its nanostructures. We have discussed the theoretical formalisms and understanding behind such phenomena with relevant experimental observations. We have discussed the various interaction regimes and their crucial role in determining the magnetic properties in this class of systems. The magnetism in graphene bipartite lattice can arise from sublattice inequality, lack of sublattice clarity, zigzag edge states in nanoribbon or nanoflake structures, charge carrier doping, elemental substitutions, external perturbations like proximity effects etc. Furthermore, we have discussed the recent developments of two-dimensional magnetic materials beyond graphene, e.g., Si-doped blue phosphorene, TMDs and van der Waals crystals.

Apart from monolayer graphene, the band structure engineering also endows electrically biased bilayer graphene with itinerant ferromagnetism [283]. The unusual flat-band topology and emergent magnetism due to enhanced Coulomb correlation in twisted bilayer graphene [284–287] have shown a tremendous surge in fundamental research. Moreover, topological frustration in nanographene can break the bipartite symmetry, thereby stabilizing magnetic domains [288, 289]. In addition to these, the formation of lateral and vertical heterostructures of different atomically thin materials has been proposed to achieve various emergent phenomena and to tailor their efficiency for a wide spectrum of applications. A profound understanding of the magnetic properties in these low-dimensional nano-architectures will be critical for their use in next-generation high-density ultra-low power electronic and spintronic devices, sustainable energy and quantum information applications. This can potentially revolutionize the quantum technologies.

Acknowledgments

The authors thank IISER Tirupati for Intramural Funding and acknowledge the Science and Engineering Research Board (SERB), Department of Science and Technology (DST), Government of India, for Early Career Research (ECR) award grant (ECR/2016/000283) and Core Research Grant (CRG/2021/001731).

References

[1] Jiles D 2015 *Introduction to Magnetism and Magnetic Materials* (Boca Raton, FL: CRC Press)

[2] Buschow K H J and Boer F R 2003 *Physics of Magnetism and Magnetic Materials* 7 (New York: Kluwer Academic/Plenum Publishers)

[3] Spaldin N A 2010 *Magnetic Materials: Fundamentals and Applications* (Cambridge: Cambridge University Press)

[4] Miller J S and Drillon M 2002 *Magnetism, Nanosized Magnetic Materials* (Hoboken, NJ: Wiley) p 403

[5] Schollwöck U, Richter J, Farnell D J and Bishop R F (ed) 2008 *Quantum Magnetism* 645 (Berlin: Springer)

[6] Sachdev S 2008 Quantum magnetism and criticality *Nat. Phys.* **4** 173–85

[7] Zhou Y, Kanoda K and Ng T K 2017 Quantum spin liquid states *Rev. Mod. Phys.* **89** 025003

[8] Mila F 2003 Quantum frustrated magnets: from theory to experiments *Int. J. Mod. Phys.* B **17** 5021–30

[9] Arnold C S, Dunlavy M and Venus D 1997 Magnetic susceptibility measurements of ultrathin films using the surface magneto-optic Kerr effect: optimization of the signal-to-noise ratio *Rev. Sci. Instrum.* **68** 4212–6

[10] Hellman F *et al* 2017 Interface-induced phenomena in magnetism *Rev. Mod. Phys.* **89** 025006

[11] Vasiliev A, Volkova O, Zvereva E and Markina M 2018 Milestones of low-D quantum magnetism *NPJ Quantum Mater.* **3** 1–13

[12] Geim A K and Novoselov K S 2007 The rise of graphene *Nat. Mater.* **6** 183

[13] Novoselov K S, Geim A K, Morozov S V, Jiang D E, Zhang Y, Dubonos S V, Grigorieva I V and Firsov A A 2004 Electric field effect in atomically thin carbon films *Science* **306** 666–9

[14] Novoselov K S, Geim A K, Morozov S V, Jiang D, Katsnelson M I, Grigorieva I, Dubonos S and Firsov A 2005 Two-dimensional gas of massless Dirac fermions in graphene *Nature* **438** 197–200

[15] Novoselov K S, Jiang D, Schedin F, Booth T J, Khotkevich V V, Morozov S V and Geim A K 2005 Two-dimensional atomic crystals *Proc. Natl Acad. Sci. USA* **102** 10451–3

[16] Neto A C, Guinea F, Peres N M, Novoselov K S and Geim A K 2009 The electronic properties of graphene *Rev. Mod. Phys.* **81** 109

[17] Avsar A, Ochoa H, Guinea F, Özyilmaz B, Van Wees B J and Vera-Marun I J 2020 Colloquium: spintronics in graphene and other two-dimensional materials *Rev. Mod. Phys.* **92** 021003

[18] Ando T 2005 Theory of electronic states and transport in carbon nanotubes *J. Phys. Soc. Jpn.* **74** 777–817

[19] Ando T 2008 Physics of graphene zero-mode anomalies and roles of symmetry *Prog. Theor. Phys. Suppl.* **176** 203–26

[20] Zhang Y, Tan Y W, Stormer H L and Kim P 2005 Experimental observation of the quantum Hall effect and Berry's phase in graphene *Nature* **438** 201–4

[21] Han W, Kawakami R K, Gmitra M and Fabian J 2014 Graphene spintronics *Nat. Nanotechnol.* **9** 794–807

[22] Roche S *et al* 2015 Graphene spintronics: the European Flagship perspective *2D Mater.* **2** 030202

[23] Bolotin K I, Sikes K J, Jiang Z, Klima M, Fudenberg G, Hone J, Kim P and Stormer H L 2008 Ultrahigh electron mobility in suspended graphene *Solid State Commun.* **146** 351–5
[24] Balandin A A, Ghosh S, Bao W, Calizo I, Teweldebrhan D, Miao F and Lau C N 2008 Superior thermal conductivity of single-layer graphene *Nano Lett.* **8** 902–7
[25] Dutta S and Pati S K 2010 Novel properties of graphene nanoribbons: a review *J. Mater. Chem.* **20** 8207–23
[26] Wakabayashi K and Dutta S 2012 Nanoscale and edge effect on electronic properties of graphene *Solid State Commun.* **152** 1420–30
[27] Wakabayashi K 2018 Graphene nanotechnology *Materials Nanoarchitectonics* (New York: Wiley), pp 109–23
[28] Ando T, Nakanishi T and Saito R 1998 Berry's phase and absence of back scattering in carbon nanotubes *J. Phys. Soc. Jpn.* **67** 2857–62
[29] Ando T and Nakanishi T 1998 Impurity scattering in carbon nanotubes – absence of back scattering *J. Phys. Soc. Jpn.* **67** 1704–13
[30] Katsnelson M I, Novoselov K S and Geim A K 2006 Chiral tunnelling and the Klein paradox in graphene *Nat. Phys.* **2** 620–5
[31] Luk'yanchuk I A and Kopelevich Y 2004 Phase analysis of quantum oscillations in graphite *Phys. Rev. Lett.* **93** 166402
[32] Park J G 2016 Opportunities and challenges of 2D magnetic van der Waals materials: magnetic graphene? *J. Phys.: Condens. Matter* **28** 301001
[33] Tombros N, Jozsa C, Popinciuc M, Jonkman H T and Van Wees B J 2007 Electronic spin transport and spin precession in single graphene layers at room temperature *Nature* **448** 571–4
[34] Ingla-Aynés J, Guimarães M H, Meijerink R J, Zomer P J and van Wees B J 2015 24-μm spin relaxation length in boron nitride encapsulated bilayer graphene *Phys. Rev.* B **92** 201410
[35] Drögeler M, Franzen C, Volmer F, Pohlmann T, Banszerus L, Wolter M, Watanabe K, Taniguchi T, Stampfer C and Beschoten B 2016 Spin lifetimes exceeding 12 ns in graphene nonlocal spin valve devices *Nano Lett.* **16** 3533–9
[36] Makarova T and Palacio F (ed) 2006 *Carbon Based Magnetism: An Overview of the Magnetism of Metal Free Carbon-Based Compounds and Materials* (Amsterdam: Elsevier)
[37] Allemand P M, Khemani K C, Koch A, Wudl F, Holczer K, Donovan S, Grüner G and Thompson J D 1991 Organic molecular soft ferromagnetism in a fullerene C60 *Science* **253** 301–2
[38] Takahashi M, Turek P, Nakazawa Y, Tamura M, Nozawa K, Shiomi D, Ishikawa M and Kinoshita M 1991 Discovery of a quasi-1D organic ferromagnet, p-NPNN *Phys. Rev. Lett.* **67** 746
[39] Tamura M, Nakazawa Y, Shiomi D, Nozawa K, Hosokoshi Y, Ishikawa M, Takahashi M and Kinoshita M 1991 Bulk ferromagnetism in the β-phase crystal of the p-nitrophenyl nitronyl nitroxide radical *Chem. Phys. Lett.* **186** 401–4
[40] Matte H R, Subrahmanyam K S and Rao C N R 2009 Novel magnetic properties of graphene: presence of both ferromagnetic and antiferromagnetic features and other aspects *J. Phys. Chem.* C **113** 9982–5
[41] Tuček J, Błoński P, Ugolotti J, Swain A K, Enoki T and Zbořil R 2018 Emerging chemical strategies for imprinting magnetism in graphene and related 2D materials for spintronic and biomedical applications *Chem. Soc. Rev.* **47** 3899–990

[42] Lieb E H 1989 Two theorems on the Hubbard model *Phys. Rev. Lett.* **62** 1201
[43] Hubbard J 1963 Electron correlations in narrow energy bands *Proc. R. Soc.* A **276** 238–57
[44] Hubbard J 1964 Electron correlations in narrow energy bands III. An improved solution *Proc. R. Soc.* A **281** 401–19
[45] Mielke A 1991 Ferromagnetic ground states for the Hubbard model on line graphs *J. Phys. A: Math. Gen.* **24** L73
[46] Tasaki H 1998 The Hubbard model-an introduction and selected rigorous results *J. Phys. Condens. Matter* **10** 4353
[47] Szabo A and Ostlund N S 2012 *Modern Quantum Chemistry: Introduction to Advanced Electronic Structure Theory* (North Chelmsford, MA: Courier Corporation)
[48] Bruus H and Flensberg K 2004 *Many-Body Quantum Theory in Condensed Matter Physics: An Introduction* (Oxford: Oxford University Press)
[49] Feldner H, Meng Z Y, Honecker A, Cabra D, Wessel S and Assaad F F 2010 Magnetism of finite graphene samples: mean-field theory compared with exact diagonalization and quantum Monte Carlo simulations *Phys. Rev.* B **81** 115416
[50] Fernández-Rossier J and Palacios J J 2007 Magnetism in graphene nanoislands *Phys. Rev. Lett.* **99** 177204
[51] Gunlycke D, Areshkin D A, Li J, Mintmire J W and White C T 2007 Graphene nanostrip digital memory device *Nano Lett.* **7** 3608–11
[52] Pisani L, Chan J A, Montanari B and Harrison N M 2007 Electronic structure and magnetic properties of graphitic ribbons *Phys. Rev.* B **75** 064418
[53] Kuroda S I and Shirakawa H 1987 Electron-nuclear double-resonance evidence for the soliton wave function in polyacetylene *Phys. Rev.* B **35** 9380
[54] Thomann H, Dalton L K, Grabowski M and Clarke T C 1985 Direct observation of Coulomb correlation effects in polyacetylene *Phys. Rev.* B **31** 3141
[55] Sorella S and Tosatti E 1992 Semi-metal-insulator transition of the Hubbard model in the honeycomb lattice *Europhys. Lett.* **19** 699
[56] Mohn P 2006 *Magnetism in the Solid State: An Introduction* 134 (Berlin: Springer Science & Business Media)
[57] Palacios J J, Fernández-Rossier J and Brey L 2008 Vacancy-induced magnetism in graphene and graphene ribbons *Phys. Rev.* B **77** 195428
[58] Hohenberg P and Kohn W 1964 Inhomogeneous electron gas *Phys. Rev.* **136** B864–71
[59] Kohn W and Sham L J 1965 Self-consistent equations including exchange and correlation effects *Phys. Rev.* **140** A1133
[60] Martin R M 2020 *Electronic Structure: Basic Theory and Practical Methods* (Cambridge: Cambridge University Press)
[61] Dutta S and Wakabayashi K 2015 Magnetization due to localized states on graphene grain boundary *Sci. Rep.* **5** 1–9
[62] Ugeda M M, Brihuega I, Guinea F and Gómez-Rodríguez J M 2010 Missing atom as a source of carbon magnetism *Phys. Rev. Lett.* **104** 096804
[63] Barzola-Quiquia J, Esquinazi P, Rothermel M, Spemann D, Butz T and Garcia N 2007 Experimental evidence for two-dimensional magnetic order in proton bombarded graphite *Phys. Rev.* B **76** 161403
[64] Esquinazi P, Spemann D, Höhne R, Setzer A, Han K H and Butz T 2003 Induced magnetic ordering by proton irradiation in graphite *Phys. Rev. Lett.* **91** 227201

[65] González-Herrero H, Gómez-Rodríguez J M, Mallet P, Moaied M, Palacios J J, Salgado C, Ugeda M M, Veuillen J Y, Yndurain F and Brihuega I 2016 Atomic-scale control of graphene magnetism by using hydrogen atoms *Science* **352** 437–41
[66] Yazyev O V and Helm L 2007 Defect-induced magnetism in graphene *Phys. Rev.* B **75** 125408
[67] McCreary K M, Swartz A G, Han W, Fabian J and Kawakami R K 2012 Magnetic moment formation in graphene detected by scattering of pure spin currents *Phys. Rev. Lett.* **109** 186604
[68] Yazyev O V 2010 Emergence of magnetism in graphene materials and nanostructures *Rep. Prog. Phys.* **73** 056501
[69] Boukhvalov D W, Katsnelson M I and Lichtenstein A I 2008 Hydrogen on graphene: electronic structure, total energy, structural distortions and magnetism from first-principles calculations *Phys. Rev.* B **77** 035427
[70] Gao F, Zhang Y, He L, Gao S and Brandbyge M 2021 Control of the local magnetic states in graphene with voltage and gating *Phys. Rev.* B **103** L241402
[71] Nafday D, Kabir M and Saha-Dasgupta T 2016 Controlling adatom magnetism on bilayer graphene by external field *Phys. Rev.* B **93** 045433
[72] Nair R R, Tsai I L, Sepioni M, Lehtinen O, Keinonen J, Krasheninnikov A V, Castro Neto A H, Katsnelson M I, Geim A K and Grigorieva I V 2013 *Nat. Commun.* **4** 2010
[73] Uchoa B, Kotov V N, Peres N M R and Neto A C 2008 Localized magnetic states in graphene *Phys. Rev. Lett.* **101** 026805
[74] Krasheninnikov A V, Lehtinen P O, Foster A S, Pyykkö P and Nieminen R M 2009 Embedding transition-metal atoms in graphene: structure, bonding, and magnetism *Phys. Rev. Lett.* **102** 126807
[75] Hong X, Cheng S H, Herding C and Zhu J 2011 Colossal negative magnetoresistance in dilute fluorinated graphene *Phys. Rev.* B **83** 085410
[76] Sepioni M, Nair R R, Rablen S, Narayanan J, Tuna F, Winpenny R, Geim A K and Grigorieva I V 2010 Limits on intrinsic magnetism in graphene *Phys. Rev. Lett.* **105** 207205
[77] Sepioni M, Nair R R, Tsai I L, Geim A K and Grigorieva I V 2012 Revealing common artifacts due to ferromagnetic inclusions in highly oriented pyrolytic graphite *Europhys. Lett.* **97** 47001
[78] Nair R R, Sepioni M, Tsai I L, Lehtinen O, Keinonen J, Krasheninnikov A V, Thomson T, Geim A K and Grigorieva I V 2012 Spin-half paramagnetism in graphene induced by point defects *Nat. Phys.* **8** 199–202
[79] Yazyev O V and Louie S G 2010 Topological defects in graphene: dislocations and grain boundaries *Phys. Rev.* B **81** 195420
[80] Podlivaev A I and Openov L A 2015 Dynamics of the Stone-Wales defect in graphene *Phys. Solid State* **57** 820–4
[81] Dutta S and Pati S K 2010 Edge reconstructions induce magnetic and metallic behavior in zigzag graphene nanoribbons *Carbon* **48** 4409–13
[82] Bae S *et al* 2010 Roll-to-roll production of 30-inch graphene films for transparent electrodes *Nat. Nanotechnol.* **5** 574–8
[83] Lahiri J, Lin Y, Bozkurt P, Oleynik I I and Batzill M 2010 An extended defect in graphene as a metallic wire *Nat. Nanotechnol.* **5** 326–9
[84] Biró L P and Lambin P 2013 Grain boundaries in graphene grown by chemical vapor deposition *New J. Phys.* **15** 035024

[85] Yeh S S, Chang W Y and Lin J J 2017 Probing nanocrystalline grain dynamics in nanodevices *Sci. Adv.* **3** e1700135
[86] Hsieh K, Kochat V, Zhang X, Gong Y, Tiwary C S, Ajayan P M and Ghosh A 2017 Effect of carrier localization on electrical transport and noise at individual grain boundaries in monolayer MoS_2 *Nano Lett.* **17** 5452–7
[87] Červenka J, Katsnelson M I and Flipse C F J 2009 Room-temperature ferromagnetism in graphite driven by two-dimensional networks of point defects *Nat. Phys.* **5** 840–4
[88] Balasubramanian K *et al* 2019 Reversible defect engineering in graphene grain boundaries *Nat. Commun.* **10** 1–9
[89] Kochat V, Tiwary C S, Biswas T, Ramalingam G, Hsieh K, Chattopadhyay K, Raghavan S, Jain M and Ghosh A 2016 Magnitude and origin of electrical noise at individual grain boundaries in graphene *Nano Lett.* **16** 562–7
[90] Akhukov M A, Fasolino A, Gornostyrev Y N and Katsnelson M I 2012 Dangling bonds and magnetism of grain boundaries in graphene *Phys. Rev.* B **85** 115407
[91] Hsieh K *et al* 2021 Spontaneous time-reversal symmetry breaking at individual grain boundaries in graphene *Phys. Rev. Lett.* **126** 206803
[92] Babar R and Kabir M 2019 Ferromagnetism in nitrogen-doped graphene *Phys. Rev.* B **99** 115442
[93] Błoński P, Tucek J, Sofer Z, Mazanek V, Petr M, Pumera M, Otyepka M and Zboril R 2017 Doping with graphitic nitrogen triggers ferromagnetism in graphene *J. Am. Chem. Soc.* **139** 3171–80
[94] Tuček J, Błoński P, Sofer Z, Šimek P, Petr M, Pumera M, Otyepka M and Zbořil R 2016 Sulfur doping induces strong ferromagnetic ordering in graphene: effect of concentration and substitution mechanism *Adv. Mater.* **28** 5045–53
[95] Lazar P, Zbořil R, Pumera M and Otyepka M 2014 Chemical nature of boron and nitrogen dopant atoms in graphene strongly influences its electronic properties *Phys. Chem. Chem. Phys.* **16** 14231–5
[96] Edwards D M and Katsnelson M I 2006 High-temperature ferromagnetism of sp electrons in narrow impurity bands: application to CaB_6 *J. Phys. Condens. Matter* **18** 7209
[97] Liu Y *et al* 2016 Elemental superdoping of graphene and carbon nanotubes *Nat. Commun.* **7** 10921
[98] Ito Y, Christodoulou C, Nardi M V, Koch N, Kläui M, Sachdev H and Müllen K 2015 Tuning the magnetic properties of carbon by nitrogen doping of its graphene domains *J. Am. Chem. Soc.* **137** 7678–85
[99] Li J, Li X, Zhao P, Lei D Y, Li W, Bai J, Ren Z and Xu X 2015 Searching for magnetism in pyrrolic N-doped graphene synthesized via hydrothermal reaction *Carbon* **84** 460–8
[100] Santos E J, Sánchez-Portal D and Ayuela A 2010 Magnetism of substitutional Co impurities in graphene: realization of single π vacancies *Phys. Rev.* B **81** 125433
[101] Berashevich J and Chakraborty T 2010 Sustained ferromagnetism induced by H-vacancies in graphane *Nanotechnology* **21** 355201
[102] Yang C K 2010 Graphane with defect or transition-metal impurity *Carbon* **48** 3901–5
[103] Zhou J, Wang Q, Sun Q, Chen X S, Kawazoe Y and Jena P 2009 Ferromagnetism in semihydrogenated graphene sheet *Nano Lett.* **9** 3867–70
[104] Khurana G, Kumar N, Kotnala R K, Nautiyal T and Katiyar R S 2013 Temperature tuned defect induced magnetism in reduced graphene oxide *Nanoscale* **5** 3346–51

[105] Tang T, Liu F, Liu Y, Li X, Xu Q, Feng Q, Tang N and Du Y 2014 Identifying the magnetic properties of graphene oxide *Appl. Phys. Lett.* **104** 123104
[106] Das S R and Dutta S 2021 Edge state induced spintronic properties of graphene nanoribbons: a theoretical perspective *Carbon Nanomaterial Electronics: Devices and Applications* (Cham: Springer), pp 165–98
[107] Wakabayashi K, Sasaki K I, Nakanishi T and Enoki T 2010 Electronic states of graphene nanoribbons and analytical solutions *Sci. Technol. Adv. Mater.* **11** 054504
[108] Wakabayashi K, Takane Y, Yamamoto M and Sigrist M 2009 Electronic transport properties of graphene nanoribbons *New J. Phys.* **11** 095016
[109] Dutta S and Wakabayashi K 2012 Anomalous energy-gap behaviour of armchair BC_3 ribbons due to enhanced π-conjugation *J. Mater. Chem.* **22** 20881–4
[110] Dutta S and Wakabayashi K 2013 Edge state induced metallicity in zigzag BC_3 ribbons *J. Mater. Chem.* C **1** 4854–7
[111] Chen Z, Lin Y M, Rooks M J and Avouris P 2007 Graphene nano-ribbon electronics *Physica* E **40** 228–32
[112] Han M Y, Özyilmaz B, Zhang Y and Kim P 2007 Energy band-gap engineering of graphene nanoribbons *Phys. Rev. Lett.* **98** 206805
[113] Tapasztó L, Dobrik G, Lambin P and Biro L P 2008 Tailoring the atomic structure of graphene nanoribbons by scanning tunnelling microscope lithography *Nat. Nanotechnol.* **3** 397–401
[114] Jiao L, Zhang L, Wang X, Diankov G and Dai H 2009 Narrow graphene nanoribbons from carbon nanotubes *Nature* **458** 877–80
[115] Kosynkin D V, Higginbotham A L, Sinitskii A, Lomeda J R, Dimiev A, Price B K and Tour J M 2009 Longitudinal unzipping of carbon nanotubes to form graphene nanoribbons *Nature* **458** 872–6
[116] Li X, Wang X, Zhang L, Lee S and Dai H 2008 Chemically derived, ultrasmooth graphene nanoribbon semiconductors *Science* **319** 1229–32
[117] Sprinkle M, Ruan M, Hu Y, Hankinson J, Rubio-Roy M, Zhang B, Wu X, Berger C and De Heer W A 2010 Scalable templated growth of graphene nanoribbons on SiC *Nat. Nanotechnol.* **5** 727–31
[118] Cai J *et al* 2010 Atomically precise bottom-up fabrication of graphene nanoribbons *Nature* **466** 470–3
[119] Liu J, Li B W, Tan Y Z, Giannakopoulos A, Sanchez-Sanchez C, Beljonne D, Ruffieux P, Fasel R, Feng X and Müllen K 2015 Toward cove-edged low band gap graphene nanoribbons *J. Am. Chem. Soc.* **137** 6097–103
[120] Ruffieux P *et al* 2016 On-surface synthesis of graphene nanoribbons with zigzag edge topology *Nature* **531** 489–92
[121] Houtsma R K, de la Rie J and Stöhr M 2021 Atomically precise graphene nanoribbons: interplay of structural and electronic properties *Chem. Soc. Rev.* **50** 6541–68
[122] Yano Y, Mitoma N, Ito H and Itami K 2019 A quest for structurally uniform graphene nanoribbons: synthesis, properties, and applications *J. Org. Chem.* **85** 4–33
[123] Nakada K, Fujita M, Dresselhaus G and Dresselhaus M S 1996 Edge state in graphene ribbons: nanometer size effect and edge shape dependence *Phys. Rev.* B **54** 17954
[124] Son Y W, Cohen M L and Louie S G 2006 Energy gaps in graphene nanoribbons *Phys. Rev. Lett.* **97** 216803

[125] Fujita M, Wakabayashi K, Nakada K and Kusakabe K 1996 Peculiar localized state at zigzag graphite edge *J. Phys. Soc. Jpn.* **65** 1920–3
[126] Wakabayashi K, Fujita M, Ajiki H and Sigrist M 1999 Electronic and magnetic properties of nanographite ribbons *Phys. Rev.* B **59** 8271
[127] Wakabayashi K, Okada S, Tomita R, Fujimoto S and Natsume Y 2010 Edge states and flat bands of graphene nanoribbons with edge modification *J. Phys. Soc. Jpn.* **79** 034706
[128] Wakabayashi K, Sigrist M and Fujita M 1998 Spin wave mode of edge-localized magnetic states in nanographite zigzag ribbons *J. Phys. Soc. Jpn.* **67** 2089–93
[129] Kobayashi Y, Fukui K I, Enoki T and Kusakabe K 2006 Edge state on hydrogen-terminated graphite edges investigated by scanning tunneling microscopy *Phys. Rev.* B **73** 125415
[130] Kobayashi Y, Fukui K I, Enoki T, Kusakabe K and Kaburagi Y 2005 Observation of zigzag and armchair edges of graphite using scanning tunneling microscopy and spectroscopy *Phys. Rev.* B **71** 193406
[131] Niimi Y, Matsui T, Kambara H, Tagami K, Tsukada M and Fukuyama H 2005 Applied surface science *The 9th Int. Symp. on Advanced Physical Fields* 241 43–8
[132] Fujita M, Igami M and Nakada K 1997 Lattice distortion in nanographite ribbons *J. Phys. Soc. Jpn.* **66** 1864–7
[133] Miyamoto Y, Nakada K and Fujita M 1999 First-principles study of edge states of H-terminated graphitic ribbons *Phys. Rev.* B **59** 9858
[134] Harigaya K and Enoki T 2002 Mechanism of magnetism in stacked nanographite with open shell electrons *Chem. Phys. Lett.* **351** 128–34
[135] Kumazaki H and Hirashima, D. S 2008 Local magnetic moment formation on edges of graphene *J. Phys. Soc. Jpn.* **77** 044705
[136] Palacios J J, Fernández-Rossier J, Brey L and Fertig H A 2010 Electronic and magnetic structure of graphene nanoribbons *Semicond. Sci. Technol.* **25** 033003
[137] Yamashiro A, Shimoi Y, Harigaya K and Wakabayashi K 2003 Spin-and charge-polarized states in nanographene ribbons with zigzag edges *Phys. Rev.* B **68** 193410
[138] Hikihara T, Hu X, Lin H H and Mou C Y 2003 Ground-state properties of nanographite systems with zigzag edges *Phys. Rev.* B **68** 035432
[139] Dutta S, Lakshmi S and Pati S K 2008 Electron–electron interactions on the edge states of graphene: a many-body configuration interaction study *Phys. Rev.* B **77** 073412
[140] Dutta S and Wakabayashi K 2012 Tuning charge and spin excitations in zigzag edge nanographene ribbons *Sci. Rep.* **2** 1–7
[141] Dutta S and Wakabayashi K 2014 Spin and charge excitations in zigzag honeycomb nanoribbons: effect of many body correlation *Jpn. J. Appl. Phys.* **53** 06JD01
[142] Dutta S and Wakabayashi K 2013 Interacting spins and holes in zigzag edge nanographene *AIP Conf. Proc.* **1566** 153–4
[143] Son Y W, Cohen M L and Louie S G 2006 Half-metallic graphene nanoribbons *Nature* **444** 347–9
[144] Rudberg E, Sałek P and Luo Y 2007 Nonlocal exchange interaction removes half-metallicity in graphene nanoribbons *Nano Lett.* **7** 2211–3
[145] Dutta S and Pati S K 2008 Half-metallicity in undoped and boron doped graphene nanoribbons in the presence of semilocal exchange-correlation interactions *J. Phys. Chem.* B **112** 1333–5

[146] Das S R and Dutta S 2019 Spin filtering and rectification in lateral heterostructures of zigzag-edge BC_3 and graphene nanoribbons: implications for switching and memory devices *ACS Appl. Nano Mater.* **2** 5365–72

[147] Wu M, Wu X, Gao Y and Zeng X C 2009 Materials design of half-metallic graphene and graphene nanoribbons *Appl. Phys. Lett.* **94** 223111

[148] Zhang L, Chen J, Zhang L, Xu F, Xiao L and Jia S 2021 Gate controllable optical spin current generation in zigzag graphene nanoribbon *Carbon* **173** 565–71

[149] Kan E J, Li Z, Yang J and Hou J G 2007 Will zigzag graphene nanoribbon turn to half metal under electric field? *Appl. Phys. Lett.* **91** 243116

[150] Hod O, Peralta J E and Scuseria G E 2007 Edge effects in finite elongated graphene nanoribbons *Phys. Rev.* B **76** 233401

[151] Kudin K N 2008 Zigzag graphene nanoribbons with saturated edges *ACS Nano* **2** 516–22

[152] Delplace P, Ullmo D and Montambaux G 2011 Zak phase and the existence of edge states in graphene *Phys. Rev.* B **84** 195452

[153] Zak J 1989 Berry's phase for energy bands in solids *Phys. Rev. Lett.* **62** 2747

[154] Cao T, Zhao F and Louie S G 2017 Topological phases in graphene nanoribbons: junction states, spin centers, and quantum spin chains *Phys. Rev. Lett.* **119** 076401

[155] Liu F, Yamamoto M and Wakabayashi K 2017 Topological edge states of honeycomb lattices with zero berry curvature *J. Phys. Soc. Jpn.* **86** 123707

[156] Kameda T, Liu F, Dutta S and Wakabayashi K 2019 Topological edge states induced by the Zak phase in A_3B monolayers *Phys. Rev.* B **99** 075426

[157] Magda G Z, Jin X, Hagymási I, Vancsó P, Osváth Z, Nemes-Incze P, Hwang C, Biro L P and Tapaszto L 2014 Room-temperature magnetic order on zigzag edges of narrow graphene nanoribbons *Nature* **514** 608–11

[158] Tao C *et al* 2011 Spatially resolving edge states of chiral graphene nanoribbons *Nat. Phys.* **7** 616–20

[159] Song S, Su J, Telychko M, Li J, Li G, Li Y, Su C, Wu J and Lu J 2021 On-surface synthesis of graphene nanostructures with π-magnetism *Chem. Soc. Rev.* **50** 3238–62

[160] Wu J, Pisula W and Müllen K 2007 Graphenes as potential material for electronics *Chem. Rev.* **107** 718–47

[161] Ćirić L, Djokić D M, Jaćimović J, Sienkiewicz A, Magrez A, Forro L, Šljivančanin Ž, Lotya M and Coleman J N 2012 Magnetism in nanoscale graphite flakes as seen via electron spin resonance *Phys. Rev.* B **85** 205437

[162] Deyo S and Hershfield S 2021 Magnetism in graphene flakes with edge disorder *Phys. Rev.* B **104** 014404

[163] Wang W L, Meng S and Kaxiras E 2008 Graphene nanoflakes with large spin *Nano Lett.* **8** 241–5

[164] Yazyev O V, Wang W L, Meng S and Kaxiras E 2008 Comment on graphene nanoflakes with large spin: broken-symmetry states *Nano Lett.* **8** 766–6

[165] Wang Y, Xue N, Li R, Wu T, Li N, Hou S and Wang Y 2019 Construction and properties of Sierpiński triangular fractals on surfaces *Chem. Phys. Chem.* **20** 2262–70

[166] Mo Y, Chen T, Dai J, Wu K and Wang D 2019 On-surface synthesis of highly ordered covalent Sierpinski triangle fractals *J. Am. Chem. Soc.* **141** 11378–82

[167] Zhang Y, Zhang X, Li Y, Zhao S, Hou S, Wu K and Wang Y 2020 Packing sierpinski triangles into two-dimensional crystals *J. Am. Chem. Soc.* **142** 17928–32

[168] Wang H, Zhang X, Jiang Z, Wang Y and Hou S 2018 Electronic confining effects in Sierpiński triangle fractals *Phys. Rev.* B **97** 115451
[169] Kempkes S N, Slot M R, Freeney S E, Zevenhuizen S J, Vanmaekelbergh D, Swart I and Smith C M 2019 Design and characterization of electrons in a fractal geometry *Nat. Phys.* **15** 127–31
[170] Clar E and Mackay C C 1972 Circobiphenyl and the attempted synthesis of 1:14, 3:4, 7:8, 10:11-tetrabenzoperopyrene *Tetrahedron* **28** 6041–7
[171] Hauptmann S 2010 The Aromatic Sextet: Von E. Clar; John Wiley & Sons LTD, London, New York, Sydney, Toronto 1972; 128 Seiten mit zahlreichen Formelbildern; Format 13 × 20 cm; broschiert £ 1, 50 *Z. Chem.* **13** 200
[172] Kang J, Wu F and Li J 2012 Spin filter and molecular switch based on bowtie-shaped graphene nanoflake *J. Appl. Phys.* **112** 104328
[173] Wang W L, Yazyev O V, Meng S and Kaxiras E 2009 Topological frustration in graphene nanoflakes: magnetic order and spin logic devices *Phys. Rev. Lett.* **102** 157201
[174] Herrera-Reinoza N, dos Santos A C, de Lima L H, Landers R and de Siervo A 2021 Atomically precise bottom-up synthesis of h-BNC: graphene doped with h-BN nanoclusters *Chem. Mater.* **33** 2871–82
[175] Li J, Sanz S, Corso M, Choi D J, Peña D, Frederiksen T and Pascual J I 2019 Single spin localization and manipulation in graphene open-shell nanostructures *Nat. Commun.* **10** 1–7
[176] Li J, Sanz S, Castro-Esteban J, Vilas-Varela M, Friedrich N, Frederiksen T, Peña D and Pascual J I 2020 Uncovering the triplet ground state of triangular graphene nanoflakes engineered with atomic precision on a metal surface *Phys. Rev. Lett.* **124** 177201
[177] Mishra S, Fatayer S, Fernández S, Kaiser K, Pena D and Gross L 2022 Nonbenzenoid high-spin polycyclic hydrocarbons generated by atom manipulation *ACS Nano* **16** 3264–71
[178] Mishra S *et al* 2021 Large magnetic exchange coupling in rhombus-shaped nanographenes with zigzag periphery *Nat. Chem.* **13** 581–6
[179] Mousavi F M and Farghadan R 2021 Electrical control of Ruderman–Kittel–Kasuya–Yosida exchange interaction in zigzag edge MoS_2 nanoflakes *J. Phys. Chem. Solids* **158** 110242
[180] Mishra S *et al* 2020 Topological frustration induces unconventional magnetism in a nanographene *Nat. Nanotechnol.* **15** 22–8
[181] Mishra S, Beyer D, Berger R, Liu J, Gröning O, Urgel J I, Mullen K, Ruffieux P, Feng X and Fasel R 2020 Topological defect-induced magnetism in a nanographene *J. Am. Chem. Soc.* **142** 1147–52
[182] Cui D, Perepichka D F, MacLeod J M and Rosei F 2020 Surface-confined single-layer covalent organic frameworks: design, synthesis and application *Chem. Soc. Rev.* **49** 2020–38
[183] Dong R *et al* 2018 A coronene-based semiconducting two-dimensional metal-organic framework with ferromagnetic behavior *Nat. Commun.* **9** 1–9
[184] Zhou Y and Liu F 2020 Realization of an antiferromagnetic superatomic graphene: Dirac Mott insulator and circular dichroism Hall effect *Nano Lett.* **21** 230–5
[185] Shastry B S, Krishnamurthy H R and Anderson P W 1990 Instability of the Nagaoka ferromagnetic state of the $U=\infty$ Hubbard model *Phys. Rev.* B **41** 2375
[186] Tasaki H 1998 From Nagaoka's ferromagnetism to flat-band ferromagnetism and beyond: an introduction to ferromagnetism in the Hubbard model *Prog. Theor. Phys.* **99** 489–548
[187] Nagaoka Y 1966 Ferromagnetism in a narrow, almost half-filled s band *Phys. Rev.* **147** 392

[188] Mattis D C 2003 Eigenvalues and magnetism of electrons on an artificial molecule *Int. J. Nanosci.* **2** 165–70
[189] Nielsen E and Bhatt R N 2007 Nanoscale ferromagnetism in nonmagnetic doped semiconductors *Phys. Rev.* B **76** 161202
[190] Dehollain J P, Mukhopadhyay U, Michal V P, Wang Y, Wunsch B, Reichl C, Wegscheider W, Rudner M S, Demler E and Vandersypen L M 2020 Nagaoka ferromagnetism observed in a quantum dot plaquette *Nature* **579** 528–33
[191] Baringhaus J, Edler F and Tegenkamp C 2013 Edge-states in graphene nanoribbons: a combined spectroscopy and transport study *J. Phys. Condens. Matter* **25** 392001
[192] Dutta S, Manna A K and Pati S K 2009 Intrinsic half-metallicity in modified graphene nanoribbons *Phys. Rev. Lett.* **102** 096601
[193] Pizzochero M, Barin G B, Cernevics K, Wang S, Ruffieux P, Fasel R and Yazyev O V 2021 Edge disorder in bottom-up zigzag graphene nanoribbons: implications for magnetism and quantum electronic transport *J. Phys. Chem. Lett.* **12** 4692–6
[194] Semenov Y G, Kim K W and Zavada J M 2007 Spin field effect transistor with a graphene channel *Appl. Phys. Lett.* **91** 153105
[195] Hallal A, Ibrahim F, Yang H, Roche S and Chshiev M 2017 Tailoring magnetic insulator proximity effects in graphene: first-principles calculations *2D Mater.* **4** 025074
[196] Yang H X, Hallal A, Terrade D, Waintal X, Roche S and Chshiev M 2013 Proximity effects induced in graphene by magnetic insulators: first-principles calculations on spin filtering and exchange-splitting gaps *Phys. Rev. Lett.* **110** 046603
[197] Swartz A G, Odenthal P M, Hao Y, Ruoff R S and Kawakami R K 2012 Integration of the ferromagnetic insulator EuO onto graphene *ACS Nano* **6** 10063–9
[198] Wei P *et al* 2016 Strong interfacial exchange field in the graphene/EuS heterostructure *Nat. Mater.* **15** 711–6
[199] Wang Z, Tang C, Sachs R, Barlas Y and Shi J 2015 Proximity-induced ferromagnetism in graphene revealed by the anomalous Hall effect *Phys. Rev. Lett.* **114** 016603
[200] Mendes J B S, Santos O A, Meireles L M, Lacerda R G, Vilela-Leão L H, Machado F L A, Rodríguez-Suárez R L, Azevedo A and Rezende S M 2015 Spin-current to charge-current conversion and magnetoresistance in a hybrid structure of graphene and yttrium iron garnet *Phys. Rev. Lett.* **115** 226601
[201] Maryenko D, Mishchenko A S, Bahramy M S, Ernst A, Falson J, Kozuka Y, Tsukazaki A, Nagaosa N and Kawasaki M 2017 Observation of anomalous Hall effect in a non-magnetic two-dimensional electron system *Nat. Commun.* **8** 1–7
[202] Leutenantsmeyer J C, Kaverzin A A, Wojtaszek M and Van Wees B J 2016 Proximity induced room temperature ferromagnetism in graphene probed with spin currents *2D Mater.* **4** 014001
[203] Singh S, Katoch J, Zhu T, Meng K Y, Liu T, Brangham J T, Yang F, Flatté M E and Kawakami R K 2017 Strong modulation of spin currents in bilayer graphene by static and fluctuating proximity exchange fields *Phys. Rev. Lett.* **118** 187201
[204] Movva H C, Lovorn T, Fallahazad B, Larentis S, Kim K, Taniguchi T, Watanabe K, Banerjee S K, MacDonald A H and Tutuc E 2018 Tunable Γ–K valley populations in hole-doped trilayer WSe_2 *Phys. Rev. Lett.* **120** 107703
[205] Wang Q H, Kalantar-Zadeh K, Kis A, Coleman J N and Strano M S 2012 Electronics and optoelectronics of two-dimensional transition metal dichalcogenides *Nat. Nanotechnol.* **7** 699–712

[206] Qian X, Liu J, Fu L and Li J 2014 Quantum spin Hall effect in two-dimensional transition metal dichalcogenides *Science* **346** 1344–7
[207] Das S R, Wakabayashi K, Yamamoto M, Tsukagoshi K and Dutta S 2018 Layer-by-layer oxidation induced electronic properties in transition-metal dichalcogenides *J. Phys. Chem.* C **122** 17001–7
[208] Yamamoto M, Dutta S, Aikawa S, Nakaharai S, Wakabayashi K, Fuhrer M S, Ueno K and Tsukagoshi K 2015 Self-limiting layer-by-layer oxidation of atomically thin WSe_2 *Nano Lett.* **15** 2067–73
[209] Das S R, Wakabayashi K, Tsukagoshi K and Dutta S 2020 Ab-initio investigation of preferential triangular self-formation of oxide heterostructures of monolayer WSe_2 *Sci. Rep.* **10** 1–6
[210] Das S R and Dutta S 2020 Inter-layer interaction induced electronic properties in partially oxidized transition metal dichalcogenides *AIP Conf. Proc.* **2265** 030723
[211] Ingla-Aynés J, Herling F, Fabian J, Hueso L E and Casanova F 2021 Electrical control of valley-Zeeman spin–orbit-coupling–induced spin precession at room temperature *Phys. Rev. Lett.* **127** 047202
[212] Bhimanapati G R *et al* 2015 Recent advances in two-dimensional materials beyond graphene *ACS Nano* **9** 11509–39
[213] Das S, Robinson J A, Dubey M, Terrones H and Terrones M 2015 Beyond graphene: progress in novel two-dimensional materials and van der Waals solids *Annu. Rev. Mater. Res.* **45** 1–27
[214] Novoselov K S, Andreeva D V, Ren W and Shan G 2019 Graphene and other two-dimensional materials *Front. Phys.* **14** 13301
[215] Novoselov K S and Neto A C 2012 Two-dimensional crystals-based heterostructures: materials with tailored properties *Phys. Scr.* **2012** 014006
[216] Novoselov K S, Mishchenko O A, Carvalho O A and Castro Neto A H 2016 2D materials and van der Waals heterostructures *Science* **353** aac9439
[217] Yamamoto M, Suizu R, Dutta S, Mishra P, Nakayama T, Sakamoto K, Wakabayashi K, Uchihashi T and Awaga K 2015 Self-assembled honeycomb lattice in the monolayer of cyclic thiazyl diradical BDTDA (=4,4′-bis(1,2,3,5-dithiadiazolyl)) on Cu(111) with a zero-bias tunneling spectra anomaly *Sci. Rep.* **5** 18359
[218] Neto A C and Novoselov K 2011 New directions in science and technology: two-dimensional crystals *Rep. Prog. Phys.* **74** 082501
[219] Gibertini M, Koperski M, Morpurgo A F and Novoselov K S 2019 Magnetic 2D materials and heterostructures *Nat. Nanotechnol.* **14** 408–19
[220] Adhikary S, Dutta S and Mohakud S 2021 Antiferromagnetic spin ordering in two-dimensional honeycomb lattice of SiP_3 *Nanoscale Adv.* **3** 2217–21
[221] Liu H, Neal A T, Zhu Z, Luo Z, Xu X, Tománek D and Ye P D 2014 Phosphorene: an unexplored 2D semiconductor with a high hole mobility *ACS Nano* **8** 4033–41
[222] Zeng J, Cui P and Zhang Z 2017 Half layer by half layer growth of a blue phosphorene monolayer on a GaN (001) substrate *Phys. Rev. Lett.* **118** 046101
[223] Zhang J L *et al* 2016 Epitaxial growth of single layer blue phosphorus: a new phase of two-dimensional phosphorus *Nano Lett.* **16** 4903–8
[224] Zhu Z and Tománek D 2014 Semiconducting layered blue phosphorus: a computational study *Phys. Rev. Lett.* **112** 176802
[225] Yazyev O V and Katsnelson M I 2008 Magnetic correlations at graphene edges: basis for novel spintronics devices *Phys. Rev. Lett.* **100** 047209

[226] Jung J, Pereg-Barnea T and MacDonald A H 2009 Theory of interedge superexchange in zigzag edge magnetism *Phys. Rev. Lett.* **102** 227205
[227] Karmakar S and Dutta S 2021 Strain tunable photocatalytic ability of BC_6N monolayer *Comput. Mater. Sci.* **202** 111002
[228] Adhikary S, Mohakud S and Dutta S 2021 Engineering anisotropic Klein tunneling in black phosphorene through elemental substitution *Phys. Status Solidi* B **258** 2100071
[229] Yonemori I, Dutta S, Nagashio K and Wakabayashi K 2021 Thickness-dependent Raman active modes of SnS thin films *AIP Adv.* **11** 095106
[230] Mak K F, Lee C, Hone J, Shan J and Heinz T F 2010 Atomically thin MoS_2: a new direct-gap semiconductor *Phys. Rev. Lett.* **105** 136805
[231] Chhowalla M, Shin H S, Eda G, Li L J, Loh K P and Zhang H 2013 The chemistry of two-dimensional layered transition metal dichalcogenide nanosheets *Nat. Chem.* **5** 263–75
[232] Jariwala D, Sangwan V K, Lauhon L J, Marks T J and Hersam M C 2014 Emerging device applications for semiconducting two-dimensional transition metal dichalcogenides *ACS Nano* **8** 1102–20
[233] Li H, Wu H, Yuan S and Qian H 2016 Synthesis and characterization of vertically standing MoS_2 nanosheets *Sci. Rep.* **6** 21171
[234] Lv R, Robinson J A, Schaak R E, Sun D, Sun Y, Mallouk T E and Terrones M 2015 Transition metal dichalcogenides and beyond: synthesis, properties, and applications of single-and few-layer nanosheets *Acc. Chem. Res.* **48** 56–64
[235] Yamamoto M, Einstein T L, Fuhrer M S and Cullen W G 2013 Anisotropic etching of atomically thin MoS_2 *J. Phys. Chem.* C **117** 25643–49
[236] Yamamoto M, Nakaharai S, Ueno K and Tsukagoshi K 2016 Self-limiting oxides on WSe_2 as controlled surface acceptors and low-resistance hole contacts *Nano Lett.* **16** 2720–7
[237] Ataca C and Ciraci S 2011 Functionalization of single-layer MoS_2 honeycomb structures *J. Phys. Chem.* C **115** 13303–11
[238] Zhang J, Soon J M, Loh K P, Yin J, Ding J, Sullivian M B and Wu P 2007 Magnetic molybdenum disulfide nanosheet films *Nano Lett.* **7** 2370–6
[239] Pan H and Zhang Y W 2012 Edge-dependent structural, electronic and magnetic properties of MoS_2 nanoribbons *J. Mater. Chem.* **22** 7280–90
[240] Li Y, Zhou Z, Zhang S and Chen Z 2008 MoS_2 nanoribbons: high stability and unusual electronic and magnetic properties *J. Am. Chem. Soc.* **130** 16739–44
[241] Deng X, Li Z and Yang J 2020 One-dimensional magnetic order stabilized in edge-reconstructed MoS_2 nanoribbon via bias voltage *J. Phys. Chem. Lett.* **11** 7531–5
[242] Gong C and Zhang X 2019 Two-dimensional magnetic crystals and emergent hetero-structure devices *Science* **363** eaav4450
[243] Wang M C *et al* 2020 Prospects and opportunities of 2D van der Waals magnetic systems *Ann. Phys.* **532** 1900452
[244] Li H, Ruan S and Zeng Y J 2019 Intrinsic van der Waals magnetic materials from bulk to the 2D limit: new frontiers of spintronics *Adv. Mater.* **31** 1900065
[245] Li X and Yang J 2014 $CrXTe_3$ (X=Si, Ge) nanosheets: two dimensional intrinsic ferromagnetic semiconductors *J. Mater. Chem.* C **2** 7071–6
[246] Onsager L 1944 Crystal statistics. I. A two-dimensional model with an order-disorder transition *Phys. Rev.* **65** 117
[247] Hohenberg P C 1967 Existence of long-range order in one and two dimensions *Phys. Rev.* **158** 383

[248] Mermin N D and Wagner H 1966 Absence of ferromagnetism or antiferromagnetism in one-or two-dimensional isotropic Heisenberg models *Phys. Rev. Lett.* **17** 1133

[249] Lines M E 1967 Comparative studies of magnetism in $KNiF_3$ and K_2NiF_4 *Phys. Rev.* **164** 736

[250] Bergmann G 1978 Transition from Pauli paramagnetism to band ferromagnetism in very thin Ni films *Phys. Rev. Lett.* **41** 264

[251] Bander M and Mills D L 1988 Ferromagnetism of ultrathin films *Phys. Rev.* B **38** 12015

[252] Bruno P 1991 Magnetization and Curie temperature of ferromagnetic ultrathin films: the influence of magnetic anisotropy and dipolar interactions *MRS Online Proc. Libr. (OPL)* **231** 299–310

[253] Huang B *et al* 2017 Layer-dependent ferromagnetism in a van der Waals crystal down to the monolayer limit *Nature* **546** 270–3

[254] Gong C *et al* 2017 Discovery of intrinsic ferromagnetism in two-dimensional van der Waals crystals *Nature* **546** 265–9

[255] Deng Y *et al* 2018 Gate-tunable room-temperature ferromagnetism in two-dimensional Fe_3GeTe_2 *Nature* **563** 94–9

[256] Lado J L and Fernández-Rossier J 2017 On the origin of magnetic anisotropy in two dimensional CrI_3 *2D Mater.* **4** 035002

[257] Guo Y, Yuan S, Wang B, Shi L and Wang J 2018 Half-metallicity and enhanced ferromagnetism in Li-adsorbed ultrathin chromium triiodide *J. Mater. Chem.* C **6** 5716–20

[258] Li S *et al* 2020 Magnetic-field-induced quantum phase transitions in a van der Waals magnet *Phys. Rev.* **10** 011075

[259] Jiang S, Shan J and Mak K F 2018 Electric-field switching of two-dimensional van der Waals magnets *Nat. Mater.* **17** 406–10

[260] Jiang S, Li L, Wang Z, Mak K F and Shan J 2018 Controlling magnetism in 2D CrI_3 by electrostatic doping *Nat. Nanotechnol.* **13** 549–53

[261] Sivadas N, Okamoto S and Xiao D 2016 Gate-controllable magneto-optic Kerr effect in layered collinear antiferromagnets *Phys. Rev. Lett.* **117** 267203

[262] Wang K, Hu T, Jia F, Zhao G, Liu Y, Solovyev I V, Pyatakov A P, Zvezdin A K and Ren W 2019 Magnetic and electronic properties of $Cr_2Ge_2Te_6$ monolayer by strain and electric-field engineering *Appl. Phys. Lett.* **114** 092405

[263] Chabungbam S, Mohakud S and Dutta S 2020 Magnetoelasticity in $CrXTe_3$ (X = C, Si) van der Waals heterobilayers *ACS Appl. Electron. Mater.* **2** 3171–7

[264] Anderson P W 1950 Antiferromagnetism. Theory of superexchange interaction *Phys. Rev.* **79** 350

[265] Goodenough J B 1955 Theory of the role of covalence in the perovskite-type manganites [La, M (II)] Mn O 3 *Phys. Rev.* **100** 564

[266] Goodenough J B 1958 An interpretation of the magnetic properties of the perovskite-type mixed crystals $La_{1-x}Sr_xCoO_{3-\lambda}$ *J. Phys. Chem. Solids* **6** 287–97

[267] Kanamori J 1959 Superexchange interaction and symmetry properties of electron orbitals *J. Phys. Chem. Solids* **10** 87–98

[268] Geertsma W and Khomskii D 1996 Influence of side groups on 90 superexchange: a modification of the Goodenough–Kanamori–Anderson rules *Phys. Rev.* B **54** 3011

[269] Chabungbam S and Sen P 2017 Computational design of a robust two-dimensional antiferromagnetic semiconductor *Phys. Rev.* B **96** 045404

[270] Casto L D *et al* 2015 Strong spin-lattice coupling in $CrSiTe_3$ *APL Mater.* **3** 041515

[271] Chen X, Qi J and Shi D 2015 Strain-engineering of magnetic coupling in two-dimensional magnetic semiconductor $CrSiTe_3$: competition of direct exchange interaction and super-exchange interaction *Phys. Lett.* A **379** 60–3

[272] Zhuang H L, Kent P R C and Hennig R G 2016 Strong anisotropy and magnetostriction in the two-dimensional Stoner ferromagnet Fe_3GeTe_2 *Phys. Rev.* B **93** 134407

[273] Wang Z, Sapkota D, Taniguchi T, Watanabe K, Mandrus D and Morpurgo A F 2018 Tunneling spin valves based on Fe_3GeTe_2/hBN/Fe_3GeTe_2 van der Waals heterostructures *Nano Lett.* **18** 4303–8

[274] Lee J U, Lee S, Ryoo J H, Kang S, Kim T Y, Kim P, Park C H, Park J G and Cheong H 2016 Ising-type magnetic ordering in atomically thin $FePS_3$ *Nano Lett.* **16** 7433–8

[275] Bonilla M, Kolekar S, Ma Y, Diaz H C, Kalappattil V, Das R, Eggers T, Gutierrez H R, Phan M H and Batzill M 2018 Strong room-temperature ferromagnetism in VSe_2 monolayers on van der Waals substrates *Nat. Nanotechnol.* **13** 289–93

[276] Kim S Y *et al* 2018 Charge-spin correlation in van der Waals antiferromagnet $NiPS_3$ *Phys. Rev. Lett.* **120** 136402

[277] Klingen W, Eulenberger G and Hahn H 1973 On the crystal structures of $Fe_2P_2Se_6$ and $Fe_2P_2S_6$ *J. Inorg. Gen. Chem.* **401** 97–112

[278] Taylor B, Steger J, Wold A and Kostiner E 1974 Preparation and properties of iron phosphorus triselenide, $FePSe_3$ *Inorg. Chem.* **13** 2719–21

[279] Wiedenmann A, Rossat-Mignod J, Louisy A, Brec R and Rouxel J 1981 Neutron diffraction study of the layered compounds $MnPSe_3$ and $FePSe_3$ *Solid State Commun.* **40** 1067–72

[280] Wildes A R, Rule K C, Bewley R I, Enderle M and Hicks T J 2012 The magnon dynamics and spin exchange parameters of $FePS_3$ *J. Phys. Condens. Matter* **24** 416004

[281] Yang K, Hu W, Wu H, Whangbo M H, Radaelli P G and Stroppa A 2020 Magneto-optical Kerr switching properties of $(CrI_3)_2$ and $(CrBr_3/CrI_3)$ bilayers *ACS Appl. Electron. Mater.* **2** 1373–80

[282] Klein D R *et al* 2018 Probing magnetism in 2D van der Waals crystalline insulators via electron tunneling *Science* **360** 1218–22

[283] Castro E V, Peres N M R, Stauber T and Silva N A P 2008 Low-density ferromagnetism in biased bilayer graphene *Phys. Rev. Lett.* **100** 186803

[284] Cao Y *et al* 2018 Correlated insulator behaviour at half-filling in magic-angle graphene superlattices *Nature* **556** 80–4

[285] Cao Y, Fatemi V, Fang S, Watanabe K, Taniguchi T, Kaxiras E and Jarillo-Herrero P 2018 Unconventional superconductivity in magic-angle graphene superlattices *Nature* **556** 43–50

[286] Chen G *et al* 2020 Tunable correlated Chern insulator and ferromagnetism in a moiré superlattice *Nature* **579** 56–61

[287] Sharpe A L, Fox E J, Barnard A W, Finney J, Watanabe K, Taniguchi T, Kastner M A and Goldhaber-Gordon D 2019 Emergent ferromagnetism near three-quarters filling in twisted bilayer graphene *Science* **365** 605–8

[288] Wakabayashi K and Harigaya K 2003 Magnetic structure of nano-graphite Möbius ribbon *J. Phys. Soc. Jpn.* **72** 998–1001

[289] Chung J H and Chai J D 2019 Electronic properties of Möbius cyclacenes studied by thermally-assisted-occupation density functional theory *Sci. Rep.* **9** 2907

IOP Publishing

Recent Advances in Graphene and Graphene-Based Technologies

Anoop Chandran, N V Unnikrishnan, M K Jayaraj, Reenu Elizabeth John and Justin George

Chapter 17

Graphene metamaterials

Gaurav Pal Singh and Neha Sardana

Metamaterials (MMs) are materials with negative permittivity and/or permeability; the combination of both being negative is not found in natural materials. MMs have periodic or non-periodic subwavelength structures that alter the material's response to the incoming electromagnetic (EM) waves. The active frequency region of the MMs depends upon the size of the structures. The fabrication of optical and infrared (IR) MMs is complex and requires advanced fabrication techniques such as electron beam lithography and nanoimprinting. The terahertz (THz) and microwave metasurfaces can be constructed by traditional techniques. However, 3D MMs again require sophisticated instrumentation. Hence, MMs with tunable response achieved by varying the dimension of the structures is not practical. Graphene has an excellent response across the EM spectrum, and it can be tuned by simple methods such as applying gate voltage and optical illumination. Therefore, the incorporation of graphene into MMs can improve performance and help develop tunable devices. The applications of graphene MMs are generally in the IR, THz, and microwave regions, where conventional materials do not perform well. In the current chapter, the applications of graphene-based MMs are broadly divided into imaging (absorber, lenses, and modulators), communication (antennas, transparency, and encryption), and sensing.

17.1 Introduction

Nature has provided humanity with a plethora of materials for a variety of applications. Over the years, the scientific community has turned towards nature for inspiration to develop new and improved materials. One of the inspirations behind the development of MMs is the wings of a butterfly, which has periodic structures whose dimensions range from tens to hundreds of nanometers that create a photonic bandgap which is dependent upon the spacing and size of the structures,

doi:10.1088/978-0-7503-3999-5ch17

therefore, creating the vibrant and unique patterns in the visible frequency range [1]. Also, the scales on the wings of a moth act as an acoustic MM that reduces ultrasonic echoes [2]. MMs have properties absent in Nature, a combination of negative permeability (μ) and permittivity (ε), when other materials (metals, 2D materials, etc) have been used to fabricate these subwavelength structures instead of what is naturally selected. The size of the structures decides the working frequency range of these new materials.

MMs are materials with properties that are not available in natural materials. Most materials have complex ε and μ, most conventional materials have both ε and μ positive (figure 17.1(a)), the real part of ε and μ decides the response of the material to EM waves, and the imaginary part dictates the losses taking place in the material. However, metals at high frequencies (ultraviolet (UV) and visible) range possess negative ε. Negative μ is not commonly found in natural materials, particularly beyond the microwave region [3]. Artificial magnetic materials (AMMs) are engineered to attain negative μ using subwavelength resonators in a dielectric medium. The response of the incoming EM wave is altered due to the induced electric current in the resonators. Furthermore, negative ε was initially obtained using an array of thin metal wires in a dielectric [4]. Combining the negative μ and ε was first realized by Veselago theoretically in 1968 [5]. The material obtained was later termed MM, double negative medium, and left-hand material. Therefore, using natural materials arranged in subwavelength structures in a dielectric medium can lead to properties unavailable in nature. The interaction of the EM waves with the periodic or non-periodic structures alters the response of its inherent material. Generally, researchers use periodic structures to impart the desired response; however, non-periodic structures are also being explored for future applications [6]. The generally used structures for MM applications include split ring resonators (SRRs), which act as magnetic resonators; and cut wires which act as electric resonators. But various other complex structures are used to achieve the desired response from the MM. Due to the complex structures present in the MMs, their fabrication is a complicated process. Although advanced fabrication techniques such as electron beam lithography and nanoimprinting have made the study of MMs accessible, their commercial applications are still limited.

Graphene is a revolutionary material due to its unique properties which include, excellent mechanical strength along with very high thermal and electrical conductivity. Further, the bandgap of single-layer graphene is zero, which is the primary source of its unique optical, electrical, and thermal properties. Graphene, owing to its unique band structure, has applications almost all over the EM spectrum. Furthermore, the bandgap of graphene can be tuned to achieve the desired resonance peak. The main methods of changing the bandgap of graphene are chemical doping and applying an electrical bias. The EM response of graphene can also be altered by optical pumping [7]. The tunability is an enticing property for researchers to develop active MMs. Tuning the response of conventional MMs is a complex and tedious process that involves optimizing the size and shape of the unit cell of the periodic array. However, active materials such as phase change, ferromagnetic and ferroelectric materials are increasingly being used to develop

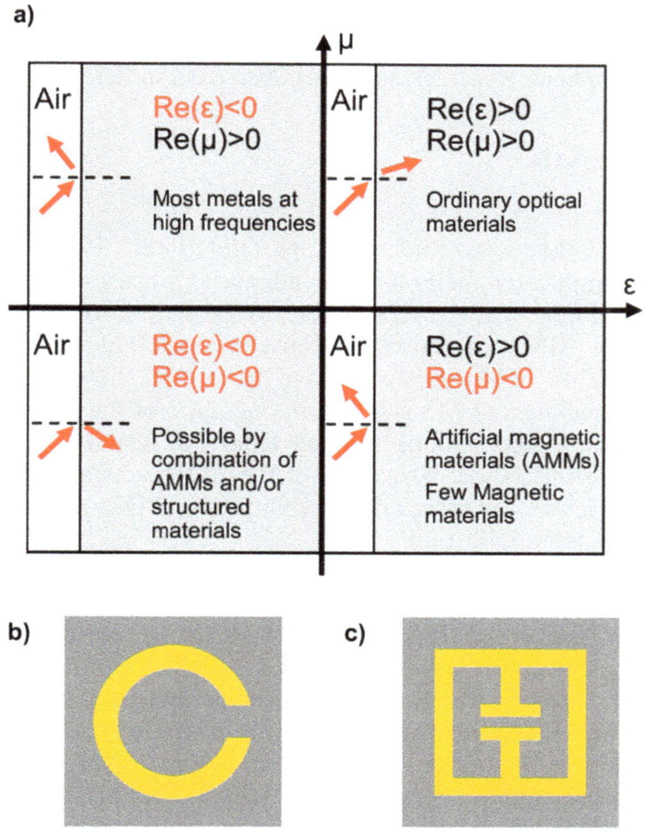

Figure 17.1. (a) The response to the various combinations of permittivity and permeability and the methods to achieve them. The basic geometry of (b) split ring resonator and (c) electrically coupled resonator.

tunable MMs [8]. A graphene-based approach to achieve a tunable response is easier to accomplish than the aforementioned techniques. Applying an electrical bias to graphene changes its chemical potential (Fermi energy), altering its conductivity; hence, the EM response of graphene is changed. Graphene-based MMs use either patterned graphene film and/or continuous graphene layer along with conventional MM structures for applications ranging from near-infrared (near-IR) to microwave frequency range. The terahertz (THz) frequency region has promising applications in communications [9], imaging [10], security [11], and sensing [12]. Conventional materials have limited usability in the THz range due to their low response [13]. However, graphene-based MMs have broadband tunability and are widely used for IR, THz, and microwave frequency range applications. The size of the structures and the spacing of the unit cells for these applications can range from nanometers (IR range) to micron range (microwave range).

The chapter will initially focus on the fabrication, design, and characterization of MMs. Followed by the basic properties and tunability of graphene, and finally, the applications of graphene-based MMs will be discussed in detail.

17.2 Metamaterials

17.2.1 Fabrication

The fabrication of MMs is a complex task, particularly 3D MMs are fabricated by either slow and complicated methods or by advanced fabrication techniques, which are still in the development stage. Hence, researchers have also focused on 2D MMs called metasurfaces (MSs) or frequency selective surfaces. There is no clearly defined thickness at which an MM can be referred to as MS; however, it is generally reported that the thickness of MS is less than one-tenth of the working wavelength. Graphene MMs have applications ranging from the near-IR to the microwave region. The dimensions of the structures of the MMs depend upon the working frequency range. Hence, the fabrication techniques of MMs are dependent on the frequency range of the application.

17.2.1.1 Optical and infrared metamaterials

The fabrication of optical and IR MMs is much harder to accomplish because of the requirement of significantly small structures (of the order of a few nanometers) due to the much higher working frequency range. Furthermore, the conventional optical materials have almost negligible magnetizability in the optical frequencies; therefore, their permeability is close to 1. However, to obtain a double negative material, control over the permeability is essential. The subwavelength structures help induce currents in the MM, which lead to the excitation of magnetic moments. This magnetic response can lead to the achievement of negative permeability. Fabricating a double negative MS in the optical frequency range is more complex. Since the combination of SRR and cut wire structures is not feasible for the nanometer range fabrication, the structure used in the MS is generally a fishnet type [14]. The induced currents in the structure create a negative permeability above the magnetic resonance frequency. To construct a bulk double negative material, the fabrication of 3D MMs is necessary. One example is the use of discrete nanoring type structures that achieved negative permittivity and permeability from 1.3 to 2.3 μm and was polarization independent [15]. The structure broke its symmetry in the 3D arrangement by creating an offset in between the unit cells. The 3D MM was fabricated by multistep electron beam lithography. The methods of fabrication of MMs for optical frequencies can include both top-down and bottom-up approaches.

17.2.1.1.1 Top-down methods

Optical lithography is a popular fabrication technique in the semiconductor industry; however, the masks required for the setup are generally fabricated using **electron beam lithography**. Although optical lithography is suitable for mass production, the development of MMs and especially 3D MMs is still in the development stage. Therefore, electron beam lithography is generally used to fabricate MSs and can be modified for 3D MMs. The resolution is lower than tens of nanometer, which is much lower than optical lithography. The basic steps are [16], deposition of a conductive substrate on which a resist (e.g., PMMA) is applied and exposed to an electron beam. The electron beam instrumentation is similar to an electron beam used for a scanning electron microscope (SEM). The pattern is scanned over the substrate and the exposed resist layer is etched away using appropriate chemicals. Next, the cavity created by the process is filled in by the selected metal. The remaining area is again chemically etched away. The aforementioned method is generally used to fabricate MSs; however, a 3D MM can be created by stacking the MS layers via deposition methods similar to evaporation. However, the technique is only feasible for a small number of layers and is prone to alignment complications [17]. The methods used for improving the electron beam lithography process include proximity correction and lift off. Proximity correction is accomplished by decreasing the incident beam dose locally to suppress the uneven spread of secondary electrons. In the lift off process, the metal film is deposited on a reversed pattern and the pattern is etched away by a solvent which takes away the excess metal film and the resist. Therefore, one step is decreased from the fabrication process. The lift off process can also be modified to develop 3D MMs by using multiple layers [18].

Nanoimprint lithography is a process that uses a master stamp which can be a template or mold, that is pressed onto the resist layer. The pattern present on the master stamp is embossed on the resist, on which a metal layer is deposited, and the remaining resist is etched away to obtain the final structures. The master stamp can be reused multiple times; however, the precision of the patterning degrades after a certain number of uses.

Focused ion beam nanostructuring is a fabrication technique that is much quicker than electron beam lithography [19]. Hence, this technique has the potential to be used for mass production. In a typical setup, gallium ions are accelerated and are incident on a metal deposited on a substrate, and the desired area is ablated away, thus, creating the desired pattern. However, the structures obtained from this technique are not as precise as electron beam lithography.

Interference lithography is a technique that is inexpensive and faster as compared to the aforementioned techniques, and can produce much larger MMs. The resist layer is exposed to two or more interfering laser beams that can create patterns of dimensions much smaller than the features present on the mask used [20]. The patterned resist is deposited with the selected metal, and the remaining resist is etched away.

17.2.1.1.2 Bottom-up methods

Bottom-up techniques have the potential to fabricate complex structures of large sizes in a short amount of time. The bottom-up approaches for nanofabrication are still in their nascent stage and will require further optimization before being ready for mass production. One of the bottom-up techniques is **self-assembly**, which is much more economical as compared to most of the top-down techniques. However, the use of the self-assembly techniques is still sparse due to the limited number of materials that can be selected, long-range disorder, and the inability to incorporate complex designs. The organization in self-assembly is achieved by interaction methods including capillary force, hydrogen bonding, and Van der Waals force. The fishnet structure required for MMs in the optical frequency range was realized by Gong *et al* [21] by using a self-assembled monolayer of polystyrene spheres which acted as the patterned mask. The mask was used to deposit multiple layers of silver electrochemically to create a 3D MM. The aforementioned technique of using polymer sphere monolayer as a mask to deposit the required material is called **colloidal lithography**. The self-assembly technique can also be used without lithography to develop 3D MMs. **Block copolymers** consisting of several covalently bonded copolymer chains can be constructed in spherical, gyroidal, and cylindrical shapes by chemical means [22]. The block copolymers are self-assembled to create a pattern that is then etched, and the required material is deposited to fabricate a 3D MM [23]. The technique is quick and economical, and can fabricate unit cells smaller than 50 nm. An advanced self-assembly technique called **DNA-templating** can also be used for fabricating optical MMs of large sizes economically [24]. Wen *et al* [25] used thiolated single strand DNA to organize gold nanoparticles in a particular design. The ligand DNA was functionalized on a gold nanoparticle which fixed onto a template DNA. The template DNA was hybridized with a support DNA which assembled itself to form the required structure. The assembled unit mainly formed trimers of nanoparticles (62%), but monomers, dimers and others were also formed. The trimer is expected to induce a magnetic dipole upon EM excitation [26]. Hence, can be used as the unit cell of the 3D MM. Other methods to achieve self-assembly are use of synthetic opals [27], chiral molecules [28], and magnetic field induced self-assembly [29].

17.2.1.2 THz and microwave metamaterials

17.2.1.2.1 Conventional methods

As explained earlier, the size of the structures present on the MMs is determined by the frequency range of its use. The fabrication of microwave and THz range MMs is much easier due to their lower working frequencies. Hence, classical manufacturing processes such as **injection molding** can produce simple designs in the millimeter scale with low melting point materials, but the technique has limited use in fabricating MMs due to the complexity of mold design and material selection [30]. Another conventional approach used for the fabrication of THz and microwave range MMs is **powder bed fusion**. Metal or polymer powder is deposited on a print bed, which is sintered using a high power laser [31]. This technique can produce complicated 2D patterns with high precision, which cannot be easily machined or injection molded. **Printed circuit board manufacturing** is also employed to develop MSs in the microwave frequency range. They usually involve computer controlled milling, soldering, and etching to create the 2D designs [32]. However, the use of the technique is limited due to the inability to create 3D structures and a large amount of wasted material. **Laser micromachining** is an advanced machining process that uses a laser to ablate holes or patterns into a metal to produce MMs [33, 34]. The patterns or holes can be of various complex shapes. The process is quick and precise; however, it is limited by the materials used because lasers cannot ablate high melting point materials. Further, MMs with 3D unit cells cannot be constructed using the aforementioned process. Also, 3D MMs with feature sizes greater than tens of millimeters can be manufactured using **casting methods** [35].

17.2.1.2.2 3D printing

3D printing is a fabrication technique in which the intended structure is developed layer by layer. Steps involved in 3D printing include creating the design to be printed on a computer, then a software slices the design into multiple layers, which are then deposited one by one. 3D printing is widely used for industrial applications [36], where the size of the components can range from micrometers to millimeters. There are more than 20 different 3D printing techniques that are recognized [43]. Complex designs with high precision and minimal waste can be attained by this technique [37]. Kronberger *et al* [38] developed an MS using a 3D printer that had a resolution of 0.1 mm. The MS worked in the microwave frequency range (below 100 GHz), and a simple array of circular holes of a few millimeters were 3D printed. The material used was a commercial brass and bronze-like powder for 3D printing applications, the exact composition of which was not reported. Specialized 3D printing methods can produce THz MMs of large areas quickly and economically. Tenggara *et al* [39] used **electrohydrodynamic jet printing** to fabricate a THz MM with feature sizes below 5 μm. SRR structures were constructed using commercial ink containing silver nanoparticles on silicon and flexible polyimide substrates. Structures in the millimeter range can also be fabricated by **fused deposition modeling**. Thermoplastics with metallic or ceramic powders can be mixed to create superior dielectric properties. The thermoplastics can be melted away by sintering to create high quality and high dielectric patterns [40]. Advances in the technique have enabled the resolution within a few tens of micrometers. **Inkjet printing** is a versatile 3D printing technique that uses multiple micron sized nozzles to deposit ink drops on a substrate. Metal nanoparticles along with a solvent can be used as the ink. After deposition, a heat treatment process is applied, evaporating the solvent and sintering the metal powder. Furthermore, when dielectric ink is used, it is generally cured using UV light because dielectrics are not able to tolerate the sintering temperature of metals [41]. Yoo *et al* [42] printed silver nanoparticles to create an MS on a paper substrate working in the microwave range. The silver structures had conductivity comparable to bulk iron. Walther *et al* [43] compared an MM fabricated by inkjet printing and lithography. The response achieved was similar; however, the inkjet printed MM had broader peaks which were deemed to be due to inhomogeneities in the structures.

The drawbacks of 3D printing are the requirement of support structures [44], anisotropy of material properties [45], and residual stresses [46]. Further, designs of very high complexity cannot be created, irregular cooling creates heterogeneous material properties and thermal stresses, and limited metals can be 3D printed.

17.2.1.2.3 Advanced methods

Direct laser writing is an advanced fabrication technique that can be used for constructing unit cells in 3D space to create a 3D MM. A connected helix structure is commonly used for the development of chiral structures. The helix arrangement can create magnetic dipoles due to the current induced in the helix. These types of structures are used for the fabrication of invisibility cloaks which are independent of polarization [47, 48]. Gansel *et al* [49] used a femtosecond laser that focused on the resist to modify the patterned region, which was carved out by chemical methods. Further, the selected metal was deposited in the carved out areas. The fabrication method in which the laser induces two photons of the same or different frequency on a material simultaneously is called **two-photon fabrication**. The laser is focused on a very small spot in the material in which chemical and/or physical change is induced. This technique was used for the fabrication of IR and THz range MMs [50, 51]. Ishikawa *et al* [52] focused a mode locked titanium sapphire laser on a silver ion solution via an objective lens. The solver ion complex was reduced to silver nanoparticles, which aggregated to form the required pyramidal structure. The crystal growth was suppressed by using a surfactant molecule n-decanoylsarcosine sodium (NDSS) [53]. The resolution achieved was less than 120 nm with reasonably high conductivity of silver. However, the two-photon fabrication is limited by the material selection and can be generally used for connected structures only. Also, the resolution limit of the technique is about 50 nm.

Optical lithography was used to create a chiral MM by Zhang *et al* [54] to act as a double negative material for circularly polarized light in the THz frequency range. The chiral effect was induced by the optical illumination of a 3D structure. The structure was a combination metal and silicon. Under illumination with energy more than the bandgap of silicon, the region containing silicon becomes conducting, which creates the necessary condition to achieve negative refractive index. The aforementioned structure had some regions where the resolution required was in a few hundreds of nanometers. Both connected and disconnected 3D MMs can be constructed by optical lithography.

17.2.2 Design

17.2.2.1 Negative permeability

An array of SRRs was one of the first structures which was used to create an AMM that had a negative μ [3]. The unit cell structure of an SRR (figure 17.1(b)) consists of an open ring generally fabricated out of a non-magnetic metal on a dielectric substrate. The size and periodicity are usually very small as compared to the active wavelength range. The effective μ of an SRR is [55]:

$$\mu_{\text{eff}}(\omega) = 1 + \frac{F\omega^2}{\omega_0^2 - \omega^2 - \mathrm{i}\gamma\omega} \tag{17.1}$$

where ω is the frequency, ω_0 represents resonant frequency, F takes into account the geometry of the structure, and γ is the damping factor. The negative real part of μ can be attained when the ratio of ω_0/γ is high enough. The SRR system can be envisioned as an inductor (L) and capacitor (C) circuit. ω_0 can be approximated as $\sqrt{1/LC}$, in which the gaps can be thought of as capacitive sources, and the split ring structures are depicted as inductance sources. The incident EM wave induces a current in the split ring, leading to the enhancement or decay of the field, depending upon the applied frequency. The SRRs can be of various shapes including, edge coupled, spiral, Hilbert, etc, to tune the resonant frequency and the operating range of the AMM [56]. The drawback of the SRR structure is that for MSs, the incident magnetic field of EM wave should be perpendicular to the surface to attain optimal magnetic coupling.

17.2.2.2 Negative permittivity

The study of artificial plasma (having negative ε) has been prevalent for a long time [57, 58], and one of the first artificial electric materials was presented by Pendry *et al* [4] in 1996, which used an array of metallic wires in a dielectric medium. To achieve a negative ε for MSs, similar to the structures for achieving negative μ, an electrically coupled resonator is used. The generally used structure of an electrically coupled resonator unit cell is shown in figure 17.1(c). As opposed to the SRR structure, the electrically coupled resonators display the optimal performance when the incident EM wave is incident normal to the surface, although the electric field has to be orthogonally polarized to the gap in the case of figure 17.1(c). Hence, the magnetic moment is nullified due to the structure, and the structure couples only with the electric field. The effective ε of the structure is [59, 60]:

$$\varepsilon_{\text{eff}}(\omega) = \varepsilon_\infty + \frac{F\omega^2}{\omega_0^2 - \omega^2 - \mathrm{i}\gamma\omega} \tag{17.2}$$

where ε_∞ represents the ε of the dielectric substrate on which the structures are fabricated, and F takes into account the geometry of the structure. Similar to the SRRs, ω_0 and γ are the resonant frequency and the resistive damping factor, respectively. When the ω_0/γ ratio is high enough, the real part of ε will be negative near the resonant frequency.

17.2.2.3 Metasurfaces

The MSs contain subwavelength periodic or non-periodic structures fabricated on a substrate that have thickness much lower than the targeted wavelength. However, the thickness is kept more than the skin depth of the material. Although the response of the MSs is complex due to the modification in both phase and amplitude of the incident wave, the equivalent circuit modeling approach can simplify the designing process. As previously explained, the LC circuits can be used to represent the unit cell of a structured periodic array. The structures which are generally metallic can be depicted to create inductive and resistive sources, while the gaps can be envisioned as capacitive sources. The MSs used for some common applications including high pass, low pass, band pass, and band stop filters are shown in figure 17.2. The unit cell of the array is represented as a circuit, which has a particular response. Altering the size and periodicity of the unit cell causes the alteration of the resonant frequency

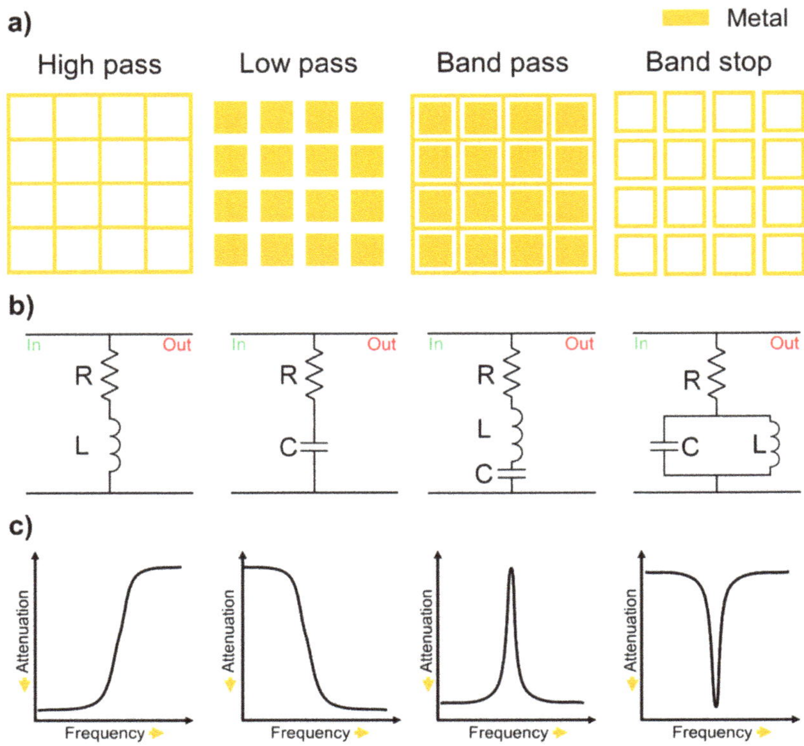

Figure 17.2. (a) High pass, low pass, band pass, and band pass MS filters along with the (b) equivalent circuits and (c) frequency response.

and the width of the peak/valley. Furthermore, creating more complex shapes can result in multimode resonance, which is desirable for specific applications. The characteristics required for a good MS are a broadband tunable range, high performance at various incidence angles, and polarization independent performance.

17.2.3 Characterization

The characterization of MMs is important to establish their performance and working range. Further, it also helps to improve the understanding of the physics of the interaction of the targeted EM waves with the MM structure. The induced fields due to the inductive and capacitive nature of the MM structure lead to modification of the reflected and/or transmitted wave. The characterization of optical frequency MMs is different from the THz and microwave frequencies because of the much higher working frequency range.

17.2.3.1 Optical and infrared

The measurements of the properties of optical and IR MMs is generally performed by techniques like angle-resolved spectroscopy measurements [61], ellipsometry [62], and femtosecond laser interferometry [63]. The main elements of a spectroscopic measurement include, a light source, polarizers, lenses, the MM sample, and detector. The instrumentation can vary depending upon the operating frequency range, optical or IR. The sample can be placed on a rotating stage, or the setup around the sample can be rotated to measure the angular sensitivity of the MM. Similarly, the incident polarization can be varied to measure the polarization sensitivity. The polarization and incidence angle response are critical to effectively gauge the performance of the MM. Ellipsometry measures the optical properties of a thin material. The basic working of ellipsometry is that an EM wave of known polarization is incident on the material, the reflected beam passes through a series of optical elements, which splits the reflected light to measure the variation in polarization. The amplitude ratio and phase difference of the reflected beam are measured. These measurements are used to calculate the optical constants of SRRs [64], fishnet MMs [65], and other types of MMs. Similar to the spectroscopic measurements, the instrumentation of the setup has to change depending upon the frequency of EM wave.

17.2.3.2 THz and microwave

The characterization of THz and microwave MMs is performed using two methods, resonant and non-resonant. Although the resonant method is very precise, its working frequency range is very narrow, and it usually works only for a single peak. The non-resonant method overcomes the aforementioned drawbacks; however, a compromise on the accuracy of the measurements has to be made. The characterization can be performed in reflection or transmission mode, depending upon the sample and the parameters that need to be measured. The impedance and the velocity specifications of the wave can help calculate the important material properties, including permittivity and permeability [66]. The two main techniques for non-resonant characterization method are waveguide measurement and free space measurement.

In **waveguide measurement**, the MM sample is placed in direct contact with waveguide adaptors in between two ports which are linked to a vector network analyzer (VNA) [67, 68]. Further, a waveguide transition is also present behind the sample. Due to the presence of the waveguide, the frequency range of this type of characterization is limited. The VNA is used to analyze both the transmitted and reflected waves. Prior to the measurement, the calibration reference plane is transformed to measure the scattering parameters at the edges of the sample. The main drawback of the waveguide measurement is that the sample has to be in direct contact with the waveguide. Therefore, after the measurement, the complex and often expensive MM sample may not be usable again.

The **free space measurement** requires no direct contact with the setup, it is more suitable for high loss materials, and it has a broadband working range compared to the waveguide measurement. The setup can be modified to work in transmitted and reflected modes. The transmitted mode requires a transmitting and a receiving antenna, but the measurement in the reflection mode can be accomplished by only one antenna for both transmitting and receiving the signal. The MM sample is located at the far field region of the antennas in a custom built sample holder. The measurements are made by aligning reference planes of the antennas at the front and back of the sample, respectively. To decrease the noise and interference in the measurements, often absorbing surfaces in the shape of pyramids are used. The free space measurement setup can be modified to improve the accuracy by placing the sample on a conducting plate between two waveguide port radiators, surrounded by absorbing surfaces [69]. The measurements are performed in empty space before including the sample to eliminate the effect of the surrounding. For more complex measurements involving different parameters, the number of transmitters and detectors, as well as the number of measurements have to be increased compared to the aforementioned examples [70].

The **errors** involved in the measurement include, systematic errors that are due to the multiple reflections at the antennas, the diffraction at the edges of the sample, impedance mismatch in the sample and the free space around it, cross talk, etc. The systematic errors can be eliminated by performing proper calibration. Moreover, the diffraction error can be reduced by using two similar plano convex lenses before and after the sample that focus the wave in the central region. The effect of multiple reflections can be suppressed by using a time domain grating with a free space calibration technique called TRL (thru, reflect, line) [71]. Also, errors can be caused when the reflected wave from the backside of the sample destructively interferes with the wave traveling in the forward direction. Other errors include drift errors, which are due to the variation in the atmospheric conditions, the main contributors to which are temperature and humidity. Also, random errors can be present, which can be mitigated by taking multiple measurements.

Following the measurements by the characterization setup, the parameters have to be extracted using algorithms such as the Nicolson–Ross–Weir algorithm [72] and the transmission line theory.

17.2.3.3 Simulation

The advancements of computation technology have made the theoretical characterization of MMs a simpler and economical task. The simulation software requires the input of the MM geometry into the software, which is then divided into small regions; this process is called meshing. Further, boundary conditions and material properties are applied to the system. Then, Maxwell's equations are solved by iterative or direct methods. The main simulation methods are finite difference time domain (FDTD) and finite element method (FEM), which are generally performed on commercial software COMSOL [73] and CST microwave studio [74]. Several output parameters can be extracted from the simulations, including the far field reflection, transmission, and absorption plots. Further, the near field electric and magnetic field distributions can also be visualized. Simulations can be used to characterize materials across the EM spectrum. Simulation is an effective tool to predict the response of the MM prior to fabrication, hence, a huge amount of capital and time can be saved.

17.3 Graphene

The previous chapters in the book have already gone through the remarkable and unique properties of graphene and the various methods to fabricate it. Briefly, in its purest form, graphene is a hexagonal structured 2D material consisting of carbon atoms. In the reciprocal space, monolayer graphene's valence and conduction bands are conical, and their tips overlap at the K point in the hexagonal lattice. Hence, the bandgap of monolayer graphene is zero, which is the major reason behind its extremely high electrical and thermal conductivity and its good response across the EM spectrum. However, the properties of graphene are also dependent on the number of layers and defects present, which are dictated by the fabrication method used. The complexity of fabrication limits the commercial application of graphene. However, graphene is widely used to improve the response of MMs, which have applications from the IR to the microwave frequency range. The bandgap of graphene can be manipulated to tune its response by two main methods, chemical doping and the application of electrical bias. The application and changing of electrical bias is much easier than altering the doping of graphene, which requires complicated fabrication steps. Therefore, most researchers tune the response of graphene via electrical bias by applying a gate voltage. Furthermore, another method of altering the response of graphene is by photodoping the graphene layer and the subsequent substrate by optical illumination. The electrical bias and photodoping is explored in greater detail in the next part of the section.

Since graphene-based MMs are still in the research and development stage, the fabrication is optimized for high purity rather than mass production. Hence, in most of the MMs, graphene is fabricated using chemical vapour deposition (CVD) because it offers low defects, high yield, and good accessibility in research labs worldwide. Furthermore, the CVD fabricated graphene can be easily transferred to a selected substrate with minimal defects. CVD can also be combined with conventional fabrication processes such as lithography.

17.3.1 Tunability via electrical bias

The property which entices researchers to develop graphene MMs or graphene-based MM is its broadband response and easy to apply high electrical tunability. Graphene's chemical potential (Fermi energy) is altered by applying a gate voltage V_g. The change in Fermi energy affects the conductivity, changing its response to the incoming EM waves. The intraband and interband transitions contribute to the conductivity of graphene. The conductivity of graphene is calculated using the Kubo formula [75]

$$\sigma_{\text{intra}} = -j\frac{e^2 k_B T}{\pi\hbar^2(\omega - j2\Gamma)}\left[\frac{\mu_C}{k_B T} + 2\ln\left(e^{-\frac{\mu_C}{k_B T}} + 1\right)\right] \quad (17.3)$$

$$\sigma_{\text{inter}} = -j\frac{e^2}{4\pi\hbar}\log\left[\frac{2\,|\,\mu_C\,| - (\omega - j2\Gamma)\hbar}{2\,|\,\mu_C\,| + (\omega - j2\Gamma)\hbar}\right] \quad (17.4)$$

where e, k_B, and $\hbar$ are the electron charge, Boltzmann constant, and reduced Planck's constant, respectively. Further, T, ω, Γ and μ_C are the working temperature, angular frequency, scattering rate, and chemical potential (Fermi energy), respectively. The main contributor to the conductivity in the THz and microwave frequency range at room temperature is the intraband term. The relation between the chemical potential and the applied gate voltage is given by the equation [76]:

$$\mu_c \approx \hbar v_F \sqrt{\frac{\pi \varepsilon_r \varepsilon_0 V_g}{e t_s}} \tag{17.5}$$

where v_F, t_s, ε_r, and ε_0 are the Fermi velocity ($\approx 10^6$ m s^{-1}), substrate thickness, permittivity of substrate, and vacuum permittivity, respectively. The above equations are valid for monolayer graphene because as the number of layers increase, the interaction between the EM waves plays a part in determining its overall response. The application of gate voltage to alter the response of graphene is seen multiple times in the next section, where applications of graphene-based MM are discussed. The application of gate voltage is generally accomplished using a field effect transistor (FET)-type fabrication. The voltage is applied between two contacts which are deposited on the graphene layer. Following the graphene layer, a semiconducting dielectric layer and a metallic grounding layer are present.

17.3.2 Tunability via photodoping

Photodoping is another route to attain the tunability of graphene that can be used as a standalone method or can be integrated with electrical biasing. The efficiency of tunability increases as a combination of electrical bias and photodoping is used [77, 78]. The usual method to achieve tunability via photodoping is optical illumination of the graphene layer deposited on a semiconducting substrate. The excess carriers generated in the semiconductors due to the optical energy received are injected into graphene. Hence, the response of the graphene layer to the incoming EM waves is altered. The transmission of the graphene is expressed as [79]:

$$t(\omega) = \frac{1}{1 + N Z_0 \sigma(\omega)/(1 + n_s)} \tag{17.6}$$

where Z_0 and n_s are the vacuum impedance and effective refractive index of the substrate, respectively. N and $\sigma(\omega)$, are number of layers of graphene and its conductivity, respectively.

Since tunability through the photodoping method is not highly prevalent in literature, the work of Li *et al* [78] is used as an example to explain its execution. A graphene-based modulator was developed that showed enhanced modulation (83%) on the simultaneous application of electrical bias and optical excitation. The study was performed in the THz range (0.4 THz to 2 THz). The p-type graphene monolayer was fabricated on a copper substrate via CVD. Two layers were transferred on an n-type silicon substrate of thickness 510 μm. Further, electrical contacts were deposited on graphene, and the silicon substrate was placed on a

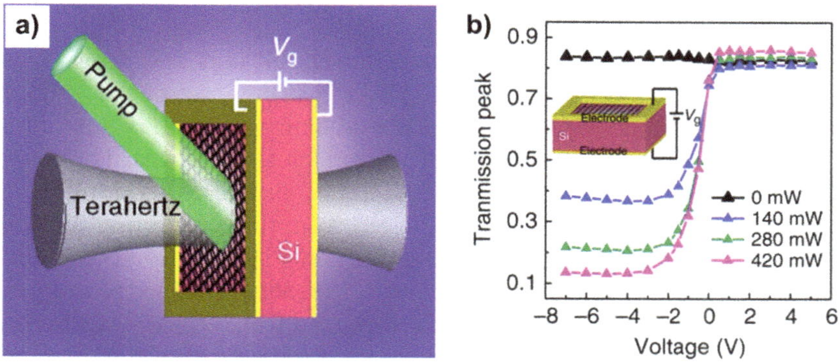

Figure 17.3. (a) Schematic representation of the graphene and silicon-based modulator. (b) Its transmission response on application of gate voltage with optical illumination. Reproduced [78], under a Creative Commons Attribution 4.0 International License, Copyright 2015, Nature Publishing Group.

grounding layer made out of a conducting material so that electrical bias could be applied. The optical excitation was executed by a continuous wave laser of wavelength 532 nm with power ranging from 140 mW to 420 mW, which was incident on the top graphene layer (figure 17.3(a)). Most of the incoming THz waves were transmitted on the application of positive bias and blocked when negative bias was applied. Upon illumination, a large number of carriers were generated in the silicon layer, and only a few carriers were generated in graphene due to its small optical absorption [80]. Therefore, the excess carriers diffused into the graphene layer that led to the creation of a depletion layer. However, the bandgap of graphene was still zero due to the recombination of the excess electrons; but its conductivity increased. When a positive bias was applied, the depletion layer thickness decreased, causing current flow and decreasing the electrons diffusing into graphene, therefore, increasing the transmission of the THz waves. Upon applying a negative bias, the depletion layer thickened and increased the electrons diffusing into graphene. Thus, its conductivity increased, causing an attenuation of the THz waves. It was observed that the modulation effect enhanced significantly when the electrical bias and optical illumination were applied simultaneously, which is shown in figure 17.3(b).

17.4 Application of graphene-based metamaterials

In the current chapter, the applications of graphene MMs are in the IR, THz, and microwave frequency range and are broadly divided into three categories, imaging, communication, and sensing. In some cases, the application may overlap into two or more categories. Before diving into the applications, it is important to establish the units used for various frequency/wavelength ranges. In the case of IR range, the wavelength is generally expressed in μm (1 μm = (speed of light × 10^6) Hz). For the THz range, the frequency is expressed in THz (1 THz = 10^{12} Hz). For microwave range, the frequency is expressed in GHz (1 GHz = 10^9 Hz).

17.4.1 Imaging

The lack of response of conventional materials in the IR and THz range emphasizes the importance of the development of MMs. Further, graphene itself has an excellent response in this frequency range. IR imaging is vital as a non-destructive evaluation technique for industrial as well as research applications. The study of the thermal response of electronics is of immense importance. Furthermore, IR imaging is used for stealth and security applications in scenarios where optical techniques cannot be used. THz imaging is gaining importance in security applications, e.g., screening of luggage and passengers at airports. Further, THz spectroscopy can help to achieve a greater understanding of the physics of materials. THz imaging has created a lot of buzz in research laboratories. Graphene-based MMs are being developed to be used as detectors, modulators, lenses, etc.

17.4.1.1 Absorbers

Absorbers can be used as detectors for imaging systems, radar absorbing surfaces, surfaces to absorb unwanted EM energy, anti-reflective surfaces, solar absorber, sensors, etc. There are several studies on the graphene-based MM as perfect absorbers. Some of them are presented in this section and are classified according to the frequency range of their application.

17.4.1.1.1 IR range applications

Safaei *et al* [82] demonstrated a mid-IR absorber (8–12 μm) with a broadband absorbance of about 2 μm. The wavelength range selected for the study is of utmost importance for thermal imaging. When the incidence wavelength was 10 μm, an absorbance of 60% was calculated theoretically, and 45% was displayed experimentally. A patterned (nanohole array with hole diameter 330 nm) graphene monolayer was grown using CVD and nanoimprinted, on which an ion gel layer was deposited. Another dielectric layer was placed below the graphene layer, followed by a gold reflector layer. The thickness of the dielectric was selected such that a constructive interference is created between the scattered and the incident electric field; hence, the system acted as an optical cavity. The strong absorption was attributed to the coupling of the surface plasmons wave with the incoming EM wave. Presence of the nanohole array leads to activation of both localized surface plasmon resonance and surface plasmon polaritons.

Lin *et al* [81] developed a tunable solar absorber with excellent thermal conductivity and an overall efficiency of 68.9%. The graphene-based MM was selected to overcome the problem of limited materials with high thermal conductivity and low thermal loss. Graphene structure consisted of square holes of size 0.59 μm and period 0.8 μm, and thickness 30 nm. The graphene structure was grown on a metallic substrate, and the overall depth of the holes was 1 μm. The basic steps of fabrication and the SEM image of the MM are shown in figures 17.4(a)–(d). The metal used was copper, and it was structured using laser printing. The copper structure was coated with graphene oxide that was converted to a 30 nm thick graphene layer upon illumination using a mercury lamp. The absorber showed high

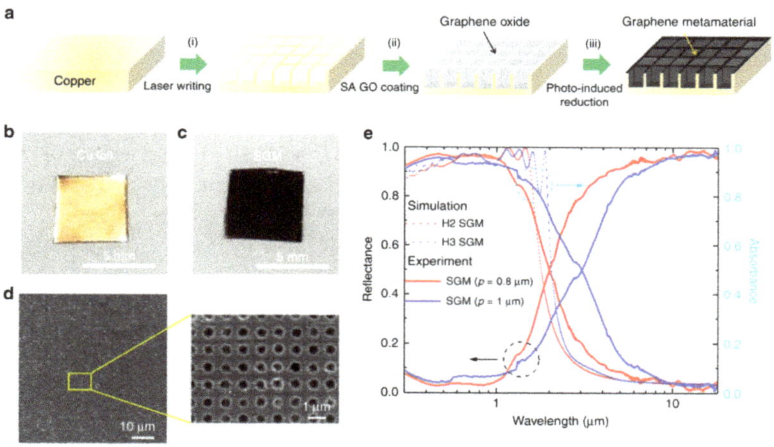

Figure 17.4. (a) Steps required for the fabrication of the graphene-based MM absorber. Images of (b) copper foil and (c) graphene-based MM. (d) SEM image of the surface of the MM. (e) The simulation and experimental reflectance and absorbance plots. For the experimental plot, the hole width (w) is kept constant at 0.57 μm, and p is the period of the array. For the simulation plot, H2 SGM represents the MM having w 0.59 μm, p 0.8 μm and depth 1 μm and H3 SGM represents the MM having w 0.79 μm, p 1 μm, and depth 2 μm. Reproduced [81] under a Creative Commons Attribution 4.0 International License, Copyright 2020, Nature Publishing Group.

absorbance from 0.28 μm to 1.6 μm, while almost zero transmission was observed in that range (figure 17.4(e)). It was suggested that the high absorbance was caused by the combined effect of surface plasmon resonance and the ability of holes to trap the light. Beyond 1.6 μm, the light wavelength was too large to reach in the holes, thus increasing the reflectance drastically beyond 1.6 μm. It was revealed through simulations that as the depth of the hole was increased beyond 1 μm, the effect on absorbance was minimal. To understand the practical working of the absorber, its performance was observed via simulation under varying angle of incidences and polarizations. It was observed that the efficiency of the absorber was almost constant for transverse electric (TE) polarization for incident angles 0°–60°. However, the efficiency decreased beyond 20° for transverse magnetic (TM) polarization. It was deduced that decreased electric field intensity as the field became orthogonal to the walls of the metallic structure was the main reason for the aforementioned phenomenon. Simulations were performed using the FDTD approach, the experimental reflectance plot was in close agreement with the simulations. The heat captured by the structured layer was effectively transported to the metal layer below, confirming the practicality of the absorber.

Mostaan *et al* [83] simulated a broadband MS absorber consisting of three-layered graphene cut in the form of an array whose unit cell was in the shape of a split rectangle. The graphene layers were on top of a dielectric substrate; gold contacts and grounding layer facilitated the application of electrical biasing. The three layers of graphene were chosen instead of one to increase the absorption of the MS by increasing the EM interaction. However, a further increase in number of layers led to splitting of the broadband absorbance into multiple peaks. It was reported that using silver, aluminum, or platinum instead of gold did not make any significant changes to the performance of the MS. The performance of a single graphene sheet with split rectangular shaped unit cells and with no splitting in the rectangular structure was compared; the sheet with split ring structures showed a much broader absorbance (30 μm with absorbance greater than 80%) in the far-IR region. The Fermi energy of 0.6 eV displayed the best absorbance. Increasing the Fermi energy led to splitting the absorbance band, and decreasing the Fermi energy caused a significant decrease in absorbance.

17.4.1.1.2 THz range applications

One of the mechanisms for perfect absorbance of MSs is their subwavelength structures acting as a Fabry–Pérot cavity, which allows only EM waves of certain frequencies to pass through them [84]. Hence, the EM wave should be in resonance with the structure. The resonance is achieved by multiple reflections in the interior of the cavity. Therefore, altering the structure can lead to responses ranging from broadband absorption to single or multiple sharp transmission/reflection peaks. Other methods such as electromagnetically induced transparency (EIT) are also responsible for the suppressed transmission/reflection, usually caused by the destructive interference of the bright mode and dark mode of structures. The Fabry–Pérot cavity effect was deemed the cause of dual band absorbance of a graphene-based MM developed by Qi *et al* [85]. The MM consisted of a dielectric layer with cross-shaped grooves forming an array that was placed on top of a graphene layer. A dielectric layer and a grounding plate followed them to apply the bias voltage. The absorber displayed two broadband peaks in the THz frequency range with bandwidths 97.8% and 31%. The absorber was polarization mode and angle (0°–80°) insensitive due to the symmetry of the structure. The absorbance of the MM increased when Fermi energy was increased from 0 eV to 1 eV. Also, a slight shift of the peak towards higher frequency was observed. Similarly, Huang *et al* [86] designed a graphene MM based absorber whose structure acted as a Fabry–Pérot cavity. The unit cell consisted of a graphene layer with cross oval shaped holes placed on a dielectric and a grounding layer. A broadband absorbance (more than 80%) was achieved from 1.2 THz to 1.8 THz. The cavity effect of the structure combined with the surface plasmon excitation of the graphene layer which resulted in the large absorbance of the MM. Also, the absorber was polarization independent. The Fermi energy was changed from 0.3 eV to 0.7 eV; as expected, the absorbance shifted to higher frequencies. Furthermore, the absorbance decreased because the resonance frequencies deviated away from the optimal surface plasmon resonance frequency.

Huang *et al* [87] designed a wide-angle tunable MM absorber for THz frequencies. The MM consisted of square patch graphene with metallic wires placed on a dielectric layer, followed by a grounding metal. The absorbance plot displayed two peaks with over 90% absorbance; this was attributed to the activation of the second order surface plasmon polaritons. The peaks shifted to a higher frequency as the Fermi energy was increased. The low frequency peak shifted from 0.85 THz to

1.01 THz, and high frequency peak shifted from 2.84 THz to 3.37 THz. Due to the asymmetric structure, the polarization modes affected the response of the MM. However, the TE mode showed good performance when the polarization angle was varied from 0° to 45°, and the angle of incidence was below 50°.

17.4.1.1.3 Microwave range applications

Huang *et al* [88] demonstrated a wideband absorber for microwave frequencies, 8 GHz to 18 GHz, fabricated on a flexible substrate. Over 90% absorbance was observed from 10.4 GHz to 19.7 GHz when the overall thickness of the prototype was 2 mm (figure 17.5(d)). An array with a complex pattern was printed using ink made of graphene nanoflakes on a flexible silicone substrate and a grounding metal (figures 17.5(a)–(c)). The simulation and experimental reflection coefficient plots were in close agreement. A VNA along with two antennas were used to measure the free space response of a flat absorber sheet and an absorber sheet bent over a cylinder. Although the performance of the bent sheet decreased, it still displayed a wideband absorbance. The device was intended to be used as a radar absorbing surface.

Zhang *et al* [89] developed an optically transparent and flexible MS. A broadband absorption (more than 90%) was observed in the microwave range of 7–18 GHz. The structure consisted of an ionic liquid electrolyte between two graphene layers with periodic square holes, forming the first layer, followed by a polyethylene glycol terephthalate (PET) substrate of thickness 3 mm and a thin grounding layer of indium tin oxide (ITO). The CVD fabricated graphene contained n-type doping and was transferred to the PET substrate using etching and lamination. Subsequently, the graphene layer was patterned using laser beam cutting. The graphene electrolyte graphene layers acted as a capacitor; as biasing was applied, the electrolyte would be polarized. The high mobility of graphene creates the response in the microwave range. The reflection of the electrolyte layer was minimal due to the low mobility of the ions. Two antennas and a network analyzer characterized the response of the

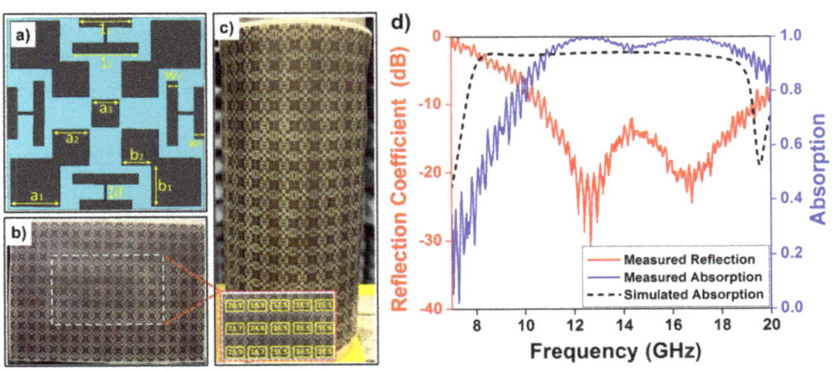

Figure 17.5. (a) The pattern of the unit cell of the graphene-based absorber. (b) Fabricated sheet. (c) The absorber sheet wrapped around a cylinder. (d) The reflection coefficient and absorption plots and the comparison of experimental and simulated absorption. Reproduced [88], under a Creative Commons Attribution 4.0 International License, Copyright 2016, Nature Publishing Group.

absorber. As the bias voltage was increased, the sheet resistance decreased. Further, the response changed from a broadband to a dual band when the voltage was varied from 0 V to 6 V. Simulation results were used to explain the phenomenon. It was observed that two resonant modes were present; in the low frequency mode, the electric field localized at the edge graphene layer and surface currents flow in the opposite direction compared to the ITO layer. While in the high frequency mode, the electric field was localized near the square pattern edges, but the surface current flowed in the same direction as the ITO layer. Further, the energy was dissipated due to the ohmic losses in graphene. The absorber could be used as a radar absorbing surface where transparency is required, for example, windows of cockpits.

Zhang *et al* [90] fabricated a tunable broadband absorber that worked in the microwave frequency range using a design consisting of an electrolyte layer (50 μm thick) sandwiched between two monolayer graphene layers (fabricated using CVD). The electrolyte used was a diaphragm paper soaked with an ionic liquid, and the whole structure was covered with polyvinyl chloride. The graphene electrolyte structure acted as a capacitor. It was calculated that the change of Fermi energy from 0.2 eV to 1.0 eV caused absorption decrease from 100% to 60%. Further, the absorption was deemed frequency independent, which is not observed in the optical or THz range. The absorber was also polarization independent and showed good performance for incidence angles less than 80°. Experimentally, the bias voltage applied was from 0 V to 4 V, which varied the Fermi energy from 0 eV to 0.17 eV. The absorption was measured by a network analyzer setup in a waveguide measurement method. The absorptivity increased from 50% to 100% as the gate voltage was increased from 1 V to 3 V.

17.4.1.2 Lenses

MMs are being progressively used for the manipulation of EM waves beyond the optical frequency range. The size and weight of the lenses can be reduced by the use of MMs due to the presence of subwavelength structures. Li *et al* [91] used a graphene-based lens to focus EM waves in the IR frequency range. The graphene layer was placed on a dielectric and a silicon substrate. The substrate thickness was decreased as the distance from the center increased, such that a cavity was formed at the center. The Fermi energy was selected to be greater than 0.175 eV because the losses in the graphene layer were low beyond this range.

Graphene can be used to make a lens that has a tunable working frequency and/or focal length. Zainud-Deen *et al* [92] designed a tunable THz lens using a graphene-based MM, whose structure consisted of a graphene array layer cut in an SRR-type unit cell situated on top of a SiO_2 substrate and another graphene layer below it. The substrate also contained air gaps of thickness 1 μm. The lens was simulated from 0.74 THz to 0.94 THz, the MM was able to focus EM waves into a directed beam. A maximum gain of 19 dB was achieved at Fermi energy 0.5 eV. The frequency range could be tuned by varying the Fermi energy. Liu *et al* [93] developed a THz lens with tunable focal length using gold structures on which a graphene layer was deposited via CVD. The gold structures were deposited on a silicon substrate and were non-periodic; each structure had a different length and rotation. The structures were designed such that the incoming waves focus on the focal point, and upon the variation of the Fermi energy, the structures should still be able to focus the wave (although at a different point). The focal length of the lens was increased from 10.46 mm to 12.24 mm as bias voltage was increased from 0 V to 2 V. Hamzavi-Zarghani *et al* [94] simulated a graphene-based lens to focus a circularly polarized THz wave whose working range was electrically tunable. The lens could both converge and diverge an incoming EM wave, depending upon the arrangement of the unit cells. The unit cell comprised a dielectric substrate sandwiched between two graphene SRRs whose gap was oriented at the same place. The unit cells were placed in a row, and each unit cell was phase shifted at a particular angle to focus the EM wave. At the incident frequency of 21 THz, the focal length obtained was 28 μm. For a particular arrangement, if a right-handed wave converged, then a left-handed wave would be diverged. Yao *et al* [95] developed a THz lens with both tunable working frequency and tunable focal length. The MS consisted of graphene nanoribbons arranged in a 1D array covered by ion gel. Further, they were placed on a dielectric substrate and a grounding layer. The structure acted as a Fabry–Pérot cavity which increased the interaction of the incoming EM waves with the graphene structure. The plasmonic modes of graphene were activated upon strong absorption, and the modes were dictated by the Fermi energy. Hence, the response of the lens was tunable. Due to the geometry of the structure, only odd modes were excited because the even modes interfered destructively. The working frequency range of the lens was from 4 THz to 6 THz. The lens's focal length could be changed from 50 μm to 150 μm by tuning graphene's Fermi energy.

The lenses using conventional materials are restricted in their capability due to the necessity of working in the diffraction limitation. Hyperlenses transform the evanescent waves into propagating waves to go beyond this limit. The most practical use of hyperlenses is in the THz and microwave range by the use of MMs, which can offer unnatural abilities like negative refractive index. However, in optical systems hyperlenses are not very common due to high losses and short resonance frequency range. Andryieuski *et al* [96] developed a graphene-based hyperlens which could resolve two sources separated by one fifth of the operating wavelength. The lens consisted of a TOPAS substrate embedded with graphene strips with features in tens of nanometers and periods in hundreds of nanometers. The simulation without the graphene strips could not resolve the two sources. The hyperlens was studied with simulations using CST Microwave Studio and COMSOL; the observations were similar in both platforms. The simulation of the hyperlens was performed around 6 THz. The hyperlens has applications in THz imaging and power concentration. It was reported that fabrication using metallic strips instead of graphene was not practically possible due to the extremely thin width required.

17.4.1.3 Modulators

Graphene-based modulators are being explored due to their high switching speeds because the modulators using conventional materials have low speeds beyond the optical region. Graphene-based modulators can be used to develop high frame rate video systems.

17.4.1.3.1 IR range applications

Fan *et al* [97] demonstrated a high-speed modulator for IR waves in free space. A patterned CVD grown monolayer graphene layer was fabricated using electron beam lithography on an aluminum oxide substrate, which was grounded by a metallic film. The experimental and simulation absorption plots were in close agreement. As the gate voltage was varied from 0 V to 100 V, the absorption peak shifted from 7.45 μm to 7.95 μm, however, absorption of greater than 90% was maintained. The change in Fermi energy caused a change in the working frequency and the focal length. At a gate voltage of 100 V, the modulation achieved was approximately 85% in the 7.5 μm frequency range. Furthermore, the modulation speed achieved was about 2.6 GHz. The modulator protype of area 100 × 120 μm^2 was practically tested using a setup comprising of a blackbody radiation source which passed through a chopper and onto the modulator that reflected the signal to an IR camera. When the graphene modulator was not used, a 5 kHz signal could not be captured by the IR camera because the signal was much faster than the camera's frame rate. However, the image of the chopper blade could be captured upon the introduction of the graphene modulator. Painelli *et al* [98] simulated a graphene-based MM modulator working in the mid-IR frequency range. Alternating layers of monolayer graphene and SiO_2 were used. As the number of graphene monolayers was increased from one to six, the working wavelength of the modulator blue shifted by 3.6 μm at Fermi energy of 0.8 eV. The six monolayer geometry had a band pass type reflectance plot with the central wavelength of about 3 μm. The working frequency was red shifted by 2.3 μm as the Fermi energy decreased from 0.8 eV to 0.2 eV. The modulator showed minimal variation in performance upon incident angle variation in the TM polarization mode, while the TE mode displayed dependence on the angle of incidence.

17.4.1.3.2 THz range applications

Zhang *et al* [99] designed a modulator that could generate linearly polarized waves whose azimuth angle could be selected from 0° to 90° by altering the Fermi energy. Further, the device also worked as a perfect absorber when the geometry of the unit cell was changed. First, considering the absorber, the unit cell consisted of cross-shaped gold resonator placed on top of two layers of cross-shaped graphene separated by a thin SiO_2 layer. Followed by a dielectric and a grounding gold layer. The dimensions of the structure were optimized by simulations; the length and width of the gold cross were 92 μm and 15 μm, respectively. Graphene cross had length equal to the period of the unit cell (120 μm) and the width was 5 μm. The electrical biasing (0.0 eV to 0.5 eV) could tune the working frequency of the modulator by 15%. Absorbance greater than 95% was attained in the THz frequency range (the simulation study was performed from 0.8 THz to 1.3 THz). Further, the absorber was polarization independent. Next, the cross-shaped geometry of the gold structures in the unit cell were changed to cut wire geometry, followed by a cut wire shaped graphene layer, thin dielectric layer, cross-shaped graphene layer, thin dielectric layer, cut wire graphene layer orthogonal to the first cut wire graphene layer, cut wire gold layer orthogonal to the first gold layer, dielectric layer and finally

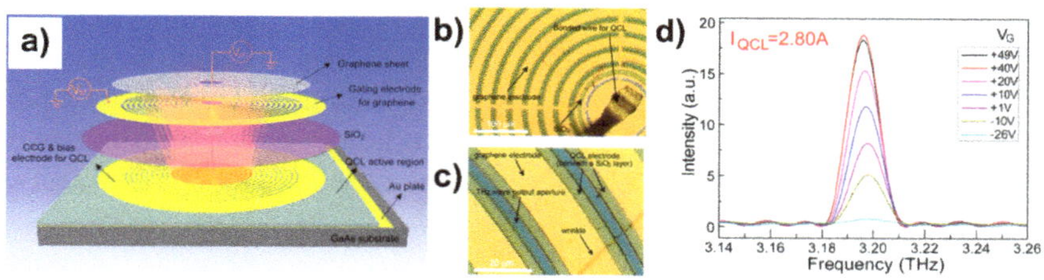

Figure 17.6. (a) The schematic representation of the circular concentric grating laser integrated modulator; the electrical pumping of the quantum cascade region takes place in the orange area. (b) Optical image and (c) zoomed in image of the area near the center of the laser integrated modulator. (d) Modulation of the THz wave by the laser integrated modulator with the variation in gate voltage V_G. Reproduced [100] under Creative Commons Attribution 4.0 International License, Copyright 2015, ACS Publications.

the grounding gold layer. The above alterations were performed to convert the absorber into a polarization modulator working in the reflection mode. The working of the polarization modulator was confirmed by performing a simulation in which the azimuth angle of the 0.87 THz wave was tuned from 0° to 90° by varying the electrical bias of the x- and y-axis graphene cut wire layers. Hence, the reflected wave could be linearly polarized in the x- or y-axis by adjusting the electrical bias. Li *et al* [77] developed a highly efficient modulator which used a combination of electrical bias and optical photodoping. Even at very low voltage ($\approx$1 V), a modulation depth of 60% was achieved. An SRR structure of aluminum was fabricated on a silicon substrate via lithography. A p-type graphene monolayer grown by CVD was transferred to the SRR structure. To apply the electrical bias, contacts and grounding layer were also incorporated. The MS was optically illuminated by a continuous wave laser which had a wavelength of 532 nm and power used was between 140 mW and 420 mW. The modulation depth in the THz frequency range was increased when gate voltage along with laser illumination was applied. The mechanism for the enhanced modulation phenomenon was similar to the one explained in section 17.3.2. Liang *et al* [100] developed a THz quantum cascade laser integrated modulator with a modulation depth of 100%. The structure of the MM consisting of graphene and circular concentric grating modulator is shown in figures 17.6(a)–(c). The thin layers of the structures helped to decrease the parasitic capacitance and overall resistance. The grating facilitates the laser beam to pass vertically through the graphene layer, which was electrically tunable. The electric field was concentrated in graphene near the output aperture. The measurements were performed when the laser operated at around 3.2 THz in the pulsed mode (pulse width 500 ns) and repetition rate 10 kHz. The intensity increased considerably as gate voltage was varied from −26 V to 49 V.

17.4.2 Communication

THz communication is a rapidly emerging field because THz waves are more directional than microwaves and are not easily scattered in the atmosphere, as in the case of IR waves. Further, MMs can be used to develop smaller and more efficient antennas. EIT-based MMs are being explored to develop slow light devices, which are extremely useful in reducing the noise and energy consumption. EIT-based devices can also be used for stealth applications. Cloaking is one of the applications of MMs, which has gained a lot of attention from the research communities over the years. Another interesting application of graphene-based MMs is holography, which can be used for fast and secure communications.

17.4.2.1 Antenna

Amanatiadis *et al* [101] designed an antenna consisting of a dielectric substrate and silicon layer with a graphene resonator. It was deduced that at the working frequency of 1.67 THz, the antenna length and efficiency could be optimized by an epsilon near zero material. However, natural materials do not display epsilon near zero behavior, and hence, a graphene-based MM layer was chosen to replace the dielectric layer. The MM layer consisted of a complex structure made out of graphene as the unit cell, which was placed vertically in the form of an array in a SiO_2 matrix. The structure's resonant frequency was 1.67 THz, and the radiation efficiency was improved by four times compared to the antenna without the MM layer. Radwan *et al* [102] simulated a patch antenna working in the THz frequency range, which used a graphene SRR array. The SRR was placed in a SiO_2 layer sandwiched between silicon and gallium arsenide (GaAs) layers. Further, a thin SiO_2 layer was situated on top of the silicon layer, and a tunable graphene radiating patch was placed on it. The simulation results confirmed that the performance of the antenna improved as the SRR structure was integrated into the antenna. When the Fermi energy of the graphene radiating patch was varied from 0.1 eV to 0.5 eV, the bandwidth of the antenna increased. When the Fermi energy of the radiating patch was kept constant at 0.2 eV, and the Fermi energy of the SRR layer was changed from 0.0 eV to 0.5 eV, the bandwidth increased from 0.08 THz to 0.5 THz, which improved the performance by 25% and 44% as compared to the antenna without SRR structure. The simulation was performed from 0.5 THz to 3.5 THz.

Degl'Innocenti *et al* [103] developed an antenna to modulate the amplitude of a quantum cascade laser that worked in THz frequency. Presence of subwavelength structures increased interaction of light with the structures of the antenna. The graphene monolayer was fabricated on copper using the CVD method. The graphene layer was moved to a silicon/SiO_2 substrate by etching and was patterned by optical lithography. Further, an SRR shaped gold array was created using electron beam lithography and evaporation techniques. The experimental reflectivity was measured to study the effect of gate voltage. As the voltage was varied from 0 V to 120 V, the reflectivity decreased. Plasmonic excitation of the metallic antennas coupled with the response of the graphene layer. Further, to test the speed of the modulator, a quantum cascade laser operating at 2.05 THz was used. The frequency selected was near the plasmonic response for the antenna which had an arm length of 26 μm. For repetition rate of 100 kHz, the modulation depth for the antenna was first increased, and after around 90 V, it decreased. The maximum modulation depth achieved was about 8.6%. Furthermore, the modulation depth variation at different speeds was measured. The modulation depth remained almost constant until about 5.5 MHz, after which it decreased.

Poorgholam-Khanjari [104] developed a modified Vivaldi antenna working from 0.5 THz to 2 THz. The Vivaldi antenna was accompanied by two parasitic elliptical structures pointing out from both the sides of the antenna to improve the gain. The structure was placed on a quartz substrate with an area of 500 × 400 μm^2. To enhance the performance and directivity, a single graphene layer was proposed to be placed in front of the antenna. High gain and directivity are essential for medical applications. Although the graphene layer improved the performance and added reconfigurability to the system, a large decrease in gain was observed. To overcome the aforementioned limitation, four layers of alternating graphene and silicon were placed in front of the antenna which acted as MM lens. Further, the Fermi energy was optimized to attain a return loss under −10 dB, it was reported that 0.7 eV was the optimal Fermi energy. The electric field distribution for Fermi energy 0.1 eV was studied. As the frequency was increased, the gain increased, and the bandwidth was reduced. The MM based structure was considered to detect tissues infected with cancer. The two antenna systems were placed on each side of the tissue under testing. The detection was based on the refractive index of the variation of the cancerous tissue as compared to healthy skin tissue. The simulation was performed for three types of samples, skin tissue, skin tissue along with cancer tissue, and cancer tissue. The reflection response contained multiple modes. However, at the first resonance, which was around 1.1 THz, the reflection of skin tissue, cancer tissue, and skin tissue were −58 dB, −45 dB, and −28 dB, respectively. Hence, the calculations could be used to detect cancerous tissues.

17.4.2.2 Transparency

Farhat *et al* [105] used multilayer graphene layers to cloak a dielectric sphere in the far-IR frequency range. The cloaking frequency could be altered by tuning the bias voltage of the layers. Initially, cloaking performance of single-layer graphene was studied. The graphene layer was able to significantly suppress the scattering cross-section of the sphere, which was visualized using the electric field distribution. The equation to minimize the scattering cross- section resulted in two solutions, one with negative permittivity and the other with positive but near zero permittivity. The negative permittivity condition resulted in better cloaking performance, although the near zero permittivity condition had greater tunability. Next, the number of graphene layers around the sphere were increased, and each layer could be electrically biased separately. The multilayer graphene cloak displayed a multiband response. The number of bands was equal to the number of graphene layers surrounding the object. Study with the single-layer graphene involved suppressing the frequency which was able to excite the dipole mode, higher frequencies will be able to excite multipoles even in simple geometries. Hence, the ability to tune multiple bands is essential for future applications. The proposed cloaking can be experimentally demonstrated by fabricating cylindrical graphene sheets [106].

Li *et al* [107] developed a modulator based on EIT that was tunable via photodoping. The unit cell of the array comprised cut wire and two square

SRR-type structures. An infrared laser was used for the optical illumination, the modulation achieved at 0.81 THz was 81.68% when 1000 mW power was used. The photodoping method was also used by Deng *et al* [108] to achieve tunable EIT in a graphene-based MM in the THz frequency (1–1.5 THz). Unit cell of the array consisted of unequal double bar geometry fabricated out of gold on a GaAs substrate, and graphene monolayer grown using CVD was added on top of the structured gold array using wet transfer technique. The optical illumination source used had a wavelength of 800 nm, pulse width 100 fs, and repetition rate 1 kHz, and was incident normally on the MM. Along with the effect of the gold array, the response to the optical illumination was dependent on the interaction of the band structures of graphene and the GaAs substrate. The two asymmetric bars led to two dipole resonances and created an EIT type window between the two resonances caused by the destructive interference of the responses in the near field. The width of both the bars was 6 μm, and the lengths were 36 μm and 46 μm. The EIT effect was also observed without the graphene layer. The distance between the bars also played a vital role in determining the response when a graphene layer was included, increasing the distance led to a red shift. Furthermore, the effect of the pump fluence on the graphene-based MM was studied. It was observed that as the pump fluence was increased, the EIT effect was suppressed and a major decrease in transmission was reported. The aforementioned effect was due to the absorption of the THz waves by the increased carriers generated due to optical illumination and suppression of the interference of the dipole resonances. The suppression was also observed in the electric field distribution.

Kindness *et al* [109] developed a modulator based on the EIT phenomenon. The unit cell of the array comprised of two C-shaped metal resonators along with graphene, as shown in figure 17.7(a). The substrate was a p-type silicon with a 300 nm SiO_2 layer, and was placed on a grounding layer. Electron beam lithography and thermal evaporation were employed to deposit the metallic array on top of the substrate. Graphene fabricated via CVD was transferred to the metal layer, and electron beam lithography and etching were used to pattern it. Initially, only the metallic components were considered; the larger resonator interacts much more strongly with the incoming THz wave because the electric field component has much more length to interact with, as compared with the smaller resonator. Hence, the smaller resonator is represented as the dark mode, and the larger one is considered the bright mode (figures 17.7(b) and (c)). Further, the capacitive coupling causes the individual plasmonic modes to be converted into two hybridized modes, one at lower frequency (bright mode at 1.49 THz) and the other at higher frequency (dark mode at 1.74 THz). Upon introduction of graphene with DC conductivity of 0.3 mS, the bright and dark modes in the resonators were still intact, however, the direction of current in the smaller resonator reversed at higher frequency (figures 17.7(d) and (e)). Hence, the graphene losses were minimal at low conductivity. But when the conductivity was 1.6 mS, the dark mode was suppressed due to the presence of graphene, as shown in figure 17.7(f). The variation of the transmission spectrum to changing conductivity is presented in figure 17.7(g).

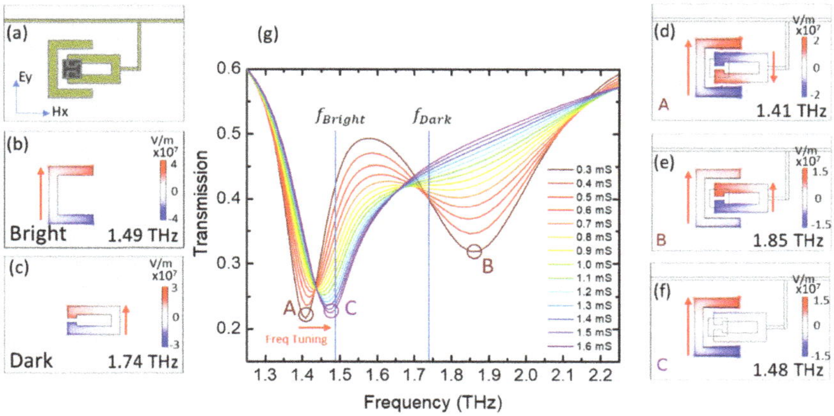

Figure 17.7. Simulation results for the modulator. (a) Schematic illustration of the unit cell, and the electric and magnetic field directions. The electric field in the *z*-direction for the individual resonators (without coupling) in the (b) bright and (c) dark modes. The red arrow represents the current direction in the resonator. The electric field in *z*-direction when the conductivity of graphene was (d and e) 0.3 mS and (f) 1.6 mS. (g) Transmission plot of the simulation for the modulators for various graphene conductivities. Reproduced with permission [109], John Wiley & Sons, copyright 2018, WILEY-VCH Verlag GmbH & Co. KGaA, Weinheim.

17.4.2.3 Encryption

Holography is a method by which the constituents of an image can be stored and then reproduced using the phase and intensity information [110]. Holography can be used to securely send information in the form of data, images, or videos [111]. Momeni *et al* [112] designed an MS using an array that was able to send encrypted THz beams through free space. The MS was reprogrammable due to the presence of graphene. The unit cell was constructed out of a Jerusalem cross-shaped graphene layer on a multilayer dielectric and a metallic grounding layer. The application of gate voltage allowed the unit cells to turn 'on' or 'off' the reflected THz wave. Therefore, the MS could be reprogrammed to conveniently change the security code. The 2D inverse Fourier transform was used to study the effect of the unit cells on the overall response of the MS. The MS was able to convert the original data wave into a parity beam that added extra data to mask the actual data which needed protection. Further, the parity beam received by the detector required decoding to reveal the actual information. The data (in the form of 1 and 0) was transferred to the MS and replicated in each row of the coding matrix. The data was phase encoded using memory registers to enable the MS to transmit parity beams instead of beams containing the actual data. Rajabalipanah *et al* [113] used an MS constructed out of graphene nanoribbons to securely send messages. The holography-based method used the phase shift of the wave when an x polarized, or y polarized struck the surface of the MS. The phase shift was due to the delay by the surface plasmons excited on graphene. When the Fermi energy was increased, the resonance frequency decreased. Hence, the reflection phases of the polarization were tuned by altering the Fermi energy of the x- and y-direction individually. A 45° phase shift was produced for the polarization in the x- and y-direction to create a 3 bit hologram.

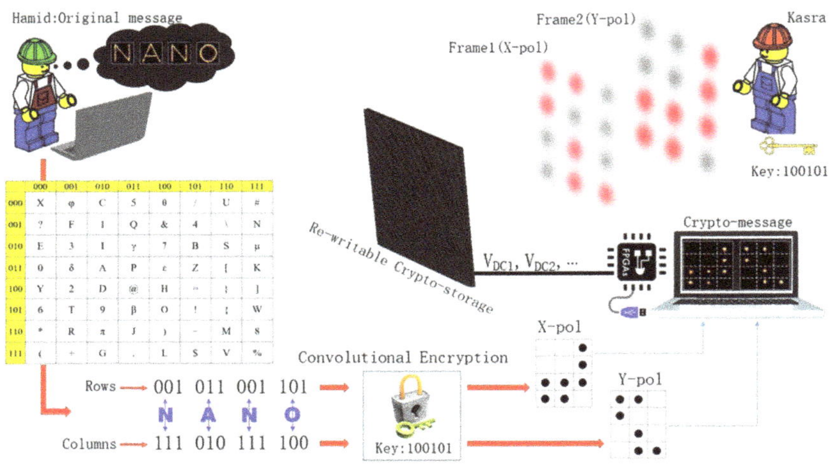

Figure 17.8. Working of the holography-based encryption system using graphene MS. As an example, a four letter word was coded into a table that had a unique key. An additional security layer was added by convolutionally encrypting the data. Due to the control over the two polarizations, two words simultaneously could be sent securely. Reproduced [113] under Creative Commons Attribution 4.0 International License, Copyright 2020, De Gruyter.

This technique decreased the interaction of the neighboring unit cells and hence, improved the performance. The electrical bias in each cell was individually controlled to feed the 3 bit data to the MS. The location of the unit cell was mapped by using two codes for the *x*- and *y*-axis. The two polarizations were able to create different images individually by employing the weighted Gerchberg–Saxton algorithm [114]. Each unit cell could be turned 'on' or 'off' by appropriately applying electrical bias; hence, the MS was reprogrammable. The working of the device was explained by sending a four-letter word securely. The location of the letters was mapped in a jumbled 8 × 8 table, as shown in figure 17.8. Therefore, the word could only be correctly identified if the user had the key to the table. Furthermore, another security layer was added to the device by encrypting the word using a convolution algorithm which needed generator polynomials to decode the data. The ability to control two polarizations individually allowed two messages to be sent simultaneously, which was not possible using conventional methods. Hence, the device was able to send data in a fast and secure manner.

17.4.3 Sensing

The high absorbance and/or reflection and tunability of graphene-based MMs makes them highly attractive for sensing applications. Most of the sensors developed are refractive index based and do not require any particular labeling. THz sensing and spectroscopy are gaining importance as the development of MMs has progressed. Yi *et al* [115] simulated a graphene MM which had two absorption peaks in the IR range (22.5 μm and 74.5 μm) with maximum absorption more than 99%. The MM comprised the following layers from top to bottom, periodic graphene layer, SiO_2 layer, a gold layer and a silicon substrate. The highest sensitivity obtained was 17.2 μm/RIU. Further, the absorption was electrically tunable. Similarly, Huang *et al* [116] developed a tunable MM absorber working in the mid-IR regions. The bandwidth obtained was 17.7 THz (82.3 THz to 94.0 THz), with absorption more than 90%. The unit cell consisted of a Jerusalem cross structure made out of gold, below which a graphene layer was present of the same pattern with slightly greater widths. The above layers were placed on top of a dielectric and a grounding gold layer. It was reported that owing to the symmetry of the unit cell, the response to TE and TM polarization was similar. Using the contours of the electric fields of the structure, it was shown that opposite charges were collected near the gap between the metal and graphene structures, and surface currents were in the opposite directions for both TE and TM modes (although, positions were different but they still cancelled each other out). Further, the enhanced absorption was explained using the two-mode waveguide coupling model. The MM absorber was tuned by varying the Fermi energy of the graphene layer, the peak absorption shifted from 85.2 THz to 93.6 THz as the Fermi energy was varied from 0.2 eV to 0.6 eV. The sensitivity to change in permittivity of the surrounding medium was established, the frequency shift varied almost linearly with the change in surrounding permittivity.

Sadafi *et al* [117] designed a THz sensor that used a SiO_2 substrate on which the unit cell was built. The cell comprised a gold SRR with two gaps that were filled by graphene strips, and the cell was grounded by a gold layer. The simulations were performed on CST Microwave Studio. The Q factor of the absorption peak obtained was 50, and the peak moved towards higher frequencies when Fermi energy was changed from 0.1 eV to 0.2 eV. The sudden rise in absorbance was caused by the destructive interference of the responses of gold and graphene structures. Dimensions of the unit cell were optimized using the simulation results. The sensitivity obtained was about 0.6 THz/RIU, which was comparable to the sensors available in literature. Lee *et al* [118] used a patterned gold structure with graphene strips in the nano range to detect single-stranded DNAs in the concentration range of a few nmol mm^{-2}. The transmission peak in the THz range was shifted due to the binding of the DNA molecules with the π π stacking of the graphene nano strips. He *et al* [119] simulated a sensor with maximum sensitivity of 0.18 THz/RIU in the THz frequency range. A MM with a graphene layer consisting of SRR and two wired slots whose length was greater than the SRR was used as the unit cell. The destructive interference between the two structures resulted in a sharp reflection peak (maximum Q factor 86.6) which was referred to as electromagnetically induced reflection. The Fermi energy of graphene was used to alter the response of the MM. The shift in reflection peak was from 0.107 THz to 0.180 THz as the Fermi energy of the graphene layer was changed from 0.5 eV to 0.2 eV. Qi *et al* [120] developed a tunable refractive index sensor with a maximum absorptivity of 49.2%. The unit cell comprised a graphene layer shaped in the form of elliptical rings, the array was situated on top of a SiO_2 substrate. The features of the unit cells were in the range of a few microns. The maximum absorption increased from 10.7% to 49.2%, and the resonant wavelength blue shifted from 107.03 μm to 53.11 μm, as the Fermi energy was increased from 0.2 eV to 0.8 eV. Further, bandwidth decreased as the Fermi energy was increased. Using the electric field distribution, it was reported that when Fermi energy was increased, the electric field near the edges of the elliptical ring increased because of localized surface plasmon resonance effect. The geometric parameters and the periodicity of the array were optimized using the simulation results. The effect of angle of incidence was also studied, as the angle was increased from 0° to 60°, although the intensity of absorbance decreased, the resonant frequency did not change significantly. To confirm the refractive index sensitivity of the sensor, the refractive index of the surrounding was increased from 1.0 to 1.4. Upon the increase of refractive index, the absorbance decreased, and a significant red shift was observed. The maximum sensitivity recorded was 14.11 μm/RIU.

17.5 Conclusions

Simultaneous negative permittivity and permeability possessed by MMs enable them to be used in unique applications, including cloaking, holography, hyperlenses, perfect absorbers and various others. MMs have subwavelength periodic or non-periodic structures which can modify the response of the incoming EM waves. Therefore, apart from the material used, the shape and size of the structures play a vital role in determining the response of the MMs. The active frequency range of the MM depends upon the size of the structures selected.

The fabrication of optical MMs and MSs is complex and often requires advanced fabrication techniques such as electron beam lithography and nanoimprint lithography. Further, self-assembly is being explored for its potential for mass production. The THz and microwave MSs fabrication is relatively simpler due to the size of the structure being in the micro to millimeter range. Hence, commercial fabrication methods such as injection molding, laser micromachining, and 3D printing can be used. However, the fabrication of 3D MMs in the THz requires advanced techniques such as two-photon fabrication and optical lithography.

Owing to the tunable electrical conductivity and EM response across the frequency spectrum, a plethora of graphene-based MMs are being explored. Most conventional MMs are considered to have a narrow spectral range [121]; the incorporation of graphene can help achieve the broadband range due to its long lifetime. Furthermore, the high switching speed of graphene-based MMs due to high carrier mobility and fast carrier relaxation time of graphene is important for applications in imaging and communication. The addition of graphene can help tune the response of MMs much more easily than by altering the subwavelength structures. The tunability using graphene can be achieved by changing its chemical structure, applying an electrical bias, and photodoping. Generally, applying an electrical bias and/or photodoping is selected for graphene-based MMs, the conductivity of graphene is altered due to the variation of Fermi energy. The electrical bias is applied by using a FET type structure; the graphene layer is placed on top of a semiconducting substrate, followed by a metallic grounding layer. Contacts are added on the graphene layer to apply a gate voltage, and thereby changing its Fermi energy. To effectively apply the photodoping, graphene is placed on a semiconducting substrate which enables excess carriers to be transferred into the graphene layer upon optical illumination.

The current chapter divided the applications of graphene-based MMs into three categories, imaging, communications and sensing. The majority of applications were in the IR, THz, and microwave frequency range. The lack of response of conventional materials in the aforementioned frequency range has caused research groups to explore graphene-based MMs. THz imaging is an emerging area for security and spectrometry applications. IR-based thermal imaging systems are essential to study the heat distribution for industrial applications, especially the electronics industry. Graphene-based MMs can be used as detectors, lenses to manipulate the EM waves, and high speed modulators. Graphene-based MMs can also be used as hyperlenses which can achieve subwavelength resolution that is not possible using conventional materials [96]. Further, the broadband absorbance of the perfect absorbers can be used as radar absorbing surfaces (particularly in the microwave frequency range), surfaces to shield electronics from unwanted EM energies, and as solar absorbers. THz waves have the advantage of being more directional than microwave waves and less prone to attenuation in the atmosphere as compared to IR waves. Therefore, THz based systems are being explored for future communication devices. The EIT phenomenon displayed by MMs can be used for stealth applications. The cloaking of a 3D object was demonstrated theoretically in the IR range [105]. The development of EIT-based devices can help in the advancement of slow light technology, which can drastically reduce the energy and noise in communication systems. Graphene-based MMs can be used to develop smaller, lightweight, and more efficient antennas. Another recent application of graphene-based MMs is the fast and secure transfer of data using THz waves [112]. Also, an MS capable of sending encoded messages using holography was presented [113]. Sensors with high refractive index sensitivity without using any labels can be developed using graphene MMs. A sensor utilizing a Vivaldi antenna with a graphene-based lens was demonstrated to detect cancerous tissues [104].

MMs have gained a lot of attention because of their unique properties and potential for futuristic applications. Simulation and theoretical methods have helped to understand the working of MMs. The development of advanced fabrication techniques has led to an increase in the testing of MMs for real-world applications, which is essential to take into account losses that are not explained by theoretical results. The inclusion of graphene can improve the performance of MMs, and tune their response by applying gate voltage and optical illumination. Therefore, graphene-based MMs are an exceptional prospect for everyday life applications.

References

[1] Dolan J A, Wilts B D, Vignolini S, Baumberg J J, Steiner U and Wilkinson T D 2015 Optical propertijues of gyroid structured materials: from photonic crystals to metamaterials *Adv. Opt. Mater.* **3** 12–32

[2] Neil T R, Shen Z, Robert D, Drinkwater B W and Holderied M W 2020 Moth wings are acoustic metamaterials *Proc. Natl Acad. Sci. USA* **117** 31134–41

[3] Pendry J B, Holden A J, Robbins D J and Stewart W J 1999 Magnetism from conductors and enhanced nonlinear phenomena *IEEE Trans. Microwave Theory Tech.* **47** 2075–84
[4] Pendry J B, Holden A J, Stewart W J and Youngs I 1996 Extremely low frequency plasmons in metallic mesostructures *Phys. Rev. Lett.* **76** 4773
[5] Veselago V G 1968 The electrodynamics of substances with simultaneously negative values of ϵ and μ *Sov. Phys. Usp.* **10** 509
[6] Mao P, Liu C, Song F, Han M, Maier S A and Zhang S 2020 Manipulating disordered plasmonic systems by external cavity with transition from broadband absorption to reconfigurable reflection *Nat. Commun.* **11** 1538
[7] Ryzhii V, Ryzhii M and Otsuji T 2007 Negative dynamic conductivity of graphene with optical pumping *J. Appl. Phys.* **101** 083114
[8] Vendik I B *et al* 2012 Tunable metamaterials for controlling THz radiation *IEEE Trans. Terahertz Sci. Technol.* **2** 538–49
[9] O'Hara J F, Ekin S, Choi W and Song I 2019 A perspective on terahertz next-generation wireless communications *Technologies* **7** 43
[10] Mittleman D M 2018 Twenty years of terahertz imaging *Opt. Express* **26** 9417–31
[11] Cheng Y, Wang Y, Niu Y and Zhao Z 2020 Concealed object enhancement using multi-polarization information for passive millimeter and terahertz wave security screening *Opt. Express* **28** 6350–66
[12] Ren A, Zahid A, Fan D, Yang X, Imran M A, Alomainy A and Abbasi Q H 2019 State-of-the-art in terahertz sensing for food and water security—a comprehensive review *Trends Food Sci. Technol.* **85** 241–51
[13] Federici J and Moeller L 2010 Review of terahertz and subterahertz wireless communications *J. Appl. Phys.* **107** 6
[14] Zhang S, Fan W, Panoiu N C, Malloy K J, Osgood R M and Brueck S R J 2005 Experimental demonstration of near-infrared negative-index metamaterials *Phys. Rev. Lett.* **95** 137404
[15] Kanté B, Park Y S, O'Brien K, Shuldman D, Lanzillotti-Kimura N D, Wong Z J, Yin X and Zhang X 2012 Symmetry breaking and optical negative index of closed nanorings *Nat. Commun.* **3** 1180
[16] Enkrich C, Wegener M, Linden S, Burger S, Zschiedrich L, Schmidt F, Zhou J F, Koschny T and Soukoulis C M 2005 Magnetic metamaterials at telecommunication and visible frequencies *Phys. Rev. Lett.* **95** 203901
[17] Capolino F 2017 *Applications of Metamaterials* (Boca Raton, FL: CRC Press)
[18] Shalaev V M, Cai W, Chettiar U K, Yuan H K, Sarychev A K, Drachev V P and Kildishev A V 2005 Negative index of refraction in optical metamaterials *Opt. Lett.* **30** 3356–8
[19] Dolling G, Wegener M and Linden S 2007 Realization of a three-functional-layer negative-index photonic metamaterial *Opt. Lett.* **32** 551–3
[20] Feth N, Enkrich C, Wegener M and Linden S 2007 Large-area magnetic metamaterials via compact interference lithography *Opt. Express* **15** 501–7
[21] Gong B, Zhao X, Pan Z, Li S, Wang X, Zhao Y and Luo C 2014 A visible metamaterial fabricated by self-assembly method *Sci. Rep.* **4** 1–7
[22] Bates F S and Fredrickson G H 2000 Block copolymers-designer soft materials *Phys. Today* **52** 32
[23] Vignolini S, Yufa N A, Cunha P S, Guldin S, Rushkin I, Stefik M, Hur K, Wiesner U, Baumberg J J and Steiner U 2012 A 3D optical metamaterial made by self-assembly *Adv. Mater.* **24** OP23–7

[24] Bidault S, García de Abajo F J and Polman A 2008 Plasmon-based nanolenses assembled on a well-defined DNA template *J. Am. Chem. Soc.* **130** 2750–1
[25] Wen Y, McLaughlin C K, Lo P K, Yang H and Sleiman H F 2010 Stable gold nanoparticle conjugation to internal DNA positions: facile generation of discrete gold nanoparticle–DNA assemblies *Bioconjugate Chem.* **21** 1413–6
[26] Watanabe-Tamaki R, Ishikawa A, Tanaka T, Zako T and Maeda M 2012 DNA-templating mass production of gold trimer rings for optical metamaterials *J. Phys. Chem.* C **116** 15028–33
[27] Luo C, Johnson S G, Joannopoulos J D and Pendry J B 2002 All-angle negative refraction without negative effective index *Phys. Rev.* B **65** 201104
[28] Baev A, Samoc M, Prasad P N, Krykunov M and Autschbach J 2007 A quantum chemical approach to the design of chiral negative index materials *Opt. Express* **15** 5730–41
[29] Alù A, Salandrino A and Engheta N 2006 Negative effective permeability and left-handed materials at optical frequencies *Opt. Express* **14** 1557–67
[30] Wu X, Su Y and Shi J 2019 Perspective of additive manufacturing for metamaterials development *Smart Mater. Struct.* **28** 093001
[31] Bonatti C and Mohr D 2017 Large deformation response of additively-manufactured FCC metamaterials: from octet truss lattices towards continuous shell mesostructures *Int. J. Plast.* **92** 122–47
[32] Askari M *et al* 2020 Additive manufacturing of metamaterials: a review *Addit. Manuf.* 101562
[33] Lin Y, Yao H, Ju X, Chen Y, Zhong S and Wang X 2017 Free-standing double-layer terahertz band-pass filters fabricated by femtosecond laser micro-machining *Opt. Express* **25** 25125–34
[34] Bhadresha R, Sukham J, Pattanayak A, Rana G, Deshmukh P, Duttagupta S P, Sarwade N, Jacob G and Prabhu S S 2015 THz bandpass filter based on sub-wavelength holes in free-standing metal thin-films *2015 40th Int. Conf. on Infrared, Millimeter, and Terahertz waves (IRMMW-THz)* (Piscataway, NJ: IEEE), pp 1–2
[35] Babaee S, Shim J, Weaver J C, Chen E R, Patel N and Bertoldi K 2013 3D soft metamaterials with negative Poisson's ratio *Adv. Mater.* **25** 5044–9
[36] Ngo T D, Kashani A, Imbalzano G, Nguyen K T and Hui D 2018 Additive manufacturing (3D printing): a review of materials, methods, applications and challenges *Composites* B **143** 172–96
[37] Lin D, Nian Q, Deng B, Jin S, Hu Y, Wang W and Cheng G J 2014 Three-dimensional printing of complex structures: man made or toward nature? *ACS Nano* **8** 9710–5
[38] Kronberger R and Soboll P 2016 New 3D printed microwave metamaterial absorbers with conductive printing materials *2016 46th European Microwave Conf. (EuMC)* (Piscataway, NJ: IEEE), pp 596–9
[39] Tenggara A P, Park S J, Yudistira H T, Ahn Y H and Byun D 2017 Fabrication of terahertz metamaterials using electrohydrodynamic jet printing for sensitive detection of yeast *J. Micromech. Microeng.* **27** 035009
[40] Lee Y, Lu X, Hao Y, Yang S, Ubic R, Evans J R and Parini C G 2007 Rapid prototyping of ceramic millimeterwave metamaterials: simulations and experiments *Microwave Opt. Technol. Lett.* **49** 2090–3
[41] Saleh E *et al* 2017 3D inkjet-printed UV-curable inks for multi-functional electromagnetic applications *Addit. Manuf.* **13** 143–8

[42] Yoo M, Kim H K, Kim S, Tentzeris M and Lim S 2015 Silver nanoparticle-based inkjet-printed metamaterial absorber on flexible paper *IEEE Antennas Wirel. Propag. Lett.* **14** 1718–21
[43] Walther M, Ortner A, Meier H, Löffelmann U, Smith P J and Korvink J G 2009 Terahertz metamaterials fabricated by inkjet printing *Appl. Phys. Lett.* **95** 251107
[44] Jin Y A, He Y and Fu J Z 2015 Support generation for additive manufacturing based on sliced data *Int. J. Adv. Manuf. Technol.* **80** 2041–52
[45] Tian Y, McAllister D, Colijn H, Mills M, Farson D, Nordin M and Babu S 2014 Rationalization of microstructure heterogeneity in INCONEL 718 builds made by the direct laser additive manufacturing process *Metall. Mater. Trans.* A **45** 4470–83
[46] Mercelis P and Kruth J P 2006 Residual stresses in selective laser sintering and selective laser melting *Rapid Prototyp. J.* **12** 254–65
[47] Kawata S, Sun H B, Tanaka T and Takada K 2001 Finer features for functional microdevices *Nature* **412** 697–8
[48] Ergin T, Stenger N, Brenner P, Pendry J B and Wegener M 2010 Three-dimensional invisibility cloak at optical wavelengths *Science* **328** 337–9
[49] Gansel J K, Thiel M, Rill M S, Decker M, Bade K, Saile V, von Freymann G, Linden S and Wegener M 2009 Gold helix photonic metamaterial as broadband circular polarizer *Science* **325** 1513–5
[50] Formanek F, Takeyasu N, Tanaka T, Chiyoda K, Ishikawa A and Kawata S 2006 Selective electroless plating to fabricate complex three-dimensional metallic micro/nanostructures *Appl. Phys. Lett.* **88** 083110
[51] Rill M S, Plet C, Thiel M, Staude I, Von Freymann G, Linden S and Wegener M 2008 Photonic metamaterials by direct laser writing and silver chemical vapour deposition *Nat. Mater.* **7** 543–6
[52] Ishikawa A, Tanaka T and Kawata S 2006 Improvement in the reduction of silver ions in aqueous solution using two-photon sensitive dye *Appl. Phys. Lett.* **89** 113102
[53] Cao Y Y, Takeyasu N, Tanaka T, Duan X M and Kawata S 2009 3D metallic nanostructure fabrication by surfactant-assisted multiphoton-induced reduction *Small* **5** 1144–8
[54] Zhang S *et al* 2012 Photoinduced handedness switching in terahertz chiral metamolecules *Nat. Commun.* **3** 942
[55] O'brien S and Pendry J B 2002 Magnetic activity at infrared frequencies in structured metallic photonic crystals *J. Phys. Condens. Matter* **14** 6383
[56] Bait-Suwailam M M 2019 Electromagnetic field interaction with metamaterials *Electromagnetic Fields and Waves* (London: IntechOpen)
[57] Rotman W 1962 Plasma simulation by artificial dielectrics and parallel-plate media *IRE Trans. Anntenas Propag.* **10** 82–95
[58] Ulrich R 1967 Far-infrared properties of metallic mesh and its complementary structure *Infrared Phys.* **7** 37–55
[59] Liu R, Cui T J, Huang D, Zhao B and Smith D R 2007 Description and explanation of electromagnetic behaviors in artificial metamaterials based on effective medium theory *Phys. Rev.* E **76** 026606
[60] Landy N I, Bingham C M, Tyler T, Jokerst N, Smith D R and Padilla W J 2009 Design, theory, and measurement of a polarization-insensitive absorber for terahertz imaging *Phys. Rev.* B **79** 125104

[61] Hoffman A J, Alekseyev L, Howard S S, Franz K J, Wasserman D, Podolskiy V A, Narimanov E E, Sivco D L and Gmachl C 2007 Negative refraction in semiconductor metamaterials *Nat. Mater.* **6** 946–50
[62] Yen T J, Padilla W J, Fang N, Vier D C, Smith D R, Pendry J B, Basov D N and Zhang X 2004 Terahertz magnetic response from artificial materials *Science* **303** 1494–6
[63] Dolling G, Wegener M, Soukoulis C M and Linden S 2007 Negative-index metamaterial at 780 nm wavelength *Opt. Lett.* **32** 53–5
[64] Rockstuhl C, Lederer F, Etrich C, Zentgraf T, Kuhl J and Giessen H 2006 On the reinterpretation of resonances in split-ring-resonators at normal incidence *Opt. Express* **14** 8827–36
[65] Oates T W H, Wormeester H and Arwin H 2011 Characterization of plasmonic effects in thin films and metamaterials using spectroscopic ellipsometry *Prog. Surf. Sci.* **86** 328–76
[66] Ghodgaonkar D K, Varadan V V and Varadan V K 1989 A free-space method for measurement of dielectric constants and loss tangents at microwave frequencies *IEEE Trans. Instrum. Meas.* **38** 789–93
[67] Weir W B 1974 Automatic measurement of complex dielectric constant and permeability at microwave frequencies *Proc. IEEE* **62** 33–6
[68] Kadaba P K 1984 Simultaneous measurement of complex permittivity and permeability in the millimeter region by a frequency-domain technique *IEEE Trans. Instrum. Meas.* **33** 336–40
[69] Ziolkowski R W 2003 Design, fabrication, and testing of double negative metamaterials *IEEE Trans. Antennas Propag.* **51** 1516–29
[70] Singh A K, Abegaonkar M P and Koul S K 2021 *Metamaterials for Antenna Applications* (Boca Raton, FL: CRC Press)
[71] Baker-Jarvis J, Vanzura E J and Kissick W A 1990 Improved technique for determining complex permittivity with the transmission/reflection method *IEEE Trans. Microwave Theory Tech.* **38** 1096–103
[72] Nicolson A M and Ross G F 1970 Measurement of the intrinsic properties of materials by time-domain techniques *IEEE Trans. Instrum. Meas.* **19** 377–82
[73] Cai W, Chettiar U K, Kildishev A V and Shalaev V M 2007 Optical cloaking with metamaterials *Nat. Photon.* **1** 224–7
[74] Hasan M, Faruque M R I, Islam S S and Islam M T 2016 A new compact double-negative miniaturized metamaterial for wideband operation *Materials* **9** 830
[75] Hanson G W 2008 Dyadic Green's functions and guided surface waves for a surface conductivity model of graphene *J. Appl. Phys.* **103** 064302
[76] Gómez-Díaz J S and Perruisseau-Carrier J 2013 Graphene-based plasmonic switches at near infrared frequencies *Opt. Express* **21** 15490–5504
[77] Li Q *et al* 2015 Dual control of active graphene–silicon hybrid metamaterial devices *Carbon* **90** 146–53
[78] Li Q, Tian Z, Zhang X, Singh R, Du L, Gu J, Han J and Zhang W 2015 Active graphene–silicon hybrid diode for terahertz waves *Nat. Commun.* **6** 7082
[79] Dawlaty J M, Shivaraman S, Chandrashekhar M, Rana F and Spencer M G 2008 Measurement of ultrafast carrier dynamics in epitaxial graphene *Appl. Phys. Lett.* **92** 042116
[80] Weis P, Garcia-Pomar J L, Hoh M, Reinhard B, Brodyanski A and Rahm M 2012 Spectrally wide-band terahertz wave modulator based on optically tuned graphene *ACS Nano* **6** 9118–24

[81] Lin K T, Lin H, Yang T and Jia B 2020 Structured graphene metamaterial selective absorbers for high efficiency and omnidirectional solar thermal energy conversion *Nat. Commun.* **11** 1389
[82] Safaei A, Chandra S, Vázquez-Guardado A, Calderon J, Franklin D, Tetard L, Zhai L, Leuenberger M N and Chanda D 2017 Dynamically tunable extraordinary light absorption in monolayer graphene *Phys. Rev.* B **96** 165431
[83] Mostaan S M A and Saghaei H 2021 A tunable broadband graphene-based metamaterial absorber in the far-infrared region *Opt. Quantum Electron.* **53** 96
[84] Zheng Y, Gao J, Zhou Y, Cao X, Yang H, Li S and Li T 2017 Wideband gain enhancement and RCS reduction of Fabry–Perot resonator antenna with chessboard arranged metamaterial superstrate *IEEE Trans. Antennas Propag.* **66** 590–9
[85] Qi L, Liu C and Shah S M A 2019 A broad dual-band switchable graphene-based terahertz metamaterial absorber *Carbon* **153** 179–88
[86] Huang M L, Cheng Y Z, Cheng Z Z, Chen H R, Mao X S and Gong R Z 2018 Design of a broadband tunable terahertz metamaterial absorber based on complementary structural graphene *Materials* **11** 540
[87] Huang M, Cheng Y, Cheng Z, Chen H, Mao X and Gong R 2018 Based on graphene tunable dual-band terahertz metamaterial absorber with wide-angle *Opt. Commun.* **415** 194–201
[88] Huang X, Pan K and Hu Z 2016 Experimental demonstration of printed graphene nanoflakes enabled flexible and conformable wideband radar absorbers *Sci. Rep.* **6** 38197
[89] Zhang J, Li Z, Shao L and Zhu W 2021 Dynamical absorption manipulation in a graphene-based optically transparent and flexible metasurface *Carbon* **176** 374–82
[90] Zhang J, Wei X, Premaratne M and Zhu W 2019 Experimental demonstration of an electrically tunable broadband coherent perfect absorber based on a graphene–electrolyte–graphene sandwich structure *Photon. Res.* **7** 868–74
[91] Li Y, Kong F and Li K 2016 Graphene-based infrared lens with tunable focal length *Prog. Electromagn. Res.* **155** 19–26
[92] Zainud-Deen S H, Malhat H A and Mabrouk A M 2017 Graphene based metamaterial lens for terahertz applications *2017 Japan–Africa Conf. on Electronics, Communications and Computers (JAC-ECC)* (Piscataway, NJ: IEEE), pp 148–51
[93] Liu W *et al* 2018 Graphene-enabled electrically controlled terahertz meta-lens *Photon. Res.* **6** 703–8
[94] Hamzavi-Zarghani Z, Yahaghi A and Matekovits L 2018 Tunable lens based on graphene metasurface for circular polarization *2018 Int. Conf. on Electromagnetics in Advanced Applications (ICEAA)* (Piscataway, NJ: IEEE), pp 644–7
[95] Yao W, Tang L, Wang J, Ji C, Wei X and Jiang Y 2018 Spectrally and spatially tunable terahertz metasurface lens based on graphene surface plasmons *IEEE Photon. J.* **10** 1–8
[96] Andryieuski A, Lavrinenko A V and Chigrin D N 2012 Graphene hyperlens for terahertz radiation *Phys. Rev.* B **86** 121108
[97] Fan K, Suen J, Wu X and Padilla W J 2016 Graphene metamaterial modulator for free-space thermal radiation *Opt. Express* **24** 25189–201
[98] Pianelli A, Kowerdziej R, Dudek M, Sielezin K, Olifierczuk M and Parka J 2020 Graphene-based hyperbolic metamaterial as a switchable reflection modulator *Opt. Express* **28** 6708–18

[99] Zhang Y, Feng Y, Zhu B, Zhao J and Jiang T 2014 Graphene based tunable metamaterial absorber and polarization modulation in terahertz frequency *Opt. Express* **22** 22743–52

[100] Liang G *et al* 2015 Integrated terahertz graphene modulator with 100% modulation depth *ACS Photonics* **2** 1559–66

[101] Amanatiadis S A, Karamanos T D and Kantartzis N V 2017 Radiation efficiency enhancement of graphene THz antennas utilizing metamaterial substrates *IEEE Antennas Wirel. Propag. Lett.* **16** 2054–7

[102] Radwan A, D'Amico M, Gentili G G and Verri V 2015 Reconfigurable THz metamaterial antenna based on graphene *2015 9th European Conf. on Antennas and Propagation (EuCAP)* (Piscataway, NJ: IEEE), pp 1–5

[103] Degl'Innocenti R *et al* 2016 Fast modulation of terahertz quantum cascade lasers using graphene loaded plasmonic antennas *ACS Photonics* **3** 464–70

[104] Poorgholam-Khanjari S and Zarrabi F B 2021 Reconfigurable Vivaldi THz antenna based on graphene load as hyperbolic metamaterial for skin cancer spectroscopy *Opt. Commun.* **480** 126482

[105] Farhat M, Rockstuhl C and Bağcı H 2013 A 3D tunable and multi-frequency graphene plasmonic cloak *Opt. Express* **21** 12592–603

[106] Ko J W, Kim S W, Hong J, Ryu J, Kang K and Park C B 2012 Synthesis of graphene-wrapped CuO hybrid materials by CO_2 mineralization *Green Chem.* **14** 2391–4

[107] Li D, Ji Z and Luo C 2020 Optically tunable plasmon-induced transparency in terahertz metamaterial system *Opt. Mater.* **104** 109920

[108] Deng Y, Zhou Q, Zhang P, Jiang N, Ning T, Liang W and Zhang C 2021 Heterointerface-enhanced ultrafast optical switching via manipulating metamaterial-induced transparency in a hybrid terahertz graphene metamaterial *ACS Appl. Mater. Interfaces* **13** 13565–75

[109] Kindness S J *et al* 2018 Active control of electromagnetically induced transparency in a terahertz metamaterial array with graphene for continuous resonance frequency tuning *Adv. Opt. Mater.* **6** 1800570

[110] Huang L, Zhang S and Zentgraf T 2018 Metasurface holography: from fundamentals to applications *Nanophotonics* **7** 1169–90

[111] Hafiz A and Qamar S 2020 Electromagnetically induced holographic imaging using monolayer graphene *Opt. Express* **28** 1970–86

[112] Momeni A, Rouhi K, Rajabalipanah H and Abdolali A 2018 An information theory-inspired strategy for design of re-programmable encrypted graphene-based coding meta-surfaces at terahertz frequencies *Sci. Rep.* **8** 6200

[113] Rajabalipanah H, Rouhi K, Abdolali A, Iqbal S, Zhang L and Liu S 2020 Real-time terahertz meta-cryptography using polarization-multiplexed graphene-based computer-generated holograms *Nanophotonics* **9** 2861–77

[114] Zalevsky Z, Mendlovic D and Dorsch R G 1996 Gerchberg–Saxton algorithm applied in the fractional Fourier or the Fresnel domain *Opt. Lett.* **21** 842–4

[115] Yi Z, Liang C, Chen X, Zhou Z, Tang Y, Ye X, Yi Y, Wang J and Wu P 2019 Dual-band plasmonic perfect absorber based on graphene metamaterials for refractive index sensing application *Micromachines* **10** 443

[116] Huang H, Xia H, Xie W, Guo Z, Li H and Xie D 2018 Design of broadband graphene-metamaterial absorbers for permittivity sensing at mid-infrared regions *Sci. Rep.* **8** 4183

[117] Sadafi M M, Karami H and Hosseini M 2021 A tunable hybrid graphene-metal metamaterial absorber for sensing in the THz regime *Curr. Appl Phys.* **31** 132–40

[118] Lee S H, Choe J H, Kim C, Bae S, Kim J S, Park Q H and Seo M 2020 Graphene assisted terahertz metamaterials for sensitive bio-sensing *Sens. Actuators* B **310** 127841

[119] He X, Zhang Q, Lu G, Ying G, Wu F and Jiang J 2016 Tunable ultrasensitive terahertz sensor based on complementary graphene metamaterials *RSC Adv.* **6** 52212–18

[120] Qi Y, Zhang Y, Liu C, Zhang T, Zhang B, Wang L, Deng X, Bai Y and Wang X 2020 A tunable terahertz metamaterial absorber composed of elliptical ring graphene arrays with refractive index sensing application *Results Phys.* **16** 103012

[121] Rout S and Sonkusale S 2017 *Active Metamaterials: Terahertz Modulators and Detectors* (Cham: Springer International Publishing)

IOP Publishing

Recent Advances in Graphene and Graphene-Based Technologies

Anoop Chandran, N V Unnikrishnan, M K Jayaraj, Reenu Elizabeth John and Justin George

Chapter 18

The way ahead for graphene-based technologies

P B Pillai, A Achari and R R Nair

Nearly two decades of academic research has dramatically expanded the boundaries of our scientific knowledge about graphene and unveiled some of its exceptional material properties. The next stages of development are expected to be led by industries designing innovative products based on graphene, equipped with the caliber to address some of the real-world problems. This chapter identifies the main application sectors where graphene is expected to play a key role and reviews some of the outstanding challenges to address. We also discuss the outcomes of the material standardization initiatives across the globe along with a roadmap detailing the anticipated timeline of future developments in graphene technology.

18.1 Introduction

Within a short span after its first isolation [1], graphene-based materials are showing clear signs of transformation from academic research to a true industrial material capable of disrupting dozens of market sectors from energy to health care. Some of the main roadblocks ahead in the commercial development of graphene-related materials (GRMs) products are the immature material supply and process challenges in transforming this atomically thin material to commercial products by preserving most of its superlative properties. Global suppliers are now cranking up the production of GRMs at a current rate of over 2.5 kilo tonnes per year. However, these over-aggressive production scale-up attempts have compromised the fine controls in the manufacturing process, essential to produce superior quality graphene. Despite these overwhelming challenges imposed by the material variability issues in the supply sector, commercialization of graphene has evolved significantly through several joint ventures between industries and academic institutions under the roof of major government investment programs for, e.g., the European graphene flagship initiative. As a result, several graphene products have already made their

doi:10.1088/978-0-7503-3999-5ch18

market entry and clear market and technological potential of several application areas have been identified. This chapter present an overview of prospects and challenges of main GRMs application areas such as composites, barriers and coatings, membranes, beyond complementary metal-oxide–semiconductor (CMOS) and biomedical applications that present significant technological and market opportunities in the short–medium term commercialization drive. The chapter also discuss global initiatives towards standardization of GRMs to improve the reliability and reproducibility of GRM products. Towards the end, a roadmap for graphene is presented with an anticipated timeline to achieve various technology milestones.

18.2 Global research initiatives

Through several worldwide initiatives, significant research investments have been made to boost the technology readiness level (TRL—a nine-point indicator to describe the maturity of a new technology, the higher the TRL number, the closer the technology to its commercialization) of GRM technology. Among these, the most notable one is Europe's ambitious €1 billion graphene flagship (GF) program, kickstarted in 2013 with a spending plan for 10 years to drive GRM technologies to their full commercialization and with an overarching aim of making Europe a powerhouse of graphene-based technologies. In 2020, the GF program identified 11 spearhead projects to push the TRL of several GRM products such as water purification membranes and fast charging sustainable batteries to flexible sensors to autonomous driving [2]. In 2021, the European Commission launched an additional €20 million investment towards the GF experimental pilot line [3] to advance the technologies in 2D electronics, optoelectronics and sensing towards commercialization. A number of application areas identified by GF overlap with the priority areas of Europe's next €100-billion science fund [4], which may open up a new funding landscape for several emerging GRM applications including spintronics and radio frequency devices. With a similar intention to focus on the graphene market, the South Korean government, with an investment budget of USD 107.7 million, introduced a five-year technology roadmap to promote the commercialization of GRMs technologies [5]. While representing over 33% of the global graphene market, China also competes in the commercialization drive for graphene by identifying GRMs industrialization as a priority in the Development Plan for New Materials as part of the Government's 13th Five-year Plan Period (2016–20) [6]. Beside these public sector investments, there is a growing investment in graphene from industrial organizations also. For example, Skeleton Technologies—the market leaders in energy storage technology in partnership with Karlsruhe Institute of Technology will have developed a graphene-based SuperBattery by 2023 with an ultra-short charging time of 15 s and signed a €1 billion letter of intent with a leading automotive manufacturer [7]. Another major private sector initiative for graphene is by South Korean fiber producer Huvis by signing a memorandum of understanding for mass production of graphene fibers and related products [8]. A joint effort between Nationwide Engineering with University of Manchester is targeting to commercialize graphene-enhanced concrete technology [9]. The list is not comprehensive, but expanding on a daily basis.

These global initiatives are intensively focusing on rapid upscaling of the GRM technology and GRM market size expected to outgrow USD 1 billion by 2027 [10].

18.3 Graphene-based products in the market

The 2017 version of the technology innovation roadmap (TIR) of the EU GF has grouped the main application areas that possess a significant technological potential into four mainstream fields of application [11, 12]. Figure 18.1 summarizes the application areas containing a broad category of products with maximum current market potential identified by the TIR [12].

Some of the major product developments from these categories are summarized here. Affordable lightweight materials solutions with high mechanical strength, flexibility, temperature tolerance and lightning strike performance for aviation and terrestrial transport sector are awaiting market entry, whilst, a number of products are already commercially available, e.g., GRM incorporated bike tyre from Vittoria, car parts by Ford Motors and thermal interface pads used in Huawei Mobile phones. Besides, a number of graphene-based products have recently been introduced or await their market entry, such as GXT-LUBE–a graphene-based synthetic additive for the treatment of engine oil, GXT–electrostatic discharge bags for electronic devices, GXT paint-customized commercial paint with graphene solution, graphene-based electrically conductive fabrics for human–machine interfacing by Grafren AB, that are expected to achieve the full roll-to-roll production goal shortly. Other notable developments in graphene towards commercialization are Grapheal's TestNpass, a digital biosensor for the antigenic detection of SARS-CoV-2, chip-based sensors developed by Nanomedical Diagnostics (USA) to evaluate activity assay of drug molecules using Agile R100 Graphene Biosensor and wireless powered brain interface that is under development by Inbrain Neuroelectronics (table 18.1).

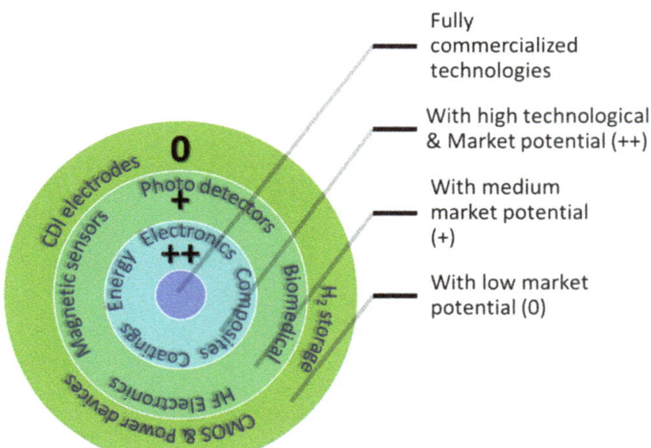

Figure 18.1. Technological and market potential categorization of various GRM technologies by EU graphene flagship TIR results.

Table 18.1. A list of GRM products available in the market or under commercial development.

Product details	Features	Manufacturer	Product reference
Graphene power break	• Graphene-composite battery for fast charging applications • 3.5× faster charging than commercial power bank	Real graphene USA	https://realgraphene.com/product/g-lite
Graphene-enhanced chronographs	• Graphene for extremely lightweight chronographs	Richard Midle	https://www.tlchardmi1 1e.oom/colleetions/rm·50-03·tourbillon-chronographsmclaren-f1
Thermal interface pads	• Graphene heat spreader for efficient SSD cooling • Heat dissipation enhanced for comibination of graphene and copper foil	TEAMGROUP	https://www.teamgroup.com/en
Graphene biosensor	• Agile R100 utilizes graphene- enhanced direct binding of challenging proteins • Excellent small molecule detection capability enabled by FET-based detection mechanism	Nanomedical Diagnostics	https://nanomedical.com/
Graphene-based engine oil additive	• A graphene-based synthetic additive for the treatment of engine oil • Atomically thin layers in the oil offer superior lubrication	Graphene-XT, Italy	https://www.graphene-xt.com/en
Bike tyre	• Performance boost specifically for speed wet grip, durability and puncture resistance	Vittoria	https://www.vittoria.com/us/en/technology/graphene
Graphene-based fabrics	• G-HEATEX textile heading elements	Grafren AB	https://grafren.se

Graphene-based biosensors	• Digital biosensor for the antigenic detection of SARS-CoV-2	Grapheal	https://www.grapheal.com/about-our-solution
3D graphene sensors	• 3D graphene sensing electrode	Emberion	https://www.intergratedgraphene.com/shop/p/gii-sens-electrodes
Graphene enforced helmets	• High impact resistance is at extreme ambient temperatures • Superior thermoregulatory features for maximum comfort	Momodesig	https://en.momodesing.com/products/fgtr-graphene-1-10
Thermal pads in smartphone	• Heat sink and interface with excellent thermal conductivity	Huawei	https://consumer.huawei.com/en/community/details/topicld_5678/
Tennis racquet, ski (GrapheneXT)	• Light weight • Fast swing speed and power	Heat	https://www.heat.com/en/sports/racquetball/tectnoloy/graphene-xt
Flexible touch screen	• Flexible touch screen for mobile phone	Changqing Graphene Technology	https://www.cqmxi.com/archives1201.html
Lithium battery electrode	• Embeded nano-silicon into graphene platelets • Triple anode specific capacity compared to traditional graphite anode (from 325 mAh g^{-1} to 1.250 mAh g^{-1})	CalBattery	https://calbattery.com/the-clbattery-solution/
Graphene nanoplatelet dispresions paints/coatings	• Excellent water-based anti-corrosion protection	Applied graphene materials	https://www.appliedgraphenematerials.com/Application/paints-and-coadings/
Neural interfaces for brain restoration	• Metal-free brain interface electrodes • Great potentieal for miniaturization	Inbarin-neuroreleronics	https://www.inbarin-neuroreleronics.com/

18.4 Prospective applications and deadlocks

As market leaders started perceiving the application potential of GRM products, the product revolution has mainly started with European market niches such as energy storage and composites. It is anticipated that product development now will spread to other areas such as flexible devices, filtration membranes, and display electronics. In the following section we try to identify the compelling opportunities for product innovation and outstanding challenges in these emerging application areas.

18.4.1 2D materials in composites and additives

18.4.1.1 Introduction

2D materials are generally used in composites to enhance the inherent properties of the host bulk materials such as polymers and ceramics. The enhancement can be in the mechanical, conducting or barrier properties of the host material. Besides, graphene and analogous 2D materials (GRMs) are also used as additives to fluids to enhance certain aspects of their application, such as to act as an adsorbent for unwanted materials in a liquid. Regarding composites, the general approach of using graphene-like materials is as an additive or as a coating. 2D materials such as graphene can be used in these applications in various forms such as few/multi-layered graphene, graphite platelets, graphene nanoribbons, or as reduced graphene oxide (rGO). Due to its existence in many different forms, the distinction between various nano and bulk forms of graphene are not very clear for many of the applications. However, this versatility also gives 2D materials a wide application range. 2D materials can be used as additive or coating for many applications that benefit from enhancement of material properties, some examples of which are noted below.

The global demand to reduce greenhouse gas emission and reduction in carbon footprint has forced the transport industries to reduce greenhouse gas emission and look for alternative manufacturing routes to build lightweight vehicles. This shift in focus has created demand for lightweight and strong materials for different manufacturing processes. In addition to being durable, these materials must be affordable, meet the existing safety standards and be functional. Polymer-based composites of 2D materials can fulfill these demands. In the automotive sector, polymer-based composites are predicted to replace the existing technologies by 2030. Wind power and aerospace engineering are pioneers in the use of such lightweight polymer composites. In aerospace, polymer-based composites are regularly used in wings and fuselage [13] in planes such as Airbus A350XWB, Boeing 787 Dreamliner to reduce greenhouse gas emission and increase efficiency.

GRM-polymer composites can combine various properties of 2D materials and polymers into a single material. This allows fabrication of lightweight materials and integration of functionalities. For example, 2D materials can bring many different properties that are not found in common polymeric materials such as high mechanical strength, thermal conductivity, electrical properties, barrier, flame retardancy, anti-friction etc (figure 18.2). 2D material incorporation can be highly utilized in polymer-based 3D printing, where additives can provide a lot of benefits

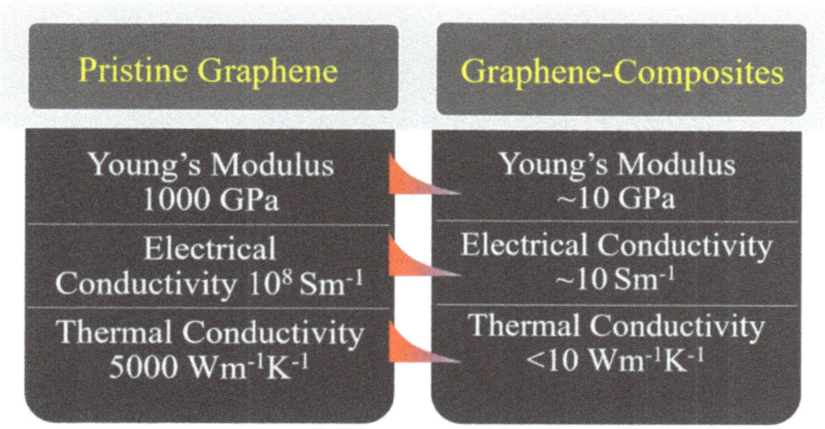

Figure 18.2. A comparison of properties of pristine graphene and graphene-based composites [14].

like higher strength and thermal/electrical conductivity [14]. GRMs can benefit from a strong growth in this sector (20%–30%) for consumables [15], ~10% for the 3D printing market [16].

In addition to being used as coating materials, GRM can be used as an additive to the metal and polymer composites to enhance related properties. Metal matrix composites are used to enhance functionalities such as lightweight metals for transport applications, improved metals for batteries and supercapacitors or to improve the corrosion resistance. On the other hand, ceramic materials can also benefit from the emerging properties of the GRMs. For example, for refractory materials, enhancement of material property can result in thermally stable composites with superior mechanical strength at high temperatures.

18.4.1.2 Opportunities:

Main application of bulk 2D materials as composites can be categorized in two main divisions: additive to bulk solids in form of composites and additive to fluids. There are several different application subdivisions under each of these two broad categories, as noted in figure 18.3.

These application areas have a broad variety of possible markets. Basically, any market that depends on or gains from material developments can use the improved composites. Looking at the composites market more broadly, it was predicted that in terms of the value of end products, the global market will be approximately $90 billion in size by 2020 [17]. The global market for functional composites is still developing, and the majority of composites are employed to utilize their superior mechanical strength and light weight. For instance, the market for conductive polymers is still in its infancy. Consumer products and electronics account for 25% of end users, followed by building, construction, storage, and pipeline (23%) and transportation (22%). Thermal (29%), electrical (23%), and barrier (19%) are the primary functions [18]. The growing popularity of GRM composites in materials research and application development is evident from a steep increase in the number

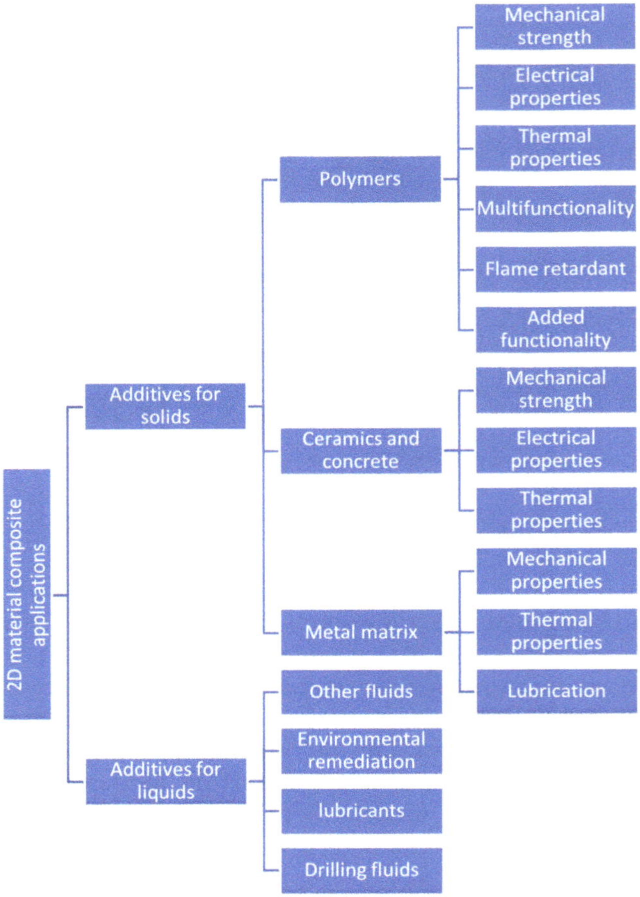

Figure 18.3. 2D materials composites and their applications. Adapted from [11].

of scientific publications and filed patents featuring 2D material composites over the past decade (figure 18.4).

18.4.1.3 Challenges

Despite the great hype surrounding graphene/2D material enhanced composites, challenges exist for its market takeover. The market targeted by these composites is already very mature, dominated by a multitude of other composites. There also exists a well-established market of composites with various nanoadditives in different hosts. Among them, the major challenger to graphene is carbon nanotubes (CNTs) due to their enhanced properties similar to that of the former. This material marketed much before graphene is better understood as a result of many decades of intense research. Thus, despite the advantages that graphene materials have due to their multifunctionality, it appears that many other composites will be competing with them in the market for at least a few more decades. Thus, for graphene-based

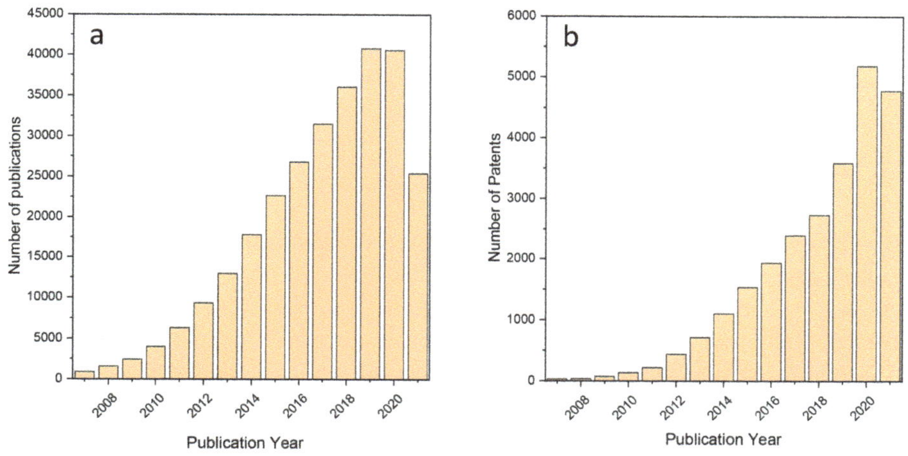

Figure 18.4. (a) Number of publications containing the phrase graphene/2D material composite from 2007, Source: Web of Science. (b) Number of patents containing the phrase graphene/2D material composites from 2007, source: Google Patents.

materials to dominate the market, a clear advantage compared to the competing materials should exist.

Another important challenge is the sourcing of materials which is subject to significant restrictions. Currently, the immense need for graphene marketization is not fully met due to the limited volume production. This is one of the biggest challenges for the graphene industry and it should be overcome. Along with that, standardized testing and quality control are crucial. It is also important to have a second source or supplier to provide the material of the same quality. Finally, a material can only be quickly adopted if it can be easily incorporated into the current production processes, and requirement of new equipment would substantially delay the implementation.

18.4.1.4 Challenges: outlook

One of the major challenges associated with the marketization of graphene-based products is that although enhanced properties such as higher Young's modulus, better thermal and electrical conductivities etc are demonstrated for them in the lab, they are not properly studied or tested for various environments and scales suitable for the desired applications (figure 18.2). There are other issues such as mutually conflicting changes in properties leading to a deteriorated overall performance. For instance, in certain composites, while the mechanical strength increases, their elongation and energy absorption before breaking diminish, which is bad for impact performance. It is important to determine which 2D material combination/configuration along with filler concentration and host gives the maximum performance for the desired application. A comparison of the cost and performance of these 2D materials-based products against those of their market rivals is essential based on which systematic improvements can be made to enable market domination.

The early products only encourage the use of graphene in a few simple specialized applications, which can only be viewed as a foreshadowing of what is to come in the following years.

18.4.2 Barriers and coatings

18.4.2.1 Introduction

Barriers and coatings are integral to the human civilization from its earliest stages. The main application of coatings through the ages has been in the protection and decoration of materials. The advancement of human civilization has further broadened their use. The excellent impermeable nature of graphene stems from its unique atomic structure coupled with robust mechanical strength comparable to steel, and is an enabling factor in designing anti-corrosion coatings and barrier films. In general, the preference for graphene and graphene-based materials is due to their immense scope to provide multifunctional properties in coatings in addition to their atomic thickness, flexibility, and chemical inertness. 2D materials are usually incorporated into a binder and can be applied as single- to few-layers large sheets or as flakes/platelets in paints or inks. Though application of paints/coatings in which the 2D materials are in single/few-layers form is more advantageous, it is difficult to mass produce. On the other hand, incorporation of 2D material in paints/coating as flakes/platelets is easily achieved. Though not to the extent of single/few-layer flakes, such type of incorporation of 2D material also achieves some interesting properties. In particular, the high mechanical stability of graphene can boost the surface coating's wear resistance, which, along with the high thermal and electrical conductivity, may be advantageous in some applications. Though the intrinsic properties of graphene promise high performance, realization of these properties in real-world applications is difficult. Such as for application in paint, basic technological aspects like achieving single-layer dispersion or possible interface interactions between polymers and 2D sheets might prove detrimental. These aspects still need to be studied in detail for better understanding in order to unlock the huge promise and realization of the full potential of 2D materials in such applications. Keeping in mind the complexity and diversity of modern coating systems, graphene and related 2D materials might find their application in improving specific functions and replacing conventional layers in a coating stack. The interest in graphene and other 2D materials in the coating industry can be extremely enhanced when the technology to replace the function of several layers of polymeric materials can be reliably done with few-layer thick 2D material paint. This will not only reduce overall material consumption by replacing several polymer layers with a single layer of 2D material-based coating but will also decrease effort and time necessary in such processes. In addition to replacing functionalities of polymeric layers, graphene and other 2D materials have also other interesting properties that are not commonly found in organic polymers such as electromagnetic properties, which can be used to tune the transparency or UV blocking of the coating. Potentially interesting fields include electroactive coatings, such as the ones used for electrochromic applications, where 2D inks might offer an intriguing option. Additionally, tunable surfaces are

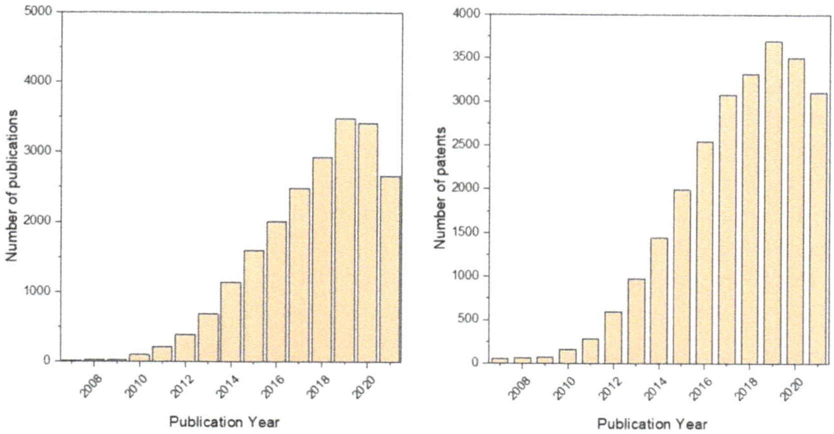

Figure 18.5. (a) Number of publications containing the phrase graphene/2D material surface and coatings from 2007, Source: Web of Science. (b) Number of patents containing the phrase graphene/2D material coatings/barriers from 2007, source: Google Patents.

conceivable, such as transparent coatings with tunable wettability and electrical conductivity. The versatility increases with the ability to build multilayer structures using various 2D materials.

If we consider graphene/GRM-based coating as a topic, the number of research articles and patent applications that contain the phrases graphene and coating are increasing each year showing increasing interest in this area (figure 18.5).

18.4.2.2 Market opportunities

Major industries experience significant financial losses due to corrosion. In the Asia Pacific area alone, the market for coatings and paints was anticipated to be greater than USD 70 billion in 2020 and is projected to reach USD 100 billion in 2028 [19]. The global market for anti-corrosion coatings alone will top USD 20 billion by 2027 [20], with Asia accounting for the majority of sales, followed by the US and Europe. Construction, aerospace, automotive, oil and gas, and marine industries are supposed to be the largest markets for anti-corrosion coatings.

Due to its unique characteristics, graphene is explored as an additive or pigment in coatings. By fine tuning the properties these coatings can be used for a plethora of applications. Utilizing the material's hydrophobicity [21], the substrate surface can be endowed with self-cleaning, antibacterial [22], and anti-biofouling [23] capabilities. If adequate orientation and density of the particles are attained during the coating process, it can build up superior barrier layers because of its chemical inertness and high aspect ratio. Furthermore, it has been demonstrated that GO, when applied directly to metal substrates, may effectively reduce the oxygen permeability of polyethyleneimine [24], provide a water vapour barrier, or function as an anti-corrosion coating [25]. Thus due to the large range of potential uses these coatings have, the market opportunities for graphene-based coatings continue to be enormous.

18.4.2.3 Market threats

In most coating applications, the 2D materials are either applied as sheets that are single/multi-layered/flakes in paints or as inks that are typically mixed with a binder. The first method helps to achieve better functionality, yet is expensive and tricky. The biggest issue is that it is not suitable for mass production. The second method on the other hand is simple and less costly, yet limited by low transparency. Apart from these complex choices, the incoherent reporting of the properties of graphene-based coatings also remains a threat to marketing. For example, while there are reports hailing the exceptionally good anti-corrosion properties of graphene [26] some other reports contradict these [27]. Such conflicting results might be a result of different forms of graphene with poor specifications, low quality and inconsistencies in parameters like size, shape etc being sold in market, which is one of the greatest challenges associated with graphene supply. A large difference in the observed properties in this context appears less surprising because it adds another level of complexity to estimating true corrosion protection characteristics of graphene when combined with various application concepts, non-optimized substrate materials, and testing conditions.

For virtually all coatings, resistance to wear and scratching is necessary. It is necessary to demonstrate this mechanical stability under strain and over time for graphene-based coatings to be commercially feasible. The wear and scratch resistance must also be tested in different environments where the coating would be employed. Studying the interaction of coating with various substrates must be studied to prevent delamination as some coatings would not be compatible with certain substrates. These are important attributes to be specified once the coating is in the market.

18.4.2.4 Outlook

2D material-based coatings and paints cover a major variety of potential applications of graphene and other 2D materials. Especially regarding graphene, there have been many early adopters for its use in paints and as coatings. However, the cost/benefit trade-off is a huge barrier to achieve the potential of 2D materials in coatings and paints. There is often an additional barrier to adopt new technologies like graphene in a market that is already matured with established coating materials. To have graphene and other 2D materials lead the market, several challenges such as cost/benefit, reproducibility and quality of coatings, reliability, and wear, environmental, health and safety concerns need to be addressed.

18.4.3 Membranes for separations and water treatment

18.4.3.1 Introduction

Membranes play a crucial role in human civilization. Their importance is rising day by day fueled by current needs and demands of energy efficiency, environmental needs, and water scarcity. Membrane-based separation techniques are very useful in chemical industries for economical as well as environmental reasons. Chemical separations account for about half of US industrial energy expenditure and 10%–15% of total energy expenditure of the country [28, 29]. Membrane-based separation techniques that do not require heat can save up to 90% of costs involved in distillation-based separations, which constitute a major part of the separation process, in turn saving 80% of total energy consumption for separation processes [29].

On the other hand, water scarcity is one of the most urgent challenges to be addressed because of worldwide shortage of clean drinking water. At present the most efficient membrane-based high performance water treatment technologies involve polymer-based reverse osmosis, micro/ultra/nanofiltration, membrane distillation and electrodialysis. However, all these areas suffer from several drawbacks such as membrane fouling, over scaling, high operation and investment costs. Research exploration has significantly advanced the membrane technology for such purposes, but over time the more stringent regulation for emerging organic contaminants in the water makes implementation challenging. Graphene and related 2D materials (GRMs) can provide a valid solution for such challenges [30]. Such materials possess some of the most desirable properties in membrane technology, such as mechanical strength, low dimensionality, chemical inertness, tunable porosity and hydrophilicity, chemical versatility and antimicrobial properties. In general, combining these properties gives 2D materials even further flexibility in mitigating the issues commonly encountered in polymer-based membrane solutions. The ability to functionalize GRMs offers an additional handle to tune their properties to our specific purposes. There is considerable industrial interest in tuning the hydrophilicity and hydrophobicity of water filtration membranes where functionalized GRMs can offer an attractive alternative.

Tunable properties of GRMs also lend themselves nicely towards removal of emerging contaminants, micropollutants, antibiotic resistive bacteria and process chemicals. GRM- based membrane solutions can be coupled with existing membrane modules for extra functionality. Graphene-based coatings can turn an ultra- or microfiltration membrane to nanofiltration membranes due to their restricted interlayer space. GRM-based filtration techniques can also be adapted to work for emerging needs in separation processes such as rejection of specific inorganic materials such as Pb or Hg or organic materials such as hormones. The compelling advantages of GRM-based membranes are unveiled from the growing number of publications and patent reports (figure 18.6), but their transition to the commercial sector is rather slow due to rivaling mature polymer-based membranes.

18.4.3.2 Opportunities

The desalination technology market is estimated to grow at a compound annual growth rate (CAGR) of 9.5% until 2028 [31]. The estimated number of operational desalination plants around the world are more than 15 000 supplying more than 90 million m^3 of water daily [32]. Africa and the Middle East are the regions where the most significant growth for desalination was observed as these areas contain

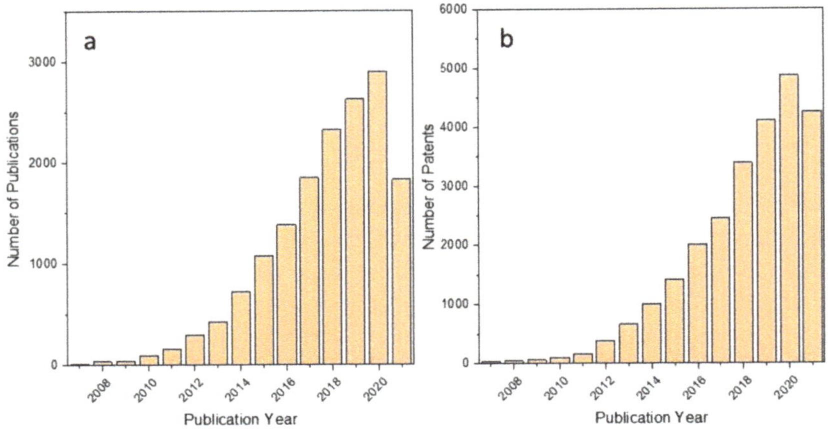

Figure 18.6. (a) Number of publications containing the phrase graphene/2D material membranes from 2007, Source: Web of Science. (b) Number of patents containing the phrase graphene/2D material membranes from 2007, source: Google Patents.

nearly half the desalination plants in the world. Other growth markets are the US, Latin America, and India [33]. In addition to desalination, there are regions where the freshwater available is highly contaminated and requires further purification steps before it can be available for drinking. The reduction of microbe levels and other pollutants are an integral part of water treatment.

Another potential market for GRM-related application is separation of industrially relevant compounds. About 10%–15% of the energy used worldwide is used in industrial separation operations. If applied to the US petroleum, chemical, and paper manufacturing industries only, methods to purify chemicals that are more energy-efficient might reduce annual carbon dioxide emissions by 100 million tonnes and energy costs by US$4 billion [34]. Other techniques might allow for the exploitation of new materials sources, such as the removal of metals from seawater. The global filtration and separation market, which was valued at USD 89 072.0 million in 2018, is anticipated to increase at a CAGR of 3.3% from 2019 to 2026 to reach USD 116 123.0 million [35].

18.4.3.3 Challenges

One of the major challenges for the successful application of membranes based on 2D materials is the demand to establish their durability. Enhanced durability and low maintenance are essential features for a new technology to become acceptable in the market. However, despite rapid testing, one will typically utilize technology in a conservative industry that has already proven to be reliable in practical applications. Currently, the low price of fossil fuels has also reduced the demand for lesser energy consumption technologies.

One of the most important applications of membranes, the desalination also presents certain challenges for 2D material-based membranes. Reverse osmosis is currently the most popular and cost-efficient desalination technology. There are also a number of older thermal desalination-based units. Other recently developed materials with potential features include thin film composites, silver nanoparticles, silica and carbon nanotubes. For capturing the market, there must be obvious advantages of 2D materials over these technologies and the current state of the art in terms of usefulness or performance. Application of 2D material membranes for pollution removal from freshwater also faces challenges because of strict cost constraints and requirements for durability and health and safety.

18.4.3.4 Outlook

If the area of graphene/2D materials as/in membranes is seen as a promising topic and is further pushed towards real-world application, the following potential actions, derived from the challenges, are suggested:

(a) The membranes should clearly show on par/improved performance from the state of the art in experimental condition.
(b) Develop scalable manufacturing concepts and compatible form factors such as hollow fibre for reverse osmosis (RO) applications.
(c) Address functionalization to improve selectivity.
(d) For freshwater purification, address the concerns for health and safety.
(e) Show improved performance/clearly demonstrate advantages over other competing emerging technologies.

Membranes for separation are increasingly important for purification of gases or liquids in an industrial process. 2D materials can be used for this in a variety of ways, either as a perforated membrane on their own or as an additive or coating for regular polymer membranes. In terms of membranes made from 2D material sheets with pores, the maturity is still rather low, however, the theoretical potential is promising. This idea has seen some industrial attention e.g., Lockheed Martin [36]. As an additive or coating to existing membranes, GRMs are easy to integrate and more mature. It has been shown that the additives improve mechanical stability, anti-fouling, and water permeability. Most essential, environmental, health, and safety issues must be addressed and verify before using water filtration methods. Market-wise, it is anticipated that there would be a rise in the need for producing drinking water and purifying gases or liquids.

18.4.4 Photocatalytic applications

18.4.4.1 Introduction

Photocatalysts hold the prospects to address some of the global challenges that the world faces today, including clean water and green energy. Photocatalysis involves light absorption by the semiconductor and generation and separation of excited carriers before they recombine to generate free radicals. However, the poor conductivity of the semiconductor material and abundance of recombination centers (defect) in them result in low photocatalytic efficiency. Graphene is considered as one of the best suited co-catalysts to improve the photocatalytic performance of the traditional semiconductors such as TiO_2, ZnO, CdS, ZnS etc. Incorporation of graphene significantly enhances the photocatalytic properties due to the improved conductivity and surface area of the composite material. However, there are several challenges still existing that are primarily linked to material quality. A more detailed description of advantages and challenges of graphene-based photocatalysts is presented in the following sections.

18.4.4.2 Opportunities

With the emergence of 2D materials, the art of engineering catalytically active materials has passed several performance milestones including highest total turnover frequency for a noble-metal-free catalyst [37], higher catalytic oxygen evolution than Ir-based noble-metal catalysts [38], about 2.4 times higher mass activities than that the benchmarking IrO_2 electrocatalysts [39] and exceptionally low onset over-potential in oxygen evolution reaction (OER) performance [40]. A race to define novel catalytic functions using 2D heterostructure materials with their precisely tuned intergalleries has begun [41, 42] and with the expanding library, 2D and their heterostructure-based materials offers innovative solutions to surpass the current technology bottlenecks in catalysis. For example the 'carbon monoxide poisoning' issue in fuels cells can be potentially alleviated by inclusion of a graphene layer on Pt (111) surface [43]. In photocatalytic applications, TiO_2 is conventionally used as a material for designing photocatalytic self-cleaning surfaces and air purifications. With tunable optoelectronic properties of 2D materials and their superlattices, they can function as superior photocatalytic enhancer as evidenced in the case of a graphene-incorporated TiO_2 system where the composite material is capable of absorbing light from the entire visible region and thereby overcoming the fundamental limitations on requirement of UV light sources of traditional TiO_2-based photocatalytic coatings [44–46] and realize self-cleaning or air-cleaning cementitious coatings for indoor applications. A major limitation of current photocatalytic disinfections coating is their relative inactivity in darkness [47]. As a prospective solution, wavelength independent absorption properties of graphene are also being extensively explored to design photocatalytic applications at mid-infrared wavelengths [48–50]. In photocatalytic H_2/O_2 production, beyond enhancing the performance of existing photocatalysts such as nitrogen-doped $Sr_2Ta_2O_7$ [51] and CdS [52], graphene-based materials are currently being explored towards the development of metal-free polymer by combining them with graphitic-carbon nitride [53–55], carbon super nitrides [56, 57] and $BiVO_4$ [58, 59]. The field is evolving at a rapid pace and graphene is focused as a standard additive in many photocatalytic applications that are being developed in scientific research.

18.4.4.3 Challenges

Commercial applications using graphene-enhanced photocatalytic coatings are yet to be demonstrated, due to technological barriers such as maintaining the homogeneous dispersion of graphene in graphene/TiO_2 composite materials, reliability and durability issues and high material cost of graphene (~100–700 €/kg). An engineering campaign to reform the current preparation technologies to improve the dispersion of graphene in various semiconductors and develop standards for testing photocatalytic performance are required as the way ahead for commercializing the potential of graphene and 2D materials-based photocatalytic enhancers. Besides, there is an immediate warrant to develop new methods to produce 2D materials with controllable size, layer, edge structure, heteroatom dopants and chemistry to identify their true potential. Another prevalent issue is lack of knowledge about the role of functional groups in controlling the n and p type character in graphene derivatives and role of various defects in graphene acting as charge recombination centers [60, 61]. For enabling long-term photocatalytic operations, it is also important to fully unveil the interface properties of oxide-semiconductor/graphene to understand the temporal variations such as interface carbon doping [62–64], structural changes in graphene [65, 66] in the composites etc. Nevertheless, *in situ* characterization techniques to probe the mechanism of catalysis and accurate theoretical models to carry out calculations at realistic conditions are needed to improve material reliability and drive the technologies towards commercialization.

18.4.4.4 Outlook

A facile way to step up the photocatalytic properties of graphene is via stacking or sandwiching them with other 2D materials where the charge transfer between the individual layers leads to enhanced surface chemical reactivity which is governed by the whole stack [67, 68]. Stacking also modifies their work function [69] and offers formidable solutions to define new hole and electron collection layers for photo-generated carriers. An alternate to approach to achieve the same feet is via intercalating the interlayers of graphene with 2D arrangement of metal atoms [70] or clusters [71, 72], leading to graphene superlattices with tunable electronic properties. Similarly, utilizing the surface charge of graphene or other 2D materials, one could create a superlattice containing alternate layers of positive and negatively charged 2D sheets via their self-assembly, as demonstrated in the case of graphene oxide and titanium oxide [73]. There are believed to be numerous opportunities in the design space for novel photocatalytic materials by combining 2D materials, 2D metals and 2D metal oxides from the vast family of layered [67] and covalently bonded 2D materials [74, 75] that are yet to be explored.

18.4.5 Beyond CMOS and spintronics applications

18.4.5.1 Introduction

With the continuing miniaturization of CMOS devices, there is an urgent demand for materials that can function as superior interconnects, ultrathin barrier layers, in heat removal components and atomically thin gate insulators [76]. As the number of transistors in microchips has doubled every year, the interconnect technology in chips becomes more and more compact, leading to inexorable issues such as strong increase in resistance and electromigration. As the interconnects currently account for the major share of power consumption in ICs and due to the growing demand for 3D integration of electrical components, novel material solutions for improved interconnect technology are inevitable soon. Similarly, there is pressing need for materials with superior diffusion barriers properties at ultra-scaled dimensions to function as barrier layers to prevent the diffusion of copper into dielectric layers of the transistors [77]. Currently, TaN/Ta-based diffusion barriers are employed in IC chips that are more resistive than Cu and occupy ~35% of the interconnect cross-section area. Thus, the current-carrying cross-sections of the interconnects are effectively reduced. It is also important to keep the TaN-based diffusion barrier thickness to ~3–4 nm to preserve the reliable diffusion barrier properties [78]. These limitations present a major bottleneck in continual scaling of interconnects in future generation technology nodes. An equally demanding sector that requires urgent material innovation is in the case of thermal interface materials (TIMs), which act as a conduction path between a heat generating device (e.g., transistors) and heat dissipating device (heat sinks) in circuits. Over 50% of energy consumption of global data centers is attributed to cooling the electronic components [79]. Two essential requirements of TIMs are high thermal conductivity (κ) that minimize the thermal resistance and low elastic modulus (E) which ensures good thermal contact area by effectively filling the air gaps under compression. Current commercial TIMs suffer from their low κ of ~1–10 W m^{-2} or high E of ~1 GPa [80, 81]. Structured carbon layers are extensively investigated as a next-generation technology for TIMs due to their low thermal resistance and excellent mechanical flexibilities [82, 83] and graphene-based 2D materials are expected to take the lead in future carbon-based TIMs.

Materials-based innovations in spintronics are also actively pursued to push the limits of future memory technology where tunnel barriers are an important building block of magnetic tunnel junctions (MTJs). Key issues with performance improvement in magnetic storage devices are challenges associated with scaling the thickness of oxide-based tunnel barriers, preserving the interface quality of ferromagnetic (FM) electrodes and oxide tunnel barrier layers and undesirable elemental diffusion across the tunnel junctions [84–86]. Incorporating atomically thin 2D layers evolved as a viable strategy to overcome these limitations and graphene-based channels with their longest spin diffusion lengths and high breakdown current capacity [87, 88] are expected to play a major role in future all-spin logic devices (ASLD) [89]. Magnetic devices where the barrier layers are sandwiched between ferromagnetic electrodes requires stringent control on the interface resistance to achieve 100% spin injection goal. The family of 2D materials offers a diverse spectrum of materials such as 2D ferromagnetic materials, 2D tunnel barriers, 2D spin transport channels with countless options to realize 2D heterostructure of these materials to realize complex spintronic and computing devices.

18.4.5.2 Opportunities

High conductivity and superior barrier properties of atomically thin graphene layers makes them an ideal fit to reform the future interconnect technology. Multilayer graphene holds great promise as CMOS interconnects with its superior current-carrying capacity and electromigration resistance [90–93]. Multilayer graphene interconnects with low resistivity via intercalation doping have shown to outperform the cobalt and ruthenium interconnects technology alternate to copper-based interconnects, and circuits with doped graphene interconnects operate with a

4× smaller delay than these metal-based interconnects at sub 20 nm line widths [91]. It is also shown via modelling that replacing copper with multilayer graphene can improve the operation of the processor cores with 9% higher speed or 16% less energy consumption and twice longer electromigration lifetime [92]. Graphene-based materials can play a vital role as diffusion barriers in circuits with their impenetrable ideal barrier energy of 30.6 eV [65] to prevent the diffusion of copper into active device components. Large-grain single-layer graphene and 1 nm graphene tri-layers have already proved superior to 2 nm TaN-based current state-of-the-art diffusion barriers [94, 95]. More importantly, 1 nm thick rGO layers are shown to retain their Cu diffusion barrier property under a thermal budget of 400 °C, which is sufficient for the back-end-of-line (BEOL) process [96] and offers robust adhesion to copper layers. Multilayer graphene and single-layer graphene offers resistivities (<100 μΩ cm) much less than 2–4 nm TaN layers (260–290 μΩ cm) is another major advantage in realizing ultrathin 2D materials-based diffusion barriers [95].

Graphene-based materials are attractive for thermal management systems in integrated circuits and 3D system-in-package chips owing to their excellent heat conduction properties stem from their unique sp^2 bonding, providing an efficient heat transfer by lattice vibrations [97, 98]. Currently, thermal interface properties of graphene are investigated by mainly incorporating graphene in polymer epoxy matrix [99–101]. Orientation of the graphene laminates in the polymer matrix is shown to have decisive role in governing the overall thermal conductivity of the composite. A number of strategies such as magnetic field directed assembly of magnetic nanoparticle anchored graphene [102], use of substantially thinner TIMs such as 'paper TIMS' [99–105], grafting graphene to polymer molecules such as polymethyl methacrylate and using them as fillers in epoxy matrix [106, 107] and by binding metal nanoparticles to graphene platelets to prevent their agglomeration [108] are used to improve the thermal dissipation properties of graphene. Proof of concept applications with graphene serving as heat spreader are demonstrated in gallium nitride transistors operating at 13 W mm^{-1} where the hotspot temperature was reduced by 20 °C [109] and numerical modelling revealed that in silicon on-insulator (SOI) structures by employing graphene heat sinks, the hotspot temperature is reduced by 70 K [110]. Recent demonstration of graphene-based TIMs with enhanced through-plane thermal conductivity of 143 W m^{-1} K^{-1}, exceeding that of many metals, and a low compressive modulus of 0.87 MPa, comparable to that of silicone, offers formidable solutions to realize next-generation graphene-based TIMs [111]. Currently, graphene-based thermal interface technology is commercially implemented in Huawei's Mate 20 and P30 Pro mobile phones [112] and is expected to play a major role in advanced circuits for the 5G era.

The ability to propagate electron spin information over tens of micrometer distance at room temperature stems from the weak spin–orbit coupling in graphene and presents a remarkable potential for commercial realization [113, 114]. Inducing a strong spin–orbit coupling via forming a heterostructure with other 2D materials and manipulating the spin using applied electric fields [115–117] is also an equally compelling functionality. Novel designs to manipulate the spin injection properties of these heterostructure devices can be realized by inserting a hexagonal boron nitride (h-BN) as tunnel barrier [118, 119]. Spin-filter magnetic tunnel junctions (sf-MTJs) based on van der Waals (vdW) heterostructure composed of graphene contacts and atomically thin four-layer chromium triiodide (CrI_3) revealed their remarkable potential for novel device options with an order of magnitude higher tunnelling magnetoresistance (TMR) than conventional MgO-based MTJs [120]. MTJs have paved the way for spin-transfer-torque magnetic random-access memories (STT-MRAM) that are already commercialized by Everspin [121]. As a next level of innovation, the writing efficiency and energy consumption of STT-MRAM can be significantly improved via introducing a perpendicular magnetic anisotropy (PMA) in ferromagnetic layers, and it is shown that graphene can dramatically enhance the PMA in ferromagnetic/graphene heterostructures [122]. To improve the performance gap in terms of high intrinsic energy consumption and delay in STT-RAM, spin Hall effect MRAMs (SHE-MRAMs) are also proposed where a charge current flowing in the spin Hall layer in contact with a recording magnetic layer exerts a spin torque on the recording layer [123–125]. To realize such ultra-compact memory devices, graphene-based heterostructure devices exhibit great scientific potential in the light of recent demonstration of room temperature gate-tunable spin-to-charge interconversion driven by SHE [126, 127].

18.4.5.3 Challenges

Prevalent edge scattering at scaled dimensions (~20 nm) and finite bandgap in nanoribbons have limited the conductivity of multilayer (ML) graphene interconnects [128, 129] and this diminishes their advantage as energy-efficient interconnects. The high CVD growth temperature of graphene of ~1000°C required to thermally dissociate the precursor molecules is incompatible with BEOL interconnect technologies [90]. The adhesion strength of graphene with interconnect materials such as copper is another major concern as a strong adhesion is vital to survive the harsh chemical–mechanical–polishing steps in chip fabrication stages to remove the unwanted copper in circuits. To be fully compatible with state-of-the-art BEOL technology, it is essential to develop the CVD growth of graphene directly on dielectric layers, while CVD growth of graphene is currently achieved on metal seeding layers that involves demanding wet etching and transfer steps, posing additional challenges to wafer-scale BEOL process. Moreover, the growth quality of graphene is highly dependent on the single-crystalline quality of metal substrate to avoid grain boundaries [130, 131] that primarily act as diffusion paths for Cu. Moreover, the incumbent technologies such as metal alloys for diffusion barrier applications or CMOS compatibility of Cu in interconnect technology present a significant entry barrier for graphene-based materials in CMOS related applications. There need to be a clear benefit in terms of performance or cost efficiency of graphene-based interconnects, that is yet to be demonstrated.

Despite their excellent in-plane thermal conductivities, another major hurdle in the widespread commercial uptake of graphene-based laminates/papers in thermal interface applications is their comparable though-plane thermal conductivity with their commercial counterparts due to their Van der Waals arrangement [132].

Besides, several factors such as the small flake size, high degree of structural defects, residual oxygen content, random orientation of individual flakes etc associated with industrial scale liquid exfoliation and self-assembly routes to graphene-based films are found to dramatically affect their thermal conductivities and elastic properties [133]. To further improve the through-plane conductivity, vertically aligned graphene platelets were demonstrated with superior though-plane thermal conductivity (615 W m^{-1} K^{-1}), however, their significantly high compressive modulus (~0.5 GPa) presents high contact thermal resistance in thermal interface applications [134]. Use of graphene as filler molecules in epoxy resins is also fraught with challenges such as lack of control of orientation of flakes in the matrix resulting in high thermal contact resistance between fillers [135], undesirable increase in electrical conductivity at high filler loadings [136] and poor thermal conductivity of polymer host matrices requiring high concentration of filler materials etc. Moreover, a direct growth of semiconductor layers for power or optoelectronic devices such as GaN or GaAs needs to be realized on graphene to function as heat spreaders which is a major challenge due to chemical inertness of graphene [137].

A major requirement to realize commercial spintronic applications of graphene is highly reproducible strategies to manufacture electronic quality graphene. For precise interface engineering using 2D heterostructure, CMOS compatible deposition and batch fabrication techniques are yet to be developed. Several discrepancies exist between theoretical predictions and experimental results in the case of spin relaxation time, energy dissipation in 2D materials, which indicates lack of accurate theoretical models [138, 139]. Another impediment is in the case of realizing pin-hole and surface state free metal-oxide barrier layers on graphene to enable spin injection and detection applications [139, 140]. Encapsulation of graphene using h-BN to improve the interface quality resulted in only a marginal improvement due to incomplete shielding of the polymer residues in fabrication stages by thin h-BN layers causing spin scattering in graphene [141]. Besides, magnetic 2D materials with high Curie temperature are still underdeveloped [142] and spin Hall conductivity of proximitized graphene often decays with temperature [126, 143].

18.4.5.4 Outlook

For the continual device scaling, along with next-generation semiconductors/device geometries, it is essential to have a significant innovation in interconnect materials. The diminishing advantages of graphene nanoribbon due to edge scattering problems can be addressed using intercalation doping in nanoribbons and doped graphene interconnects with superior performance and reliability over current copper-based interconnect technology is already demonstrated [144]. The current-carrying capacities of these doped graphene interconnects can be further improved by depositing on high thermal conductivity dielectrics, like hexagonal boron nitrides [144]. Catalyzed edge annealing on Cu(111) substrate [145], controlled hydrogen plasma etching [146] and growth of graphene nanoribbons in pre-patterned h-BN trenches with precise zig-zag edges [147, 148] are developed as promising strategies to improve the edge structure of graphene ribbons. A recently unveiled vdW doping strategy of graphene using monolayer tungsten oxyselenide offers ultra-clean doping strategies to achieve sheet resistance below 50 Ω sq^{-1} [149]. Towards BEOL, compatible plasma enhanced CVD (PECVD)-based synthesis methods for electronic quality graphene are also being developed with carrier mobility as high as 9000 $cm^2\ V^{-1}\ s^{-1}$ for graphene synthesized on AuNi using acetylene at 600 °C [150] and growth of graphene on dielectric substrates was also realized with career mobility in the range 550–1600 $cm^2\ V^{-1}\ s^{-1}$ [151]. A foremost requirement in developing superior diffusion barrier layers using ultrathin graphene layers is to establish a grain-boundary-free growth of graphene. A promising CVD growth method for single-crystal graphene layers exceeding the size of 1.5 inch by using local carbon precursor feeding on optimized $Cu_{85}Ni_{15}$ alloy substrates [152] and large-area fold-free single-crystal graphene via tuning the growth temperature [153] has recently been demonstrated. Research data on the barrier properties of these defect-free single-crystalline 2D layers via time-dependent dielectric breakdown tests [154, 155] need to be developed to bench mark their performance. These developments in terms of BEOL-compatible low temperature growth, strategies to tune the edge structures and intercalation doping methods to improve the breakdown characteristics may drive the technology towards commercial level implementation of graphene-based interconnects for continual device scaling.

Commercial graphene-based thermal interface pads are now produced by several manufacturers including Smart High Tech AB [156] and Incubation alliance Inc. [157]. Graphene monolith TIMs exhibit best performance in terms of their high thermal conductivity, while graphene–polymer composite TIMs offer best processability, but suffer from their low thermal conductance. To improve their reliability and heat radiation properties, the influence of binders or alignment promoters used in the matrix need to be assessed before developing their commercial applications. Rapid progress in the synthesis methods for high quality graphene materials may improve the though-plane thermal conductivity bottleneck of graphene-based TIMs. It is also of paramount importance to optimize the contact thermal resistance and elastic modulus of the graphene-based TIMs to enable their widespread applicability in commercial heat spreaders.

The future developments in graphene-based MTJs will depend on realizing the best combination of vdW materials with graphene and their direct growth on ferromagnetic contacts. Several factors need to be addressed towards this goal such as chemical state and crystallinity of FM, conductivity mismatch between FM and graphene channels, 2D materials thickness or functionalization and effect of intercalation of FM at 2D material interfaces, etc. Novel ferromagnetic materials with high T_c well over room temperature [142] and voltage control of the spin–orbit coupling in 2D FMs that forms magnetic and spin valves with graphene layers may open up additional spintronic opportunities [158, 159]. It is also predicted that a device-consisting of a bilayer graphene sandwiched by a 2D ferromagnet and a 2D transition metal dichalcogenide layer, called 'ex-so-tic' vdW heterostructure, is capable of swapping the exchange and spin–orbit coupling in a single device [160] and expected to play a key role in spintronics and quantum metrology. Realization of super-clean interfaces with controllable interlayer twist between magnetic vdW layers (twistronics) may also uncover countless options to tune the spin–orbit coupling in graphene [161].

18.4.6 Biomedical applications

18.4.6.1 Introduction

Extensive exploration of biomedical prospects of CNTs in the past decades uncovered several potential application areas where graphene can be incorporated and make transformative changes. Most of the clinical and pharmaceutical applications of graphene were built on the experience with CNTs in biomedical applications. However, the considerable advantages of graphene over CNTs such as versatile functionalization options, planar structure and low-cost synthesis methods etc are expected to result in rapid development in potential areas including biosensors, drug delivery and biomolecule separation. Despite some developments on assessing the biocompatibility of CNTs and graphene, these materials are still far from obtaining regulatory approval to be used in clinical applications. However, the fast-paced growth of graphene technology and demand of new materials in biomedical research to improve the efficiency of sensors, therapies and diagnosis methods maintains a huge interest in unlocking novel biomedical applications using graphene. The following sections discuss their potential advantages, barriers to overcome and their outlook.

18.4.6.2 Opportunities

The suitability of graphene in major commercial sectors of the biomedical industry such as drugs and pharmaceuticals, medical devices and equipment and medical research and testing has been widely explored over the past years [162–164]. The intrinsic near infrared (NIR) absorption properties of graphene-based materials are expected to be utilized in photodynamic (PD) and photothermal (PT) therapies in cancer treatment [165, 166]. The remarkably high $^1O_2$ generation efficiency of graphene oxide will also be exploited in the development of simultaneous imaging and perform *in vivo* PD therapy. Oxidation degree in graphene can be chemically or thermally modified to tune the NIR absorption properties and distinct polar/non-polar drug loading capacity can be achieved to design new treatment protocols. Three-dimensional assemblies of graphene and polymers are most pursued in nanomedicine as a multifunctional scaffold as substrate for the growth and differentiation of cells, or as implants to replace tissues in living beings [167–169].

18.4.6.3 Challenges

One of the main outstanding challenges to tackle is the toxicity issues arising primarily due to the incomplete removal of contaminants from the synthesis stage of graphene-based materials. These contaminants are mainly toxic metal ions from CVD growth/transfer process or from the modified Hummer's process [170–173]. Moreover, safety issues related to local heating, inhomogeneous drug release efficiency at changing pH conditions in the intracellular environments [174], lack of assessment of their long-term toxicity and challenges with their recovery after the targeted therapeutic treatments etc [163] pose major roadblocks for using them for human clinical trials. Lack of understanding on the controllable functionalization of graphene with biolinkers also limits their application in medical devices.

Though antimicrobial activity of graphene is well-known [175, 176], there are contradictory reports on the enhanced microbial growth in the presence of graphene [177, 178], which means the antimicrobial properties strongly linked to experimental conditions or material properties and detection mechanism on the antimicrobial properties need to be further developed. Similarly, it is also important to apprehend the haemolytic activity of hydrophilic graphene moieties that lead to thrombotoxicity [179], which also seems to be strongly linked to material properties [180].

18.4.6.4 Outlook

As a primary step, industrial scale production capability of 'bio-grade' quality graphene should be developed which is devoid of any heavy metal traces and other harmful synthesis debris. Simultaneously, a clear understanding on the toxicological and pharmacokinetic profiles of graphene [181] with different flake size, shape, oxidation state, dispersion quality, concentration and the corresponding variations in their physicochemical properties should also be developed and the critical nanosafety criteria based on these clinical research outcomes should be constantly updated. Enzyme-aided biodegradation of graphene [182] and enhanced biocompatibility of graphene via surface functionalization [183, 184] open up possible avenues to circumvent some of the challenges associated with cytotoxicity of graphene. A coordinated effort by involving physicists, chemists, material and clinical scientists is required to explore the biomedical potentials of graphene-based materials.

18.5 Standardization problems and new recommendations

A nagging challenge faced by graphene application development firms is ensuring a steady and assured supply of graphene from manufacturers following stringent quality control measures [173, 185–189]. As a next-generation material technology propelled to address several grand challenges for society including clean water, renewable energy, sustainable food processing and health care applications, lack of adequate material quality for application development not only impedes the early development of such applications, but at the same time poses a severe safety risk. As per the definition provided by the International Organization for Standardization (ISO), a stack of graphene crystals with more than ten layers at room temperature ($T \approx 300$ K) can be treated as bulk graphite [190]. A recent quality assessment study performed using graphene samples from more than 60 suppliers across the globe revealed that the quality of graphene produced is suboptimal to realize many of the proposed applications and a final product containing at least 50% graphene content is scarce [187]. Besides, the global market of graphene covers a wide range of products such as graphene flakes, graphene oxide, rGO and graphene films. Each of these product categories possess individual signature properties and exhibit significant differences in their lateral size, thickness, defect density and material properties. The structural features of these materials are also strongly linked to synthesis procedures, making it challenging to source identical product quality across various suppliers. These limitations in materials availability and quality have limited the commercial applicability of graphene for niche applications such as composites, lubricants and coatings etc. To unlock the commercial potential of graphene in diverse areas, it is of paramount importance to ensure the widespread availability of high-quality graphene-based supplies. A firm footing towards this important requirement is to establish a meticulous standardization scheme for graphene-based supplies. To minimize the variability and improve the material quality, a new international standard on measurement technology 'ISO/TS 21356-1 Structural characterization of graphene from powders and liquid dispersions' has been published [191]. This standard includes the measurement techniques required to assess the quality of graphene material manufactured and provides a methodical flow chart to enable the efficient decision making process and associated measurands to be considered (figure 18.7). The European Union though its 1 Billion Euro science initiative (Graphene flagship) has formed a graphene flagship standardization committee (GFSC) and currently hosts more than 35 projects to form new guidelines on nomenclature, metrology, characterization and quality control to graphene suppliers and buyers [192]. National Physical laboratory of UK in participation with University of Manchester has also published a guide using accepted metrology to assess the quality of industrially supplied graphene samples [193]. To promote the rapid commercialization of graphene, China has published a number of standards for industries covering material measurements, methods, application and terminology that are summarized in a recent short communication by Zhu *et al* [194]. With a number of standards now globally available for graphene, it is also important to establish coherent benchmarking criteria to help the supply chains to assess, design and deliver reliable graphene products with minimum variability in performance to

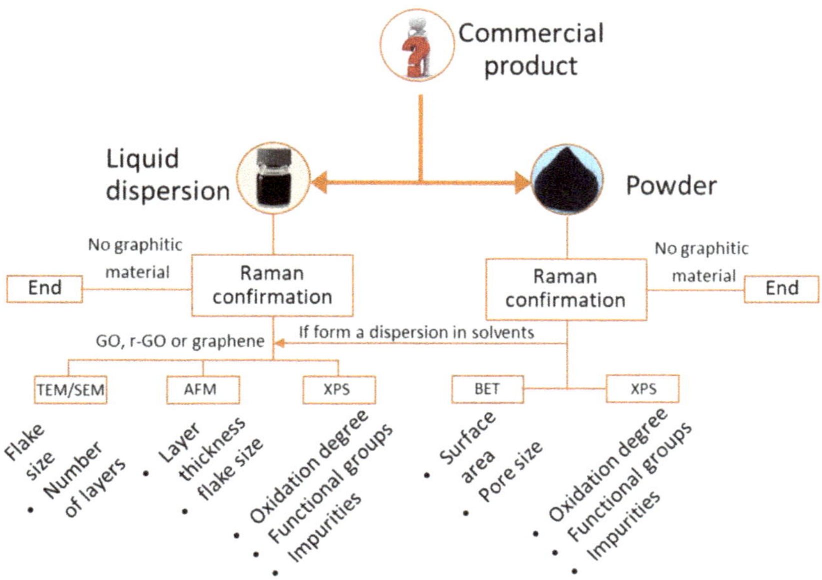

Figure 18.7. A flow chart of the measurement techniques to assess the quality of graphene-based materials and main material properties to be considered (based on ISO/TS 21356-1 published standard for graphene suspensions/powders).

end users. Such globally accepted standards are essential to eliminate the current stumbling blocks in materials synthesis and characterization and drive the graphene-based technology towards a fully-fledged commercialization. These efforts may form the basis for a full ISO standard for graphene by 2025.

18.6 A roadmap for GRM technology

Ever since the first successful isolation of graphene in 2004, there have been rapid developments in terms of understanding their physical properties, large-scale synthesis, developing plethora of applications and integration into systems. Beyond silicon, graphene meets the criteria as a legacy material to enable universal applications with the caliber to make disruptive changes in various sectors of daily life by its unmatched material properties, such as being transparent as plastic, but seamlessly conducting heat and electricity, with superior flexibility and mechanical strength, solution processability, highest carrier mobilities, high intrinsic capacitance and specific surface area, photocatalytic effects, chiral superconductivity and tunable spin–orbit coupling etc. The surge in graphene has led to the development of several other 2D materials with metallic, semiconducting, and insulating properties. Combining graphene with this large family of 2D materials with tailored properties may unveil a cornucopia of new applications and push the boundaries of several existing technologies. Despite these superlatives, major hurdles in realizing cost-effective material scale-up techniques while preserving material quality has diminished the high expectations created by academic research. However, graphene finds its way into our daily life via thriving sectors such as composites, plastics, anti-corrosion coating and barrier films, thermal interface materials etc, where clear signs of market adoptions by leaders in technology are evident [195]. Graphene is evolving as an industrial material with flake production capacity exceeding thousands of tonnes per year. As a guidance to potential stakeholders before the practical supply chain begins, the Fraunhofer Graphene Roadmap Team as part of the European graphene flagship programme has initiated an evolving technology and innovation roadmap (TIR) for GRMs by compiling the latest status and prospects of GRM industrialization [12].

The large corporation at original equipment manufacturers level in the supply chain now plays only a passive role in the GRM industry as the market potential of GRMs is still unclear due to outstanding challenges in cost, commoditized supply and standardization. GRM suppliers hence struggle to scale up their production due to the lack of market demand and thus funding. The fate of GRM commercialization is entangled in this dilemma and a coordinated effort between scientist and application developers for validation, standardization and regulation process is essential to keep up the momentum and facilitate the market penetration of new products that have the potential to solve real-world problems. Moreover, as the initial GRM products originate from SMEs or spin-offs, a trusted relationship and information exchange between them is vital to be promoted to develop graphene as an industrial material and uncover its realistic market potential. To clear the path towards the development high reward applications (photovoltaics, spintronics,

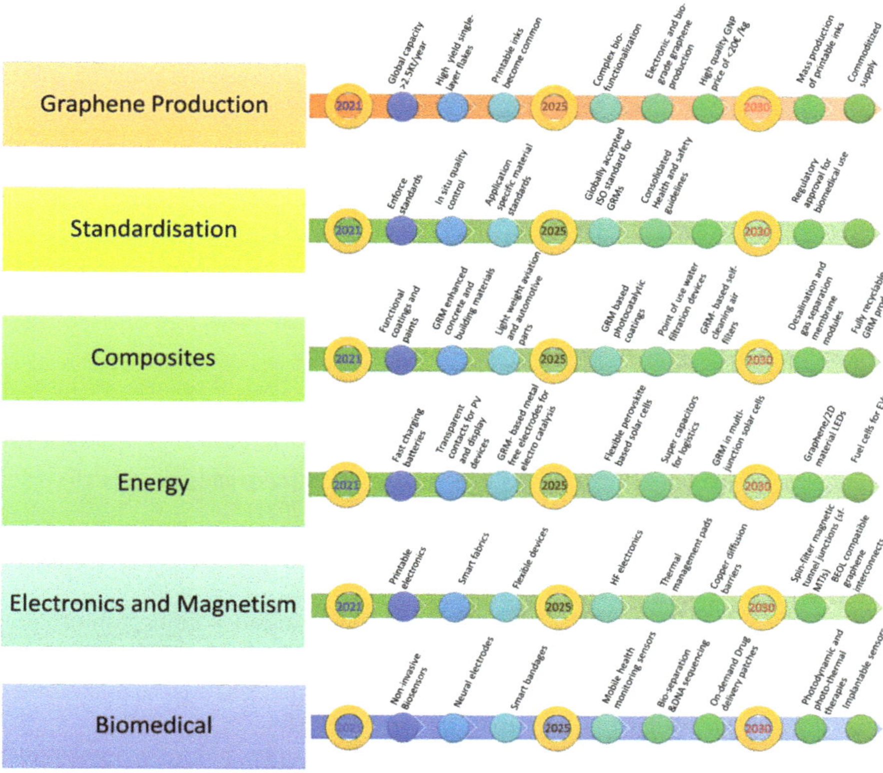

Figure 18.8. A roadmap for production, standardization and application development for GRM technology (based on EU graphene flagship's industrialization roadmap 2020).

neural and biomedical sensors etc), the market share of the initial niche products such as coatings, rubber/polymer composites, concrete etc should significantly step up. Market leaders will eventually step in for large-scale commercialization of the technology and established material suppliers may fill the gap between supply and demand. The spintronics and biomedical applications of GRMs may be fully developed by 2030 once the ISO standard is established along with commercial supply of electronic and bio-grade graphene (figure 18.8). The rise of GRM technology from these global efforts will lead us towards a way to use carbon in a sustainable way and make several other technologies carbon neutral or negative.

References

[1] Novoselov K S, Geim A K, Morozov S V, Jiang D E, Zhang Y, Dubonos S V, Grigorieva I V and Firsov A A 2004 Electric field effect in atomically thin carbon films *Science* **306** 666–9

[2] Graphene flagship annual report., 2020. Available at: https://graphene-flagship.eu/media/dwweq5sw/graphene-flagship-annual-report-2020_final.pdf (accessed 14 November 2021)

[3] Nature Materials 2021 Graphene on the pilot line *Nat. Mater.* **20** 573

[4] Schiermeier Q 2021 How Europe's €100-billion science fund will shape 7 years of research *Nature* **591** 20–1
[5] Graphene Commercialization Promotion Technology Roadmap – Korea Policy Briefing, 2015. Available at: https://korea.kr/news/pressReleaseView.do?newsId=156054402 (accessed 14 November 2021)
[6] Global and China Graphene Industry Report, 2020–2026., 2021. Available at: https://reportlinker.com/p04539093/Global-and-China-Graphene-Industry-Report.html?utm_source=PRN (accessed 14 November 2021).
[7] Skeleton Technologies, 2020. Available at: https://graphene-info.com/skeleton-technologies-working-graphene-superbattery-evs-signs-1-billion-loi (accessed 15 November 2021).
[8] HUVIS, 2021. Available at: https://huvis.com/eng/prCenter/NewsView.asp?news_seq=592 (accessed 15 November 2021).
[9] Graphene@Manchester solves concrete's big problem, 2021. Available at: https://manchester.ac.uk/discover/news/greener-and-cheaper-graphenemanchester-solves-concretes-big-problem/ (accessed 15 November 2021)
[10] Graphene Market Size Forecast, 2020. Available at: https://grandviewresearch.com/press-release/global-graphene-market (accessed 15 November 2021)
[11] Meister M, Braun A, Husing B, Schmoch U and Reiss T 2017 Graphene and other 2D materials Technology and Innovation Roadmap Version 3 Available at: http://www.isi.fraunhofer.de (accessed 6 November 2021)
[12] Döscher H, Schmaltz T, Neef C, Thielmann A and Reiss T 2021 Graphene Roadmap Briefs (No. 2): industrialization status and prospects 2020 *2D Mater.* **8** 022005
[13] IATA – Technology Roadmap (2020). Available at: https://iata.org/en/programs/environment/technology-roadmap/ (accessed 18 November 2021)
[14] Samorì P, Kinloch I A, Feng X and Palermo V 2015 Graphene-based nanocomposites for structural and functional applications: using 2-dimensional materials in a 3-dimensional world *2D Mater.* **2** 030205
[15] Fullerex (2014) Fullerex Report. Available at: https://cdn.ymaws.com/www.thegraphenecouncil.org/resource/resmgr/Files/Bulk_Graphene_Pricing_Report.pdf
[16] Emergence of 3D Printing Materials, 2016. Available at: https://reportlinker.com/p03621330/Emergence-of-3D-Printing-Materials-.html (accessed 18 November 2021)
[17] Global Composites Market 2015–2020, 2015. Available at: https://businesswire.com/news/home/20150429005414/en/Research-and-Markets-Global-Composites-Market-2015–2020—Type-Manufacturing-Technology-Resin-Type-Application-Analysis (accessed 1 November 2021).
[18] Functional Composites Market, 2021. Available at: https://marketsandmarkets.com/Market-Reports/functional-composite-market-164726061.html (accessed 1 November 2021)
[19] Paints and Coatings Market Size, Share, Report, 2021. Available at: https://fortunebusinessinsights.com/industry-reports/paints-and-coatings-market-101947 (accessed 8 November 2021)
[20] Global Market Insights Inc. Report, 2021. Available at: https://prnewswire.com/news-releases/the-anti-corrosion-coatings-market-is-projected-to-exceed-20-billion-by-2027-says-global-market-insights-inc-301249785.html (accessed 18 November 2021)
[21] Ghauri F A, Raza M A, Baig M S and Ibrahim S 2017 Corrosion study of the graphene oxide and reduced graphene oxide-based epoxy coatings *Mater. Res. Express* **4** 125601

[22] Yousefi M, Dadashpour M, Hejazi M, Hasanzadeh M, Behnam B, de la Guardia M, Shadjou N and Mokhtarzadeh A 2017 Anti-bacterial activity of graphene oxide as a new weapon nanomaterial to combat multidrug-resistance bacteria *Mater. Sci. Eng.* C **74** 568–81

[23] Armendáriz-Ontiveros M M, García A G, De Los Santos Villalobos S and Weihs G F 2019 Biofouling performance of RO membranes coated with iron NPs on graphene oxide *Desalination* **451** 45–58

[24] Liu H, Kuila T, Kim N H, Ku B C and Lee J H 2013 In situ synthesis of the reduced graphene oxide–polyethyleneimine composite and its gas barrier properties *J. Mater. Chem.* A **1** 3739–46

[25] Gomasang P, Kawahara K, Yasuraoka K, Maruyama M, Ago H, Okada S and Ueno K 2019 A novel graphene barrier against moisture by multiple stacking large-grain graphene *Sci. Rep.* **9** 3777

[26] Prasai D, Tuberquia J C, Harl R R, Jennings G K and Bolotin K I 2012 Graphene: corrosion-inhibiting coating *ACS Nano* **6** 1102–8

[27] Schriver M, Regan W, Gannett W J, Zaniewski A M, Crommie M F and Zettl A 2013 Graphene as a long-term metal oxidation barrier: worse than nothing *ACS Nano* **7** 5763–8

[28] Humphrey J L and Keller G E 1997 *Separation Process Technology* (New York: McGraw-Hill) pp 113–51

[29] Oak Ridge National Laboratory 2005 *Materials for Separation Technologies: Energy and Emission Reduction Opportunities* (Oak Ridge, TN: Oak Ridge National Laboratory)

[30] Liu G, Jin W and Xu N 2016 Two-dimensional-material membranes: a new family of high-performance separation membranes *Angew. Chem. Int. Ed.* **55** 13384–97

[31] Global Water Desalination Forecast 2020–2028, 2021. Available at: https://adroitmarketresearch.com/industry-reports/water-desalination-market (accessed 9 November 2021)

[32] Jones E, Qadir M, van Vliet M T, Smakhtin V and Kang S M 2019 The state of desalination and brine production: a global outlook *Sci. Total Environ.* **657** 1343–56

[33] Global Water Desalination Market Outlook & Forecast Report 2021–2026,. 2021. Available at: https://businesswire.com/news/home/20211028005641/en/Global-Water-Desalination-Market-Outlook-Forecast-Report-2021-2026—ResearchAndMarkets.com (accessed 9 November 2021)

[34] Brueske S, Kramer C and Fisher A 2015 Bandwidth Study on Energy Use and Potential Energy Savings Opportunities in US Petroleum Refining (No. DOE/EE-1230). Energetics

[35] Filtration & Separation Market Size, Share and Growth Forecast – 2026,. 2020. Available at: https://alliedmarketresearch.com/filtration-and-separation-market-A05976 (accessed 9 November 2021).

[36] Perforene Membranes, 2021. Lockheed Martin Perforene Membranes. Available at: https://lockheedmartin.com/en-us/products/perforene-graphene-membrane.html (accessed 27 November 2021)

[37] Feng K, Zhong J, Zhao B, Zhang H, Xu L, Sun X and Lee S T 2016 $Cu_xCo_{1-x}O$ nanoparticles on graphene oxide as a synergistic catalyst for high-efficiency hydrolysis of ammonia–borane *Angew. Chem. Int. Ed.* **55** 11950–54

[38] Ma T Y, Dai S, Jaroniec M and Qiao S Z 2014 Graphitic carbon nitride nanosheet–carbon nanotube three-dimensional porous composites as high-performance oxygen evolution electrocatalysts *Angew. Chem.* **126** 7409–13

[39] Huang J *et al* 2015 CoOOH nanosheets with high mass activity for water oxidation *Angew. Chem.* **127** 8846–51

[40] Liu Y, Xiao C, Lyu M, Lin Y, Cai W, Huang P, Tong W, Zou Y and Xie Y 2015 Ultrathin Co_3S_4 nanosheets that synergistically engineer spin states and exposed polyhedra that promote water oxidation under neutral conditions *Angew. Chem.* **127** 11383–7

[41] Xiong P, Ma R, Sakai N, Nurdiwijayanto L and Sasaki T 2018 Unilamellar metallic MoS_2/graphene superlattice for efficient sodium storage and hydrogen evolution *ACS Energy Lett.* **3** 997–1005

[42] Xiong P *et al* 2019 Interface modulation of two-dimensional superlattices for efficient overall water splitting *Nano Lett.* **19** 4518–26

[43] Yao Y *et al* 2014 Graphene cover-promoted metal-catalyzed reactions *Proc. Natl Acad. Sci. USA* **111** 17023–8

[44] Akhavan O and Ghaderi E 2009 Photocatalytic reduction of graphene oxide nanosheets on TiO_2 thin film for photoinactivation of bacteria in solar light irradiation *J. Phys. Chem.* C **113** 20214–20

[45] Lee J S, You K H and Park C B 2012 Highly photoactive, low bandgap TiO_2 nanoparticles wrapped by graphene *Adv. Mater.* **24** 1084–8

[46] Li X, Shen R, Ma S, Chen X and Xie J 2018 Graphene-based heterojunction photocatalysts *Appl. Surf. Sci.* **430** 53–107

[47] Cornu C J, Colussi A J and Hoffmann M R 2001 Quantum yields of the photocatalytic oxidation of formate in aqueous TiO_2 suspensions under continuous and periodic illumination *J. Phys. Chem.* B **105** 1351–4

[48] Hu X, Liu X, Tian J, Li Y and Cui H 2017 Towards full-spectrum (UV, visible, and near-infrared) photocatalysis: achieving an all-solid-state Z-scheme between Ag_2O and TiO_2 using reduced graphene oxide as the electron mediator *Catal. Sci. Technol.* **7** 4193–205

[49] Yan L, Gu Z, Zheng X, Zhang C, Li X, Zhao L and Zhao Y 2017 Elemental bismuth–graphene heterostructures for photocatalysis from ultraviolet to infrared light *ACS Catal.* **7** 7043–50

[50] Xia Y, Cheng B, Fan J, Yu J and Liu G 2020 Near-infrared absorbing 2D/3D $ZnIn_2S_4$/N-doped graphene photocatalyst for highly efficient CO_2 capture and photocatalytic reduction *Sci. China Mater.* **63** 552–65

[51] Mukherji A, Seger B, Lu G Q and Wang L 2011 Nitrogen doped $Sr_2Ta_2O_7$ coupled with graphene sheets as photocatalysts for increased photocatalytic hydrogen production *ACS Nano* **5** 3483–92

[52] Jia L, Wang D H, Huang Y X, Xu A W and Yu H Q 2011 Highly durable N-doped graphene/CdS nanocomposites with enhanced photocatalytic hydrogen evolution from water under visible light irradiation *J. Phys. Chem.* C **115** 11466–73

[53] Xiang Q, Yu J and Jaroniec M 2011 Preparation and enhanced visible-light photocatalytic H_2-production activity of graphene/C_3N_4 composites *J. Phys. Chem.* C **115** 7355–63

[54] Zhang X, Yuan X, Jiang L, Zhang J, Yu H, Wang H and Zeng G 2020 Powerful combination of 2D g-C_3N_4 and 2D nanomaterials for photocatalysis: recent advances *Chem. Eng. J.* **390** 124475

[55] Rahman M Z, Kibria M G and Mullins C B 2020 Metal-free photocatalysts for hydrogen evolution *Chem. Soc. Rev.* **49** 1887–931

[56] Kim I Y, Kim S, Jin X, Premkumar S, Chandra G, Lee N S, Mane G P, Hwang S J, Umapathy S and Vinu A 2018 Ordered mesoporous C_3N_5 with a combined triazole and

triazine framework and its graphene hybrids for the oxygen reduction reaction (ORR) *Angew. Chem.* **130** 17381–6
[57] Qi S, Fan Y, Wang J, Song X, Li W and Zhao M 2020 Metal-free highly efficient photocatalysts for overall water splitting: C_3N_5 multilayers *Nanoscale* **12** 306–15
[58] Ng Y H, Iwase A, Kudo A and Amal R 2010 Reducing graphene oxide on a visible-light $BiVO_4$ photocatalyst for an enhanced photoelectrochemical water splitting. The *J. Phys. Chem. Lett.* **1** 2607–12
[59] Soltani T, Tayyebi A and Lee B K 2018 Efficient promotion of charge separation with reduced graphene oxide (rGO) in $BiVO_4$/rGO photoanode for greatly enhanced photoelectrochemical water splitting *Sol. Energy Mater. Sol. Cells* **185** 325–32
[60] Li X, Yu J, Wageh S, Al-Ghamdi A A and Xie J 2016 Graphene in photocatalysis: a review *Small* **12** 6640–96
[61] Mondal A, Prabhakaran A, Gupta S and Subramanian V R 2021 Boosting photocatalytic activity using reduced graphene oxide (RGO)/semiconductor nanocomposites: issues and future scope *ACS Omega* **6** 8734–43
[62] Tachikawa T, Tojo S, Kawai K, Endo M, Fujitsuka M, Ohno T, Nishijima K, Miyamoto Z and Majima T 2004 Photocatalytic oxidation reactivity of holes in the sulfur-and carbon-doped TiO_2 powders studied by time-resolved diffuse reflectance spectroscopy *J. Phys. Chem.* B **108** 19299–306
[63] Di Valentin C, Pacchioni G and Selloni A 2005 Theory of carbon doping of titanium dioxide *Chem. Mater.* **17** 6656–65
[64] Treschev S Y, Chou P W, Tseng Y H, Wang J B, Perevedentseva E V and Cheng C L 2008 Photoactivities of the visible-light-activated mixed-phase carbon-containing titanium dioxide: the effect of carbon incorporation *Appl. Catal.* B **79** 8–16
[65] Zhao G *et al* 2017 The physics and chemistry of graphene-on-surfaces *Chem. Soc. Rev.* **46** 4417–49
[66] Sutar D S, Kushwaha N, Appani S K and Major S S 2020 Energy level alignment of graphene oxide and its derivatives with ZnO *J. Electron. Spectrosc. Relat. Phenom.* **243** 146953
[67] Geim A K and Grigorieva I V 2013 Van der Waals heterostructures *Nature* **499** 419–25
[68] Deng D, Novoselov K S, Fu Q, Zheng N, Tian Z and Bao X 2016 Catalysis with two-dimensional materials and their heterostructures *Nat. Nanotechnol.* **11** 218–30
[69] Kang J, Tongay S, Zhou J, Li J and Wu J 2013 Band offsets and heterostructures of two-dimensional semiconductors *Appl. Phys. Lett.* **102** 012111
[70] Premlal B, Cranney M, Vonau F, Aubel D, Casterman D, De Souza M M and Simon L 2009 Surface intercalation of gold underneath a graphene monolayer on SiC (0001) studied by scanning tunneling microscopy and spectroscopy *Appl. Phys. Lett.* **94** 263115
[71] Cranney M, Vonau F, Pillai P B, Denys E, Aubel D, De Souza M M, Bena C and Simon L 2010 Superlattice of resonators on monolayer graphene created by intercalated gold nanoclusters *Europhys. Lett.* **91** 66004
[72] Fu Q and Bao X 2017 Surface chemistry and catalysis confined under two-dimensional materials *Chem. Soc. Rev.* **46** 1842–74
[73] Cai X, Ozawa T C, Funatsu A, Ma R, Ebina Y and Sasaki T 2015 Tuning the surface charge of 2D oxide nanosheets and the bulk-scale production of superlatticelike composites *J. Am. Chem. Soc.* **137** 2844–7

[74] Sreepal V *et al* 2019 Two-dimensional covalent crystals by chemical conversion of thin van der Waals materials *Nano Lett.* **19** 6475–81

[75] Zheng Z, Yao J, Li J and Yang G 2020 Non-layered 2D materials toward advanced photoelectric devices: progress and prospects *Mater. Horiz.* **7** 2185–207

[76] Hoefflinger B 2011 ITRS: the international technology roadmap for semiconductors *Chips 2020* (Berlin: Springer) pp 161–74

[77] Gupta T 2009 Diffusion and barrier layers *Copper Interconnect Technology* (New York: Springer) pp 111–59

[78] Witt C, Yeap K B, Leśniewska A, Wan D, Jordan N, Ciofi I, Wu C and Tokei Z 2018 Testing the limits of TaN barrier scaling *2018 IEEE Int. Interconnect Technology Conf. (IITC) June 2018* (Piscataway, NJ: IEEE) pp 54–6

[79] Masanet E, Shehabi A, Lei N, Smith S and Koomey J 2020 Recalibrating global data center energy-use estimates *Science* **367** 984–6

[80] Engelmann G, Senoner T, van Hoek H and De Doncker R W 2015 A systematic comparison of various thermal interface materials for applications with surface-mounted (DirectFET TM) MOSFETs *2015 IEEE 11th Int. Conf. on Power Electronics and Drive Systems (June 2015)* (Piscataway, NJ: IEEE) pp 112–7

[81] Cui Y, Qin Z, Wu H, Li M and Hu Y 2021 Flexible thermal interface based on self-assembled boron arsenide for high-performance thermal management *Nat. Commun.* **12** 1284

[82] Moore A L and Shi L 2014 Emerging challenges and materials for thermal management of electronics *Mater. Today* **17** 163–74

[83] Hansson J, Nilsson T M, Ye L and Liu J 2018 Novel nanostructured thermal interface materials: a review *Int. Mater. Rev.* **63** 22–45

[84] Ikeda S, Hayakawa J, Lee Y M, Matsukura F, Ohno Y, Hanyu T and Ohno H 2007 Magnetic tunnel junctions for spintronic memories and beyond *IEEE Trans. Electron Devices* **54** 991–1002

[85] Reiss G, Schmalhorst J, Thomas A, Hütten A and Yuasa S 2008 Magnetic tunnel junctions *Magnetic Heterostructures* (Berlin: Springer) pp 291–333

[86] Bhatti S, Sbiaa R, Hirohata A, Ohno H, Fukami S and Piramanayagam S N 2017 Spintronics based random access memory: a review *Mater. Today* **20** 530–48

[87] Murali R, Yang Y, Brenner K, Beck T and Meindl J D 2009 Breakdown current density of graphene nanoribbons *Appl. Phys. Lett.* **94** 243114

[88] Zomer P J, Guimarães M H D, Tombros N and Van Wees B J 2012 Long-distance spin transport in high-mobility graphene on hexagonal boron nitride *Phys. Rev.* B **86** 161416

[89] Behin-Aein B, Datta D, Salahuddin S and Datta S 2010 Proposal for an all-spin logic device with built-in memory *Nat. Nanotechnol.* **5** 266–70

[90] Hoofman R J O M, Verheijden G J A M, Michelon J, Iacopi F, Travaly Y, Baklanov M R, Tökei Z and Beyer G P 2005 Challenges in the implementation of low-k dielectrics in the back-end of line *Microelectron. Eng.* **80** 337–44

[91] Jiang J, Chu J H and Banerjee K 2018 CMOS-compatible doped-multilayer-graphene interconnects for next-generation VLSI *2018 IEEE Int. Electron Devices Meeting (IEDM) December 2018* (Piscataway, NJ: IEEE) pp 34–5

[92] Li L and Wong H S P 2018 Integrating graphene into future generations of interconnect wires *2018 IEEE Int. Electron Devices Meeting (IEDM) December 2018* (Piscataway, NJ: IEEE) pp 5.5.1–4

[93] Lo C L, Helfrecht B A, He Y, Guzman D M, Onofrio N, Zhang S, Weinstein D, Strachan A and Chen Z 2020 Opportunities and challenges of 2D materials in back-end-of-line interconnect scaling *J. Appl. Phys.* **128** 080903

[94] Nguyen B S, Lin J F and Perng D C 2014 1-nm-thick graphene tri-layer as the ultimate copper diffusion barrier *Appl. Phys. Lett.* **104** 082105

[95] Li L, Chen X, Wang C H, Lee S, Cao J, Roy S S, Arnold M S and Wong H S P 2015 Cu diffusion barrier: graphene benchmarked to TaN for ultimate interconnect scaling *2015 Symp. on VLSI Technology (VLSI Technology) June 2015* (Piscataway, NJ: IEEE) pp T122–3

[96] Bong J H, Yoon S J, Yoon A, Hwang W S and Cho B J 2015 Ultrathin graphene and graphene oxide layers as a diffusion barrier for advanced Cu metallization *Appl. Phys. Lett.* **106** 063112

[97] Balandin A A 2011 Thermal properties of graphene and nanostructured carbon materials *Nat. Mater.* **10** 569–81

[98] Lewis J S, Perrier T, Barani Z, Kargar F and Balandin A A 2021 Thermal interface materials with graphene fillers: review of the state of the art and outlook for future applications *Nanotechnology* **32** 142003

[99] Shahil K M and Balandin A A 2012 Graphene–multilayer graphene nanocomposites as highly efficient thermal interface materials *Nano Lett.* **12** 861–7

[100] Liang C, Qiu H, Han Y, Gu H, Song P, Wang L, Kong J, Cao D and Gu J 2019 Superior electromagnetic interference shielding 3D graphene nanoplatelets/reduced graphene oxide foam/epoxy nanocomposites with high thermal conductivity *J. Mater. Chem.* C **7** 2725–33

[101] Huang X, Zhi C, Lin Y, Bao H, Wu G, Jiang P and Mai Y W 2020 Thermal conductivity of graphene-based polymer nanocomposites *Mater. Sci. Eng. R: Rep.* **142** 100577

[102] Yan H, Tang Y, Long W and Li Y 2014 Enhanced thermal conductivity in polymer composites with aligned graphene nanosheets *J. Mater. Sci.* **49** 5256–64

[103] Zhu H, Li Y, Fang Z, Xu J, Cao F, Wan J, Preston C, Yang B and Hu L 2014 Highly thermally conductive papers with percolative layered boron nitride nanosheets *ACS Nano* **8** 3606–13

[104] Song N, Jiao D, Ding P, Cui S, Tang S and Shi L 2016 Anisotropic thermally conductive flexible films based on nanofibrillated cellulose and aligned graphene nanosheets *J. Mater. Chem.* C **4** 305–14

[105] Yao Y, Zeng X, Sun R, Xu J B and Wong C P 2016 Highly thermally conductive composite papers prepared based on the thought of bioinspired engineering *ACS Appl. Mater. Interfaces* **8** 15645–53

[106] Eksik O, Bartolucci S F, Gupta T, Fard H, Borca-Tasciuc T and Koratkar N 2016 A novel approach to enhance the thermal conductivity of epoxy nanocomposites using graphene core–shell additives *Carbon* **101** 239–44

[107] Li A, Zhang C and Zhang Y F 2017 RGO/TPU composite with a segregated structure as thermal interface material *Composites* A **101** 108–14

[108] Chen L, Zhao P, Xie H and Yu W 2016 Thermal properties of epoxy resin based thermal interfacial materials by filling Ag nanoparticle-decorated graphene nanosheets *Compos. Sci. Technol.* **125** 17–21

[109] Yan Z, Liu G, Khan J M and Balandin A A 2012 Graphene quilts for thermal management of high-power GaN transistors *Nat. Commun.* **3** 827

[110] Subrina S, Kotchetkov D and Balandin A A 2010 Thermal management with graphene lateral heat spreaders: a feasibility study *2010 12th IEEE Intersociety Conf. on Thermal*

and Thermomechanical Phenomena in Electronic Systems June 2010 (Piscataway, NJ: IEEE) pp 1–5

[111] Dai W *et al* 2019 Metal-level thermally conductive yet soft graphene thermal interface materials *ACS Nano* **13** 11561–71

[112] Graphene-Info, 2019. Available at: https://graphene-info.com/huawei-mate-p30-pro-adopts-graphene-based-heat-management-film (accessed 27 October 2021).

[113] Dlubak B *et al* 2012 Highly efficient spin transport in epitaxial graphene on SiC *Nat. Phys.* **8** 557–61

[114] Guimarães M H, Zomer P J, Ingla-Aynés J, Brant J C, Tombros N and van Wees B J 2014 Controlling spin relaxation in hexagonal BN-encapsulated graphene with a transverse electric field *Phys. Rev. Lett.* **113** 086602

[115] Avsar A *et al* 2014 Spin–orbit proximity effect in graphene *Nat. Commun.* **5** 4875

[116] Wang Z, Ki D K, Chen H, Berger H, MacDonald A H and Morpurgo A F 2015 Strong interface-induced spin–orbit interaction in graphene on WS_2 *Nat. Commun.* **6** 8339

[117] Dankert A and Dash S P 2017 Electrical gate control of spin current in van der Waals heterostructures at room temperature *Nat. Commun.* **8** 16093

[118] Kamalakar M V, Dankert A, Bergsten J, Ive T and Dash S P 2014 Enhanced tunnel spin injection into graphene using chemical vapor deposited hexagonal boron nitride *Sci. Rep.* **4** 6146

[119] Kamalakar M V, Dankert A, Kelly P J and Dash S P 2016 Inversion of spin signal and spin filtering in ferromagnet|hexagonal boron nitride-graphene van der Waals heterostructures *Sci. Rep.* **6** 21168

[120] Song T *et al* 2018 Giant tunneling magnetoresistance in spin-filter van der Waals heterostructures *Science* **360** 1214–8

[121] Slaughter J M *et al* 2012 High density ST-MRAM technology *2012 Int. Electron Devices Meeting (December 2012)* (Piscataway, NJ: IEEE) pp 29–33

[122] Yang H, Vu A D, Hallal A, Rougemaille N, Coraux J, Chen G, Schmid A K and Chshiev M 2016 Anatomy and giant enhancement of the perpendicular magnetic anisotropy of cobalt–graphene heterostructures *Nano Lett.* **16** 145–51

[123] Miron I M, Garello K, Gaudin G, Zermatten P J, Costache M V, Auffret S, Bandiera S, Rodmacq B, Schuhl A and Gambardella P 2011 Perpendicular switching of a single ferromagnetic layer induced by in-plane current injection *Nature* **476** 189–93

[124] Liu L, Pai C F, Li Y, Tseng H W, Ralph D C and Buhrman R A 2012 Spin-torque switching with the giant spin Hall effect of tantalum *Science* **336** 555–8

[125] Hao Q and Xiao G 2015 Giant spin Hall effect and switching induced by spin-transfer torque in a $W/Co_{40}Fe_{40}B_{20}/MgO$ structure with perpendicular magnetic anisotropy *Phys. Rev. Appl.* **3** 034009

[126] Safeer C K, Ingla-Aynés J, Herling F, Garcia J H, Vila M, Ontoso N, Calvo M R, Roche S, Hueso L E and Casanova F 2019 Room-temperature spin Hall effect in graphene/MoS_2 van der Waals heterostructures *Nano Lett.* **19** 1074–82

[127] Benítez L A, S Torres W, Sierra J F, Timmermans M, Garcia J H, Roche S, Costache M V and Valenzuela S O 2020 Tunable room-temperature spin galvanic and spin Hall effects in van der Waals heterostructures *Nat. Mater.* **19** 170–5

[128] Yang Y and Murali R 2010 Impact of size effect on graphene nanoribbon transport *IEEE Electron Device Lett.* **31** 237–9

[129] Goharrizi A Y, Pourfath M, Fathipour M, Kosina H and Selberherr S 2011 An analytical model for line-edge roughness limited mobility of graphene nanoribbons *IEEE Trans. Electron Devices* **58** 3725–35

[130] Xu X *et al* 2017 Ultrafast epitaxial growth of metre-sized single-crystal graphene on industrial Cu foil *Sci. Bull.* **62** 1074–80

[131] Huang M, Deng B, Dong F, Zhang L, Zhang Z and Chen P 2021 Substrate engineering for CVD growth of single crystal graphene *Small Methods* **5** 2001213

[132] Zhang J, Shi G, Jiang C, Ju S and Jiang D 2015 3D Bridged carbon nanoring/graphene hybrid paper as a high-performance lateral heat spreader *Small* **11** 6197–204

[133] Fu Y *et al* 2019 Graphene related materials for thermal management *2D Mater.* **7** 012001

[134] Zhang Y F, Han D, Zhao Y H and Bai S L 2016 High-performance thermal interface materials consisting of vertically aligned graphene film and polymer *Carbon* **109** 552–7

[135] Su Y, Li J J and Weng G J 2018 Theory of thermal conductivity of graphene-polymer nanocomposites with interfacial Kapitza resistance and graphene-graphene contact resistance *Carbon* **137** 222–33

[136] Chen J and Gao X 2019 Directional dependence of electrical and thermal properties in graphene-nanoplatelet-based composite materials *Results Phys.* **15** 102608

[137] Chung K, Lee C H and Yi G C 2010 Transferable GaN layers grown on ZnO-coated graphene layers for optoelectronic devices *Science* **330** 655–7

[138] Roche S and Valenzuela S O 2014 Graphene spintronics: puzzling controversies and challenges for spin manipulation *J. Phys.* D **47** 094011

[139] Roche S *et al* 2015 Graphene spintronics: the European Flagship perspective *2D Mater.* **2** 030202

[140] Han W, Pi K, McCreary K M, Li Y, Wong J J, Swartz A G and Kawakami R K 2010 Tunneling spin injection into single layer graphene *Phys. Rev. Lett.* **105** 167202

[141] Gurram M, Omar S, Zihlmann S, Makk P, Schönenberger C and Van Wees B J 2016 Spin transport in fully hexagonal boron nitride encapsulated graphene *Phys. Rev.* B **93** 115441

[142] Kabiraj A, Kumar M and Mahapatra S 2020 High-throughput discovery of high Curie point two-dimensional ferromagnetic materials *NPJ Comput. Mater.* **6** 35

[143] Guo Y, Zhou S and Zhao J 2021 Two-dimensional intrinsic ferromagnets with high Curie temperatures: synthesis, physical properties and device applications *J. Mater. Chem.* C **9** 6103–21

[144] Jiang J, Kang J, Cao W, Xie X, Zhang H, Chu J H, Liu W and Banerjee K 2017 Intercalation doped multilayer-graphene-nanoribbons for next-generation interconnects *Nano Lett.* **17** 1482–8

[145] Safron N S, Choi J W, Kim M, Shin N, Gopalan P and Arnold M S 2015 Zigzag faceting and width refinement of graphene nanoribbons and nanoperforations via catalyzed edge-annealing on Cu (111) *Solid State Commun.* **224** 76–80

[146] Zhang X *et al* 2013 Experimentally engineering the edge termination of graphene nanoribbons *ACS Nano* **7** 198–202

[147] Chen L *et al* 2017 Oriented graphene nanoribbons embedded in hexagonal boron nitride trenches *Nat. Commun.* **8** 14703

[148] Wang H S *et al* 2021 Towards chirality control of graphene nanoribbons embedded in hexagonal boron nitride *Nat. Mater.* **20** 202–7

[149] Choi M S *et al* 2021 High carrier mobility in graphene doped using a monolayer of tungsten oxyselenide *Nat. Electron.* **4** 731–9

[150] Woo Y S *et al* 2013 Low temperature growth of complete monolayer graphene films on Ni-doped copper and gold catalysts by a self-limiting surface reaction *Carbon* **64** 315–23
[151] Wei D, Lu Y, Han C, Niu T, Chen W and Wee A T S 2013 Critical crystal growth of graphene on dielectric substrates at low temperature for electronic devices *Angew. Chem.* **125** 14371–6
[152] Wu T *et al* 2016 Fast growth of inch-sized single-crystalline graphene from a controlled single nucleus on Cu–Ni alloys *Nat. Mater.* **15** 43–7
[153] Wang M *et al* 2021 Single-crystal, large-area, fold-free monolayer graphene *Nature* **596** 519–24
[154] Wong T K 2012 Time dependent dielectric breakdown in copper low-k interconnects: mechanisms and reliability models *Materials* **5** 1602–25
[155] Lo C L, Catalano M, Smithe K K, Wang L, Zhang S, Pop E, Kim M J and Chen Z 2017 Studies of two-dimensional h-BN and MoS_2 for potential diffusion barrier application in copper interconnect technology *NPJ 2D Mater. Appl.* **1** 42
[156] GT – Smart High Tech AB, 2021. Available at: https://sht-tek.com/product/gt/ (accessed 30 October 2021).
[157] INCUBATION ALLIANCE, INC. (2021). Available at: https://incu-alliance.co.jp/english/graphenesheet/ (accessed 30 October 2021)
[158] Cardoso C, Soriano D, García-Martínez N A and Fernández-Rossier J 2018 Van der Waals spin valves *Phys. Rev. Lett.* **121** 067701
[159] Zollner K, Gmitra M and Fabian J 2018 Electrically tunable exchange splitting in bilayer graphene on monolayer $Cr_2X_2Te_6$ with X = Ge, Si, and Sn *New J. Phys.* **20** 073007
[160] Zollner K, Gmitra M and Fabian J 2020 Swapping exchange and spin-orbit coupling in 2D van der Waals heterostructures *Phys. Rev. Lett.* **125** 196402
[161] David A, Rakyta P, Kormányos A and Burkard G 2019 Induced spin-orbit coupling in twisted graphene–transition metal dichalcogenide heterobilayers: twistronics meets spintronics *Phys. Rev.* B **100** 085412
[162] Chung C, Kim Y K, Shin D, Ryoo S R, Hong B H and Min D H 2013 Biomedical applications of graphene and graphene oxide *Acc. Chem. Res.* **46** 2211–24
[163] Reina G, González-Domínguez J M, Criado A, Vázquez E, Bianco A and Prato M 2017 Promises, facts and challenges for graphene in biomedical applications *Chem. Soc. Rev.* **46** 4400–16
[164] Ghosal K and Sarkar K 2018 Biomedical applications of graphene nanomaterials and beyond *ACS Biomater. Sci. Eng.* **4** 2653–703
[165] Hui Y Y, Chang H C, Dong H and Zhang X (ed) 2019 *Carbon Nanomaterials for Bioimaging, Bioanalysis, and Therapy* (New York: Wiley)
[166] Nurunnabi M and McCarthy J (ed) 2019 *Biomedical Applications of Graphene and 2D nanomaterials* (Amsterdam: Elsevier)
[167] Shin S R, Li Y C, Jang H L, Khoshakhlagh P, Akbari M, Nasajpour A, Zhang Y S, Tamayol A and Khademhosseini A 2016 Graphene-based materials for tissue engineering *Adv. Drug Delivery Rev.* **105** 255–74
[168] Cheng X, Wan Q and Pei X 2018 Graphene family materials in bone tissue regeneration: perspectives and challenges *Nanoscale Res. Lett.* **13** 289
[169] Chun H J, Park C H, Kwon I K and Khang G (ed) 2018 *Cutting-Edge Enabling Technologies for Regenerative Medicine* vol 1078 (Singapore: Springer)

[170] Ambrosi A, Chua C K, Khezri B, Sofer Z, Webster R D and Pumera M 2012 Chemically reduced graphene contains inherent metallic impurities present in parent natural and synthetic graphite *Proc. Natl Acad. Sci. USA* **109** 12899–904

[171] Wong C H A, Sofer Z, Kubešová M, Kučera J, Matějková S and Pumera M 2014 Synthetic routes contaminate graphene materials with a whole spectrum of unanticipated metallic elements *Proc. Natl Acad. Sci. USA* **111** 13774–9

[172] Barbolina I, Woods C R, Lozano N, Kostarelos K, Novoselov K S and Roberts I S 2016 Purity of graphene oxide determines its antibacterial activity *2D Mater.* **3** 025025

[173] Lin L, Peng H and Liu Z 2019 Synthesis challenges for graphene industry *Nat. Mater.* **18** 520–4

[174] McCallion C, Burthem J, Rees-Unwin K, Golovanov A and Pluen A 2016 Graphene in therapeutics delivery: problems, solutions and future opportunities *Eur. J. Pharm. Biopharm.* **104** 235–50

[175] Zou X, Zhang L, Wang Z and Luo Y 2016 Mechanisms of the antimicrobial activities of graphene materials *J. Am. Chem. Soc.* **138** 2064–77

[176] Lu X, Feng X, Werber J R, Chu C, Zucker I, Kim J H, Osuji C O and Elimelech M 2017 Enhanced antibacterial activity through the controlled alignment of graphene oxide nanosheets *Proc. Natl Acad. Sci. USA* **114** E9793–801

[177] Liu S, Zeng T H, Hofmann M, Burcombe E, Wei J, Jiang R, Kong J and Chen Y 2011 Antibacterial activity of graphite, graphite oxide, graphene oxide, and reduced graphene oxide: membrane and oxidative stress *ACS Nano* **5** 6971–80

[178] Ruiz O N, Fernando K S, Wang B, Brown N A, Luo P G, McNamara N D, Vangsness M, Sun Y P and Bunker C E 2011 Graphene oxide: a nonspecific enhancer of cellular growth *ACS Nano* **5** 8100–7

[179] Singh S K, Singh M K, Nayak M K, Kumari S, Shrivastava S, Grácio J J and Dash D 2011 Thrombus inducing property of atomically thin graphene oxide sheets *ACS Nano* **5** 4987–96

[180] Sasidharan A, Panchakarla L S, Sadanandan A R, Ashokan A, Chandran P, Girish C M, Menon D, Nair S V, Rao C N R and Koyakutty M 2012 Hemocompatibility and macrophage response of pristine and functionalized graphene *Small* **8** 1251–63

[181] Fadeel B *et al* 2018 Safety assessment of graphene-based materials: focus on human health and the environment *ACS Nano* **12** 10582–620

[182] Kurapati R, Russier J, Squillaci M A, Treossi E, Ménard-Moyon C, Del Rio-Castillo A E, Vazquez E, Samorì P, Palermo V and Bianco A 2015 Dispersibility-dependent biodegradation of graphene oxide by myeloperoxidase *Small* **11** 3985–94

[183] Yang K, Feng L, Hong H, Cai W and Liu Z 2013 Preparation and functionalization of graphene nanocomposites for biomedical applications *Nat. Protoc.* **8** 2392–403

[184] Martín C, Kostarelos K, Prato M and Bianco A 2019 Biocompatibility and biodegradability of 2D materials: graphene and beyond *Chem. Commun.* **55** 5540–6

[185] Park S 2016 The puzzle of graphene commercialization *Nat. Rev. Mater.* **1** 16085

[186] Bøggild P 2018 The war on fake graphene *Nature* **562** 502–3

[187] Kauling A P, Seefeldt A T, Pisoni D P, Pradeep R C, Bentini R, Oliveira R V, Novoselov K S and Castro Neto A H 2018 The worldwide graphene flake production *Adv. Mater.* **30** 1803784

[188] Kong W, Kum H, Bae S H, Shim J, Kim H, Kong L, Meng Y, Wang K, Kim C and Kim J 2019 Path towards graphene commercialization from lab to market *Nat. Nanotechnol.* **14** 927–38

[189] Clifford C A, Martins Ferreira E H, Fujimoto T, Herrmann J, Hight Walker A R, Koltsov D, Punckt C, Ren L, Smallwood G J and Pollard A J 2021 The importance of international standards for the graphene community *Nat. Rev. Phys.* **3** 233–5

[190] ISO – ISO/TS 80004-13:2017 – Nanotechnologies, 2017. Available at: https://iso.org/standard/64741.html (accessed 5 October 2021)

[191] ISO – ISO/TS 21356-1:2021—Nanotechnologies—structural characterization of graphene—part 1: graphene from powders and dispersions, 2021. Available at: https://iso.org/standard/70757.html (accessed 7 October 2021)

[192] Graphene standardisation – Graphene Flagship (no date). Available at: https://graphene-flagship.eu/innovation/industrialisation/standardisation/ (accessed 3 November 2021)

[193] Characterisation of the structure of graphene. Measurement Good Practice Guide. 145, 2017. Available at: https://eprintspublications.npl.co.uk/8654/1/mgpg145.pdf (accessed 5 November 2021)

[194] Zhu Y, Qu B, Andreeva D V, Ye C and Novoselov K S 2021 Graphene standardization: the lesson from the east *Mater. Today* **47** 9–15

[195] Barkan T 2019 Graphene: the hype versus commercial reality *Nat. Nanotechnol.* **14** 904–6

www.ingramcontent.com/pod-product-compliance
Ingram Content Group UK Ltd.
Pitfield, Milton Keynes, MK11 3LW, UK
UKHW051910100926
8556IPUK00005B/827

9 780750 33997